Methods in Applied Soil Microbiology and Biochemistry

Methods in Applied Soil Microbiology and Biochemistry

Edited by

Kassem Alef

Institute of Ecological Chemistry and Geochemistry, University of Bayreuth, Germany.

and

Paolo Nannipieri

Department of Soil Science and Plant Nutrition, University of Florence, Italy.

ACADEMIC PRESS

Harcourt Brace & Company, Publishers

London San Diego New York Boston Sydney Tokyo Toronto

ACADEMIC PRESS LIMITED
24–28 Oval Road
LONDON NW1 7DX

US Edition published by
ACADEMIC PRESS INC.
San Diego, CA 92101

This book is printed on acid free paper

A catalogue record for this book is available from the British Library

ISBN 0–12–513840–7

Typeset by J&L Composition Ltd, Filey, North Yorkshire
Transferred to digital print on demand, 2005
Printed and bound by Antony Rowe Ltd, Eastbourne

Preface

The rapid advances in industrialization and technology together with changes in agricultural management have brought not only advantages for human society but also enormous environmental problems. As important fields in the environmental sciences, soil microbiology and biochemistry have now to deal with new aspects of, and to offer reasonable solutions for questions related to, agricultural practice, soil microbial ecology, ecotoxicology and different aspects of biotechnology. Both soil microbiology and biochemistry also deal with different aspects of new microbiological technologies such as the bioremediation of contaminated soils. A knowledge of the methodology and its theoretical basis is essential for successful work in this field and for beneficial cooperation with related environmental sciences. This book provides details of up-to-date methods in and fundamental knowledge of applied soil microbiology and biochemistry.

With respect to bioremediation of soils, this book bridges the gap between general and applied soil microbiology and biochemistry. It presents an integrated discussion of concepts, theories and methods. The experimental section offers systematic schemes of analyses for the different soil biological, chemical and physical parameters. Particular emphasis is laid upon a uniform, simple and clear presentation of the different methods.

This book is essential reading for scientists and students of environmental and natural sciences and engineering, for industry, and for consulting companies, as well as for authorities such as the Ministry of Environment and The Offices of Environmental Protection.

We wish to thank Dr Tessa Picknett (Academic Press) and Professor Dr O. Hutzinger (Institute of Ecological Chemistry and Geochemistry, University of Bayreuth, Germany) for their help and cooperation.

The editors wish to express their appreciation to their many colleagues for helpful criticisms and kind suggestions.

Dr K. Alef is grateful to his wife, Wafa, and son, Mido, for their patience, encouragement and assistance during the preparation of the book. Professor Dr P. Nannipieri expresses his gratitude to his wife, Manuelita, and his daughters, Ilaria and Elisa, for their help and patience during the preparation of this book.

K ALEF
P. NANNIPIERI
APRIL 1995

Contents

Contributors

Dr Kassem Alef, Institute of Ecological Chemistry and Geochemistry, University of Bayreuth, 95440 Bayreuth, GERMANY.

Professor Paolo Nannipieri, Dipartimento della Scienza del Suolo, e Nutrizione della Pianta, Piazzale delle Cascine 28, 50144 Firenze, ITALY.

Dr Gabiala Aden, Institute of Soil Science & Forestry, Buesgenweg 2, 37077 Göttingen, GERMANY.

Dr Bert Albers, Institute of Soil Ecology, GSF, Neuherberg, D–85758 Oberschleißheim, GERMANY.

Professor Dr Stuart Bamforth, Dept of Ecology, Evolution and Organismal Biology, 310 Dinwiddie Hall, Tulane University, New Orleans LA 70118–5698, USA.

Dr Habil Fritz Beese, Institute of Soil Ecology, GSF Neuherberg, 85758 Oberschleißheim, GERMANY.

Dr Giro Benckieser, Institute of Microbiology, University of Giessen, Senckenberg Str 3, 35390 Giessen, GERMANY.

Dr Hans-Michael Berstermann, Wartig Chemieberatung GmbH, Ketzerbach 27, 35094 Lahntal-Sterzhausen, GERMANY.

Dr Jaap Bloem, Dienst Landbouwkindig Onderzoek, Institut voor Bodemruchtbaarheid (IB–DLO), Postbus 30003, 9750 RA Haren, THE NETHERLANDS.

Dr Popko R Bolhuis, Dienst Landouwkundig Onderzoek, Institut woor Bodemvruchtbaarheid IB–DLO), Postbus 30003, 9750 RA Haren, THE NETHERLANDS.

Dr Philip C Brookes, Rothamsted Experimental Station, Harpenden, Herts AL5 2JQ, UK.

Professor Dr Gerhard Brümmer, Institute of Soil Science, University of Bonn, Nuß Alle 13, D–53115 Bonn, GERMANY.

Dr Rainer Brumme, Institute of Soil Science und Forestry, Büsgenweg 2, 37077 Göttingen, GERMANY.

Dr Peter Burauel, Institute of Radioagronomy, Research Center Jülich, Postfach 1913, 52425 Jülich, GERMANY.

Dr Harald Burmeier, Wodward/Clyde International Hauptstr. 45a, 30974 Wennigsen, GERMANY.

Dr Carlo Ciardi, Institute of Soil Chemistry, CNR Via Corridoni 78, 56100 Pisa, ITALY.

Dr Henrik Christensen, Institute of Biotechnology, Lundtoftevey 100, Bld 217, 2800 Ljngby, DENMARK.

Dr Soren Christensen, Department of Population Biology, Universitetsparken 15, 2100 Copenhagen, DENMARK.

Professor Johanna Döbereiner, EMBRAPA, National Center for Soil Biology Research, km 47 Seropedica, 23851–970 Rio de Janeiro, BRAZIL.

Dr Helmut Deschauer, Institute of Soil Science, University of Bayreuth, 95440 Bayreuth, GERMANY.

Dr Reinhard Eisermann, XENEX, Gesellschaft zur biotechnologischen Schadstoffsanierung mBH, 58642 Iserlohn-Letmathe, Stenglingser Weg 4–12, GERMANY.

Dr Sawsan Attalah Oran, Dept of Biological Sciences, Faculty of Science, The University of Jordan, Amman, JORDAN.

Dr Jürgen C. Forster, Institute of Soil Science, University of Bayreuth, 95440 Bayreuth, GERMANY.

Professor Dr William T Frankenberger Jr, Dept of Soil & Environmental Science, University of California, Riverside CA 92521, USA.

Professor Dr Fritz Führ, Institute of Radioagronomy, Research Center Jülich, Post 1913, 52425 Jülich, GERMANY.

Dr Uwe Gauglitz, BASF AG, Division of Environmental Protection, 67056 Ludwigshafen, GERMANY.

Dr Gaared Genouw, Lab of Microbial Ecology, Coupure Links 653, 9000 Gent, BELGIUM.

Dr Habil Anton Hartmann, Institute of Soil Ecology, GSF Neuherberg, 85758 Oberschleißheim, GERMANY.

Dr Thomas Held, Trischler & Partner GmbH, Umwelttechnik, Berliner Allee 6, 64295 Darmstadt, GERMANY.

Dr Kjeld Ingvorsen, Institute of Biological Sciences, Dept of Microbial Ecology, University of Aarhus, Bldg 540, Ny Munkegade, 8000 Aarhus C, DENMARK.

Dr Rainer Georg Jörgensen, Institute of Soil Science, University of Göttingen, Siebold Str. 4, 3400 Göttingen, GERMANY.

Professor Dr Ingrid Koegel-Knabner, Institute of Geography & Soil Ecology, University of Bochum, Postfach 102148, D-44780 Bochum, GERMANY.

Professor Dr Kirstin Lindström, Dept of Applied Chemistry & Microbiology, Division of Microbiology, POB 27 SF–00014, University of Helsinki, FINLAND.

Dr Hans-Joachim Lorch, Institute of Microbiology, University of Giessen, Senckenberg Str. 3, 35390 Giessen, GERMANY.

Dr Inge Lorenz, Wartting Chemieberatung GmbH, Ketzerbach 27, 35094 Lahntal-Sterzhausen, GERMANY.

Dr Karin Mathes, Dept of Biological Sciences, University of Bremmen, PO Box 33 0404, 28334 Bremen, GERMANY.

Dr Wolfgang Müller-Markgraf, Linde AG, Process Engineering & Contracting Division, Dr Carl-von-Linde Str 6–14, 82049 Hollriegelskreuth ba, München, GERMANY.

Professor P. Nannipieri, Dipartimenti di Scienza, del Suolo e Nutrizione Della Planta, Universitá, degli Studi, Piazzale dell Cascine 15, 50144 Firenze, ITALY.

Professor Dr Johanes C G Ottow, Institute of Microbiology, University of Giessen, Senckenberg Str. 3, 35390 Giessen, GERMANY.

Dr Harald Platen, Bio-DATA GmbH, Labor fur Umwelt-Analyse Phillip-Str. 4, 35440 Linden, GERMANY.

Dr Ludwig Ries, Federal Environmental Agency, Dept II 3.4, PO Box 330022, 14191 Berlin, GERMANY.

Dr Gerd Rippen, Trischler u Partner GmBH, Umwelttechnik, Berliner Allee 6, 64295 Darmstadt, GERMANY.

Dr Nils Risgaard-Petersen, Institute of Biological Sciences, Dept of Microbial Ecology, University of Aarhus, Bldg 540 Ny Munkegade, 8000 Aarhus C, DENMARK.

Dr Soren Rysggard, National Environmental Research Institute, Dept of Freshwater Ecology, Vejlsovej 25 8600, Silkeborg, DENMARK.

Dr Amio Saano, Dept of Applied Chemistry & Microbiology, Division of Microbiology, PO Box 27, SF–00014, University of Helsinki, FINLAND.

Dr Bernd Schieffer, Warttig Chemieberatung GmbH, Ketzerbach 27, 35095 Lahntal-Sterzhausden, GERMANY.

Dr Graham Sparling, Soil Science & Plant Nutrition, University of Western Australia, Nedlands, Perth WA 6009, AUSTRALIA.

Dr Wilhelm Steffen, Institute of Radioagronomy, Research Center, Jülich, Postfach 1913, 52425 Julich, GERMANY.

Dr M. A. Tabatabai, Dept of Soil & Environmental Science, University of California, Riverside CA 92521, USA.

Dr Kai U Totsche, Bayreuth Institute of Terrestial Ecosystem Research, Dept of Soil Physics, University of Bayreuth, 95447 Bayreuth, GERMANY

Dr Carmen Trazar-Cepeda, CSIC, Saniago de Compostela, SPAIN.

Dr Meint R. Veninga, Dienst Landbouwkundig Onderzoek, Institut voor Bodemruchtbaarheid (IB–DLO), Postbus 30003, 9750 RA Harn, THE NETHERLANDS.

Dr Herman A. Verhoef, Free University of Amsterdam, Biological Lab, PO Box 7161, 1007 Amsterdam, THE NETHERLANDS.

Professor Dr Welly Verstraete, Lab of Microbial Ecology, Coupure Links 653, 9000 Gent, BELGIUM.

Dr Luc van Vooren, Lab of Microbial Ecology, Coupure Links 653, 9000 Gent, BELGIUM.

Dr Jim W L van Vuurde, Research Institute for Plant Protection, PO Box 9060, NL-6700 GW Wageningen, THE NETHERLANDS

Dr Gerhard Welp, Institute of Soil Science, University of Bonn, Nuß Allee 13, 53115 Bonn, GERMANY.

Dr Henre van der Werf, Lab of Microbial Ecology, Coupure Links 653, 9000 Gent, BELGIUM.

Dr Johanna Wieringa, Dienst Lanbouwkundig Onderzoek, Institut voor Bodemruchtbaarheid (IB-DLO), Postbus 3003, 9750 RA Haren, THE NETHERLANDS.

Dr J. M. van der Wolf, Research Institute for Plant Protection, PO Box 9060, N>–6700 GW Wageningen, THE NETHERLANDS

Dr Jörg Zausig, Institute of Soil Science, University of Bayreuth, 95440 Bayreuth, GERMANY.

Dr Lazlo Zelles, Institute of Soil Ecology, GSF, Neuherberg, 85758 Oberschleißheim, GERMANY.

Introduction

1

Soil is a highly complex system characterized by a variety of biological, chemical and physical processes, which are markedly influenced by environmental factors. Microorganisms inhabit soil and, together with exocellular enzymes and the soil mesofauna and macrofauna conduct all known metabolic reactions. Microorganisms play a key role in the decomposition of soil organic matter and nutrient cycling, and therefore microbial activity is most important for maintenance of *soil fertility.* The integrity of the metabolic capacity of the soil microflora is a fundamental requirement for any concept of soil protection, soil bioremediation and recultivation. Thus soil microbiology and biochemistry are now considered as important fields of the **environmental sciences**, and reliable methods are required to obtain information on the interactions between microbial populations and environmental factors in soil.

The present book deals with methods in the microbiology, biochemistry and biotechnology of soil. These methods are essential for quantifying microbial processes in *agricultural, forest* and *contaminated* soils. To present the different aspects of this subject and to avoid confusion, several important terms routinely encountered should be defined and discussed.

Measurements of metabolic activities (aerobic and anaerobic) performed under laboratory conditions on sieved soil samples without plants and visible animals express only the contributions to the metabolic activities of living microorganisms inhabiting soil (Nannipieri et al, 1990). Therefore it is preferable to term the metabolic activity under such conditions **microbial** or **microbiological activity**. In contrast, the term **biological activity** implies the contribution to overall metabolic activity of all organisms in soil including microorganisms, fauna and plants.

Another aspect to be carefully considered in interpreting soil microbial and biochemical measurements is that these assays are generally performed under optimal assay conditions in the presence of substrate to obtain high activity rates (Burns, 1982; Nannipieri et al, 1990). These measurements express the potential activity in the soil. Assays carried out under laboratory conditions but in the absence of substrate are usually termed actual activity. These two terms, **potential activity** and **actual activity**, are frequently misinterpreted in the literature. Strictly speaking, all microbial or enzyme activities estimated under laboratory conditions (e.g. sieved soil, optimal pH, water potential and temperature, substrate, etc.) express a potential activity. According to this definition, *short- or long-term* estimations, such as *basal respiration*, performed in the absence of substrate but under optimal conditions for microorganisms express the *potential activity.* Short-term measurements (up to 6 h), such as *arginine ammonification* and *substrate-induced respiration*, are also considered as *potential activities.* To avoid any misunderstanding, we propose to use **basic potential activity** for measurements performed in the absence of substrate and **potential activity** for measurements in the presence of substrate. Changes in the relative composition of soil microflora are expected in long-term estimations while the physiological state of microorganisms may change in both long- and short-term assays.

The term **actual activity** comprises all estimations performed under natural conditions and probably this term can only be applied to

Methods in Applied Soil Microbiology and Biochemistry
ISBN 0-12-513840-7

field measurements without the addition of external substrates.

In the literature, activities in soil have been separated into *general* and *specific* activities. The term **general activity** includes criteria, such as respiration, ATP content, heat output and dehydrogenase activity, which reflect the activity of metabolic processes conducted by all, or at least a large group of, microorganisms. In contrast, the term **specific activity** only comprises the metabolic activity (nitrification, nitrogen fixation, etc.) conducted by some microbial groups or the activity of specific reactions catalysed by cell-free enzymes.

It may be difficult to decide whether a criterion is specific or not. For example, it is well established that not all microorganisms utilize glucose when added to soil (Wardle and Parkinson, 1990). In this case, should the glucose-induced respiration be considered a general or a specific criterion? The use of synthetic substrates for quantifying some microbial activities in soil can create problems. For example, it is not known whether substrates like TTC (2,3,5-triphenyl tetrazolium chloride) and INT (2(p-iodophenyl)-3-(p-nitrophenyl)-5-phenyl tetrazolium chloride), which are used to estimate dehydrogenase activity, can be taken up by all microorganisms inhabiting soil. Soil respiration and dehydrogenase activity (listed as general activities) must be considered as specific activities when anaerobic conditions are taken into account. The subdivision into general or specific criteria is not always practicable and should, therefore, be avoided.

It is also difficult to classify the methods according to their role in single nutrient cycles for several reasons:

1. Criteria, such as ATP, adenylate energy charge, dimethylsulphoxide reduction, dehydrogenase activity, the hydrolysis of fluoresceine diacetate, and the catalase activity, etc., reflect the transformation of different nutrients.
2. Many assays use synthetic substrates, such as *p*-nitrophenylsulphate [a substrate used to estimate the activity in soil of arylsulphatase, an enzyme supposed to play a fundamental role in sulphur mineralization according to Tabatabai and Bremner (1970)], that do not exist in nature and thus the measured activity may not represent the investigated nutrient cycle.
3. Nearly all laboratory methods do not give information on the actual rate of nutrient transformations in soil. Such information can be more realistically obtained by using field methods, such as lysimeter experiments, decomposition of cellulose *in situ* and monitoring the fate of labelled compounds, etc.
4. The arrangements of methods according to the nutrient cycles would also not discriminate between laboratory and field methods, and between aerobic and anaerobic methods.

The activity of any enzyme in the soil is due to enzymes that may have a different location. They can be associated not only with proliferating or non-proliferating cells, but also with dead cells or cell debris, or be immobilized on clays and humic colloids (Burns, 1982). The most serious problem in interpreting determinations of enzyme activities is to decide which combination of activities has been determined experimentally (Nannipieri et al, 1990). The distinction between the intracellular and extracellular components would be important. It seems reasonable to think that under certain circumstances (short-term assays, using dry soils and bacteriostatic agents) that the measured activity is due almost completely to the extracellular rather than the intracellular component (e.g. estimation of urease activity). However, with the present assays, it is very difficult to verify this hypothesis. Treatment by bacteriostatic agents like toluene present some problems and their mode of action can not be unequivocally interpreted because their effects vary with the soil, assay conditions and the enzyme (Ladd, 1985).

In this book, a distinction has been made between the terms **microbial activity** and **enzyme activity** because enzymes in soil are not only originated from microorganisms, but also from plants and animals. Their activities are not always bound to active cells. Furthermore, the effect of environmental factors (organic pollutants, heavy metals, cultivation and treatments of soils, etc.) on enzyme activities is mostly different from that on active microbial populations.

The present book also deals with the new technology of *Bioremediation of soils*, which indicates how microorganisms can be used to clean up chemically contaminated soils. To present this topic, some terms should also be defined. Unnatural compounds are known as **xenobiotics**, in contrast to **biogenic** or naturally occurring compounds.

The term **biodegradation** refers to the microbial transformation of organic compounds. Complete degradation of chemicals to harmless inorganic molecules is termed **mineralization.** Under these conditions, the microorganisms utilize the substrate as a source of carbon and energy with the formation of additional microbial biomass. The incomplete degradation of chemicals is known as **co-metabolism**, if the microbial growth is supported by an alternative source of carbon and energy. Incomplete degradation of the xenobiotic may lead to the accumulation of metabolites.

A chapter is also presented dealing with several aspects of project design, geostatistics and mathematical tools related to quality control and quality assurance in applied soil microbiology and biochemistry.

The great interest in environmental quality, soil fertility and soil protection has resulted in the development of methods for determining *aerobic* and *anaerobic microbial activities* and in the use of new biochemical and molecular biological techniques for monitoring microbial *community structure.* Up-to-date methods for estimating *microbial biomass* and *enzyme activities* are also presented. Methods for the *enrichment, isolation* and *enumeration* of microbial populations are covered extensively in this book.

Special attention has been given to *field methods.* Applications of lysimeters and procedures to quantify nitrogen mineralization and decomposition of organic matter under natural conditions are discussed.

Procedures dealing with *soil sampling, soil sterilization*, and the *physical* and *chemical analysis of soil* are also included. These methods are indispensable for microbiological and biochemical investigations in soils.

A separate chapter covers several aspects of *bioremediation* including guidelines, the technology of remediation, and microbial and chemical analysis.

The purpose of the book is to provide modern and authoritative prescriptions of laboratory and field methods for the microbiological characterization of soils as well as for biotechnological soil decontamination. We hope to guide the reader through the methodology of applied soil microbiology and biochemistry and to point out the necessity of interdisciplinary cooperation to allow a rapid development in this field.

References

Burns RG (1982) Enzyme activity in soil: location and a possible role in microbial ecology. Soil Biol Biochem 14: 423–427.

Ladd JN (1985) Soil enzymes. In: Soil Organic Matter and Biological Activity. Vaughan D, Malcolm RE (eds). Martinus Nijhoff Jr. W. Junk Publishers, pp. 175–221.

Nannipieri P, Ceccanti B, Grego S (1990) Ecological significance of the biological activity in soil. In: Soil Biochemistry, vol 6. Bollag J-M, Stotzky G (eds). Marcel Dekker, New York, pp. 293–355.

Tabatabai MA, Bremner JM (1970) Arylsulphatase activity of soils. Soil Sci Soc Am Proc 34: 225–229.

Wardle DA, Parkinson D (1990) Response of the soil microbial biomass to glucose and selective inhibitors, across a soil moisture gradient. Soil Biol Biochem 22: 825–834.

Quality control and quality assurance in applied soil microbiology and biochemistry

2

Quality – project design – spatial sampling

K. Totsche

An introduction

Modern soil microbiology and biochemistry is confronted with new and different challenges derived not only from agricultural practices (e.g. water and soil treatment/management/quality), but also from environmental protection (e.g. environmental impact statements, ecotoxicology, waste management, trend and risk assessment, intervention analysis), soil bioremediation and research. Whoever searches for answers to problems associated with soil microbiology and biochemistry must be concerned with the principal question of quality control and quality assurance.

How accurate is the decision, how good is the inference drawn?

Suppose one has to screen an industrial site for different pollutants. A number of questions are pertinent to conducting such a survey:

- Is there a specific contamination (survey, screening)?
- What levels can be determined (means, maximum, minimum)?
- Are these levels hazardous (with respect to ecotoxicological and human toxicological limit values)?
- What is the size of the contaminated site (regionalization problem)?
- What influence exerts heterogeneity on the distribution?

Methods in Applied Soil Microbiology and Biochemistry
ISBN 0-12-513840-7

- What type of clean-up methods are appropriate?
- Is the selected clean-up method efficient?

The related consequences exemplify the need for an accurate, reliable and valid methodology to assure the highest achievable level of quality. The principal statements of statistical quality control have been reported by Besterfield (1990), Burr (1979), Grant and Leavenworth (1988), Prier et al (1975), Rinne and Mittag (1989), Rosander (1985) and Wadsworth et al (1986).

Although developed mainly for industrial production and health care, the methodologies introduced are useful as a guideline and starting point for the understanding and the development of quality control and quality assurance methods in applied soil microbiology. The application of quality control and quality assurance enables us to cope with the increased responsibility we have for people and the environment.

But what is quality?

Quality is a certain property of a process or result, of which it is an inherent characteristic. This process or result may be a laboratory analysis, survey, scientific experiment or validation of a hypothesis. However, quality can only be examined if the process and/or result is documented in detail, i.e. the following are provided: information on data sampling, data analysis, data processing, the applied methodology, the mathematical tools used, the errors and their supposed error sources, the occurred problems and difficulties, etc. Otherwise, the data presented within a process or result can not be compared or interpreted.

Quality is examined by the determination of the levels of quality categories or parameters. To avoid confusion, definitions presented in this chapter are those used commonly in the literature on statistical quality control and classical statistics. We use the definitions in a more general way, which does not contradict the classical definition of the categories.

These categories are:

- Precision: a process/result is called precise, when the corresponding error characterizing the process deviates only slightly from its mean. In other words, precision is a measure of the deviation of repeated measurements of one quantity.
- Accuracy: does the process or result provide information on the actual value of the target parameter? If the average of a random variable characterizing the process is approaching zero, we call the process accurate. In other words, accuracy is a measure of the deviation of the measured value from the real, but unknown value.
- Reliability: does the process or result provide a qualitative and quantitative estimation of random error and variability? Therefore, reliability is the consequence of high precision combined with high accuracy.
- Representativity: does the process/result come from a sample of a given population that is representative of the whole population?
- Validity: does the process/result show less sensitivity to variable experimental conditions? (To avoid confusion, the validity term used with mathematical modelling is a measure of whether a mathematical model is able to simulate real-world results when using independently measured real-world parameters. Here, validity is a category of quality. It stands for the robustness of a process or result against any variability introduced by the experimenter or by the experimental conditions.)
- Objectivity: is the applied methodology independent of the experimenter?
- Reproducibility: does the documentation of a process or result provide sufficient information on the applied methodology?
- Relevance: does the process/result provide satisfactory answers to the question under investigation?
- Applicability: is the applied methodology transferable with respect to financial and personal resources?
- Efficiency: is the process/result achievable at the same level of quality with another methodology with less effort and expense?

A certain level of quality is achieved by optimizing these quality parameters. The constraints of this optimization procedure are temporal, financial and personnel expenses, which can be spent on raising quality. Nevertheless, when considering quality, all these categories have to be taken into account and evaluated carefully.

The complementary aspects of quality are quality control, quality assurance and quality improvement. Quality control summarizes all the methods and tools that deal with the evaluation or estimation of quality categories. Quality control methods are used to assess the particular levels of the quality categories in order to ascertain whether a particular product or services match the quality requirements. Quality assurance is the application of state of the art methodologies and tools that have been checked for quality requirements and therefore guarantee the highest achievable level of quality of a process or result.

Quality improvement summarizes the development and application of new methods, tools and techniques in order to raise the present level of quality. This point will not be discussed in detail because this discipline is the object of current scientific research.

In the most general sense, quality control and quality assuring methods are the means by which the question for goodness of decision can be qualitatively and quantitatively answered at the highest achievable level of objectivity.

In the following chapter, a methodology will be introduced that enables practitioners and scientists to cope with the quality requirements of soil microbiology and biochemistry.

From sampling to data interpretation (project design strategies)

Independent of the project being conducted (scientific experiment, research study, survey of contaminated sites, environmental impact statement), the design of a scientifically based plan should be the first step. The objective of the design is to plan an appropriate experiment so that the data obtained can provide valid and objective conclusions. Since all data are subject to variation, which is the consequence of characteristics of the system and experimental error, the only way to gain valid results is by applying statistical methodology. Since the statistical analysis is highly sensitive and closely related to the method of collecting data, the experimenter has to choose the appropriate composition of experimental and statistical tools used carefully to study the respective topic.

Empirical studies and surveys – in the following termed "projects" – performed under well-defined environmental conditions can be classified with respect to their objectives:

- Inventory: providing information on a system and its components. In this context, projects are well-defined plans for the collection and analysis of data of a given system with given properties.
- Response assessment: providing information on the reaction of a system and its components to changing conditions. In this context, projects are well-defined manipulations of one or more factors in order to gain knowledge about the effect of this manipulation (the response) on another variable.
- Hypothesis verification: providing information in order to reject or accept a hypothesis with respect to the reaction of a system to changing conditions. In this context, projects are well-defined manipulations of a set of variables.
- Correlations research: providing information on relations between a set of variables. In this context, projects are well-defined measurements of a set of variables that are not manipulated by any experimental action.

Independent of the type of project to be performed, there is a need for a meaningful and appropriate project design.

The zero step: project design

Project design resembles all activities leading to the construction of a scientifically based project plan. A project plan is the formulation of a set of sequential procedures in order to provide information suitable to answer a certain scientific or research problem. The plan describes a multistep process with internal regulatory elements and is set up during the project design (Fig. 2.1). Each individual step will be discussed in detail in the following section.

The problem statement – setting up a scientific hypothesis

Obviously, a project has to start with a problem. For example, let us suppose that a microbiologist wants to study the effect of a certain contaminant on the microbial mineralization of nitrogen under variable environmental condi-

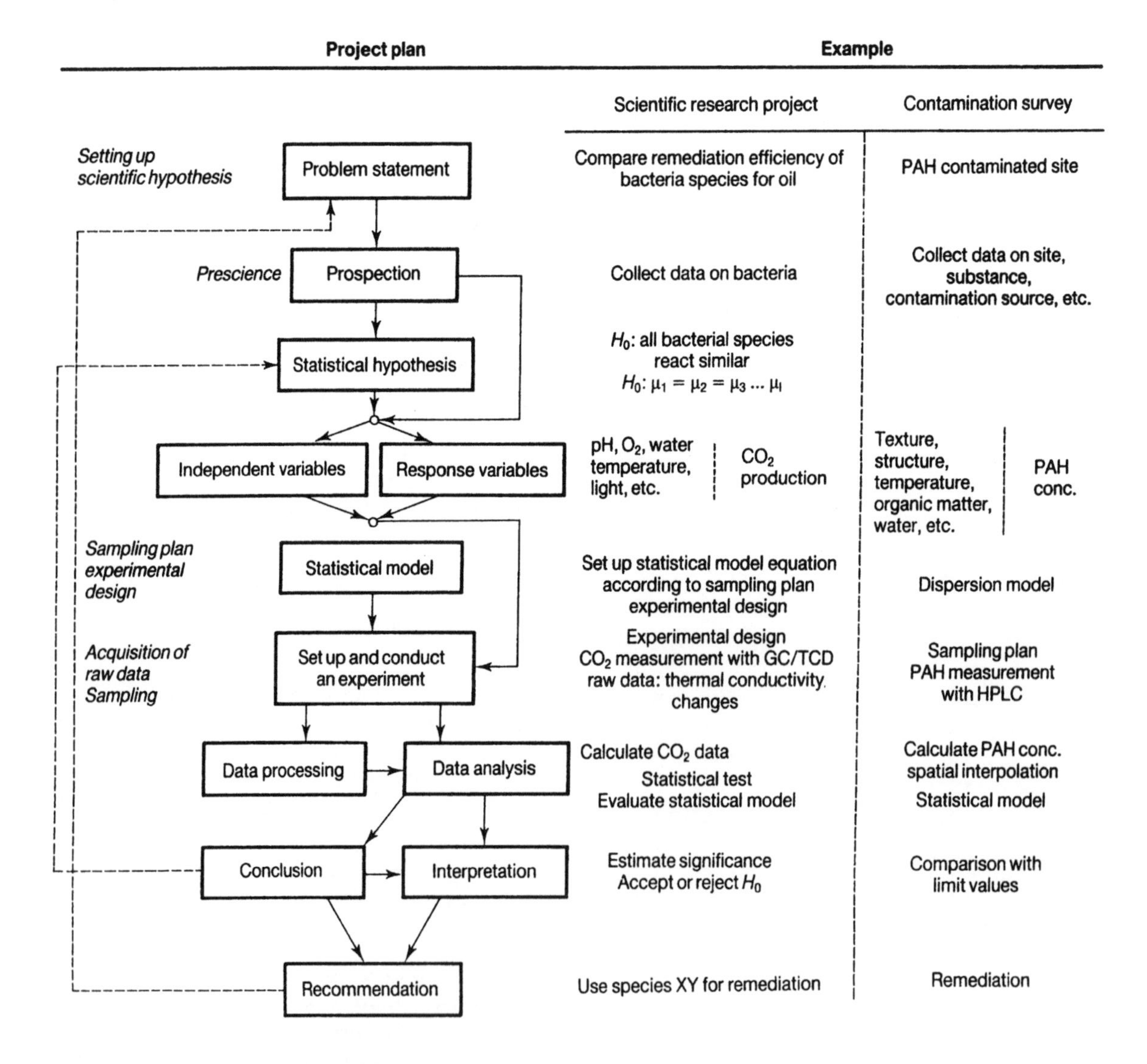

Figure 2.1. Project design plans.

tions. The experimenter has to think about an appropriate problem statement and has to define the conditions required to perform valid experiments, i.e. the experimenter has to formulate a scientific or research hypothesis to set the parameters of the problem. Furthermore, for a better understanding of the problem and its implications, the discussion and formulation of different, alternative scientific hypotheses are useful.

Therefore, the experimenter has to define in detail the possible statements of the problem. For the abovementioned example, these are:

- What is the influence: inhibition, increase or no effect on mineralization?
- What are the environmental conditions, how variable are they and what is their relevance to the problem?

The scientific or research hypothesis could be, for example, that a certain xenobiotic inhibits the microbiological mineralization of nitrogen under constant environmental conditions in soils.

Setting up a statistical hypothesis

Only in a few cases experimenters are lucky enough to have identical scientific and statistical hypotheses. Usually, statistical hypotheses are formulations testing a scientific hypothesis. They should be formulated according to logical requirements, i.e. first test condition A, then response B. For the above example the appropriate statistical hypothesis (H_0) would be: if a certain xenobiotic is present in soils, microbiological mineralization is inhibited. A statistical alternative formulation hypothesis (alternative hypothesis H_1) would then be: if a certain xenobiotic is present in soil, the microbiological mineralization is not inhibited.

Prospecting

Prospecting is the collection of a priori available data relevant to the project prior to conducting the experiments. The results of the prospecting provide the prescience of the experiment. A lot of information is already available before any experiments are started. The collection and analysis of this data can help to:

- Modify the hypothesis
- Modify the experimental set up
- Assess limitations of the experiments
- Estimate the relevance and need for surveys or experiments
- Reduce financial and personnel investment (cost).

The data may be collected by analysing, e.g.

- Historical data documenting the past use of sites (industrial, dwelling)
- Data and information on chemicals
- Information on physical, chemical and biological processes
- Geological, biological, hydrological and/or pedological information
- Meteorological data.

Prospecting provides the means to raise efficiency and economy of a certain project.

Selection and control of independent variables

Independent variables are the factors that modify the response of a system. They can influence the system quantitatively and qualitatively. In the previously mentioned example, the independent factor is the level of concentration of the xenobiotic added to the system, but there are a lot of other independent factors that have to be known and, if needed, controlled. The term environmental condition summarizes a more or less indefinite number of independent variables, say water content, organic C level, O_2-pressure, texture, microbiological species, etc. In order to conduct a well-defined experiment, probably all of these factors have to be controlled and therefore measured. Thus the experimenter has to formulate ranges and tolerances for the factors that guarantee the requirement of constant environmental conditions, i.e. the experimenter has to set up the experimental conditions.

Selection and control of dependent variables (response)

The dependent variable is used to measure the influence of one or more factors. In the above example it would be nitrogen mineralization, but, if the problem is "does a certain xenobiotic influence soil bacteria? ", influence in this context could be, for example:

- Changing the community structure
- Inhibiting a special physiological metabolic process of a specific group of bacteria
- Inducing mutageneous effect.

Each type of "influence" needs an individual dependent variable. The conclusion drawn is very sensitive to the selection of the dependent variable. If the experimenter focuses on the inhibition of physiological processes caused by the xenobiotic, for example, but actually the xenobiotic causes a change in the community structure, the question would probably be inappropriate. This illustrates that the experimenter has to take care with the selection of the dependent variable with respect to the scientific problem. If a certain dependent variable is chosen, the experimenter may think about the way to measure the variable, i.e. the response and how accurate these measurements are.

Setting up a statistical model (response function)

If we were not only interested in whether a certain factor influences a response variable (qualitative relation) but also in the quantitative relationship of one or more factors, we have to set up a statistical model (design model equation, regression model equation, see later). The method of assessing the regression model and the properties of the regression function will be discussed in detail in the section entitled "Spatial sampling and analysis". The design model equations are introduced later in this section. At this point, the reader should notice that a statistical model is not a physical description of the underlying processes but a phenomenological description of an observed relationship subjected to variation. As a consequence, a statistical model can be used as a measure of the response of a variable to a number of factors. Statistical models are valid and inexpensive tools to establish whether it is worthwhile to invest further experimental effort for understanding the physical processes causing the observed relationship.

Setting up and conducting an experiment

In order to choose an appropriate experimental design, the experimenter has to think about the magnitude of the acceptable risk. The higher the risk, the lower the number of samples to be taken or replications to be carried out, respectively. Unfortunately the lower the risk, the higher the number of samples and therefore the cost. During the experiment the experimenter should carefully document each step of the experimental process in order to verify whether:

- The experiment proceeds according to plan
- Dependent variables are under control to maintain the experimental condition

Any findings that do not follow the experimental plan should be documented whether or not they appear to be important or irrelevant with respect to the scientific or statistical hypothesis. Frequently the data analysis is carried out after the completion of the data collection phase of the experiment. If there are any irregularities within these data, one can try to identify the causes for the erroneous data with the extra findings documented in the "experimental control chart".

Data processing and analysis

A number of experimental set ups do not measure the response variable directly but measure a property of a dependent variable that is known to be related to the response variable by a well-defined mathematical functional relationship. All spectroscopic methods, for example, give this type of measurement. Therefore, the measured data have to be transformed in a mathematical way. The direct method is to use the relationship of the dependent variable to the measured property and calculate the value of the dependent variable. The other way is to use a calibration technique. Therefore, the experimenter uses well-defined systems with different but known levels of the response variable for setting up a calibration curve. The calibration curve will then be used to determine the level of the response variable in the system under investigation. Experimenters should be very careful in supplying calibration curves. All independent variables influencing the measured property of the response variable have to be known and controlled. Once the data on the response variable are known, the data analysis is performed. A large number of numerical packages (SPSS, SAS, CoStat, etc.) are available for the analyis of spatial and temporal statistical data. The experimenter may be cautious in the choice of the program package used for the data analysis: some numerical procedures for the statistical data analysis are optimized for discrete data. The reason for this is that these packages were developed for social and medical science and research, which deal with a good number of discrete data. In applied soil microbiology and biochemistry, we more frequently have to face continuous rather than discrete data. The experimenter should be aware of this difference and should consult an expert to decide whether a chosen program package is suitable and adequate for analysing the data. The sections entitled "Spatial sampling and analysis" and "Mathematical and statistical tools" discuss in detail a number of adequate statistical methods used for the analysis of statistical data.

The statistical model should now be subjected to a critical discussion with respect to the validity and relevance of the underlying assumptions and its applicability.

Conclusions and interpretations

The final step of the proposed project design plan is the interpretation of the data. The statistical aspect of data interpretation is the acceptance or rejection of the statistical hypothesis. The acceptance/rejection will be qualified by the level of confidence chosen by the experimenter and by a detailed description of the experiment and its condition, say a detailed documentation of steps 1–7. In the case of a different statistical and scientific hypothesis, the experimenter has to evaluate the consequence of the result of the statistical test for the scientific hypothesis. The experimenter should emphasize the practical significance of the results, for example, the observed difference between the means of two differently treated experimental plots is statistically proven but may only be of academic interest, especially if the difference is gradual.

The data interpretation has a large subjective component: the way the interpretation is done. It is the experimenter who, on the basis of his or her personal knowledge, experience and social and ethic awareness, has to formulate appropriate recommendations for taking action. It is the experimenter's duty to conduct a final critical discussion with regard to the fitness of the applied methodology with respect to the particular problem. It is the experimenter's responsibility to present the findings in a comprehensive manner and document in detail the complete project design, so that the findings can be checked and the methodology is reproducible. The proposed project design should be used as a guideline on how to conduct scientific or research studies whilst simultaneously increasing accuracy. A priori, the design will probably neither take into account all the possible outcomes nor estimate the influence of the independent variable. A posteriori, the results may show that a variable assumed to be unimportant showed a prominent influence and other considered variables are of less significance. Consequently, we face an iterative process: the results may show that further investigations are needed in order to answer the initial problem.

Experimental designs and sampling strategies – principal activities

After an appropriate hypothesis has been formulated and a proper statistical procedure has been chosen, experimental action starts by collecting data that are suitable to answer the problem in a statistical objective manner. The way of collecting data is preconditioned by the problem statement and the statistical methodology chosen to sample and analyse the observations.

Four rules for experimental design to collect data can be specified: randomization, replication, blocking and error control.

Randomization

When applying statistical methods, care should be taken that the observations are independent and randomly distributed variables. Randomization is the way to determine both the order of the experimental runs and the experimental conditions or material at random. In this way any kind of researcher-dependent bias can be avoided. Another advantage of randomization is the averaging out of extraneous factors that may possibly be present. By studying the effect of some xenobiotic on the microbiological mineralization of nitrogen in soils, for example, the selected experimental units may show a gradual decrease in a factor that is relevant to the mineralization rate, say water content. If we assign by chance the treatment with the xenobiotic to the higher water content and the treatment without the xenobiotic to the lower water content, we may achieve invalid experimental results. Randomization provides the means to damp this effect.

Replication

Replication signifies that the basic experimental treatments are performed repeatedly. In the abovementioned example, the two basic treatments would be the trial with the xenobiotic and that without the xenobiotic. If ten units are treated with the xenobiotic and ten without the xenobiotic, we have ten replicates of each treatment. With replication we obtain an estimate on the experimental error. This estimation is a prerequisite for the decision whether a difference in the observation is statistically significant. Further, if we use the sample mean (x) as an estimate of the effect of a treatment in the experiment, replication allows us to obtain a more precise estimate of this effect.

Blocking

Blocking is the assignment of a collection of experimental units to blocks that show less heterogeneity with respect to independent variables compared to the complete set of these units. Consequently, a block is a fraction of the experimental site, which is more homogeneous than the complete experimental site. The units are commonly called plots. In this way the precision of an experiment is increased.

Error control

Error signifies the standard error defined by

$$\sqrt{(\sigma^2/r)} \qquad (2.1)$$

where σ is the standard deviation and r is the number of replications.

The objective of error control or local control is to minimize the error variance. Error control is achieved by increasing the homogeneity of the experimental conditions. Another way of error control is the method of blocking mentioned previously. An increase in the number of replications does not affect the error variance itself but leads to a more stable estimate of the error variance, i.e. a decrease of confidence intervals. This rule is an integral component of each of the experimental designs introduced.

Analysis of variance (ANOVA) is a powerful tool that is used as the statistical methodology to identify differences between appropriately defined sample populations. The fundamentals of ANOVA design have been discussed by Eisenhart (1947) and by Dunn and Clark (1974). All more complex and sophisticated experimental designs evolve from three basic ANOVA designs, the completely randomized design, the randomized block design and the Latin square design. Recently, procedures have been developed to investigate the influence of one or more independent variables on two or more dependent variables. These procedures are called MANOVA (multivariate analysis of variance) techniques. As mathematics and numerics do not differ in principle from ANOVA designs, we restrict ourselves to discussion of the most common ANOVA designs.

The reader should note that each design requires its own method of randomization, i.e. the way of allotting the treatments to the blocks, and that each design has its own method of determining the minimum sample size (number of replications) and dealing with missing observations. What they all have in common is that the order by which the observations are taken (the response is measured) is determined at random. Therefore, they are called completely randomized designs.

The aim of the following discussion will not be to present a formal description of the experimental designs in detail, but to give a short but comprehensive overview of the various designs and the type of problems for which they were developed. For a detailed statistical analysis of the designs, the reader should consult, e.g. Anderson and McLean (1974), Clarke (1980), Kirk (1982) and Montgomery (1984).

Completely randomized and building block designs

Completely randomized design

As indicated earlier, the objective of blocking is to control the variability caused by erroneous sources systematically. Suppose, for example, that an experimenter wants to check whether four species of bacteria can use a different organic contaminant as the carbon source. The degradation may be checked by measuring the CO_2 production with respect to time. The bacteria are cultured on a medium containing the organic contaminant and 12 replications are taken. The one and only factor considered in this experiment will be the bacteria species. A completely randomized single factor design would be to assign randomly each of the $12 \times 4 = 48$ runs to one experimental unit (the contaminated agar plates) and to observe the production of CO_2 of each of the units. This allotment reflects the simplest form of an experimental ANOVA design and is known as the completely randomized design (CRD) (Table 2.1).

The dependent variable is the CO_2 production time and the independent variable is the bacteria species, called treatment A. In this particular case, the levels of the treatment are identical for the four different bacteria

Table 2.1. Completely randomized design. A schematic representation of an experimental design to elucidate the inhibition of CO_2 production of four bacteria species ($a_1 \ldots a_4$) is shown. Y is the response variable that designates the production of CO_2 in one individual replicate.

a_1	a_2	a_3	a_4
Y_{11}	Y_{12}	Y_{13}	Y_{14}
Y_{21}	Y_{22}	Y_{23}	Y_{24}
.	.	.	.
$Y_{12,1}$	$Y_{12,2}$	$Y_{12,3}$	$Y_{12,44}$

1 = replicates; 2 = treatments.

species, a_1, a_2, a_3 and a_4, (a_j in general). The amount of CO_2 produced by treatment level a_j is denoted with $Y_{i,j}$. The statistical hypothesis that we will test may be

$$H_0: \mu_1 = \mu_2 = \mu_3 = \mu_4, \quad (2.2)$$

where the treatment level means μ_1, μ_2, μ_3, μ_4 (the means for the production of CO_2 of each bacteria species) are the same. We have 48 agar plates and will assign these randomly to the bacteria species with the restriction that each bacteria species (level of treatment) receives 12 agar plates (12 replicates; $i = 1, \ldots, 12$).

Obviously, the production of CO_2 is not only affected by the bacteria species, but also by the replicates used and by the environmental conditions. The knowledge obtained on the response is formulated by the experimental design model equation. The statistical model equation for this particular case (completely randomized design) would be

$$Y_{i,j} = \mu + a_j + e_{i,j}$$
$$(j = 1, \ldots, 4;\ i = 1, \ldots, 12) \quad (2.3)$$

where $Y_{i,j}$ is the value of CO_2 production of treatment level j and replicate i, μ the grand population mean, constant for all 48 experimental units, a_j the treatment effect of level j, $a_j = \mu_j - \mu$, and $e_{i,j}$ the error effect, the variation in the score response due to extraneous environmental conditions.

Completely randomized block design

If the culture media used differ slightly in their properties, then the experimental units contribute to the variability in the production of CO_2 of the different bacteria species. Consequently, the experimental error is composed of the random error and the variability of the experimental units. The following designs are developed to minimize the error effect by isolating the sources of variation of this type.

To minimize the experimental error in this context, the variability between the experimental units must be removed from the experimental error. This could be accomplished, for example, by using the blocking procedure. With respect to the CO_2 production experiment, this would mean assigning culture media from the same batch of raw material to one block. This block would be more homogeneous with respect to the nuisance variable than a collection of culture media from different batches of raw material. The improvement to the completely randomized design, for example, would be to form 12 blocks (equal to 12 replicates) and to assign randomly each of the four treatment levels (bacteria species) to each block (Table 2.2).

This design is called a completely randomized block design. With this type of design, we can test the following hypothesis:

$$H_0: \mu_1 = \mu_2 = \mu_3 \text{ (treatment population means are equal)} \quad (2.4)$$

$$H_0: \mu_1 = \mu_2 = \ldots = \mu_{1\ 2} \text{ (block population means are equal)} \quad (2.5)$$

This design guarantees more homogeneous experimental conditions. Obviously, it increases the accuracy of the comparison of the bacteria species by reducing the variability between the experimental unit (note that the block is more homogeneous than the entire population of experimental units). Within each block, the order in which the four bacteria are tested is determined randomly. The experimental design model equation for the completely randomized block design will be

Table 2.2. Completely randomized design. The blocks are composed of culture media prepared with the same raw batch material. $a_1 \ldots a_4$ are bacteria species. Y is the response variable that designates the production of CO_2 of one individual replicate (see Table 2.1).

	a_1	a_2	a_3	a_4	*Block means*
Block 1	$Y_{1,1}$	$Y_{1,2}$	$Y_{1,3}$	$Y_{1,4}$	$Y_{1,.}$
Block 2	$Y_{2,1}$	$Y_{2,2}$	$Y_{2,3}$	$Y_{2,4}$	$Y_{2,.}$
.	.	.	.	.	.
.	.	.	.	.	.
.	.	.	.	.	.
Block 12	$Y_{12,1}$	$Y_{12,2}$	$Y_{12,3}$	$Y_{12,44}$	$Y_{12,.}$
Treatment level means	$Y_{.,1}$	$Y_{.,2}$	$Y_{.,3}$	$Y_{.,4}$	Grand mean

$$Y_{i,j} = \mu + a_j + p_i + e_{i,j} \quad (2.6)$$

where μ is the grand population mean, a_j the treatment effect of level j, p_i the block effect of block i due to variation of batch raw material, and $e_{i,j}$ the error effect, the remaining effect after subtracting the block effect, the treatment effect and the grand mean.

Latin square design

Suppose a soil microbiologist is interested in the effect of four different formulations of a fertilizer on the microbial activity in forest soils. This may be studied with respect to forest soil acidification: fertilizers are applied to increase not only plant growth but also the capacity of the soil to neutralize the acidity. In this example we may face the situation that every formulation is mixed from a batch of raw materials that is not large enough to support replicate mixings. Consequently, the experimenters have to use different batches of raw material. Furthermore, the formulations may be prepared by different operators. Obviously, there are two extraneous factors to be averaged out, the operators and the batches of raw material. The proper design for this problem would be to let each operator mix the formulations exactly once and to test each formulation exactly once in each batch of raw material. Such a design is called Latin square design (LSD) and is shown in Table 2.3.

LSDs are used to control and eliminate two extraneous sources of variability simultaneously by blocking systematically in two

Table 2.3. Latin square design. A schematic representation of an experiment design to measure the effect of four different fertilizer formulations on microbial activity of four different species of bacteria in forest soil is shown. Extraneous factors are four batches of raw material for fertilizer formulations and four different operators to conduct the experiment. Each column represents one individual operator, each row represents one individual batch of raw material. $a_1 \ldots a_4$ represent the different bacteria species. Y is the response variable (CO_2 production).

	Operator 1	*Operator 2*	*Operator 3*	*Operator 4*	*Row means*
Raw batch 1	a_1 $Y_{1,1,1,1}$ $Y_{2,1,1,1}$ ⋮ $Y_{12,1,1,1}$	a_2 $Y_{1,2,1,2}$ $Y_{2,2,1,2}$ ⋮ $Y_{12,2,1,2}$	a_3 $Y_{1,3,1,3}$ $Y_{2,3,1,3}$ ⋮ $Y_{12,3,1,3}$	a_4 $Y_{1,4,1,4}$ $Y_{2,4,1,4}$ ⋮ $Y_{12,4,1,4}$	$Y_{..1.}$
Raw batch 2	a_2 $Y_{1,2,2,1}$ $Y_{2,2,2,1}$ ⋮ $Y_{12,2,2,1}$	a_3 $Y_{1,3,2,2}$ $Y_{2,3,2,2}$ ⋮ $Y_{12,3,2,2}$	a_4 $Y_{1,4,2,3}$ $Y_{2,4,2,3}$ ⋮ $Y_{12,4,2,3}$	a_1 $Y_{1,1,2,4}$ $Y_{2,1,2,4}$ ⋮ $Y_{12,1,2,4}$	$Y_{..2.}$
Raw batch 3	a_3 $Y_{1,3,3,1}$ $Y_{2,3,3,1}$ ⋮ $Y_{12,3,3,1}$	a_4 $Y_{1,4,3,2}$ $Y_{2,4,3,2}$ ⋮ $Y_{12,4,3,2}$	a_1 $Y_{1,1,3,3}$ $Y_{2,1,3,3}$ ⋮ $Y_{12,1,3,3}$	a_2 $Y_{1,2,3,4}$ $Y_{2,2,3,4}$ ⋮ $Y_{12,2,3,4}$	$Y_{..3.}$
Raw batch 4	a_4 $Y_{1,4,4,1}$ $Y_{2,4,4,1}$ ⋮ $Y_{12,4,4,1}$	a_1 $Y_{1,1,4,2}$ $Y_{2,1,4,2}$ ⋮ $Y_{12,1,4,2}$	a_2 $Y_{1,2,4,3}$ $Y_{2,2,4,3}$ ⋮ $Y_{12,2,4,3}$	a_3 $Y_{1,3,4,4}$ $Y_{2,3,4,4}$ ⋮ $Y_{12,3,4,4}$	$Y_{..4.}$
Column means	$Y_{...1}$	$Y_{...2}$	$Y_{...3}$	$Y_{...4}$	
Treatment level means	$Y_{.1..}$	$Y_{.2..}$	$Y_{.3..}$	$Y_{.4..}$	Grand mean Y

1 = replicates; 2 = treatment; 3 = raw batch material; 4 = operator.

directions. With this in mind, we can interpret the rows and columns of the LSD as two restrictions on randomization. In general, a LSD for k factors ($k \times k$ square), is a symmetric matrix containing k rows and k columns. Each cell of the resulting k^2 matrix contains the k letters which correspond to the treatments (formulations) and each letter occurs once and only once in each row and in each column. In practice, this means that the levels of the one nuisance variable (batches of raw materials) are assigned to the rows of the square and the levels of the second nuisance variable (operators) are assigned to the columns of the square. The levels of treatment are assigned to the cell of the square. The randomization here is more complicated than in completely randomized designs. The number of rows and columns have to be equal to the number of treatment levels, in our case the four formulations of the fertilizer. With the Latin square design, we are able to test three hypotheses:

$$H_0\text{: } \mu_{1..} = \mu_{2..} = \mu_{3..} \text{ (treatment level population means are equal)} \quad (2.7)$$

$$H_0\text{: } \mu_{.1.} = \mu_{.2.}\ \mu_{.3.} \text{ (row population means are equal)} \quad (2.8)$$

$$H_0\text{: } \mu_{..1} = \mu_{..2} = \mu_{..3} \text{ (column population means are equal)} \quad (2.9)$$

The Latin square design model equation is

$$Y_{i,j,k,l} = \mu + a_j + b_k + g_l + e\ (i = 1, \ldots, n;\ j = 1 \ldots p;\ k = 1 \ldots, p;\ l = 1, \ldots p) \quad (2.10)$$

where μ is the grand population mean, a_j the treatment level effect, b_k the row effect due to nuisance variable 1, g_l the column effect due to nuisance variable 2 and e the composite error effect.

The reader may note that the squared composite error effect of the LSD is usually smaller compared with the randomized block design (RBD). Thus, as a consequence of the isolation of two nuisance variables, we accomplished a more powerful and efficient experimental design. Generally, the more efficient and powerful the design, the more restrictive are the statistic assumptions, and the more sophisticated the designs and the analysis of data.

The origin of the name Latin square design arises from an ancient riddle which deals with the question of how the letters of the Latin alphabet can be allotted to a square matrix, so that each letter appears only once in each row and each column.

Graeco Latin square design

Graeco Latin square designs (GLSDs) are used to control and eliminate three extraneous sources of variability simultaneously, by systematically blocking in three directions. A GLSD is achieved by superimposing two p^2 LSDs. The one LSD will use Latin letters for denoting the treatments, the second LSD will use Greek letters to denote the treatments. As in the LSD, the Greek letters must occur only once in each row and each column, and each Greek letter only once with each Latin letter. Such an allotment is called orthogonal and the resulting design is called Graeco Latin square. With the GLSD we are able to study four factors (rows, columns, Greek letters, Latin letters) simultaneously with only k^2 runs.

Incomplete block designs

We are not usually able to conduct experiments with all treatments run within each block. This may be because we do not have enough experimental and personal facilities, the size or capacity of the experimental units is not sufficient (remember the previous example with the agar plates) or there is simply a lack of money. A lot of effort has been invested in developing a valid and statistical methodology to reduce the number of replicates without losing accuracy in answering statistical hypothesis and identifying parameters of statistical models. These efforts resulted in randomized block designs, which did not have every treatment in each block. They are therefore called incomplete block designs (IBDs).

Balanced incomplete block design

If all treatments are equally important, as a prerequisite, then the treatment combinations in each block should be selected in a balanced manner. With this in mind, the experimenter should take care that any pair of treatments occurs the same number of times as any other treatment. Let there be k treatments. Every block may hold l ($l < k$) treatments. In general, a balanced incomplete block design (BIBD) is achieved by taking (l/k) blocks and allotting a different combination of treatments to every block. Mostly balances can be achieved with less than (l/k) blocks. For tables of BIBD the more interested reader may refer to Cochran and Cox (1957).

Youden squares, lattice designs and partially balanced incomplete block designs

All these designs have been developed to reduce further the number of treatment combinations, blocks and replications. Youden squares can be interpreted as symmetrical balanced incomplete block designs. They allow two sources of variation to be eliminated simultaneously. A Youden square design may be constructed by eliminating one row, one column or a diagonal of a Latin square design. The experimenter has to be careful not to eliminate more than one row, column or diagonal. The resulting square may have lost its balance, i.e. every treatment appears exactly once in each block and therefore the Youden square is not valid any more. Another advantage of the Youden square is its ability to eliminate the position effect. Since the treatments occur only once with each position and only once within each block, the positions are orthogonal to the blocks and the treatments. Lattice designs allow the number of replications to be reduced further. One of the main disadvantages of BIBD is that the number of replications is usually high (especially if there is a larger number of treatment combinations). In lattice designs, the number of replications of the treatments are flexible. Cochran and Cox (1957) give a detailed description of how to construct lattice designs. Partially balanced incomplete block designs no longer require that all treatment combinations occur the same number of times in each block. These designs require only a minimum number of replications and treatment combinations.

Factorial designs

Factorial designs provide the means to study the effect of two or more factors (treatments) simultaneously at two or more levels. They are based on completely randomized building block designs. Whether a variable is assumed to be a nuisance variable or a factor is mainly subject to the experimenter. As Kirk (1982) points out, a "nuisance variable is included in a design for the purpose of improving the efficiency and the power of the design. A treatment (factor) is included because it is related to the scientific hypothesis that an experimenter wants to test". Besides the literature dealing with proper experimental design, the reader should consult Mulaik (1972) for detailed discussions of the foundations of factorial designs.

Suppose a microbiologist wants to study the influence of temperature and water availability on the nitrogen fixation by a bacterium in agricultural soils. In this case we face two treatments (factors): soil moisture and temperature. In studying this problem, the experimenter has to think about the levels of the factors. If he chooses a priori discrete levels of the factor, the factorial experiment is said to be a fixed model factorial experiment. If the levels of both factors are determined at random, we speak of a random model type factorial experiment. If, for some good reason, the levels of one factor are determined at random, say the factor water content – one of the most variable factors under field conditions, which should, therefore, be determined at random – and the levels of the other factors are chosen as fixed levels, say temperature – assume that, below a certain temperature threshold, there is no fixation and, above this temperature, the factor may obey the Michaelis–Menten equation – we speak of a mixed model. The types of model used with factorial designs have their specific advantages:

- Fixed models have the limitations that the conclusions drawn are only valid for the

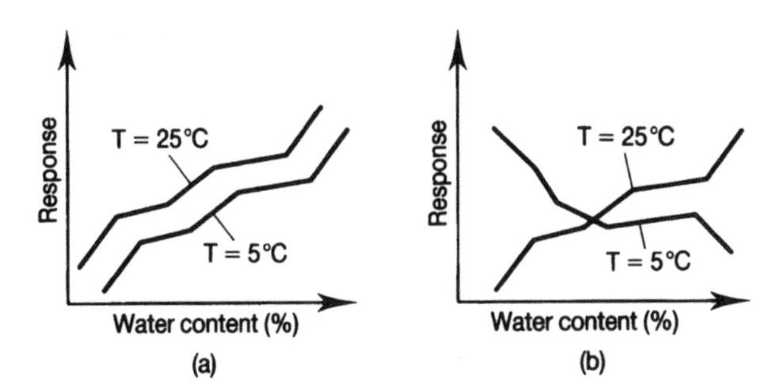

Figure 2.2. Schematic representation of interaction of factors. (a) No interaction is present; (b) interaction is present.

levels of factors for which they were checked but they are conducted with relatively little effort with respect to the combinations.

- Random models have the advantage that the conclusions drawn are valid over ranges of factor levels but require a greater effort compared to the fixed models.
- Mixed models, as one may assume, combine the advantages of both the fixed and the random models, and should be used in cases where details of one factor are known.

By using factorial designs, the experimenter is able to distinguish between effects that are changes in response to changes in the levels of one primary factor (main effect) and interactions. By interaction we mean that the response to the levels of a factor (A) depends on the level of a second factor (B) (Fig. 2.2).

Using the example mentioned previously, the factor or treatment A may be temperature and may be studied at two different levels, say a_1 = 5°C and a_2 = 25°C. Factor B is water content and may be studied at three different levels, say b_1 = 95% saturation, b_2 = 50% saturation and b_3 = 20% saturation. The resulting factorial design consists of six possible treatment combinations, i.e. a_1b_1, a_2b_1, a_1b_2, a_2b_2, a_1b_3, a_2b_3. There will be 48 experimental units and the six-treatment combination will be assigned randomly to these units with the restriction that each particular treatment combination is replicated eight times (Table 2.4).

Table 2.4. Factorial design. A schematic representation of an experimental design to measure the simultaneous effect of two or more independent variables (factors) at different levels is shown. *a* represents the factor temperature and *b* is soil moisture. Each column represents a specific factor-level combination, which is replicated 12 times. The indices of factors *a* and *k* indicate the levels of the respective factor. *Y* represents the response variable of individual factor combination (see Tables 2.1 and 2.3). The first index of *Y* represents the replicate, the second the level of factor *a* and the third the level of factor *b*.

a_1b_1	a_2b_1	a_1b_2	a_2b_2	a_1b_3	a_2b_3
$Y_{1,1,1}$	$Y_{1,2,1}$	$Y_{1,1,2}$	$Y_{1,2,2}$	$Y_{1,1,3}$	$Y_{1,2,3}$
$Y_{2,1,1}$	$Y_{2,2,1}$	$Y_{2,1,2}$	$Y_{2,2,2}$	$Y_{2,1,3}$	$Y_{2,2,3}$
.	.	.	.	.	.
.	.	.	.	.	.
.	.	.	.	.	.
$Y_{12,1,1}$	$Y_{12,2,1}$	$Y_{12,1,2}$	$Y_{12,2,2}$	$Y_{12,3,3}$	$Y_{12,2,3}$

With this factorial design we can test the following hypotheses:

$$H_0\text{: } \mu_{1a} = \mu_{2a} \text{ (treatment A population means are equal} \quad (2.11)$$

$$H_0\text{: } \mu_{1b} = \mu_{2b} = \mu_{3b} \text{ (treatment B population means are equal} \quad (2.12)$$

$$H_0\text{: } \mu_{j,k} - \mu_{j,'k} - \mu_{j,k'} + \mu_{j',k'} = 0 \text{ (all AB interactions are 0)} \quad (2.13)$$

The hyothesis on the interaction is specific for factorial designs.

The factorial design model equation is

$$Y_{i,j,k} = \mu + a_j + b_k + (ab)_{j,k} + e_i\,(j,k) \quad (i=1,\ldots,n;\ j=1,\ldots,p;\ k=1,\ldots q) \quad (2.14)$$

where μ is the grand population mean, a_j the treatment effect of level a_j, b_k the treatment effect of level b_k, $(ab)_{j,k}$ the interaction effect level for a_j and b_k, and $ei_{(j,k)}$ the error effect for subject i in treatment combination j,k.

Summarizing, we may consider the following advantages of factorial experiments:

- Efficiency: factorial experiments are more efficient than single factor experiments, i.e. they allow conclusions to be drawn with the same precision as single factor experiments but need smaller numbers of observations.
- Analysis: factorial experiments allow interactions between two factors to be identified. If the experimenter is interested in the main effect, he or she has to vary the levels of one factor while keeping the levels of the other factors fixed. They also allow valid conclusions to be drawn over a range of experimental conditions.

In the past, a large number of different factorial designs were developed and the statistical analysis for the general cases of k factors were derived. We will not discuss these in detail. The particular factorial designs most widely used are the $2k$ and $3k$ factorial designs. In the $2k$ factorial design there are only two levels of k factors present, say level 0 and 1, or low and high. The $3k$ factorial design allows for three levels of each factor. The $2k$ and $3k$ factorial designs are useful in the initial stage of experimental studies, when there are many factors that may have an effect and there is no data on the interactions of the factors.

For cases where it is impossible to perform a complete replication of the factorial design in one block, a special technique has been developed in analogy to the incomplete block design, which is called confounding. Confounding therefore is a method to design blocks with sizes smaller than the number of treatment combinations. When using a confounding technique, we lose information about the higher order interactions. They are indistinguishable from block effects and said to be confounded with the blocks. The higher order interactions have been lost, but the main effects and lower order interactions can now be estimated with a higher precision. Up to now we had to face the limitation that the factors had to have the same number of levels. This restriction is somewhat inconvenient, for example, if one wants to study the biodegradation of oil by microbes as influenced by light and oxygen availability. One possible scenario could be different oxygen partial pressure, say five levels, but with only two levels of light, light present or absent.

This type of factorial experiment is covered by asymmetrical factorial design experiments. Naturally, confounding techniques can be used with these types of experiment.

Split plot factorial designs have been developed for cases where one factor might require larger experimental units than other factors. For example, an agroforestry experiment wants to study the effect of different ways of cultivation (deep ploughing, hacking, etc.) in combination with the application of different fertilizers. In this case there are at least two factors: the cultivation technique and the fertilizer. It is not too difficult to apply different levels of fertilizer on to small neighbouring plots, but it is very difficult and inconvenient to apply different cultivation techniques to small neighbouring plots. To meet this requirement, split plot designs have been developed. In this, a larger plot (main plot) has first been set up to meet the factors requiring more space for the main treatment, and then this larger plot is divided into smaller sub-plots to meet the requirements of the other factors.

Criteria for selecting an appropriate design

The most important task for experimenters is to decide which design is the most appropriate for a particular problem. In general, there is no "decision recipe" which, on the basis of the experimenters' prescience and scientific expertise, will lead to a unique solution of these problems. The following list of criteria, which has been partially adapted from Kirk (1982), will, when followed carefully, offer some help in the decision process. However, the criteria have to be verified with the project design plan we introduced in the section entitled "From sampling to data interpretation" with respect of their suitability within the whole project plan.

Will the chosen experimental design provide the information to achieve the objective?

- Does the chosen experimental design provide representative data?
- What treatment levels should be used and should they be fixed a priori or at random from a treatment level population?
- Are the parameters measured really influencing the effect (factor of nuisance variable)?
- Are interaction effects assumed to play a prominent role?
- Are higher order interactions of any relevance (confounding)?
- Are all treatments and treatment levels of equal importance?

Will the proposed sample population be sufficient to answer the problem with accuracy and reliability?

- Do the experimental units represent a random sample of the grand population?
- Does blocking lead to an increase of homogeneity within blocks?
- Have all the interacting parameters been checked?
- Is there any data available on heterogeneity of influencing parameters?
- Does the experimental set up provide for an observation of the units under different levels?
- Is the application of the experimental design independent of the experimenter?
- Is the experiment applied reproducible?
- Are the risks of committing type I and type II errors considered appropriately?

Is the proposed experimental design sufficient and economic?

- Are sampling plans the means to reduce sample size?
- Is efficiency improved by blocking or by randomly assigning a large number of experimental units to the treatments?
- Does control of nuisance parameters increase efficiency or has the sample size to be increased?
- Can regression techniques increase efficiency with respect to parameters, which influence dependent variables?
- Can efficiency be increased by reducing the time used to plan a more sophisticated experimental design, when conducting a simpler design with a higher number of replicates?

The missing value problem

One of the main problems occurring with experimental designs is the lack of observations (data) during data acquisition. The reasons for this phenomenon are numerous: carelessness of the experimenter, a defect in analytical units, and defects in and unavoidable damage to experimental units. Missing observations in an experiment introduce serious problems for the analysis and therefore for the interpretation of the data: we no longer face the situation where every treatment occurs within every block (non-orthogonality). Two principal approaches are common in dealing with the missing value problem. The first approach uses an estimation of the missing value for the following analysis of variance and the second consists of an exact analysis of the missing observation. We will not discuss the missing value problem further here but further details can be found in Dodge (1985).

Spatial sampling and analysis

Sampling is the proper method of collecting specimens in order to obtain usable data for statistical analysis. Sampling is the interface between problem formulation and data analysis. Obviously, the statistical data analysis and data interpretation must depend heavily on the sampling method used and vice versa. Therefore, a lot of effort should be invested in the construction of an appropriate sampling plan in order to obtain representative, reliable and indicative data.

Unfortunately, our daily experience with sampling water and soil exemplify the need for quality assurance tools with respect of the applied sampling method. Objective of a "proper" sampling design should be the development of a sampling procedure, suitable to answer the statistical hypothesis with respect to the problem statement. Typical distributions of costs between project planning, sampling, chemical analysis and data interpretation may serve as an indicator for deficient realization of a proper project design plan, illustrated by the distribution of investments for each step in the project design plan. As we learned, quality assurance starts with the formulation of "quality aims" and leads to the development of practical sampling methods by adapting general models. The objective of quality assurance is the derivation of the methodology, which enables the practitioner to set up a reproducible, site- and substance-specific optimal sampling plan taking into account any available data, the problem statement, and the practicability with respect to time, cost and labour.

The sampling problem has to be faced not only by the surveyor who wants to gain knowledge on the spatial and temporal distribution pattern of a number of certain parameters, but also by the experimenter who studies the effect of a number of different factors at different levels on a certain parameter. Neither surveyor nor experimenter is capable or willing to analyse the entire survey site or experimental unit. Therefore, both are forced to draw a limited number of specimens and consider the representativity and reliability of their sample population. The most general, closed and comprehensive theory on sampling particulate matter was presented first by Dr Pierre Gy. Gy's theory on sampling particulate matter introduces a unique methodology on how to select and draw a correct, accurate and precise sample, how to estimate the related errors, and how to calculate the "real" value of the parameter under consideration. Pitard (1992) presented an overview of Gy's theory addressed primarily to the practitioners in the sampling field.

In the following we will discuss two principal sampling strategies: rigid (fixed) sampling plans and algorithm-orientated sampling plans. The criteria for the selection of a particular sampling plan, whether rigid or algorithm orientated, are derived from the prescience and the problem statement. Scholz et al (1993) provide a list to help determine which sampling plans should be applied for soil surveys and discuss different sampling plans. They recommend the use of rigid plans to assess the relevant parameters of contamination, and algorithm-orientated sampling plans to verify the origin and distribution of contaminants in the soil environment.

A number of publications deal with the problem of proper sampling, e.g. Cochran (1977), Gy (1982, 1989) and Provost (1984).

Rigid sampling plans

Rigid sampling plans are used for the detection and determination of relevant parameters, for example, contaminant concentration, maximum or minimum concentrations, or ranges of concentrations. Rigid sampling plans originate from general screening plans like random or square sampling plans. These should be adequately adapted and modified with respect to the problem statement. The rigid sampling plans are subclassified into proportionate and non-proportionate sampling plans.

Proportionate sampling plans

Proportionate means that each equal-sized parcel (allotment) is sampled the same number of times as every other parcel of a screen.

Rectangular proportionate sampling plans

These sampling plans are the simplest with respect to the practical procedure (Fig. 2.3). The screen elements are quadrates. The sampling spots are the intersections of a rectangular screen of the survey site. They are widely

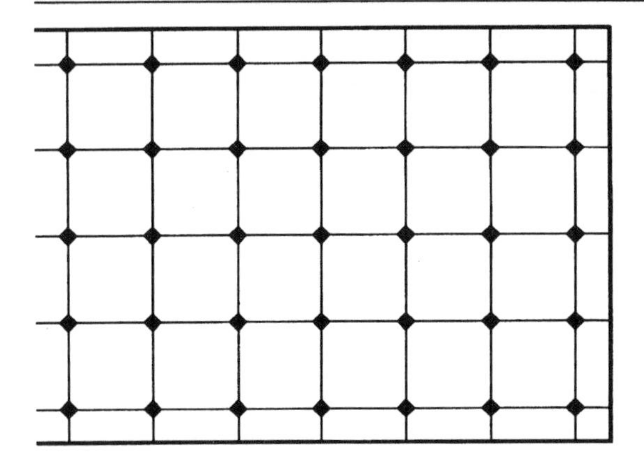

Figure 2.3. Rectangular sampling plan. A schematic representation of an orthogonal, rectangular sampling plan is shown. Screen intersection points indicate sampling spots (solid diamonds).

used and recommended for use in new situations where previous data are not available.

Non-rectangular proportionate sampling plans

The rectangular screen is not the only proportionate screen (Fig. 2.4). Scholz et al (1993) proved the optimality of the screen of non-rectangular proportionate sampling plans for a number of different problem statements. The screen elements are equilateral triangles.

Non-proportionate sampling plans

Random sampling plans

Random sampling plans (Fig. 2.5) are often used. The term "random" indicates that each

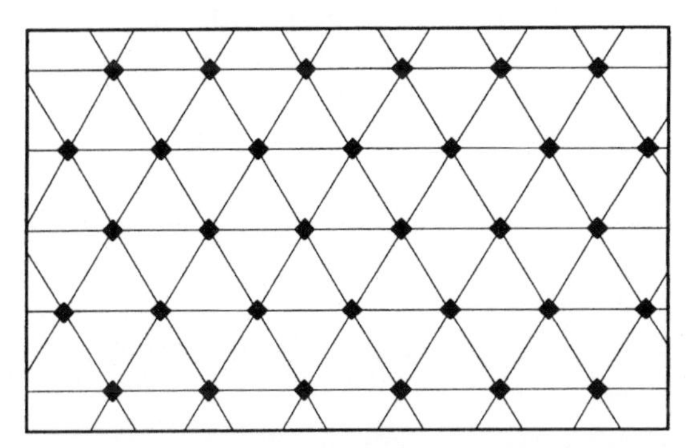

Figure 2.4. Non-rectangular sampling plan. A schematic representation of a non-orthogonal, rectangular sampling plan is shown. Screen intersection points indicate sampling spots (solid diamonds).

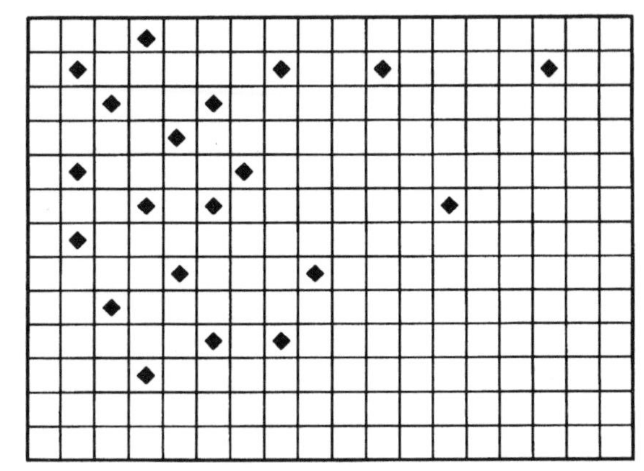

Figure 2.5. Random sampling plan. A schematic representation of a randomized, orthogonal sampling plan is shown. A screen is used to divide the area into fractions of equalized units. According to a randomization procedure, a number of units are chosen for taking samples (indicated by solid diamonds).

spot of a survey site or experimental unit is equally likely to be sampled, and that each sample is independent from the previous and successive sample. Consequently, it is possible that the same spot may be sampled more than once. Scholz et al (1993) classified random sample plans as "suboptimal" and disadvantageous for the following reasons:

- Random sample spots have to be fixed by using suitable maps
- Irregular topography (elevation) increases the difficulty in identifying random sample spots
- They result in a different density of sampling spots.

Efforts have been made to cope with the problems of random sampling plans. One is to divide the entire survey site into different plots and to assign the sample spots randomly within these plots (stratified random sampling). The other is to fix the sample spot *y* coordinates at random along the *x* axis and the *x* coordinates at random along the *y* axis within a subdivided survey site (unaligned random sampling).

Polar sampling plans

Polar sampling plans (Fig. 2.6) should be applied in situations where point sources are

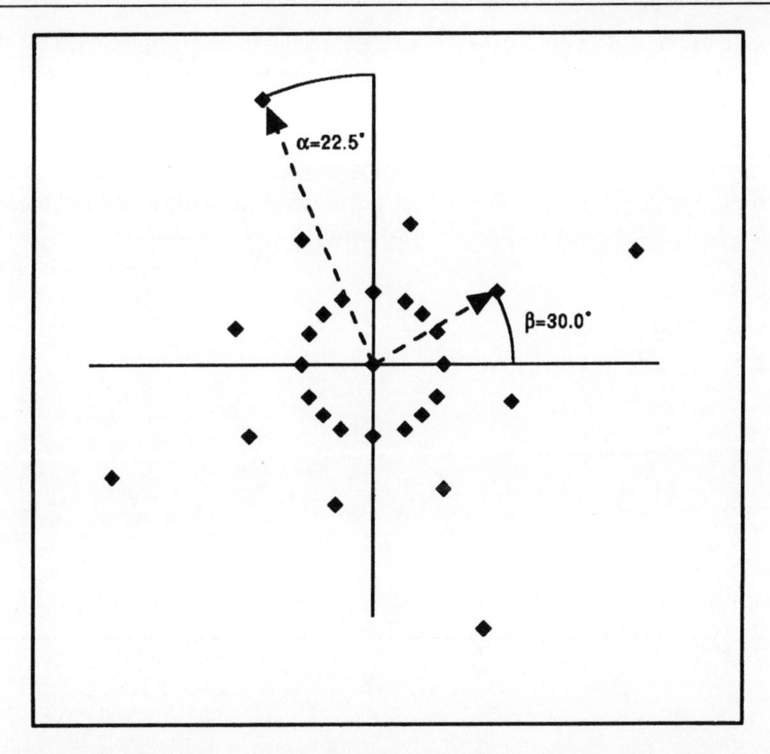

Figure 2.6. Polar sampling plan. A schematic representation of the construction of a polar sampling plan is shown. Three concentric circles make up this sampling scheme. A modification is achieved by selecting samples not according to the direction of the radius, but by rotating each circle by a certain angle with respect to the previous circle. The sampling spots are indicated by solid diamonds.

assumed to be the origin of a contamination. They not only provide information on the contamination level (maximum concentration) in the centre but also information on the distribution pattern. The polar sampling plans should be limited to the cases where heterogeneity plays a minor role and where there is no preferred direction of distribution. For these cases the assumption that the spreading is equally likely in all directions holds true and the concentration may gradually decline with distance to the centre.

Linear sampling plans

Linear sampling plans apply with situations where, for example, leakages along pipes may be identified.

Diagonal sampling and related plans

This type of sampling plan is widely used for agricultural and survey issues because of their easy application. Scholz et al (1993) discuss a number of points to be considered with the application of diagonal sampling plans:

- They should only be applied when proportionate distribution of the parameter homogeneity of sample site is required
- There should be more than one sampling diagonal, the supplemental sampling
- The diagonal should cross or parallel the main sampling diagonal
- There should be equidistant sample points on each diagonal.

Algorithm-orientated sampling plans

In order to identify sources for contamination, the distribution dynamic, the distribution pattern and/or the fate of a substance in the environment, the solitary application of a rigid sample plan is not sufficient. Therefore, specific plans have to be developed that can cope with the complexity of these types of problem. The term algorithm-orientated sampling plan is simply the general principle of project design plan we introduced in the previous section in this chapter on project design strategies. With the application of these strategies in combination with statistical experimental design theory and a specific sampling plan, the experimenter or surveyor is provided with a powerful, state of the art methodology to cope with a variety of different problem statements.

Mathematical and statistical tools

K. Mathes
L. Ries

In applied soil microbiology and biochemistry, statistical methods that are suited to describing, analysing and evaluating the information and quality of field data are important. The purpose of this section is to present a guide to the principles and options for statistical analysis methods in such studies or experiments. To assist in the selection of suitable methods, it is not intended to give a complete review of recent statistical literature, but some of it will be organized and made accessible.

Some principles for the selection of statistical analysis methods will be discussed and concepts relevant to the following sections will be introduced. Statistical analysis is used to provide criteria to assist in decision-making when uncertainty is introduced by random errors. Firstly, the distinction between random error and bias must be kept in mind: bias is the systematic error, this means it is the magnitude and drift of the deviation from the value intended to measure. Statistical methods can only cope with random errors. "No amount of statistics will reveal whether the pipette used throughout an experiment was wrongly calibrated" (Green 1979). Secondly, statistical analysis cannot eliminate the uncertainty resulting from random errors; however, it can quantify it. In general, increasing the sample size or number of experimental units improves the precision of the results, but at some sample sizes the improvements are no more important in relation to the efforts necessary to answer the question under consideration.

Principles for the selection of a suitable statistical analysis method

The problems encountered and the questions asked can be organized into the same framework independent of whether one studies, e.g. microbial activities, biomass, litter decomposition, or the content of organic or inorganic matter in the soil. An important characteristic of this kind of data is that they are mostly quantitative (interval) and therefore may be described by continuous distributions. Other types of data are ranks, preferences (ordinal) or categories that cannot be ordered by the inequalities greater than or less than (nominal). The measurement scale is one criterion that guides the choice of the appropriate statistical analysis method.

How does one find the methods suitable for the question under consideration? This will be introduced by examples, e.g. we wish to determine how quickly litter decomposes on a field site at 10–20 cm soil depth. n litter bags are placed randomly on the site at the respective soil depth. After 2 weeks the bags are retrieved, the decomposition rate of each is determined and the mean value is calculated. In order to answer the question we must estimate an interval within which we are reasonably confident that the true but unknown mean μ lies (see section entitled "Estimating the unknown population mean"). In some cases it may be sufficient to display the standard deviation in connection with the estimated mean. If the sampling and measuring procedure described above is repeated in an appropriate manner every 2 weeks over 5 years, a time series of the mean decomposition rate and the respective confidence limits are obtained. In doing this, we are further interested in the variation of the decomposition rate in time. Mostly, the aim is to identify seasonal, cyclical, trend and irregular patterns (see section entitled "Time series analysis"). Another problem may be to estimate the spatial distribution of the litter decomposition rate of the site. In order to achieve this, bags are dispersed as discussed in the section on spatial sampling and points not measured are interpolated by the methods described in the section on evaluation of spatial data sets. Statistics that utilize spatial and temporal data simultaneously are currently in development and the reader should refer to Cressie (1991) and Haas (1993). Up to now our focus has been on univariate data gained in a survey study of one site or region.

Methods for establishing differences between field sites are reviewed in the section entitled "Establishing differences between field sites in the univariate case". The litter bag example can be used to determine whether the decomposition rate differs between two sites. n litter bags are dispersed randomly at each site at the same soil depth, they are retrieved after a month, and the data obtained. The confidence interval is estimated for the difference of the decomposition rate or a statistical test is applied to see whether there is a significant difference between decomposition rates at the two locations. If treatment effects need to be identified, such "unreplicated experiments" are not sufficient. The aim now may be to decide whether litter that contains residues of an insecticide will decompose more slowly than untreated litter of the same kind at the same soil depth. n bags of untreated litter are placed at random within a 10 m^2 plot (U) and n bags of treated litter are placed at random within a second "identical" plot (T). This kind of experimental design is called pseudoreplication (Hurlbert 1984) because the treatments are segregated and not interspersed. The only hypothesis testable by this is that untreated litter at location U decays at a different rate than treated litter at location T. The supposed "identicalness" of the two plots almost certainly does not exist and the experiment is not controlled for the possibility that the seemingly small initial dissimilarities between the two plots will have an influence on decomposition rate. Nor is it controlled for the possibility that an uncontrolled extraneous influence or chance event during the experiment could increase the dissimilarity of the two plots. If the test result is generalized to the 10–20 cm soil depth of a certain class of fields, two sets of litter bags must be distributed in some randomized fashion over all, or a random sample, of these fields. Sampling designs that can be used are described in the section on experimental designs and sampling strategies.

In applying the correlation and regression methods reviewed in the section entitled "Quantifying relationships between responses and predictors", the aim is to quantify the relationship between responses and predictors. We consider first the bivariate case where methods are generally applied to a sample of bivariate observations (x_i, y_i) $(i = 1,2, \ldots, n)$. For example x_i may represent different temperatures at which n_i litter bags are incubated in the laboratory and we wish to analyse the relationship between temperature and litter decomposition.

Mostly a linear relationship is assumed. Concerning the example our hypothesis may be that

$$\text{Litter decomposition} = a + b \times \text{temperature} \qquad (2.15)$$

and our aim is to estimate the parameters a and b as well as to judge the precision and certainty of these estimates. If the relationship between a response variable and more than one predictor variable is to be identified and estimated, multiple regression methods must be applied (see the section entitled "Quantifying relationships between responses and predictors").

In the litter bag example, our hypothesis now may be that

$$\text{Litter decomposition} = a + b \times \text{temperature} + c \times \text{moisture} \qquad (2.16)$$

and the influence of the two predictor variables, temperature and moisture, on litter decomposition is analysed simultaneously. The statistical methods of regression and correlation analysis are also used in calibration problems, time series analysis and gradient analysis. For example, the non-parametric calibration or inverse regression, a problem in which one uses regression to estimate *x* corresponding to later observed *y* is the subject of a paper by Knafl et al (1984).

In the section "Structuring of multivariate data sets by descriptive and explanatory analysis" multivariate descriptive methods are introduced. Multivariate means that more than one "response" variable of interest is considered by the described ordination and clustering procedures. Ordination will produce a reduced number of variables. Clustering methods will produce groups with differing but internally homogeneous biological characteristics. In general, neither ordination nor clustering techniques use knowledge about the source of each sample. Thus, one can examine the outcome for evidence that samples within a site are placed nearer to each other in the ordination, or cluster more often in the same groups, than would be expected by chance. For example, the aim may be to investigate the homogeneity of the soil nitrogen balance. This means that different variables have to be analysed simultaneously. Then clustering or ordination methods may be applied in order to judge the homogeneity of different plots or sites by looking at all variables of the nitrogen balance simultaneously.

In summary, besides organizing and summarizing data, multivariate methods can be used to analyse the spatial heterogeneity and to generate hypotheses from the observed spatial pattern. Whereas ordination and clustering techniques do not make assumptions as to how the underlying populations are distributed, this information is essential for the selection of appropriate estimation and test methods. It should be emphasized that any inferential statistical analysis method assumes certain things about the data. This influences the estimation and test results, and the most efficient statistical method should be selected.

An efficient method will be as conservative, powerful and robust as possible. It is conservative if it has a low probability (α) of making an error of the first kind – concluding that there were biological effects of the impact or treatment when in fact there were none. If it is powerful, it will have a low probability (β) of making the error of the second kind – concluding there were no effects when in fact there were. If it is robust, the error levels will not be seriously affected even if the data does not satisfy the assumptions. Especially microbiological field data often do not satisfy the assumptions of classical statistical analysis. In this case, there are basically three possibilities for continuing:

- to use non-parametric (or distribution-free) statistical methods
- to transform the raw data in order to reduce the violations and their consequences to an acceptable level
- to proceed as if the assumptions were fulfilled but in the knowledge that this will affect the tests of significance and estimation of parameters.

The way the choice is made will depend on the particular problem and kind of data. Information to assist in deciding which possibility is the most appropriate is provided in the individual sub-sections. However, it should be mentioned that in recent years there has been an increasing interest in non-parametric methods because the assumptions needed for their validity are so broad that they apply to data that vary widely in their characteristics. This applies particularly to field or laboratory data that are collected in surveys, but perhaps less to controlled experimental data. It is clearly intuitive that the more relevant information you use, the better a testing or estimation procedure should be in terms of its efficiency, but "if there is any doubt about distributional assumptions, we should not lose much efficiency, if any, by using the most appropriate non-parametric test" (Sprent 1990).

Furthermore, non-parametric methods are usually the only methods available for use with data that specify only ranks or proportions and not quantitative observational values. It must be stressed that weaker assumptions do not mean that non-parametric methods are assumption free. What can be deduced depends on what assumptions can validly be made. Another criterion for categorizing the statistical methods is the kind of sampling in time. Sequential statistics exist (Sen 1981; Bauer et al 1986), but they are not often applied in soil microbiology and biochemistry. The principal advantage of sequential sampling is that, in the actual study, sampling continues until a decision is made so that prior estimation of the adequate sample number is unnecessary. "On the one hand, sequential sampling is very flexible because designs can be based on virtually any sampling distribution, such as the normal, Poisson, binomial, or negative binomial. However, a given design is useful only so long as the sampling distribution on which it is based does not change" (Green 1979).

Analysis of univariate data gained by descriptive field sampling

Estimating the unknown population mean

In soil microbiology and biochemistry, to estimate the true mean value of a variable, such as enzyme activity, the sample units are often mixed. In doing so, no information concerning the dispersion of the values of the single sample units is available, and therefore it is not possible to evaluate the accuracy and certainty of the point estimate. As long as it is not known how many sample units are necessary to estimate the mean with sufficient precision, the sample units should not be mixed. Inferring mean values for particular variables within a given experimental plot or field site implicitly means interpolating values for all points not measured. As long as assumptions regarding sample independence and normality are met, methods for estimating the mean (method of moments: $\bar{x} = 1/n \times \Sigma x_i$) and its confidence interval are readily available (Kachigan 1986; Rasch 1987; Köhler et al 1992). Confidence intervals give an indication of the accuracy and certainty of the point estimate by providing limits for it. Showing the point estimate, the coefficient of variation ($s/\bar{x}$ as estimate for the population ratio σ/μ) and the confidence interval as well as the sample size used is the most clear and meaningful presentation of the results and their quality.

Unfortunately, distributions of many chemical, physical and microbiological properties of soils appear to be skewed to the right, and hence are better approximated by the lognormal distribution. Parkin et al (1988) assessed the efficiency of three methods to estimate the mean, variance and coefficient of variation (CV) of lognormal data. Three test lognormal populations were used in the evaluation with coefficients of variation that span the range seen for many soil variables (CVs of 50%, 100% and 200%). They found Finney's method (1941) was the best for estimating the mean and variance of lognormal data when the coefficient of variation of the underlying lognormal frequency distribution exceeds 100%; below this value the extra computational effort required to implement Finney's technique provides little, relative to the method of moments. In order to obtain confidence limits, the lognormal data can be transformed to normality and the respective parametric estimation method can be used. Another possibility of handling non-normal data is to apply non-parametric estimation methods. Sprent (1990) and Büning and Trenkler (1978) describe methods for constructing confidence intervals for the median. If the sample comes from a population with a symmetrical distribution, the mean and median coincide. In this case we have a non-parametric method to estimate the confidence limits for the population mean also. The situation is more difficult if we cannot assume a symmetrical distribution. The location parameter "median" is based on ranks and known as a robust estimator of location but it is often very difficult to interpret the biological meaning of the median. Therefore, if none of the methods described above is applicable, we recommend data transformations in order to get a normal or at least symmetrical distribution. Then we can calculate the confidence limits in the transformed units.

Up to now the aim has been to estimate the mean of the variable of concern. Another aim may be to describe the underlying distribution. Then, besides measures of dispersion and the location parameters "mean" and "median", the "mode" is of special importance if multimodal distributions are possible. By displaying all these three location parameters an impression is obtained of whether a unimodal distribution can be assumed and whether it is skewed or symmetrical. This will help in deciding how to estimate the population mean and its confidence limits.

Evaluation of spatial data sets

An important point in the evaluation of soil data sets is the spatial analysis and display of the data. In this field the framework of geostatistics has created valuable tools for general use.

What is geostatistics?

In the present context the application of geostatistics means to use spatial correlation in a data set under investigation for sampling, measurement network design and spatial prediction (interpolation). Note that all examples in this section are shown for areal data sets. However, geostatistical methods generally can also be used for higher dimensional problems (see Haas 1993).

Before giving a detailed explanation, a short introduction will be given about the development of geostatistics, which is a relatively new discipline in statistics: the French mathematician G. Matheron (1971) became known worldwide for his "Theory of regionalized variables and its applications". However, the roots of geostatistical methods go back to the 1940s, when engineers, mathematicians, physicists and biometricians worked out independently from each other the basics for this statistical model. A concise introduction is given by Cressie (1990). For the original literature, see Gandin (1963), Kolmogorov (1961), Krige (1951), Matern (1960), Wiener (1949) and Yaglom (1962).

The geostatistical approach enables the soil scientist to analyse the data of a sampling in the following way:

1. For the process under investigation, the maximum distance (called the range) for which spatial autocorrelation exists should be determined. This can provide the answer to the question concerning the maximum distance of two sampling points.
2. Whether a sampling consists of redundant data or has a lack of samples should be determined. In this way the geostatistical analysis enables the sampling design to be optimized (see Fig. 2.7, γ_4 and γ_3).
3. In a further step the analysis of spatial autocorrelation can be used for a spatial prediction of the data under investigation. That means that this technique can be used for predicting (respectively interpolating) maps with the soil data.

Steps 1 and 2 can be done with a relatively simple morphological analysis of the empirical function of the spatial covariance, called variogram. The variogram characterizes the spatial variances of the data set dependent on distance and direction. Thus variogram analysis is an elementary tool for statistical analysis. In Fig. 2.7 a short example for a morphology of variograms is given in the following way.

With typical forms of empirical variograms, geostatistical problems can be classified as follows:

- γ_1 shows no variation in distance zero and therefore a good degree of autocorrelation for spatial interpolation.
- γ_2 shows a sufficient degree of autocorrelation for interpolation, however, the interception, called the nugget effect ($C_0 > 0$), indicates an additional (random) source of variance either in the samples, on which can hint at errors in the applied sampling

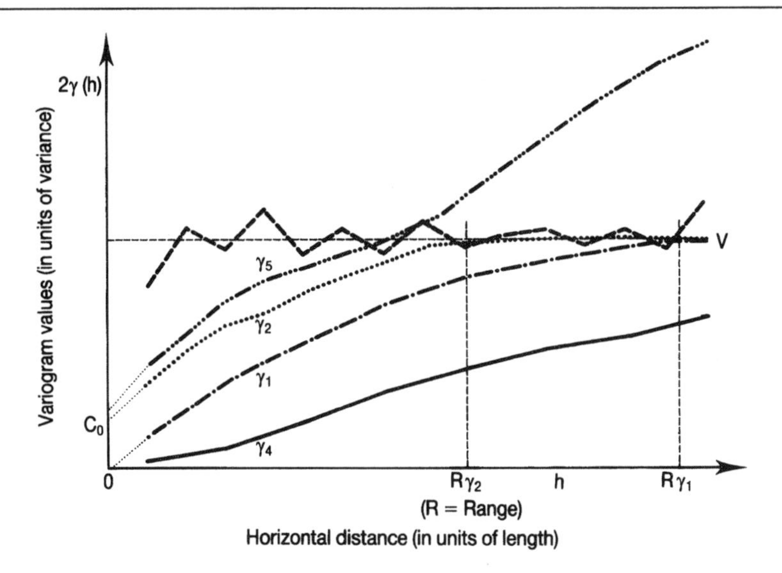

Figure 2.7. Short morphology of variograms.

technique, or the laboratory or measurement method. This also can mean that small-scale variation exists in the data.

- γ_3 points to a plain nugget effect, which indicates no autocorrelation in the data. This is called a random variogram. It means that no spatial covariance can be used for interpolation. In future, measurement points will have to be taken closer together.
- γ_4 is an overdetermined variogram. This occurs when redundant measurement stations exist. In this case the density of measurement points can be reduced.
- γ_5 can indicate additional spatial trend in the data.
- Range, another valuable property of the variogram is that the maximum distance of spatial dependency can be determined. This distance is called the range, *R*. It can be estimated when the variogram meets the horizontal line of the sample variance of the data set.

A practical introduction to variogram analysis, including the morphology of empirical variograms is presented by Akin and Siemes (1988). The definition of the variogram (Cressie 1986) and the instruction about the general structure of a variogram (Cressie 1987) provide information about the theoretical background. Finally, the introductions to variogram analysis by David (1979) (with sources for computer programs) and Clark (1979) should be noted.

The task of **spatial prediction** which is mentioned in step 3 can be fulfilled under two different constraints as follows.

First case: for a spatial prediction, the stable mean of the values is of major importance. This is the case when, for example, volume integrals have to be determined. This approach can be used when the data reproduce continuous processes (e.g. diffusion) in the soil. The triangular surface in Fig. 2.8 gives the impression of data with a typical form of continuous variation. In nature such data can be found, for example, when data for chemical substances in a homogeneous still water sediment have to be used for spatial prediction. More precisely, such data could be generated by a spherical diffusion process with spatial random distribution (which is only one example for several different possibilities of continuous variation). The technique for spatial prediction of this kind of data is known in geostatistics as **Kriging.**

Second case: investigated processes in soil consist of structural discontinuities and local variations without or besides a continuous spatial variation. Therefore, the technique of Kriging is inappropriate because of its strong

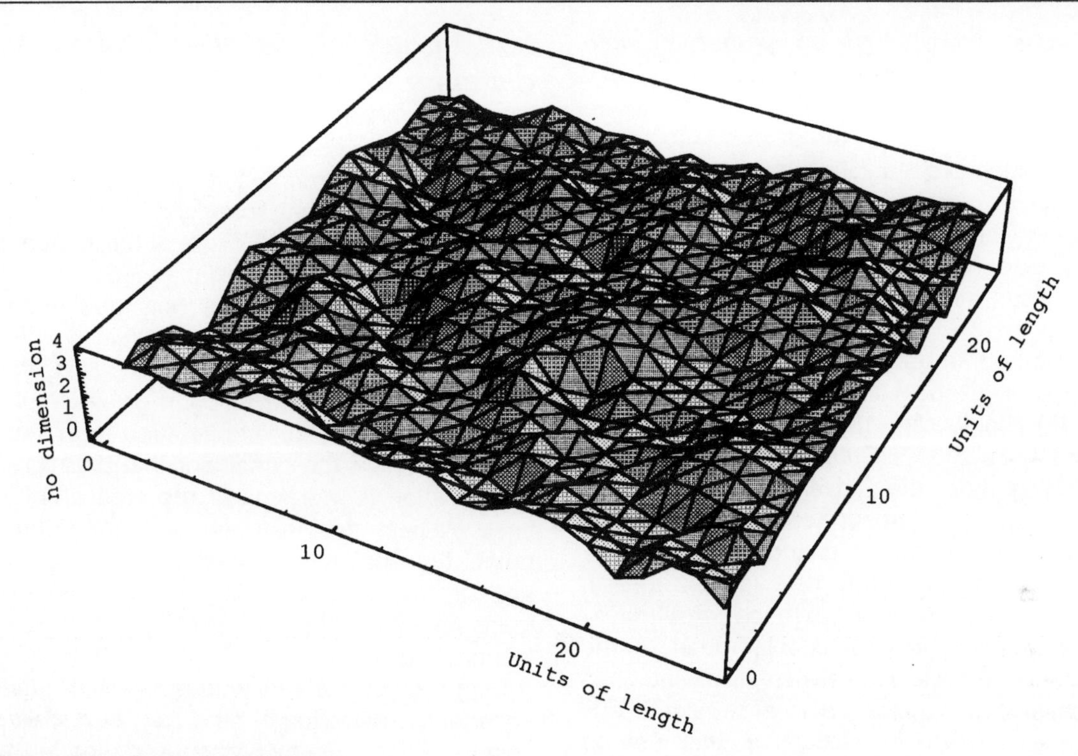

Figure 2.8. Synthetic test data.

smoothing effect. With the use of Kriging these local variations in the data are treated like errors and taken away in a predicted map. In this case the application of IDW+LOGRAN can help the soil scientist. The IDW technique, which means inverse distance weighted interpolation with moving averages, has the general ability to provide better spatial differentiation in interpolated results, especially in the case of sparse samplings. Together with the local gradient analysis (LOGRAN) technique for selecting more homogeneous samples for spatial prediction IDW+LOGRAN, it possesses the algorithmic potential for detection and display of higher local variations in a spatial prediction. The triangular surface plot of copper concentrations in soil shown in Fig. 2.10 gives an impression of the discontinuities that often occur in environmental data. It can be stated that, in environmental sciences, spatial processes with originally strong local variations and structural discontinuities are nothing extraordinary.

Geostatistical analysis and spatial prediction

In practice, a geostatistical analysis is a stepwise process for evaluation and interpolation:

1. An empirical function, which shows the dependency of spatial covariance to distance, is calculated from the data set under investigation. This function is called the **empirical variogram.**
2. By regression a continuous function is approximated to the empirical variogram. This function is called the **theoretical variogram.** In the case of direction-independent (isotropic) data, one overall variogram is modelled. With anisotropic data for several directions, several specific variograms have to be found.
3. The theoretical variogram is used for the calculation of distance- (and direction-) dependent interpolation weights for interpolation.

4. Spatial predictions are performed with Kriging, using steps 1–3.

Kriging

Interpolation procedures, which calculate the weights as shown and simultaneously minimize the error variance of the interpolation are called Kriging predictions, named after the African mining engineer Krige (1951) who developed basic ideas and applications for this method. In general statistical terminology Kriging is the best linear unbiased predictor (BLUP) (Goldberger 1962) applied on spatial data (spatial BLUP) (Cressie 1990).

Kriging has the advantage of creating smooth interpolated surfaces with a relatively high stability. It avoids distortions caused by spatial clustering of the data (screen effect). This provides the right approach for all questions that look for the stable mean of the interpolated data. Also, when mass integrals of measured substances in area-related soil volumes have to be calculated, this method should be used. Kriging requires normal distributed data or data sets that have been transformed to normality. Because of the basic idea of the method, minimum and maximum peaks of data are smoothed. In fact Kriging works like a low band filter (higher frequencies of variations in the original measurement data are taken away in interpolated values).

This model works well as long as spatially smooth and continuous processes (e.g. diffusion in soil or in water sediment) have to be modelled by spatial prediction. But, on the other hand, there are numerous cases where a higher degree of locality in spatial predictions is requested in scientific research, for example, the monitoring of high substance concentrations, high temperatures, high and low groundwater speeds, or high population numbers of vertebrates in a soil area. All these examples have in common a specific interest in the tails of the distribution of the predicted values. In this case, another method for spatial interpolation and investigation can be applied.

Local gradient analysis with inverse distance weighted interpolation

The local gradient analysis (LOGRAN) together with inverse distance weighted moving averages (IDW) forms a solution in the following way. An algorithm scans the spatial variation of the sampled data. Right at the locations where a spatial prediction is planned, the local gradient analysis determines local gradients in the form of normalized numerical values, expressing the degree of variation in this point. Then a mapping function gives circular areas of influence for interpolation from local gradients under the constraint that the higher the variation is, the smaller the area of influence will become. Finally, for each preselected point, the spatial prediction is done by the use of the specific radius for selection of measurement points in the neighbourhood (subsamples).

With the LOGRAN technique, areas of influence for the selection of data can be modelled using the dependency of local variation in the measured data. With increasing variation in the data, the size of the area of influence becomes reduced. By using the variable interpolation radii, the input data for spatial prediction become more homogeneous compared to selection by constant interpolation radii (or constant ellipses in the case of anisotropy of the data), which are generally used for Kriging or IDW. Spatial predictions under use for the varying areas of influence described have a higher degree of locality, which will be shown later in the methodical comparison of Kriging and IDW+LOGRAN. IDW+LOGRAN consists of the following steps:

1. At first, an initial approximation with simple IDW is calculated for the desired final square grid geometry. In this situation the relatively high methodical stability of IDW interpolation is an advantage (see Eckstein, 1989, concerning the stability of the IDW interpolation). This is of special importance because the following modelling of circular areas of influence depends on the quality of this first step.
2. Then for each grid square the local gradient analysis computes an individual value of the

local gradient (which is normalized between 0 and 1). The local gradient gives a numerical value for the degree of variation between this point and its surrounding neighbourhood.

3. Considering the empirical variogram, a maximum interpolation radius is found and, from the empirical distribution of the Euclidean pairwise distances of measurement stations, a minimum interpolation radius is determined.
4. By using a mapping function from local gradient to radius for each location, an individual interpolation radius is found under the constraint that the higher the gradient is, the smaller the area of influence (radius) for interpolation becomes.
5. Finally, the IDW interpolation is applied again using varying areas of influence (radii) for interpolation (IDW+LOGRAN).

Before showing the results of a methodical comparison in the next section, the basic principles of IDW interpolation have to be explained with reference to the relevant literature.

The IDW interpolation does not minimize the variance of the prediction error like Kriging. It uses a constant function for calculation of interpolation weights, whereas Kriging uses a varying function as a result of the regression of the empirical variogram. Nonetheless, both methods are unbiased, and produce balanced positive and negative prediction error sums. The means of the predicted values are practically the same.

Both methods are exact interpolators (in a location with a known value, the predicted value is identical to the measured value) and both methods are linear. They calculate the predicted value from linear combinations of interpolation weights and the values of neighbouring points. So IDW could also be expressed as spatial unbiased linear predictor (**spatial LUP**). The IDW procedure is described in detail by Ripley in his book about spatial statistics. An easy to read introduction to applied geostatistics with little formal requirement has been written by Isaaks and Srivastava (1989) and *Statistics for Spatial Data* (Cressie 1991) is an up to date standard textbook with more theoretical background. The method of local gradient analysis is explained in detail by Ries (1993) and applications are shown in *Chemometrics and Intelligent Laboratory Systems* by the same author (Ries 1995).

Methodical comparison of Kriging and inverse distance weighted moving averages (IDW and IDW+LOGRAN)

The following methodical comparison is theoretically made using direction-independent (isotropic) data for which interpolation radii are applied. For methodical comparison these data have been applied to a method called predictive sample reuse (PSR), which is also known as "leaving one out method" or cross-validation. Given a set of n spatial data, the first to the n-th data point is left out. Then at the location of the data point that has been left out the value of this data point is predicted using the rest of the values with the method under comparison. In this way n calculations are made. Finally n data pairs of the measured value Z and the predicted value $\hat{Z}$ exist. Then the absolute difference $(\hat{Z} - Z)$ as a percentage of Z is computed and defined as percentage prediction error. In this section all figures of the PSR validation show graphs of the cumulative percentage prediction error.

Further, it should be noted that, despite the fact that all the examples shown in the following paragraph are structured in square grids, the methods compared can generally be used for each kind of spatially irregular distributed data also. The different characteristics of spatial prediction presented will not change.

At the beginning of the methodical comparison, the prediction techniques are tested with theoretically ideal data for Kriging. The triangular surface shown in Fig. 2.8 shows synthetic spatial data with continuous variation. These data have been generated with the random coin method according to Sironvalle

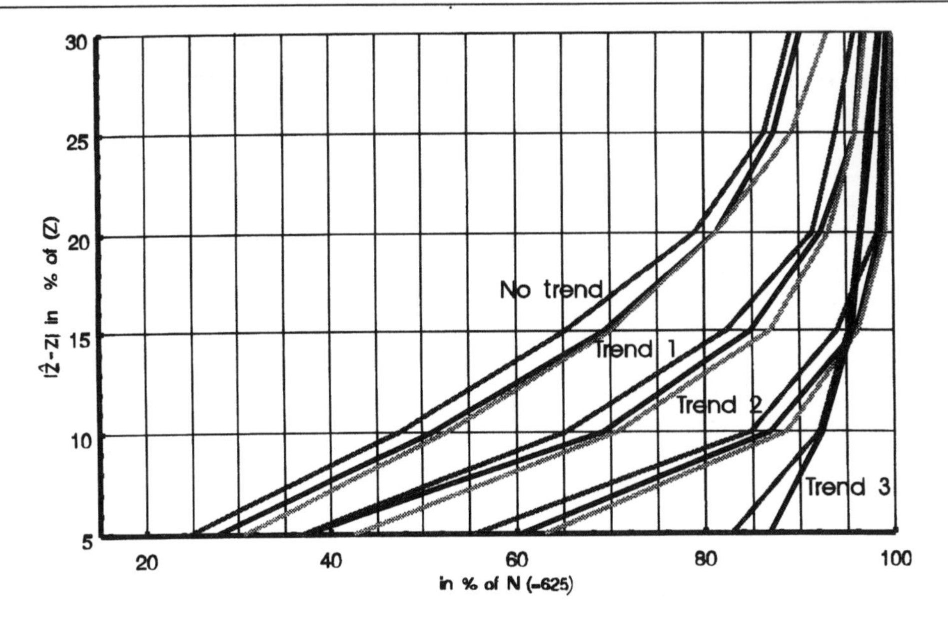

Figure 2.9. PSR validation of Kriging (light grey), IDW (grey) and IDW+LOGRAN (black).

(1980) and represent a spherical variogram with a range of five units of distance. The data obey the stationarity assumption of the method of ordinary point Kriging, which means that the values may not contain an additional trend component.

Figure 2.9 shows a methodical comparison in the form of PSR validation applied to the data of Fig. 2.8, which represent the case termed "no trend". In the cases "trend 1", "trend 2" and "trend 3", increased trend components have been added to the data of Fig. 2.8. Trend 1 has a maximum additive trend component equal to the mean of the data from Fig. 2.8. Trend 2 has a maximum additional trend component which is equal to the triple mean of the data without trend. And the maximum additional trend component of Trend 3 equals the 100 times mean of the data without trend. However, the fact that, for the application of ordinary Kriging for theoretical reasons no trend in the data is allowed, the prediction errors of Kriging (graph: light grey), IDW (grey) and IDW+LOGRAN (black) show a similar relation in all cases with and without trend. Generally the IDW+LOGRAN prediction shows a reduced prediction error, close to the error of Kriging.

Kriging is considerably better only in preventing larger amounts of errors (especially with "no trend"). In the case of trend 3, the numerical algorithm used for Kriging was not able to handle the high additional trend component. It should be noted that the graphs of the prediction errors generally are quite close together. This is correct, since Kriging applied to the ideal test data (which is shown in this figure) is only a few percentage points better than the usual IDW (see also Isaaks and Srivastava 1989). Generally, in all the following figures of PSR validation the different graphs are shown in the following grey shades: Kriging, light grey; IDW, grey; IDW+LOGAN, black.

After the excursion to theoretically ideal data for ordinary Kriging, all the following examples deal with data for heavy metal concentrations in the first mineral soil horizon.

Figure 2.10 shows a structure that is typical for real world environmental data. One part of the measured data shows a relatively continuous spatial variation while a certain part of the data originates high local variations with extreme values. The n=100 soil samples have been collected in a 14 to 8, 25 m^2 grid, where 12 sampling points have been excluded in

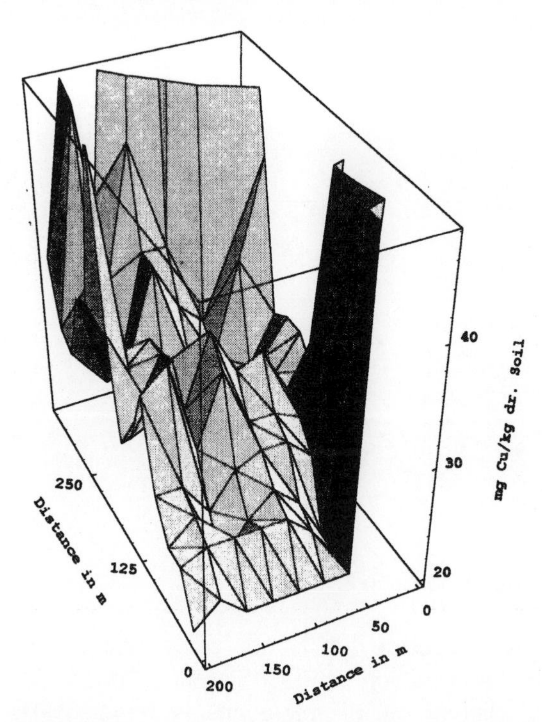

Figure 2.10. Triangular surface plot of copper in soil, site C.

the upper right corner. The minimum of the measured values is over 20 and the maximum is under 130 mg Cu/kg of dried soil matter.

Figure 2.11 compares the cumulative percentages of the prediction error ($Z' - Z$) in Z of the predictive sample reuse method. Here Z' is the predicted value and Z is the measured value. According to these results the advantage of IDW + LOGRAN over the simple IDW prediction becomes obvious with a reduced percentage of the previously introduced PSR prediction error. It should be noted that real world data with a fairly continuous characteristic of spatial variation does not always show an advantage of IDW+LOGRAN compared to Kriging in terms of a continuously reduced PSR prediction error. However, in every case the application of the local gradient analysis results in a considerably less smoothed spatial prediction, which can also be shown in Fig. 2.12, where a triangular surface plot of lead concentration values in upper soil. The data show a strong anisotropic characteristic: the measurement values possess a relatively high spatial autocorrelation along the lines. Crossing the lines the autocorrelation decreases considerably but the spatial variation increases highly. The n = 65 values are in a 5

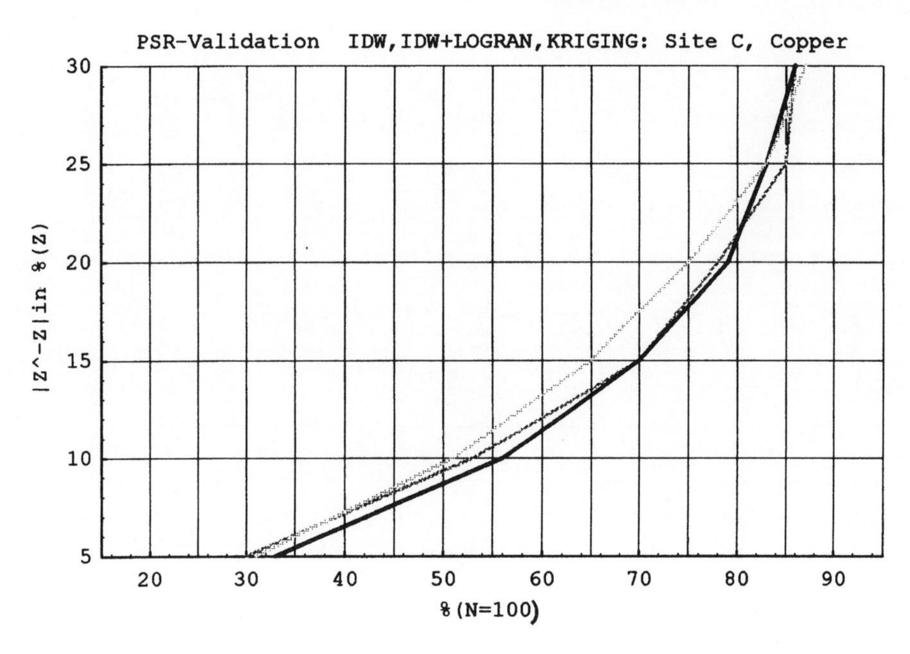

Figure 2.11. PSR validation for copper, Site C.

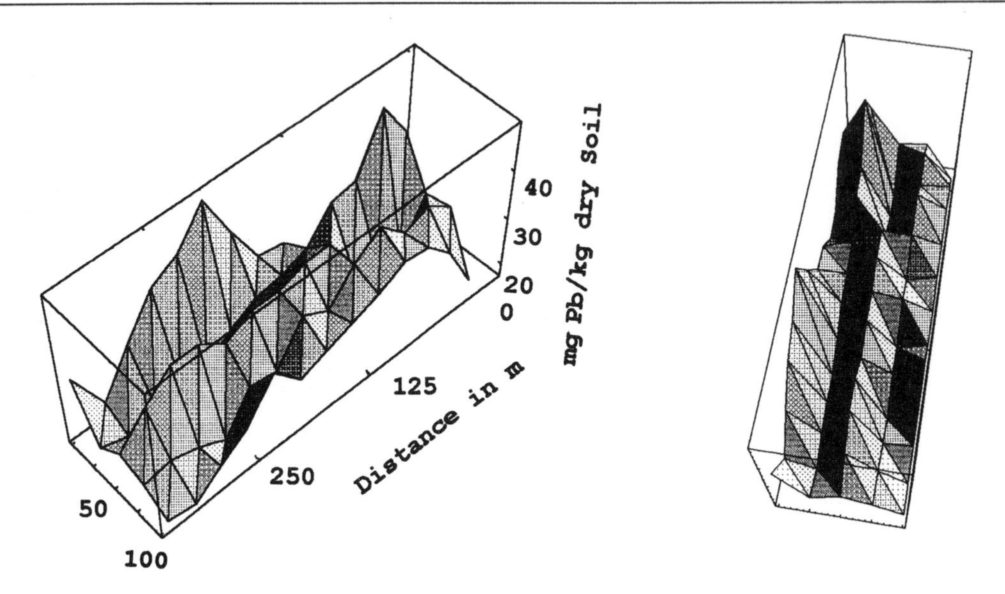

Figure 2.12. Triangular surface of lead concentrations in upper soil, site E.

to 13, 25 m^2 grid structure. The range of the measured values is between 18 and 49 mg Pb/kg of dried soil matter.

In Fig. 2.13 (PSR validation of Kriging, IDW and IDW+LOGRAN with lead data from site E) IDW+LOGRAN appears to have a considerable advantage compared to Kriging. The main reason for this is that the data set shows a clear anisotropy. It is not independent from direction, which is the theoretical reason for circular areas of influence. However, even with the theoretically insufficient circular areas of influence, the variable modelling of prediction input results in a smaller prediction error.

In order to compare the mappings of the

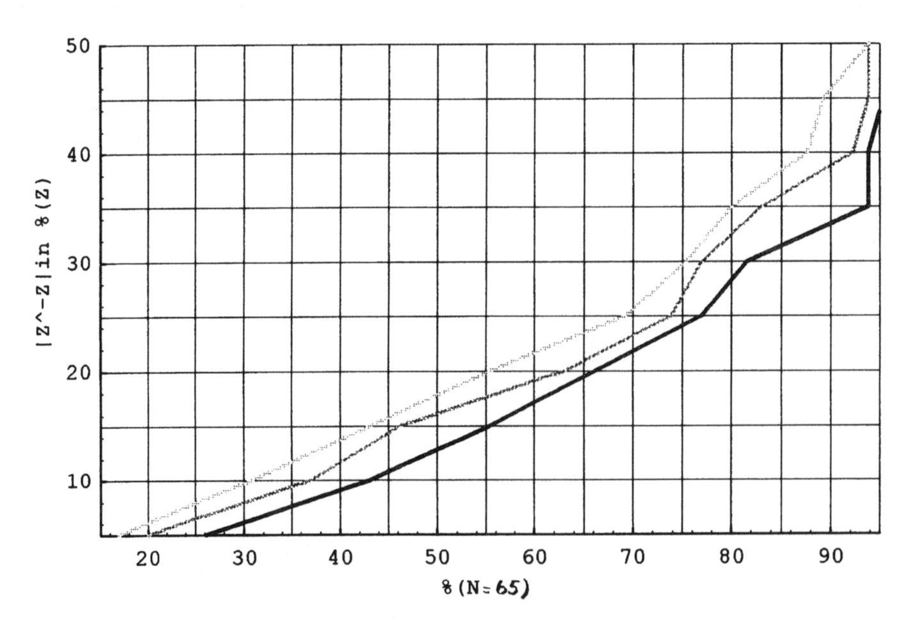

Figure 2.13. PSR validation of lead concentration in upper soil, site E.

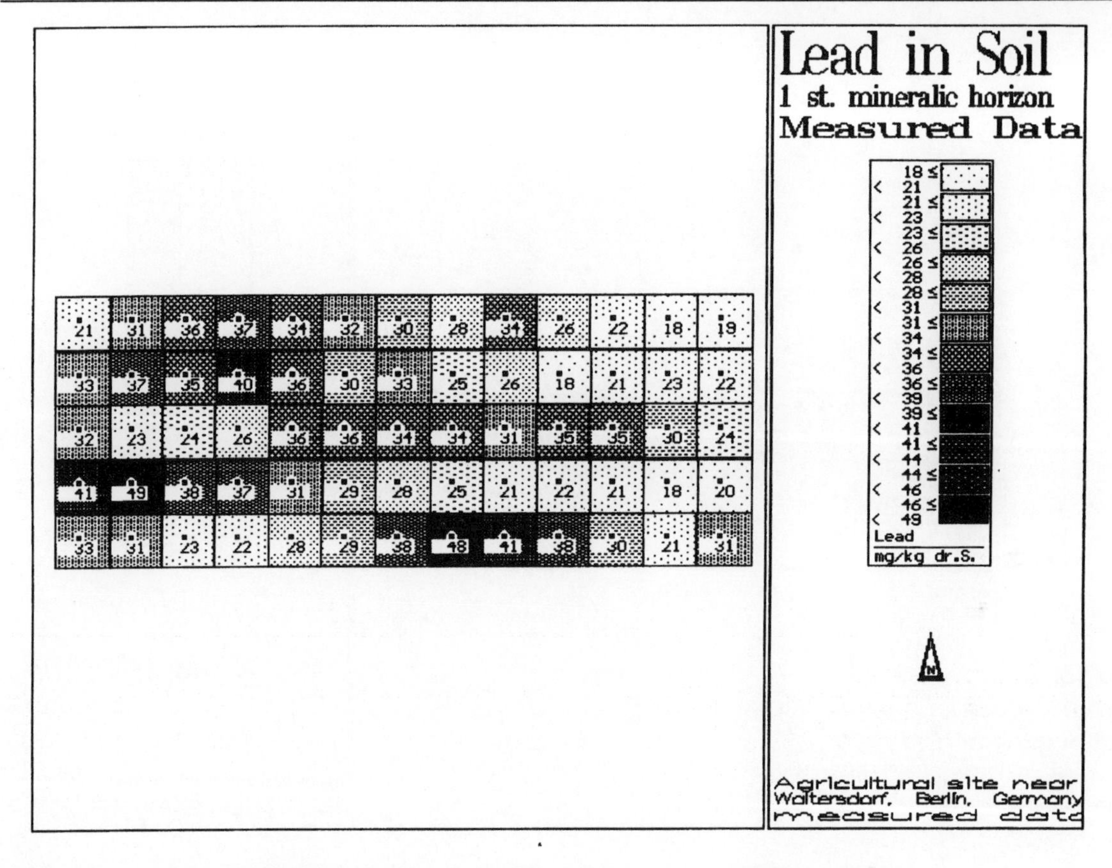

Figure 2.14. Grey scale diagram of lead concentrations in upper soil horizon, site E.

prediction results, the values measured are given in Fig. 2.14 in the centres of the grid cells. For comparison Fig. 2.15 shows the predicted concentrations in the soil of site E with the method IDW+LOGRAN. The location of the measured values is shown by a black point. The place of a predicted value is indicated by a white point.

The predicted lead concentrations in soil (site E) according to the method of Kriging are shown in Fig. 2.16. The grid geometry and shift to the predicted grid is identical to the parameters in the previous figure, where IDW+LOGRAN prediction results are given. Two aspects become obvious: the strong smoothing effect of Kriging and the effect of constant areas of influence (here radius = range) that are used by Kriging for interpolation. In fact, Kriging results still give clearer predictions when constant ellipses along the lines of the lead concentration structure are taken, but in order to do this an expert has to make extra intellectual and manual input. Despite these additional operations and also when the minimum number of points for prediction is reduced, the local variation in autocorrelation will be smoothed to a comparatively high degree. In the case of IDW+LOGRAN, however, an algorithm computes the variable areas of influence directly. Furthermore, there are also elliptic or irregular areas of influence that can be computed. Finally, an important property of IDW+LOGRAN has to be pointed out: with local gradient analysis, more homogeneous samples for spatial prediction can be modelled.

Figure 2.17 shows the reduction of standard deviation in samples for spatial prediction with variable interpolation radii (IDW+LOGRAN) compared to samples taken with a constant interpolation radius (IDW). The previously shown examples of copper in site C and lead in site E are included among other examples.

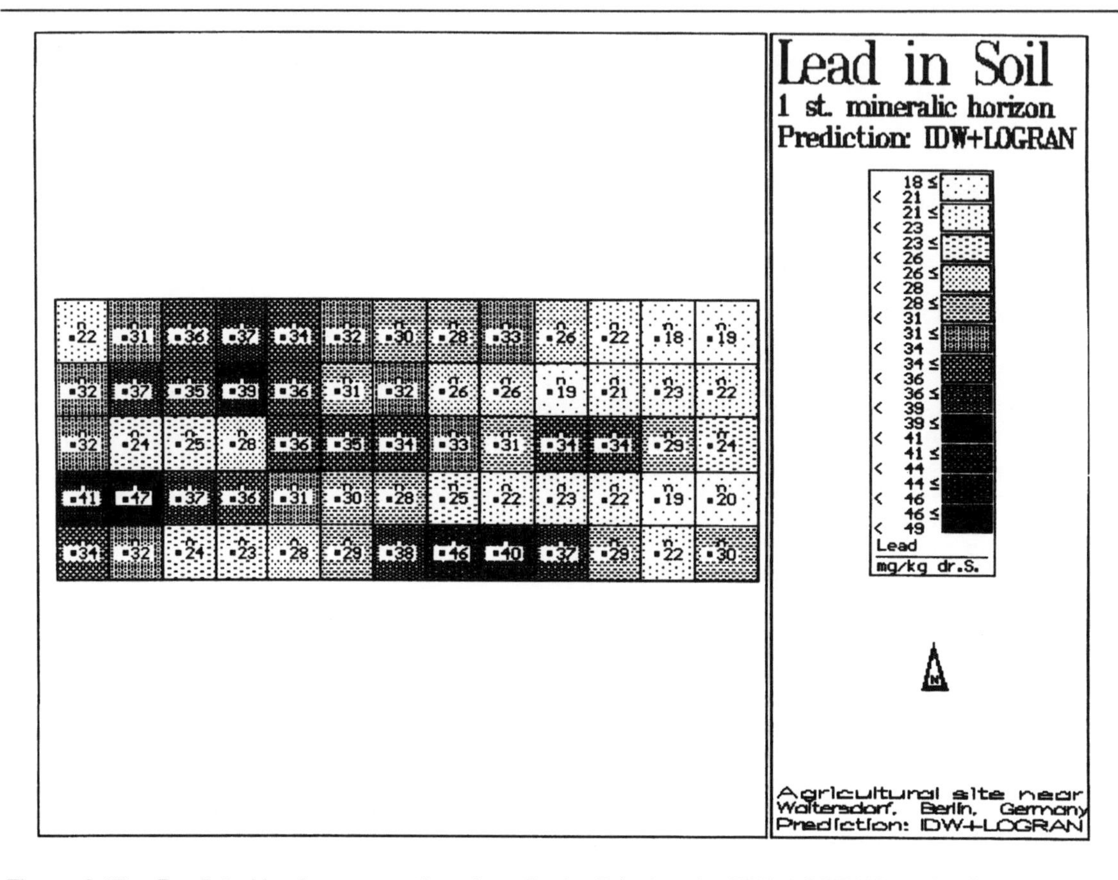

Figure 2.15. Predicted lead concentrations in soil, site E (using the IDW+LOGRAN method).

For this comparison, the same parameters except for interpolation radii have been taken. The results show that the mean standard deviation of prediction samples, taken with variable radii are typically reduced by 40% compared to samples that are selected with a constant radius (radius = range) for interpolation. This means that this prediction input is more homogeneous.

Software availability

Software for Kriging is available from the United States Environmental Protection Agency with the (public domain) program GEO–EAS (see Englund and Sparks 1990).

Time series analysis

If a variable is measured repeatedly over time, the resulting sequence of values is referred to as a time series. The regularity or systematic variation of a time series can differ in degree. On the one hand, the variation can be regular, so that no sophisticated techniques are needed to discern the pattern — it will be obvious to anyone. On the other hand, the variation may be so irregular as to defy description by any known means. We are left, then, with intermediate situations in which the patterns of variation are evident but not so obvious as to preclude statistical analysis. In addition to seasonal and cyclical patterns of variation, a time series can show an overall trend. Seasonal variation occurs within well-defined periods defined in terms of physical or biological phenomena. In the case of cycles, on the other hand, unless some theory exists that can predict the period of a hypothesized cycle, the period must be abstracted from the data itself. If the variations are described, they can be used to adjust past values of the variable in question. That is,

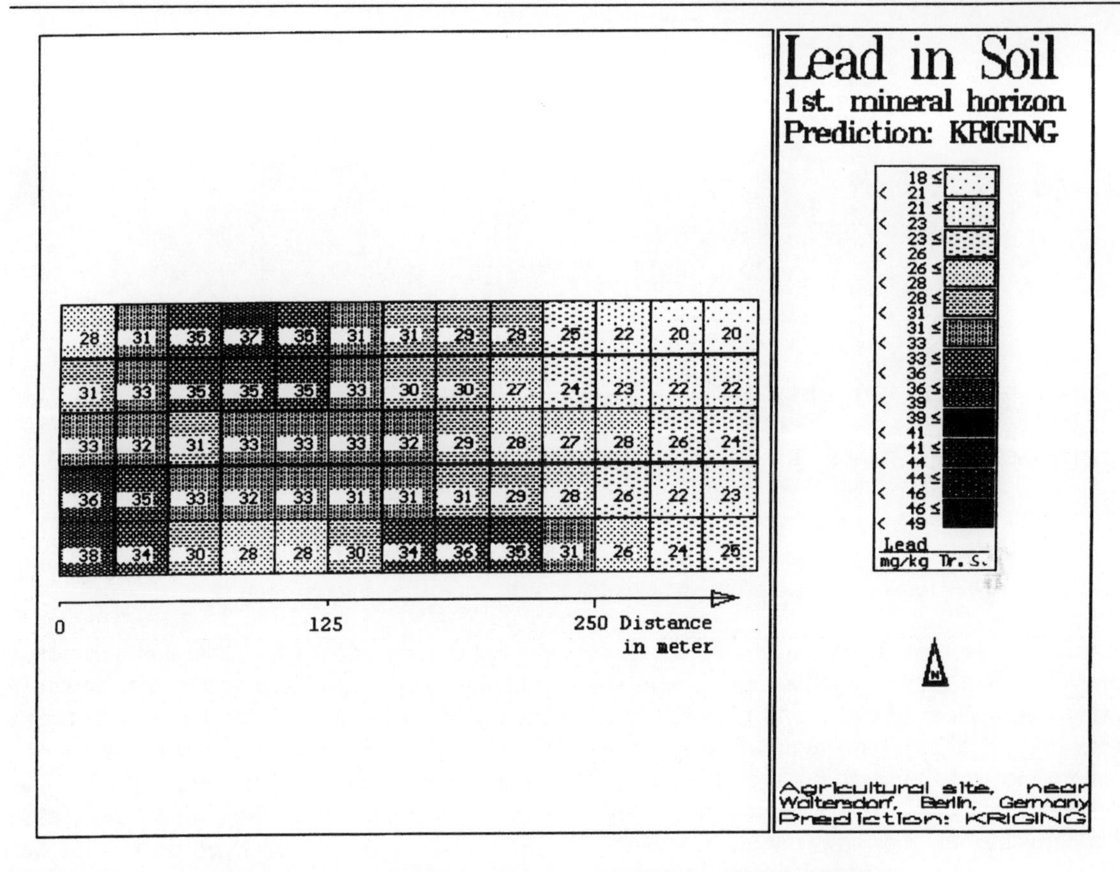

Figure 2.16. Predicted lead concentration in soil, site E (using the Kriging method).

they can be used to remove the respective source of variation in a time series in order to display the remaining source of variation. After seasonal, cyclical and trend variations have been identified in a time series, the residual variation is often referred to as irregular. Such descriptive methods, which can be used to display as well as to adjust seasonal, cyclical and trend patterns, were put together by Kachigan (1986), Chatfield (1989), and Schlittgen and Streitberg (1987). In descriptive analysis "anything goes" and therefore such methods are easily applicable. On the other hand, generalizing beyond time series data is not an easy task but nevertheless of substantial importance for ecological studies. Here, the characteristic situation is that we are often confronted with "unreplicated experiments" and are supposed to identify when variables begin to vary outside the normal range, thereby indicating that the ecosystem is perturbed or stressed. Thus, we are concerned with inferential statistics in order to detect whether a time series changes, whether it is a change in trend, in oscillatory behaviour or the impact of "unusual" events. The classic approach is the decomposition of a series into trend, oscillatory and irregular components (Schlittgen and Streitberg 1987; Chatfield 1989). However, as pointed out by Jassby and Powell (1990), decomposition of a time series is not always possible and flexibility is an indispensable feature of any attempt to investigate changing time series. The variety of approaches to analysing time series is reviewed in a special feature of *Ecology* (1990)*. The goal of this

* *Ecology* (1990) published a special feature on "Statistical analysis of ecological response to large-scale perturbations". Reprints of this 32-page special feature are available. Order reprints from the Ecological Society of America, Arizona State University, Tempe, AZ 85287, USA.

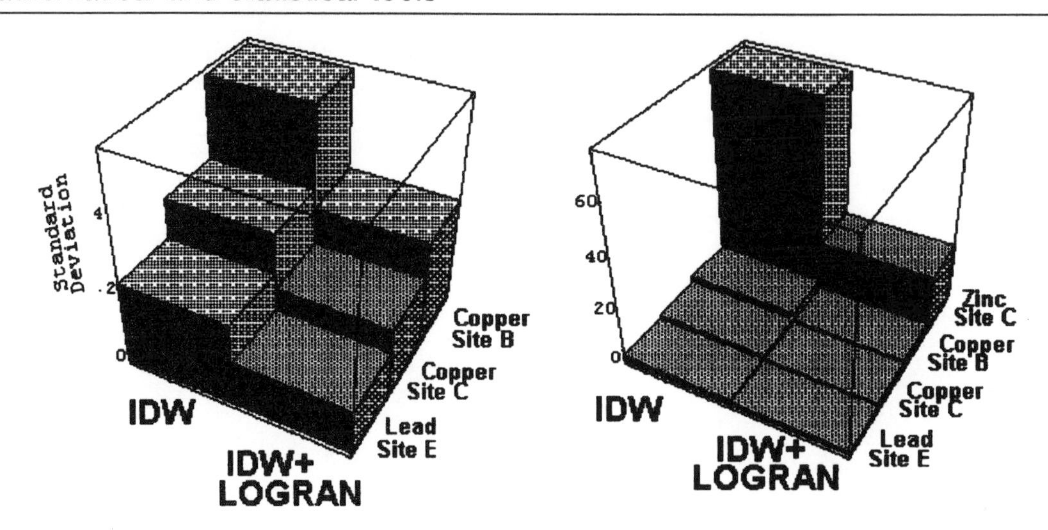

Figure 2.17. The standard deviation in samples for spatial prediction with variable interpolation radii (IDW+LOGRAN) is reduced compared to samples taken with a constant interpolation radius (IDW).

special feature was to inform ecologists of a range of statistical approaches useful for interpreting ecological responses to perturbations occurring on ecosystem, regional and global scales. Most of the methods represented will also be applicable in soil microbiology and biochemistry. In the first paper, Carpenter (1990) presents an overview of statistical approaches for large-scale planned and unplanned experiments. The second paper, by Jassby and Powell (1990), focuses on trend detection and inference in time series data. Reckhow (1990), in the third paper, addresses opportunities in Bayesian statistics, and points out key differences between Bayesian statistics and the frequentist statistics.

Establishing differences between field sites in the univariate case

Introduction

For univariate measures, the aim is usually to allow interval estimation and tests. As pointed out in the section entitled "Estimating the unknown population mean", a prerequisite for applying such inferential statistical methods is that information concerning the dispersion of the variable of interest is available. Hypothesis formulation is a second prerequisite for the application of statistical tests. Such tests have meaning only when they are made against an a priori null hypothesis that can never be proved correct but can be rejected knowing that there are risks of being wrong in doing so. The null hypothesis may be that a given impact (i.e. a contamination by a toxic substance) does not increase the activity of a specified enzyme. In this case, the null hypothesis must be rejected, if increased activity is observed on the contaminated plots. The test of a null hypothesis results in a decision to accept or reject, based on some risk of being wrong in that decision. The probability α that one will reject the null hypothesis when it is in fact true is commonly set by convention at $\alpha = 0.05$. However, for any given sampling and analysis design, lowering the level α will increase the level β and vice versa. The only way to reduce one error level without increasing the other is to improve the design, i.e. by increasing the number of samples or experimental units. Small sample sizes and small to moderate effects result in low statistical power. With low power the probability that any study will fail to detect a significant result is high even when the true effect is not zero. In principle, the possibility of showing significant treatment effects that do exist is a function of the magnitude of α, the distribution of the response variable, the magnitude of the treatment effect, the number of experimental units receiving each treatment and the sample size. Another problem is pseudoreplication, which results from the use of inferential statistics to test for treatment effects with data from experiments where either treatments are not replicated (though samples may be), or replicates are not statistically independent. In spite of these problems, statistical tests are often applied and therefore will be reviewed here also. In addition, estimation methods for displaying the difference between sites will be presented.

The comparison of two means

As mentioned in the introduction, many problems in soil microbiology and biochemistry lead to the comparison of two means. In the first step we have to decide whether we are only interested in an indication that there is some difference between the sites or instead that we are interested in the magnitude of the difference in order to judge its biological importance. In the latter case we have to calculate the confidence interval for the difference of the means. Then, the problems encountered and the strategy to solve them are very similar to those described in the section entitled "Estimating the unknown population mean". The first step is to check which kind of distribution we can assume. If we have good reasons to believe that our data is normal, we can easily estimate the difference and the confidence limits for this difference (Kachigan 1986; Lorenz 1988). If we can transform the data in such a way that normality can be assumed for the transformed data, we can calculate the confidence limits in the transformed units. At this point, Clark and Green (1988) emphasize backtransformation because statistical significance of a between-site difference is less important than its biological significance. If normality can not be achieved, we can choose robust estimation procedures for the difference of the medians (in the case of symmetrical distributions, median = mean). Such non-parametric estimation methods for independent samples as well as non-independent paired measures are put together by Büning and Trenkler (1978) and Sprent (1990). Even in the case of pseudoreplication the calculation and presentation of confidence intervals is possible, but then, if the value zero is not included in the interval, we can not infer statistical signficance of the treatment effect.

In order to apply statistical tests, the same questions as mentioned above must be answered before the suitable test can be selected. If our raw or transformed data are normal, the *t*-test can be applied (Kachigan 1986; Lorenz 1988; Köhler et al 1992). A non-parametric and hence robust test against deviations from the normal distribution is the *U*-test for independent samples and the Wilcoxon test for non-independent samples (Sachs 1984; Sprent 1990). Especially when small sample sizes were used, it is very difficult – if not impossible – to decide whether a normal distribution can be assumed or not. Then the application of the *U*-test is a useful alternative, even if the distribution of the population is in fact normal. However, if the assumptions are violated because the underlying populations are severely skewed, or have extremely long tails in both directions, the *U*-test is the more powerful alternative (Pagenkopf 1976). Nevertheless, we should be cautious about judging significant results biologically if the distribution is not symmetrical. Then, we have tested the median!

If the inherent variability of the variable is very high, it should be kept in mind that it may be very difficult to demonstrate the difference between control and treatment plots unless there are very large effects. Conquest (1983) has analysed the relationship between the coefficients of variation, minimum detectable distances between means, and the statistical power of a two-sample *t*-test. For selected small sample sizes, the results are presented graphically in this paper. These can help to assess the statistical adequacy of an experiment before it is started.

The comparison of more than two means

The problems to be solved differ from those discussed only as far as more than two treatments should be evaluated in terms of significant differences. Most of the principles outlined there are also valid here. To handle the comparison of say *m* treatments or *m* variations of one factor, an approach known as analysis of variance, often abbreviated to the acronym ANOVA is available. This can be used to show whether or not any two of the *m* means are significantly different. If the result of the ANOVA is significant, it is possible to conclude that at least one of the population means differs from one or more of the others. Later multiple comparison procedures must be applied in order to determine how many and which two means are really significantly different. To review the broad body of approaches in the ANOVA, the problems encountered, the strategies for solving them and the possible mistakes made would clearly fill many pages. The methods are described in the standard statistical literature, e.g. Kachigan (1986) as far as normality can be assumed, in Sprent (1990) for non-parametric methods and Köhler et al (1992) for normal as well as rank data. However, it should be pointed out that one of the most difficult subjects is the selection of the suitable multiple comparison procedure and such non-parametric methods are more restricted than their parametric counterparts.

The application of ANOVA in environmental studies is discussed by Green (1979), Hurlbert (1984), and Eberhardt and Thomas (1991). All of these articles contain very good compilations of the use of ANOVA in applied environmental studies but differ in their statements concerning the use of inferential methods in "unreplicated experiments". Whereas Green (1979) suggests that it is valid to use inferential statistics to test for environmental impacts in such studies, Hurlbert (1984) emphasizes that, although most statistical methods can be applied to either experimental or observational studies, their proper use in the former case requires that several conditions be met concerning the physical conduct of the experiment. We agree with his conclusion that in many cases accuracy would be better served by not applying inferential statistics at all. In the case of pseudoreplication, the best thing to do is to develop graphs and tables that clearly show both the approximate mean values and the variability of the data on which they are based.

Quantifying relationships between responses and predictors

Regression and correlation analysis in the bivariate case

A key difference between correlation and regression is that correlation is concerned solely with the strength of a linear (or linear in ranks) relationship, whereas regression problems deal with the precise form of a linear (or non-linear) relationship. In order to detect such relationships and to display their certainty and accuracy, it is better to apply both methods simultaneously. The first step is to analyse whether a linear or non-linear relationship can be assumed. Here, a plot of y_i against x_i is very helpful in indicating the kind of pattern. If no linear relationship can be assumed, a suitable data transformation should be attempted to satisfy the linearity assumption. Some linearizing transformations are shown by Sachs (1984). If a linear relationship: $y_i = a + b \times x_i$ is met, usually the equation for the best-fitting line by the method of least squares as well as the product moment correlation coefficient of Pearson is calculated (Kachigan 1986; Köhler et al 1992). However, for the resulting correlation coefficient and regression equation to be properly interpreted, a number of assumptions must be met. Often, these are not satisfied or cannot be checked, especially if small sample sizes are used. In this case, we recommend the application of non-parametric estimation and test procedures. Two non-parametric measures of correlation based on ranks are the correlation coefficients of Spearman and Kendall. Both coefficients share with those of Pearson the properties that they lie between −1 and +1, and that they take the value +1 when the ranks of x and y agree completely, and − 1 when they are precisely opposite. If the ranks show no obvious relationship, values near zero may be expected. When using these correlation coefficients, linearity of the quantitative data is not necessary but a monotonic relationship is assumed. Furthermore, for these coefficients there are tests of the null hypothesis which prove that the respective populations are independent. In regression analysis, robust estimation of the slope of the regression line is possible (method of Theil) and tests as well as interval estimation procedures are available (Büning and Trenkler 1978; Sprent 1990).

However, by transforming quantitative data into ranks, on the one hand it is possible to achieve robust methods but on the other hand the information contained in the data is reduced. Therefore, other techniques, called bootstraps, were developed, which use the original data and need no distributional assumptions at hand. The application of such a method to determine the precision of the correlation estimate is shown by Diaconis and Efron (1983).

Nevertheless, correlation and regression analysis is concerned with associations among random variables, and does not allow causal interpretations for such relationships, playing rather a descriptive and hypothesis-generating role.

Multiple regression and correlation

The multiple linear regression equation will be recognized as similar to the bivariate regression equation but, instead of a single predictor variable x, there are several predictor variables $x_1, x_2, \ldots, x_p$. The respective equation does not represent a straight line as in the bivariate case but rather planes in a multi-dimensional space. If p is greater than two, it is impossible to portray it graphically and it is even more difficult to check violations of the linearity assumption by graphical techniques. In applying multiple linear correlation methods, we want to know the total explanatory power of a set of predictor variables combined. The multiple correlation coefficient also measures the correlation between two variables, which is just that one between the response variable and the variable derived as a weighted combination of several others. The partial correlation coefficient between two variables when the common variance of other variables is extracted can also be considered. Furthermore, the partial regression coefficients can provide the rank order of importance of the predictor variables but they can only show the relative importance of the various predictor variables, not the absolute contribution. However, in order to judge the results, confidence limits should be estimated and/or tests should be applied. In order to achieve this, many assumptions must be met, just as in the bivariate case. In addition, the multiple correlation coefficient has some problems of interpretation. The most severe is that, if the sample size is not large in relation to the number of predictor variables, an artificially inflated multiple correlation coefficient is obtained. Therefore, in cases of small sample size, multiple correlation and regresssion should not be used.

Parametric methods for dealing with multiple and partial correlation and regression problems are provided by Sachs (1984), Flury and Riedwyl (1983), and Kachigan (1986). Partial correlation coefficients corresponding to Kendall's coefficient are discussed by Conover (1980). Various extensions of Theil's method to multiple regression have been considered and Sprent (1990) proposes that of Agee and Turner (1979).

However, it must be pointed out that relationships between microbiological data and environmental factors are rarely linear and often not even monotonic.

Structuring of multivariate data sets by descriptive and explanatory analysis

Introduction

Multivariate procedures such as ordination and clustering are useful for exploratory descriptive purposes such as reducing massive, complex data sets to "what is going on". Now the data consists either of more than one response or criterion variable only or one or more predictor variables have been measured additionally. However, descriptive multivariate statistical methods can describe a supposed structure that is nothing more than random variation and it is easy to be convinced that things to be seen in the data are in the data. An ordination analysis will produce continuous structures and a cluster analysis will produce discontinuous clusters from any data set. It has been a common assumption in the past that cluster analysis and ordination are methods in competition. We emphasize that it is often a good strategy to do both and compare the results.

Ordination analysis

Referring to criterion variables only, it is possible to arrange the sampling units in relation to one or more coordinate axes so that their relative positions to the axes and to each other provide maximum information about their similarities. By identifying the sample units in a collection that are most similar to those based on coordinate position, it is possible to search for underlying factors that might be responsible for the observed pattern. On the other hand, if predictor variables are measured, the aim is to elucidate the relationship between microbiological parameters and environmental factors. This is often done in the context of gradient analysis. In community ecology especially, such methods are used extensively. Two reviews of ordination methods showing their possibilities, limitations and applications are worth mentioning (Green 1979; Ludwig and Reynolds 1988) because most of the knowledge can be transferred to the respective problems in soil microbiology and biochemistry. They show the progression of the use of ordination methods from the use of polar ordinations to the more recent emphasis on principal components analysis, detrended correspondence analysis, and non-metric multidimensional scaling. When the data matrix has non-linear relationships, it is recommended that non-metric multidimensional scaling is used. This method relies only on the rank order of similarities. Clark and Green (1988) consider non-metric multidimensional scaling (Kruskal and Wish 1978) to be one of the most powerful ordination methods.

Cluster analysis

Cluster analysis is a classification technique for placing similar entities such as microbiological samples or sample units into groups or clusters. Most often, cluster analysis models are used to place similar samples into such clusters, which are arranged in a hierarchical tree-like structure called a dendrogram (Ludwig and Reynolds 1988). Many clustering procedures exist, based on different algorithms related to different definitions of a group and different measures to quantify similarity or dissimilarity. The use of these procedures calls for subjective decisions. First, there is no general or optimal rule for the selection of similarity measures (Bock 1980). Second, to identify specific groups from a dendrogram, one needs to decide when dividing should be stopped. Although it can be helpful to use such methods as an aid in interpreting data, no strong emphasis should be placed on results of a single analysis. In cases where the data set is large, somewhat complex in nature and with no obvious pattern, the results of the various clustering procedures can often vary, in some cases even substantially. In the last few years especially, cluster analysis models based on graph theoretical approaches have been developed (Godehardt 1990).

There are three main reasons why these models are more suited to analysing problems encountered in soil microbiology and biochemistry. Firstly, non-hierarchical graph theoretical cluster analysis provides probability models from which tests for the hypothesis of homogeneity within a data set ("randomness" of the clusters found) can be derived for many situations. Secondly, in some cases it may be better not to compute one single similarity between any pair of samples or sample units but more say *t*-similarities, i.e. the structure of a set of mixed data can more appropriately be described by a superposition of *t* graphs, so-called "completely labelled multigraphs". Lastly, the subjective decisions necessary for clustering multivariate data become more transparent to the users. They are directly involved in the decisions required by the method as to which dissimilarity measure should be used for which dimensions, or whether single or *k*-linkage classification would be appropriate. Some of the other clustering methods tend to be a "black box" where data is fed into a computer program and the user is largely detached from the analysis.

References

Agee WS, Turner RH (1979) Application of robust regression to trajectory data reduction. In: Robustness in Statistics. Learner RL, Wilkinson GN (eds). Academic Press, London.

Akin H, Siemes H (1988) Praktische Geostatistik. Springer, Berlin.

Anderson VL, McLean RA (1974) Design of Experiments. Marcel Dekker, New York.

Bauer P, Scheiber V, Wohlzogen FX (1986) Sequentielle statistische Verfahren. In: Biometrie. Lorenz RJ, Vollmar J. (eds). Gustav Fischer, Stuttgart.

Besterfield DH (1990) Quality Control, 3rd edn. Prentice Hall, Englewood Cliffs, NJ.

Bock H-H (1980) Clusteranalyse – Überblick und neuere Entwicklungen. OR Spektrum 1: 211–232.

Büning H, Trenkler G (1978) Nichtparametrische Statistische Methoden. Walter de Gruyter, Berlin.

Burr IB (1979) Elementary Statistical Quality Control. Marcel Dekker, New York.

Carpenter SR (1990) Large-scale perturbations: opportunities for innovation. Ecology 71: 2038–2043.

Chatfield C (1989) The Analysis of Time Series: An Introduction. Chapman and Hall, London.

Clark I (1979) Practical Geostatistics. Applied Science Publishers, London.

Clark KR, Green RH (1988) Statistical design and analysis for a "biological effects" study. Marine Ecol–Progress Ser 46: 213–226.

Clarke GM (1980) Statistics and Experimental Design, 2nd edn. Arnold, London.

Cochran WG (1977) Sampling Techniques, 3rd edn. Wiley, New York.

Cochran WG, Cox GM (1957) Experimental Design, 2nd edn. Wiley, New York.

Conover WJ (1980) Practical Nonparametric Statistics, 2nd edn. Wiley, New York.

Conquest LL (1983) Assessing the statistical effectiveness of ecological experiments: utility of the coefficient of variation. Int J Environmental Studies 20: 209–221.

Cressie NAC (1986) Kriging nonstationary data. J Am Stat Assoc 81: 625–634 (Applications).

Cressie N (1987) Variogram. In: Encyclopedia of Statistical Sciences vol 9. Kotz, Johnson (eds). Wiley, New York, 489–490.

Cressie N (1990) The origins of Kriging. Math Geol 22: 239–252.

Cressie NAC (1991) Statistics for Spatial Data. Wiley, New York.

David M (1979) Geostatistical Ore Reserve Estimation. Elsevier, Amsterdam.

Diaconis P, Efron B (1983) Statistik per Computer: der Münchhausentrick. Spektrum der Wissenschaft 56–71.

Dodge Y (1985) Analysis of Experiments with Missing Data. Wiley, New York.

Dunn OJ, Clark VA (1974). Applied Statistics: Analysis of Variance and Regression. Wiley, New York.

Eberhardt LL, Thomas JM (1991). Designing environmental field studies. Ecol Monographs 61: 53–73.

Eckstein BA (1989) Evaluation of spline and weighted average interpolation algorithms. Computers Geosci 15: 79–94.

Eisenhart C (1947) The assumptions underlying the Analysis of Variance. Biometrics 3: 1–21.

Englund EJ, Sparks A (1990) GEO–EAS (Geostatistical Environmental Assessment Software) User's Guide, Version 1.2. United States Environmental Protection Agency Environmental Monitoring Systems Laboratory (for MS–DOS operating-system), Las Vegas.

Finney DJ (1941) On the distribution of a variate whose logarithm is normally distributed. Roy Stat Soc Lond J Suppl 7: 155–161.

Flury B, Riedwyl H (1983) Angewandte multivariate Statistik – Computergestützte Analyse mehrdimensionaler Daten. Gustav Fischer, Stuttgart.

Gandin LS (1963) Objective Analysis of Meteorological Fields. Gimiz, Leningrad. (Translation: Israel Program for Scientific Translations, Jerusalem, 1965.)

Godehardt E (1990) Graphs as structural models. The application of graphs and multigraphs in cluster analysis. In: Advances in System Analysis, 2nd edn. Möller DPF (ed.). Vieweg, Braunschweig.

Goldberger AS (1962) Best linear unbiased prediction in the generalized linear regression model. Am Stat Assoc J 369–375.

Grant E, Leavenworth RS (1988) Statistical Quality Control, 6th edn. McGraw Hill, New York.

Green RH (1979) Sampling Design and Statistical Methods for Environmental Biologists. Wiley, New York.

Gy PM (1982) Sampling of Particulate Materials. Elsevier, Amsterdam.

Gy PM (1989) Heterogeneity – Sampling – Homogenization. Elsevier, Amsterdam.

Haas TC (1993) Spatio temporal Kriging within a moving window. In: Ecoinforma Proceedings, vol. 4. Ries L, Fiedler H, Wagner G, Hutzinger O (eds). Ecoinforma Press, Bayreuth, pp. 241–262.

Hurlbert SH (1984) Pseudoreplication and the

design of ecological field experiments. Ecol Monographs 54: 187–211.

Isaaks EH, Srivastava RM (1989) Applied Geostatistics. Oxford University Press, Oxford.

Jassby AD, Powell TM (1990) Detecting changes in ecological time series. Ecology 71: 2044–2052.

Kachigan SK (1986) Statistical Analysis – An interdisciplinary introduction to univariate and multivariate methods, 2nd edn. Radius Press, New York.

Kirk RE (1982) Experimental Design: Procedures for the behavioural sciences, 2nd edn. Brooks/Cole Publishing Co., Belmont, California.

Knafl G, Sacks J, Spiegelman C, Ylvisaker D (1984) Nonparametric calibration. Technometrics 26: 233–241.

Köhler W, Schachtel G, Voleske P (1992) Biostatistik – Einführung in die Biometrie für Biologen und Agrarwissenschaftler, 2nd edn. Springer, Berlin.

Kolmogorov AN (1961) The local structure of turbulence in an incompressible fluid at very large Reynolds numbers. Doklady Akademii Nauk SSR (1941) 30: 301–305. Reprinted in: Friedlander SK, Topping L (eds) Turbulence: Classic papers on statistical theory. Interscience Publishers, New York, pp. 151–155.

Krige DG (1951) A statistical approach to some basic mine valuation problems on the Witwatersrand. J Chem Metall Min Soc S Africa 52: 119–139.

Kruskal JB, Wish M (1978) Multidimensional Scaling. Sage Publications, Beverley Hills, CA.

Lorenz RJ (1988) Grundbegriffe der Biometrie. In: Biometrie, 2nd edn. Lorenz RJ, Vollmar J (eds). Gustav Fischer, Stuttgart.

Ludwig JA, Reynolds JF (1988) Statistical Ecology: A primer on methods and computing. Wiley, New York.

Matern B (1960) Spatial variation. Meddelanden fran Statens Skogsforskningsinstitute 49: 144.

Matheron G (1971) The Theory of Regionalized Variables and its Applications. Centre de Morphologie Mathématique de Fontainebleau, Fontainebleau.

Montgomery DC (1984) Design and Analysis of Experiments. 2nd edn. Wiley, New York.

Mulaik SA (1972) The Foundations of Factor Analysis. McGraw Hill, New York.

Pagenkopf J (1976) Güte und Effizienz einiger nichtparametrischer Tests bei kleinen Stichproben. Dissertation, Universität Hamburg.

Parkin TB, Meisinger JJ, Chester ST, Starr JL, Robinson JA (1988) Evaluation of statistical estimation methods for lognormally distributed variables. Soil Sci Soc Am J 52: 323–329.

Pitard FF (1992) Pierre Gy's Sampling Theory and Sampling Practice. CRC Press, Boca Raton, FL.

Prier JE, Bartola JT, Friedman H (eds) (1975) Quality Control in Microbiology. University Park Press, Tokyo.

Provost LP (1984) Statistical methods in environmental sampling, In: Environmental Sampling for Hazardous Wastes. Scheitzer GE, Santolucito JA (eds), pp. 79–96. ACS Symposium Series 267. ACS, Washington, DC.

Rasch D (1987) Biometrie: Einführung in die Biostatistik, 2nd edn. Harri Deutsch, Frankfurt am Main.

Reckhow KH (1990) Bayesian Inference in nonreplicated ecological studies. Ecology 71: 2053–2059.

Rinne H, Mittag HJ (1989) Statistische Methoden der Qualitätssicherung, 3rd edn. Hanser, München.

Ries L (1993) Areas of influence for IDW-interpolation. Catena (Journal) 20: 199–205.

Ries L (1994) Applications of local gradient analysis on Gereal data. Chem Intelligent Lab Syst (submitted).

Ripley BD (1981) Spatial Statistics, Wiley, New York.

Rosander AC (1985) Quality and Reliability. Marcel Dekker, New York.

Sachs L (1984) Angewandte Statistik, 6th edn. Berlin, Springer.

Schlittgen R, Streitberg BHJ (1987) Zeitreihenanalyse, 2nd edn. R Oldenbourg, München.

Scholz RW, Nothbaum N, May Th W (1993) Starre und hypothesengeleitete Probenahmepläne – Grundlagen, Strategieen und Fallbeispiele, vol 4. In: Biomonitoring und Umweltprobebanken, Umweltdatenbanken und Informationasysteme, Ökometrie, Qualitätssicherung. Ries L, Fiedler H, Wagner G, Hutzinger O (eds). pp. 229–240. Ecoinforma Press, Bayreuth, Germany.

Sironvalle MA (1980) The random coin method: solution of the problem of the simulation of a random function in the plane. Math Geol 12: 25–32.

Sen PK (1981) Sequential nonparametrics. Invariance principles and statistical inference. Wiley, New York.

Sprent P (1990) Applied Nonparametric Statistical Methods, 2nd edn. Chapman and Hall, London.

Wadsworth HM, Stephens KS, Godfrey AB (1986) Modern Methods for Quality Control and Improvement. Wiley, New York.

Wiener N (1949) Extrapolation, Interpolation, and Smoothing of Stationary Time Series. MIT Press, Cambridge, MA.

Yaglom AM (1962) An Introduction to the Theory of Stationary Random Functions. Prentice Hall, Englewood Cliffs, NJ. Republished by Dover, New York, in 1973.

Soil sampling, handling, storage and analysis

3

Soil sampling and storage

J.C. Forster

Microbial life in soil has been studied intensively and a multitude of parameters are available for its characterization. However, the problems related to sample representativity and storage history have often been ignored. Both can influence analytical results markedly.

Representative sampling

The organic and especially the microbiological properties of soil are often extremely heterogeneous in the field. Although this has been recognized by soil surveyors, only a few papers have dealt with representative sampling from the microbiologist's point of view. Suttner (1990) had as much as 100 sampling points per 1 ha plot, and Beck (1986) suggested a similar number of individual samples for an even smaller test plot.

Öhlinger et al (1986) measured protease activity from fresh samples and phosphatase activity from air-dried samples on 1 ha plots and found that 10 individual samples had to be combined for enzyme activity measurements on an agricultural plot (0–25 cm) to give a maximum 20% deviation from the "true" means ($P < 0.05$). The respective numbers in a pasture plot (0–10 cm) were five samples for protease (±10%; $P < 0.05$) and 25 samples for phosphatase (±20%; $P < 0.05$). Larger numbers of individual samples did not improve the accuracy significantly.

As a rule, as many individual samples as possible (depending on available time and working capacity) should be taken and analysed separately not only to even out the heterogeneity but also to determine the mesoscale and macroscale variability. If the variability of soil parameters is already known, the number of samples that needs to be taken may be calculated to obtain a given precision with a specified probability (Petersen and Calvin 1986) by the equation

$$n = t_\alpha^2 s^2 / D^2 \qquad (3.1)$$

where t_α is Student's t at $(n-1)$ degrees of freedom and probability a, s is the variance and D is the specified precision limit.

It must not be overlooked, however, that absolute representativity can not be achieved. For detailed sampling plans, the reader is referred to Chapter 2 on quality control. The dimensions of the soil sample should be large enough to include all forms of microscale variability due to soil aggregation and penetration by plant roots. Nevertheless, conventional bulk soil sampling does not take into account the spatial arrangement of microorganisms and enzymes on and in structural soil units that can give rise to a series of misinterpretations of laboratory data with regard to field conditions.

Sampling depth is normally the plough layer in agricultural soils (0–25 cm) or the most intensively rooted A horizon under grassland (0–10 cm). When forest soils or whole soil profiles are investigated, samples should be

Methods in Applied Soil Microbiology and Biochemistry
ISBN 0–12–513840–7

taken representatively from their genetic horizons. However, if an extrapolation of data to unit area and soil depth is desired, sampling of soil segments to a fixed depth is preferable due to the convenience of data handling.

Seasonal variations exert significant influence on activity and biomass data. This is a result of combined temperature, moisture and vegetation/rhizosphere effects. Repeated sampling throughout a year is thus the only way to obtain complete information on a soil and averaged data may then be used to compare soils from contrasting locations. For a single measurement, sampling time should not be chosen immediately after a period of freezing or drying, especially not if the dissipation of chemicals has to be assessed in the laboratory (Anderson 1987). Edwards and Cresser (1992) state that the availability of various nutrients increased after a series of freeze/thaw cycles imposed on different soil types, including peat, as a consequence of changes in the arrangement of mineral and organic particles thus liberating fresh reactive sites. Newly available nutrients include solubilized organic carbon and organic carbon released from the lysis of plant and microbial cells, leading to a flush of CO_2 and inorganic nitrogen production when the soil thaws (Soulides and Allison 1961; Christensen and Tiedje 1990). Skogland et al (1988) also measured a respiratory burst after freezing and thawing of soil due to the mineralization of cellular material from killed bacteria.

The highest extractable carbon flushes have to be expected in the driest months, while extractable carbon in unfumigated samples is positively related to the field moisture content (Ross 1990). Inputs of mineral fertilizer or fresh or composted organic matter would produce increases in microbial activity and biomass (Nannipieri et al 1990).

Sampling device

The choice of sampling procedure and device depends on the purpose of the subsequent investigation; undisturbed (so-called volume) and disturbed (bulk) samples have to be distinguished. Disturbed samples may be collected from topsoil horizons by using a spade or hand shovel, or special hand samplers. Care has to be taken that equivalent proportions of soil are taken throughout the whole sampling depth, as there might be a gradient of the property to be measured even within a limited vertical distance. This prerequisite may be met by the use of a metal or plastic frame (e.g. 20 $\times$ 20 cm^2 basal area and 10 cm height), which is driven into the soil and whose content is sampled quantitatively.

A hammer-driven Pürckhauer auger (available from Forstkultur GmbH, Schlüchtern, Germany) is useful for sampling profiles to a depth of 1–2 m of soils that contain only minor amounts of coarse fragments. Edelman augers (Eijelkamp, Giesbeek, The Netherlands) are drilled into the soil by hand; special augering tips are available for clay or stony soils, respectively. While a core of *c.* 20 cm depth is removed, there is some danger of mixing with topsoil material when cores from the subsoil are taken.

Motor-driven auger equipment may be advisable if large numbers of samples or samples from deep soil profiles have to be investigated.

Sample transport

Samples should be transferred to the laboratory as soon as possible after sampling to avoid dramatic changes in microbial populations and activities. If possible, samples should be cooled or at least kept in heat-isolated protective cases to prevent exposure to elevated temperatures. Undisturbed samples must not be subjected to shaking or vibration.

Composite samples and subsampling

Large numbers of samples are required if heterogeneous areas are investigated; however, laboratory capacity may be limited so that the number of samples to be analysed has to be reduced. Composite samples should consist of equal amounts (weight or volume) of individual samples from the same horizon or depth that are mixed thoroughly in a container with a hand shovel. One or a few portions of soil are removed at random and kept for further examination; if possible, the rest is discharged at the place of sampling.

Storage of samples

When large numbers of samples are taken on one sampling date, and when methodology or manpower do not permit immediate processing and analysis in the laboratory, samples have to be stored under conditions that impose as little change as possible on them.

The choice of the storage conditions should take into account: (1) the object of interest, i.e. whether dead, moribund, resting or metabolically active microorganisms are being examined (Anderson 1987); (2) which parameter is under study, e.g. whether biomass or activity data should be collected (Skujins 1967); and (3) whether microbially influenced processes are being investigated, e.g. the dissipation of pesticides or the dynamics of inorganic nitrogen.

No general rule can be derived from data in the literature because storage effects have been investigated by a multitude of authors only for single parameters in arbitrarily sampled soil horizons (Ross 1970, 1972; Jager and Bruins 1975; Powlson and Jenkinson 1976b; Ross et al 1980; Salonius 1983; Bottner 1985; Sparling et al 1986; West et al 1986, 1987; Shen et al 1987; Malkomes 1989, 1991; Zelles et al 1991). It has been shown, however, that inherent soil properties (e.g. particle size distribution), initial water content, and time and temperature of storage exert different influences on individual activity or biomass parameters. Thus, air-dried soil is not recommended for microbiological measurements. Samples should be stored preferably at 2–4°C for as short a time as possible until examination; a storage time of up to 4 weeks seems acceptable. Freezing at −20°C should be preferred for longer storage. A pre-incubation of 24–48 h may be useful, especially if physiological measurements are planned. It is most important to treat all samples equally to ensure comparability of results.

Sterilization of soil and inhibition of microbial activity

K. Alef

For studying the transport of chemicals in soil, estimation of microbial biomass or enzyme activity, the soil needs to be sterilized and microbial activities need to be inhibited. Several methods for the sterilization and inhibition of cell proliferation in soil are presented and discussed.

Sterilization by autoclaving

Principle of the method

The method is based on the incubation of soil samples in an autoclave at 121°C and 1.1 atm.

Materials and apparatus

Autoclave
Glass beaker (the volume depends on the soil required)
Aluminium foil

Chemicals and solutions

Distilled water

Procedure

Pour distilled water into the autoclave as described in the manufacturer's instructions. Place the soil sample in a glass beaker and cover with aluminium foil. Put the beaker in the autoclave and sterilize for 30 min at 120°C and 1.1 atm (the autoclaving time starts when the autoclave shows the desirable temperature and pressure). For soil samples of more than 500 g, a sterilization time of about 1 hour is recommended. At the end of sterilization, open the autoclave (only when the autoclave shows the value of natural air pressure) and allow the soil to cool at room temperature.

A fractional sterilization can also be carried out by cooling the autoclaved soil samples overnight in the autoclave. The sterilization procedure is then repeated once or twice. Soil samples should be stored under sterile conditions.

Discussion

The fractional sterilization is used when strong soil contamination with fungal spores is expected.

Sterilization by moist heat destroys the soil structure significantly; thus studies on the transport of chemicals can not be performed. Furthermore, autoclaving results in the liberation of high concentrations of ammonium and organic substances, such as amino acids.

Soil fumigation

(Powlson and Jenkinson 1976a)

Principle of the method

This technique is based on the fumigation of soil samples with chloroform. It is also used for the estimation of microbial biomass.

Materials and apparatus

Desiccator (30.5 cm i.d.)
Glass beaker (the volume depends on the soil required)
Anti-bumping granules
Vacuum pump
Water pump
Filter paper

Chemicals and solutions

Alcohol-free chloroform
Distilled water

Procedure

Soil portions (250 g or more) are placed in glass beaker and put in a large desiccator (30.5 cm i.d.) lined with moist paper. The desiccator contained a beaker with alcohol-free $CHCl_3$ and a few anti-bumping granules. The desiccator is evacuated until the $CHCl_3$ boils vigorously. Finally, the tap is closed and the desiccator is left in the dark at 25°C for 18–24 h. The beaker with the $CHCl_3$ and the paper are then removed and $CHCl_3$ vapour is removed from the soil by repeated evacuation in the desiccator. Six 3-min evacuations, three with a water pump and three with a high-vacuum oil pump, are usually sufficient to remove the smell of $CHCl_3$ from the soil; two additional evacuations are then given with the high vacuum pump.

Discussion

In the same manner ethylene oxide can be used to sterilize soils. Ethylene oxide is explosive and toxic for humans, therefore special precautions must be taken in its handling.

Fumigation also destroys the fine structure of soils and is not suitable for chemical transport studies.

Irradiation of soils

(Powlson and Jenkinson 1976b)

Principle of the method

The method is based on the irradiation of soil samples at a gamma irradiation unit.

Materials and apparatus

Gamma irradiation unit
Polythene bags
Dry ice

Procedure

The soils are irradiated in sealed polythene bags, each containing the amount of moist soil required for the experiment. Irradiation is carried out at a gamma irradiation unit. Soil samples are given a dose of 2.5 Mrad, at a dose rate of about 2 Mrad h^{-1}. Soils are kept frozen during transport to and from the irradiation unit. Further transfer of irradiated soils must be carried out under sterile conditions.

Discussion

No disturbance of soils occurs using this method.

Inhibition of microbial activity by azide, cyanide and toluene

Azide, cyanide and toluene are the most widely used microbial inhibitors in soil microbiology (Alef and Kleiner 1989). Azide and cyanide at concentrations of 1–10% strongly inhibit microbial activity. In soil suspensions, a concentration of 5–10% toluene is sufficient.

Principle of the method

The method is based on mixing soil samples with solutions of azide, cyanide or toluene.

Materials and apparatus

Glass beaker or reaction flasks (the volume depends on the soil required)
Aluminium foil or rubber septa
Glass bottles with different volumes

Chemicals and reagents

Azide solution (10%)

Dissolve 10 g of sodium azide in 70 ml of distilled water and bring up with distilled water to 100 ml. Store in closed bottles.

Cyanide solution (10%)

Dissolve 10 g of sodium cyanide in 70 ml of distilled water and bring up with distilled water to 100 ml. Store in closed bottles.

Toluene

Procedure

Place the soil sample or soil suspension in a glass beaker or flask, add azide, cyanide or toluene solution to obtain final concentrations of 1–10% in response to azide and cyanide, and 5–10% in response to toluene. Close the flask with a rubber septum or the beaker with aluminium foil, and incubate the soil sample at the required temperature. The addition of inhibitors several hours before starting the measurements is recommended. In leaching experiments azide, cyanide or toluene concentrations should remain constant in the mobile phase. To check the intensity of the inhibition, estimation of microbial count or other microbial parameters (e.g. respiration) in the treated soils is recommended.

Discussion

No significant disturbance of soil structure occurs by using this method.

Azide and cyanide are toxic; therefore special precautions must be taken in their handling.

If required, prepare more concentrated azide and cyanide solutions, or increase the amount of toluene added to soil.

It should be pointed out that a high azide concentration (10%) causes the removal of sodium ions from the soil in leaching experiments.

Determination of soil pH

J. Forster

All (bio)chemical transformations in soil are influenced by the proton activity in the soil solution, which is usually given as the pH value. Community structure and activity of soil microbes are also governed by soil pH. It is thus one of the basic features that a microbiologist needs to know when working with soil samples. Protons are hydrated in solution and occur only as H_3O^+ (hydronium) ions. Most of the protons are dissociated H^+which originates from places of variable charge (organic carboxyl and hydroxyl groups, surface hydroxyl groups from Fe and Al hydroxides). A pH measurement in water includes easily dissociated protons, while the addition of salt (usually 0.01 M $CaCl_2$ or 1 M KCl) may also mobilize tightly bound H^+, e.g. from the Al hexaquo complexes.

Principle of the method

Potentiometric determination of the H^+ activity in a soil suspension in water or salt solution.

Materials and apparatus

pH meter, equipped with a glass electrode and a built-in reference electrode

Chemicals and solutions

Buffer solutions, pH 4 and 7.
Salt solutions (if desired, 0.01 M $CaCl_2$)

Dissolve 219.08 g calcium chloride hexahydrate in 600 ml distilled water, dilute to 1000 ml and mix well.

1 M KCl

Dissolve 74.56 g potassium chloride in 600 ml distilled water, dilute to 1000 ml and mix well.

Procedure

Weigh 10 g of air-dried and sieved (< 2 mm) soil into a plastic beaker; after addition of 25 ml distilled water or salt solution, and stirring for *c.* 1 min, the pH is measured in the supernatant after 1 h of standing and a second short stirring.

Discussion

Adjustments of the pH meter/glass electrode have to be made with buffers at pH 7 and pH 4 before measuring the samples; if the expected sample pH will exceed this range, use appropriate buffers, e.g. at a pH of 10. Repeat adjustments periodically while measuring large numbers of samples.

The solution temperature is essential for activity measurements; modern pH meters include a temperature sensor and control.

Measurements of organic soil samples require a wider soil:solution ratio; a 1:10 v/v ratio has been found convenient (Kögel 1987). Sewage sludges are examined in their original state or after the addition of a little water without stirring (DIN 38414, German Institute of Standards 1981).

Proper pH measurements require a minimum of ionic strength in solution; the electrical conductivity of the solution should be > 150 $\mu S\ cm^{-1}$ to avoid drifting of the measured pH value.

Standing times may vary between 30 min and 24 h or overnight (DIN 19684, German Institute of Standards 1973); however, they should be held constant for all samples.

Store the electrode in 1 M KCl to avoid desiccation of the glass membrane. Reactivation of used electrodes may be achieved by a short (10–30 s) treatment with dilute (e.g. 0.1 M) hydrofluoric acid.

Measurement of oxygen partial pressures in soil aggregates

J. Zausig

In structured soils, microorganisms are usually situated at the surface of or inside the single aggregates which form the bulk soil. Oxygen diffusion to the respiring organisms takes place in the fine intra-aggregate pores and is induced only by the oxygen concentration gradient, which is a result of the respiration process. Thus research on the activity of aerobic microorganisms and their environment and oxygen supply in well-structured soils has to focus on aggregate properties and on the processes of O_2 consumption and O_2 transport.

Principle of the method

Amperometric measurement of O_2 partial pressure *in situ* with a Clark-type microelectrode. If the polarization voltage of the platinum cathode is constant and falls within a range −580 to −780 mV, the current in the measuring circuit depends only on the O_2 concentration.

Materials and apparatus

The construction of Clark-type oxygen-sensitive microelectrodes was outlined by Revsbech and Ward (1983). Stepniewski et al (1991) constructed similar microelectrodes and describe a motor-equipped microdrive, which allows the continuous measurement of pO_2 profiles in soil aggregates.

Construction of O_2 sensitive microelectrodes:

The platinum cathode and the reference electrode (Ag/AgCl) are placed in an aqueous electrolyte (1 M KCl) in an outer glass casing. This is made from a soda-lime glass capillary (ϕ 5.0/7.0 mm) by shaping it with a pointed flame and using a microforge to form the very tip. The outer casing is sealed at the tip with a silicone rubber membrane which is extremely permeable to oxygen (RTV 108, General Electric).

The cathode is made from a 0.05 m long piece of platinum wire (ϕ 0.1 mm), which is etched in saturated KCN while 3 V AC is applied. Then the Pt wire is rinsed in

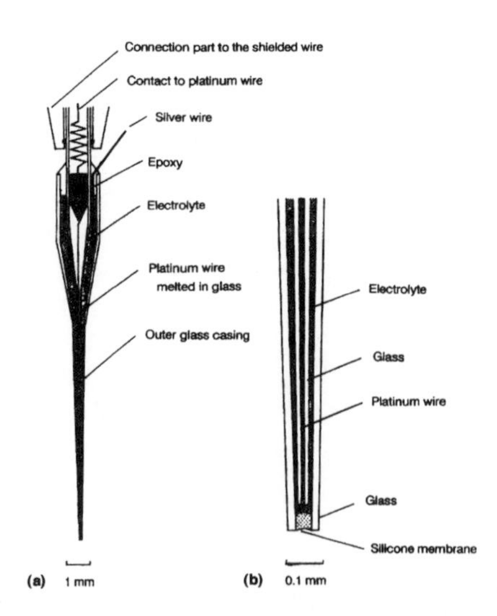

Figure 3.1 Design of an oxygen-sensitive microelectrode: (a) total electrode; (b) detail of the electrode tip

concentrated HCl, H_2O (millipore quality), and ethanol and then melted inside a thin glass capillary (Reedglas 8530 or 8533, Schott Glaswerke, Landshut, Germany). The tip of the cathode is exposed by grinding on a rotating disc and thereby etched in saturated KCN with 3 V AC applied to obtain a small recess. The cathode is pushed into the outer casing with a micromanipulator and fixed in the right position with a fast-curing polyester. The electrode is filled with methanol and deaerated in a desiccator by vacuum. The methanol is then replaced by 1 M KCl. An AgCl-covered silver wire is installed in the electrolyte as a reference electrode. Finally the top of the electrode is sealed completely with polyester. A shielded cable is used for electric contact to the voltage source and nanoammeter (Fig. 3.1).

Further apparatus:

- O_2-sensitive microelectrode
- Voltage source for constant negative voltage supply to the Pt cathode, nanoammeter (NA meter), O_2, N_2 and air pressure for calibration in gas-saturated water

Procedure

The cathode is polarized with a −580 to −780 mV constant voltage and the current in the measuring circuit is measured with a NA meter and recorded with a chart recorder.

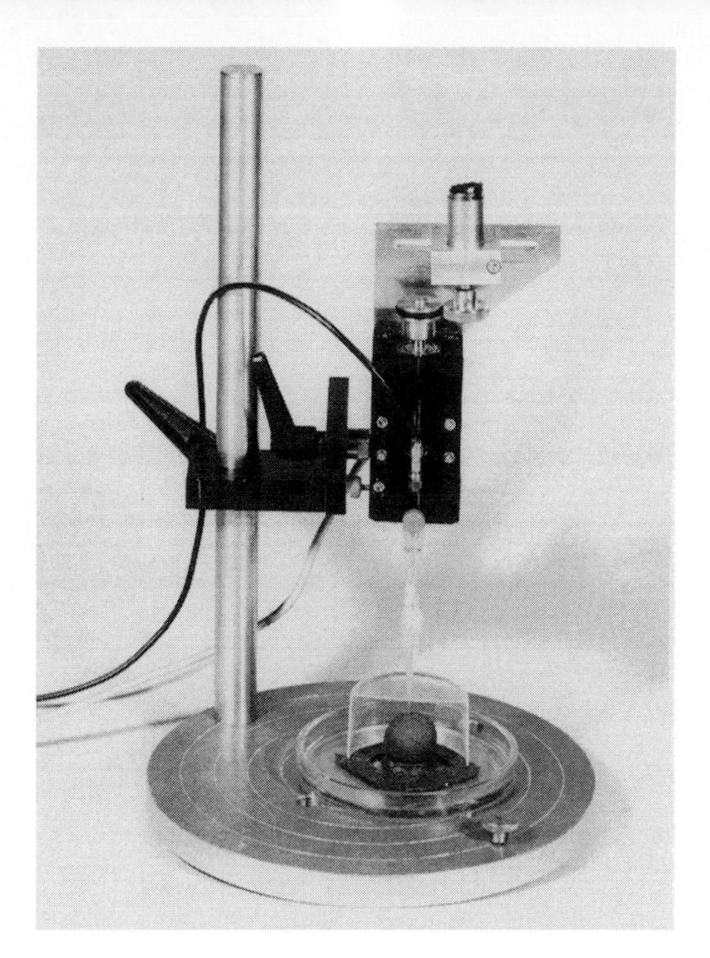

Figure 3.2 Photograph showing apparatus for measurement of oxygen partial pressure in a soil aggregate.

Calibration:

A current with an electrode in nitrogen-saturated water gives a zero value and in air-saturated water gives a maximum value (oxygen partial pressure = 20.5 kPa at T = 20°C and normal air pressure). At a voltage range of −580 to −780 mV polarization voltage, the current output was shown to be linear depending on oxygen partial pressure (Baumgärtl and Lübbers 1983). Typical signals are 0.1–0.5 nA in nitrogen-saturated water (nitrogen) and 2–7 nA in air-saturated water (air) depending on the diameter of the sensing electrode tip. The maximum current should be 10–70 times the zero current.

Measurement:

Measurements should be done at the same pressure and temperature as the calibration. The oxygen concentration at a microsite is lowered by the measurement itself (O_2 reduction at the surface of a Pt electrode). This is why the electrode should not be kept longer in one position than is necessary to reach equilibrium. The O_2-sensitive micro-electrode is mounted to a micromanipulator and pushed stepwise through the soil aggregate (Fig. 3.2). After the O_2 reduction current has stabilized, the electrode can be pushed deeper for another measurement.

Calculation

$$pO_2 \text{ at depth } x = \frac{I_x - I_{nitrogen}}{(I_{air} - I_{nitrogen})} pO_{2\ air\ (T,\ P)} \qquad (3.2)$$

where I_x is the current at depth x.

Discussion

The use of an electric motor with velocities lower than 0.1 mm min^{-1}) attached to the micromanipulator allows continuous radial profiles of pO_2 distribution to be measured.

Aggregates should not contain coarse sand or stones to avoid damage to the electrode tip.

Aggregates should not be drier than −10 kPa soil water suction because soil mechanical strength increases and electrodes may break.

A range of soil water potentials between −10 kPa soil water suction and water saturation may be monitored. At lower soil water potentials, the oxygen supply to aerobic microorganisms is not restricted because sufficient pathways for oxygen diffusion exist.

Insufficient oxygen supply and low O_2 concentrations in soil aggregates are due to (1) low O_2 diffusion due to low pore continuity and/or a water-filled pore space, and (2) oxygen consumption by aerobic microorganisms or abiotic O_2 consumption.

Organic carbon

J.C. Forster

Organic carbon is essential to microorganisms not only as a nutrient source, but also as a physical soil conditioner that influences soil aggregation and water characteristics. In turn, soil microorganisms, together with the soil enzymes, hold a key position in the processes of humification and mineralization of organic substrates, which lead to the production of persistent humus, degradable organic compounds and carbon dioxide. There is often a direct relationship between the percentages of total organic carbon and microbial biomass carbon found in soils that belong to the same climatic zone, and a relationship also exists with soil enzyme activities (O'Toole et al 1985; Suttner and Alef 1988). Organic carbon percentages are between 40 and 50% in peat soils and organic layers of forest soils, and usually < 1% in desert soils.

Total organic carbon

A large number of methods – manual or automatic – is available for the estimation of this basic soil parameter. All include the complete or partial oxidation of carbon compounds and detection of evolved CO_2. The latter need a conversion factor to calculate total organic carbon.

Nelson–Somers method

(Nelson and Somers 1975)

Principle of the method

Wet potassium dichromate digestion and titrimetric measurement of unreacted dichromate.

Materials and apparatus

Metal-free ball mill
Analytical balance
Digestion block and tubes, and drying oven
Burette (20–50 ml, 0.1 ml scale)
Magnetic stirrer
Glassware: 100 ml Erlenmeyer flasks, 100 ml and 1000 ml volumetric flasks, pipettes

Chemicals and solutions

Potassium dichromate solution (1.00 M)

Dissolve 49.024 g dry potassium dichromate in 600 ml distilled water; dilute to 1000 ml.

Sulphuric acid, concentrated (95–97%)

Ferrous ammonium sulphate, 0.20 M

Dissolve 78,390 g ferrous ammonium sulphate in 50 ml concentrated sulphuric acid, and dilute to 1000 ml with distilled water.

Indicator solution

Dissolve 0.1 g *N*-phenylanthranillic acid and 0.1 g sodium carbonate in 60 ml of distilled water; make up to 100 ml and mix well.

Procedure

Weigh 0.1–0.5 ± 0.001 g dried (40°C) and ground soil (depending on total carbon content but no more than 8 mg) into a block digestor tube; the exact sample weight (W) is noted. Then add 5.00 ml of potassium dichromate solution and 7.5 ml concentrated sulphuric acid. Place the tube in a preheated block at 150°C (drying oven) for 30 min, remove and allow to cool. Transfer the digest to a 100 ml Erlenmeyer flask and rinse with little water if necessary. Add 0.3 ml of indicator solution. Place the flask on a magnetic stirrer and titrate with Ferrous ammonium sulphate solution; a colour change from violet to green indicates the endpoint. Record the titre (T_s) and correct for the blank (T_o).

Calculation

$$\% \text{ organic carbon} = \frac{(T_s - T_o) \times 0.2 \times 0.3}{W} \quad (3.3)$$

where T_s and T_o are measured in millilitres and W is the sample weight in grams; the factors 0.2 and 0.3 originate from the molarity and amount in millilitres of the indicator solution added.

Discussion

While the method is well adapted to the rapid, routine carbon analysis of mineral soils, it is generally not recommended for organic soils (Nelson and Somers 1975).

Interferences are reported from large quantities of chloride, manganese and iron. The chloride interference can be eliminated by the addition of silver sulphate (Ag_2SO_4) to the oxidizing reagent (Nelson and Somers 1975).

Soil storage after air drying does not affect the results of carbon analysis (Nelson and Somers 1975).

This method is also suitable for light fraction analysis of microbial biomass C (Anderson and Ingram 1989).

Modified Walkley-Black method

(Jackson 1958)

Principle of the method

Colorimetric determination after wet potassium dichromate digestion.

Materials and apparatus

Metal-free ball mill
Dispensers (10 ml and 20 ml)
Glassware: 100 ml Erlenmeyer flasks, 1000 ml volumetric flasks

Chemicals and solutions

Potassium dichromate, 5% w/v

Dissolve 50 g in 600 ml distilled water; make to 1000 ml and mix well.

Sulphuric acid, concentrated (95–97%).

Barium chloride, 0.4% w/v

Dissolve 4 g $BaCl_2$ in 600 ml distilled water; make to 1000 ml and mix well.

Stock standard solution (50 mg C ml^{-1})

Dissolve 12.5 g dry glucose in 60 ml distilled water; make to 100 ml and mix well.

Calibration

Prepare working solutions: transfer 0, 100, 200, 300, 400 and 500 µl of the stock solution into labelled 100 ml Erlenmeyer flasks which now contain 0, 5, 10, 15, 20 and 25 mg C. Dry at 105°C. Perform the oxidation procedure as outlined above and read the absorbance of the resulting solutions.

Procedure

Weigh 1 ± 0.001 g dried, thoroughly mixed and ground soil into a labelled 100 ml Erlenmeyer flask. Note the exact weight (W). Add 10 ml of potassium dichromate solution and shake gently to dissolve soil and standards completely. Add 20 ml concentrated sulphuric acid with a dispenser and shake gently. Allow to cool, then add 50 ml of barium chloride solution and swirl to mix thoroughly. Allow to stand for 8 h, overnight, or centri-fuge. Pipette an aliquot of the clear (!) super-natant solution into a photometer cuvette, and record the absorbance at 600 nm.

Calculation

Plot a graph of absorbance against standard concentration. Determine sample and blank concentrations.

$$\% \text{ organic carbon} = \frac{(K_s - K_o) \times 0.1}{W \times 0.74} \quad (3.4)$$

where K_s is the sample concentration and K_o is the mean of the blanks, and W is the sample weight in grams. A factor of 0.74 is applied to correct for incomplete digestion, a factor of 0.1 is for conversion from grams per kilogram to per cent.

Conductometric method

(Wösthoff Carmhomat)

Principle of the method

Thermal dissociation of organic matter in an oxygen stream at $T = 1000°C$ and introduction of evolved CO_2 into a NaOH reservoir; the conductivity decrease of NaOH (through displacement of OH^- by HCO_3^{2-}) is recorded and converted into carbon concentration.

Materials and apparatus

Wösthoff Carmhomat 8-ADG or similar apparatus
Sample containers, preglowed at 1000°C

Chemicals and solutions

Sodium hydroxide (0.05 M)
Calcium carbonate, analytic grade (for calibration)
Seasand, carbon-free
Oxygen tank

Procedure

For calibration, weigh 1 ± 0.001 g of calcium carbonate into prepared sample containers. Push the sample container into the combustion oven and adjust the display to "120" (which means 120 mg C) with the respective control. Repeat the procedure with four different standards to ensure constant results. Process a control standard: weigh in 0.500 ± 0.001 g of calcium carbonate; a displayed value of "60" (mg C) is expected.

Weigh a suitable amount of dried (40°C) sample (< 500 mg C) into a prepared container. Cover with a thin layer of carbon-free seasand to prevent popping when high carbon levels (> 10%) are expected. Push the sample into the combustion oven and read the measured amount of carbon on the display.

Calculation

$$\%\ \text{organic carbon} = \frac{\text{Displayed value}}{\text{Sample weight} \times 10} \qquad (3.5)$$

where the displayed value is in milligrams of carbon and the sample weight is in grams.

Discussion

Change the sodium hydroxide in the conductivity cell every day before calibration

Different automatic analysers use thermal conductivity (e.g. Carlo Erba 1500 automatic nitrogen analyser) or infrared detection (e.g. Rosemount CSA 302) after combustion of the sample in a stream of oxygen. Often there is the possibility of simultaneous or sequential (after modification of the apparatus) measurement of a second element (nitrogen or sulphur). Tabatabai and Bremner (1983) reported on the use of the Leco 70-second carbon analyser; additional commercial instruments and references are listed in their work.

Dissolved organic carbon

Organic carbon measurements as described above reflect the sum of all forms of organic matter in soils, regardless of whether they are present as stable, immobile macromolecules or as highly charged molecules that are easily mobilized and transported in the soil solution.

The latter have the ability to co-transport other charged or uncharged species, e.g. metals, pesticides, etc. At the same time, they may act as a carbon source for microorganisms that become dissolved organic carbon (DOC) sources themselves when walls and contents of lysed cells are released into the soil solution. The fumigation–extraction method makes use of this effect for the determination of microbial biomass (Vance et al 1987). The water-extractable C content has been successfully related to the denitrification capacity of soils (Burford and Bremner 1975). A DOC analysis thus provides information on the amount and structure of "active" or "available" carbon.

Estimation of DOC involves an extraction step followed by instrumental analysis. "Extraction" can be performed either by sampling soil solution in the field with freely draining lysimeters or suction cups or by preparing a water or salt extract from, preferably, fresh soil in the laboratory. New ceramic P80 suction cups cause qualitative and quantitative alterations of percolated DOC solutions, while field-equilibrated cups did not change concentration and composition of DOC (Guggenberger and Zech 1992). The authors recommend the use of suction cups where DOC is not subjected to significant fluctuations. As by definition DOC excludes organics > 0.45 µm in size, membrane filtration is required prior to analysis. The problems related to filter material have generally been realized; while sorption of DOC on glass fibre filters has been observed (Abdel-Moati 1990), Norman (1993) reports sustaining DOC leakage from prepacked filter units and polycarbonate filters. Only acid-washed polysulphone membranes (Gelman Supor 200, 0.2 µm pore diameter) mounted in an acid-washed polycarbonate filter holder gave satisfying results. Total organic carbon (TOC) measurements (unfiltered samples or filtered with a larger pore size) are useful when larger fractions, e.g. bacterial cells, should be included.

It is still unclear which method would be most suitable to extract DOC solution from soil samples. Water is a common extractant, but salt solutions or mineral acids have also been used. Batch extracts at a 1:10 soil:solution ratio, centrifugates of saturated soil pastes and percolation extracts of organic layers have been investigated. Fresh soils should be processed shortly after sampling, after a short time of storage at 4°C, or after freezing and storage over longer periods. Air or oven drying will lead to the mobilization of microbial carbon. Extraction time strongly influences the amount of DOC; increasing mobilization at the beginning of the extraction procedure might be followed by re-adsorption of DOC to the soil matrix (Kaupenjohann 1989).

The quantitative analysis of dissolved organic matter includes compounds with high structural diversity. To reduce chemical heterogeneity, and to characterize the original DOC extract, Leenheer and Huffman (1976) and Leenheer (1981) have developed a DOC fractionation method. This includes a separation according to charge characteristics performed on a combination of nonionic as well as cation and anion exchange resins. The procedure leads to a maximum of six DOC fractions (hydrophobic or hydrophilic neutrals, acids, and bases).

Modern DOC analysers operate according to contrasting principles: thermal oxidation or UV digestion are combined with infrared (IR) detection or colorimetric procedures. Instrumental problems of DOC determination have been discussed extensively in a recent workshop report (Hedges and Lee 1993).

Determination of dissolved organic carbon by ultraviolet digestion/ colorimetric analysis

The procedure described below follows the application notes for the Alpkem RFA-300 air-segmented continuous flow system.

Principle of the method

Samples are first acidified to convert inorganic carbon to CO_2 which is removed by sparging with a high velocity stream of nitrogen. The sample is then mixed with a potassium persulphate digestion reagent and subjected to UV radiation which generates free radicals through the photolysis of persulphate; a complete oxidation of organic compounds (except perhalogenated species) may be expected (Peyton 1993). The generated CO_2 is dialysed through a silicone rubber membrane into a weakly buffered phenolphthalein indicator. The decrease in the colour of the indicator is proportional to the carbon concentration.

Materials and apparatus

Alpkem RFA-300 TOC application A303-S361

Chemicals and solutions

Sulphuric acid, 1 N

Add 2.8 ml of sulphuric acid to 60 ml of distilled water; dilute to 100 ml with distilled water and mix well.

Digestion reagent

Dissolve 1.21 g potassium persulphate and 3.4 g of sodium tetraborate in 60 ml distilled water. Heat gently to facilitate dissolution, cool to room temperature and dilute to 100 ml with distilled water.

Add 0.1 ml of Triton X-100 to 100 ml distilled water and mix well.

Phenolphthalein indicator

Dissolve 10.6 g sodium carbonate in 600 ml distilled water; dilute to 1000 ml and mix well (sodium carbonate, 0.1 M). Dissolve 8.4 g sodium bicarbonate in 600 ml distilled water; dilute to 1000 ml and mix well (sodium bicarbonate, 0.1 M). Mix 50 ml of the sodium carbonate with 100 ml of the sodium bicarbonate solution. Prepare a stock phenolphthalein solution by dissolving 1.0 g in 60 ml anhydrous methanol; dilute to 100 ml and mix well. Prepare the working phenolphthalein indicator by diluting 0.6 ml of stock phenolphthalein and 10 ml of the carbonate–bicarbonate buffer to 1000 ml with distilled water. Mix well and add exactly 0.5 ml of Triton X-100 (1:1 in isopropyl alcohol). Mix thoroughly and store in a tightly sealed, amber glass bottle.

Calibration

Dissolve 2.1254 g of potassium biphthalate in 60 ml distilled water, add 0.5 ml of 1 N sulphuric acid, and dilute to 100 ml with distilled water to give a 10,000 mg C l^{-1} stock solution. Dilute by a factor of 100 to obtain an intermediate calibrant containing 100 mg C l^{-1}. Pipette 0.1, 1.0, 2.0, 4.0, 6.0, 8.0 and 10.0 ml of the intermediate calibrant into 100 ml volumetric flasks and make up to volume with distilled water; the nominal concentration of the working calibrants is 0.1, 1.0, 2.0, 4.0, 6.0, 8.0 and 10.0 mg C l^{-1}.

Procedure

Follow the instructions given in Alpkem application note A303-S361 (Alpkem Corporation, Clackamas, USA).

Discussion

Analysis should be performed as soon as possible after sampling to exclude changes in DOC concentration due to microbial activity or precipitation. Sample preservation should be at 4°C; acidification may result in the precipitation of undissociated compounds and is not recommended here.

Contamination of water with DOC is a frequent source of error; check that the water has been cleaned by passage through exchange resins for DOC concentration. Never store distilled water in new plastic containers.

Automatic instruments for DOC/TOC are available from Technicon and Skalar (UV/colorimetry), Heraeus LiquiTOC (UV/IR),

Dohrman (thermal oxidation/IR), Shimadzu and others.

The Alpkem device has an hourly throughput of 20 samples.

The recoveries of a wide variety of chemically different compounds was close to 100% with the combustion/IR technique, while the UV/colorimetric device detected less than 50% of nitroaniline, thus indicating lower sensitivity for complex aromatics. A different UV/IR device, however, recovered *c.* 100% of the latter; it is therefore recommended that the performance of a system with standard compounds is checked.

Determination of dissolved organic carbon by high temperature catalytic oxidation/non-dispersive infrared detection (HTCO/NDIR)

Principle of the method
This method is based on the use of a Dohrmann DC-90 analyser combined with a model 903 autosampler or similar apparatus. After removal of the dissolved inorganic carbon (DIC) by the addition of phosphoric acid and sparging with CO_2-free oxygen gas, the aqueous sample is injected into a sample loop, heated to 900°C and oxidized completely under CO_3O_4 catalysis. After the elimination of halogens, water, and corrosive gases, CO_2 is determined by NDIR.

Materials and apparatus

Dohrmann DC-90 total carbon analyser
Oxygen tank (O_2 purity > 4.5, CO_2-free)
Amber bottles for storage of standard solutions

Chemicals and solutions

Phosphoric acid, concentrated (*c.* 85%)
Aqueous phosphoric acid carrier (0.2% H_3PO_4):

Dissolve 3 ml H_3PO_4 in 1000 ml distilled water

Deionized or distilled water, < 0.2 mg DOC l^{-1}

Calibration

Stock DOC standard solution, 2000 mg C l^{-1}

Dissolve 0.425 g potassium dihydrogen phthalate in 60 ml distilled water, add 0.1 ml concentrated H_3PO_4 and make up to 100 ml; mix well and store in the dark at 2–4°C.

Working DOC single standard solution, 10 mg C l^{-1}

Dilute 1.25 ml of the stock DOC standard solution to 250 ml and mix well.

DOC quality control (QC) solutions

Prepare a 1000 mg C l^{-1} stock QC solution monthly by dissolving 0.5313 g potassium dihydrogen phthalate in 150 ml distilled water; add 0.25 ml concentrated H_3PO_4 and make up to 250 ml; mix well. Prepare QC working solutions by diluting 50, 1000 and 3000 ml of the stock QC solution to 100 ml with distilled water including 0.05 ml concentrated H_3PO_4.

Procedure

Set up the instrument as outlined in the operation manual. Adjust the oxygen pressure to 2.0 bars. Allow the IR detector to warm up for at least 2 h; it is best to leave the detector on permanently. Calibrate the instrument with five 10 mg C l^{-1} standards. Check the linearity of the calibration with standards containing 0.50, 10.0 and 30.0 mg C l^{-1}; results are acceptable within ± 0.1, 0.5 and 1.5 mg C l^{-1}, respectively [Environmental Protection Agency (EPA) 1987].

Analyse a sample batch containing at least three standards and one blank solution. Blanks should contain less than 0.1 mg C

l^{-1}, however, due to the high detection limit it is often difficult to decide whether there is a contamination of the used distilled water.

Calculation

Read the DOC concentration from the instrument output if the calibration was made with a single standard. If the non-linear response is observed in the concentration range of interest, erase the instrument calibration, and analyse a set of standards containing 0.500, 1.000, 5.000, 10.00 and 30.00 mg C l^{-1}. Plot a calibration curve, and read or calculate sample concentrations from the first-order regression equation.

Discussion

The detection limit for the DC-90 is *c.* 1 mg C l^{-1}; a value of 0.8 mg C l^{-1} has been reported (EPA 1987) for the DC-80 analyser which uses UV-catalysed persulphate oxidation.

Organic particles up to 0.5 mm size are oxidized quantitatively.

Samples with salt contents of < 3% may be measured reliably.

The concentration range is dependent on the size of the sample loop. The loop sizes and respective DOC concentration ranges available for the Dohrmann DC-90 are as follows: 40 μl (100–2500 mg l^{-1}); 200 μl (10–700 mg l^{-1}); 500 μl (0.2–20 mg l^{-1}); the accuracy of loop filling is ± 2%.

Standard preparation should be repeated monthly.

Proper sample handling and cleaning of storage bottles is essential for accurate results. Bottle necks and caps should be treated most carefully.

Composition of soil organic matter

I. Kögel-Knabner

The chemical structure of plant litter and humic substances in soil is very complex. However, it is possible to obtain a characterization if the organic material is grouped into several compound classes (Ryan et al 1990): (1) extractable and saponifiable lipids; (2) hydrolysable proteins; (3) crystalline and non-crystalline polysaccharides; (4) lignin. These compounds are found in plant litter as well as in the soil organic matter. The analytical assay differs with respect to the first step, which can be an extraction or depolymerization/extraction procedure. This is followed by a second step that includes the separation and/or determination of the released compounds. The compounds may be quantified by gravimetric determination (lipids), colorimetric procedures (total carbohydrates), or chromatographic separation and determination (lignin oxidation products).

The procedures described below are applicable to plant litter, forest floor materials and mineral soils. Sample treatment often leads to changes in the chemical structure of the residue, which most probably influences the behaviour of these compounds in the following step, e.g. the harsh conditions necessary for the hydrolysis of polysaccharides will alter the lignin compounds in the residue. During hydrolysis with 72% sulphuric acid a small amount of lignin is solubilized, the so-called acid-soluble lignin, and the lignin residue has undergone a partial hydrolysis of the β-O-4 linkages (Leary et al 1986). We therefore do not favour a proximate carbon fraction analysis which is based on sequential extraction/depolymerization procedures. All the procedures described below are applicable to bulk soil samples, but also to fractions obtained from chemical or physical fractionation procedures. In the recent decade, the progress in solid-state ^{13}C nuclear magnetic resonance (NMR) spectroscopy has allowed the bulk chemical composition of soil organic matter to be characterized without extraction or purification steps. The use of solid-state ^{13}C NMR spectroscopy in soil and sediment chemistry has been described in detail by Wilson (1987).

Determination of total carbohydrates

(Pakulski and Benner 1992)

Carbohydrates are the main constituents of plants, representing between 50% and 75% of their dry weight. In mineral soils they account for 5–25% of the total organic matter. They represent the major carbon and energy source for soil microorganisms.

Principle of the method

Several possibilities exist for the quantitative determination of carbohydrates in soil or in plant litter, depending on the information required. The method described here is a two-step selective hydrolytic depolymerization reaction followed by the colorimetric analysis of the released monomers. The type of hydrolysis procedure determines the different fractions of carbohydrates that are released. Total carbohydrates are determined after hydrolysis with 12 M

H_2SO_4/1 M H_2SO_4, which effectively release crystalline (cellulose) and non-crystalline polysaccharides, such as plant hemicelluloses and microbial polysaccharides. For the hydrolysis of non-crystalline polysaccharides, which include plant-derived hemicelluloses, but also most microbial polysaccharides, a treatment with dilute H_2SO_4, dilute HCl or trifluoroacetic acid (TFA) may be used.

The determination of the monomers released can be done by analysis of individual monomers by gas chromatography or high performance liquid chromatography (HPLC). The amount of total carbohydrates in the hydrolysate can also be determined by colorimetric assays, such as the phenol sulphuric reagent, the anthrone reagent (Dubois et al 1956) or the MBTH (3-methyl-2-benzothiazolinone hydrazone hydrochloride) reagent. In contrast to chromatographic methods, the spectrophotometric determination of carbohydrates is rapid, sensitive, and requires a minimum of equipment and manipulation beyond hydrolysis. The MBTH assay is a combination of a series of well-established chemical procedures. Hexoses and pentoses are reduced to sugar alcohols with borohydride in aqueous solution. The sugar alcohols containing two 1,2-diol structures are oxidized by periodic acid yielding 2 moles of formaldehyde per mole of sugar alcohol. MBTH is used to determine formaldehyde in aqueous solution. Two moles of formaldehyde are formed from hexoses and pentoses and sugar alcohols, and one mole from deoxyhexoses, containing only one 1,2-diol structure.

Materials and apparatus

Hot plates
Glass vessels for hydrolysis with screw caps
Test tubes with rubber stoppers or Teflon-lined caps

Chemicals and solutions

12 M H_2SO_4
0.1 M HCl
1 M HCl
3 M NaOH
0.36 M HCl
0.025 M periodic acid
0.25 M sodium arsenite
2 M HCl
2.5% (w/v) $FeCl_3$
Acetone p.a. (analytical grade)
Potassium borohydride, 20 g KBH_4 l^{-1}; prepare fresh every day
MBTH reagent

The MBTH reagent is prepared by heating 2.76 g MBTH (3-methyl-2-benzothiazolinone hydrazone hydrochloride, Aldrich) in 100 ml 0.1 M HCl for 2 h. After cooling, undissolved hydrazone and precipitated MBTH are removed by filtration through a glass fibre filter and the filtrate is diluted with 100 ml 0.1 M HCl. The reagent can be stored for at least 5 days in an amber bottle.

Procedure

Hydrolysis of polysaccharides with concentrated sulphuric acid. A total of 50 mg of a sample with a high organic carbon content (litter, plant materials) or 250–1000 mg of a mineral soil sample is suspended in 1.5 ml 12 M H_2SO_4 in a closed vial and shaken for 16 h at room temperature. The solution is then diluted with 17.5 ml distilled water to result in 1 M H_2SO_4 and hydrolysed at 100°C for 5 h. After cooling, the hydrolysate is filtered, neutralized with 3 M NaOH (pH 6–7) and made up to a final volume of 100 ml.

Hydrolysis of non-crystalline polysaccharides. A total of 100 mg of a sample with a high organic carbon content (litter, plant materials) or 250–1000 mg of a mineral soil sample is suspended in 50 ml 1 M HCl in a screw-capped vial (volume 250 ml) under N_2. For hydrolysis the vials are kept in a laboratory oven at 100°C for 5 h. After cooling, the suspension is filtered over Whatman GF/F glass fibre filters and the residue is washed with 180 ml of water. The filtered solution is neutralized with 3 M NaOH (pH 6–7) and made up to a final volume of 250 ml. Alternatively, for the hydrolysis of non-crystalline polysaccharides, trifluoroacetic acid or 0.63 M HCl (Fengel and Wegener 1979, 1984; Beyer et al 1993) can be used.

Colorimetric determination of carbohydrates by MBTH. For spectrophotometric analysis the hydrolysates have to be diluted between 1:25 and 1:100, depending on the wide range of carbon contents of plant material, litter layers and soils.

One millilitre of neutralized hydrolysate is incubated with 0.05 ml of cold freshly prepared solution of potassium borohydride for 4 h in darkness to reduce the aldehyde groups of the sugars to alcohols. For routine analysis, the reduction step is practically performed overnight. Before oxidation, excess borohydride must be destroyed with 0.05 ml 0.36 M HCl. A 0.1 ml portion of 0.025 M periodic acid is added after 10 min to oxidize the alditols to formaldehyde. Ten minutes later, the reaction is stopped with 0.1 ml of a 0.25 M solution of sodium arsenite and the solution is left for 10 min. Then 0.2 ml of 2 M HCl are added. Then 0.4 ml MBTH reagent are added and the tightly stoppered tubes are heated in a boiling water bath. The samples are then cooled to room temperature in a water bath and oxidized with 0.2 ml 2.5% $FeCl_3$ solution. The solution is mixed with 1 ml acetone, and the absorbance is read at 635 nm in a 1 cm quartz cell against a distilled water–acetone (2:1) reference.

The analysis is carried out in triplicate in capped glass tubes (except during the borohydride destruction). To correct for turbidity, free aldehydes and reagents, three control samples are treated exactly the same way as the samples except for the periodate oxidation. Formation of formaldehyde is avoided by adding 0.2 ml of a premixed solution (1:1) of periodic acid and sodium arsenite solution. The difference between the sample and the control represents the formaldehyde liberated by the oxidation of the sugar alcohols. The carbohydrate contents of the hydrolysates can then be calculated as glucose equivalents.

Discussion

The assay is insensitive to a series of amino acids and other organic compounds. The procedure for the controls corrects for any substance reacting directly with MBTH and for turbidity. Consequently, even coloured soil hydrolysates can be analysed without purification steps. Possibly interfering compounds are those yielding formaldehyde upon periodate oxidation. These are compounds containing 1,2-diol or 1-hydroxy-2-amino structures, e.g. glycol, glycerol, and serine. However, the amount of glycerol and serine in soils is low and therefore their contribution to the total amount of carbohydrate determined by this procedure can be neglected.

The determination of individual sugars in the hydrolysates can be carried out with gas chromatographic separation after reduction and derivatization. Suitable procedures were described by Cowie and Hedges (1984) or Sugahara et al (1992).

The colorimetric determination of carbohydrates with MBTH can also be applied to determine other fractions, such as free water-soluble sugars or total sugars in hydrolysates of water extracts.

Characterization of the amount and degree of lignin biodegradation by CuO oxidation

Lignin is considered to be one of the major precursors of aromatic carbon in humic compounds. Lignin is composed of cross-linked phenyl-propane monomers, commonly referred to as guaiacyl/vanillyl, syringyl and *p*-hydroxyphenyl units. The relative proportions of each monomeric unit in the lignin of a particular plant species depend on its phylogenetic origin (Sarkanen and Ludwig 1971). The lignin of hardwoods such as beech consists of about equal proportions of guaiacyl and syringyl monomers, softwood lignin is composed mainly of guaiacyl units, whereas the lignin of grasses consists of about equal amounts of guaiacyl, syringyl and *p*-hydroxyphenyl units.

Principle of the method

The method includes oxidation of lignin with CuO and 2 M NaOH at 170°C under N_2 for 2 h. The lignin of wood yields the ketone, aldehyde and acid phenols of the vanillyl (acetovanillon, vanillin, vanillic acid), syringyl (acetosyringon, syringaldehyde, syringic acid) and *p*-hydroxyl (*p*-hydroxy-benzoic acid, *p*-hydroxy-benzaldehyde) type. Lignins of non-woody tissues additionally yield *p*-coumaric acid and ferulic acid, which are partly bound via ester linkages.

Materials and apparatus

Pressure bomb with Teflon vessels
Solid–liquid extraction system equipped with C18 reversed-phase columns or cartridges
HPLC with a binary gradient system, equipped with a UV detector or gas chromatograph (GC) equipped with a flame ionization detector (FID)
Rotary evaporator
Centrifuge with glass vessels
Hot plate

Chemicals and solutions

CuO p.a.
Fe $(NH_4)_2(SO_4)_2 \times 6H_2O$
NaOH, 2 M (stored under N_2)
N_2
6 M HCl
Acidified water (1 ml concentrated HCl in 1 l H_2O)
Ethyl acetate
Methanol

For HPLC:

Acetonitrile p.a.
Phosphate buffer 50 mM, brought up to pH 2 with H_3PO_4

For GC:

Phenyl acetic acid (100 mg l^{-1} in methanol)
Bovine serum albumin (BSA)

Procedure

CuO oxidation (Hedges and Ertel 1982). Each vessel of the pressure bomb is loaded with 50 mg of organic material (litter, plant material) or 100–500 mg mineral soil material and 50 mg Fe $(NH_4)_2$ $(SO_4)_2$ $6H_2O$, 250 mg CuO and 15 ml 2 M NaOH. Nitrogen is bubbled through the solution for 5 min and the vessel is sealed. The pressue bomb is heated to 170°C and the reaction temperature is maintained for 2 h. After cooling, the suspension is centrifuged for 25 min at 3000 rev min^{-1}. The supernatant is saved and the sediment washed with water and centrifuged as before. The combined supernatants are acidified to pH 2 with 6 M HCl and kept at room temperature in the dark for 1 h. The suspension is centrifuged, decanted and saved, and the residue washed with acidified water and centri-fuged again. Before sample application, the disposable columns are conditioned with methanol and water as described by the supplier. The combined, acidified supernatant recovered from centrifugation is added, applied by vacuum to the column and the sample passed through. After drying of the column in a stream of N_2 for several minutes, lignin oxidation products are selectively eluted in four volumes of 0.5 ml ethyl acetate. The ethyl acetate is removed with a rotary evaporator and the residue is dissolved in 1 ml of methanol. Samples should be kept in the dark to pre-vent isomerization, e.g. of cinnamic acids.

Separation and quantification of CuO oxidation products.

The generated lignin-derived phenols are separated and quantified by HPLC or gas chromatography. For HPLC separation a binary gradient system equipped with a UV detector is necessary (Kögel and Bochter 1985). The individual conditions will depend on the HPLC system and the C18 reversed-phase column used for separation. Commonly, a concave gradient is used, with a mixture of phosphate buffer (pH 2) and acetonitrile, starting from 10% acetonitrile and ending at 90% acetonitrile. The samples can be injected to the HPLC column without being cleaned up further. For gas chromatographic analysis, it is necessary to derivatize the lignin oxidation products and to add an internal standard (Hedges and Ertel 1982; Guggenberger et al 1993). The lignin oxidation products in methanol as obtained from the solid–liquid extraction procedure are transferred to a reacti-vial and dried under a gentle stream of N_2. A certain amount of the internal standard phenyl acetic acid (100–200 µl) is added and the solvent is again removed by drying under N_2. The residue is dissolved in 100–200 µl BSA and allowed to stand at room temperature for about 1–2 h. The gas chromatographic separation is conventionally achieved with a non-polar fused silica column, such as OV1 or SE 54, 25 m and a temperature programme from 100°C (3 min) to 250°C (10 min) at a heating rate of 10°C min^{-1} (Guggenberger et al 1994).

Discussion

The yields and composition of phenols are different for lignins derived from different types of plants (Sarkanen and Ludwig 1971). Gymnosperm wood lignin produces mainly vanillyl type oxidation products and angiosperm wood lignin produces syringyl units in addition to vanillyl units. High yields of cinnamyl units are characteristic for non-woody angiosperm and gymnosperm tissues (Ertel et al 1984). Syringyl phenols are released with 90% efficiency by alkaline CuO oxidation compared to 30% efficiency for vanillyl units (Ertel et al 1984). The absolute yields then depend on the proportions of syringyl and vanillyl units in the plant tissue investigated. The mass ratio of acid/aldehyde for the vanillyl units (Ac/Al)V can be used to determine the degree of oxidative decomposition of lignin within a sample (Ertel and Hedges 1984, 1985).

Further information on lignin composition and degradative processes in soils and sediments can be obtained by analysing the dimeric oxidation products in addition to the monomers as described by Goni and Hedges (1992). The same CuO oxidation procedure has also been used to determine the amount of cutin-derived material in a soil or sediment sample (Goni and Hedges 1990).

Total amino sugars

(Scheidt and Zech 1990)

Amino sugars are almost exclusively derived from soil microorganisms. The major sources for amino sugars in soils are bacteria that produce both glucosamine and galactosamine, and fungi that mainly produce galactosamine (Parsons 1981). Minor amounts of mannosamine have been found in soils.

Principle of the method

The analysis of amino sugars is done after a conventional hydrolysis with 6 м HCl, which is also common for the hydrolysis of proteins and the determination of amino acids. The colorimetric determination of total amino sugars as described here is done over a period of 3 days. About 15 samples can be analysed within this period. The analysis of individual amino sugars requires the use of chromatographic separation techniques, as described by Kögel and Bochter (1985).

Materials and apparatus

Spectrophotometer
Laboratory oven
Screw cap vials
Mixer for test tubes
Evaporating flasks

Chemicals and solutions

Concentrated HCl
6 м HCl
5 м NaOH
Ethanol p.a.
Acetylacetone p.a.
1 м Na_2CO_3
Diethyl ether p.a.
p-Dimethylaminobenzaldehyde
$HgCl_2$
D-Glucosamine
D-Galactosamine
Ehrlich reagent

Prepare a mixture of 400 mg *p*-dimethylaminobenzaldehyde, 15 ml ethanol and 15 ml concentrated HCl. The reagent is stable for several days if stored in the refrigerator. The reagent can no longer be used when the yellow-green colour is lost.

Acetyl acetone solution

Use 1.5 ml acetyl acetone and add 1 м Na_2CO_3 to a final volume of 50 ml. This reagent is stable only for several hours and should therefore be prepared fresh before the spectrophotometric determination of amino sugars.

Procedure

Hydrolysis. Use 50–200 mg of a dry (65°C) ground soil sample, depending on its total organic carbon content, in screw cap vials. Add 20 ml 6 м HCl and shake gently (about 100 min^{-1}) on a horizontal shaker for 4 h. Then hydrolyse for 4 h at 100°C in a laboratory oven and allow to cool. At this point the samples can be stored overnight in the refrigerator.

Spectrophotometric determination of amino sugars. The hydrolysed samples are filtered with cellulose filters in 50 ml evaporating flasks and the residue is washed with distilled water. The solution is evaporated to dryness (essential) at a temperature of < 40°C. This procedure requires about 1 h. The dried residue can be stored in the capped containers overnight if necessary.

Add 10 ml of distilled water to the evaporating flask and dissolve the residue completely. This probably requires the use of an ultrasonic bath. Use 2 ml of this solution for the analysis and pipette in a test tube. Add 1 ml of acetyl acetone solution, close the tube with a rubber stopper and mix on a test tube mixer. The solution is then heated for 40 min in a water bath (80–95°C). Subsequently the solution is removed from the water bath and allowed to stand at room temperature to cool. Then 1 ml of Ehrlich reagent and 1 ml ethanol are added to the cooled solution and mixed. The colour of the solution changes to red. The test tubes are left open and are again heated in the water bath at a temperature of 60–70°C for 5–10 min. Then 1 ml 5 м NaOH is added to the solution, which changes colour to yellow after thorough mixing. Add 2 ml diethyl ether and mix again intensively. The yellow colour is found in the ether phase. The ether phase is transferred to glass cuvettes with a pasteur pipette and closed with a Teflon stopper. The amount of amino sugars is determined at 467 nm against an ether reference cuvette. The spectrophotometric analysis of the ether phase should be done within 30 min and 2 h from the ether addition to the test tube.

Calibration curve. Prepare solutions containing between 0 and 150 mg D-glucosamine l^{-1}. These standards (or mixtures of D-glucosamine and D-galactosamine) are treated similarly to the hydrolysates.

Discussion

The amount of total amino acids can be determined as described below in the same hydrolysates.

Extractable lipids

Soil lipids comprise a wide range of individual components soluble in organic solvents but differing in chemical structure and physico-chemical properties. A detailed characterization of lipids from various origins is given by Holloway (1984) and Dinel et al (1990). Close correlations were obtained between the lipid fraction and the microbial activity of soils (Capriel et al 1990).

Principle of the method

The lipid fraction is extracted with a mixture of chloroform and methanol (Bligh and Dyer 1959). The chloroform phase, which contains the lipids, is dried and weighed.

Materials and apparatus

Erlenmeyer flasks, 50 ml, with glass stoppers
Horizontal shaker
Büchner funnel and filtering flasks
Separatory funnels
Rotary evaporator
Desiccator
Evaporating flasks, 100 ml

Chemicals and solutions

$CHCl_3$ p.a.
Methanol p.a.
Mixture of $CHCl_3$/methanol (1/2 v/v)
0.1 M KCl

Procedure

Soils are dried and ground, and 6.00 g of the mineral soil (or 2.00 g of the plant, litter or peat materials) are placed in glass-stoppered Erlenmeyer flasks (50 ml) and moistened with 8 ml distilled water for 6–8 h. A mixture of $CHCl_3$ and methanol (30 ml) is added and the suspension is shaken under N_2 for 16 h at room temperature. For purification of the extracted lipids 10 ml $CHCl_3$ are added to the monophasic solvent system and shaking is continued for another hour. Then 30 ml of 0.1 M aqueous KCl are added and shaking is continued for 15 min. The mixture is then filtered under suction through glass microfibre filters on a Büchner funnel. Wash the residue with a small amount of $CHCl_3$. The filtrate is transferred to a separation funnel and allowed to separate (15 min) into a methanolic aqueous phase (upper layer) and a chloroform phase (lower layer). The weight of the evaporating flask is determined. The methanol–water layer is discarded and the $CHCl_3$ layer containing the purified lipids is collected and evaporated to dryness on a rotary evaporator at 30–35°C. The dried residue is weighed and the amount of lipids is then obtained by difference. The amount of lipid is given as a percentage of dry soil or as percentage of soil organic matter.

Discussion

The extraction with chloroform–methanol is preferred to the extraction with other solvent mixtures or petroleum ether (Heng and Goh 1981) because of higher yields. The procedure described above extracts the free lipids occurring in soil. Additionally, other paraffinic materials exist in soil, which can not be determined by wet chemical methods. For these bound aliphatic materials, other techniques, such as solid-state ^{13}C NMR spectroscopy or analytical pyrolysis are necessary (Hatcher et al 1981; Kögel-Knabner et al 1992).

Hydrolysable proteins

Principle of the method

The α-amino group of amino acids is determined by a colorimetric procedure after release of amino acids from proteins by hydrolysis with 6 M HCl. Compounds interfering with the colorimetric determination by ninhydrin (NH_3, amino sugars) are destroyed (Stevenson and Cheng 1970).

Materials and apparatus

Hydrolysis vessels
Evaporating flasks
Volumetric flasks, 100 ml
Eppendorf caps
Laboratory oven
Centrifuge
Vortex mixer
Water bath
Spectrophotometer

Chemicals and solutions

Cellulose filters
6 M HCl/1 M formic acid (10/1)
Phenolphthalein (0.1%, w/v) in ethanol p.a.
5 M NaOH
Ethanol/H_2O (1/1 v/v)
Concentrated acetic acid
Sodium acetate × 3 H_2O
Sodium citrate
Dimethylsulphoxide (DMSO) p.a.
Ninhydrin
Hydrindantin
$SnCl_2$ × 2 H_2O p.a.
Glycine p.a.
Nitrogen
Sodium citrate

29.4 g in 250 ml distilled water (store in refrigerator)

4 M sodium acetate buffer

Add 544 g sodium acetate × $3H_2O$ to 400 ml distilled water and heat to dissolve. After cooling add 100 ml concentrated acetic acid, adjust pH to 5.5 with NaOH or acetic acid, and add distilled water to a final volume of 1 l.

Ninhydrin reagent

Add 1.000 g ninhydrine and 70 mg hydrindantin to 50 ml DMSO treated with N_2 for 5 min. Treat with N_2 again. Add 25 ml sodium acetate buffer, 25 ml H_2O, treat with N_2 again, add 80 mg $SnCl_2$ × 2 H_2O, treat with N_2 for 20 min and close the bottle tightly. The colour of the reagent is yellow; a pink colour indicates that the reagent can not be used any more. The bottle should be treated with N_2 for 10 min each time after it has been opened.

Procedure

Use 100–200 mg organic material (litter, plant residues) or up to 1 g dried and ground mineral soil. Weigh soil in a hydrolysis vessel and add 5 ml of the 6 M HCl/1 M acetic acid mixture, and bubble N_2 through the suspension for 5 min. Close the hydrolysis vessel. Hydrolysis is achieved at 110°C for 12 h in a laboratory oven. After cooling the suspension is filtered and the filtrate is collected in an evaporating flask. The filtrate is evaporated to dryness with a rotary evaporator at 40°C. The residue is washed with distilled water, the filtrate collected in the same evaporating flask and again evaporated to dryness to remove traces of acid. The residue is dissolved in 5 ml bidistilled water. An ultrasonic bath can be used for this procedure if necessary. Add several drops of phenolphthalein (0.1% in ethanol) and titrate with 5 M NaOH until a pink colour shows. Subsequently the solution is kept at 110°C in the laboratory oven for 45 min. During this period of time, NH_4 will escape; the pink colour does not change. The solution is then reduced to dryness at the rotary evaporator again, redissolved in water (ultrasonic bath) and transferred to a 100 ml volumetric flask. After thorough mixing, 1.5 ml of this solution is transferred to Eppendorf caps. They are centrifuged for 12 min at 5000 rev min^{-1} and the solution is transferred to fresh Eppendorf caps. For colorimetric determination, use 25 µl of the solution containing the hydrolysed amino acids and pipette in a fresh Eppendorf cap. Add 25 µl of sodium citrate solution, 100 µl of ninhydrine reagent and mix with a vortex mixer. The Eppendorf caps are then heated for 20 min in a boiling water bath and subsequently cooled in cold water (flowing) for 10 min. Add 1.25 ml ethanol/water mixture (1/1) to the cap and mix again. If a precipitate is found, centrifuge at 8000 rev min^{-1} for 10 min. The absorbance of the solution is read at 570 nm against a reference solution that is free of amino acids.

Calibration

A calibration curve is obtained with solutions of glycine in concentrations of 10, 20, 50, 70 and 100 mg l^{-1}. The result is given as micrograms of α-amino-N per gram soil. The amount of total protein per gram of soil or per gram of organic matter can be calculated by multiplying this result by a factor of 6.25.

Discussion

The amount of total amino sugars can be determined as described above in the same hydrolysates.

The determination of individual amino acids requires a chromatographic separation. This is conventionally done by ion chromatography (amino acid analyser), but is also possible via gas chromatography or HPLC.

Physical fractionation of soil organic matter

Soil scientists have used physical separation techniques to study humification processes (Greenland and Ford 1964; Turchenek and Oades 1979). The turnover of soil organic matter (SOM) can be described by using soil fractions, provided the isolated fractions are related to structural or functional compartments of soil (Cambardella and Elliott 1993). The concept supporting physical fractionation of SOM emphasizes the role of soil minerals in SOM stabilization and turnover (Martin and Haider 1986; Oades 1988; Baldock et al 1992; Christensen 1992). The separates obtained from physical fractionation can be further investigated for their chemical structural composition and also for microbial biomass content (Jocteur Monrozier et al 1991) or other soil biological parameters, such as carbon or nitrogen mineralization potential (Cambardella and Elliott 1993).

Physical fractionation includes density and particle size separation of primary organo-mineral complexes in soils. Fractionation of soil according to particle size or density yields organo-mineral fractions with distinctly different properties in terms of organic matter turnover and chemistry (Kögel-Knabner and Ziegler 1993; Guggenberger et al 1994). The physical fractionation techniques are based on the general concept that the fresh or only slightly decomposed plant remains are associated with the light or coarse fractions and the humified materials with the heavy or small fractions. Several procedures described in the literature use a combination of particle size and density separation (Baldock et al 1991; Elliott and Cambardella 1991; Christensen 1992).

A major problem is associated with the dispersion of the particle size fractions, which is commonly carried out using ultrasonic treatment. Standardization of ultrasonic dispersion techniques is still lacking. The determination of the power output of the ultrasonic probe is recommended as a check for routine operation. Detailed instructions for the optimization of the ultrasonic treatment and the measurements necessary are given by Christensen (1992).

Density fractionation

Principle of the method

A common method for isolating soil organic matter fractions according to density involves the use of a mixture of high density organic solvents, such as $CHBr_3$ (or $CClBr_3$). The density liquids described here are mixtures of $CHBr_3$ and ethanol, which can be adjusted to different densities. Often density liquids of 1.6, 2.0 and 2.4 g cm^{-3} are used (Turchenek and Oades 1979; Brückert 1982).

Materials and apparatus

For density separation:

- Laboratory hood
- Glass centrifuge vessels, 80 ml volume
- Glass funnels
- Centrifuge
- Ultrasonic probe
- Erlenmeyer flasks
- Drying oven under a hood
- Graded cylinders, 500 ml
- Pipette, 10 ml
- Glass beaker
- Brown glass bottles for storage of reagents

For recovery of $CHBr_3$:

- Separatory funnel
- Distillation bridge
- Oil bath
- Heating plate
- Erlenmeyer flasks

Chemicals and solutions

$CHBr_3$, p.a.
Ethanol, p.a.
Acetone, p.a.
Na_2SO_4, dried at 105°C for 16 h
Filter paper, 125 mm diameter

Procedure

Preparation of density liquids. The calculation of the percentage of ethanol in the ethanol–$CHBr_3$ mixture is done as follows (Richter et al 1975):

$$\text{vol\% ethanol} = (100\, d_s - 100\, d)/(d_s - d_E) \quad (3.6)$$

where d_s = density of $CHBr_3$; d_E = density of ethanol; d = density to be prepared (usually 1.6, 2.0 and 2.4 g cm^{-3}).

Check for density and correction: 10 ml of the liquid are pipetted in a beaker and weighed under a laboratory hood. This procedure has to be repeated several times. The density has to be corrected if the actual density of the liquid differs by more than 0.02 g cm^{-3}. The correction is done by replacing a specific amount of the liquid with ethanol if the density is too high or with $CHBr_3$ if the density is too low, according to

$$V\,(\%) = w/(d - d_{s/E}), \quad (3.7)$$

where w = real weight of 10 ml of liquid – calculated weight; d = real density; $d_{s/E}$ = density of $CHBr_3$ or ethanol, respectively.

It is important to determine the density for every bottle of reagent before use (see below). The density given by the manufacturer can differ significantly from the actual density measured, which is extremely important if $CHBr_3$ is recovered after use.

Fractionation procedure. The fractionation procedure starts with the separation of the light fraction and proceeds to the separation of heavy fractions. The whole fractionation procedure has to be done under a laboratory hood. Extreme care must be exercised during the whole procedure, as $CHBr_3$ is a severe poison.

Four grams of organic material or 8 g mineral soil are weighed into a glass centrifuge tube and 60 ml of the density liquid (1.6 g cm^{-3}) are added. The suspension is subjected to ultrasonic dispersion for 30 s (40 W, 20 kHz) and subsequently centrifuged for 20 min at 3500 rev min^{-1}. The floating material is decanted in a funnel and the filtrate is recovered and added to the soil residue again. The suspension is dispersed by ultrasonic vibration for 15 s and again centrifuged for 20 min at 3500 rev min^{-1}. The procedure is repeated several times until no material is floating on the liquid surface (usually two or three replicate treatments). The light fraction is washed three times with 10 ml ethanol. The heavy fraction is recovered from the centrifuge tubes with ethanol, washed, filtered and washed again. Both fractions are dried in a drying oven (under a laboratory hood) at 65°C overnight. The heavy fraction is then further separated by adding 30 ml of the liquid with density 2.0 g cm^{-3} in a centrifuge tube. Only low volumes (30 ml) of high-density liquids (2.0 and 2.4 g cm^{-3}) may be used. The soil residue is again treated with ultrasound (15 s) and centrifuged as described before. After separation of the material with a density of 1.6–2.0 g cm^{-3}, the material is dried as described previously and subjected to the same procedure with a liquid of density 2.4 g cm^{-3}. After drying, all fractions are weighed so that the yield and distribution of fractions can be calculated.

Recovery of $CHBr_3$. As $CHBr_3$ is rather expensive, a separation and recovery step can be added. The $CHBr_3$/ethanol mixture is separated with H_2O in a separatory funnel for some minutes. Ethanol is found in the aqueous phase. The heavy phase ($CHBr_3$) is transferred to another separatory funnel and the ethanol/water phase is discarded. As the latter contains traces of $CHBr_3$, an appropriate disposal procedure has to be applied. The separation procedure for the heavy phase is repeated twice with water. The $CHBr_3$ phase is collected in an Erlenmeyer flask and dried by adding Na_2SO_4. The Na_2SO_4 is filtered and the $CHBr_3$ is distilled at about 140–145°C in an oil bath, which is kept at 180°C. The first fraction is discarded and the $CHBr_3$ is collected when the boiling point has stabilized, and a clear and colourless liquid is obtained.

Discussion

The use of organic solvents leads to a loss of extractable lipids from all fractions. The losses of total SOM for the whole fractionation procedure by ethanol/$CHBr_3$ mixtures are lower than for other solvents; mean losses were 5%, and 10% was not exceeded in work reported by Beudert et al (1989). Alternatively, inorganic high-density liquids, such as $ZnCl_2$, NaJ or sodium polytungstate have been used. However, Ertel and Hedges (1985) reported that $ZnCl_2$ led to high losses (20%) of total C, and concluded this compound is inappropriate for soil or sediment fractionation. The use of NaJ leads to high losses of organic C because of changes in the cationic composition of the soil exchange complex, especially the soil organic matter. Sodium polytungstate has been used successfully for combined particle size and density fractionation (Cambardella and Elliott 1993). The major problem associated with this procedure is that the fractions obtained are contaminated with sodium polytungstate, which can not be removed completely from the density separates.

Particle size fractionation

The study of soil organic matter, for a long time, has concentrated on bulk soil samples or on the classical acid–base solubility concept (Schnitzer 1991). Unfortunately, this traditional approach is lacking a clear differentiation of organic matter, which represents a continuum of degradation states and degrees of association with the mineral soil matrix. Investigations on the micromorphology, chemical structure and biochemical turnover of soil size separates showed that there is a clear pattern that is related to the particle size class (Turchenek and Oades 1979; Elliott and Cambardella 1991; Baldock et al 1992; Christensen 1992). Particle size fractionation has thus proved to be a most powerful tool in research on the distribution and cycling of elements that are mainly linked to SOM, such as carbon, nitrogen and sulphur (Anderson et al 1981). The dynamics of heavy metals and xenobiotics (pesticides, aromatic and halogenated hydrocarbons) might also be reflected by their distribution over the particle size spectrum of soils.

Principle of the method
(Christensen 1992)

Ultrasonic dispersion of a soil sample in water at a low soil:water ratio, and separation of particle size classes by sieving, sedimentation and centrifugation.

Materials and apparatus

Ultrasonic disintegrator, 300–500 W power input
Glass beakers, 200 ml
Ice bath
Magnetic stirrer
Analytical sieves, mesh width 2.00 and 0.05 mm
Wet-sieving apparatus
Sedimentation flasks, e.g. Atterberg flask, 1 l
Plastic buckets for the collection of soil suspensions, 10 l
Freeze-dryer

Chemicals and solutions

$MgCl_2 \times 10\ H_2O$, reagent grade

Procedure

Weigh a portion of fresh sieved (< 2 mm) soil equivalent to 20.00 g dry weight into a 200 ml glass beaker. Add 100 ml distilled water and a Teflon-coated stirring rod, and place the beaker in the ice bath on a magnetic stirrer. Install the tip 0.5–1 cm below the water surface (depending on the instructions of the manufacturer) and sonicate for 15 min at maximum output under permanent stirring. Rinse the sonicator tip to remove adhering particles and pass the soil suspension over a 0.05 mm sieve; use a wet-sieving apparatus for completing the separation of the sand fraction. Collect the suspended fine fraction < 0.05 mm in buckets and transfer it to a set of sedimentation flasks. Collect the sieve fraction in a glass beaker, and separate organic particles from sand grains by floating and decanting over a funnel and filter paper. Dry the organic and mineral sand fraction at 65°C in an oven.

Bring the sedimentation flask to volume with distilled water, stopper and shake overhead for 1 min by hand. Allow to settle for an appropriate period of time under constant temperature and drain off the supernatant clay suspension (< 2 μm) into a bucket. Repeat this procedure until the supernatant is clear after the respective settling time. Precipitate the suspended clay by adding portions of $MgCl_2$ under stirring to dissolve the salt properly. Decant the clear supernatant solution of the clay and residual silt fractions in the bucket and sedimentation flask, respectively. Transfer the silty sediment to a 1 l flask with distilled water and freeze-dry. Wash the clay fraction by repeated shaking in a centrifuge flask with distilled water, centrifuge, transfer to a 1 l flask and freeze-dry.

Discussion

The procedure yields one sand (2–0.05 mm), silt (0.05–0.002 mm) and clay fraction (< 0.002 mm) only. Although a larger number of size classes was introduced by many authors, only little additional information is usually obtained (McKeague 1971; Catroux and Schnitzer 1987; Christensen 1992).

The finest (clay) fraction contains dissolved organic matter and inorganic nitrogen, which has consequences on measurements and interpretation of the elementary and structural composition.

The degree of dispersion is highly dependent on the actual power output of the ultrasonic tip, which may be measured by the temperature increase in a reference experiment (North 1977; Christensen 1992). The tip should be carefully maintained and adjusted to keep the power output constant. An energy input of between 300 and 500 J ml^{-1} has been found sufficient for the disruption of microaggregates (Gregorich et al 1988). It is advisable to compare the particle size distribution with that obtained from conventional particle size analysis to check for completeness of dispersion (Christensen 1991).

The light silt and clay size fraction that represents largely uncomplexed organic matter may be isolated by density fractionation.

Although ultrasonic dispersion has no direct chemical effects, it is suspected of inducing organic matter redistribution and even structural changes through high-energy cavitation (Suslick 1989).

Soil nitrogen

J.C. Forster

Microorganisms are highly involved in the cycling of nitrogen in soil, because they carry out nitrogen fixation, nitrification, denitrification, and nitrogen mineralization – immobilization turnover. Total nitrogen analysis, the C:N ratio and the mineral N fractions (mainly ammonium and nitrate) provide insight into the N supply to microflora and plants, and thus reflect an aspect of the microbiological status of a soil.

Determination of total nitrogen

(Keeney and Nelson 1982)

Digestion of soil samples

Principle of the method

Total nitrogen estimation usually consists of a sulphuric acid digestion (the Kjeldahl procedure) and subsequent titrimetric or colorimetric analysis. As the complete breakdown of organic compounds is desired, hydrogen peroxide may be added. Selenium acts as a catalyst, while alkali sulphates are added to raise the boiling point. Ammonium measurements with ion-sensitive electrodes are possible; an excellent discussion of this method is given by Keeney and Nelson (1982).

Materials and apparatus

Digestion blocks and quartz glass tubes
Drying oven

Chemicals and solutions

Sulphuric acid, concentrated (95–97%)
Selenium catalyst; selenium-free catalysts are also available and have proven to be of equal value
Lithium sulphate
Hydrogen peroxide (30%)
Digestion mixture

Add 0.42 g selenium powder and 14 g lithium sulphate to 350 ml 30% hydrogen peroxide and mix well. Slowly add 420 ml concentrated sulphuric acid under water or ice cooling. This solution is stable for 4 weeks at 2°C.

Procedure

About 0.2 ± 0.001 g ground soil or plant material are weighed into a digestion tube. The exact weight is recorded. Add 4.4 ml digestion mixture to each tube. Prepare at least six blank digests for the correction of results. Digest at 360°C for 2 h; alternatively, an overnight digestion at 200°C is practicable. After cooling, transfer to a 100 ml volumetric flask with about 50 ml distilled water and allow to cool again. Make up to volume, mix well and allow to settle until there is a clear solution.

Discussion

No accurate results can be expected for compounds containing N–N and N–O linkages unless a pretreatment is included. Nitrate and nitrite are rendered digestable by the same pretreatments.

Sodium or potassium sulphate are appropriate in place of Li_2SO_4 to raise the temperature during digestion.

Digestion temperature is related to the alkali sulphate concentration; while high concentrations increase the reaction rate, they lead to a solidification of the digest when higher than 1 g ml^{-1}.

Analysis by steam distillation and titration

(Keeney and Nelson 1982)

Principle of the method

Ammonium in digests is transformed to free ammonia under excess alkali and collected in boric acid. Titration is carried out with hydrochloric acid to a pH endpoint of 4.5.

Materials and apparatus

Steam distillation apparatus
Burette (10 ml) graduated at 0.01 ml intervals
Glassware: 100 ml and 1000 ml volumetric flasks

Chemicals and solutions

Standard ammonium sulphate (100 mg NH_4-N l^{-1})

Dissolve 0.4717 g dry ammonium sulphate in 600 ml distilled water; dilute to 1000 ml and mix well.

0.01 M hydrochloric acid

Add 113.9 g (98.2 ml) concentrated HCl (32%) to 800 ml distilled water in a 1000 ml volumetric flask and make up to volume to give 1 M HCl. Dilute a 10 ml aliquot to 1000 ml.

Alkali mixture

Dissolve 500 g sodium hydroxide and 25 g sodium thiosulphate in distilled water. Cool and dilute to 1000 ml with distilled water.

Boric acid indicator solution

Dissolve 20 g boric acid in 600 ml distilled water. Prepare an indicator solution from 0.099 g bromocresol green and 0.066 g methyl red in 100 ml ethanol. Add 20 ml of the indicator to the boric acid solution. Add 0.1 M NaOH until a purple colour appears, and make up to volume with distilled water.

Procedure

Transfer a 50 ml aliquot of the digest to the distillation flask and add 25 ml of the alkali solution. Use 5 ml boric acid indicator solution in a suitable flask graduated at 30 ml to collect the distillate. Commence distillation immediately; stop the reaction when 25 ml distillate are collected. Titrate the distillate with 0.01 M HCl; the endpoint is indicated by a colour change from green to pink.

Calculation

1 ml 0.01 M HCl is equal to 0.14 mg NH_4-N.

$$\%\ \text{total N} = \frac{\text{Titre (ml)}}{50 \times \text{sample weight (g)}} \quad (3.8)$$

Discussion

Distilled water is reported possibly to contain some ammonium. For this reason, use of deionized water or water treated with an ion exchange resin is preferred.

Colorimetric analysis

Principle of the method

Ammonium reacts with salicylate and hypochlorite in a buffered alkaline solution in the presence of sodium nitroprusside to form the salicylic acid analogue of indophenol blue. The blue–green colour produced is measured at 660 nm. A complexing agent is added to remove interfering polyvalent cations.

Materials and apparatus

Spectrophotometer equipped with 1 cm cuvettes
Glassware: 1000 ml volumetric flasks

Chemicals and solutions

Color reagent

Dissolve 34 g sodium salicylate, 25 g sodium citrate and 25 g sodium tartrate in 750 ml distilled water. Add 0.12 g sodium nitroprusside, dissolve and make up to 1000 ml.

Alkaline hypochlorite solution

Dissolve 30 g sodium hydroxide in 750 ml distilled water. Add 10 ml sodium hypochlorite solution (> 5% available chloride) and make up to 1000 ml.

Calibration

Dissolve 4.719 g dry ammonium sulphate in 400 ml distilled water in a 1000 ml volumetric flask and make up to volume (= 1000 mg l^{-1} stock solution).

Pipette 0, 0.5, 1, 1.5, 2.0 and 2.5 $\pm$ 0.001 ml of the stock solution into 100 ml volumetric flasks to give working standards of 0, 5, 10, 15, 20 and 25 mg NH_4-N l^{-1}.

Compensate for the digest matrix as indicated above and make up to volume.

Procedure

Transfer a 0.1 ml aliquot of sample or standard solution to a test tube. Add 5.0 ml of color reagent, mix well and allow to stand for 15 min. Add 5.0 ml of alkaline hypochlorite solution and mix well. The colour develops within 1 h and is stable only for several hours. Read the absorbance on a spectrophotometer at 660 nm.

Calculation

Plot a calibration curve from the absorbance readings versus standard concentrations, and determine sample and blank concentrations. Calculate the corrected sample concentration (*C*) by the difference of sample minus blank concentrations.

$$\% \text{ nitrogen} = \frac{C \times 0.1}{\text{Sample weight}} \quad (3.9)$$

where *C* is measured in milligrams per litre, sample weight in grams and a factor of 0.1 is used for conversion into percentage nitrogen.

Discussion

The method is similar to the original indophenol blue procedure but substitutes the toxic phenol dye by its salicylate analogue.

Automated colorimetric procedures are available from a number of suppliers, e.g. Alpkem (RFA-300 application A303-S021-00), Skalar, etc., which allow a sample throughput of 120 h^{-1}.

Inorganic nitrogen

Ammonium and nitrate in the soil solution are the nitrogen sources of plants. Inorganic N is liberated by mineralization of organic compounds or added to soils as fertilizers. The concentration of mineral N forms presents a great seasonal variation. Only repeated or permanent measurements can give insight into the mineral N status of a soil by (1) numerous soil sampling dates during a year, or (2) the installation of suction cups to monitor the respective concentrations continuously.

Potassium chloride at 2 M is often used for the extraction of nitrate and ammonium N (nitrite does not contribute significantly). The colorimetric analysis of NH_4-N is usually done by the phenate (indophenol blue) or salicylate methods. Nitrate (plus nitrite) is measured by colorimetry after reduction to nitrite or by direct methods as described below.

Extraction of mineral nitrogen

Principle of the method

Batch extraction with a potassium salt at high ionic strength to remove free and absorbed NH_4^+, NO_3^-, and NO_2^- from soil. Fixed ammonium from the interlayer space of clay minerals is not recovered.

Materials and apparatus

Horizontal shaker
Polyethylene bottles (250 ml) for sample extraction and collection of the filtrate
Polyethylene funnels
Whatman 2V filter paper

Chemicals and solutions

Potassium chloride, 2 M

Dissolve 149.12 g KCl in 600 ml distilled water and make up to 1000 ml with distilled water; mix well.

Procedure

Weigh fresh soil equivalent to 10 g dry mass into a 250 ml polyethylene bottle and add 100 ml 2 M potassium chloride solution. Shake for 1 h at 120 min^{-1} and filter the soil suspension into another 100–250 ml polyethylene bottle.

Discussion

Measurements should be performed immediately; alternatively, the filtrates should be stored at temperatures lower than −18°C.

Colorimetric determination of ammonium

See earlier section on colorimetric analysis.

Colorimetric nitrate analysis

(Keeney and Nelson 1982)

Principle of the method

The Griess–Ilosvay method for the determination of nitrate involves the reduction to nitrite by cadmium and subsequent staining by the formation of an azo dye. As the method includes nitrite that is present initially, an extra nitrite determination without Cd reduction has to be run if significant amounts of NO_2^- are expected; the nitrate concentration is then calculated by the difference between the two determinations.

Materials and apparatus

Spectrophotometer and 1 cm cuvettes.

Glass column, 1 cm i.d., 30 cm length, fitted with a glass frit and Teflon stopcock at the outlet, and a liquid reservoir (> 75 cm^3) on the top. The outlet can be connected to a 100 ml volumetric flask via flexible tubing and a glass tube, which penetrates a two-hole rubber stopper. A vacuum source and regulating valve are connected similarly.

Chemicals and solutions

Copperized Cd reagent

50 g of coarse-powdered or granular Cd metal are pretreated with 6 M HCl for 1 min; the acid is decanted and the residue is washed twice with distilled water. Two successive treatments with 2% w/v $CuSO_4 \times 5\ H_2O$ are followed, with distilled water washings before and, thoroughly, after the second treatment. When all the blue and light grey colour has disappeared from the washing water, the reagent is filled into the glass column.

Concentrated ammonium chloride solution

Dissolve 100 g NH_4Cl in 500 ml distilled water and mix well.

Dilute ammonium chloride solution

50 ml of the concentrated ammonium chloride solution are diluted to 2 l with distilled water.

0.5 g sulphanilamide are dissolved in 100 ml 2.4 M HCl and stored at 4°C (diazotizing reagent).

0.3 g of [*N*-(1-naphthyl)-ethylenediamine]hydrochloride are dissolved in 100 ml 0.12 M HCl and stored in an amber bottle at 4°C (coupling reagent).

Calibration

Nitrate stock standard solution (1000 mg l^{-1})

Dissolve 7.218 g dry potassium nitrate (analytical grade) in 600 ml distilled water, dilute to 1000 ml and mix well.

Nitrate working standard solutions

Pipette 0, 0.20, 0.40, 0.60, 1.00 and 2.00 ml of the stock standard solution into 100 ml volumetric flasks and make to volume with distilled water. The resulting standard concentrations are 0, 2, 4, 6, 10 and 20 mg NO_3-N l^{-1}.

Procedure

Fill the glass column with dilute NH_4Cl to one third of its height followed by Cd granules up to a height of 20 cm. Note that air bubbles are undesirable because of oxidation effects. Wash with dilute NH_4Cl by tenfold the pore volume of the Cd column at a flow rate of 8 ml min^{-1}.

Add 1 ml concentrated NH_4Cl before treating a soil extract and lower the liquid level to the top of the Cd column. Pass 75 ml dilute NH_4Cl over the column at a flow rate of 110 ml min^{-1} until the top of the column is reached and collect the effluent in a flask attached to the rubber stopper. Again add 1 ml concentrated NH_4Cl and an aliquot of soil extract (normally between 2 and 5 ml) containing a maximal amount of 20 mg NO_3-N. Attach a new 100 ml volumetric flask and drain the liquid until the top of the column is reached. Wash the top of the glass column with 2 ml dilute NH_4Cl and drain as before. Pass 75 ml of dilute NH_4Cl over the column at 110 ml min^{-1} and collect everything in the volumetric flask.

Pipette 2 ml of the diazotizing solution to the contents of the volumetric flask. Add 2 ml of the coupling reagent after 5 min and make to volume with distilled water. Mix well and allow to stand for 20 min. Read absorption at 540 nm against a reagent blank. Treat working standard solutions like soil extracts using 1 ml aliquots throughout and determine absorbance as described above.

Calculation

Plot a calibration curve of absorbance versus concentration, and read or calculate the apparent NO_3-N concentration of the extract that is equal to the true concentration if the same aliquots from extracts and standards were used. The concentration, c_e, in the original soil is given by

$$c_e = \frac{c_a \cdot V_s \cdot V_e}{V_a W} \quad (3.10)$$

where c_a is the apparent concentration of the extract taken from the calibration curve, V_s, V_e and V_a are the volumes of the standard aliquot, the total extract and the extract aliquot, respectively, and W is the soil dry weight used.

Discussion

Store the copperized Cd column under dilute NH_4Cl when not in use.

Several reduction columns may be run simultaneously to increase sample throughput. It is advisable to check if the results from different columns are identical by running separate calibration curves. If results are comparable, one calibration curve may be sufficient.

Column regeneration is only necessary after several hundred samples and is preceded by a decrease of absorbance per concentration unit.

Column performance may be checked by comparing the results of a unit amount of NO_3^- to the same amount of NO_2^-.

Interferences may be eliminated by diluting the extract.

Direct colorimetric determination of nitrate

Principle of the method

Salicylic acid is nitrated in an alkaline solution; light absorbance is measured at 410 nm.

Materials and apparatus

Spectrophotometer equipped with a 1 cm cuvette
Test tube shaker

Chemicals and solutions

4 M sodium hydroxide

Dissolve 160 g NaOH cakes in 600 ml distilled water; make up to 1000 ml and mix well.

5% salicylic acid

Dissolve 5 g salicylic acid in 95 ml concentrated sulphuric acid (95–97%).

Calibration

1000 mg NO_3-N l^{-1} stock solution

Dissolve 7.223 g dry potassium in distilled water and make up to volume in a 1000 ml volumetric flask.

Pipette 0, 200, 400, 600, 800 and 1000 μl of the stock solution into 100 μl volumetric flasks, and make up to volume to give the 0, 2, 4, 6, 8 and 10 mg NO_3-N l^{-1} working standards.

Procedure

Pipette 500 μl of each standard and sample into a test tube. Add 1 ml salicylic acid solution, mix well immediately and allow to stand for 30 min. Add 10 ml of 4 M NaOH and leave for 1 h for colour development. Read absorbance at 410 nm on a spectrophotometer.

Calculation

Plot the standard curve from absorption measurements versus concentration. Determine sample and blank concentrations. After correction for the mean of the blanks, the original concentration in soil is calculated by

$$NO_3\text{-N (mg kg}^{-1}) = \frac{C \cdot V}{W} \quad (3.11)$$

where C is the corrected NO_3-N concentration in sample solutions, V is the final solution volume and W is the equivalent dry weight of soil. For the above 2 M KCl extraction procedure, the NO_3-N concentration in soil is equal to $C \times 10$.

Discussion

In the case of coloured soil extracts, an additional blank has to be substracted where 1 ml of concentrated sulphuric acid without salicylic acid has been added.

Although there are some ill-defined prerequisites by Keeney and Nelson (1982), the method is recommended as a standard procedure by Anderson and Ingram (1989).

Nitrate determination by ion chromatography

(Tabatabai and Dick 1979)

Principle of the method

A soil extract is injected into a low pressure liquid chromatography system. The sample passes through a precolumn, a separator column and a suppressor column, and the anions contained in the analyte are detected in a conductivity cell (Small 1978; Fritz et al 1982). Anion identification is based on retention time. Separation is due to the affinity of ions to a low capacity anion exchange resin contained in the first two columns. The suppressor column is packed with a strong acidic cation exchange resin where cations are adsorbed and H^+ is released into the solution, which leads to a decrease in conductivity. Packed-bed, fibre and membrane suppressors are available. The first has to be regenerated off-line, whereas the other two are regenerated continuously during the analysis. Moreover, their lower dead-volume makes them preferable. Autosampler and data processing systems supply the operator with the possibility of running 24 h analyses with a throughput of about five samples per hour.

Materials and apparatus

Commercial ion chromatograph, e.g. Dionex Model 10, Model 2010i or equivalent.

Chemicals and solutions

Eluent solution

3.0 mM $NaHCO_3$/1.8 mM Na_2CO_3 (Tabatabai and Dick 1979) or 0.75 mM $NaHCO_3$/2.0 mM Na_2CO_3

Regenerant solution

0.025 M H_2SO_4

Calibration

Nitrate stock standard solution (1000 mg NO_3^- l^{-1})

Dissolve 1.6305 g potassium nitrate in 600 ml distilled water; make to 1000 ml and mix well.

Nitrate intermediate standard solution (10 mg NO_3^- l^{-1})

Dilute 10 ml of the stock standard solution to 1000 ml and mix well.

Nitrate working standard solutions

Pipette 200 µl and 1, 5, 10 and 30 ml of the intermediate standard solution into 100 ml volumetric flasks, and make up to volume with distilled water. The resulting standard concentrations are 0.02, 0.10, 0.50, 1.00 and 3.00 mg l^{-1}.

Procedure

Set up the ion chromatograph instrument according to the operation manual. Note that different separator and suppressor columns may need different working conditions. Adjust the detector to the appropriate working concentration range. Inject samples manually or by autosampler, using 5–10 times the sample loop volume for flushing. Analyse a blank solution within every batch of samples.

Calculation

If a data processing unit is not available, plot a calibration curve from peak heights (or peak area) versus standard concentrations after correction for the blank. Read the sample concentrations from the diagram or calculate them from the first-order regression equation.

Discussion

The method detection limit for the Dionex 2010i ion chromatography system is 0.005

mg NO_3^- l^{-1}, the optimum concentration range is 0.1–5.0 mg l^{-1} (EPA 1987).

Membrane filtration of sample solutions (45 μm pore size) is recommended to prevent system malfunction or damage. For the same reason, eluents should be filtered (< 20 μm; EPA 1987).

No substances that interfere with retention times similar to nitrate have been found in soil solutions or extracts.

Overlapping with neighbouring peaks due to high ion concentrations can be resolved by dilution or spiking of samples.

Ion chromatography is similarly useful for the determination of sulphate and phosphate in soil solutions and extracts; simultaneous measurements are possible due to the chromatographic separation of species in the separator column. Analysis time is then a function of the complexity of the analyte and hence one of resolution requirements.

Colorimetric determination of nitrite

(Keeney and Nelson 1982)

This method is analogous to colorimetric nitrate determination by the modified Griess–Ilosvay procedure except that the reduction step is omitted.

Principle of the method

Nitrite in soil extracts forms a diazonium salt with sulphanilamide, a primary aromatic amine in acidic solution. The absorbance at 540 nm is measured after coupling with *N*-(1-naphthyl)-ethylenediamine.

Materials and apparatus

Spectrophotometer and 1 cm cuvettes

Chemicals and solutions

Sulphanilamide solution

See the earlier section on colorimetric nitrate analysis

[*N*-(1-naphthyl)-ethylenediamine]-hydrochloride solution

See the earlier section on colorimetric nitrate analysis

Calibration

Nitrite stock standard solution (1000 mg NO_3-N l^{-1})

Dissolve 4.925 g dry sodium nitrite in 600 ml distilled water; make to 1000 ml and mix well. This solution is stable for at least 6 months if stored in a refrigerator.

Nitrite intermediate standard solution (10 mg NO_3-N l^{-1})

Dilute 10 ml of the stock standard solution to 1000 ml with distilled water and mix well.

Nitrite working standard solutions

Pipette 0, 50, 100, 200, 400 and 600 μl into 50 ml volumetric flasks, and dilute with 45 ml distilled water.

Procedure

Dilute 2 ml soil extract with 45 ml distilled water in a 50 ml volumetric flask. Add 1 ml of sulphanilamide solution, mix well, and allow to stand for 5 min. Add 1 ml of the coupling reagent, mix well and allow colour development for 20 min. Make to volume with distilled water, mix well and read the absorbance at 540 nm against a reagent blank. Treat the working standard solutions in the same way.

Calculation

Plot a calibration curve from standard absorbances versus concentrations; read sample concentrations from the plot or calculate from the first-order regression equation. The soil nitrite concentration is given by the equation

$$c_e = \frac{c_a \cdot f_e \cdot V_e}{W} \tag{3.12}$$

where c_a is the apparent concentration of the extract taken from the calibration curve, V_e is the total volume of the soil extract, W is the soil dry weight used, and f_e is the dilution factor of the measuring solution, which is 25 if 2 ml of extract and 50 ml final volume are used as above.

Discussion

The nitrite concentration in extracts remains stable for several weeks at 4°C (Keeney and Nelson 1982).

Most cations and anions, and extracted organic matter do not interfere with the colorimetric determination, while Hg^{2+} and Cu^{2+} lead to higher and lower results, respectively, due to the decomposition of the diazonium salt in the latter case.

Each series of analyses should include a control to correct for the colour of solutions.

Soil phosphorus

J.C. Forster

Soil phosphorus is made up of an inorganic (bound or dissolved) and an organic fraction with varying percentages between 5% and 95% of the total for each. Only a small part appears in the soil solution (< 0.01–1 mg l^{-1}). The soil phosphorus cycle is almost restricted to the soil, water and plant compartments, as it lacks wet or dry inputs from and gaseous output to the atmosphere, and an output to headwater streams or the groundwater. However, the latter may occur if there were soils with low adsorption capacity and high water permeability, or erosion of soils heavily fertilized with P applications. Soil microbes are involved in the mineralization of P from organic debris. The dissolution of inorganic P is promoted by the emission of carbon dioxide and organic acids by microorganisms. Extracellular phosphatases are produced by microorganisms and plant roots, and contribute to the mineralization of organic P. Mycorrhizal hyphae are most effective in soil P uptake due to their ability to penetrate even small organic particles and soil aggregates; the minimum P level for uptake is much lower than that of plant roots, thus plant–mycorrhiza symbiosis leads to an enhanced P supply to plants.

Among the multitude of methods for characterizing different P forms in soil, a rapid and simple procedure for total P, and the resin method for labile P are presented. In a later chapter of this book, a method is presented for the estimation of microbial P in fumigated soil samples. Speciation of organic and inorganic P forms may be performed on suitable soil extracts by ^{31}P NMR spectroscopy.

Estimation of total phosphorus

(Bowman 1988)

Fusion or extraction procedures have been used for the extraction of total P; their aim is to dissolve such different forms of P, as are the organic P esters, mixed Ca phosphates, P in the lattice of alumino-silicates, and P bound to or occluded in iron and aluminium oxides. This makes a laborious combination of extraction steps or use of carbonate fusion necessary, which is often time consuming, expensive and even dangerous to handle. The presented method needs only 30 min and has been found to be as effective as carbonate fusion, which has been the reference method for the last decades.

Principle of the method

Soil is treated with concentrated sulphuric acid and hydrogen peroxide to dissolve P in organic and non-silicate inorganic forms. P in the silicate lattice is released by hydrofluoric acid treatment.

Materials and apparatus

Hotplates or heated sand bath
Teflon beakers, 100 ml

Chemicals and solutions

Concentrated sulphuric acid (95–97%)
Hydrogen peroxide (30%)
Concentrated hydrofluoric acid (40%)

Procedure

Weigh 0.5 g finely ground, well-mixed soil into a 100 ml Teflon beaker. Use 0.25 g for sandy and high organic matter soils. Add 5 ml concentrated H_2SO_4 and swirl gently. Add 3 ml 30% H_2O_2 in 0.5 ml portions. Swirl vigorously (caution: foaming may lead to an overflow of soil material in high organic matter samples). When reaction with H_2O_2 has subsided, add 1 ml hydrofluoric acid in 0.5 ml portions and swirl gently. Place the beaker on a hot plate at 150°C for 10–12 min to eliminate excess H_2O_2. After slight cooling, wash down the sides of the beaker with about 15 ml distilled water. Mix and cool to room temperature. Transfer the beaker contents quantitatively to a 50 ml volumetric flask, passing it through a quantitative filter paper. Make two additional washings of the beaker each with 10 ml distilled water, filtrate and make up the extract to volume.

Extraction of organic phosphorus

(Bowman 1989, modified)

Ignition–extraction and direct extraction methods are available; the most widespread Saunders–Williams ignition–extraction method has been found to overestimate organic P in highly weathered soils (Condron et al 1990). In contrast, combined acid–alkali extractions gave reasonable results for soils of different degrees of weathering, and the Bowman method is recommended here for its additional advantage of requiring little equipment and manpower.

Principle of the method

Soil is extracted with concentrated sulphuric acid followed by sodium hydroxide; inorganic and total P are determined by colorimetry, the latter after digestion of the extracts with perchloric acid. Inorganic P is calculated by their difference. Water additions to concentrated sulphuric acid lead to elevated temperature and enhanced dissolution of organophosphates (Bowman 1989).

Materials and apparatus

High-speed centrifuge with 100 or 250 ml tubes
Analytical balance
100 ml volumetric flasks
Horizontal shaker with holding device for centrifuge tubes

Chemicals and solutions

Concentrated sulphuric acid (95–97%)
0.5 M sodium hydroxide

Dissolve 20.00 g of NaOH in 1000 ml distilled water; cool with water to avoid excessive heat development.

Concentrated perchloric acid (72%)

Procedure

Weigh 2.000 g dry, finely ground soil into a centrifuge tube. Add 3 ml concentrated sulphuric acid and, successively, 5 ml of distilled water while gently shaking for mixing. Cool and add 40 ml distilled water; shake for 2 h. Centrifuge and filter the supernatant through a 45 µm pore size membrane. Add 100 ml 0.5 M NaOH, include the membrane filter and shake for another 2 h. Centrifuge and filter the extract.

Discussion

The method may be modified for total P analysis in highly weathered soils as follows: make a first NaOH extraction with 50 ml at ambient temperature, and add a second hot NaOH extraction at 80–90°C in a shaking water bath. Use capped glass tubes.

The unmodified Bowman procedure recovered about 79% of total P determined by the sodium carbonate fusion method (Condron et al 1990).

Extraction of labile phosphorus

(Resin-P method; Sibbesen 1978)

Phosphorus that is thought to be available to plants is usually extracted by salt or acid solutions. Although there is some relationship between extracted P and the P supply and P content of plants, the P fraction thus extracted is comparatively ill-defined. In contrast, a well-defined phosphate fraction that is in equilibrium with soil solution may be determined by isotopic dilution or, at less cost, by resin extraction. Both lead to similar results.

Principle of the method

Soil is equilibrated with anion exchange resin in a batch experiment. After elution of the resin with hydrochloric acid, P is measured in the eluate by the molybdenum blue method as described below.

Chemicals and solutions

Strongly basic anion exchange resin, e.g. Dowex 1 × 8, > 30 mesh.

Preparation of resin bags

Polyester (Estal Mono PE 400 µm) or Nylon (Weisse and Eschrich polyamide 53 µm) sreening is folded to give a 4 cm × 3 cm bag; edges are sealed in a cold flame except one, which is left open to fill in the resin. The bag is filled with anion exchange resin (0.4 g) using a spatula and the remaining edge is sealed.

Conditioning and regeneration of the resin

Transfer the resin bags to a glass beaker containing 0.5 M $NaHCO_3$ and soak for 1 h under repeated stirring. Repeat the treatment in fresh 0.5 M $NaHCO_3$ and follow by two washings with distilled water. Store the bicarbonate solution from treatment number 2 and use for number 1 of a following batch of resin bags.

1 M hydrochloric acid

Add 98.2 ml (= 114 g) concentrated HCl (32% = 10.18 M) to 800 ml distilled water and make up to volume in a 1000 ml volumetric flask.

0.5 M sodium bicarbonate, pH 8.5

Dissolve 42 g sodium bicarbonate in 600 ml distilled water in a 1000 ml volumetric flask. Adjust to pH 8.5 with 2 M HCl or 2 M NaOH, and make up to volume.

Procedure

Weigh 1.0 ± 0.01 g ground soil into a 50–100 ml polyethylene flask. Add 40 ml of water and one resin bag, cap the flask and shake gently overnight (about 10 h). Remove the resin bag, wash free of soil with distilled water and shake for 1 h with 20 ml of 0.5 M HCl. Determine the phosphorus content by the molybdenum blue method described below; use 0.5 M HCl or, if samples have to be diluted, a suitable dilution thereof for the standards.

Discussion

For prolonged storage, place the resin bags in distilled water and add a spatula tip of sodium azide (NaN_3), preferably in a refrigerator (Anderson and Ingram 1989). Bags may also be stored in 0.5 M HCl without azide addition (Sibbesen 1978).

Quantification of phosphorus in soil extracts

(John 1970)

The molybdenum blue method is still preferable for P determinations because of its rapidity and simplicity. The procedure presented below is a modification of the Murphy and Riley (1962) method.

Principle of the method

Orthophosphate reacts with molybdenum and antimony in an acid medium to form a phosphoantimonyl–molybdenum complex. This complex is subsequently reduced with ascorbic acid to form a mixed valence complex. Light absorption is measured at 880 nm.

Materials and apparatus

Spectrophotometer equipped with 1 cm cuvette
Test tube shaker

Chemicals and solutions

Antimony potassium tartrate
Ammonium molybdate
Concentrated sulphuric acid (95–97%)
Ascorbic acid
Colour reagent, stock solution

Dissolve 20 g ammonium molybdate and 0.5 g ammonium potassium tartrate in 600 ml distilled water. Slowly add 250 ml concentrated sulphuric acid, and dilute and make up to 1000 ml in a volumetric flask. Keep this solution in the dark at 2–4°C for 4 weeks maximum.

Colour reagent, working solution

Dissolve 1.5 g ascorbic acid in 100 ml of stock solution and mix well. This solution should be prepared daily.

Procedure

Pipette a suitable aliquot (= V_a; 0.1–1 ml) of soil extract in a test tube. The final P concentration in the test solution should be 0.01–0.6 mg l^{-1}. Pipette 0, 20, 40 and 60 µl of the P standard stock solution into test tubes. Add 1 ml of the working colour reagent. Add an amount of soil extractant to the standards equivalent to that in the sample aliquots. Make up to 10 ml total volume with a suitable amount of distilled water, shake well and allow to stand for at least 30 min. Read the adsorption on a spectrophotometer at 882 nm.

Calculation

Plot a calibration curve from measured absorption versus concentration. Determine sample and blank concentrations and subtract the mean of the blanks from sample values to give the corrected sample solution concentration (C_s). The concentration in the original soil sample

$$C_p \text{ (mg P kg}^{-1}) = \frac{C_s \cdot 10 \cdot V_e}{V_a \cdot W} \quad (3.13)$$

where V_a is the volume of the aliquot in millilitres, V_e is the volume of the extract in millilitres and W is the sample dry weight in grams; the final volume of the analyte is 10 ml.

Discussion

Colour development is influenced by acidity, arsenate and silicate concentrations, and by compounds which interfere in the redox conditions during analysis. Special modifications are thus indispensable to avoid serious errors. Interference due to ferric (800 mg l^{-1}), fluoride (100 mg l^{-1}) and chloride (130 g l^{-1}) ion is within 5% error (John 1970). Arsenate is not tolerable in the measuring solution. Interference of fluoride at > 0.05 M can be overcome by the addition of boric acid (0.1% in final solution).

A straight linear relationship between standard concentration and absorption is observed in the range between 0 and 0.6 mg l^{-1}.

Colour development is stable after 30 min at acid concentrations between 0.25 and 0.65 M, and independent of temperature between 10 and 60°C.

Automated colorimetric applications are available from Alpkem (Clackamas, Oregon, USA), Skalar (De Breda, the Netherlands), Technicon, and others.

New techniques for P analysis include inductively coupled plasma-atomic emission spectrometry and ion chromatography.

^{31}P nuclear magnetic resonance spectroscopy of soil extracts

The colorimetric analysis of soil extracts reflects the total (detectable) quantity of extracted P, irrespective of the individual chemical structure of P compounds. The ^{31}P NMR spectroscopy offers a powerful tool to distinguish organic and inorganic P forms, to which different functions in the P cycle of soils may be attributed. Although P concentrations in soil are comparatively low, this possibility is offered by the nearly 100% natural abundance of the ^{31}P isotope that possesses a spin quantum number of 1/2. While the P uptake of plants is only in the inorganic orthophosphate form, soil P further consists of pyrophosphate, polyphosphates, organic orthophosphate monoesters (mainly inositol) and diesters (phospholipids, nucleotides), teichoic acids and phosphonates (Newman and Tate 1980; Tate and Newman 1982; Condron et al 1985, 1990).

Principle of the method

Dissolution of soil P by sonication treatment in concentrated sodium hydroxide solution and measurement by solution-^{31}P NMR spectroscopy.

Materials and apparatus

Ultrasonic generator
Ultracentrifuge and centrifuge tubes (30–50 ml volume)

Chemicals and solutions

0.5 M sodium hydroxide solution

Dissolve 20 g NaOH pellets in 600 ml distilled water; cool the volumetric flask in a water or ice bath. Make up to 1000 ml with distilled water when the solution is at room temperature.

Reference solution

Add 1 ml D_20 to 10 ml concentrated H_3PO_4; transfer *c.* 3 ml of this solution to a NMR tube.

Procedure

Weigh 6.7 g finely ground air-dried soil into a centrifuge tube and add 20 ml 0.5 M NaOH. Sonicate for 3 min under ice cooling to avoid heating of the sample, which might lead to the hydrolysis of organophosphate esters. Centrifuge and concentrate the supernatant solution to about 2 ml under a stream of nitrogen at 40°C. Add 1 ml of D_20 and transfer the solution to a 1 cm diameter NMR tube.

Recording conditions for NMR spectra are dependent on the P content in the soil extract and on the type of NMR spectrometer used. Therefore, exemplary data are given for the spectra reported in Forster and Zech (1992). They used a Jeol FX-90Q spectrometer at an observation frequency of 36.2 MHz; further parameters were as follows: proton-decoupling, 90° pulse angle, 1 s pulse delay, 0.4 s acquisition time, 4 kHz spectral width, data collection by 10,000 to 30,000 pulses.

Interpretation of results

^{31}P NMR spectra show well-resolved peaks in places representing chemical shifts, *d*, relative to the reference solution (85% H_3PO_4). Peak assignments are presented in Table 3.1, following the data given by Newman and Tate (1980) and Condron et al (1985, 1990). Quantification is carried out via manual or computer integration of peak areas; results should be given as the percentage distribution of P among the different species or/and as their absolute amounts if an internal standard has been included in the analysis. Data on the recoveries of organic and inorganic P are helpful in the discussion of results.

Table 3.1. Peak assignments in ^{31}P NMR spectroscopy relative to 85% H_3PO_4.

Chemical shift (δ)	*Peak assignment*
19.8	Alkyl-phosphonic acids (phosphonates)
18.3	Alkyl-phosphonic esters (phosphonolipids)
5.3	Inorganic orthophosphate
3.5–5.3	Orthophosphate monoesters
3.6	Choline phosphate
5.3, 4.4, 4.0, 3.9	*Myo*-inositol hexaphosphate
0.36–0.95	Teichoic acids
−0.8	Orthophosphate diesters: phospholipids, DNA
−5.5	Pyrophosphate
−21.4	Polyphosphates

Discussion

An upfield shift of the relative peak positions may be observed with increasing hydroxyl concentration in the soil extract; the reverse is true for more dilute NaOH extracts (Alt 1990, personal communication); shift errors may also occur due to the broadness and inexact peak position of the internal standard, H_3PO_4 (Wilson 1990).

Only half of the total P has been found in most NaOH extracts of different soils (Zech et al 1985). Condron et al (1985, 1990) obtained higher recoveries by a sequential alkali extraction procedure followed by ultrafiltration.

Orthophosphate and monoesters are usually the dominant P forms in NaOH extracts. Phosphonates were only found in cool and moist soils, and a close relationship with annual precipitation was found. Diesters were absent only in dry soils (Tate and Newman 1982).

Choline phosphate, orthophosphate diesters and phosphonates correlated well with biomass C and ATP content (Tate and Newman 1982).

The NMR experiment produces quantitative data that fit well with wet-chemical P determinations (Tate and Newman 1982; Forster and Zech 1993).

Orthophosphate monoesters are relatively persistent and tend to accumulate in soils, while diesters (and phosphonates) may be decomposed easily to replenish the available soil P pool (Tate and Newman 1982; Hinedi et al 1988). This was also indicated by a decreasing diester:mono-ester ratio with increasing soil depth in Humic Cambisols from north west Spain (Trasar-Cepeda et al 1989).

Solid-state NMR has not produced reasonable spectra for ^{31}P so far (Condron et al 1990a and b).

Soil sulphur

J.C. Forster

Sulphur exists in soil in solid or dissolved inorganic forms as sulphides (reduced) or sulphate (oxidized), in soil organic matter, especially in proteins with a natural S:N ratio of 1:36, in the soil air as H_2S or SO_2, or in minor trace gases. Organic sulphur comprises more than 95% in most terrestrial soils, whereas mineral S (except S from fertilizers and sulphate S in Aridisols) is only present in minor amounts. Oxidation of inorganic sulphides to sulphates is brought about by lithotrophic bacteria, a process well-known from "cat-clay" acid sulphate soils that appears after the drying of tidal marshes. Sulphate reduction is found in gleyic soils, which may be recognized by the evolution of H_2S.

Sulphur is added to deficient soils in the form of sulphate salts. Dry and wet input as a component of acid rain has a similar fertilizing effect as that which has been observed for nitrogen. There is no satisfactory method available for the determination of reduced sulphur forms because of the ease of oxidation during analysis (Tabatabai 1982). Even more uncertainty exists about the nature of organic S compounds. The S-containing amino acids cystine and methionine were found to comprise between 10% and 26% of total soil sulphur (Stevenson 1956; Freney et al 1972). A common organic S fractionation procedure is based on the reducibility by HI (ester sulphates) and Raney nickel (S-containing amino acids).

Total sulphur analysis

Procedures for determining total S in soil usually involve the conversion to SO_4^{2-} and subsequent determination by gravimetry, turbidimetry, colorimetry or ion chromatography. The most widely accepted oxidation procedure is fusion with sodium bicarbonate for total element analysis in the presence of sodium peroxide, which produces complete oxidation. However, the procedure is tedious and hazardous and uses expensive platinum crucibles. To avoid these specific disadvantages, low-temperature ashing with $NaHCO_3$ and Ag_2O has been recommended by Steinbergs et al (1962). Acid digestion with $HClO_4$/HNO_3 (Arkley 1961) is more rapid, although there is a danger of explosion and fire when using $HClO_4$. Tabatabai and Bremner (1970) introduced the alkaline NaOBr digestion procedure, which is presented here because it is cheap, rapid and safe.

Principle of the method
(Tabatabai and Bremner 1970)

Alkaline digestion with NaOBr at 250°C.

Materials and apparatus

Ceramic crucibles
Hotplates or sand bath
Ultrasonic tank

Chemicals and solutions

Sodium hypobromite solution (NaOBr)

Add 3 ml Br_2 dropwise to 100 ml 2 M NaOH in a 250 ml Erlenmeyer flask. Always prepare a fresh solution before the digestion procedure.

Procedure

Weigh 0.100 ± 0.0001 g finely ground dry soil or 0.500 ± 0.0001 g oven-dry (65°C) organic material into a ceramic crucible (or a digestion flask if S analysis by the Johnson and Nishita method is desired). Add 3 ml hypobromite solution and allow to stand for 5 min. Place on a hot sand bath or a hotplate at 250–260°C and evaporate to dryness. Allow to stand on the hot sand bath for a further 30 min. Cool to room temperature, add 3 ml distilled water and heat for 30 s. Sonicate to dissolve the residue and cool to ambient temperature. Transfer quantitatively to a 10 ml volumetric flask. If the digest is to be analysed by the methylene blue procedure of Johnson and Nishita (1952), add 1 ml distilled water and heat for 30 s. Shake or sonicate to dissolve the residue, and cool to ambient temperature. Perform colorimetric analysis as described below. Repeat water addition twice and make the solution to volume in the volumetric flask. If there are particles in the solution that could interfere in the subsequent S analysis, centrifuge or filter the digest before measurement.

Sulphur analysis in digests, and salt or aqueous solutions

(Bardsley and Lancaster 1962)

The methylene blue method of Johnson and Nishita (1952) has long been the most widespread because of its high sensitivity and accuracy. Its lack of interferences makes it superior to turbidimetric or colorimetric procedures (Tabatabai 1982); the sulphur is converted to H_2S, which is collected in a receiving flask. However, this method is very tedious and time consuming, and therefore different methods are preferred unless interferences occur. The colorimetric methylthymol blue (MTB) procedure is available on automated flow analysis systems (e.g. Alpkem RFA™ methodology[C] A303-S031, Alpkem Corporation, Clackamas, Oregon, USA), which involves the liberation of free MTB from Ba-MTB complexes by the formation of $BaSO_4$ at a pH between 2.5 and 3.0. Free MTB is determined by spectrophotometry at 460 nm; interfering polyvalent cations are removed by on-line ion exchange. Sulphur may also be determined by inductively coupled plasma–atomic emission spectroscopy (ICP-AES; Novozamsky et al 1986) or energy-dispersive X-ray fluorescence analysis (EDXRA).

Principle of the method

The turbidimetric method of Bardsley and Lancaster (1962) given here is rapid and simple, however, if precise results are required, another method should be used.

Materials and apparatus

Spectrophotometer

Chemicals and solutions

Acid seed reagent

Dilute 40.0 ml of a 1000 mg l^{-1} SO_4-S standard to 1000 ml with distilled water. Mix 50.0 ml of this solution with 50.0 ml concentrated HCl (37%).

Procedure

Pipette a suitable aliquot of the S-containing digest or extract, standards and blank into a test tube, and dissolve to 10 ml to give between 2 and 40 mg S l^{-1} in the final solution. Add 1 ml acid seed reagent and mix well. Add 0.5 g $BaCl_2 \times 2\ H_2O$ crystals (20–60 mesh) and allow to stand for exactly 1 min. Mix until the crystals are dissolved, and read the absorbance after 2–8 min on a photometer equipped with a 1 cm cell at 420 nm.

Calculation

Plot a calibration curve from the photometer readings versus concentrations of the standards, and read the sample concentrations from the curve. The S concentration (in mg kg^{-1}) of the sample is calculated by

$$C_s = C_m \,.\, V/W_s \qquad (3.14)$$

where C_m is the measured S concentration in solution, V is the volume of the extract or digest and W_s is the original weight of the sample.

Discussion

Analytical errors may be due to the formation of insoluble salts of barium and other polyvalent cations.

Sulphur analysers for total sulphur determination

Commercial S analysers have long been considered unsuitable for the total S analysis of soil samples (Tabatabai and Bremner 1983). Poor recoveries and low reproducibility of results were observed. David et al (1989), however, reported that a LECO™ SC-132 analyser gave satisfactory results if accelerators were admixed to the samples and if calibration was made with plant or soil standard material. Their method was also recommended by the (US) Environmental Protection Agency (1990).

Principle of the method
(David et al 1989)
Combustion of a sample/accelerator mixture at 1370°C under an atmosphere of > 99.5% O_2. Liberated sulphur dioxide is determined by an infrared detector.

Materials and apparatus

LECO™ SC-132 sulphur analyser
S-free combustion ceramic boats
Iron powder (LECO™ part 501-078)
LECOCEL™ (LECO™ part 763-266)
Standard material: LECO™ orchard leaves, LECO™ soil

Chemicals and solutions

Magnesium perchlorate, $Mg(ClO_4)_2$, used as desiccant for the gas drying tube of the analyser.

Procedure

Weigh 0.800 ± 0.001 g of air-dried soil sample into a ceramic boat. Add 0.5 g iron powder and 1.5 g LECOCEL™ and mix thoroughly. Run the total S analysis as outlined in the instruction manual (LECO™ 1983). Use 0.300 ± 0.001 g of LECO™ orchard leaves or 0.800 g of LECO™ soil as calibrants. Calibrate the instrument at least once a day or before measuring a sample batch, and use different (three at least) standards or weights to obtain a calibration curve with different amounts of S covering the range expected in samples.

Discussion

Freeze-dried samples gave results close to fresh, air-dried material, while oven drying at 65°C reduced recoveries of S significantly (David et al 1989). Air-dried samples should be kept at 2–4°C before analysis (EPA 1990).

The automated procedure compared well with an alkaline oxidation/Johnson–Nishita procedure (David et al 1989).

1.00 g of V_2O_5 iron powder accelerator (LECO™ Part No. 501-636) is recommended by the EPA (1990).

Sample size has to be reduced, if soils with high organic content are measured, to prevent an explosion. High amounts of organic matter may also lead to a delayed or incomplete combustion (EPA 1990).

Up to five analyses are required for the condi-tioning of the instrument after system main-tenance or after a longer measurement break.

Replace the desiccant if the first few centimetres in the drying tube are becoming wet; a baseline shift might be observed simultaneously.

Keeping the analyser close to its recommended working temperature will impose less strain on the combustion tube and refractory liner.

The detection limit is about 0.003% S (EPA 1990); David et al report a value of 0.22 μmol S g^{-1}, which is equal to 7 mg S kg^{-1}.

Soil iron

J.C. Forster

Iron in soils exists in the form of crystalline and amorphous iron oxides, hydroxides and oxihydroxides, e.g. hematite H α-Fe_2O_3, ferrihydrite $Fe(OH)_3$ and goethite H α-FeOOH; in silicate minerals, e.g. biotite, amphiboles, pyroxenes and olivines, and as iron sulphides (pyrite FeS_2); chelated to colloidal and dissolved organic matter; adsorbed to mineral surfaces in exchangeable form at pH < 3. Total Fe content in soils is between < 1 (e.g. desert soils formed from quartz sand, E horizons of Spodosols) and > 10% (e.g. Oxisols from basalt, B horizons of Spodosols). The microbial influence on iron transformation in soils may be grouped roughly into: (1) direct involvement in the reduction–oxidation processes; and (2) direct transformation of iron from the organic (chelated) to the inorganic form, and vice versa. For example, chelating organic compounds (siderophores), which mediate the transport of ferric iron (Fe^{3+}), are produced by microorganisms. Microbial action thus exerts multiple influence on soil formation, water quality, and natural formation or biotechnological exploration of iron ores. The ability to reduce Fe^{3+} has been utilized to test the influence of xenobiotics on microorganisms in H soils (pH > 4.5; Welp and Brümmer 1985).

The analytical approach to soil iron by wet chemistry either distinguishes between different degrees of crystallinity or determines the plant available iron fraction; it is even possible to separate ferrous (Fe^{2+}) from ferric iron. Total iron is best determined by atomic absorption spectroscopy after carbonate fusion or HF digestion, which is presented below, because it is more rapid and because there is no danger of losing elements. Iron determinations are also possible by colorimetry using an *o*-phenanthroline reagent (Olson 1965). A method for the quantification of Fe in X-ray amorphous (oxi-)hydroxides is included. A speciation and (semi-)quantification of crystalline iron oxides is possible by X-ray diffractometry, differential thermal analysis and Mössbauer spectroscopy (Bigham et al 1978).

Total iron by hydrofluoric acid digestion

Principle of the method

(Jackson 1958)

Soil minerals are digested in concentrated hydrofluoric acid; concentrated perchloric acid is added to destroy organic matter, and the sample is finally dissolved in sulphuric acid.

Materials and apparatus

Teflon beakers, 50 ml volume, with a Teflon lid
Hotplates or sand bath

Chemicals and solutions

Concentrated hydrofluoric acid (48%) (Caution: do not use glassware for manipulation!)
Concentrated perchloric acid (72%)
Concentrated sulphuric acid (95–97%)
6 M hydrochloric acid
Concentrated nitric acid (65%)

Procedure

Weigh a sample of about 500 mg into a 50 ml Teflon beaker; record the exact weight. Carry out the following steps under a fume hood: wet the soil with a few drops of H_2SO_4, and add 5 ml HF and 0.5 ml $HClO_4$. Place the beaker on a hotplate or in a sand bath, cover with the Teflon lid to 90%, and heat to 180°C; evaporate the contents to dryness. (Caution: Teflon beakers are resistant to temperatures up to 260°C only.)

Cool the beaker and add 2 ml distilled water and a few drops of $HClO_4$ to destroy the remaining organic matter; heat and evaporate to dryness. Cool the beaker and add 5 ml 6 M HCl and 5 ml distilled water. Heat gently until the solution boils. Cool and transfer to a 50 ml volumetric flask if the solution is clear and free of particles. Wash the beaker twice with little distilled water and collect the washings in the volumetric flask. Make up to volume with distilled water.

Discussion

Organic soils are heated with 3 ml concentrated HNO_3 and 1 ml $HClO_4$ until the appearance of white $HClO_4$ fumes prior to step 2. Cool before adding HF!

If complete dissolution of the sample in HCl cannot be achieved, evaporate to dryness and repeat the procedure starting from step 2.

Amorphous iron by acid ammonium oxalate

Principle of the method
(Schwertmann 1964)

Ferric iron in amorphous hydroxides is reduced to Fe^{2+} by oxalic acid at pH 3 and kept in solution by chelation with oxalate anions; hydroxyl anions liberated in the dissolution of hydroxides are neutralized. The extraction is performed in the dark to prevent the reduction of crystalline iron by oxalic acid (Schwertmann 1964).

Materials and apparatus

Horizontal shaker

Chemicals and solutions

0.2 M oxalic acid

Dissolve 25.22 g in 600 ml distilled water in a 1000 ml volumetric flask and make up to volume.

0.2 M ammonium oxalate

Dissolve 28.42 g in 600 ml distilled water in a 1000 ml volumetric flask and make up to volume.

Extracting solution

Add oxalic acid solution to the ammonium oxalate solution in a glass beaker until pH 3.00 is reached.

Procedure

Weigh 1 g air dry, sieved (< 2 mm) soil into a 250 ml polyethylene bottle. In the dark: add 100 ml of the extracting solution and shake for 2 h. Filter through qualitative filter paper and collect the filtrate in a polyethylene bottle. For measurement on an atomic absorption spectrophotometer, dilute at least by a factor of 10 with distilled water to prevent clogging of the burner capillary.

Iron determination by atomic absorption spectrometry

Principle of the method

The element under concern is dissolved in a digesting or extracting solution that is sucked through a capillary and sprayed into an air–acetylene flame. The element is converted to its atomic form and thus is capable of absorbing light at characteristic wavelengths. The elementary concentration in a sample solution is calculated by comparing its absorption data to a calibration curve made from a blank and several standard measurements.

The iron determination by atomic absorption spectrometry is highly sensitive (the detection limit is < 0.01 mg l^{-1} in the air–acetylene flame); costs are low compared to ICP-AES or X-ray fluorescence analysis. The measurement is almost free of interferences, and a large number of samples can be processed within a short space of time.

Materials and apparatus

Atomic absorption spectrometer

Chemicals and solutions

Commercial iron stock solution, 1.000 ± 0.002 g Fe l^{-1} (e.g. Merck 9972 Titrisol)

Standards

Dilute a 1.000 g stock solution to 1000 ml in a volumetric flask and mix well (= working stock solution).

Pipette appropriate amounts of the working stock solution into 100 ml volumetric flasks to give the desired standard concentrations including a blank; at the most sensitive wavelength 248.3 nm, the optimum working range is between 0.2 and 5.0 mg Fe l^{-1}.

Add appropriate amounts of the matrix solution which has been used to digest or extract the samples; care for eventual dilution of the original sample solutions. Make up to volume with distilled water, cap and mix well.

Procedure

Measure blank and standard absorptions according to the operation instructions of your instrument using an oxidizing air–acetylene flame. Determine sample concentrations from the calibration curve or take over the printout from the connected microcomputer.

Calculation

The iron concentration in the original sample C_s is calculated by

$$C_s = (C_m - C_{bl}) \cdot f \cdot V_e / W_s \qquad (3.15)$$

where C_m and C_{bl} are the measured Fe concentrations in the sample and blank solutions in milligrams per litre, f is the dilution factor applied to the sample solution, V_e is the volume of the extractant in millilitres and W_s is the sample weight in grams.

Iron determination by colorimetry

(Olson 1965)

Principle of the method

Iron in solution is reduced with hydroxylamine hydrochloride and forms a specific red ferrous tri-(1,10)-phenanthroline complex with *o*-phenanthroline. Fe^{2+} and Fe^{3+} may be distinguished by measuring one aliquot each with and without reductant and calculating Fe^{3+} by their difference; this can be done properly if reducing agents are absent in the sample solution and if photoreduction by direct light is avoided (Olson 1965).

Materials and apparatus

Spectrophotometer

Chemicals and solutions

5 M ammonium acetate

Dissolve 385.40 g NH_4OAc in 600 ml distilled water in a 1000 ml volumetric flask; make up to volume and mix well.

10% hydroxylamine hydrochloride ($NH_2OH.HCl$)

Add 90 ml distilled water to 10 g hydroxylamine hydrochloride.

o-Phenanthroline reagent

Dissolve 0.30 g *o*-phenanthroline monohydrate in water by heating to 80°C; cool and make up to 100 ml.

6 M hydrochloric acid

Add 500 ml concentrated HCl (37%) to 450 ml distilled water in a 1000 ml volumetric flask and make up to volume.

1 M sodium acetate

Dissolve 136.08 g sodium acetate trihydrate in 600 ml distilled water and make up to volume.

Standards

100 mg 1^{-1} stock Fe solution

Dissolve 0.7022 g of ferrous ammonium sulphate hexahydrate [$Fe(NH_4)_2(SO_4)_2.6 H_2O$] in 100 ml concentrated sulphuric acid (95–97%); warm if necessary, and dilute to 1000 ml final volume.

Pipette 0, 0.5, 1.0, 1.5, 2.0 and 2.5 ml of the stock Fe solution into 50 ml volumetric flasks to give 0, 1, 2, 3, 4 and 5 ml Fe Hl^{-1} in the final solutions.

Add matrix solution according to the samples.

Add 1 ml of 10% hydroxylamine hydrochloride and shake the flask.

Add 1 ml of the *o*-phenanthroline reagent.

Add 1 M NaOAc dropwise until a bright orange to red colour appears, and again add 1.5 ml of 1 M NaOAc.

Dilute to volume with distilled water and mix well.

Read light absorbance against a distilled water blank on a spectrophotometer at 508 nm.

Procedure

Pipette a suitable aliquot of sample solution into a 50 ml volumetric flask; e.g. an aliquot size of 500 ml would be appropriate in the case of a HF digestion of 0.500 g sample containing 5% Fe and a final volume of 50 ml digest. Apply hydroxylamine addition and the following steps of the standard preparation procedure. Read light absorbance as indicated above.

Calculation

Plot a calibration curve from concentration versus absorption of the standards and determine sample concentrations. The iron concentration in the original sample C_s is calculated by

$$C_s = (C_m - C_{bl}) . f. V_e / W_s \quad (3.16)$$

where C_m and C_{bl} are the measured Fe concentrations in the sample and blank solutions, f is the dilution factor applied to the sample solution, V_e is the volume of the extractant and W_s is the sample weight.

Heavy metals

J.C. Forster

The inhibitory effect of heavy metals (HMs) on microbial activity has been demonstrated in mineral and organic soil horizons (Wilke 1986; Laskovski et al 1994). Some HMs, however, are part of soil enzymes (Tyler 1981) and therefore play an important role in the metabolism. Heavy metals may exist as cations (Cd, Co, Cr, Cu, Hg, Ni, Pb, Zn) or anions (Mo, Se, As) in soils. They may be incorporated into silicates or oxides, complexed to the soil organic matter, or adsorbed to exchange sites of permanent or variable charge. Heavy metal contamination in soils is frequently judged according to the total content (Anonymous 1992), however, because of the variety of binding and transport mechanisms that are specific to each individual element and because of the many soil properties influencing those mechanisms, mobile fractions have to be considered to obtain a realistic view of the actual contamination of the soil, the risk of groundwater pollution (Brümmer et al 1986) and of inhibition of the soil microflora (Angle et al 1993).

Total heavy metal contents by aqua regia digestion

(DIN 38414-S7, 1983; McGrath and Cuncliffe 1985)

The aqua regia soluble fraction represents nearly the total heavy metal content of soils, except part of those contained in silicates and Al, Cr, Fe and Ti oxides. Although the overall contamination of a site is reflected, it gives no information on possible sources or ecologically meaningful concentrations.

Principle of the method

Dissolution of a soil sample in a concentrated nitric acid:hydrochloric acid mixture (1:3). Determination of heavy metals by inductively coupled plasma, or by flame and graphite furnace atomic absorption spectroscopy.

Materials and apparatus

Metal-free ball mill
Analytical sieve, mesh size 0.1 mm
Drying oven
Desiccator
Reaction flasks, 250 ml
Reflux cooler
Absorption device (Fig. 3.3)
Glass beads with a rough surface
Oil bath
Funnels and 100 ml volumetric flasks
Ashless filter paper (e.g. Schleicher and Schüll 1505)
Instrument for heavy metal measurements (ICP (inductively coupled plasma) or graphite furnace AA spectrometers)

Chemicals and solutions

Concentrated hydrochloric acid (37%), analytical grade.
Concentrated nitric acid (65%), analytical grade.
0.5 M nitric acid
Dilute 35 ml concentrated nitric acid to 1000 ml with water of bidistilled or comparable quality.

Procedure

Rinse all materials with warm 0.5 M HNO_3 and then with bidistilled water. Weigh 3.00 g oven-dried (40°C), finely ground (95% < 0.1 mm) soil into a reaction flask. Moisten with a little (2–3 ml) bidistilled water, and successively add 21 ml concentrated hydrochloric acid and 7 ml nitric acid. Add 10 ml 0.5 M HNO_3 to the absorption device, and connect the cooler to the device and reaction flask. Allow it to stand at room temperature overnight, then boil for 2 h. The condensation zone should not exceed one third of the height of the cooler. Allow to cool to room temperature and combine the contents of the absorption device with the reaction mixture. Rinse the device and cooler with approximately 5 and 10 ml 0.5 M HNO_3, respectively.

Transfer the clear digest directly to a 100 ml volumetric flask, rinse the reaction flask with 0.5 M HNO_3 and make up to volume with bidistilled water. If sample turbidity prevents subsequent measurement, centrifuge or filtrate through 0.45 µm membranes. Perform HM analyses by ICP or atomic absorption spectroscopy according to the instructions given by the manufacturer.

Discussion

Foaming in organic samples is avoided by adding acids dropwise.

Instead of bidistilled water, distilled water cleaned by a further purification device may be used.

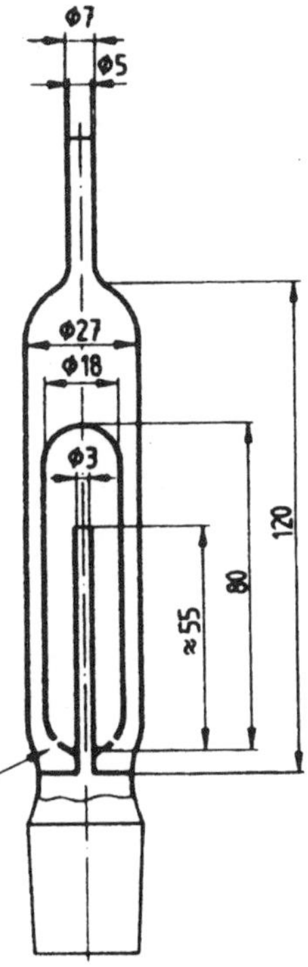

Φ7
Φ5
Φ27
Φ18
Φ3
120
80
≈55
4 holes
φ 1 - 2

Figure 3.3 Absorption device to be plugged on to the cooler of the aqua regia digestion apparatus (from DIN 1983); all dimensions are given in millimetres.

Heavy metal speciation

(Zeien and Brümmer 1989, 1991)

A procedure is presented that differentiates seven forms of heavy metal binding in soils. Extractants were optimized for selectivity and absence of interferences in the instrumental analysis. Problems associated with an adverse order of extractants and with redistribution processes were overcome. Important conclusions can be drawn on the availability of single heavy metals to microorganisms and plants, and thus on the toxic potential of a distinct soil.

Principle of the method

Successive extraction of soil samples with solutions of increasing exchange strength and acidity. The procedure includes the following extractants:

1 M ammonium nitrate (NH_4NO_3) for mobile, exchangeable (= unspecifically adsorbed) HMs and soluble organic complexes.

1 M ammonium acetate (NH_4OAc, pH 6.0) for HMs that are specifically sorbed, occluded near oxide surfaces, bound in organic complexes of moderate strength, or in carbonates.

0.1 M hydroxylamine hydrochloride (NH_2OH–HCl) + 1 M NH_4OAc pH 6.0 for HMs bound in manganese oxides.

0.025 M NH_4-EDTA pH 4.6 for organic complexes of increased strength.

0.2 M ammonium oxalate pH 3.25 for HMs bound to amorphous Fe oxides.

0.1 M ascorbic acid and 0.25 M ammonium oxalate pH 3.25 for HMs bound in crystalline Fe oxides.

Concentrated perchloric acid and nitric acid for total residual HMs, e.g. in silicates.

Materials and apparatus

Polypropylene centrifuge tubes and caps, 100 ml, prerinsed with 0.5 M HNO_3 and bidistilled water
Horizontal shaker or equivalent
Centrifuge
Funnels and 100 ml polythene bottles, prerinsed with 0.5 M HNO_3 and bidistilled water

Chemicals and solutions

1 M HN_4NO_3

Dissolve 80.04 g in 1000 ml bidistilled water.

1 M NH_4OAc pH 6.0

Dissolve 77.08 g in 900 ml bidistilled water, adjust the pH to 6.0 with acetic acid (50%) and make up to 1000 ml with bidistilled water.

0.1 M NH_2OH–HCl and 1 M NH_4OAc pH 6.0

Dissolve 6.95 g NH_2OH–HCl and 77.08 g NH_4OAc in 900 ml bidistilled water, adjust the pH to 6.0 with acetic acid (50%), and make up to 1000 ml with bidistilled water.

0.025 M NH_4-EDTA pH 4.6

Dissolve 7.31 g NH_4-EDTA (Titriplex II, Merck, Germany) in 900 ml bidistilled water, adjust to pH 4.6 with dilute ammonia, and make up to 1000 ml with bidistilled water.

0.2 M NH_4-oxalate pH 3.25

Dissolve 28.42 di-ammonium-oxalate monohydrate and 25.21 g oxalic acid dihydrate in 900 ml bidistilled water, adjust to pH 3.25 with dilute ammonia, and make up to 1000 ml with bidistilled water.

0.1 M ascorbic acid and 0.2 M NH_4-oxalate pH 3.25

Dissolve 17.61 g L-ascorbic acid in 900 ml of oxalate solution, readjust the pH to 3.25 with dilute ammonia, and make up to 1000 ml with oxalate solution.

$HClO_4$ (70–72%) and HNO_3 (65%), analytical grade.

Procedure

Fraction I Weigh 2.00 g air-dried (25°C) soil into a polypropylene centrifuge tube, add 50 ml of ammonium nitrate solution and shake for 24 h at 180 min^{-1} at 25°C. Centrifuge at 2500 rev min^{-1} for 15 min at 20°C and pass the supernatant through an ashless filter paper into a polythene bottle. Add 0.5 ml HNO_3 (65%) to stabilize the filtrate.

Fraction II Redissolve the sediment with 50 ml of ammonium acetate solution; add dilute HCl (0.1 M) to neutralize for carbonates at contents of less than 5%. Perform shaking, centrifugation and decanting as indicated above. Redissolve the sample in 25 ml of ammonium nitrate, shake for 10 min, centrifuge and combine with the preceding extract. Stabilize with 0.5 ml concentrated HNO_3.

Fraction III Redissolve the sample in 50 ml of hydroxylamine/ammonium acetate solution. Perform shaking for 30 min, centrifugation and decanting as indicated above. Redissolve the sample twice in 25 ml of 1 M NH_4OAc solution pH 6.0, shake for 10 min, centrifuge and combine with the preceding extract. Stabilize with 0.5 ml concentrated HNO_3.

Fraction IV Redissolve the sample in 50 ml of 0.025 M NH_4-EDTA solution pH 4.6. Perform shaking for 90 min, centrifugation and decanting as indicated above. Redissolve the sample in 25 ml of 1 M

NH_4OAc solution pH 6.0, adjust the pH to 4.6 with concentrated acetic acid, shake for 10 min, centrifuge and combine with the preceding extract.

Fraction V Redissolve the sample in 50 ml of 0.2 M NH_4-oxalate solution pH 3.25. Perform shaking for 60 min in the dark, centrifugation and decantation as indicated above. Redissolve the sample in 25 ml of 0.2 M NH_4-oxalate solution pH 3.25, shake for 10 min in the dark, centrifuge and combine with the preceding extract.

Fraction VI Redissolve the sample in 50 ml of 0.1 M ascorbic acid and 0.2 M NH_4-oxalate solution pH 3.25. Keep at 96 ± 3°C on a hot water bath for 30 min, allow to cool, and perform centrifugation and decanting as indicated above. Redissolve the sample in 25 ml of 0.2 M NH_4-oxalate solution pH 3.25, shake for 10 min in the dark, centrifuge and combine with the preceding extract.

Fraction VII Transfer the sediment to Teflon beakers with a total of 15 ml concentrated $HClO_4$ and 15 ml concentrated HNO_3 under a fume hood. Place the beakers on a sand bath and successively heat to 80 and 120°C for 1 h each. Increase the temperature to 170°C and fumigate the acids until the soil colour changes to a light grey. Cool and redissolve the residue in a total of 20 ml 5 M HNO_3, transfer the solution to a 100 ml volumetric flask and make up with bidistilled water. Filter this solution into a 100 ml polythene bottle.

Perform measurements on an ICP-AES, an ICP-atomic absorption spectrometer (AAS), a flame photometric AAS, a graphite furnace AAS or a hydride generation-AAS, respectively, according to the manufacturer's instructions.

Discussion

The fractionation procedure was developed for aerobic soils with less than 5% $CaCO_3$.

The method presented includes single steps from procedures published by Shuman (1985), Chao (1972), Tessier et al (1979) and Sposito et al (1982).

The main forms of binding of single elements are (Shuman 1985, Zeien and Brümmer 1989, 1991):

Cd, in fractions I, II, and III.
Co, in fraction III and partly in fraction V.
Cu, in fractions IV, V and VII.
Fe, in fractions V, VI and VII.
Mn, in fractions III, IV and VII.
Ni, in fractions V, VI and VII.
Pb, in fractions III, IV and V.
Zn, in fractions V, VI and VII, and in fractions I and II, depending on the soil pH.

The method presented uses homogenized soil and therefore does not account for the spatial differentiation in aggregated soils. The physico-chemical heterogeneity of aggregate surface and core has been demonstrated (Kaupenjohann 1989); differences in pH, redox potential, amount of organic substances, etc. suggest large differences in the binding of HMs.

Soil physical analysis

J.C. Forster

Determination of the gravimetric water content and soil dry mass

Principle of the method

The weight of a soil subsample is determined in its "field-moist" state and again after oven-drying, and the difference attributed to loss of water is calculated and related to the soil moist weight.

Materials and apparatus

Drying oven
Wide-mouthed heat-resistant sample jars or glass beakers, 100–250 ml
Desiccator, supplied with P_2O_5 or silica gel desiccant

Procedure

Weigh about 20 g field-moist soil into a suitable drying container (wide-mouthed heat-resistant sample jar) and record the weight to the nearest 0.01 g. Avoid thick layers of soil to ensure complete drying. Place the container in a drying oven at 105°C overnight. Remove the container and cool in a desiccator. Reweigh the container with the dry sample.

Calculation

$$\% \text{ water content} = [(\text{moist weight} - \text{dry weight})/\text{moist weight}] \times 100 \qquad (3.17)$$

$$\% \text{ dry mass} = (\text{dry weight}/\text{moist weight}) \times 100 \qquad (3.18)$$

Discussion

Organic soils and soil horizons should be dried at 60–65°C to avoid loss of volatile organic compounds (EPA 1990).

Water is not completely removed from clay soils at 105°C (Hendrickx 1990).

Drying may also be achieved in a microwave oven within 6 to 20 min. (Hendrickx 1990).

For conversion of gravimetric into volumetric water content, a measurement of the soil bulk density is needed.

Field moisture determinations may be conducted with neutron scattering, gamma-ray attenuation, electrical resistance blocks or time-domain reflectrometry techniques.

Determination of the soil bulk density

Bulk density is the dry weight per unit volume of soil; it is useful for converting a weight-based result to a volume-based number. Bulk densities of A horizons are around 1.2 g cm^{-3}, whereas mineral soil may reach up to 2.0 g cm^{-3}; organic layers and volcanic ash soils may be between 0.1 and 0.5 g cm^{-3}.

Measurements are made as described in the earlier section in this chapter entitled "Sterilization of soil and inhibition of microbial activity" except that core samples contained in stainless steel rings with a defined volume are used. Samples with an intermediate water content should be used to avoid deformation. At least five replicates are needed to allow for field variability. Soil from steel rings may be collected quantitatively in a suitable container and treated as above; this might be of value if a large number of rings are collected and combined in the field. This method is unsuitable for stony or very loose soils where excavation methods are preferable. These include the removal and weighing of a unit of soil the volume of which is determined by replacing the hole with sand or a water-filled rubber balloon (Blake and Hartge 1986). The clod method (USDA/SCS 1984) is in use with soils that yield stable clods that are covered with a thin water-repellant wax mantle and weighed under air (fresh and oven-dry) and water. The bulk density is calculated under the assumption of distinct individual densities of soil particles, pore water and applied wax material (EPA 1990).

Determination of the water-holding capacity of soils

Microbiological measurements are often made after adjusting the water content to a constant value for all soils or, preferably, to the individual water-holding capacity or a defined percentage thereof. This ensures similar conditions for the microbes as concerns the availability of water, which is crucial for their growth and metabolic activity.

Principle of the method

An excess amount of water is percolated through a known amount of field-moist soil; the volume of the percolate is determined and the volume of water stored in the soil is calculated.

Materials and apparatus

Polyethylene funnels and folded filter paper, e.g. Whatman 2V or Schleicher & Schüll 597 1/2.
Collecting flasks

Procedure

Place duplicate 20 g field-moist soil samples in a funnel and filter paper mounted on a suitable, preweighed collecting flask. Record the exact soil weight to the nearest 0.01 g. Add 100.00 ± 0.01 g distilled water in small portions and allow to stand overnight while covering the funnel with aluminum foil to prevent evaporation. Gently tap the funnel to the neck of the flask to move adhering water drops into the flask. Weigh the collecting flask to the nearest 0.01 g. Run duplicate blanks, including a funnel and filter paper without soil. Determine the water content of the soil samples as outlined in the earlier section on sterilization of soil and inhibition of microbial activity.

Calculations

$$\% \text{ water-holding capacity} = [(100 - W_p) + W_i] / dwt \times 100 \qquad (3.19)$$

where W_p is the weight of the percolated water in grams, W_i is the initial amount of water in grams contained in the sample, and *dwt* is the soil dry weight in grams.

Determination of the soil water potential

Principle of the method

The water potential concept was developed in 1907 by Buckingham to describe the energy status of the soil solution phase. Water-filled, porous ceramic cups (tensiometers) equipped with a pressure-sensitive device were soon found to be suitable for the measurement of the soil water tension; they have been proved to be useful in studies of soil water transport and soil water–plant relationships

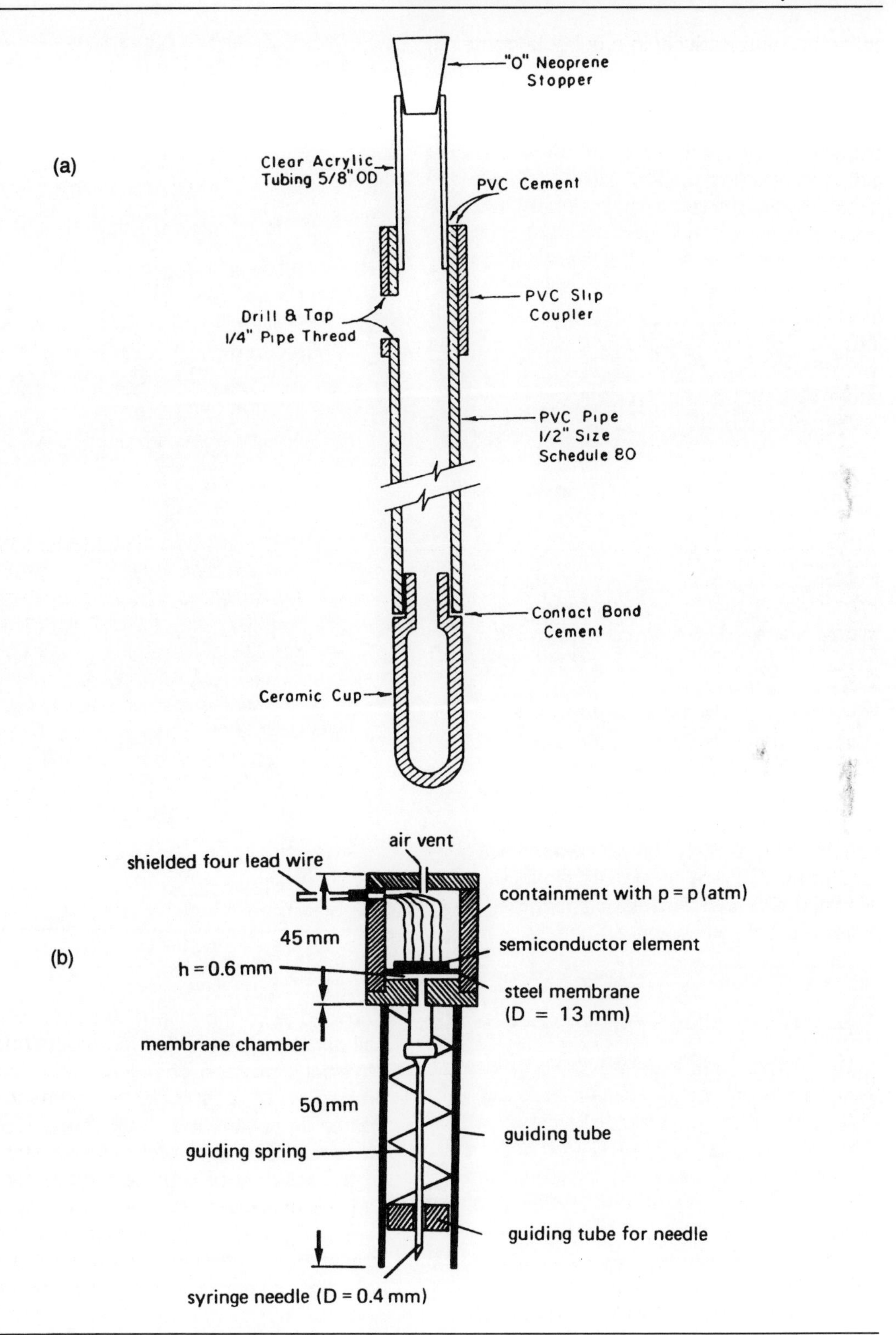

Figure 3.4. Diagram of a tensiometer with a rubber stopper (a) (Henderson and Rogers 1963) and a mounted pressure transducer (b) (Marthaler et al 1983).

(Cassel and Klute 1986). The soil water potential may also be used to monitor the humidity conditions as an important factor in microbial transformation processes in soil because a characteristic water potential–water content relationship exists for every soil layer. Soil pores are assumed to be filled with soil solution when their equivalent diameter is smaller than the present corresponding soil water potential.

The water in suction cups is assumed to be in equilibrium with soil water (which is actually not the case with respect to all thermodynamic aspects), so it may flow into or out of the cup if the soil is wetting or drying. The pressure within the tensiometer thus reflects the soil water potential.

Materials and apparatus

Measurements are made with vacuum gauges, water- or mercury-filled manometers, or electrical pressure transducers. The connection between tensiometer and pressure transducer is achieved by a water-filled tube or an injection needle that penetrates a septum stopper mounted on the top of the tensiometer tube (Fig. 3.4) and taps the air reservoir above the tensiometer water column. Readings can be made periodically or by automatic registration with a data logger, which makes the installation and simultaneous use of many tensiometers possible.

Field installation and maintenance

Tensiometers are installed in the field in a horizontal or vertical position in holes prepared with a soil auger. For horizontal installation, a soil pit is first dug and bore holes with diameters slightly wider than that of the tensiometer are made. A soil or quartz powder suspension is then slaked into the hole to ensure a close contact between the suction cup and bore hole surface. On sloping sites the tensiometer should be directed horizontally or diagonally towards the slope to avoid draining effects. These are particularly dangerous in vertical tensiometer installation, however, they might be overcome by a narrow bore hole and thorough installation.

It should be remembered that porous ceramic suction cups are not suitable for soil water tensions higher than 800 cm due to air entry and dissolution. Cup conductance is usually around 3×10^{-5} $cm^2\ s^{-1}$ and may range up to 1×10^{-3} cm^2 s^{-1}; response time is between 40 and 60 s with mercury-filled manometers and 5 s for a vacuum gauge (Cassel and Klute 1986). Maintenance primarily includes regular inspections for air accumulation that are necessary to maintain correct measurements. Unreliable readings may be identified by the history of suction values and by comparison with parallel measurements from different instruments. To remove excess air, open the tensiometer cap and refill the barrel with deaerated water. Create a suction (30 s) with a hand-operated vacuum pump to remove air bubbles. Add deaerated water again to fill the barrel completely and recap the instrument. Refilling will have to be done more frequently under hot and dry than under cold and wet conditions. It may be advis-able to “retire” tensiometers for longer periods where the suction limit of 800 hPa is exceeded.

Particle size analysis

Knowledge of the size distribution of primary soil particles is essential for understanding the physical, physico-chemical and biological properties of a soil. If the primary mineral matrix is examined, soil texture analysis involves a series of pretreatment steps, such as the removal of organic matter, carbonates and sesquioxides, to eliminate any type of cementing agent. The subsequent dispersion step alone is sufficient if the natural size distribution of soil components has to be preserved. Analysis means the separation of particles by sieving and sedimentation techniques, according to their size and density.

Pretreatments

Removal of carbonate and soluble salts

Materials and apparatus

Centrifuge with 250 ml bottles
Glassware: 1000 ml volumetric flask

Chemicals and solutions

1 M sodium acetate solution

Dissolve 136.08 g sodium acetate trihydrate in 200 ml distilled water and make to 1000 ml with distilled water.

Silver nitrate and barium chloride solution for Cl^- and SO_4^{2-} tests.

Procedure

Weigh a portion of soil (between 10 and 60 g depending on clay content) in a 250 ml centrifuge bottle. Then add 100 ml of water and 10 ml 1 M sodium acetate (pH 5), shake and centrifuge (10 min, 1500 rev min^{-1}), and discard the clear supernatant. Wash twice with 50 ml distilled water and discard the clear supernatant after centrifugation. In cases of turbidity, which may be due to high amounts of soluble salts or gypsum, further washing is recommended. Salts may be tested by $AgNO_3$ for Cl^- and $BaCl_2$ for SO_4^{2-}.

Discussion

Sodium acetate is recommended instead of hydrochloric acid because it does not affect the crystalline lattice of clay minerals.

Heating accelerates the carbonate dissolution reaction.

Alkaline salts can cause decomposition of H_2O_2 and have therefore to be removed before the removal of organic matter.

Oxidation of organic matter

Materials and apparatus

Water or sand bath
1000 ml glass beakers, glass covers and glass rods

Chemicals and solutions

Hydrogen peroxide, 30%, technical grade
Ethyl alcohol or amyl alcohol, technical grade

Procedure

Weigh or transfer soil (from centrifuge bottles, see above) into 1000 ml glass beakers. Place the beakers on a water or sand bath at $T = 60°C$. Add about 100 ml distilled water. Add 5 ml H_2O_2, stir and cover for several minutes. Add alcohol dropwise, if necessary to minimize foaming, or cool the glass beakers in cold water. Repeat H_2O_2 addition until organic matter is destroyed completely, judged by the rate of reaction (foaming) and the bleaching of black or brown colour. Avoid liquid volumes greater than 100 ml to preserve a narrow soil:solution ratio, which is important for an effective treatment. Transfer the sample to a 250 ml centrifuge bottle.

Discussion

In contrast to a commonly held opinion, even samples that contain less than 2% organic matter should be treated with an oxidizing agent: (1) to give comparable results, and (2) because organic binding agents are overproportionately efficient at low concentrations.

A number of alternative oxidizing agents have been used successfully, e.g. sodium hypochlorite, sodium hypobromite and potassium permanganate. Perchloric acid is not useful because layer lattice clays are destroyed.

Removal of organic matter might be improved if light particular matter is floated off the sample in a preceding step after a short ultrasonic treatment.

An ultrasonic pretreatment improves the accessibility for and thus the effectivity of the oxidizing agent.

Manganese dioxide decomposes H_2O_2 and thus has to be complexed or removed before organic matter oxidation.

The following iron oxide removal procedure may be performed in the glass beakers. The sample must not be transferred to the centrifuge bottles.

Removal of iron oxides

Principle of the method

Soil samples are treated with a buffered solution of a reducing plus a chelating agent. Iron oxides and hydroxides are dissolved and the Fe (and Al) removed.

Materials and apparatus

Water or sand bath, or hotplates
Centrifuge and glass beakers (250 ml volume is desirable)

Chemicals and solutions

Citrate bicarbonate buffer.

0.3 M sodium citrate

Dissolve 88 g of trisodium citrate in 1000 ml distilled water (solution 1).

1 M sodium bicarbonate

Dissolve 8.4 g of $NaHCO_3$ in 100 ml distilled water (solution 2).

Mix 800 ml of solution 1 and 100 ml of solution 2 in a 1000 ml glass beaker (solution 3).

Sodium dithionite, technical grade.

Saturated sodium chloride solution

Dissolve 400 g of reagent grade NaCl in 1000 ml distilled water and filter the solution through suitable filter paper to remove the undissolved salt (the solubility of NaCl at 20°C is 358 g per kg of water).

10% sodium chloride solution

Dissolve 100 g of reagent grade sodium chloride in 1 kg distilled water.

Procedure

Add solution 3 to give a total liquid volume of 150 ml, and shake the glass beaker or centrifuge tube. Slowly add 3 g of sodium dithionite to prevent foaming. Place the container on a water bath at 80°C for 20 min with intermittent stirring. Remove the container from the water bath, add 10 ml of saturated NaCl, mix, centrifuge and discard the supernatant. Add a second portion of sodium dithionite if the soil is not bleached. Wash the sample with 50 ml of solution 3 and 20 ml of saturated NaCl. Wash the sample twice with 50 ml of 10% NaCl and twice with distilled water. If the solution is not clear, centrifuge at high speed or add acetone, warm and centrifuge. Add 150 ml distilled water, shake and check pH which should be > 8 because of complete sodium saturation.

Discussion

Oxides should not be removed in oxidic soils because primary mineral grains would be destroyed (El-Swaify 1980).

Dispersion

Materials and apparatus

Horizontal shaker
1000 ml polyethylene bottles

Chemicals and solutions

Sodium hexametaphosphate solution, 50 g l^{-1}

Dissolve 50 g of sodium hexametaphosphate in 1000 ml distilled water.

Procedure

Transfer washed soil after pretreatments into a 1000 ml round bottom flask and dry on a freeze-dryer. Weigh 10 g of freeze-dried soil into a 1000 ml polyethylene bottle. Cautiously add 200 ml distilled water to prevent loss of sample dust and 10 ml of sodium hexametaphosphate (HMP) solution. Shake overnight on a horizontal shaker.

Discussion

Other commonly used dispersants are: sodium pyrophosphate ($Na_4P_2O_7$) at concentrations between 0.025 M and 0.4 M, for soil samples with dominant Al and Fe organic cementing; 2% sodium carbonate (Na_2CO_3), if subsequent clay mineral analysis has to be performed; sodium oxalate ($Na_2C_2O_4$), if incubation or degradation studies have to be carried out on size separates and P fertilization is undesirable.

For later hydrometer measurements, use 100 ml hexametaphosphate solution and soil weight as follows: 10–20 g of fine-textured soils (clays, silts); 40 g of loamy soils; and 60–100 g of coarse sands.

Sieve analysis

Materials and apparatus

Set of appropriate sieves, 100 mm diameter, mesh size usually 2000 to 50 μm
Plastic funnel
Rubber wiper
1000 ml acrylic cylinders
Glass beakers, 250 ml
Drying oven
Analytical balance

Procedure

Place a plastic funnel on the acrylic cylinder. Place the set of sieves on the plastic funnel with the mesh size decreasing from top to bottom. Transfer the dispersed sample on to the top sieve and pass the smaller particles through by gentle use of the rubber wiper and rinsing with distilled water. Take off the top sieve and transfer the respective size fraction into a preweighed 250 ml glass beaker. Continue with the fraction on the following sieve. Take care that the water volume in the cylinder does not exceed 1000 ml. Allow the soil particles to settle in the beakers and decant the supernatant water. Place the beakers in a drying oven at T = 105°C overnight, cool the samples in a desiccator and reweigh on an analytical balance.

Calculation

% by weight of a size fraction =

$$\frac{(W_f - W_o) \times 100}{W_s} \qquad (3.20)$$

where W_f is the weight of the glass beaker plus dried soil fraction, W_o is the weight of the sole glass beaker and W_s is the total weight of the dispersed sample.

Discussion

Instead of freeze-drying, washed samples may also be transferred to suitable glass beakers, dried at 65°C, and the dried soil gently crushed in a mortar, homogenized and weighed into the polyethylene beakers for dispersion.

Pipette analysis

Principle of the method

The analysis of fine soil particles (< 50 μm) is no longer possible by sieve analysis but needs a different procedure. The pipette and hydrometer methods are the most popular, and are rapid and simple. Both use the fact that the sinking velocity is dependent on (amongst other things) particle equivalent size, which is described in Stokes' law. While the pipette method involves sampling of an aliquot of a soil suspension after a defined range of time at constant temperature, the hydrometer method is based on the change of the buoyancy force due to the settling of soil particles.

Materials and apparatus

Pipette device
Plastic or glass beakers, 50 or 100 ml volume, preweighed
Drying oven
Water bath or constant temperature location
Rubber stoppers for 1000 ml cylinders

Procedure

Make up the 1000 ml cylinder containing the soil fraction < 50 μm to volume with distilled water. Put the cylinder into a water bath or take it to a constant temperature room. Let the sample equilibrate for several hours or overnight. Place a rubber stopper on the cylinder and gently shake the cylinder end-over-end for 1 min. Position the cylinders in a row, avoiding further vibration. Bring the pipette rack behind the cylinder after the appropriate time interval. Lower the pipette to 10 cm depth and withdraw an aliquot of the soil suspension starting exactly after the calculated settling time. Bring the pipette out of the cylinder, wipe away adhering drops and transfer the contents to a preweighed beaker. Rinse the pipette with distilled water and add this to the sampled aliquot in the beaker. Allow the particles to settle and decant excess water; evaporate the sample to dryness at 105°C overnight.

Continue with size fractions < 20, < 5 and < 2 μm at appropriate settling times.

Calculation

The percentage (by weight) of a size fraction is calculated according to the equation:

% size fraction =

$$\frac{(w_i - w_{d,i}) - (w_{i-1} - w_{d,i-1}) \times 100}{w_t} \quad (3.21)$$

where w_i and w_{i-1} are the weights of fraction i and the following smaller fraction i-1 including the weights $w_{d,i}$ and $w_{d,i-1}$ of the respective glass beakers, and w_t is the total weight of the dispersed soil.

Discussion

Silt (5–20 and 2–5 μm) and total clay (< 2 μm) fractions may be analysed after an appropriate settling time, starting with the coarsest fraction.

Settling times for water and HMP dispersed soil at 25°C are given in Table 3.2; data for different conditions are listed in Gee and Bauder (1986).

Hydrometer methods

Standard Bouyoucos-type hydrometer method

(Gee and Bauder 1986)

This represents an alternative procedure for estimating the distribution of fine particles (< 50 μm). Unlike the pipette method, there is no subsampling involved, so repeated measurements on the same sample are possible.

Table 3.2 Sedimentation times for 2, 5 and 20 µm fractions at a particle density of 2.6 g cm^{-3}, a 10 cm sampling depth, for a 0.5 g l^{-1} sodium hexametaphosphate solution.

Temperature (°C)	*Viscosity (10^{-3} kg cm^{-1} s^{-1})*	*2 µm*	*5 µm*	*20 µm*
18	1.055	8 h 25 min	1 h 20 min 39 s	5 min 02 s
20	1.004	8 h 00 min	1 h 16 min 48 s	4 min 48 s
22	0.957	7 h 38 min	1 h 13 min 12 s	4 min 36 s
24	0.913	7 h 17 min	1 h 09 min 54 s	4 min 24 s
26	0.872	6 h 57 min	1 h 06 min 48 s	4 min 12 s
28	0.834	6 h 39 min	1 h 03 min 54 s	4 min 00 s
30	0.799	6 h 22 min	1 h 01 min 12 s	3 min 48 s

Principle of the method

The density of a soil suspension is measured by introducing a glass body with additional weight (e.g. lead balls). Two principles for measurement involving different ways of reading the results have to be distinguished:

1. Reading the upper edge of the meniscus appearing at the scale which is fixed directly to the glass body (so-called Bouyoucos type hydrometer; Gee and Bauder 1986).

2. Lowering the glass body to constant depth whereafter the result is read on a mobile scale (De Leenheer et al 1955).

Materials and apparatus

Hydrometer, Bouyoucos type, e.g. ASTM 152 H; the latter is usually calibrated in grams per litre at 20°C.
Sedimentation cylinders, 1000 ml volume, liquid level at ± constant height from inner bottom.
Electric or manual stirring device.

Procedure

For calibration, use 100 ml sodium hexametaphosphate (50 g l^{-1}) solution made up to 1000 ml with distilled water at room temperature in a sedimentation cylinder. Mix well and introduce the hydrometer. Read the blank, R_L, at the upper edge of the meniscus. Transfer the dispersed soil suspension to the sedimentation cylinder and make to volume with distilled water. Mix thoroughly with a plunger for 1 min, using strong upward strokes to dislodge sediment and finishing with a few smooth strokes to decrease turbulence of the solution. Lower the hydrometer into the solution about 10 s before the reading, and read R at 0.5, 1, 3, 10, 30, 60, 90, 120 and 1440 min as given above. Clean the hydrometer after every reading, and take blank readings periodically.

Calculation

The concentration of a soil fraction in the suspension at a given time is $C = R - R_L$, where R is the uncorrected reading of the sample in grams per litre and R_L is the blank reading. The sum of the percentage, P, of a soil fraction at a given time interval is calculated by $C/C_0 \times 100$, where C_0 is the oven-dry weight of the soil sample. The respective average mean diameter in suspension, X (µm), is calculated by the equation

$$X = \theta t^{-1/2} \quad (3.22)$$

where t is the settling time in minutes, θ is given by

$$\theta = 1000\,(Bh')^{-1/2} \quad (3.23)$$

B and h' are given by the equations

$$B = 30\,\eta/[g(\rho_s - \rho_l)] \quad (3.24)$$

$$h' = -0.164R + 16.3 \quad (3.25)$$

where θ is the sedimentation parameter (µm $min^{1/2}$), h' the effective hydrometer depth (cm), η the viscosity of the blank solution (poise, g cm^{-1} s^{-1}), g the gravitational constant (cm s^{-2}), ρ_s the soil particle density (g cm^{-3}) and ρ_l the solution density (g cm^{-3}); ρ_l and η can be approximated by the following equations:

$$\rho_1 = \rho_0\,(1 + 0.63\,C_s) \quad (3.26)$$

$$\eta = \eta_0\,(1 + 4.25\,C_s) \quad (3.27)$$

where ρ_0 is the solution density at temperature t (g ml^{-1}), η_0 the water density at temperature t [g ml^{-1}] and C_S the concentration of HMP (g ml^{-1}). η values at a given temperature have been tabulated by Gee and Bauder (1986).

Discussion

To ensure reproducible measurements, use 10–20 g of fine-textured, 40 g medium-textured and 60 g or more of sandy soil for the hydrometer method.

Make all measurements at constant temperature. Only use temperature-equilibrated distilled water.

End-over-end shaking of stoppered cylinders is an alternative for stroking with a plunger to mix the soil suspension.

Ethanol or amyl alcohol may be used dropwise to eliminate foam on the measuring solution.

Error analyses show that the reading of the result on the scale is the major source of error of the hydrometer method (Gee and Bauder 1979), while that of the pipette method is the sampling and weighing (precision of clay determination: ± 1%). De Leenheer et al (1955) state a five times better reproducibility of the chain hydrometer versus the pipette method.

Particle density is usually assumed to be 2.65 g cm^{-3}, which is that of quartz, while clays may amount to even more than 2.7 g cm^{-3}. Mineral particles bound to organic matter may have densities below 2.5 g cm^{-3}.

A nomograph is available (Sur and Singh 1976) for the determination of the mean particle diameter, X, from hydrometer readings.

The chain hydrometer procedure does not include a direct measurement of the 20–50 µm fraction, which is instead calculated by the difference of 100 minus the sum of the percentages of the other size fractions. Hence, it accounts for the sum of measurement errors of the other fractions. An increase of the error was observed with increasing organic matter percentage (De Leenheer et al 1955).

Simplified Bouyoucos-type hydrometer procedure

(Sur and Kukal 1992)

Day's hydrometer procedure (1965) may be replaced by the method of Sur and Kukal (1992) who showed that, if readings were made at t = 4.5 and 120 min, the clay content was almost identical to that measured by the former, much more laborious method. However, the applicability of their procedure was only tested on the hyperthermic families of Typic Ustipsamments and Ustochrepts.

Calculation

Prepare samples, calculate C, P and X, and make determinations of sand, silt and clay percentages as outlined above.
Hydrometer readings, however, are predicted by the equation:

$$R_j = a/b - (a/b - R_i)(t_j/t_i)^B \quad (3.28)$$

where R_j is the hydrometer reading predicted for time t_j (g l^{-1}), R_i is the initial hydrometer reading at time t_i (g l^{-1}), a and b are empirical constants taken from the relationship (Sur and Singh 1976)

$$\theta = m(a - bR)^{-1/2}$$

where m is another empirical, temperature-dependent constant. m, a and b depend on the measuring conditions and on the hydrometer used. B is the slope of the regression line for the relation between log $(a/b - R_i)$ and log (t_j/t_i).

Chain hydrometer method

(De Leenheer et al 1955)

Here the hydrometer is lowered to a constant depth below the water table by the use of small weights and a fine chain. The weights compensate for the additional buoyancy force of the solution, which depends on the presence of soil particles leading to an increase in the solution density compared to the blank dispersing. Calibration is again performed on a temperature-equilibrated blank solution made from the dispersing agent (e.g. sodium hexametaphosphate) and filled up to 1000 ml in a sedimentation cylinder. The hydrometer is lowered into the solution and the complete scale rack plus holder is moved vertically until the needle just touches the solution surface. The indicator remains on the zero point of the scale during this procedure.

Measurements are started by lowering the hydrometer into the soil suspension after an appropriate settling time. Then sufficient weights are added to make the needle approach the water table within a distance between 1 and 2 cm. After adjusting the chain to zero, the scale rack is moved relative to the holder until the needle just touches the liquid surface. The result is read on the scale that refers to the respective weight.

Calculation

The sedimentation time, t, is equivalent to the time required for a particle fraction to pass the distance between the needle tip (the solution surface) and the middle of the hydrometer. The density of the soil suspension, q_s, is given by the equation

$$q_s = (w_h + w_w + w_c) / v \quad (3.29)$$

where w_h is the weight of the hydrometer, w_w is the weight of the additional weights, w_c is the effective weight of the chain and v is the hydrometer volume.

References

Abdel-Moati AR (1990) Adsorption of dissolved organic carbon (DOC) on glass fibre filters during particulate organic carbon (POC) determination. Water Res 24: 763–764.

Alef K, Kleiner D (1989) Rapid and sensitive determination of microbial activity in soils and soil aggregates by dimethylsulfoxide reduction. Biol Fertil Soils 8: 349–355.

Anderson DW, Saggar S, Bettany JR, Stewart JWB (1981) Particle size fractions and their use in studies of soil organic matter: I. The nature and distribution of forms of carbon, nitrogen, and sulfur. Soil Sci Soc Am J 45: 767–772.

Anderson JM, Ingram JSI (1989) Tropical Soil Biology and Fertility: A handbook of methods. CAB International, Wallingford.

Anderson JPE (1987) Handling and storage of soils for pesticide experiments. In: Pesticide, Effects on Soil Microflora. Sommerville L, Greaves MP (eds). Taylor & Francis, London, pp. 45–60.

Angle JS, Chaney RL, Rhee D (1993) Bacterial resistance to heavy metals related to extractable and total metal concentrations in soil and media. Soil Biol Biochem 25: 1443–1446.

Anonymous (1992) Klärschlammverordnung (AbfKlärV). Bundesgesetzblatt 21: 912–934.

Arkley TH (1961) Sulfur Compounds of Soil Systems. Ph.D. thesis, University of California, Berkeley, CA.

Baldock JA, Currie GJ, Oades JM (1991) Organic matter as seen by ^{14}C NMR and pyrolysis tandem mass spectrometry. In: Advances in Soil Organic Matter Research: The impact on agriculture and the environment. Wilson WS (eds) The Royal Society of Chemistry, Special Publication No. 90, Cambridge, pp. 45–60.

Baldock JA, Oades JM, Waters AG, Peng X, Vassallo AM, Wilson MA (1992) Aspects of the chemical structure of soil organic materials as revealed by solid-state ^{13}C NMR spectroscopy. Biogeochem 16: 1–42.

Bardsley CE, Lancaster JD (1962) Determination of reserve sulfur and soluble sulfate in soils. Soil Sci Soc Am Proc 24: 265–268.

Baumgärtl H, Lübbers W (1983) Microcoaxial needle sensor for polarographic measurement of local O_2 pressure in the cellular range of living tissue. Its construction and properties. In: Polarographic Oxygen Sensors. Gnaiger E, Forstner H (eds). Springer Verlag, Berlin, pp. 37–65.

Beck T (1986) Aussagekraft und Bedeutung enzymatischer und mikrobiologischer Methoden bei der Charakterisierung des Bodenlebens von landwirtschaftlichen Böden. Veröff Landwirtsch-Chem Bundesanstalt Linz/Donau 18: 75–100.

Beudert G, Kögel-Knabner I, Zech W (1989) Micromorphological, wet-chemical and ^{13}C NMR spectroscopic characterization of density fractionated forest soils. Sci Total Environ 81/82: 401–408.

Beyer L, Wachendorf C, Koebbemann C (1993) A simple chemical extraction procedure to characterize soil organic matter (SOM): 1. Application and recovery rate. Commun Soil Sci Plant Anal 24: 1645–1663.

Bigham JM, Golden DC, Bowen LH, Buol SW, Weed SB (1978) Iron oxide mineralogy of well-drained Ultisols and Oxisols: I. Characterization of iron oxides in soil clays by Mössbauer spectroscopy, X-ray diffractometry, and selected chemical techniques. Soil Sci Soc Am J 42: 816–825.

Blake GR, Hartge KH (1986) Bulk density. In: Methods of Soil Analysis, Part 1, Physical and Mineralogical Methods. Klute A (ed). Agronomy 9(1), ASA, SSSA, Madison, WI, pp. 363–375.

Bligh EG, Dyer WJ (1959) A rapid method of total lipid extraction and purification. Can J Biochem Physiol 37: 911–917.

Bottner F (1985) Response of microbial biomass to alternate moist and dry conditions in a soil incubated with ^{14}C and ^{15}N-labelled plant material. Soil Biol Biochem 17: 329–337.

Bowman RA (1988) A rapid method to determine total phosphorus in soils. Soil Sci Soc Am J 52: 1301–1304.

Bowman RA (1989) A sequential extraction procedure with concentrated sulfuric acid and dilute base for soil organic phosphorus. Soil Sci Soc Am J 53: 362–366.

Brückett S (1982) Analysis of the Organo-mineral Complexes of Soils. In: Constituents and Properties of Oils. Bonneau M, Souchier B (eds), Academic Press, London.

Brümmer GW, Gerth J, Herms U (1986) Heavy metal species, mobility and availability in soils. Z Pflanzenern Bodenkd 149: 382–398.

Burford JR, Bremner JM (1975) Relationship between the denitrification capacities of soil and total water soluble and readily decomposable organic matter. Soil Biol Biochem 7: 389–394.

Cambardella CA, Elliott ET (1993) Methods for physical fractionation and characterization of soil organic matter fractions. Geoderma 56: 449–457.

Capriel P, Beck T, Borchert H, Härter P (1990) Relationship between soil aliphatic fraction extracted with supercritical hexane, soil microbial biomass, and soil aggregate stability. Soil Sci Soc Am J 54: 415–420.

Cassel DK, Klute A (1986) Water potential: tensiometry. In: Methods of Soil Analysis, Part 1, Physical and Mineralogical Methods. Klute A (ed). Agronomy 9(1), ASA, SSSA, Madison, WI, pp. 563–596.

Catroux G, Schnitzer M (1987) Chemical, spectroscopic, and biological characterization of organic matter in particle size fractions separated from an Aquoll. Soil Sci Soc Am J 51: 1200–1207.

Chao TT (1972) Selective dissolution of manganese oxides from soils and sediments with acidified hydroxylamine hydrochloride. Soil Sci Soc Am Proc 36: 764–768.

Christensen BT (1991) Physical fractionation of soil and organic matter turnover. In: Decomposition and Accumulation of Organic Matter in Terrestrial Ecosystems. N van Breemen (ed). Research Priorities and Approaches. Ecosystems Research Report No. 1, Commission of the European Communities, Brussels, pp. 14–26.

Christensen BT (1992) Physical fractionation of soil and organic matter in primary particle size and density separates. In: Advances in Soil Science, vol 20. Stewart BA (ed). Springer, New York, pp. 1–90.

Christensen S, Tiedje JM (1990) Brief and vigorous N_2O production by soil at spring thaw. J Soil Sci 41: 1–4.

Condron LM, Goh KM, Newman RH (1985) Nature and distribution of soil phosphorus as revealed by a sequential extraction method followed by ^{31}P nuclear magnetic resonance analysis. J Soil Sci 36: 199–207.

Condron LM, Frossard E, Tiessen H, Newman RH, Stewart JWB (1990a) Chemical nature of organic phosphorus in cultivated and uncultivated soils under different environmental conditions. J Soil Sci 41: 41–5.

Condron LM, Moir JO, Tiessen H, Stewart JWB (1990b) Critical evaluation of methods for determining total organic phosphorus in tropical soils. Soil Sci Soc Am J 54: 1261–1266.

Cowie GL, Hedges JI (1984) Determination of neutral sugars in plankton, sediments, and wood by capillary gas chromatography of equilibrated isomeric mixtures. Anal Chem 56: 497–504.

David MB, Mitchell MJ, Aldcorn D, Harrison RB (1989) Analysis of sulfur in soil, plant and sediment materials: sample handling and use of an automated analysis. Soil Biol Biochem 21: 119–124.

Day PR (1965) Particle fractionation and particle size analysis. In: Methods of Soil Analysis, Part 1, Physical and Mineralogical Methods. Black CA (ed) Agronomy 9(1), ASA, Madison, WI, pp. 545–567.

De Leenheer L, Van Ruymbeke M, Maes L (1955) Die Kettenaräometer-Methode für die mechanische Bodenanalyse. Z Pflanzenernähr Düng Bodenkd 68: 10–19.

Deutsches Institut für Normung (German Institute of Standards) (1983) DIN 38 414 S 7. Beuth, Berlin.

Dinel H, Schnitzer M, Mehuys GR (1990) Soil lipids: origin, nature, content, decomposition, and effect on soil physical properties. In: Soil Biochemistry, vol 6. Bollag J-M, Stotzky G (eds). Marcel Dekker, New York, pp. 397–429.

Dubois M, Gilles KA, Hamilton JK, Rebers PA, Smith F (1956) Colorimetric method for determination of sugars and related substances. Anal Chem 28: 350–356.

Edwards AC, Cresser MS (1992) Freezing and its effect on chemical and biological properties of soil. Adv Soil Sci 18: 56–79.

Elliott ET, Cambardella CA (1991) Physical separation of soil organic matter. Agric Ecosyst Environ 34: 407–419.

El-Swaites SA (1980) Physical and Mechanical properties of oxisols. In: Soils with Variable Charge. Theng BKG (ed). New Zealand Society of Soil Science, Lower Hutt, pp. 303–324.

Environmental Protection Agency (1987) Handbook of Methods for Acid Deposition Studies: Laboratory Analyses for Surface Water Chemistry. US EPA, Washington, DC.

Environmental Protection Agency (1990) Handbook of Methods for Acid Deposition Studies: Laboratory Analyses for Soil Chemistry. US EPA, Washington, DC.

Ertel JR, Hedges JI (1984) The lignin component of humic substances: distribution among soil and sedimentary humic, fulvic, and base-insoluble fractions. Geochim Cosmochim Acta 48: 2065–2074.

Ertel JR, Hedges JI (1985) Sources of sedimentary humic substances: vascular plant debris. Geochim Cosmochim Acta 49: 2097–2107.

Ertel JR, Hedges JI, Perdue EM (1984) Lignin signature of aquatic humic substances. Science 223: 485–487.

Fengel D, Wegener G (1979) Hydrolysis of polysaccharides with trifluoroacetic acid and its application to rapid wood and pulp analysis. Adv Chem Ser 181: 145–158.

Fengel D, Wegener G (1984) Wood: Chemistry, Ultrastructure, Reactions. De Gruyter, Berlin.

Forster JC, Zech W (1993) Phosphorus status of a soil catena under Liberian evergreen rain forest: results of ^{31}P NMR spectroscopy and phosphorus adsorption experiments. Z Pflanzenern Bodenkd 156: 61–66.

Freney JR, Stevenson FJ, Beavers AH (1972) Sulfur containing amino acid in soil hydrolysates. Soil Sci 114: 468–476.

Fritz JS, Gjerde DT, Pohlandt C (1982) Ion Chromatography. Hüthig, Heidelberg.

Gee GW, Bauder JW (1986) Particle-Size Analysis.

In: Methods of Soil Analysis, Part 1, Physical and Mineralogical Methods. Klute A (ed). Agronomy 9(1), ASA, SSSA, Madison, WI, pp. 383–411.

Goni MA, Hedges JI (1990) Potential applications of cutin-derived CuO reaction products for discriminating vascular plant sources in natural environments. Geochim Cosmochim Acta 54: 3073–3081.

Goni MA, Hedges JI (1992) Lignin dimers: structures, distribution, and potential geochemical applications. Geochim Cosmochim Acta 56: 4025–4043.

Greenland DJ, Ford GW (1964) Separation of partially humified organic material from soil by ultrasonic dispersion. Trans 8th Int Congr Soil Sci 3: 137–148.

Gregorich EG, Kachanoski RG, Voroney RP (1988) Ultrasonic dispersion of aggregates: distribution of organic matter in size fractions. Can J Soil Sci 68: 395–403.

Guggenberger G, Zech W (1992) Sorption of dissolved organic carbon by ceramic P 80 suction cups. Z Pflanzenern Bodenkd 155: 151–155.

Guggenberger G, Zech W, Schulten H-R (1994) Formation and mobilization pathways of dissolved organic carbon: evidence from chemical structural studies of organic carbon fractions in acid forest floor solutions. Org Geochem 21: 51–66.

Guggenberger G, Christensen BT, Zech W (1994) Land use effects on the composition of organic matter in soil particle size separates. I. Lignin and carbohydrate signature. Europ J Soil Sci 45: 449–458.

Hatcher PG, Schnitzer M, Dennis LW, Maciel GE (1981) Aromaticity of humic substances in soils. Soil Sci Soc Am J 45: 1089–1094.

Hedges JI, Ertel JR (1982) Characterization of lignin by gas capillary chromatography of cupric oxide oxidation products. Anal Chem 54: 174–178.

Hedges JI, Lee C (1993) Measurement of dissolved organic carbon and nitrogen in natural waters. Proc NSF/NOAA/DOE Workshop, Seattle, WA, Marine Chem 41.

Henderson DW, Rogers EP (1963) Tensiometer with plastic materials. Soil Sci Soc Am Proc 27: 239–240.

Hendrickx JMH (1990) Determination of hydraulic soil properties. In: Process Studies in Hillslope Hydrology. Anderson MG, Burt TP (eds). Wiley, Chichester, pp. 43–92.

Heng S, Goh KM (1981) A rapid method for extracting lipid components from forest litter especially adapted for ecological studies. Commun Soil Sci Plant Anal 12: 1283–1292.

Hinedi ZR, Chang AC, Lee RWK (1988) Mineralization of phosphorus in sludge-amended soils monitored by phosphorus-31-nuclear magnetic resonance spectroscopy. Soil Sci Soc Am J 52: 1593–1596.

Holloway PJ (1984) Surface lipids of plants and animals. In: Handbook of Chromatography, Lipids, vol. 1. Mangold H (ed). CRC Press, Boca Raton, FL., pp. 347–380.

Jackson ML (1958) Soil Chemical Analysis. Constable & Co. Ltd, London.

Jager G, Bruins EH (1975) Effect of repeated drying at different temperatures on soil organic matter decomposition and characteristics, and on the soil microflora. Soil Biol Biochem 7: 153–159.

Jocteur Monrozier L, Ladd JN, Fitzpatrick RW, Foster RC, Raupach M (1991) Components and microbial biomass content of size fractions in soil of contrasting aggregation. Geoderma 50: 37–62.

John MK (1970) Colorimetric determination of phosphorus in soil and plant materials with ascorbic acid. Soil Sci 109: 214–220.

Johnson CM, Nishita H (1952) Microestimation of sulfur in plant materials, soils, and irrigation waters. Anal Chem 24: 736–742.

Kaupenjohann M (1989) Chemischer Bodenzustand und Nährelementversorgung immissionsbelasteter Waldböden. Bayreuther Bodenkundl Ber 11: 1–202.

Keeney DR, Nelson DW (1982) Nitrogen–inorganic forms. In: Methods of Soil Analysis, Part 2, Chemical and Microbiological Methods. Page AL, Miller DR, Keeney DR (eds). Agronomy 9(2), ASA, SSSA, Madison, WI, pp. 643–698.

Kögel I (1986) Estimation and decomposition pattern of the lignin components in forest soils. Soil Biol Biochem 18: 589–594.

Kögel I, Bochter R (1985) Characterization of lignin in forest humus layers by high-performance liquid chromatography of cupric oxide oxidation products. Soil Biol Biochem 17: 637–640.

Kögel-Knabner I, Ziegler F (1993) Carbon distribution in different component of forest soils. Geoderma 56: 515–525.

Kögel-Knabner I, De Leeuw JW, Hatcher PJ (1992) Nature and distribution of alkyl carbon in forest soil profiles: implications for the origin and humification of aliphatic biopolymers. Sci Total Environ 117/118: 175–185.

Laskovski R, Maryanski M, Niklinska M (1994) Effect of heavy metals and mineral nutrients on forest litter respiration rate. Environ Pollut 84: 97–102.

Leary GJ, Newman RH, Morgan GR (1986) A carbon-13 nuclear magnetic resonance study of chemical processes involved in the isolation of Klason lignin. Holzforschung 40: 267–272.

LECO Corporation (1983) Instruction Manual SC-132. LECO Corporation, MI.

Leenheer JA (1981) Comprehensive approach to preparative isolation and fractionation of dissolved organic carbon from natural waters and wastewaters. Environ Sci Technol 15: 578–587.

Leenheer JA, Huffman EWD (1976) Classification of organic solutes in water by using macroreticular resins. J Res US Geol Survey 4: 737–751.

Malkomes H-P (1989) Einfluß der Lagerung von Bodenproben auf den Nachweis von Herbizid-Effekten auf mikrobielle Aktivitäten. Zbl Mikrobiol 144: 389–398.

Malkomes H-P (1991) Einfluß variierter Temperatur und Feuchte auf mikrobielle Aktivitäten im Boden unter Laborbedingungen. Z Pflanzenern Bodenkd 154: 325–330.

Marthaler HP, Vogelsanger W, Richard F, Wierenga PJ (1983) A pressure transducer for field tensiometers. Soil Sci Am J 47: 624–627.

Martin JP, Haider K (1986) Influence of mineral colloids on turnover rates of soil organic carbon. In: Interactions of Soil Minerals with Natural Organics and Microbes. Huang PM, Schnitzer M (eds). SSSA, Madison, WI, pp. 283–304.

McGrath SP, Cuncliffe CH (1985) A simplified method for the extraction of the metals Fe, Zn, Cu, Ni, Cd, Pb, Cr, Co, and Mn from soils and sewage sludges. J Sci Food Agric 36: 794–798.

McKeague JM (1971) Organic matter in particle-size and specific gravity fractions of some Ah horizons. Can J Soil Sci 51: 499–505.

Murphy J, Riley JP (1962) A modified single solution method for determination of phosphate in natural waters. Anal Chem 27: 970–976.

Nannipieri P, Ceccanti B, Grepo S (1990) Ecological significance of the biological activity in soil. In: Soil Biochemistry, vol 6. Bollag J-M, Stotzky G (eds). Marcel Dekker, New York, pp. 293–355.

Nelson DW, Somers LE (1975) A rapid and accurate method for estimating organic carbon in soil. Proc Indiana Acad Sci 84: 456–462.

Norman B (1993) Filtration of water samples for DOC studies. Mar Chem 41: 239–242.

Newman RH, Tate KR (1980) Soil phosphorus characterisation by ^{31}P nuclear magnetic resonance. Commun Soil Sci Plant Anal 11: 835–842.

North PF (1977) Towards an absolute measurement of soil structural stability using ultrasound. J Soil Sci 27: 451–459.

Novozamsky I, van Eck R, van der Lee GG, Houba VJG, Temminghoff E (1986) Determination of total sulphur and extractable sulphate in plant materials by inductively-coupled plasma atomic emission spectrometry. Commun Soil Sci Plant Anal 17: 1147–1157.

Öhlinger R, Eibelhuber A, Fischerlehner J (1986) Bodenprobennahme für Enzymaktivitätsbestimmungen. Veröff Landwirtsch-Chem Bundesanstalt Linz/Donau, 18: 255–283.

Oades JM (1988) The retention of organic matter in soils. Biogeochem 3: 35–70.

Olson RV (1965) Iron. In: Methods of Soil Analysis, Part 2, Chemical and Microbiological Methods. Black CA (ed). Agronomy 9(1), ASA, Madison, WI, pp. 966–967.

O'Toole P, Morgan MA, McGarry SJ (1985) A comparative study of urease activities in pasture and tillage soils. Commun Soil Sci Plant Anal 16: 759–773.

Pakulski JD, Benner R (1992) An improved method for the hydrolysis and MBTH analysis of dissolved and particulate carbohydrates in seawater. Mar Chem 40: 143–160.

Parsons JW (1981) Chemistry and distribution of amino sugars in soil and soil organisms. In: Soil Biochemistry, vol 5. Paul EA, Ladd JN (eds). Marcel Dekker, New York, pp. 197–227.

Petersen RG, Calvin LD (1986) Sampling. In: Methods of Soil Analysis, Part 1, Physical and Mineralogical Methods. Klute A (ed.) Agronomy 9(1), ASA, SSSA, Madison, WI, pp. 33–5.

Peyton GR (1993) The free-radical chemistry of persulfate-based total organic carbon analyzers. Mar Chem 41: 91–103.

Powlson DS, Jenkinson DS (1976a) The effects of biocidal treatments on metabolism in soil – I. Fumigation with chloroform. Soil Biol Biochem 8: 167–177.

Powlson DS, Jenkinson DS (1976b) The effects of biocidal treatments on metabolism in soil – II. Gamma irradiation, autoclaving, air-drying and fumigation. Soil Biol Biochem 8: 179–188.

Revsbech NP, Ward DM (1983) Oxygen microelectrode that is insensitive to medium chemical composition: use in an acid microbial mat dominated by Cyanidium caldarium. Appl Environ Microbiol 45: 755–759.

Richter M, Mizuno I, Aranguez S, Uriarte S (1975) Densimetric fractionation of soil organo-mineral complexes. J Soil Sci 26: 112–123.

Ross DJ (1970) Effects of storage on the dehydrogenase activity of soils. Soil Biol Biochem 2: 55–61.

Ross DJ (1972) Effect of freezing and thawing of some grassland topsoils on oxygen uptake and dehydrogenase activities. Soil Biol Biochem 4: 115–117.

Ross DJ (1990) Estimation of soil microbial C by a fumigation–extraction method: influence of seasons, soils and calibration with the fumigation–incubation procedure. Soil Biol Biochem 22: 295–300.

Ross DJ, Tate KR, Cairns A, Meyrick KF (1980) Influence of storage on soil microbial biomass estimated by three biochemical procedures. Soil Biol Biochem 12: 369–374.

Ryan MG, Melillo JM, Ricca A (1990) A comparison of methods for determining proximate carbon fractions of forest litter. Can J For Res 20: 166–171.

Salonius PO (1983) Effects of air drying on the

respiration of forest soil microbial populations. Soil Biol Biochem 15: 199–203.

Sarkanen KV, Ludwig CH (1971) Lignins. Wiley-Interscience, New York.

Scheidt M, Zech W (1990) A simplified procedure for the photometric determination of amino sugars in soil. Z Pflanzenern Bodenkd 153: 207–208.

Schnitzer M (1991) Soil organic matter – the next 75 years. Soil Sci 151: 41–58.

Schwertmann U (1964) Differenzierung der Eisenoxide des Bodens durch photochemische Extraktion mit saurer Ammoniumoxalatlösung. Z Pflanzenernähr Düng Bodenkd 105: 194–202.

Shen S-M, Brookes PC, Jenkinson DS (1987) Soil respiration and the measurement of soil microbial biomass C by the fumigation technique in fresh and in air-dried soil. Soil Biol Biochem 19: 153–158.

Shuman LM (1985) Fractionation method for soil microelements. Soil Sci 140: 11–22.

Sibbesen E (1978) An investigation of the anion-exchange resin method for soil phosphate extraction. Plant Soil 50: 305–321.

Skogland T, Lomeland S, Goksoyr J (1988) Respiratory burst after freezing and thawing of soil. Soil Biol Biochem 20: 851–856.

Skujins J (1967) Enzymes in soil. In: Soil Biochemistry. McLaren AD, Peterson GH (eds). Marcel Dekker, New York, pp. 371–414.

Small H (1978) An introduction to ion chromatography. In: Ion Chromatographic Analysis of Environmental Pollutants. Sawicki JD, Mulik E, Wittgenstein E (eds). Ann Arbor Science, Ann Arbor, MI, pp. 11–21.

Soulides DA, Allison FE (1961) Effect of drying and freezing soil on carbon dioxide production, available mineral nutrients, aggregation and bacterial population. Soil Sci 91: 291–298.

Sparling GP, Speir TW, Whale KN (1986) Changes in microbial biomass C, ATP content, soil phosphomonoesterase and phospho-diesterase activity following air-drying of soils. Soil Biol Biochem. 18: 363–370.

Sposito G, Lund LJ, Cheng AL (1982) Trace metal chemistry in arid-zone field soils amneded with sewage sludge. 1 Fractionation of Ni, Cu, Zn, Cd, and Pb, in solid phases. Soil Sci Am J 46: 260–264.

Steinbergs, A, Iismaa O, Freney JR, Barrow NJ (1962) Determination of total sulphur in soil and plant material. Anal Chim Acta 27: 158–164.

Stepniewski W, Zausig J, Niggemann S, Horn R (1991) A dynamic method to determine the O_2-partial pressure distribution within soil aggregates. Z Pflanzenern Bodenkd 154: 59–61.

Stevenson FJ (1956) Isolation and identification of some amino compounds in soils. Soil Sci Soc Am Proc 20: 201–204.

Stevenson FJ, Cheng CN (1970) Amino acids in sediments: recovery by acid hydrolysis and quantitative estimation by a colorimetric procedure. Geochim Cosmochim Acta 34: 77–88.

Sugahara K, Washio T, Akimori N, Inutsuka K, Matsuzaka Y (1992) Determination of neutral sugars in acid hydrolysates of rice straw and straw compost by high resolution gas chromatography. Soil Sci Plant Nutr 38(4): 689–697.

Sur HS, Kukal SS (1992) A modified hydrometer procedure for particle size analysis. Soil Sci 153: 1–4.

Sur HS, Singh NT (1976) A nomograph for hydrometer method of particle size analysis. Soil Sci Soc Am J 40: 457–458.

Suslick KS (1989) The chemical effects of ultrasound. Sci Am 260: 80–86.

Suttner T (1990) Zur mikrobiellen Aktivität bayerischer Böden in Abhängigkeit von der Nutzung und unter besonderer Berücksichtigung des bodenbiologischen Transformationsvermögens. Bayreuther Bodenkundl Ber 14: 125.

Suttner T, Alef K (1988) Correlations between the arginine ammonification, enzyme activities, microbial biomass, physical and chemical properties of different soils. Zbl Mikrobiol 143: 569–573.

Tabatabai MA (1982) Sulfur. In: Methods of Soil Analysis, Part 2, Chemical and Microbiological Methods. Page AL (ed). Agronomy 9(2), ASA, SSSA, Madison, WI, pp. 301–312.

Tabatabai MA, Bremner JM (1970) Comparison of some methods for determination of total sulfur in soils. Soil Sci Soc Am Proc 34: 417–420.

Tabatabai MA, Bremner JM (1983) Automated instruments for determination of total carbon, nitrogen, and sulphur in soils by combustion techniques. In: Soil Analysis. Smith KA (ed). Marcel Dekker, New York, pp. 171–194.

Tabatabai MA, Dick WA (1979) Ion chromatographic analysis of sulfate and nitrate in soils. In: Ion Chromatographic Analysis of Environmental Pollutants, vol 2. Mulik JD, Sawicki E (eds). Ann Arbor Science, Ann Arbor, MI, pp. 361–370.

Tate KR, Newman RH (1982) Phosphorus fractions of a climosequence of soils in New Zealand tussock grassland. Soil Biol Biochem 14: 191.

Tessier A, Chambell PG, Bisson M (1979) Sequential extraction procedure for the specification of particulate trace metals. Anal Chem 51: 844–851.

Trasar-Cepeda MC, Gil-Sotres F, Zech W, Alt HG (1989) Chemical and spectral analysis of organic P forms in acid high organic matter soils in Galicia (N.W. Spain). Sci Total Environ 81/82: 429–436.

Turchenek LW, Oades JM (1979) Fractionation of organo-mineral complexes by sedimentation and density techniques. Geoderma 21: 311–343.

Tyler G (1981) Heavy metals in soil biology and biochemistry. In: Soil Biochemistry, vol 5. Paul EA, Ladd JN (eds). Marcel Dekker, New York, pp. 371–414.

USDA/SCS (1984) Soil survey laboratory methods and procedures for collecting soil samples. Soil Survey Investigations Report No. 1. US Government Printing Office, Washington, DC.

Vance ED, Brookes PC, Jenkinson DS (1987) An extraction method for measuring soil microbial biomass C. Soil Biol Biochem 19: 703–707.

Welp G, Brümmer G (1985) Der Fe(III)-Reduktionstest - ein einfaches Verfahren zur Abschätzung der Wirkung von Umweltchemikalien auf die mikrobielle Aktivität in Böden. Z Pflanzenern Bodenkd 148: 10–23.

West AW, Ross DJ, Cowling JC (1986) Changes in microbial C, N, P and ATP contents, numbers and respiration on storage of soil. Soil Biol Biochem 18: 141–148.

West AW, Sparling GP, Grant WD (1987) Relationships between mycelial and bacterial populations in stored, air-dried and glucose-amended arable and grassland soils. Soil Biol Biochem 19: 599–605.

Wilke BM (1986) Einfluβ verschiedener potentieller anorganischer Schadstoffe auf die mikrobielle Aktivität von Waldhumusformen unterschiedlicher Pufferkapazität. Bayreuther Geowissenschaftliche Arbeiten 8: 1–151.

Wilson MA (1987) NMR Techniques and Applications in Geochemistry and Soil Chemistry. Pergamon Press, Oxford.

Wilson MA (1990) Application of nuclear magnetic resonance spectroscopy to organic matter in whole soils. In: Humic Substances in Soil and Crop Sciences: Selected Readings. MacCarthy CE, Clapp E, Malcolm RL, Bloom PR (eds). ASA, SSSA, Madison, WI, pp. 221–260.

Zech W, Alt HG, Zucker A, Kögel I (1985) ^{31}P-NMR-spectroscopic investigations of NaOH-extracts from soils with different land use in Yucatan (Mexico). Z Pflanzenern Bodenkd 148: 626–632.

Zeien H, Brümmer GW (1989) Chemische Extraktion zur Bestimmung von Schwermetallbindungsformen in Böden. Mitt Dtsch Bodenkundl Ges 59/I: 505–510.

Zeien H, Brümmer GW (1991) Ermittlung der Mobilität und Bindungsformen von Schwermetallen in Böden mittels sequentieller Extraktion. Mitt Dtsch Bodenkundl Ges 66/I: 439–442.

Zelles L, Adrian P, Bai QY, Stepper K, Adrian MV, Fischer A, Maier A, Ziegler A (1991) Microbial activity measured in soils stored under different temperature and humidity conditions. Soil Biol Biochem 23: 955–962.

Enrichment, isolation and counting of soil microorganisms

4

A great number of morphological and physiological types of microorganisms can be found in soils. The community structure is strongly dependent on the properties of the soil. Besides studying the physical, chemical and biochemical soil parameters, it is worthwhile to isolate and count soil microorganisms. This may give useful information on the effect of environmental factors on the microbial community at the sites studied. However, it should be pointed out that enrichment and isolation procedures generally give little information on the ecological importance of soil microorganisms. The purpose of this chapter is to provide a general introduction to laboratory techniques for sterilization and preparation of cultural media. Cultivation of aerobic and anaerobic soil microorganisms that are involved in the carbon, nitrogen and sulphur cycles, as well as modern and classical methods for counting bacteria, fungi, protozoa and cyanobacteria will also be presented.

Methods in Applied Soil Microbiology and Biochemistry
ISBN 0-12-513840-7

Nutrients, sterilization, aerobic and anaerobic culture techniques

K. Alef

Nutrients

Enrichment and isolation of soil microorganisms is based on the cultivation of these microorganisms in liquid or agar media. Microorganisms are diverse in their specific physiology and correspondingly in their specific nutrient requirements. A culture medium must contain all nutrients required for microbial growth. Therefore, a great number of different media have been proposed for cultivation of soil microorganisms.

Every medium must contain a carbon and nitrogen source, and minerals. Minerals such as potassium, magnesium, calcium, iron, manganese, cobalt, copper, molybdenum and zinc are required by nearly all microorganisms. The quantities required of manganese, cobalt, molybdenum and zinc are always very small. They are present as contaminants of the major inorganic constituents of media and are often referred as **trace elements**.

A trace elements solution may contain the following (in mg):

Fe III-citrate	1000.000
$MnCl_2.4H_2O$	10.000
$ZnCl_2$	5.000
5.000 LiCl	5.000
KBr	2.500
KI	2.500
$CuSO_4$	0.005
$CaCl_2$	1000.000
$Na_2 MoO_4. 2H_2O$	1.000
$CoCl_2$	5.000
$SnCl_2.H_2O$	0.500
$BaCl_2$	0.500
$AlCl_3$	1.000
H_3BO_3	10.000
EDTA	20.000
(Distilled water	1000 ml)

2–10 ml of the trace elements solution should be added to 1 litre of medium.

Many microorganisms require for their growth vitamins such as biotine, thiamine, cyanocobalamine and pyridoxamine.

Heterotrophic microorganisms obtain carbon principally from organic nutrients such as glucose, malate, citrate and cellulose. Organic substrates usually have a dual nutritional role; they serve at the same time as a source of carbon and energy. Photosynthetic microorganisms and bacteria obtaining energy from the oxidation of inorganic compounds typically use the most oxidized form of carbon, CO_2, as the sole or principal source of cellular carbon.

The requirement for a reduced nitrogen source is relatively common and can be met by the provision of nitrogen as ammonium or nitrate salts. Very few bacteria groups are capable of fixing molecular nitrogen. Amino acids can also be used by microorganisms as nitrogen sources. Media can be prepared by dissolving all nutrients in their inorganic form. This solution can then be supplemented with carbon and nitrogen sources, and growth factors. A medium composed of chemically defined nutrients is termed a **synthetic medium**, one that contains ingredients of unknown composition such as yeast extract, meat extract or blood, is termed a **complex medium**.

- **Meat extract** is prepared from lean meat, containing growth factors and can be used as both carbon and nitrogen sources.
- **Yeast extract** is prepared from yeast cells. It

contains vitamins, and can be used as both carbon and nitrogen sources.

- **Malt extract** is prepared from malt and consists of about 50% maltose, and glucose, dextrin and starch; it also contains vitamins.
- **Casein hydrolysat** is prepared from casein and mainly contains amino acids.
- **Soil extract** contains growth factors, trace elements, organic substances and can be added to the media. It can be prepared as follows: suspend 500 g of grassland soil in 1000 ml distilled water, and boil the suspension for several minutes. After cooling, filter the suspension. For use, dilute the suspension 1:10 or 1:50.

The hydrogen ion concentration in the medium is also an important factor that affects the growth of microbial populations. The optimal pH for bacterial growth ranges from 6.5 to 7.5, while that for fungi is between 4 and 6. To prevent excessive changes in hydrogen ion concentration during the development of microbial cultures, buffers (phosphate, Tris-HCl) are usually added to the medium.

To prepare solid media, agar is usually added at concentrations between 1.2 and 1.5%. Agar is a polysaccharide, consisting of agarose (70%) and agaro pectine (about 30%) and can be extracted from red algae.

Sterilization of solutions and media

Sterilization is a treatment that frees the treated object of all living organisms. It can be achieved by exposure to lethal physical or chemical agents or, in the special case of solutions, by filtration.

Sterilization by moist heat

This technique is suitable for sterilizing liquid media, solutions, cotton, paper and rubber. The treatment is usually carried out in an autoclave, which can be filled with steam at a pressure greater than the atmospheric value. Sterilization can thus be achieved at temperatures considerably above the boiling point of water. A sterilization for 20–30 min at 120°C is sufficient for solutions. A period of 10 min can be used for sterilizing glassware. It should be noted that the volume of the liquid medium should be less than 50% of that of the vessel used, otherwise the solution would boil over during autoclaving. Flasks are usually closed with cotton plugs and bottles with screw caps. Do not cap the bottle very tightly, otherwise it may break.

Fractional sterilization can be used and its success depends on the germination of spores that survive the initial heating. It can be carried out as follows: after autoclaving the medium or a microbial culture, allow it to stand at room temperature overnight. Repeat the autoclaving procedure on 3 successive days.

Sterilization by dry heat

Dry heat is used principally to sterilize glassware or other heat-stable solid materials. The objects are usually wrapped in aluminium foil or protected from subsequent contamination, and exposed to a temperature of 180°C for 2 h in an oven.

Sterilization by filtration

Heat-labile solutions can be sterilized by filtration through filters (< 0.3 µm) capable of retaining microorganisms. Microorganisms are retained in part by the small size of the filter pores and in part by adsorption on to the pore walls during their passage through the filter. Large viruses are also retained by filters with a pore size of 0.1–0.15 µm.

Microorganisms can be removed from gases by filtration through cotton plugs.

Sterilization by chemical treatment

Sterilization by chemical treatment is required when samples contain heat-labile substances. Ethylene oxide is usually used for this purpose. This compound boils at 10.7°C and can be added to solutions in liquid form (final concentrations of 0.5–1.0%) at a temperature ranging from 0°C to 40°C. Ethylene oxide is both explosive and toxic for humans, so special precautions must be taken for handling this compound.

Sterilization by irradiation

Disinfection of rooms, clean bench or instruments can be performed by treatment with UV rays (210–290 nm). Such treatment does not kill all living microorganisms completely. Gamma rays are more effective and can be used to sterilize different types of equipment.

Aerobic culture techniques

Enrichment techniques provide a means for isolating different microbial types from nature, by taking advantage of their specific requirements. The inoculation of a liquid or agar medium with e.g. a soil suspension and incubation at a defined temperature promotes growth of different groups of microorganisms depending on the growth conditions; such cultures are termed **mixed cultures** or mixed microbial populations.

To obtain aerobic growth conditions it is sufficient to fill 1/5th of the culture vessel (e.g. Erlenmeyer flasks) with the medium and to incubate with continuous shaking. For cultures with volumes greater than 2 l, bubbling of the medium with air with continuous stirring is recommended.

The diffusion of oxygen through the surface of the agar medium (e.g. agar dishes) also supports aerobic growth conditions.

Preparation of agar plates

Cool the autoclaved media (nutrient solution and agar) at 50–60°C. After cooling, flame the vessel and fill the Petri dishes with 25–30 ml medium. To remove the air bubbles, flame the surface of the agar medium in the dish very gently. After the medium has hardened, allow the dishes to stand (upside down) for 1–3 days at room temperature (to evaporate the condensed water), then store at 4°C.

Isolation of pure culture

The preparation of a pure culture involves the isolation of a given microorganism from a mixed natural microbial population and the maintenance of the isolated microorganism in an artificial environment. Pure cultures may be simply obtained by the **streaked plating** method. This method involves the separation and immobilization of individual organisms on or in a nutrient medium solidified with agar. A sterilized bent wire is dipped into a suitably diluted suspension of organisms (mixed culture) and then used to make a series of parallel, non-overlapping streaks on the surface of an already solidified agar plate.

The inoculum is progressively diluted with each successive streak and well-isolated colonies develop along the lines of later streaks (Fig. 4.1). Using a sterilized bent wire, a visible colony is then removed and suspended in a physiological solution (e.g. 0.9% NaCl). An inoculum from the resulting suspension is

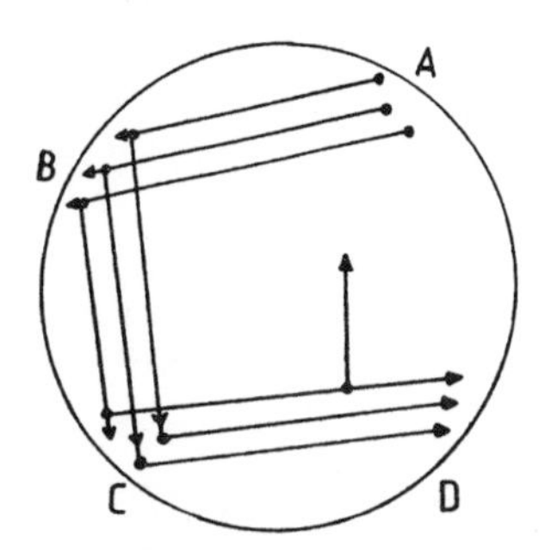

Figure 4.1. Isolation of a pure culture by the streak method (Alef 1991).

used to make a series of streaks on the surface of a new agar plate. This procedure should be repeated several times. A colony from the last plate should then be removed, and used to inoculate a sterilized agar or liquid medium. The pure culture grown must contain only one type of identical cells.

Anaerobic culture technique

(Hungate 1965)

To obtain anaerobic growth conditions for facultative anaerobes, it is sufficient to prevent the oxygen diffusion in the growth medium. This can be achieved by using growth vessels (bottles and cultivation tubes) with screw caps as follows. Fill the vessel with sterile medium (> 95%). After the inoculation, close the vessels with the screw caps (not very tightly to allow produced gases to escape) and incubate at the desired temperature. The dissolved O_2 in the medium is used by the growing microorganisms and a shift to anaerobic conditions occurs. In contrast to facultative anaerobes, strict anaerobes can not consume the dissolved O_2 in the medium. Oxygen even has a toxic effect on obligate anaerobes. For the cultivation of these bacteria, it is necessary to remove the O_2 from the medium and to reach a negative redox potential < -100 mV. Such negative redox potentials (up to -40 mV) can be obtained by the addition of redox solutions.

Removal of oxygen from gases

Slowly pass a high-purity gas (N_2, He or Ar) over a hot column (reduced Cu). If necessary, clean the gas further by passing it over a cold catalyst or Oxisorb. Sterilize the gas by passing it through a filter composed of a heat-resistant tube (20 ml glass syringe) filled with cotton and provided with a Teflon syringe needle connector that provides an easy mount for the sterile needle. Before use dry sterilize the filter.

Preparation of anaerobic media

Place the heat-stable components of the medium or solution in a flask (250–500 ml), and gently boil while sparging with O_2-free gas (e.g. N_2, He, Ar or N_2 + CO_2) through the gassing probe. Continue gassing while dispensing the medium into thick-walled containers; then apply the stopper tightly and autoclave. After slow cooling, using a syringe, add the heat-labile ingredients, reducing agents and redox indicators that have been previously filter-sterilized.

Reducing agents and redox indicators

1. Sodium thioglycolate
(Bogdahn 1985; Kasper and Tiedje 1982)

Sodium thioglycolate (5 g)
Distilled water (50 ml)

Autoclave and store under O_2-free gas. At a thioglycolate concentration in the medium of 0.05%, a redox potential of < -100 mV is achieved.

2. Titanium (III) citrate
(Zehnder and Wuhrmann 1977)

Add 5 ml of titanium III chloride (15%) to 50 ml of sodium citrate solution (0.2 M) and neutralize by adding saturated Na_2CO_3 solution, then filter-sterilize and store under O_2-free gas. At titanium citrate concentrations in the medium between 0.2 and 0.5 mM, a redox potential of < -480 mV is achieved.

3. Cysteine
(Kasper and Tiedje 1982)

A cysteine solution (1 g/100 ml distilled water) is autoclaved and stored under O_2-free gas. At a cysteine concentration of 0.025% in the medium, a redox potential of −340 mV is achieved.

4. Sodium sulphide
(Kasper and Tiedje 1982)

Dissolve 1.2 g $Na_2S.9H_2O$ in 80 ml distilled water and bring up with distilled water to 100 ml, autoclave, and store under O_2-free gas. At a sodium sulphide concentration of 0.025% in the medium, a redox potential of −225 mV is achieved.

All the redox agents mentioned can also be filter-sterilized and stored under O_2-free gas.

The following solutions can be used as redox indicators (Jacob 1970).

1. Resazurin

Dissolve 0.1 g resazurin in 100 ml distilled water and filter-sterilize. Add the solution to the growth medium to give a final concentration of 0.0001% and 0.0003%.

2. Methylene blue
(Skinner 1971)

Solution I Dilute 3 ml of a methylene blue solution (0.5% in water) with distilled water to 100 ml and store in the dark.

Solution II 0.5 g glucose and one small crystal of thymol are dissolved in 100 ml distilled water.

Solution III 1.06 g Na_2CO_3 and $NaHCO_3$ are dissolved in 100 ml distilled water.

Add solution III to solution II until a value of pH 10 is reached. This mixture should be stored under cold dark conditions. Mix the product with solution I to give a light blue colour. Boil portions of this methylene blue indicator in small beakers or test tubes until it becomes colourless, and then place it in containers to be monitored for anaerobiosis. Close the containers immediately and make the atmosphere anaerobic.

3. Phenosafranin

Dissolve 0.1 g phenosafranin in 100 ml distilled water and filter-sterilize. Add the solution to the growth medium to give a final concentration of 0.0001% and 0.0002%.

Preparation of anaerobic solid media

Place 0.15 g agar in a flat bottle (50 ml), which is continuously and very gently gassed with a N_2, He, Ar or CO_2/N_2 stream. Pipette 10 ml of the anaerobic medium (without agar, but containing both redox agent and indicator). Apply a rubber stopper very tightly to the bottle and autoclave for 20 min at 120°C. After autoclaving mix the contents of the bottle and lay it flat until the medium becomes solid.

Isolation of pure cultures of anaerobic bacteria

The isolation of strictly anaerobic bacteria by plating methods poses special problems. In the case that the microorganisms are not rapidly killed by exposure to oxygen, plates may be prepared in the usual manner and then incubated under N_2, He or Ar in closed containers. For more O_2-sensitive anaerobes, a *dilution shake culture* is preferred. A tube of melted and cooled agar is inoculated and mixed, and about one tenth of its contents is transferred to a second tube, which is then mixed and used to inoculate a third tube in a similar fashion. After 6–10 successive dilutions, the tubes are cooled rapidly and sealed, by pouring a layer of sterile petroleum jelly and paraffin on the surface, thus preventing access of air to the agar column.Isolation of pure cultures of anaerobic bacteria

In shake cultures, the colonies develop deep in the agar column. To transfer colonies, the petroleum jelly–paraffin seal is removed with a sterile needle, and the agar column is gently blown out of the tube into a sterile Petri dish by passing O_2-free gas through a capillary pipette inserted between the tube wall and the agar. The column is sectioned into discs with a sterile knife to permit the examination and transfer of colonies.

Enrichment of aerobic soil bacteria

The enrichment and isolation of different microorganisms from soils may be needed to study their role in mineralization of nutrients and microbiological decontamination of soil, because specific microorganisms are needed for the degradation of pollutants in soil (see Chapter 11).

Enrichment of physiological groups

K. Alef

Enrichment of cellulose-decomposing aerobic bacteria

(Drews 1968)

Cellulose is mainly degraded by fungi in soils. In contrast, most bacteria can only be cultivated on partially hydrolysed cellulose.

Growth medium

0.30 g NH_4Cl
0.50 g $NaNO_3$
1.00 g K_2HPO_4
0.30 g $MgSO_4.7H_2O$
0.10 g $CaCl_2$
0.05 g $FeCl_2$

Dissolve in 800 ml distilled water; adjust the pH to 7.0 and bring up with distilled water to 1000 ml.

Procedure

Place strips of filter paper moistened with 10 ml of the growth medium described above in a Petri dish, distribute a small amount of soil sample on the filter and incubate at room temperature. Visible colonies can be detected within several weeks. Bacterial colonies can then be removed and cultivated on a new agar medium with cellobiose as a carbon source.

Enrichment of Pseudomonaceae

Pseudomonaceae are gram-negative bacteria that are found in soil, water and air. They can use a wide variety of organic matter and also heterocyclic aromatic compounds as their carbon source. Under anaerobic conditions, some species are capable of dissimilatory nitrate reduction to nitrite and/or N_2. Furthermore, several groups synthesize water-soluble pigments.

A very simple method to detect *Pseudomonas* species in soils is based on the cultivation of the bacteria on agar medium and identification of cell pigments.

Growth media

Pseudomonas-agar F

10.0 g Pepton from casein
10.0 g Pepton from meat
1.5 g $MgSO_4.7H_2O$
1.5 g K_2HPO_4
12.0 g Agar
10.0 ml Glycerin

Dissolve in 800 ml distilled water; adjust the pH to 7.2 and bring up with distilled water to 1000 ml.

Pseudomonas-agar P

20.0 g Pepton from gelatine
1.4 g $MgSO_4.7H_2O$
10.0 g K_2SO_4
12.6 g Agar
10.0 ml Glycerin

Dissolve in 800 ml distilled water; adjust

the pH to 7.2 and bring up with distilled water to 1000 ml.

Kings-agar B

20.0 g Pepton
1.5 g $MgSO_4.7H_2O$
1.8 g $K_3PO_4.3H_2O$
10.0 g Agar
10.0 ml Glycerin

Dissolve in 800 ml distilled water; adjust to pH 7.2 and bring up with distilled water to 1000 ml.

Procedure

Place 99 ml of distilled water in an Erlenmeyer flask and autoclave at 120°C for 20 min. After cooling add 1.0 g soil and incubate under continuous shaking at 30°C for 30 min. Then allow to stand for about 1 h. From the supernatant, prepare a dilution series using 0.9% NaCl solution (10^{-4}–10^{-7}).

From each dilution step take 0.1 ml and disperse (spread) on the surface of the agar medium and incubate at 30–37°C for several days. The colonies of *Pseudomonas fluorescens* (Kings-agar B) can be identified by a yellow to yellow–green fluorescence. *Pseudomonas aeruginosa* colonies (*Pseudomonas*-agar P) can be identified by detecting the fluorescence at 366 nm. It should be pointed out that not only Pseudomonaceae but also many other microorganisms can grow on the media described.

Enrichment of oligotrophic bacteria

Oligotrophic bacteria include taxonomically and physiologically different microbial groups. These bacteria are capable of growing at a slow rate in media and environmental sites with very low nutrient concentrations; under these conditions, the growth of other microbial groups is not possible (Ohata and Hattori 1983; Suwa and Hattori 1984; Williams 1985; Whang and Hattori 1988; Sugimoto et al 1990).

Growth media
(Whang and Hattori 1988; and see Hattori 1976; Hattori and Hattori 1980)

10 g Pepton
10 g Meat extract
5 g NaCl

Dissolve in 800 ml water; adjust to pH 7.0–7.2 with NaOH (0.1 M) and bring up with water to 1000 ml. This stock solution must be diluted 100-, 1000- and 10,000-fold before being used. For preparation of solid media, add agar (1.5%, pure) to the medium.

The following medium can also be used for the isolation of oligotrophic bacteria.

KB-medium
(Sugimoto et al 1990)

990.0 ml Double distilled water
1.5 g K_2HPO_4
1.5 g $MgSO_4.7H_2O$
20.0 g Pepton
10.0 ml Glycerin

Adjust the pH to 7.2 with NaOH (0.1 M). For the enrichment and isolation of oligotrophic bacteria, this medium must be diluted with distilled water at least 250-fold.

The addition of the following growth inhibitors to the diluted KB-medium makes the enrichment of oligotrophic Pseudomonaceae possible:

(A) 45 mg/l medium Novobiocin
75,000 units/l medium Pencillin G
75 mg/l medium Cycloheximide

Dissolve these antibiotics in 3 ml ethanol (95%), dilute with 50 ml sterilized distilled water and add to 940 ml of autoclaved, cooled (45°C) KB-250 medium.

(B) Irgasan 25 mg/l medium

Dissolve 25 mg Irgasan in 2 ml ethanol (95%), dilute with 50 ml sterilized distilled water and add to 950 ml of autoclaved, cooled (45°C) KB-250 medium.

(C) Crystal violet-nitrofurantoin (Kruger and Sheikh 1987)

Crystal violet 1–2 mg/l medium
Nitrofurantoin 350 mg/l medium

Prepare a 0.1% crystal violet solution in distilled water and 5% nitrofurantoin in *N,N*-dimethylformamide, and store at room temperature in the dark. To 990 ml of autoclaved and cooled (45°C) KB-medium, add 2 ml of filter-sterilized crystal violet solution and mix well. Finally, add 7 ml of filter-sterilized nitrofurantoin solution to the medium. Inoculate the diluted (100–10,000) medium (containing the inhibitors) with soil (0.1 g/50 ml) and incubate at 22°C in the dark for at least 14 days.

Enrichment of nitrifying bacteria

The liberated ammonium during the ammonification of organic substances in soils can be oxidized under aerobic conditions to nitrite and then to nitrate by the nitrifying bacteria (see Chapter 5).

For the enrichment of nitrifying bacteria the following procedure is recommended.

Growth medium for ammonium-oxidizing bacteria

1.00 g $(NH_4)_2SO_4$
0.50 g K_2HPO_4
2.00 g NaCl
0.20 g $MgSO_4.7H_2O$
0.05 g $FeSO_4.7H_2O$
10.00 ml Trace elements solution (see earlier section on nutrients)
6.00 g $CaCO_3$

Dissolve in 800 ml distilled water; adjust the pH to 7.6 and bring up with distilled water to 1000 ml.

Growth medium for nitrite-oxidizing bacteria

0.1 g $NaNO_2$
6.0 g $CaCO_3$
0.500 g K_2HPO_4
0.200 g $MgSO_4.7H_2O$
0.005 g $FeSO_4.7H_2O$
0.500 g NaCl

Dissolve in 800 ml distilled water; adjust the pH to 7.6 and bring up with distilled water to pH 7.6.

Procedure

Place 50 ml of the medium in Erlenmeyer flasks (300 ml) and autoclave at 120°C for 20 min. After cooling, place 0.1–0.3 g of soil in the Erlenmeyer flask and incubate under continuous shaking at 30°C in the dark for up to 3 weeks. The nitrite and/or nitrate formed can be estimated qualitatively using Merck paper.

Enrichment of denitrifying bacteria

Many microbial groups are able to use nitrate as their only nitrogen source (assimilatory nitrate reduction). Denitrifying bacteria are aerobes that are capable of using nitrate as a terminal electron acceptor under anaerobic conditions. N_2 is the end product of this reaction (denitrification).

For the enrichment of denitrifying bacteria the following media can be used.

Growth media

Mineral medium

10.0 g Sodium acetate
20.0 g KNO_3
0.5 g K_2HPO_4
0.2 g $MgSO_4.7H_2O$
10.0 ml Trace element solution (see earlier section on nutrients)

Dissolve in 800 ml distilled water; adjust to pH 7.0 and bring up with distilled water to 1000 ml.

Complex medium

1 g Meat extract
5 g Pepton
2 g Yeast extract
15 g NaCL
10 g KNO_3

Dissolve in 800 ml distilled water; adjust to pH 7.0 and bring up with distilled water to 1000 ml.

Bouillon medium

40 g Bouillon
10 g KNO_3

Dissolve in 800 ml distilled water; adjust the pH to 7.4 and bring up with distilled water to 1000 ml.

Procedure

Pipette 15 ml of the medium into tubes with screw caps, place a Durham tube (upside down) in each tube and autoclave at 120°C for 20 min. After cooling, add 0.1–0.3 g soil, seal the tubes tightly with screw caps and incubate at 30°C in the dark up to 3 weeks. Gas formation can be detected by the Durham tubes. Nitrite and nitrate can be estimated qualitatively using Merck paper.

Isolation and identification of aerobic nitrogen-fixing bacteria from soil and plants

J. Döbereiner

Until the early 1970s the classical concept of aerobic diazotrophs, that is bacteria able to use molecular N_2 as their sole nitrogen source for growth, referred only to bacteria able to grow under atmospheric oxygen concentrations. This property, due to the extreme sensitivity of the nitrogenase enzyme, requires mechanisms that protect the enzyme from O_2, even though these bacteria require O_2 for respiration and generation of ATP. This mechanism is known as respiratory protection.

The diazotrophs of this type are from the genera *Azotobacter* and *Azomonas*. The very tenacious slime capsules of diazotrophs of the genera *Derxia* and *Beijerinckia* seems to have a mechanical oxygen protection. Members of these four genera were the only known aerobic diazotrophs until the 1970s. Better understanding of the mechanisms of nitrogenase function led then to use a new method, which permitted the isolation of several new aerobic diazotrophs. These bacteria do not possess oxygen-protection mechanisms and therefore are more efficient in terms of their carbon use for nitrogen fixation. They have been found to be the most common plant associated diazotrophs. Bacteria of this type belong to the genera *Azospirillum*, *Herbaspirillum*, *Azoarcus* and *Acetobacter*.

Isolation and identification methods

(Döbereiner 1966; Krieg and Döbereiner 1984)

Principle of the method

Due to their respiratory protection mechanisms *Azotobacter* spp. and *Azomonas* can grow and form colonies on agar plates incubated in air, even if the medium contains no nitrogen source. *Derxia* spp. and *Beijerinckia* spp. form very tenacious colonies on such plates, which mechanically protect the nitrogenase against O_2. This property makes the isolation much easier because no other bacteria can grow on such media.

Materials and apparatus

Incubator adjustable to 25–35°C
Petri dishes
Silica gel plates
Autoclave
Clean bench
Sterile glassware
Microscope

Chemicals and solutions

Medium for the isolation of *Azotobacter* and *Azomonas* spp. (LG medium)

20.00 g Sucrose
0.05 g K_2HPO_4
0.15 g KH_2PO_4
0.01 g $CaCl_2$
0.20 g $MgSO_4.7H_2O$
2 mg $Na_2MoO_4.2H_2O$
0.01 g $FeCl_2$
2.00 ml bromothymol blue (0.5% sol. in ethanol)
1.00 g $CaCO_3$
15.00 g Agar

Dissolve in 800 ml distilled water and bring up with distilled water to 1000 ml.

Medium for the isolation of *Derxia*

Same medium as for *Azotobacter*, but the sucrose is replaced by glucose or starch, and the $CaCO_3$ by 0.01 g of $NaHCO_3$ (Campelo and Döbereiner 1970).

Medium for the isolation of *Beijernckia* and solutions for preparing silica gel plates

Stock solution A: HCl diluted to a density of 1.10 (measured with a densitometer).

Stock solution B: $Na_2O_5.SiO_2$ (water glass) solution with a density of 1.060.

Winogradsky's salt solution C:

5.00 g KH_2PO_4
2.50 g $MgSO_4.7H_2O$
2.50 g NaCl
5 mg $NaMoO_4.2H_2O$
0.05 g $MnSO_4.4H_2O$
0.05 g $Fe_2(SO_4)_3$

Dissolve in 800 ml distilled water; adjust the pH to 6.5 and bring up with distilled water to 1000 ml.

A total of 500 ml of solution B is poured into 503 ml of solution A mixing constantly and then distributed immediately into Petri dishes (30 ml). These are allowed to solidify for 24–48 h and then placed into running water for 3 days until the $AgNO_3$ test for chloride is negative (no white precipitate is formed when a drop of a 1% solution of $AgNO_3$ is placed on the plates). Such plates can be stored in a closed vessel for many months. On the day they are to be used, the plates are sterilized by immersion in boiling distilled water, which also removes any residual chloride. Glucose is then added to Winogradsky's solution to give a concentration of 10%; then 2 ml portions are heated to boiling in test tubes and poured over the plates, which are allowed to dry in a clean drying oven.

Procedure

Plates containing the growth medium are inoculated with about 100 mg of finely sieved soil or with appropriate dilutions of soil. Colonies of *Azotobacter chroococcum* appear after 24 h of incubation at 30°C as white moist colonies turning dark brown after 3–5 days. *A. vinelandii* and *Azomonas* colonies are similar but do not turn dark. *A. paspali* colonies on these plates appear only after 48 h, and become yellowish in the centre due to assimilation of bromothymol blue and acidification of the medium.

Isolation of these organisms is usually possible after only one additional streaking out on the same medium. Cells of *A. chroococcum* and *A. vinelandii* are large (3 × 6 µm) occurring in pairs. *A. paspali* cells are very motile with peritrichous flagellation and become 10–15 µm long and 2 µm wide (Döbereiner 1966). Colonies of *Derxia* on these plates initially are white and small, but later become large, curled and brown. The bacteria within the colonies start fixing N_2 and growing faster, because the capsular polysaccharides of the bacteria form colonies large enough for nitrogenase to be protected against O_2.

Isolation of *D. gummosa* is easier after incubation at 35°C for 7 days, when the large tough colonies are removed and left for 2 h in test tubes with sand and sterile water to soften the consistency, and then homogenized. The suspension is then streaked out on LG medium (see above) where this organism forms two types of colonies: small colonies, which are slightly beige and do not fix N_2, and large brown colonies, which are able to start fixing N_2. When the plates are incubated under reduced pO_2 (0.05 atm) all colonies become uniformly large and brown (Campelo and Döbereiner 1970). Colonies on plates inoculated with soil or root pieces seem to have enough nitrogen to start growth to a certain size when they are able to start fixing N_2. *Derxia* cells are 1–1.2 µm by 3–6 µm and contain small refractive lipid bodies. Counting is difficult and the best estimates of their occurrence seem to be the percentage of soil grains or root pieces containing *Derxia*.

Beijerinckia are best isolated on silica gel plates. Air-dried soil is then sieved (1 mm) and 20–100 mg spread uniformly over the plates. After 4–10 days incubation at 30°C, *Beijerinckia indica* forms small raised white colonies that rapidly become larger and very tenacious. *B. fluminense* colonies are small, lightly beige and dry. They stop growing and do not become sticky. *Beijerinckia* cells are very characteristic under the microscope. In wet mounts, they appear as medium-sized rods (1 × 3 µm) with two very refractive fat globules, one at each extremity. The globules stain with Sudan black. *B. fluminense* cells are packed in zoogloea-like clusters surrounded by a membrane visible under the light microscope. For their isolation, entire colonies are dispersed in test tubes with sand and sterile water (colonies become soft after 2 h in water) and streaked out on LG medium (see *Azotobacter* isolation above) from which $CaCO_3$ was omitted. Estimates of numbers can only be made from microcolonies occurring in soil, which form 10^3–10^4 colonies on silica gel plates.

Azotobacter, Azomonas, Derxia and *Beijerinckia* are all typical soil bacteria for which few data indicate plant–bacteria associations. The only exception is the specific association of *A. paspali* with *Paspalum notatum* cv. batatais (Döbereiner 1966). This species only occurs in the rhizosphere of this grass.

Recent unpublished data from our

Table 4.1. Physiological differential characteristics of aerobically N_2-fixing bacteria (form colonies on N-free agar plates incubated under air).

	Azotobacter	A. paspali*	Azomonas	Beigerinckia	Derxia
Cells larger than 1.5 × 3 µm	+	+	+	–	–
Cysts formed	+	+	–	–	–
Surface pellicle on liquid medium	+	+	+	–	–
NO_3^- - NO_2^-	+	–	+	±	–
Use of C-source for N_2-dependent growth					
Starch	–	–	–	–	+
Glucose, fructose	+	+	+	+	+
Sucrose	+	+	+	+	–
Acid from glucose, sucrose, fructose	–	+	–	+	–
Optimum growth temperature (°C)	25–32	30–35	25–32	25–30	25–35
Optimum pH	6.5–8.0	5.5–7.8	6.5–7.5	4.5–6.0	5.5–8.0

* This species is separated due to various important differences.

laboratory showed this species occurs not only with roots but also with stems of *Paspalum notatum*. For additional differential characteristics of these genera see Table 4.1. Identification of species within these genera can be made according to Krieg and Döbereiner (1984).

Methods for isolation and identification of *Azospirillum* spp.

(Reinhold et al 1987; Khammas et al 1989; Döbereiner 1992)

Principle of the method

The discovery of the five species of this genus was due to the introduction of nitrogen-free semi-solid isolation media where these organisms, attracted by their aerotactic characteristic, move to the region within this medium where their respiration rate is in equilibrium with the oxygen diffusion rate. They form very characteristic veil-like pellicles 5 mm below the surface; then they move up close to the surface when, due to their N_2-dependent growth, more and more cells accumulate.

Materials and apparatus

Petri dishes
Incubator adjustable to 35–41°C
Autoclave
Microscope
Clean bench
Sterile glassware

Chemicals and solutions

(NFb) basic medium for the isolation of *Azospirillum* spp.

5.00 g D,L-Malic acid
0.50 g K_2HPO_4
0.20 g $MgSO_4.7H_2O$
0.10 g NaCl
0.02 g $CaCl_2.2H_2O$
2.00 ml Minor element solution
2.00 ml Bromothymol blue (0.5% solution in 0.2 M KOH)
1.64% Fe EDTA solution
1.00 ml Vitamin sol
1.75 g Agar agar

Dissolve in 800 ml distilled water; adjust the pH to 6.8 with KOH and bring up with distilled water to 1000 ml.

It is important that the various ingredients are added in the given sequence to avoid precipitation of Fe or other salts due to high pH.

Minor element solution

0.40 g $CuSO_4.5H_2O$
0.12 g $ZnSO_4.7H_2O$
1.40 g H_2BO_4
1.00 g $Na_2MoO_4.2H_2O$
1.50 g $MnSO_4.H_2O$

Dissolve in 800 ml distilled water and bring up with distilled water to 1000 ml.

Vitamin solution

10 mg Biotin
20 mg Pyridoxol–HCl
100 ml Distilled water

Potato medium

A total of 200 g of peeled fresh potatoes are cooked for 30 min in 1000 ml distilled water and then filtered through sheets of cotton. 2.5 g D,L-malic acid, 2.5 g sucrose and 15 g agar agar are then added, and the pH adjusted to 6.8.

Semi-soil medium for the isolation of *A. amazonese* (LGI)

0.200 g K_2HPO_4
0.600 g KH_2PO_4
0.002 g $CaCl_2.2H_2O$
0.200 g $MgSO_4.7H_2O$
0.002 g $Na_2MoO_4.2H_2O$
0.010 g $FeCl_3$
5.000 ml Bromothymol blue (0.5% solution in 0.2 M KOH

Table 4.2. Physiological differential characteristics of *Azospirillum* spp.

	A. brasilense	*A. lipoferum*	*A. amazonense*	*A. irakense*	*A. balopraeferans*
Pleomorphic cells	–	+	±	+	+
NO_3^-–NO_2^-	+	+	±	±	+
NO_2^-–N_2O	+	+	–	–	+
Use of C-source for N_2-dependent growth					
Glucose	–	+	+	+	–
Sucrose, maltose	–	–	+	+	–
Pectine hydrolysis	–	–	–	+	–
Optimum growth Temperature (°C)	22–37	32–37	32–37	30–33	41
Optimum pH range	6.0–7.8	5.7–6.8	5.7–6.5	6.4–6.7	6.8–8.0

5.000 g Sucrose
1.800 g Agar agar

Dissolve in 800 ml distilled water; adjust to pH 6.0 and bring up with distilled water to 1000 ml.

Procedure

For the isolation of *Azospirillum brasilense* and *A. lipoferum*, the NFb medium is inoculated with 0.1 ml of soil or root suspensions, and incubated at 35°C. Paper-like white pellicles are usually formed at the surface after 3–5 days; the cultures are streaked out on agar plates (15 g agar l^{-1}) with the same medium but containing 0.020 g of yeast extract. Colonies of these two species after 1 week are small, dry white and curled. Individual colonies must then be rechecked by transferring them into new nitrogen-free semi-solid medium and purified by streaking them out on potato medium.

Small, dry curled colonies that turn pinkish after 1 week are then selected and transferred into new semi-solid NFb vials from where the bacteria can be identified in wet mounts under the microscope.

A. brasilense cells are medium-sized (1 × 3–5 µm), very motile, curved rods with spirilloid movement. *A. lipoferum* cells initially are indistinguishable from the former but, once the medium has turned alkaline (blue colour), change into large pleomorphic forms. The physiological characteristics of these organisms are summarized in Table 4.2.

A. irakense according to Khammas et al (1989) can be isolated in the same way except that incubation is carried out at 30°C instead of 35°C.

A. halopraeferans (Reinhold et al 1987) is isolated in the same basic medium but with the following modification: pH is adjusted to 8.5, 1.2% NaCl is added and plates are incubated at 41°C.

A. amazonense is best isolated in semi-solid sucrose medium (LGI). Subsurface pellicles after 3–5 days incubation at 35°C are streaked out on agar plates (20 g agar agar) with the same LGI medium containing 0.020 g yeast extract. Small, white, curled colonies, similar to those of the other *Azospirillum* spp., are formed after 5 days and are checked again in semi-solid LGI medium for N_2-dependent growth without acid production (the medium remains green). Purity is then checked by streaking them out on potato agar containing sucrose and malate (see the above composition). *A. amazonense* colonies on this medium are white and become larger (5 mm) with a raised margin; they are very different from the other *Azospirillum* spp.

Azospirillum spp. are soil bacteria but their numbers are enriched in the rhizosphere and certain strains have been shown to be able to infect roots or even stems of various plants (Döbereiner 1992).

Isolation and identification of endophytic diazotrophs

(Cavalcante and Döbereiner 1988; Gillis et al 1989; Döbereiner 1992; Reinhold-Hurek et al 1993)

Principle of the method

In contrast to the diazotrophs described above, *Herbaspirillum* spp., *Azoarcus* spp. and *A. diazotrophicus* are obligate plant endophytes and therefore can only be isolated from their host plants. This property, which was discovered only recently, seems to be related to a much more efficient biological nitrogen fixation resulting in better plant growth, especially in the tropics, when compared to rhizosphere associations. *H. seropedicae* has been isolated from a large number of Gramineae but not from other plants (Olivares et al 1993). *H. rubrisubalbicans*, until recently known as (*Pseudomonas*) *rubrisubalbicans* (Gillis et al 1991; Pimentel et al 1991), a mild sugar cane pathogen causing mottled stripe disease in a few sensitive varieties, has only been isolated from sugar cane. So far *Azoarcus* has only been isolated from Kallar grass (*Leptochloa fusca*) in Pakistan (Reinhold-Hurek et al 1993).

Materials and apparatus

Incubator adjustable at 30–35°C
Autoclave
Microscope
Sterile glassware
Clean bench
Petri dishes

Chemicals and solutions

Semi-solid (JNFb) medium for the isolation of *Herbaspirillum* spp., *H. seropedicae* and *H. rubrisubalbicans*

5.00 g	D,L-malic acid
1.50 g	K_2HPO_4
0.20 g	$MgSO_4.7H_2O$
0.02 g	$NaCl_2.2H_2O$
2.00 ml	Minor element solution (see *Azospirillum* above)
1.00 ml	Vitamin solution (see *Azospirillum* above)
4.00 ml	Fe EDTA 1.64% solution
2.00 ml	Bromothymol blue solution (0.5% in 0.2 M KOH)
2.00 g	Agar agar

Dissolve in 800 ml distilled water; adjust the pH to 6.0 and bring up with distilled water to 1000 ml. All ingredients should be added in the given sequence to avoid undesired reactions.

Procedure

Both *Herbaspirillum* spp., *H. seropedicae* and *H. rubrisubalbicans* are best isolated by the inoculation of 10^{-2} to 10^{-6} dilutions of roots, stems or leaves of all kinds of Gramineae (*H. seropedicae*) or sugar cane (*H. rubrisubalbicans*) into vials with semi-solid JNFb medium.

Growth of *Herbaspirillum* spp. in this medium occurs in thin pellicles very similar to those of *Azospirillum* spp. but wet mounts observed under the phase-contrast microscope show much smaller cells (0.6–0.7 μm × 3–5 μm) that are usually curved and only show spiraloid movement when close to air bubbles. The two species of this genus are very similar and can only be distinguished by their differential growth on two carbon sources (Table 4.2) and 23S rRNA sequencing (Hartmann, personal communication). Isolation is done on NFb agar plates with 0.02 g yeast extract and 4.0 ml bromothymol blue. There small, moist, initially white, colonies appear which, after 1 week, become dark blue in the centre. Purification on potato agar with sucrose and malate (see above) yields small, wet, raised colonies, which become brownish in the centre.

Growth in semi-solid NFb medium and isolation procedures for *Azoarcus* spp. are the same as described for *Herbaspirillum* spp. above. Colonies on NFb medium form a non-diffusable yellowish pigment, which is more intensive with ethanol as a carbon source. *Azoarcus* cells occur singly or in pairs, 0.4–1.0 μm wide, 1.1–4 μm long and are slightly S-shaped (Reinhold-Hurek et al 1993).

Acetobacter diazotrophicus, the only species of this genus so far found to fix N_2, is isolated in semi-solid LGIP medium with the following composition: LGI medium used for *A. amazonense* (see above) with the sucrose, or better, cane sugar concentration increased to 100 g l^{-1} and the agar agar concentration increased to 2.0 g. The pH is adjusted to 5.5 by the addition of acetic acid. In this medium, 4–6 days after inoculation of 0.1 ml of stem or leaf macerates of sugar cane, the pellicle initially at the subsurface moves to the surface and becomes dark orange while the medium below becomes colourless due to the assimilation of the bromothymol blue by the bacteria. After streaking out such cultures on the same medium in agar plates (20 g agar agar l^{-1}), small moist dark orange colonies develop after 1 week. These colonies are easily recognized and are purified on potato agar (see above) containing no malate but 10% cane sugar. Dark brown moist colonies are formed after 1 week.

Table 4.3. Physiological characterization of endophytic diazotrophs.

	H. seropedicae	*H. rubrisubalbicans*	*Azoarcus* spp.	*Acetobacter diazotrophicus*
Colony pigmentation	None	None	Yellowish	None
Vibroid cells	+	+	+	–
Veil-like pellicle in semi-solid media	+	+	+	–
NO_3^-–NO_2^-	+	+	+	–
Nitrogenase activity with 10 mM NO_3^-	–	–	–	+
Use of C-sources for N_2-dependent growth				
Glucose	+	+	–	+
Sucrose	–	–	–	+
C-sources for growth with combined N				
N-Acetylglucosamine	+	–	+	
Meso-erythritol	–	+	–	
Optimum growth temperature (°C)	34	30	37–40	30
Optimum pH	5.3–8.0	5.3–8.0	6.5–6.8	4.5–6.0

Acetobacter diazotrophicus so far has only been isolated from sugar cane, sweet potatoes and cameroon grass, all sugar-rich plants that are propagated vegetatively. This indicates that this obligate endophytic bacterium is transmitted within stem cuttings of these plants. So far it has not been isolated from soil or any other plant (Cavalcante and Döbereiner 1988; Gillis et al 1989). Physiological characteristics of these four endophytic diazotrophs are summarized in Table 4.3.

Enrichment and isolation of obligate anaerobes

K. Alef

Enrichment of sulphate reducers

(Drews 1968)

Sulphate reducers are obligate anaerobic bacteria that can be found in anaerobic soils and water. They are heterotrophic and can use hexose, alcohols and organic acids as a carbon source. They reduce sulphate, sulphite or thiosulphate (as electron acceptors) to H_2S.

Chemicals and solutions

NaCl solution (0.2%)

Dissolve 0.2 g NaCl in 80 ml distilled water, bring up with distilled water to 100 ml and autoclave.

Salt solution

0.5 g K_2HPO_4
1.0 g NH_4Cl
1.0 g $CaSO_4$
2.0 g $MgSO_4.7H_2O$
5.0 g Sodium lactate (70% wt/wt)
10.0 g Agar
930.0 ml Distilled water

Adjust to pH 8.1.

Ferrous ammonium sulphate [$FeSO_4$ $(NH_4)_2.6H_2O$] solution

Steam 1% (wt/vol) $FeSO_4$ $(NH_4)_2SO_4.6H_2O$ solution for 1 h on 3 successive days.

Yeast extract solution

Dissolve 10% in water and autoclave.

Sodium thioglycolate solution (10%).

Growth medium, 1 l

Autoclave solution 1.2, cool and put into an anaerobic glove box. Filter-sterilize 50 ml of ferrous ammonium sulphate solution, 10 ml of yeast extract solution and 10 ml of sodium thioglycolate solution, and add to salt solution. Adjust final pH to 7.2–7.6. Transfer portions of the medium to sterilized tubes or flasks.

Procedure

Put soil samples in the glove box and suspend in NaCl solution. Prepare dilutions (1:10, use NaCl solution) and inoculate the tubes or flasks with 0.1–1.0 of soil suspension. After an incubation period of at least 4 weeks at room temperature, a growth of sulphate reducing bacteria is visible. To obtain pure cultures of sulphate reducers, see earlier section on isolation of pure cultures of anaerobic bacteria.

Enrichment of carbon dioxide reducers

(Braun et al 1979)

Carbon dioxide reducers are obligate anaerobic bacteria and catalyse the following reaction

$$H_2 + CO_2 \rightarrow CH_4, \text{ acetate} \qquad (4.1)$$

These bacteria can be only found in water-flooded anaerobic soils.

Chemicals and solutions

Vitamin solution (Wolin et al 1964)

1000.00 ml Distilled water
2.00 mg Biotin
2.00 mg Folic acid
10.00 mg Pyridoxine.HCl
5.00 mg Thiamine.HCl
5.00 mg Riboflavine
5.00 mg Nicotinic acid
5.00 mg Calcium pantothenate
0.01 mg Vitamin B_{12}
5.00 mg *p*-Aminobenzoic acid
1.00 mg Lipoic acid

Store at 4°C.

Mineral solution (Wolin et al 1964)

0.50 g Titripex I
6.20 g $MgSO_4.7H_2O$
0.55 g $MnSO_4.7H_2O$
1.00 g NaCl
0.10 g $FeSO_4.7H_2O$
0.17 g $CoCl_2.6H_2O$
0.13 g $CaCl_2.2H_2O$
0.180 g $ZnSO_4.7H_2O$
0.050 g $CuSO_4.7H_2O$
0.018 g $AlK(SO_4)_2.5H_2O$
0.010 g H_3BO_3
0.011 g $Na_2MoO_4.2H_2O$

Dissolve titripex I in 10 ml NaOH (0.5 M) and dilute with distilled water to 500 ml. Dissolve the salts in the solution and dilute with distilled water to 1000 ml.

Redox agent

165.5 ml Distilled water
15.5 ml NaOH (1 M)
2.5 g $Cysteine.HCl.H_2O$
2.5 g $Na_2S.9H_2O$

Dilute the NaOH with water and boil gently, while sparging with O_2-free nitrogen; cool on ice, add cysteine and Na_2S, and autoclave in 250 ml bottles. Add 1 ml of this solution to each 100 ml of nutrient solution.

Resazurine solution

See earlier section on reducing agents and redox indicators.

Dilution fluid

1.0 g NH_4Cl
0.1 g $MgSO_4.7H_2O$
0.4 g K_2HPO_4
0.4 g KH_2PO_4
2.0 ml Resazurin solution
40.0 ml Redox agent

Dissolve in 800 ml distilled water and bring up with distilled water to 1000 ml.

Bromocresol green agar

100 ml Distilled water
1 g Agar
0.01 g Bromocresol

Nutrient solution

950.00 ml Distilled water
1.00 ml Resazurin solution
1.00 g NH_4Cl
0.68 g KH_2PO_4
0.87 g K_2HPO_4
0.04 g $MgSO_4.7H_2O$
20.00 ml Vitamin solution
20.00 ml Mineral solution
2.00 g Yeast extract
10.00 g $NaHCO_3$

Adjust the pH to 7.8, boil the nutrient solution gently, while sparging with N_2/CO_2 (80%/20%) stream. After cooling, add 0.5 g $cysteine.HCl.H_2O$ and 0.25 g $Na_2S.9H_2O$. Under a continuous stream of this gas mixture, pour portions of the nutrient solution into flat bottles (100 ml), stopper the bottles and autoclave for 30 min at 120°C. After cooling, if necessary adjust the pH to 7.8.

Procedure

Suspend under anaerobic conditions (in a glove box) 1 g of soil in 10 ml of dilution fluid and prepare dilutions (1:10). Evenly streak portions (0.1 ml) of these dilutions on the agar under a continuous flow of 67% H_2 and 33% CO_2. Close the bottles and incubate at 37°C for several weeks. To distinguish between methanogens and acetogens, carefully pour 5 ml of bromocresol green agar solution over the colonies. Acid-producing colonies and their surrounding area turn yellow after about 1 h, whereas CH_4-producing colonies remain blue.

Enrichment of cellulose-decomposing *Clostridia*

(Skinner 1971)

Clostridia are obligate anaerobic spore-forming bacteria and their primary habitat is soil. They can use sugars, starch, pectin, amino acids, organic acids, proteins and cellulose as carbon and energy sources. The cellulose-decomposing bacteria are of special importance, since cellulose is an important component of plant residues.

Chemicals and solutions

Cellulose suspension

Together with a little sodium carboxy cellulose, suspend ball-milled cellulose in distilled water to give a 2% (wt/vol.) suspension.

Dilution fluid

1.0 g Ammonium sulphate
0.1 g $MgSO_4.7H_2O$
2.0 g NaCl
0.1 g $CaCl_2$
1.3 g K_2HPO_4
0.7 g KH_2PO_4

Dissolve in 800 ml distilled water; adjust the pH to 7.0, bring up with distilled water to 1000 ml and autoclave.

Salt solution A

1 g Ammonium sulphate
2 g NaCl
13 g K_2HPO_4

Dissolve in 800 ml distilled water; adjust to pH 7.0 and bring up with distilled water to 1000 ml.

Salt solution B

50.00 ml Distilled water
0.05 g $MgSO_4.7H_2O$
0.05 g $CaCl_2$

Resazurin solution

See earlier section on reducing agents and redox indicators.

Nutrient solution

500 ml Salt solution A
200 ml Cellulose suspension
250 ml Distilled water
1 ml Resazurin solution
15 g Agar
1 g Yeast extract

Make up with distilled water to 1000 ml and autoclave. After cooling (45°C), store in an anaerobic container. Dissolve 0.5 g cysteine.HCl in 50 ml salt solution B, filter-sterilize and add to the nutrient solution. Store portions of the nutrient solution in sterile anaerobic tubes at 45°C.

Procedure

Place the soil sample in the anaerobic glove box and suspend 1 g soil in 10 ml of the dilution fluid. Inoculate the warm agar tubes each with 0.1 ml suspension (several tubes per dilution), spread, solidify and incubate all tubes at 28–35°C. Growth of cellulytic bacteria is indicated by clearing of the cellulose agar.

Enrichment of fungi

K. Alef

Fungi are heterotrophic microorganisms, which have an important role in the decomposition of organic matter (e.g. cellulose and lignin) and in the mineralization of nutrient in soils.

Growth media

A considerable variety of nutrient agar media have been used for the isolation of soil fungi. The most frequently used media are presented below.

Czapek–Dox agar

30.00 g Sucrose
3.00 g Sodium nitrate
0.50 g $MgSO_4.7H_2O$
0.50 g KCl
1.00 g K_2HPO_4
0.01 g $FeSO_4$
15.00 g Agar
0.50 g (if desired) Yeast extract

Dissolve inorganic constituents separately in 800 ml distilled water; add $FeSO_4$ last; adjust the pH to 7.3 and bring up with distilled water to 1000 ml. Add sucrose before sterilization.

Malt extract agar

15.0 g Malt extract
1.0 g Pepton
13.0 g Maltose
3.0 g Dextrose
1.0 g K_2HPO_4
1.0 g NH_4Cl
15.0 g Agar

Dissolve in 800 ml distilled water, adjust the pH to 4.8 and bring up with distilled water to 1000 ml.

Dextrose–peptone agar

1000.0 ml Distilled water
10.0 g Dextrose
5.0 g Peptone
1.0 g KH_2PO_4
0.5 g $MgSO_4.7H_2O$
20.0 g Agar
3.3 ml Rose bengal (1%)
30.0 mg Streptomycin

Dissolve in 800 ml distilled water and bring up with distilled water to 1000 ml. Autoclave the medium and add streptomycin to the still liquid agar medium (45–55°C).

Procedure

Disperse 10 g of moist soil in 90 ml of diluent (with 0.2% dextrin) and shake for 15 min. Prepare a dilution series using 1 ml blowout pipettes and 0.2% dextrose as diluent. Aliquots of chosen dilutions (e.g. 10^{-5}, 10^{-6}) are plated on to the chosen nutrient agar medium. Incubate the agar plates at 20–25°C for 14 days. Colonies developing on the plates can be subcultured for subsequent studies (identification, isolation of pure cultures, etc.). It should be pointed out that the large majority of fungal colonies developing on soil dilution plates originate from spores.

For more information on media, enrichment and counting of fungi, see Chapter 4 (basic methods for counting microorganisms in soil and water; microscopic methods for counting bacteria and fungi in soil).

Basic methods for counting microorganisms in soil and water

H-.J. Lorch
G. Benckieser
J.C.G. Ottow

Soils and water may contain up to 10^8–10^{12} bacterial cells (propagules) per gram of soil or millilitre of water, if determined by microscopic techniques. Many of these bacteria are non-culturable on common or special agar media, and have never been isolated and identified under laboratory conditions because of their stenobiotic requirements with respect to the composition of the media, the presence of specific growth factors or environmental conditions (pO_2, pH, gradient conditions etc.). Thus, the total number of bacteria, actinomycetes or even moulds can not be obtained using agar media (Parkinson et al 1971; Wollum 1982). The total number of plate count bacteria usually covers 0.1–10% (exceptionally 50%) of the total propagules observed with special microscopic techniques (Schmidt and Paul 1982; Gray 1990; Lorch et al 1992).

Number and size of samples

From the field soil and its profile horizons or from the various materials in question (compost, sediment, rhizosphere soil, etc.), as many single samples as possible (at least three) should be collected. They should be freed from debris, stones and roots, bulked, carefully homogenized and stored in plastic bags (at 4–5°C or by deep freezing). Most significant changes in population densities occur during transport and subsequent days of storage (at 4–5°C). Variation among random samples collected from the same field unit, horizon, experimental plot or compost heap can be estimated by calculating the standard deviations and variance s^2 (1–4)

$$s = \sqrt{\frac{\Sigma(x - \bar{x})^2}{n - 1}} \tag{4.2}$$

where $x - \bar{x}$ is the deviation from the mean ($\bar{x}$), n is the number of samples and n-1 is the number of degrees of freedom. The variance (s^2), which it is often more convenient to use, is the square of the standard deviation.

$$s^2 = \frac{\Sigma(x - \bar{x})^2}{n - 1} \tag{4.3}$$

Calculation of $(x - \bar{x})^2$ is laborious and it is quicker to use

$$\Sigma x^2 - \frac{(\Sigma x)^2}{n} \tag{4.4}$$

to calculate this, i.e.

$$s^2 = \frac{\Sigma x^2 - (\Sigma x)^2/n}{n - 1} \tag{4.5}$$

The standard error ($s_{\bar{x}}$) is a measure of reproducibility and can be calculated from the standard deviation or the variance.

$$s_{\bar{x}} = \frac{s}{\sqrt{n}} \tag{4.6}$$

$$\text{or } s_{\bar{x}} = \sqrt{\frac{s^2}{n}} \tag{4.7}$$

If the distribution of the estimate is approximately normal and the number of samples is greater than 30, the interval $x \pm s_x$ includes approximately two thirds of the means of similarly drawn batches of the samples. This is the range in which the means of future samples batches can be expected to fall (Parkinson et al 1971; Koch 1981). Microorganisms are heterogeneously distributed in soil, mainly adsorbed to and coated by organic clay particles. Therefore, a sufficient large sample of at least 20 g fresh material should be used in serial dilutions. All results are converted to the customary dry soil or sludge basis. Drying of soil, sludge or compost reduces the number and changes the composition of the microflora.

Materials and apparatus

Dilution flasks (screw-capped, autoclavable, 250 ml) with 180 ml sterile water
Dilution flasks (screw-capped, autoclavable, 250 ml) with 90 ml quarter strength's Ringer solution
Pipettes, sterile (1.0 ml and 10 ml)
Petri dishes (7.5 cm diameter), sterile (glass or plastic)
Reagent tubes
Durham vials
Drigalski (hockey) stick (sterilized in aluminium foil)
Ringer solution 25%

NaCl, 2.25; KCl, 0.105; $CaCl_2$, 0.045; $NaHCO_3$, 0.05; citric acid, 0.034 g l^{-1}

$Na_4P_2O_7.10H_2O$ (sodium pyrophosphate as soil dispersing agent)
Ethanol (95%)
Chemicals, p.a.

See composition of media

Autoclave
Incubator
Rotary shaker
Magnetic stirrer and paddles
Dispenser (variable)

Preparation of serial dilutions

From the fresh and carefully homogenized bulk soil, 20 g (or 20 ml water, sludge or slurry, respectively) are transferred into 180 ml sterile water (250 ml flask) and mixed with sodium pyrophosphate (0.18% final concentration) to disperse the soil colloids. Pyrophosphate is added after autoclaving (15 min at 121°C) the flasks and omitted in the case of water samples. The flask (dilution 10^{-1}) is shaken (15 min on a horizontal or rotary shaker) and left to settle the soil (15 min). Ten millilitres of this dilution (10^{-1}) are pipetted into a 250 ml flask containing 90 ml of sterile quarter strength Ringer solution, hand shaken (with up and down movements) and subsequently serially diluted [up to 10^{-7} for plate count methods or up to 10^{-9} for the most probable number (MPN) technique]. One millilitre of each dilution (beginning with the highest dilution) is transferred with 1 ml (blowout) pipettes into (a) Petri dishes (three or five parallel plates per dilution) for plate counting or into (b) reagent tubes containing 7 ml of an appropriate nutrient broth (each tube with or without a Durham vial), usually with three or five replicate tubes per dilution. Replicate tubes of liquid media are inoculated and used for the determination of the MPN.

Cultural methods for enumerating soil and water organisms

Poured plate counts

Twenty millilitres of the selected sterile (15 min at 121°C) molten agar medium (45–50°C) (Table 4.4) are poured over the inoculum (1 ml) in the Petri dish and the medium is carefully mixed with the sample by gently hand-rotating each dish on the

Table 4.4. Plate count agar media used to quantify the total number of bacteria.

*Acronym for agar**	*Characteristic ingredients of the media†‡*	*References*
PGY agar	Peptone–glucose–yeast extract agar (pH 7.0)	Hirsch and Rades-Rohkohl (1983); Marxsen (1988)
AlbGY agar	Albumin–glucose–yeast extract agar (pH 7.2)	Hattori (1982); Sato et al (1984)
PG agar	Peptone–glucose agar (pH 7.2)	Kölbel-Boelke et al (1988)
PMA agar	Peptonized milk–actidione agar (pH 7.0)	Larkin (1972)
PYGG agar	Proteose peptone–yeast extract–glycerol–glycerophosphate agar (pH 7.5)	Litchfield et al (1975)
PYS agar	Peptone–yeast extract–soil extract agar (pH 6.8–7.0)	Singh-Verma (1968); Jager and Bruins (1975); Martin (1975)
DPM agar	Diluted peptone–meat extract agar, (pH 7.0–7.2)	Ohto and Hattori (1983)

* For routine examinations commercially available dehydrated media such as bacto nutrient agar or bacto tryptic soy agar (Difco Laboratories, Detroit), casein peptone–yeast extract–dextrose agar (Oxoid, Unipath GmbH, Wesel), standard meat extract–peptone plate count agar (Merck, Darmstadt), plate count agar (bioMérieux, Nürtingen) or nutrient agar (Kyohuto Seiyaku Co., Tokyo) may be used.

† In order to facilitate the enumeration of red-coloured bacterial colonies, each medium can be supplied with 0.001% aqueous triphenyltetrazoliumchloride (TTC) solution just prior to pouring the plates (Unger 1958).

‡ For ingredients, see Table 4.5.

table surface (about three times clockwise and counter-clockwise). The cooled plates (room temperature) are stacked upside down in piles and incubated (25–30°C) in darkness. Colonies are counted using a hand counter and a magnifying lens on an illuminated counting chamber at those dilutions that have developed 20–100 colonies after 5–7 days, if not indicated differently. If moulds are spreading and covering more than 15–20% of the agar surface, the plates should be rejected, because the development of bacteria may have been inhibited (Parkinson et al 1971). The development of fungi can be reduced by adding mycostatin (2 μg ml^{-1}) prior to pouring the plates (Williams and Davies 1965). For soil and water samples the incubation temperature should range between 25°C and 30°C, if not indicated differently (see survey of media composition in Table 4.5).

Spread plate counting technique

In many cases with special substrates (milk, food-derived samples, clinical materials, etc.) or specific bacteria, aliquots (1 or 0.1 ml samples) of the dilution series are spread on the poured, cooled (selective) agar and homogeneously distributed on the surface using a sterile Drigalsky glass stick (hockey stick). Flame (ethanol) the hockey stick between each use. Spread plating (3–5 replicates per dilution) is a commonly used technique for determining total viable counts (TVC) of pure cultures or specifically coloured (by pigments or dyes in the agar) aerobic bacteria that are both selectively identified and counted. This surface spread-plate method can be carried out by automatic procedures using a spiral plater (e.g. Spiral Systems Inc., Ohio). The success of this spread-plating technique depends on the organisms in question, the medium, the

Table 4.5. Survey of the composition of media used to enumerate bacteria.

Acronym	*Ingredients (in g l^{-1} distilled water), final pH and comments on preparation*†*
PGY agar	Glucose 0.25, peptone 0.25, yeast extract 0.25, K_2HPO_4 0.1, $MgSO_4.7H_2O$ 0.05, $FeSO_4.7H_2O$ 0.02, agar 15, pH 7.0
AlbGY agar	Glucose 0.1, egg albumin (Difco) 0.25, yeast extract 0.05, K_2HPO_4 0.5, $MgSO_4.7H_2O$ 0.4, $Fe_2(SO_4)_3.7H_2O$ (trace), agar 15, pH 7.2
PG agar	Glucose 0.1, peptone 1.0, K_2HPO_4 0.1, $FeSO_4.7H_2O$ 0.02, agar 15, pH 7.2
PMA agar	Peptonized milk (Difco) 1.0, actidione (= cyclohexamide) 0.1, agar 15, pH 7.0
PYGG agar	Proteose peptone 1.0, yeast extract 1.0, glycerol 5 ml, sodium glycerophosphate 0.5, agar 15, pH 7.4
PYS agar	Glucose 1.0, peptone 0.2, yeast extract 0.1, K_2HPO_4 0.4, $MgSO_4.7H_2O$ 0.05, 100 ml soil extract, agar 15, pH is adjusted to 6.8–7.0. Preparation of soil extract: 1 kg of dried sieved top soil is mixed with 1 l tap water and autoclaved (30 min at 121°C). The suspension is settled and the supernatant filtered through a gauze and subsequently through a paper filter in a Buchner funnel (with vacuum pump) and made up to 1 l.
DPM agar	Peptone 0.01, meat extract 0.01, NaCl 5.0, agar 15, pH adjusted to 7.0–7.2

* Compared to original literature, small modifications in the composition may occur.

† Fanny Angelina Hesse introduced agar agar into microbiology (Hesse 1992).

accuracy of the dilution and the care used in spreading the inoculum on the surface of plates (Krieg 1981).

Total number of bacteria

Agar media commonly used to count a broad number of bacteria (saprophytes) in soil, sediment, water, waste water or compost, should be complex in nature (including peptone or meat extract and/or yeast extract) but relatively meagre in composition (Table 4.5). Nutrient agars (NA) that are rich in carbon sources such as glucose and/or proteins (peptone) may develop colonies very rapidly (within 3 days at 30°C) but their total number usually remains considerably lower than on media with a relatively poor composition low in concentration (e.g. Standard-II-Nutrient Agar, Merck). Diluted compositions (1:10 or 1:100) should therefore be compared with the original compositions (Hattori and Hattori 1980). In general, most agar media used to count and isolate microorganisms from soil and water are too rich in ingredients. As a consequence, mainly eurybiotic (zymogenous) bacteria may develop rapidly and suppress other fastidious cells that grow relatively slowly under less appropriate plate conditions, such as high osmotic values and supply of nutrients as well as by acidification (from glucose or sucrose) or alkalization (through deamination of relatively high quantities of peptone, tryptone or meat extract). On relatively rich agar media at higher temperatures (30–35°C), a few microorganisms tend to overgrow and inhibit others, resulting in a decrease of the "total" count. In general the use of low concentrated or diluted agar media is recommended rather than relatively rich and complex media with a great variety of ingredients. Relatively simple and less concentrated media containing soil extract usually yield the highest number of colonies (Harris and Keeney 1968; Singh-Verma 1968; Jager and Bruins 1975; Martin 1975). The major disadvantages of soil extract agars are the time required for their preparation and their lack of standardization. Probably, the positive effect of soil extract on the increase of bacterial colonies should not be ascribed to growth factors in the extract but to the chelating, binding and buffering effects of extracted humic substances and other compounds that lower toxic or inhibiting effects of ingredients, ions and/or metabolites.

Total plate number of actinomycetes

If the total population of actinomycetes is determined, special agar media are required. Actinomycetes usually develop slowly (incubation 10–14 days at 25–30°C) forming small, round, tenaceous, white-headed (aerial mycelium) colonies on or partly in the agar. Most representatives of this group are neutrophiles (pH 6–7) and physiologically versatile. Therefore, selectivity of most agar media is obtained essentially by the addition of fungal antibiotics and/or bacterial inhibitors. Antifungal antibiotics such as cycloheximide, mycostatin, pimaricin or anfotericine as well as the bacteriostatic dye rose bengal have been recommended. Rose bengal (at 0.035–0.067 g l^{-1}) suppresses bacteria, and simultaneously inhibits growth and spreading of fungal colonies (Martin 1950; Ottow and Glathe 1968; Ottow 1972). In media containing rose bengal, actinomycetes are easily recognized as small, intensively pink coloured colonies that develop inside or on a slightly pink medium. Because of their slow development, plates should be incubated at 25–30°C for mesophilic and at 45–55°C for thermophilic species for at least 10–14 days (Williams and Wellington 1982). Agar media (Table 4.6) routinely used for the enumeration and/or isolation of actinomycetes (mainly *Streptomyces* spp. and *Micromonospora* spp.) from soil sediments or water are:

- Starch–casein–nitrate agar (SCN), pH 7.0–7.2 (Küster and Williams 1964).
- Rose bengal–malt extract agar (RBME), pH 6.0–6.2 (Ottow and Glathe 1968)
- Colloidal chitin–mineral salt agar (CCMS), pH 8.0 (Hsu and Lockwood 1975).

SNC-agar has been used in various modifications including the addition of antifungal substances such as actidione (cyclohexamide) and nystatin (mycostatin) each at a concentration of 50 µg ml^{-1} (Davies and Williams 1970), pimaricin (50 µg ml^{-1}; Porter et al, 1960) or anfotericine B (fungizon, 30 µg ml^{-1}; Coelho and Drozdowicz 1978), or the bacteriostatic dye rose bengal at 0.035 gl^{-1} (Ottow 1972). Most actinomycete colonies on agar plates originate from spores, other propagules as well as from pseudomycelial fragments and thus the colony forming units (per gram of dry soil) may give only little information on the location, number and characteristic growth of these organisms in the soil matrix, sludges or composts (Williams et al 1984). If moulds and actinomycetes are evaluated simultaneously on RBME- or RBSCN-agar, colony counting of actinomycetes can usually be done at higher dilutions than those used for fungi. The antibiotics are dissolved in sterile water and added to the autoclaved agar medium just prior to pouring the plates, whereas rose bengal can be autoclaved with the medium. Media containing rose bengal should not be stored (at 4–5°C) and remelted, because a pink flocculation may occur that interferes on the plates with colony counting (Ottow 1972).

Fungal plate count

The plate count method is less suitable for the enumeration of fungal populations or densities because spores as well as fragments of one and the same mycelium fragmented by the dilution agitation will develop as single colonies, which are counted. If used, the results should be expressed as colony forming units per

Table 4.6. Survey of the composition of agar media for actinomycetes.

Acronym	*Ingredients (in g l^{-1} distilled water), final pH and comments on preparation**
SCN agar	Starch 10, casein (Difco, vitamin free) 0.3, KNO_3 2.0, NaCl 2.0, K_2HPO_4 2.0, $MgSO_4.7H_2O$ 0.05, $CaCl_2$ and $FeSO_4.7H_2O$ (traces), agar 15, pH adjusted to 7.2 before autoclaving
RBSCN agar	SCN agar with rose bengal (Fluka) 0.035, pH 7.0–7.2
RBME agar	Malt extract 20, K_2HPO_4 0.5, rose bengal 0.067, agar 17–20, pH adjusted to 6.0–6.2
CCMS agar	Colloidal chitin (Callbiochem Corp.) 2.5, K_2HPO_4 0.7, KH_2PO_4 0.2, $MgSO_4.7H_2O$ 0.5, $FeSO_4.7H_2O$ 0.01, $ZnSO_4$ 0.001, $MnCl_2$ 0.001, agar 15, after autoclaving the pH is adjusted with 1 M NaOH solution to 8.0. Preparation: 20 g of the ground chitin (Waring blender) is dissolved in *c.* 200 ml concentrated HCl, filtered through a glass filter wool and poured subsequently into 1 l water (at 5–10°C). The reprecipitated chitin is washed with sterile water until pH neutral and 2.5 g (dry weight) is added to the mineral agar.

* Compared to the original composition, small modifications may occur.

gram of dry soil (CFU g^{-1}). Agar plates are more appropriate for qualitative examinations and isolation purposes. Media used to enumerate and isolate fungi may receive streptomycin (30 μg ml^{-1}), aureomycin (20 μg ml^{-1}) or chloramphenicol (100 μg ml^{-1}) after sterile filtration in order to suppress the development of rapidly growing bacteria. Some agar media for the enumeration of fungal CFU are listed in Table 4.7. If both fungi and actinomycetes are the objectives of the examination, RBSCN at pH 7.0–7.2 may be used as it allows both the enumeration of moulds and actinomycetes on the same poured plates at different dilutions (Turian and Ottow 1983). Pepton–dextrose–rose bengal–streptomycin–aureomycin agar (Martin 1950) has been very useful for counting CFU and for isolating fungi by the dilution plate method, because bacteria and actinomycetes are

Table 4.7. Common agar media used for counting and isolating fungi from soil and other natural substrates.

Acronym	*Characteristic composition of the medium**	*Comments and references*
ME agar†	Malt extract agar (pH 5.5)	Commercially available
RBME agar	Rose bengal–malt extract agar (pH 6.0)	Ottow and Glathe (1968)
RBSCN agar	Rose bengal–starch–casein–nitrate agar (pH 7.0)	Ottow (1972)
Czapek-Dox agar†	Sucrose–nitrate–mineral salt agar (pH 7.3)	Warcup (1960)
PDRB agar	Peptone–dextrose–rose bengal–streptomycin (or aureomycin) agar (pH 7.0)	Martin (1950)
Sabouraud agar†	Peptone–dextrose (or maltose) agar (pH 6.5)	Commercially available
PGBRC agar	Peptone–glucose–rose bengal–chloramphenicol agar (pH 7.0–7.2).	Jarvis (1973)

* The ingredients are specified in Table 4.10 and may have small modifications compared to their original composition.

† These media (in various modifications) are commercially available in dehydrated form from bioMérieux (Nürtingen), Difco (Detroit), Merck (Darmstadt) or Oxoid (Unipath GmbH, Wesel).

Table 4.8. Liquid media for enumerating some functional groups of bacteria by the MPN method.

Functional group	*Broth composition and acronyms**	*References and comments*
Saprophytes	Casein peptone–starch broth (CPS)	Collins (1963)
	Proteose peptone–yeast extract-glycerol–glycerophosphate broth (PYGG)	Litchfield et al (1975)
	Casein peptone–lactate–acetate–glycerol–glucose broth (CLAGG)	Lorch et al (1990)
Prototrophic bacteria	Synthetic acetate–glutamate broth (AG) with bromothymol blue indicator (blue colouring = positive)	Lorch et al (1990)
Fermenting bacteria	(Gas in Durham vials of CPS or PYGG broth from starch or glycerol, respectively = positive)	
Enterobacteriaceae	Dextrose–peptone–brilliant green broth (DP) (gas in Durham vial = positive)	Mossel et al (1974)
Coliforms	MacConkey broth (MC) (gas and acid after 48 h, 37°C = total coliforms) (gas and acid after 48 h, 44°C = faecal coliforms)	commercially available
Faecal streptococci (*S. faecalis-faecium*)	Kanamycin–aesculine–azide broth (KAA) (black colouring after 48 h, 37°C = positive)	Mossel (1977)
Aerobic spore forming bacilli (*Bacillus* spp.)	Growth in PYGG broth after pasteurization, 15 min, 80°C	Ottow et al (1984)
Pseudomonads (*P. aeruginosa-fluorescens-putida*)	Cetrimide–peptone–sulphate broth (CPSB)	Brown and Lowbury (1965); Schmider and Ottow (1981)
Nitrifiers (chemolitho-autotrophic)	Ammonium or nitrite–mineral salt–solution (MS), incubation 28 days, 25°C	Abeliovich (1987)
Denitrifiers†	Casamino acid–yeast extract broth (CY) (gas in Durham vial = positive)	Tomlinson and Hochstein (1972)
	Lactate–peptone–yeast extract broth (LPY) (gas in Durham vial = positive)	Malek et al (1974)
	Citrate–asparagine broth (CA) (gas in Durham vial = positive)	Bollag et al (1970)
	AG broth (gas in Durham vial = positive)	Lorch et al (1990)
Sulphate-reducing bacteria	Lactate–thioglycolate–sulphate broth (LTS), incubation 28 days, 25°C (black colouring = positive)	Oberzill (1970)
Fe(III)-reducing bacteria	Glucose–asparagine–Fe_2O_3 broth (GAF) (Fe(II) /detected by red colouring after addition of 2,2-dipyridyl solution)	Ottow (1969)
Fe(III)-reducing, N_2-fixing, saccharolytic clostridia	Nitrogen-free glucose–Fe_2O_3 broth (NGF) (Fe(II) detected by red colour after adding 2,2-dipyridyl solution)	Hammann and Ottow (1976)
Cellulolytic bacteria	Cellulose (filter paper strip)–mineral salt broth (CM), incubation 35 days, 28°C	Sato et al (1984)
N_2-fixing bacteria‡ (*Azopirillum* spp.)	Semi-solid malate–yeast extract–glucose broth (MYG) (positive tubes show clear pellicle formation and ethylene formation after 3 days, 30°C)	Okon et al (1977); Jagnow (1982)

* The ingredients (with modifications) are specified in Table 4.11.

† For the total number of denitrifying bacteria, use CY or LPY broth, but for the MPN of prototrophic denitrifiers, CA or AG broth are recommended.

‡ Because of the great variety of potentially N_2-fixing bacteria in the soil and rhizosphere with specific requirements, the reader is referred to Postgate (1981), Ladha et al (1983) and Blandreau (1983, 1986) for details of methods and media.

essentially suppressed. Otherwise, peptone–glucose–rose bengal–chloramphenicol agar has been successfully used for counting moulds and yeasts in foods (Jarvis 1973).

Most probable number technique

Principle of the method

Population densities of various groups of (functional) bacteria can be estimated by the MPN method without an actual count of single cells or colonies. This method is based on the presence or absence of bacteria by using an extinction dilution in which replicate tubes of special (selective) broth are inoculated with 1 ml aliquots of the serial dilution. One prerequisite of this MPN method is that microorganisms that are to be enumerated selectively must be able to cause some characteristic metabolite (e.g. gas formation) or product that can be detected easily by a specific reagent (e.g. Fe(II)-formation with 2,2-dipyridyl solution or nitrite by Griess–Illosvay reagents). Growth or positive tests on a specific product are recorded from the tubes at the end of the incubation time (usually 7–14 days, 24–30°C). The tabulated results are referred to probability tables (Meynell and Meynell 1965; De Man 1975, 1983) and the MPN of the bacterial population is expressed per millilitre (water sample) or per gram (dry soil, compost, sludge or sediment). Multiplying the results by the dilution factor will give the MPN per millilitre or gram dry matter (Koch 1981; Alexander 1982).

Procedure

One millilitre samples of a 10-fold dilution series (up to 10^{-9}) are pipetted into reagent tubes (3 or 5 parallel tubes per dilution) containing a specific liquid broth (7 ml/tube, with Durham vial, if required; Table 4.8). The tubes are carefully mixed (by hand rolling) and incubated at 20°C or 30°C (if not stated differently). At the end of the incubation period, the tubes are observed for growth (pellet formation and/or turbidity in the case of the total MPN of bacteria), for gas production (for the MPN of glucose or lactose-fermenting bacteria or for the MPN of potentially denitrifying bacteria), for a specific colour (after adding a specific reagent on ammonium, nitrite, nitrate or ferrous iron formation) or for blackening (MPN of sulphate-reducing bacteria). For the MPN of cellulose-degrading bacteria (aerobic and anaerobic), each tube is shaken vigorously (after 5 weeks at 28°C) and the rupture of the cellulose paper strip (filter paper) at the liquid surface is recorded as a positive result. If they are incubated anaerobically (in a jar or incubator), the MPN of anaerobic cellulolytic bacteria can be quantified.

Calculation of MPN and confidence limits

From the tabulated results, select: (1) the tubes in the least concentrated dilution in which all tubes have been recorded positive or in which the greatest number of tubes is positive; and (2) the number of positive tubes in the next two higher dilutions. From Table 4.9 (De Man 1983) the characteristic numbers are identified that correspond to the values observed experimentally (the number of positive results in the left row, e.g. 3–2–1 = characteristic number). The MPN is obtained by multiplying 15 (Table 4.9, second row) by the appropriate dilution. The 95% and 99% confidence limits for the MPN values are obtained from Table 4.9.

Table 4.9. Population densities of bacteria estimated by the MPN method (De Man 1983).
(a) MPN table for 3 × 1, 3 × 0.1 and 3 × 0.01 g (ml)

Number of positive results			MPN	Category when number of tests 1	2	3	5	10	Confidence limits ⩾ 95%	⩾ 95%	⩾ 99%	⩾ 99%
0	0	0	<0.30						0.00	0.94	0.00	1.40
0	0	1	0.30	3	2	2	2	1	0.01	0.95	0.00	1.40
0	1	0	0.30	2	1	1	1	1	0.01	1.00	0.00	1.60
0	1	1	0.61	0	3	3	3	3	0.12	1.70	0.05	2.50
0	2	0	0.62	3	2	2	2	1	0.12	1.70	0.05	2.50
0	3	0	0.94	0	0	0	0	3	0.35	3.50	0.18	4.60
1	0	0	0.36	1	1	1	1	1	0.02	1.70	0.01	2.50
1	0	1	0.72	2	2	1	1	1	0.12	1.70	0.05	2.50
1	0	2	1.10	0	0	0	3	3	0.40	3.50	0.20	4.60
1	1	0	0.74	1	1	1	1	1	0.13	2.00	0.06	2.70
1	1	1	1.10	3	3	2	2	2	0.40	3.50	0.20	4.60
1	2	0	1.10	2	2	1	1	1	0.40	3.50	0.20	4.60
1	2	1	1.50	3	3	3	3	2	0.50	3.80	0.20	5.20
1	3	0	1.60	3	3	3	3	2	0.50	3.80	0.20	5.20
2	0	0	0.92	1	1	1	1	1	0.15	3.50	0.07	4.60
2	0	1	1.40	2	1	1	1	1	0.40	3.50	0.20	4.60
2	0	2	2.00	0	3	3	3	3	0.50	3.80	0.20	5.20
2	1	0	1.50	1	1	1	1	1	0.40	3.80	0.20	5.20
2	1	1	2.00	2	2	1	1	1	0.50	3.80	0.20	5.20
2	1	2	2.70	0	3	3	3	3	0.90	9.40	0.50	14.20
2	2	0	2.10	1	1	1	1	1	0.50	4.00	0.20	5.60
2	2	1	2.80	3	2	2	2	1	0.90	9.40	0.50	14.20
2	2	2	3.50	0	0	0	0	3	0.90	9.40	0.50	14.20
2	3	0	2.90	3	2	2	2	1	0.90	9.40	0.50	14.20
2	3	1	3.60	0	3	3	3	3	0.90	9.40	0.50	14.20
3	0	0	2.30	1	1	1	1	1	0.50	9.40	0.30	14.20
3	0	1	3.80	1	1	1	1	1	0.90	10.40	0.50	15.70
3	0	2	6.40	3	3	2	2	2	1.60	18.10	1.00	25.00
3	1	0	4.30	1	1	1	1	1	0.90	18.10	0.50	25.00
3	1	1	7.50	1	1	1	1	1	1.70	19.90	1.10	27.00
3	1	2	12.00	3	2	2	2	1	3.00	36.00	2.00	44.00

3	1	3	16.00	0	0	0	3	3	3.00	38.00	2.00	52.00
3	2	0	9.30	1	1	1	1	1	1.80	36.00	1.20	43.00
3	2	1	15.00	1	1	1	1	1	3.00	38.00	2.00	52.00
3	2	2	21.00	2	1	1	1	1	3.00	40.00	2.00	56.00
3	2	3	29.00	3	3	3	2	2	9.00	99.00	5.00	152.00
3	3	0	24.00	1	1	1	1	1	4.00	99.00	3.00	152.00
3	3	1	46.00	1	1	1	1	1	9.00	198.00	5.00	283.00
3	3	2	110.00	1	1	1	1	1	20.00	400.00	10.00	570.00
3	3	3	>110.00									

(b) MPN table for 5 × 1, 5 × 0.1 and 5 × 0.01 g (ml)

Number of positive results			*MPN*	*Category when number of tests*					*Confidence limits*			
				1	*2*	*3*	*5*	*10*	*≥ 95%*	*≥ 95%*	*≥ 99%*	*≥ 99%*
0	0	0	<0.18						0.00	0.65	0.00	0.93
0	0	1	0.18	2	2	2	1	1	0.00	0.65	0.00	0.93
0	1	0	0.18	1	1	1	1	1	0.01	0.65	0.00	0.93
0	1	1	0.36	3	3	3	2	2	0.07	0.99	0.02	1.40
0	2	0	0.37	3	2	2	2	1	0.07	0.99	0.02	1.40
0	2	1	0.55	0	0	0	3	3	0.17	1.40	0.09	2.10
0	3	0	0.56	0	3	3	3	3	0.17	1.40	0.09	2.10
1	0	0	0.20	1	1	1	1	1	0.02	0.99	0.01	1.40
1	0	1	0.40	2	1	1	1	1	0.07	1.00	0.02	1.40
1	0	2	0.60	0	0	3	3	3	0.17	1.40	0.09	2.10
1	1	0	0.40	1	1	1	1	1	0.07	1.10	0.03	1.40
1	1	1	0.61	3	2	2	2	1	0.17	1.40	0.09	2.10
1	1	2	0.81	0	0	0	0	3	0.33	2.20	0.20	2.80
1	2	0	0.61	2	1	1	1	1	0.18	1.40	0.09	2.10
1	2	1	0.82	3	3	3	3	2	0.33	2.20	0.20	2.80
1	3	0	0.83	3	3	3	3	2	0.33	2.20	0.20	2.80
1	3	1	1.00	0	0	0	0	3	0.30	2.20	0.20	2.80
1	4	0	1.10	0	0	0	0	3	0.30	2.20	0.20	2.80
2	0	0	0.45	1	1	1	1	1	0.08	1.40	0.04	2.10
2	0	1	0.68	2	1	1	1	1	0.18	1.50	0.09	2.10
2	0	2	0.91	0	3	3	3	3	0.33	2.20	0.20	2.80

Table 4.9. Continued.

Number of positive results			MPN	Category when number of tests					Confidence limits			
				1	2	3	5	10	⩾ 95%	⩾ 95%	⩾ 99%	⩾ 99%
2	1	0	0.68	1	1	1	1	1	0.19	1.70	0.10	2.30
2	1	1	0.92	2	2	1	1	1	0.33	2.20	0.20	2.80
2	1	2	1.20	0	0	3	3	3	0.40	2.50	0.20	3.40
2	2	0	0.93	1	1	1	1	1	0.34	2.20	0.20	2.80
2	2	1	1.20	3	3	2	2	2	0.40	2.50	0.20	3.40
2	2	2	1.40	0	0	0	0	3	0.60	3.40	0.40	4.40
2	3	0	1.20	3	2	2	2	1	0.40	2.50	0.20	3.40
2	3	1	1.40	0	3	3	3	3	0.60	3.40	0.40	4.40
2	4	0	1.50	0	3	3	3	3	0.60	3.40	0.40	4.40
3	0	0	0.78	1	1	1	1	1	0.21	2.20	0.12	2.80
3	0	1	1.10	1	1	1	1	1	0.40	2.20	0.20	2.90
3	0	2	1.30	3	3	3	2	2	0.60	3.40	0.40	4.40
3	1	0	1.10	1	1	1	1	1	0.40	2.50	0.20	3.40
3	1	1	1.40	2	1	1	1	1	0.60	3.40	0.40	4.40
3	1	2	1.70	3	3	3	3	2	0.60	3.40	0.40	4.40
3	2	0	1.40	1	1	1	1	1	0.60	3.40	0.40	4.40
3	2	1	1.70	2	2	2	1	1	0.70	3.90	0.50	5.10
3	2	2	2.00	0	3	3	3	3	0.70	3.90	0.50	5.20
3	3	0	1.70	2	2	1	1	1	0.70	3.90	0.50	5.20
3	3	1	2.10	3	3	3	2	2	0.70	3.90	0.50	5.20
3	3	2	2.40	0	0	0	3	3	1.00	6.60	0.70	9.40
3	4	0	2.10	3	3	2	2	2	0.70	4.00	0.50	5.20
3	4	1	2.40	0	3	3	3	3	1.00	6.60	0.70	9.40
3	5	0	2.50	0	0	0	3	3	1.00	6.60	0.70	9.40
4	0	0	1.30	1	1	1	1	1	0.40	3.40	0.30	4.40
4	0	1	1.70	1	1	1	1	1	0.60	3.40	0.40	4.40
4	0	2	2.10	3	2	2	2	2	0.70	3.90	0.50	5.20
4	0	3	2.50	0	0	0	0	3	1.00	6.60	0.70	9.40
4	1	0	1.70	1	1	1	1	1	0.60	3.90	0.40	5.10
4	1	1	2.10	1	1	1	1	1	0.70	4.10	0.50	5.30
4	1	2	2.60	3	3	2	2	2	1.00	6.60	0.70	9.40
4	1	3	3.10	0	0	0	0	3	1.00	6.60	0.70	9.40

4	2	0	2.20	1	1	1	1	1	0.70	4.80	0.50	6.10
4	2	1	2.60	2	1	1	1	1	1.00	6.60	0.70	9.40
4	2	2	3.20	3	3	3	2	2	1.00	6.60	0.70	9.40
4	2	3	3.80	0	0	0	0	3	1.30	10.00	0.90	14.70
4	3	0	2.70	1	1	1	1	1	1.00	6.60	0.70	9.40
4	3	1	3.30	2	2	1	1	1	1.00	6.60	0.70	9.40
4	3	2	3.90	3	3	3	3	2	1.30	10.00	0.90	14.70
4	4	0	3.40	2	2	1	1	1	1.30	10.00	0.90	14.70
4	4	1	4.00	3	3	2	2	2	1.30	10.00	0.90	14.70
4	4	2	4.70	0	0	0	3	3	1.40	11.30	0.90	14.70
4	5	0	4.10	3	3	3	3	2	1.30	10.00	0.90	14.70
4	5	1	4.80	0	0	3	3	3	1.40	11.30	0.90	14.70
5	0	0	2.30	1	1	1	1	1	0.70	6.60	0.50	9.40
5	0	1	3.10	1	1	1	1	1	1.00	6.60	0.70	9.40
5	0	2	4.30	3	2	2	2	1	1.30	10.00	0.90	14.70
5	0	3	5.80	0	0	0	3	3	2.10	14.90	1.40	20.00
5	1	0	3.30	1	1	1	1	1	1.00	10.00	0.70	14.70
5	1	1	4.60	1	1	1	1	1	1.40	11.30	0.90	14.70
5	1	2	6.30	2	2	1	1	1	2.10	14.90	1.40	20.00
5	1	3	8.40	3	3	3	3	2	3.40	22.00	2.10	27.00
5	2	0	4.90	1	1	1	1	1	1.50	14.90	0.90	20.00
5	2	1	7.00	1	1	1	1	1	2.20	16.80	1.40	23.00
5	2	2	9.40	2	2	1	1	1	3.40	22.00	2.10	28.00
5	2	3	12.00	3	3	2	2	2	3.00	24.00	2.00	32.00
5	2	4	15.00	0	0	0	0	3	6.00	35.00	4.00	45.00
5	3	0	7.90	1	1	1	1	1	2.30	22.00	1.50	27.00
5	3	1	11.00	1	1	1	1	1	3.00	24.00	2.00	32.00
5	3	2	14.00	1	1	1	1	1	5.00	35.00	3.00	45.00
5	3	3	17.00	3	2	2	2	1	7.00	39.00	4.00	51.00
5	3	4	21.00	3	3	3	3	2	7.00	39.00	4.00	51.00
5	4	0	13.00	1	1	1	1	1	3.00	35.00	3.00	45.00
5	4	1	17.00	1	1	1	1	1	6.00	39.00	4.00	51.00
5	4	2	22.00	1	1	1	1	1	7.00	44.00	4.00	57.00
5	4	3	28.00	2	1	1	1	1	10.00	70.00	6.00	92.00
5	4	4	35.00	2	2	2	1	1	10.00	70.00	6.00	92.00
5	4	5	43.00	0	0	3	3	3	15.00	106.00	9.00	150.00
5	5	0	24.00	1	1	1	1	1	7.00	70.00	4.00	92.00
5	5	1	35.00	1	1	1	1	1	10.00	106.00	6.00	150.00

Number of positive results			MPN	Category when number of tests					Confidence limits			
				1	2	3	5	10	⩾ 95%	⩾ 95%	⩾ 99%	⩾ 99%
5	5	2	54.00	1	1	1	1	1	15.00	166.00	10.00	223.00
5	5	3	92.00	1	1	1	1	1	23.00	253.00	15.00	338.00
5	5	4	160.00	1	1	1	1	1	40.00	460.00	20.00	620.00
5	5	5	>160.00									

(c) Explanation of categories for results

Category	*Definition*
1	When the number of bacteria in the sample is equal to the MPN found, the result is one of those that have the greatest chance of being obtained. There is only at most a 5% chance of obtaining a result that is less likely than the least likely one in this category.
2	When the number of bacteria in the sample is equal to the MPN found, the result is one of those that have less chance of being obtained than even the least likely one in category 1, but there is only at most a 1% chance of obtaining a result that is less likely than the least likely one in this category.
3	When the number of bacteria in the sample is equal to the MPN found, the result is one of those that have less chance of being obtained than even the least likely one in category 2, but there is only at most a 0.1% chance of obtaining a result that is less likely that the least one in this category.
0	When the number of bacteria in the sample is equal to the MPN found, the result is one of those that have less chance of being obtained than even the least likely one in category 3. There is only a 0.1% chance of obtaining a result in this category, without anything being wrong.

Before starting to test it should be decided which category will be acceptable, i.e. only 1, 1 and 2, or even 1, 2 and 3. When the decision to be taken on the basis of the result is of great importance, only category 1, or at most 1 and 2 results should be accepted. Category 0 results should be considered with great suspicion.

Caution: The confidence limits given in the tables are meant only to provide some idea of the influence of statistical variations on results. There will always be other sources of variation, which may sometimes be even more important.

Table 4.10. Survey of composition of agar media used for fungi.

Acronym	*Ingredients (in g l^{-1} distilled water), final pH and comments on preparation**
ME agar	Malt extract (or biomalt extract) 10, agar 20–25 (depending on the quality to give a firm solid medium), adjust the pH to 5.5–5.6 by lactic acid solution (10%) before autoclaving
RBME agar	See Table 4.6
RBSCN agar	See Table 4.6
Czapek–Dox agar	Sucrose (= saccharose) 30, $NaNO_3$ 2.0, KCl 0.5, K_2HPO_4 1.0, $MgSO_4.7H_2O$ 0.5, $CaCl_2$ 0.1, $FeSO_4.7H_2O$ 0.01, agar 15, pH 7.2–7.3. After pH adjustment a precipitate may occur that is eliminated by filtration of the dissolved medium (glass wool) before autoclaving
PDRB agar	Dextrose 10, peptone 5.0, KH_2PO_4 1.0, $MgSO_4.7H_2O$ 0.5, rose bengal 0.035, streptomycin, 30 μg ml^{-1}, aureomycin 2 μg ml^{-1}, agar 15, pH 7.0. Preparation: all ingredients are boiled, bottled and autoclaved (15 min at 121°C); just prior to pouring plates, streptomycin and aureomycin are added from sterile aqueous solutions to the cooled (45–50°C) agar.
Sabouraud agar	Glucose 20, peptone 10, agar 17–20, pH is adjusted to 5.6 with 10% lactic acid solution before autoclaving
PGBRC agar	Glucose 10, peptone 5.0, KH_2PO_4 1.0, $MgSO_4.7H_2O$ 0.5, rose bengal 0.025, chloramphenicol 0.1, agar 15, pH 7.0–7.2. Preparation: chloramphenicol (dissolved in acetone) is added to the medium just prior to pouring plates

*Small modifications are included.

Discussion

To prepare an agar-solidified medium, first adjust the pH of the basal liquid medium to the desired value and add the agar (15–20 g l^{-1}). Different brands and grades of agar require different concentrations to achieve the required degree of firmness. The manufacturer's instructions should be consulted. The mixture is dissolved by boiling (using a hot plate, steamer or microwave oven), the pH checked (using indicator paper) and the contents dispensed with suitable dispensors into flasks or (reagent) tubes before autoclaving (15 min at 121°C). Tubes with 7 ml broth are sterile after only autoclaving for 1 min at 121°C, which saves time and the quality of the ingredients. If the pH of the agar medium is 6 or less (e.g. malt extract agar; Table 4.10) the medium should receive at least 20 g l^{-1} good quality agar before sterilization, because part of the agar will fail to solidify due to hydrolysing during autoclaving. In order to prevent agar hydrolysis by a low pH during autoclaving, the medium (with 15 g l^{-1} agar) is adjusted to a pH value above 6, subsequently autoclaved and the cooled portions (45–50°C) brought to the pH required by adding sterile lactic acid (10% solution) aseptically. The correct amount of acid should be determined previously in a separate experiment. For general details on laboratory safety, techniques for sterilization, microbial reagents, specific nutritional requirements, assay procedures, special growth media and conditions (anaerobic cultures), the reader is referred to the *Manual of Methods for General Microbiology* (Gerhardt et al 1981).

Table 4.11. Survey of liquid media used for the MPN technique.

Acronym	*Ingredients (in g l^{-1} distilled water), final pH and comments on preparation**
CPS broth	Starch 0.5, glycerol 1 ml, casein peptone 0.5, proteose peptone 0.5, K_2HPO_4 0.2, $MgSO_4.7H_2O$ 0.05, 0.01% $FeCl_3.6H_2O$ 0.2 ml; pH 7.4
PYGG broth	Proteose peptone 1.0, yeast extract 1.0, glycerol 5 ml; sodium glycerophosphate 0.5, pH 7.4
CLAGG broth	Glucose 0.5, sodium lactate 0.5, sodium citrate 0.5, sodium acetate 0.5, glycerol 0.5, casein 1.0, K_2HPO_4 3.0, $FeCl_3.6H_2O$ 0.02, pH 7.2
AG broth	Sodium acetate 1.0, L-glutamate (sodium salt) 0.5, Na_2HPO_4 3.0, KH_2PO_4 1.0, KNO_3 3.0, $MgSO_4.7H_2O$ 0.2, Fe(III)NH_4 citrate, Mn(II)-acetate, Na_2MoO_4 (traces), bromothymol blue 0.02, pH 7.1
DB broth	Dextrose 5.0, peptone 10, Na_2HPO_4 6.5, KH_2PO_4 2.0, brilliant green 0.0135, sodium dodecylsulphate 1.0, pH 7.0
MC broth	Lactose 10, peptone (from casein) 20, bile salts (dried) 5.0, bromocresol purple 0.01, pH 7.1
KAA broth	Casein peptone 20, yeast extract 5.0, sodium citrate 1.0, NaCl 5.0, Fe(III)NH_4 citrate 0.5, aesculin 1.0, sodium azide 0.15, pH 7.0
CPSB broth	Peptone from gelatin 16, casein hydrolysate 10, K_2SO_4 10, $MgCl_2$ 1.4, *N*-cetyl-*N,N,N*-trimethylammoniumbromide (cetrimide) 0.5, nalidixic acid 0.015, pH 7.1
MS solution	Stock solution: Na_2HPO_4 3.5, KH_2PO_4 0.7, $MgSO_4.7H_2O$ 0.1, $NaHCO_3$ 0.5, $CaCl_2.2H_2O$ 0.18, $FeCl_3.6H_2O$ 0.014, pH 7.8 (+ $(NH_4)_2SO_4$ 0.5 for ammonium oxidizers) (+ $NaNO_2$ 0.5 for nitrite oxidizers)
CY broth	Casamino acid 5.0, yeast extract 5.0, KNO_3 6.0, NaCl 3.5, pH 7.3
LPY broth	Calcium lactate 1.0, peptone (casein-tryptone) 10, yeast extract 0.5, KNO_3 5.0, K_2HPO_4 3.0, pH 7.0
CA broth	Sodium citrate 8.5, L-asparagine 1.0, KNO_3 3.0, KH_2PO_4 1.0, $MgSO_4.7H_2O$ 0.2, $CaCl_2.2H_2O$ 0.2, $FeCl_3.6H_2O$ 0.05, pH 7.4
LTS broth	Sodium lactate 5.0, yeast extract 1.0, K_2HPO_4 0.5, NH_4Cl 1.0, $CaCl_2.2H_2O$ 0.1, $MgSO_4.7H_2O$ 3.5, sodium thioglycolate 0.5, $Fe(NH_4)_2SO_4$ 0.1, agar 2.0, pH 7.2
GAF broth	Glucose 20, asparagine 5.0, yeast extract 0.5, K_2HPO_4 3.0, KH_2PO_4 0.8, KCl 0.2, $MgSO_4.7H_2O$ 0.2, $Fe_2O_3.3H_2O$ powder (Merck) 1.0, pH 7.0. The broth is stirred constantly (magnetic stirrer) while dispensing the iron oxide suspension in 7 ml portions; Fe(II) formation (after 7 days at 30°C) is detected by a red colour after adding about 1 ml of an 0.2% 2,2-dipyridyl solution in 10% acetic acid
NGF broth	Glucose 20, sodium acetate 5.0, K_2HPO_4 0.25, KH_2PO_4 0.25, $MgSO_4.7H_2O$ 0.05, $Na_2MoO_4.2H_2O$ (trace), $Fe_2O_3.3H_2O$ powder (Merck) 1.0, vitamin solution (commercially available) 5 ml, pH 7.0. For preparation and detection of Fe(II) formation, see GAF broth
CM broth	Cellulose (filter paper strip), yeast extract 0.02, K_2HPO_4 0.5, KH_2PO_4 0.3, $MgSO_4.7H_2O$ 0.2, $Fe_2(SO_4)_3.7H_2O$ (trace), pH 7.2. A strip of filter paper is introduced into each tube; degradation of the filter paper strip in the standing tubes after shaking the cultures by hand at 1 week intervals for 5 weeks is considered as positive

Table 4.11. Continued.

Acronym	*Ingredients (in g l^{-1} distilled water), final pH and comments on preparation**
MYG broth	D,L-Malic acid 5.0, yeast extract (Difco) 0.1, glutamic acid 0.07, K_2HPO_4 6.0, KH_2PO_4 4.0, $MgSO_4.7H_2O$ 0.2, NaCl 0.1, $CaCl_2$ 0.02, $FeCl_3$, $Na_2MoO_4.2H_2O$, $MnSO_4$, H_3BO_3, $Cu(NO_3)_2.3H_2O$, $ZnSO_4.7H_2O$ (traces), agar (Difco Noble) 1.75, pH adjusted to 6.8. The boiled medium is dispensed into tubes and sterilized (1 min at 121°C); the inoculated tubes are incubated (48 h at 30°C) and evaluated (for clear subsurface pellicle formation). The cotton plugs of positive tubes are exchanged for rubber serum stoppers and acetylene gas (12% v/v) is added. The acetylene reduction assay (ARA) is measured by gas chromatography after 24 h at 30°C and tubes containing more than 6 nmol C_2H_4 per tube are considered as nitrogenase positive

* Compared to the original composition, small modifications may occur.

Microscopic methods for counting bacteria and fungi in soil

J. Bloem
P.R. Bolhuis
M.R. Veninga
J. Wieringa

Since 1970 epifluorescence microscopy has become the major technique for direct enumeration of bacteria and fungi in soil. In principle, a known amount of homogenized soil suspension is placed on a known area of a microscopic slide, the microorganisms are stained with a fluorescent dye and numbers are counted using a microscope. Biovolumes and biomass can be estimated from lengths and widths. The frequency of dividing cells (FDC), i.e. the percentage of cells showing an invagination, can be used as an index of the *in situ* specific growth rate of soil bacteria (Bloem et al 1992a).

Metabolically active fungal hyphae can be estimated after staining with fluorescein diacetate (FDA) which becomes fluorescent after enzymatic hydrolysis (Söderström 1977).

Extensive reviews of microscopical techniques in microbial ecology have recently been given by Fry (1990), Frankland et al (1990) and Gray (1990). Fry (1990) concluded that a variety of preparation and staining methods have been used for direct counts of soil bacteria and that a consensus has not yet emerged. For example, homogenization of soil suspensions has been performed using a Waring blender at speeds of up to 23,000 rev min^{-1} for periods of 1–16 min, but also the more vigorous Ultra-Turrax tissue-homogenizer (24,000 rev min^{-1} for 10 min) has been suggested to produce finer suspensions with very even bacterial distributions. Moreover, different sonication regimes have been used. Homogenization of soil suspensions has been performed in particle-free water with and without additions such as sodium pyrophosphate, sodium hexametaphosphate (Calgon) and Tween 80. These detergents and deflocullents have been recommended for preventing reaggregation of soil particles and bacterial clumps. Bakken (1985), however, found no differences in bacterial numbers between different dilution media and distilled water.

Soil bacteria have been counted in soil smears or in soil collected on membrane filters, using several stains such as fluorescein isothiocyanate (FITC, which stains proteins), acridine orange (AO, which stains nucleic acids), ethidium bromide (EB, which stains nucleic acids) and europium chelate/fluorescent brightener differential stain (DFS). Europium chelate stains DNA and RNA, and fluorescent brightener stains polysaccharides (cell walls) (Gray 1990). Only a few quantitative comparisons of stains have been published. Wynn-Williams (1985) obtained similar counts with AO and FITC. Jenkinson and Ladd (1981) considered FITC to be one of the best stains for soil work and mentioned non-specific staining and rapid fading as disadvantages of AO. Anderson and Slinger (1975a, 1975b) reported higher counts with DFS than with FITC.

Fungal hyphae have been counted in agar films (Lodge and Ingham 1991) or on membrane filters using fluorescent brightener, phenol aniline blue (PAB) or phase contrast microscopy. West (1988) compared different methods and recommended the membrane filtration method with fluorescent brightener.

Microscopic counting of microbial numbers

works reasonably quickly (15–30 min per slide) but it is subjective because of the presence of weakly stained cells. Visual measurement of bacterial sizes and counts of dividing cells, however, are very inaccurate and time consuming (1 h per slide). In aquatic samples, measurements of bacterial numbers and volumes have been improved greatly by using video microscopy and (semi-)automatic image analysis (Sieracki et al 1985; Bjørnsen 1986). Using the system described by Bjørnsen (1986) consisting of a low-light video camera on an epifluorescence microscope and an IBAS image analysis system (Kontron), it was also possible to measure bacterial sizes in soil smears (Bloem et al 1992a). Because of the thickness of the smears (about 3 μm) and the limited depth of focus of a conventional microscope, however, it was not possible to focus all cells simultaneously. Therefore, we presently use a confocal laser-scanning microscope (CLSM, Leica Lasertechnik GmbH, Heidelberg), which offers an extended depth of focus by combining a series of optical slices into one image (Brakenhoff et al 1989). Moreover, contrast is better due to the suppression of stray light from outside the focal plane, and less bleaching of the fluorescent dye occurs due to the short illumination time. Using the CLSM combined with a Quantimet 570 image analysis system (Leica Cambridge Ltd, Cambridge), it is possible to measure cell numbers, lengths and widths and also dividing cells automatically without human intervention (Bloem et al 1994a). An automated image analysis method to determine fungal biomass in soil has recently been described by Morgan et al (1991).

In this section, we compare different methods for preparing and staining soil bacteria as reviewed by Fry (1990) and Gray (1990). Although our measurements were performed with a CLSM and automatic image analysis, the results also apply to conventional visual microscopy because the techniques are the same. Methods with which we have gained good experience are described in detail: measurement of total bacterial numbers, volumes and FDC in soil smears stained with 5-(4,6-dichlorotriazin-2-yl) aminofluorescein (DTAF), (an FITC analogue that stains proteins), and measurement of total and active fungal hyphae on membrane filters with fluorescent brightener and FDA, respectively. For methods of studying specific organisms by immunofluorescence and studying organisms in relation to soil structure (thin soil sections), see Postma et al, (1988), Altemüller and Van Vliet-Lande (1990), Postma and Altemüller (1990), and Gray (1990), and Chapter 9.

Preparation of soil suspensions

Principle of the method

Soil is homogenized in sterilized water. The soil suspension is then treated with formaldehyde to fix bacterial cells and fungal hyphae.

Materials and apparatus

Balance
Plastic beakers (100 ml) and spoon for weighing 20 g soil samples
Graduated cylinder, 250 ml
Blender (Waring)
Timer
Adjustable volume pipettes with 1 and 5 ml tips
Capped polypropylene tubes (13 ml) in a rack for storage of fixed soil suspensions
Vortex mixer

Chemicals and solutions

Fresh demineralized and filter-sterilized (0.2 μm pore size) water
Filter-sterilized formaldehyde solution, 37%, (0.2 μm pore size)

Procedure

At least three soil samples are taken per treatment, e.g. from three 2 × 2 m subplots. From each subplot 15 soil cores

(3.5 cm diameter) are bulked in a plastic bag. Usually the samples have to be stored overnight at 2°C.

Of each hand-mixed sample, 20 g of soil and 190 ml demineralized and filter-sterilized water are homogenized in a blender for 1 min at maximum speed (20,000 rev min^{-1}). Nine millilitres of the soil suspension is fixed by adding 1 ml formalin (3.7% final concentration) and mixing on the vortex. Thus, the suspension contains 90 mg soil ml^{-1}. Inhalation of and skin contact with formaldehyde must be prevented.

Discussion

Effects of homogenization on bacteria: we compared the effects of different treatments using sand (Tynaarlo, 3.1% clay, 11.4% organic matter, pH-KCl 4.3), loam (Lovinkhoeve, 20.4% clay, 2.2% organic matter, pH-KCl 7.3) and clay (Finster wolde, 45.8% clay, 5.2% organic matter, pH-KCl 7.1) soils. Characteristics of the soils have been described by Kooistra et al (1989) and Hassink (1993). Bacteria were measured by automatic image analysis in DTAF-stained soil smears. The statistical significance of effects of treatments was tested by analysis of variance and means were compared using Student's *t*-tests. All differences reported are significant at the $P < 0.05$ level.

Homogenization of sand in 0.3% (w/v) sodium pyrophosphate in a Waring blender (Waring, New Hartford, CO) was compared with sonication (MSE Soniprep 150, MSE Scientific Instruments, Crawley, West Sussex) and with mixing in a laboratory multi-purpose mixer emulsifier (Silverson Machines Ltd, Waterside, Chesham, Bucks). The Silverson mixer was used because the similar Ultra-Turrax mixer (Fry 1990) jammed in the sand suspension. Blending at maximum speed (20,000 rev min^{-1}, 20 g soil and 190 ml water) for 0.5, 1, 2 × 1 and 4 × 1 min with intermediate cooling on ice yielded similar numbers (Fig. 4.2). However, 8 × 1 min blending yielded significantly higher numbers than the other treatments. Bacterial numbers were drastically reduced by mixing at maximum speed (15,000 rev min^{-1}, tubular processing head with a diameter of 19 mm, 2 g soil and 19 ml water) for 10 min (on ice), and by 2 and 5 min sonication (maximum power, amplitude 14 µm, small probe with a diameter of 9.5 mm, 2 g soil and 19 ml water, cooled on ice). The coefficient of variation (CV) was on average 35% and was not even reduced by the most drastic homogenization treatment. The mean cell volume and the FDC of bacteria were not affected by the homogenization treatments. In an earlier experiment with the loam soil, we found no significant differences between 1 and 5 × 1 min blending.

Effects of homogenization on fungal hyphae: we did not check effects of homogenization on fungal hyphae. Using an MSE-HR blender at half speed, West (1988) found the greatest total hyphal lengths with 1 min blending, but there were no significant differences between 0.5 and 4 min blending. Faegri et al (1977) used a Waring blender for 3 × 1 min and found no decrease in fungal respiration after

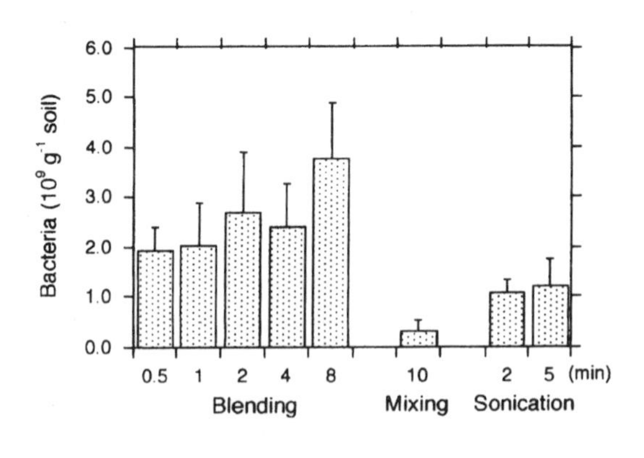

Figure 4.2 Effects of homogenization with a Waring blender, a Silverson mixer emulsifier and an MSE sonifier on bacterial numbers counted in a sandy soil. Error bars indicate standard deviation (SD) $n = 5$. Least significant difference (LSD) at $P = 0.05 = 1.23 \times 10^9$ bacteria per gram of soil.

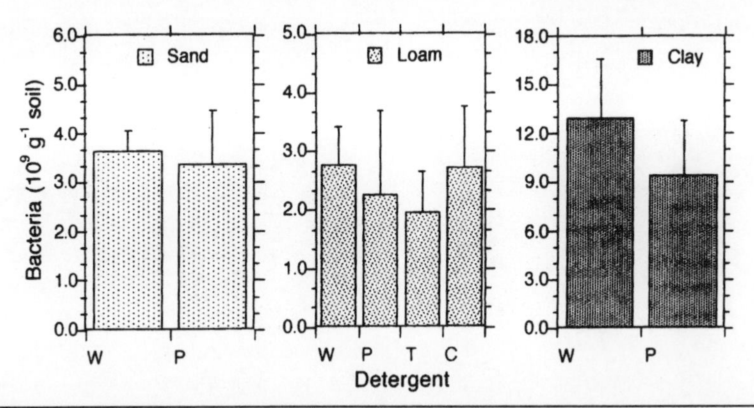

Figure 4.3 Effects of detergents on bacterial numbers counted in sand, loam and clay. W = water; P = 0.3% (w/v) sodium pyrophosphate; T = 0.3% sodium pyrophosphate and 0.5% Tween 80; C = Calgon = 0.3% sodium pyrophosphate and 0.5% sodium hexametaphosphate buffered to pH 8 with Na_2CO_3. Error bars indicate SD, $n = 5$. LDSs are: 1.23, 1.23 and 5.21×10^9 bacteria per gram of soil for sand, loam and clay, respectively.

blending. Ingham and Klein (1982) checked the effect of a Waring blender at maximum speed for 2 min on batch cultures and found reductions of about 10% and 20% for total and FDA-active fungal hyphae, respectively, compared to unblended samples. Söderström (1979) used an MSE ATO-mix at half speed (6000 rev min^{-1}) and found 2 min dispersion to be the optimum for FDA-active hyphae, whereas 5 min yielded 30% lower counts. At 2 min, 20% of pure culture FDA-active biomass added to sterile soil was still destroyed. Thus similar results were obtained with the MSE ATO-mix and the Waring blender. When large amounts of samples must be processed, it is practical to use the same homogenized soil suspension for bacterial and fungal counts. Then a Waring blender can be used for 1 min at maximum speed. Although 8 × 1 min blending may yield higher bacterial numbers (Fig. 4.2), we find it too time consuming for routine use.

Effects of detergents and storage: Using the Waring blender at maximum speed for 1 min, dilution and homogenization in 0.3% sodium pyrophosphate, 0.3% sodium pyrophosphate and 0.5% sodium hexametaphosphate buffered to pH 8.5 with Na_2CO_3 (Calgon), or 0.3% sodium pyrophosphate and 0.5% Tween 80, neither increased the counts nor reduced the variation compared to homogenization in pure water (Fig. 4.3). Similar results were obtained by Babiuk and Paul (1970) and Bakken (1985). Thus, detergents and deflocculents did not improve total direct counts in any of the soils tested.

We counted bacteria in fixed suspensions of the loam soil 0, 1, 3 and 6 weeks after fixation with 3.7% formaldehyde and found significantly lower numbers and mean cell volumes after 3 weeks storage at 2°C. It is safest to make the preparations as soon as possible after fixation and to store the slides.

Direct count of bacteria

Principle of the method

Formaldehyde-fixed soil suspension is smeared on a glass slide. After air-drying, the bacteria on the slide we stained with a fluorescent dye which binds to proteins in bacterial cell walls and cytoplasm. The bacteria are counted using an epifluorescence microscope.

Materials and apparatus

Self-adhesive plastic tape (4 cm × 25 m) with removable non-adhesive backing, normally used to cover books.
Punch (12 mm diameter) to punch holes in the tape.
Glass slides and cover slips. The microscope slides are precleaned with 70% ethanol to remove hydrophobic material from the glass surface, and are sealed with pieces of tape (4 × 2 cm) with a hole of 12 mm diameter in the centre. Alternatively, printed microscope slides (Bellco Glass Inc, Vineland NJ, or Cel-line Associates Inc, Newfield NJ can be used.
Knife to remove tape from microscope slides.
Pipette with 10 μl tips.
Plastic trays with cover.
Baths with slide holders for rinsing.
Microscopic slide files for storage of slides.
Epifluorescence microscope equipped with an ocular counting grid and a filter set for blue light (BP 450–490 nm exciter filter, 510 nm beam splitter and LP 520 nm barrier filter).

Chemicals and solutions

Buffer solution consisting of 0.05 M Na_2HPO_4 (7.8 g l^{-1}) and 0.85% NaCl (8.5 g l^{-1}), adjusted to pH 9.

Stain solution consisting of 2 mg DTAF (Sigma Chemical Company, St Louis, MO) dissolved in 10 ml of the buffer (Sherr et al 1987). Skin contact and inhalation of all stains (powder) must be avoided.

All solutions are filtered through a 0.2 μm membrane before use. The stain solution should not be stored for longer than a day.

Immersion oil.

Nail varnish.

Procedure

The fixed soil suspensions are resuspended on a Vortex mixer. Some soils require a further 3–10-fold dilution (Discussion). After 2 min of settling to remove coarse particles, 10 μl soil suspension (0.9 mg soil) is evenly smeared in the hole of the tape on a glass slide (Babiuk and Paul 1970). The water-repellent tape keeps the suspension in a defined area of 113 mm^2. The slides are placed on tissue in a plastic tray. After air-drying the tape is peeled off. The spots of dried soil film are flooded by drops of stain solution for 30 min at room temperature. To prevent drying, the tissue is moistened before and the trays are covered during staining. Next, the slides are rinsed three times for 20 min with buffer, and finally for a few seconds with water, by putting them in slide holders and passing them through four baths. After air-drying, a cover slip is mounted with a small drop of immersion oil and the edges are sealed with nail varnish. The slides can be stored at 2°C until observation.

Counts are performed using an epifluorescence microscope at 1000–1250 × magnification in a dark room (Ploem and Tanke 1987). Usually it is sufficient to count bacteria in 10 fields of view (enclosed by an ocular grid). The fields should cover the soil smear randomly, e.g. by selecting fields along two central transects at right angles (Fry 1990). The specimen must not be observed while fields are being changed. Besides total numbers, cell lengths and widths can be estimated using an ocular micrometer and dividing cells can be counted as an index of the *in situ* specific growth rate. It is convenient to use a PC with appropriate software beside the microscope for tallying counts and for instantaneous calculation of results and statistics (Bloem et al 1992b)

Calculation

Bacterial numbers per gram of soil B are calculated from

$$B = (N/X)\ (A/B)\ (1/S) \qquad (4.8)$$

where N is the number of bacteria counted, X is the number of fields of view (grids) counted, A is the area of the slide covered by sample (113 mm^2), B is the area of the field of view (e.g. $0.048 \times 0.048 = 2.304 \times 10^{-3}$ mm^2), depending on microscope and magnification, to be measured with an object stage micrometer, and S is the amount of soil smeared on the slide (usually 0.9 mg).

Biovolumes V (μm^3) can be calculated from length (L, μm) and width (W, μm) using the equation

$$V = (\pi/4)\, W^2\, (L - W/3) \tag{4.9}$$

(Krambeck et al 1981). The amount of carbon can be estimated from the biovolume using a specific carbon content of 3.1×10^{-13} g C μm^{-3} (Fry 1990).

Discussion

Preparation of slides: It is not possible to give one universal method for all soils and ideally any technique should be optimized for any soil. An important difference between different soils is the amount of soil that can be used per square centimetre of slide. A dilution which gives at least 25 cells per field of view is desirable (Bloem et al 1992b). However, too much soil per square centimetre results in severe masking of bacteria by soil particles. In this case the linear relationship between the amount of soil on the slide and the number of bacteria counted is lost. For our sand and loam soils 0.8 mg soil cm^{-2} was found to be the maximum, whereas for clay the maximum was 0.27 mg cm^{-2} (Fig. 4.4). Faegri et al (1977) reported maxima of 0.08 and 0.4 mg cm^{-2} for an organic subalpine soil (88.5% organic matter) and the A_0 horizon of an alpine podzol (80% organic matter), respectively.

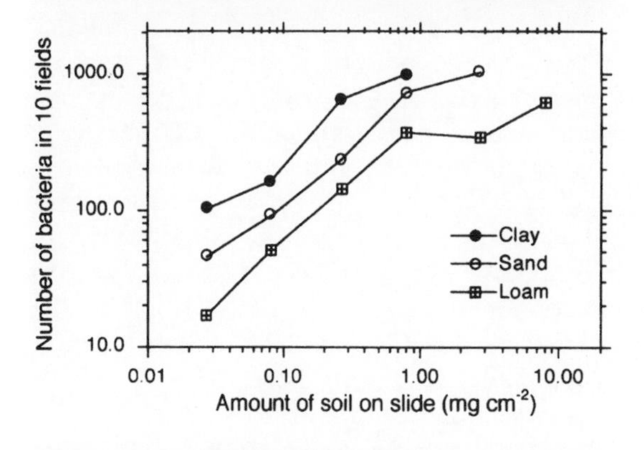

Figure 4.4 Relationship between the amount of soil per square centimetre slide and the number of bacteria counted in 10 fields of view, for clay, sand and loam. Points represent means of three replicates.

Even in sufficiently diluted soil suspensions some masking will occur. Although methods have been developed to correct for masking of bacteria by particles (Fry 1990), they do not seem to be very practical for routine counting. Babiuk and Paul (1970) added known amounts of bacteria to soils with clay contents ranging from 20% to 80% and recovered 94.5% to 101.3% of the added bacteria in soil smears stained with FITC. This indicates that the presence of soil in the preparations is not a serious problem but it gives no information about underestimation of native bacteria enclosed in soil aggregates. The observation that detergents and deflocculents did not increase the number of bacteria counted (Fig. 4.3) may indicate that underestimation due to aggregates is limited. On the other hand, homogenization treatments which completely disrupt aggregates will also destroy bacteria (Fig. 4.2). We compared 0.5, 2, 4 and 8 min settling periods for suspensions of all three soils and found no significant effect of settling time on bacterial numbers and on the variance between replicates.

Both membrane filters and soil smears have been used for counting bacteria. When we compared 10 black 0.2 μm pore-size polycarbonate filters (Nucleopore) with 10 smears of the same sandy soil, we found similar numbers on the filters ($2.78 \pm 1.4 \times 10^9$ bacteria per gram of soil, mean $\pm$ SD) and the smears ($2.37 \pm 0.47 \times 10^9$

bacteria per gram of soil). We prefer soil smears because they are very flat and show little background staining compared to filters. In smears we can measure practically all bacteria with the CLSM by scanning seven subsequent optical slices of 0.5 μm thickness. Thus, a layer of 3.5 μm can be measured automatically without intermediate focusing between fields. The filters had to be focused by hand on each new field. We also tried aluminium oxide membrane filters (Anotec Separations Ltd, Banbury, Oxon), which are flatter than polycarbonate filters (Jones et al 1989). These filters, however, are enclosed by a plastic ring which prevents the preparation of very thin slides.

In contrast to fixed soil suspensions, DTAF-stained soil smears can be stored at 2°C in the dark for at least a year. After 1 year we did not find significant differences in bacterial numbers, cell volumes, length to width ratios and FDCs of 20 soil smears of sand, loam and clay. Means after 0 and 1 year storage were: 2.24 and 2.23 $\times 10^9$ cells g^{-1} soil, 0.21 and 0.23 μm^3 $cell^{-1}$, 1.51 and 1.47 (L/W) and 5.47 and 5.47% (FDC), respectively. Since these identical values were obtained using fixed settings of laser power, pinhole and photomultiplier (= detector) sensitivity, the fluorescence intensity had not significantly decreased after storage. Filter preparations stained with DTAF or FITC could not be stored for longer than a few days because the background fluorescence increased so much that bacteria could not be distinguished anymore.

Comparison of stains: for total bacterial counts in aquatic samples AO and DAPI are the best stains. The position for soil is less clear (Fry 1990). The first and most commonly used fluorescent stains for soil bacteria are FITC (Babiuk and Paul 1970) and AO (Trolldenier 1973). According to Gray (1990), dead cells lose their ability to be stained with FITC after 2 days in soil. FITC was also used first in our laboratory but because sometimes staining failed for unknown reasons it was replaced by DFS (Anderson and Slinger 1975a, 1975b). In our experience DFS is a useful stain to count bacteria in soil. In order to avoid a high level of blanks (counts in preparations without soil added) due to precipitation of europium chelate, it is better to prepare the DFS solution 1 day before use and to filter the stain through a 0.2 μm pore size membrane immediately before use. With all stains blanks should be checked regularly. When we started to use the CLSM, it appeared that the europium chelate in DFS could not be detected by the laser-scanning microscope, because europium chelate is a phosphorescent dye (Davidson and Hilchenbach 1990). This means that light emission starts microseconds after excitation by the scanning laser beam. Because excitation and detection occur simultaneously, the signal does not reach the detector in time and only fluorescent dyes, which respond in nanoseconds, can be used. As we needed an alternative for DFS and we did not want to return to FITC, we tested the applicability of several fluorochromes for counting bacteria in soil smears.

The nucleic acid stains DAPI and Hoechst 33258 have been recommended for counting bacteria attached to particles in aquatic samples (Paul 1982). Although we tried these stains at various concentrations, pH levels and staining times, in all cases the soil (loam) stained so yellow that hardly any bacteria could be distinguished. Very good images were obtained with the protein stain DTAF, which has binding, absorption and emission properties nearly identical to FITC, but is superior with regard to purity and stability (Blakeslee and Baines 1976). DTAF has been used to label bacteria for quantification of uptake by protozoa (Bloem et al 1989). In a comparison of DTAF and DFS by visual counts of nine samples, we found no significant difference in bacterial numbers, which were on average 13% lower with DTAF. Using automatic

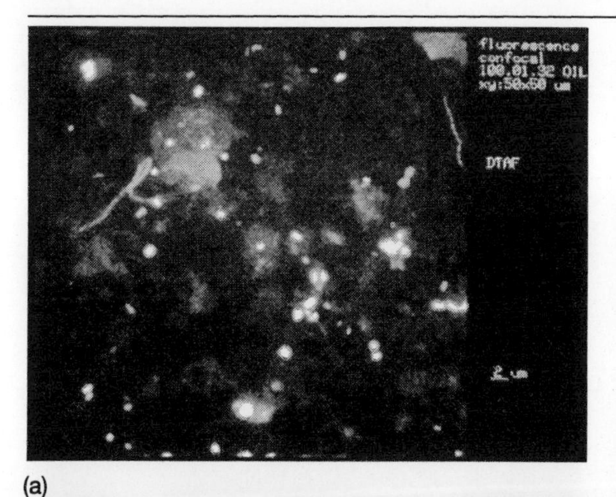

(a)

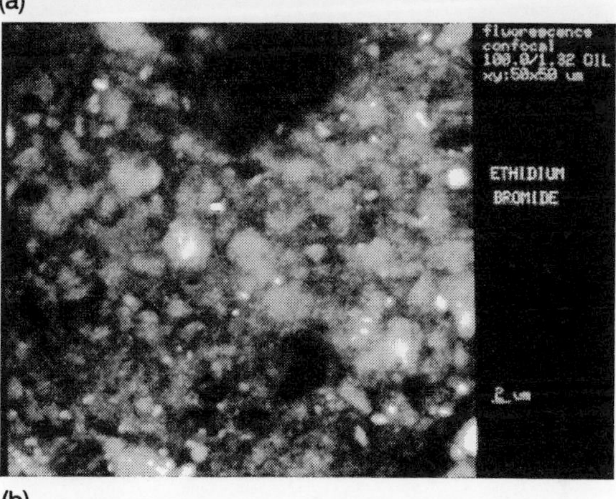

(b)

Figure 4.5 Images of soil smears stained with (a) 5-(4,6-dichlorotriazin-2-yl) aminofluorescein (DTAF) and (b) ethidium bromide (EB), obtained with a confocal laser-scanning microscope. Both smears were prepared from the same suspension of a loam soil. The small bright spots (≤2 μm diameter) are assumed to be bacteria.

image analysis we compared DTAF (200 mg l^{-1} for 30 min, as described above) with FITC (400 mg l^{-1} for 3 min, after Babiuk and Paul 1970), AO (30 mg l^{-1} for 5 min, rinsed 8 × 3 min, after Wynn-Williams 1985), EB (250 mg l^{-1} for 5 min, after Roser 1980) and propidium iodide (PI, 250 mg l^{-1} for 5 min). PI is a nucleic acid stain similar to EB. With each stain 10 smears were prepared from the same suspension of a loam soil. Images with good contrast and little background staining were obtained with FITC and DTAF (Fig. 4.5a). With AO, EB and PI, non-specific background staining of soil particles was much higher and contrast was less (Fig. 4.5b). In contrast with the other stains, AO-stained slides could not be mounted with immersion oil (several types) or glycerol because the originally non-fluorescent oil seemed to dissolve some stain resulting in a bright green background which made counting impossible. Therefore, the slides were immersed in water with the anti-fading agent 1,4-diazobicyclo-2,2,2-octane (DABCO, 5 g l^{-1} final concentration). The latter was added because the AO-stained slides bleached rapidly. Later we found that the bright green background fluorescence with AO did not occur when Cargille (type B) immersion oil was used (R.P. Cargille Laboratories Inc., Cedar Grove, NJ).

DTAF, FITC and AO were excited at 488 nm with an Omnichrome 155T argon ion laser (2–50 mW, Omnichrome, Chino, CA). A beam splitter of 510 nm and a barrier filter of 530 nm were used. EB and PI were excited at 514 nm using a beam splitter of 580 nm and a barrier filter of 590 nm. For all stains, the optimum contrast was obtained at the same laser power (95,000) and photomultiplier sensitivity (700 V). This resulted in video images with grey values between 0 (black) and 255 (white). After image enhancement, the image was segmented to a binary image (0 = black background; 1 = white object) using a fixed threshold at grey level 25. Particles in binary images were measured at a pixel size of 0.098 μm in 10 fields (46.7 × 44.2 μm) per slide. The fields were scanned using a motorized *x–y* stage controlled by the image analysis program. The scan pattern was defined by three steps of 1200 μm in the *x*-direction and four steps of 1000 μm in the *y*-direction. Thus, 12 fields were defined and the first 10 fields were measured.

Bacterial numbers were similar with AO, DTAF and FITC, but significantly higher with EB and PI (Fig. 4.6). EB and PI also yielded significantly higher (apparent) FDCs and a higher average length to width ratio.

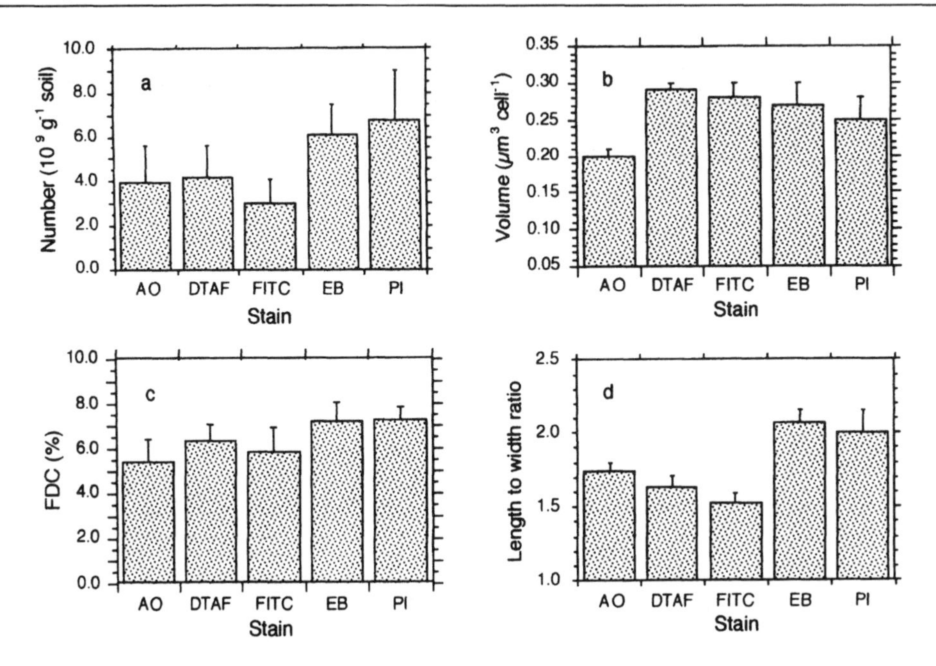

Figure 4.6 Bacterial numbers (a), mean cell volumes (b), frequencies of dividing cells (FDC) (c) and length to width ratios (d) in soil smears from one suspension of a loam soil as measured after staining with acridine orange (AO), 5-(4,6-dichlorotriazin-2-yl) aminofluorescein (DTAF), fluorescein isothiocyanate (FITC), ethidium bromide (EB) and propidium iodide (PI). Error bars indicate SD, $n = 10$. LSDs are: (a) 1.45×10^9 bacteria per gram of soil; (b) 0.019 μm^3; (c) 0.78%; (d) 0.088.

Given the high background staining, the different results obtained with EB and PI may have been due to measurement of non-bacterial particles. It is not clear why the mean cell volumes were lower with AO than with the other stains. It may be possible to match the results of different stains by counting only brighter particles when the background is high. However, this increases the subjectivity especially in visual counts. It is better to use a stain with a low background like FITC and DTAF. We recommend DTAF because of our good experiences in various soils and litter, and because of the problems with FITC in the past. These problems may have been caused by great variations in purity and labelling capacity both between manufacturers and lots (Blakeslee and Baines 1976).

Estimates of activity

For aquatic bacteria, several methods for counting active cells have been published but no single best method has emerged yet (Fry 1990). For bacteria in soil suspensions, the FDA method has been used to stain cells showing esterase activity, and the redox dye 2-(*p*-iodophenyl)-3-(*p*-nitrophenyl)-5-phenyl tetrazolium chloride (INT) has been used to stain actively respiring cells. A disadvantage of the FDA method and also of other viability probes (Morgan et al 1993) is that it does not work well with gram-negative bacteria due to the low permeability of the cell wall.

We have tried to optimize the use of the redox probe 5-cyano-2,3-ditolyl tetrazolium chloride (CTC, Rodriguez et al 1992) for bacteria in soil suspensions. This method is identical to the INT method except that the reduced intracellular formazan deposits are fluorescent, which enhances detection and sensitivity. Incubation of soil suspensions with 5 mM CTC for 4 h at 20°C indeed yielded fluorescently stained (respiring) cells, whereas prekilled blanks showed no fluorescent particles. However, when soil was enriched with tryptone soya broth, we found no increase in CTC-stained cells, while the total numbers increased from 0.64 to 7.0 × 10^9 bacteria per gram of soil in 2 days. This suggests that the growing population was not able to take up CTC. Thus, the CTC method seems to have limitations similar to those for the FDA method.

During the exponential growth between day 0 and day 2, the average cell volume increased from 0.32 to 0.44 μm^3 and the FDC increased from 4.0% to 9.6%. In contrast with the previous methods, the FDC method does not require incubation and can be determined in the same slides that are used for total counts. We have observed significant increases in FDC after rewetting of dried soil (Bloem et al 1992a), and after harvest and soil tillage in winter wheat fields (Bloem et al 1994b).

Measurement of fungi

Total hyphal length

Principle of the method

Formaldehyde fixed soil suspension is mixed with fluorescent brightener which stains polysaccharides in cell walls. The suspension is filtered through a black membrane filter. The total length of fungal hyphae on the filter surface is estimated using an epifluorescence microscope.

Materials and apparatus

Test tubes in a rack.
Adjustable volume pipettes with tips of 1 and 5 ml.
Membrane filtration unit, e.g. with an effective filter area of 201 mm^2 (16 mm diameter).
Black 0.8 μm pore size 25 mm diameter polycarbonate membrane filters (e.g. Nuclepore no. 110659).
Microscopic slides and cover slips.
Epifluorescence microscope equipped with an ocular counting grid (10 × 10) and a filter set for UV illumination (BP 340–380 nm exciter filter, 400 nm beam splitter and LP 430 nm barrier).

Chemicals and solutions

Stain solution consisting of 2 g fluorescent brightener (calcofluor white M2R, e.g. Sigma no. F 6259) per litre of demineralized, sterile filtered water.
Demineralized sterile filtered water.
Immersion oil (Cargille type B).

Procedure

The fixed soil suspension (90 mg soil ml^{-1}) is resuspended and diluted with water in three steps: 1:5, 1:5 and 1:5, resulting in 5 ml suspension containing 3.6 mg soil. Five millilitres of the fluorescent brightener solution is added (1 mg ml^{-1} final concentration) (West 1988; Morgan et al 1991). After 2 h of staining at room temperature in the dark, 4 ml of the suspension (1.44 mg soil) is filtered through a black 0.8 μm pore size polycarbonate filter. The filter is rinsed three times with 3 ml water and mounted on a slide with immersion oil. After 2 months of storage at 2°C in the dark, we observed no visible decrease in fluorescence and contrast. In two slides we counted 14 and 26 metre hyphae g^{-1} soil at time 0, and 27 and 24 metre g^{-1} after 2 months.

The slides are observed in a dark room under an epifluorescence microscope with UV illumination at 400× magnification. In one or two transects over the filter (about 50–100 fields) hyphal lengths are estimated by counting the number of intersections of hyphae with (all) the lines of the counting grid (Fry 1990).

Calculation

The hyphal length H (μm $grid^{-1}$) is calculated as:

$$H = I \pi A/2L \quad (4.10)$$

where I is the number of intersections per grid, A is the grid area (e.g. 233 × 233 = 54,289 μm^2) and L is the total length of lines in the counting grid (e.g. 22 lines × 233 μm = 5126 μm).

The total length of fungal hyphae, F (m g^{-1} soil) is calculated as:

$$F = H\, 10^{-6} (A/B) (1/S) \quad (4.11)$$

where H is the hyphal length (μm $grid^{-1}$), 10^{-6} is the conversion of μm to m, A is the area of the filter covered by sample (e.g. 201 mm^2), B is the area of the grid (e.g. 0.053 mm^2), and S is the amount of soil on the filter (e.g. 1.44 mg).

Biovolumes V (μm^3) can be calculated from length (L, μm) and width (W, μm) using the equation $V = (\pi/4)\, W^2\, (L-W/3)$.

The amount of carbon can be estimated from the biovolume using a specific carbon content of 1.3×10^{-13} g C μm^{-3} (Van Veen and Paul 1979; Bakken and Olsen 1983).

FDA-active fungal hyphae

Principle of the method

Fresh soil suspension is mixed with fluorescein diacetate (FDA). In metabolically active fungal hyphae FDA is enzymatically hydrolyzed and becomes fluorescent. The suspension is filtered through a black membrane filter. The length of FDA-active fungal hyphae on the filter surface is estimated using an epifluorescence microscope.

Materials and apparatus

Same as for total fungal hyphae except that the epifluorescence microscope must be equipped with a filter set for blue illumination (BP 450–490 nm exciter filter, 510 nm beam splitter and LP 520 nm barrier filter).

Chemicals and solutions

Same as for total fungal hyphae with the following modifications. The stain solution consists of 2 mg of FDA per ml acetone. This stock solution is stored at −20°C. Before use the stock solution is diluted 1:18 with demineralized filter-sterilized water to 111 μg ml^{-1}.

Phosphate buffer pH 7.5 made of 800 ml of solution A and 200 ml of solution B, where A is 11.9 g l^{-1} $Na_2HPO_4.2H_2O$ and B is 9.08 g l^{-1} KH_2PO_4.

Procedure

A fresh soil suspension is prepared by homogenizing 15 g soil with 190 ml filter-sterilized buffer for 1 min at maximum speed (20,000 rev min^{-1}) in a blender. The suspension (75 mg soil ml^{-1}) is diluted with buffer in three steps: 1:10, 1:5 and 1:5, resulting in 5 ml suspension containing 1.5 mg soil. Then, 0.5 ml FDA solution is added (10 μg ml^{-1} final concentration) and the suspension is mixed thoroughly (Söderström 1977). After 3 min the suspension is filtered through a black 0.8 μm pore size polycarbonate filter. The filter is mounted on a slide with immersion oil and observed immediately under an epifluorescence microscope with blue illumination.

Calculation

Same as for total hyphae.

Discussion

Dilution: the most appropriate dilution depends on the soil type. In the soils we have tested, a dilution resulting in 0.3–0.8 mg soil cm^{-2} slide area was sufficient to prevent serious masking of bacteria by soil particles (Fig. 4.3). West (1988) reported similar maximum values of 0.4 and 0.8 mg cm^{-2} for fungal counts in two silt loam soils.

Magnification: estimates of hyphal length have been performed at magnifications ranging from 100 to 1250×. Magnifications above 800× have been reported to give two-fold higher estimates than lower magnifications (Bååth and Söderström 1980; West 1988). When fungal densities are relatively low (c. 40 m g^{-1} soil), however, high magnifications increase the probability of missing hyphae and too many fields have to be counted. In such samples, Ingham and Klein (1984) found 2–7-fold higher estimates at 150–350× than at 600–1000×.

Isolation and counting of protozoa

S.S. Bamforth

Litters and soils are mosaics of microhabitats, therefore, composite samples are collected from an area, mixed well, and subsamples used for identification of species and estimations of numbers. Portions of the samples should also be measured for moisture content and pH because the presence and abundance of many species reflect these two parameters.

Many protozoa in soils are encysted. Direct counts of testacea (Couteaux 1967; Korgonova and Geltser 1977) distinguish between active, encysted forms and empty tests, furnishing lists of species richness and relative abundance (diversity), and estimation of numbers. The Petri dish and agar plate methods described below estimate species richness for the less easily extractable "naked" groups – flagellates, naked amoebae and ciliates.

Most probable number culture methods (Singh 1955; Stout 1962; Darbyshire et al 1974) are used to estimate total numbers of naked protozoa and can identify major taxa, but do not distinguish between active and encysted forms. A direct count density gradient centrifugation, with subsequent fluorescent staining, method (Griffiths and Ritz 1988) likewise estimates total numbers of protozoa but does not distinguish between taxa.

Straight counts on fresh soil suspensions (Foissner 1987; Luftenegger et al 1988) estimate active forms only for ciliates and testacea. A combined direct count with culture technique (Bamforth 1991) estimates total numbers, and proportions of active and encysted ciliates.

Since there is no universal method for soil protozoa identification and enumeration, investigators must choose the methods which best address their objectives, recognizing limitations of the techniques.

Sampling and microscopic examination

Collect 6–12 samples by spatula or cork borer (about 15 mm diameter) from 0–3 cm soil depth in a 4–16 m^2 area; preferably several such areas in a forest, grassland, agricultural field, etc. should be sampled. Mix the samples thoroughly and withdraw subsamples for studies.

All of the following methods require a compound microscope with oil immersion objective, slides and coverslips, and pipettes for transferring organisms, so these apparatus are not repeated in the methods described below. Phase-contrast or interference-contrast microscopy improves observation of many taxonomic features without the use of stains, which can produce artifacts and lysis. Observations on live specimens is often critical for identifying genera and species, especially amoebae; for locomotor patterns and the position of organelles (e.g. nucleus, contractile vacuole) in the cell, often constitute taxonomic features.

Useful taxonomic references include the following. General: Lee et al (1985); Lousier and Bamforth (1990). Flagellates: Patterson and Larsen (1991). Gymnamoebae: Page (1988); Bovee (1985). Ciliates: Foissner (1982, 1985, 1987). Testacea: Cash (1905, 1909); Cash et al (1915, 1918); Foissner (1987); Bovee (1985).

Species richness

Protozoa in general (Petri dish)

Principle of the method

The moistened sample in a Petri dish is a closed ecological system, inducing a succession of species that are recorded as they appear in 20–30 days.

Materials and apparatus

Sterile Petri dishes, 10–15 cm diameter

Chemicals and solutions

Distilled water

Procedure

Place 10–50 cm of litter or soil sample, at least 1 cm deep, in a Petri dish. Saturate, but do not flood, with distilled water, until 5–20 ml will drain off when material is gently pressed with a finger. Examine the run-off for protozoa:

Days	
2–3	Flagellates, colpodid, some other ciliates
5–6	Amoebae, small ciliates, hypotrichs
8–10	Hypotrichs, testacea
15–20	Mainly testacea
30	Mainly testacea

Discussion

More amoebae will be found by agar plate methods (see section below on amoebae) and additional small testacean species from permanent slides (see section below on permanent direct counts for Testacea). The large amount of soil used is responsible for the large number of species that appear.

Amoebae (agar plates)

Principle of the method

Many amoebae are inhibited by the large amount of water and competition of other protozoa in the preceding Petri dish method. Bacterized agar plates induce amoebae in the added soil to excyst and multiply, as well as inhibiting ciliates, which require thicker water films than the thin film on the surface of the agar.

Streak plates

Materials and apparatus

Sterile Petri dishes, 10–15 cm diameter.

Reagents and bacteria

1.5% water (non-nutrient) agar
Distilled water
Culture of *Escherichia coli*, *Klebsiella aerogenes*, or a suitable soil bacterium

Procedure

Streak two "sine waves" of a bacterial suspension on a Petri dish containing water agar. Add two straight streaks of sample suspension bisecting the bacterial streaks. Incubate at room temperature (20°C) and examine for amoebae migrating into the bacterial streaks from the 4th day and beyond.

Discussion

Amoebae should be transferred from agar to glass slides for identification, since active forms flatten out on agar. Cysts are important in identification (Page 1988).

Well plates

(Robinson 1992)

Materials and apparatus

Sterile Petri dishes, 10–15 cm diameter
10 mm cork borer or glass rod
Steel spatula
Bunsen burner
Thin L-shaped glass rod

Chemicals, solutions and bacteria

1.5% water (non-nutrient) agar
Distilled water
95% ethyl alcohol
Culture of *Escherichia coli* or *Aerobacter aerogenes* or a suitable soil bacterium

Procedure

Using alcohol-flamed instruments, cut a well 5 cm long and 1 cm wide in one side of agar in a Petri dish, and three holes with a cork borer or glass rod. Spread the agar with bacterial suspension. Fill the four wells with soil and moisten with distilled water, but do not overflow wells. Incubate at room temperature and examine at 4 days and thereafter for amoebae migrating from wells over the agar.

Discussion

Transfer amoebae to slides for identification (see earlier section on streak plates). The larger quantities of soil furnish more species.

Enumeration

Total protozoa

Most probable number

Principle of the method

The presence or absence of organisms in several individual portions in each of several consecutive dilutions of soil are determined and compared to MPN statistical tables (e.g. Fisher and Yates 1963).

Singh two-fold dilution method

Materials and apparatus

Sterile Petri dishes
Sterile glass or propylene rings

Reagents and bacteria

1.5% water (non-nutrient) agar
Soil extract: boil 300 g soil in 1000 ml distilled water for 10 min, decant, filter and autoclave. Dilute 1 part soil extract to 5 parts distilled water for use. Adjust to pH of the soil being investigated with HCl or NaOH.
Culture of *Escherichia coli* or *Aerobacter aerogenes* or a suitable soil bacterium

Procedure

Embed eight rings in agar per Petri dish for 15 dishes (= levels). Prepare an initial 1:5 soil suspension, using soil extract as the diluent and then make two-fold dilutions (1:10, 1:20, 1:40, etc.). Pipette 0.5 ml of each dilution into each of eight rings to obtain 15 levels. Incubate at room temperature and examine at 4-day intervals. After 1 week, add a bacterial suspension to the wells to induce encysted protozoa (mainly amoebae) to appear. Count number of negative rings and consult appropriate tables (Fisher and Yates 1963) to estimate numbers.

Stout (1962) modification

Materials and apparatus

(see Singh two-fold dilution method)

Reagents and bacteria

(see Singh two-fold dilution method)

Procedure

Using larger rings, embed 10 rings in agar plates per level for three levels. Prepare three ten-fold dilutions. Dilution range depends on soils: 10^{-1}–10^{-3} for nutrient poor, 10^{-4}–10^{-6} for very rich soils. Pipette 1.0 ml of each dilution into 10 rings to obtain three levels. Incubate and consult MPN tables as above in the section above.

Darbyshire et al (1974) modification

Materials and apparatus

96-well microtitre plates
Other apparatus same as in above two sections

Reagents and bacteria

(see the two sections above)

Procedure

Prepare a two-fold dilution series as in the two sections above. Pipette 0.05 ml of each dilution into eight wells of a 96-well microtitre plate. Examine and consult MPN tables as in the two sections above.

Discussion of MPN methods

The MPN methods allow distinction of flagellates, amoebae, and ciliates but, at low dilutions, drops must be removed from rings to find the protozoa, which may be obscured by soil particles. Cultural conditions may not be optimal, and protozoa may compete and prey upon each other. Dormant protozoa (especially amoebae) may excyst during the incubation period, thus obscuring interpretation of how many protozoa are encysted and how many active. Many workers have used Cutler's (1920) 2% HCl technique on subsamples to destroy active forms and incubate the surviving "encysted" forms. However, the acid also destroys cysts (Bodenheimer and Reich 1933; Foissner 1987) and stimulates some protozoa to excyst that would not in the untreated sample. Despite these criticisms, the MPN method provides an approximate estimate of numbers and the protozoan resource in soils, hence the method is widely used.

Density centrifugation

Principle of the method

Protozoan cells are separated from soil particles by density gradient centrifugation, stained with epifluorescence dyes and concentrated on a filter for microscopical examination.

Materials and apparatus

Wrist action shaker
Centrifuge
Black 0.8 μm pore size membrane filters, 25 mm diameter

Chemicals and solutions

Tris buffer, 50 mM, pH 7.5
Iodonitrotetrazolium (INT), 0.4% (w/v)
Glutaraldehyde, 25% (v/v)
Percoll column: 5 ml each of Percoll, 0.1 phosphate buffer
Diamidinophenyl indole (DAPI), 5 μg ml^{-1}

aqueous solution
Acridine orange, 33 μg ml^{-1} solution

Procedure

Shake 5 g of fresh sieved soil in 50 ml Tris buffer for 10 min on wrist shaker.
Settle 1 min and then remove 1 ml aliquot from 5 cm below the meniscus in the tube.
Incubate with 0.1 ml 4% aqueous INT for 4 h at 25°C.

Fix with 0.1 ml of 25% glutaraldehyde.
Load aliquot onto a 5 ml Percoll phosphate column in a sterile 15 ml polycarbonate centrifuge tube.

Allow to settle for 30 min.

Centrifuge at 3000*g* for 2 h.

Decant the supernatant. Stain with 1 ml DAPI.

Filter through black 25 mm diameter 0.8 mm pore size filter, using gentle suction of −7 kPa.Counterstain cells with acridine orange solution.

Mount filters on microscope slides, and observe for protozoa using epifluorescence and an oil immersion lens.

Discussion

Fixing the cells quickly after soil sampling eliminates the problems associated with MPN techniques (see earlier section on discussion of MPN methods) and density gradient centrifugation separates more protozoa from soil particles than other methods. This method does not differentiate active from encysted forms. Large ciliates may be distinguished from small flagellates but intermediate taxa cannot be determined, and generic and species identifications cannot be made. This method is more useful in agricultural than in forest soils (Griffiths and Ritz 1988).

Fresh direct counts for ciliates and Testacea

Principle of the method

Combining a small amount of well-mixed soil with a few millilitres of soil extract (to reduce osmotic stress) enables drop-by-drop observation of active ciliates and Testacea, stained in the preparations, to estimate the number of active individuals.

Chemicals and solutions

Soil extract as described earlier in the section on the Singh two-fold dilution method.
Phenolic aniline blue: 5% phenol (30 ml), 1% aniline blue (2 ml), glacial acetic acid (8 ml). Allow to stand for 1 h before using.

Procedure

To 0.2–0.4 g of fresh soil, add 3–6 ml of soil extract and examine dropwise on the day of sampling for ciliates. To another 0.1 g of soil, add 5 ml of phenolic aniline blue and allow to stand overnight. Wash out the aniline blue by centrifugation and macerate the pellet. Add 0.5 ml albumin–glycerin or 1 ml of 0.5% agar to 5 ml of the pellet suspension, and immediately examine dropwise on a microscope slide. For both groups, the amount of dilution depends upon soil type and abundance of protozoa.

Permanent direct counts for Testacea

The presence of a shell enables stained permanent preparations, showing active and encysted individuals, and empty shells of Testacea on slides. The amount of soil per slide is known, hence the method is also quantitative.

Membrane filter method

(Couteaux 1967)

Materials and apparatus

Millipore filters, 25 mm diameter, 0.45 mm pore size
Millipore filter holder
Wrist or other type of shaker

Chemicals and solutions

Bouin–Hollande fixative

6.25 g copper acetate, 250 ml distilled water, 10 picric acid. Filter. Add 25 ml of 40% formalin and 2.5 ml glacial acetic acid.

Xylidine ponceaux 2R, 1% (v/v).

Procedure

Fix 0.25 of a well-mixed soil sample in 1 ml Bouin–Hollande fixative for 24 h. Stain with 3 ml of xylidine ponceaux for 30 min. Dilute with distilled water to 250 ml and agitate for several hours on shaker to dislodge Testacea from soil particles. Filter 5 ml through Millipore filter under vacuum pressure. Air dry the filter, clear in xylene, and mount in Canada balsam or other mounting medium on to a slide. Examine a determined area of the filter.

Smear slide

(Korgonova and Geltser 1977)

Materials and apparatus

Pipettes drawn out into J-shaped spreaders
Coplin staining jars
Hot plate to warm agar

Chemicals and solutions

Agar (0.5%)
Phenolic erythrosin, 1 g in 100 ml of 5% phenol

Procedure

Place 5 g soil in 50 ml distilled water in a 250 ml flask. Soak for several hours. Shake vigorously by hand for 10 min or on shaker for several hours. Pipette 1 drop (= 0.05 ml) from the centre of the flask on to a clean slide and add one drop of (warmed) 0.5% agar. Spread with a glass rod over an 8 cm^2 area. Air-dry the slide and place it in a staining jar for 1 h. Pass the slide through three water rinses and air-dry. Examine a determined area of the slide.

Direct count/culture estimation for ciliates

Principle of the method

This method employs direct counting (see earlier section on fresh direct counts) for active ciliates in one part of the soil sample, and cultures a second part in isolated chambers, allowing each active ciliate to function, and dormant ciliates to excyst independently of other ciliates.

Materials and apparatus

96-well microtitre plate
Magnetic stirrer

Reagents and bacteria

Soil extract as described in the earlier section on Singh two-fold dilution method. Culture of *Escherichia coli* or *Aerobacter aerogenes*, or a suitable soil bacterium.

Procedure

From a well-mixed soil sample, 0.6 g is analysed. Of this, 0.2 g is mixed with 3 ml of sterile soil extract and examined drop by drop. The remaining 0.4 g is mixed with 10 ml bacterized soil extract in a small beaker on a magnetic stirrer, and 0.04 ml aliquots are transferred to 60 wells of a 96-well microtitre plate. The outer wells are filled to retard evaporation and the plate is incubated at room temperature for 6–7 days. Each well is examined for ciliates. The active count is subtracted from the microtitre count (total count) to give the number of cysts.

Isolation and counting of cyanobacteria

S. Oran

Cyanobacteria are prokaryotic microorganisms containing chlorophyll a and producing oxygen as a byproduct of their photosynthetic activity. They inhabit soil, fresh water and marine environments. Cyanobacteria represent different forms of organizations: unicellular, colonial and filamentous (Bold et al 1987). The size of the cells range from 1–6 μm. In general cyanobacteria are larger than the bacteria but smaller than most algae.

Cyanobacteria are capable of fixing nitrogen and therefore they are important for soil fertility. Cyanobacteria are used in the revitalization of soil and as partial crop fertilizers, especially for rice. It is also reported that cyanobacteria such as *Nostoc*, *Scytonema* and *Anabaena* can be employed in the reclamation of alkaline land (Hamdi 1988; Shield 1982; Stern 1988; Yanni 1991; Edward et al 1992; Fernandes et al 1993; Russel and Tranvik 1993).

Cultivation of cyanobacteria

Many of the adopted laboratory culture media methods are known to be effective for growing the unicellular, filamentous and the multicellular cyanobacteria. It is preferable to use solid media and they have proved to be more successful than liquid media, although many species can be easily maintained in liquid media. Cultures can be preserved by using methods that keep the culture viable at room temperature.

The media described in this chapter have been shown to be successful for some species of cyanobacteria. Some recommended methods for preserving cultures are also discussed by Carr et al (1973).

Selected solid media for laboratory culture

Medium A (Allen 1968), modified from Hughes et al (1958)

1.500 g $NaNO_3$
0.039 g K_2HPO_4
0.075 g $MgSO_4.7H_2O$
0.020 g Na_2CO_3
0.027 g $CaCl_2$
0.058 g $Na_2SiO_2.9H_2O$
0.001 g EDTA
0.006 g Citric acid
0.006 g Ferrous citrate
1000.000 ml Distilled water
1.000 ml Microelement solution

Adjust the pH to 7.8.

Microelements – 1 ml l^{-1} of a solution containing (g l^{-1})

2.8600 g H_3BO_4
1.8100 g $MnCl_2.H_2O$
0.2220 g $ZnSO_4.7H_2O$
0.3910 g $Na_2MoO_4.2H_2O$
0.0790 g $CuSO_4.5H_2O$
0.0494 g $Co(NO_3)_2.6H_2O$
1000.0000 ml Distilled water

To prepare solid media, autoclaved 1.5% agar (48°C) is added to the mineral salts.

Medium B (Castenholz 1970)

0.100 g Nitrilotriacetic acid

0.060 g $CaSO_4.2H_2O$
0.100 g $MgSO_4.7H_2O$
0.008 g NaCl
0.103 g KNO_3
0.689 g $NaNO_3$
0.111 g $NaHPO_4$
1.000 ml $FeCl_3$ solution (0.2905 g l^{-1})
0.500 ml Microelement solution
1000.000 ml Distilled water

Microelement solution

2.280 g $MnSO_4.H_2O$
0.500 g $ZnSO_4.7H_2O$
0.500 g H_3BO_3
0.025 g $CuSO_4.5H_2O$
0.025 g $Na_2MoO_4.2H_2O$
0.045 g $CoCl_2.6H_2O$
0.500 ml H_2SO_4 (conc.)
1000.000 ml Double distilled water

pH is adjusted to 8.2 with NaOH (1 M), which gives a final pH of 7.5–7.6 in the autoclaved, cooled and cleared medium.

This medium is used to maintain at least 100 strains of approximately 25 species of thermophilic blue–green algae at temperatures between 45°C and 70°C (Carr et al 1973). Many other media for laboratory culture are widely employed (Carr et al 1973).

Cultivation

Cultures of cyanobacteria can be preserved after growth and should not be maintained in refrigerator like other prokaryotes.

Preparation of plates or tubes

(Wollum 1982)

Materials

Milted, 180 ml of sterile medium in 250 ml Erlenmeyer flasks
Sterile Petri dishes, 12
Sterile pipettes, 10 ml

Procedure

Prepare the media as indicated above and sterilize in the autoclave at 121°C. Place flasks in a 48°C water bath until ready for use. Select a range of four dilutions that will encompass the optimum number of organisms for counting. Starting with the most dilute bottle, transfer a 1 ml portion of a freshly mixed suspension into each of three sterile Petri dishes using a sterile 1 ml pipette. Before each pipetting, draw up the soil suspension and empty the pipette into a bottle. Repeat the transfer process involving the next lower dilution. Continue the process for the next two successive lower dilutions, starting with dilutions 10^{-7}, 10^{-6}, 10^{-5}, respectively. Pour about 15 ml of the medium for the organism being studied into the plate. Immediately after pouring, carefully swirl the plates to mix their contents thoroughly. After the medium has solidified, invert the plates and place in an incubator.

Spread plates

Materials

Petri plates with appropriate medium, 12
Sterile 1 ml pipettes, marked in 0.1 ml divisions
Glass spreader (hockey stick)
Ethanol (95%)

Procedure

Prepare the media according to the directions. Sterilize in an autoclave at 121°C for 20 min and cool to a pouring temperature of 48°C. Distribute about 15 ml of media into each plate and allow the agar to solidify. Plates should be cured for 2 days, either by placing the Petri dishes in the incubator or leaving them on a laboratory bench in a relatively protected location. The plates that are not used immediately can be stored in the refrigerator, from where they should be removed sufficiently ahead of time so that spreading of the inoculum can take place when the plates have reached ambient temperature.

To use, select a range of four dilutions that will adequately characterize the organisms in the sample. Transfer three 0.1 ml aliquots to separate plates from the highest dilution. Note that a 0.1 ml aliquot from 10^{-6} corresponds to an actual dilution of 10^{-7} on the plate. Repeat the process, transferring 0.1 ml aliquots from each of the next three successive and lower dilutions on to each of the triplicate plates for each dilution. Spread the suspension on the agar surface using a sterile glass spreader (hockey stick) for each plate. In this step start with the highest dilution, and progress to the next lower dilution, continuing the sequence until all the plates have been spread. Alcohol-flame the glass spreader between each use. Invert the plates and place in an incubator.

Tubes

Materials

Tubes, 5 ml, of appropriate sterile broth, 40
Sterile 1 ml pipettes
Vortex mixer

Procedure

The same steps are used as explained above except for adding 1.0 ml aliquots to each of five tubes of each dilution step. As explained before, start with the highest dilution and continue with the next successive lower dilution until all the desired dilutions have been used. Mix each tube to ensure a uniform suspension throughout the tube. Place the tubes in an incubator.

Counting of cyanobacteria

Most probable number method

(Allen 1957)

Principle of the method

The MPN method is applied to the dilution counting technique, which is a method for isolating and quantifying soil micro-organisms. This method is of great value in the case of cyanobacteria that grow as single motile or non-motile cells, e.g. species of *Chlorococcum* (Shield 1982). The MPN method is based on the determination of the presence or absence of microorganisms in several individual portions of each of several consecutive dilutions of soil or other material (Alexander 1982).

Materials and apparatus

Incubation containers, 25, each containing 20 ml of substrate solution
Erlenmeyer flasks (200 ml), test tubes (25 mm in diameter, capped or plugged) or 57–113 ml screw-capped bottles can be used
Autoclave
Bacteriological transfer loop
Microscope
Slides and cover slides
Glass beads (2 mm)
Shaker

Chemicals and solutions

Substrate A

For a suitable substrate use one of the following nutrient solutions or another standard nutrient solution.

(A) Bristol's solution

1.0 g $NaNO_3$
1.0 g K_2HPO_4

0.3 g $MgSO_4.7H_2O$
0.1 g NaCl
Trace $FeCl_3.6H_2O$

Dissolved in 800 ml distilled water and brought up with distilled water to 1000 ml.

(B) Wilson's solution

1.00 g $Ca(NO_3)_2$
0.25 g KCl
0.25 g $MgSO_4.7H_2O$
0.25 g KH_2PO_4
0.30 g $FeCl_3.6H_2O$

Dissolved in 800 ml distilled water and brought up with distilled water to 1000 ml.

Sterile water blanks (Clarke 1965; Wollum 1982)

(A) 8 oz, screw-cap bottles with approximately 36 spherical glass beads (the purpose of the beads is to facilitate the disintegration of soil aggregates) and 95 ml of water (alternatively, 16 oz, screw-cap bottles with 190 ml of water and about 72 beads).

(B) 8 oz, screw-cap bottles with 90 ml of water and no beads. Cap all the bottles, sterilize them by autoclaving at 121°C for 15 minutes. For a single soil sample, one water blank with beads, and approximately seven without beads, are required.

Procedure

Prepare the serial dilution series. For each soil sample, make a dilution blank containing 95 ml diluent and 15–20 glass beads. Make seven 90 ml dilution blanks with the diluent. Cap the bottles, autoclave at 121°C for 20 min and cool to room temperature. Transfer 10 g of moist soil into a blank containing 95 ml of water (or diluent) glass beads (alternatively, transfer 20 g of soil to a 190 ml blank with beads). Tightly cap the bottle and shake it for 10 min in a horizontal position in a reciprocating shaker. Within 10 min of removing the bottle from the shaker, shake the bottle vigorously by hand for a few seconds and immediately transfer 10 ml from the centre of the suspension to a 90 ml water blank, using a sterile 10 ml pipette. This establishes a 10^{-2} dilution. Cap and vigorously shake this bottle and transfer 10 ml of the suspension dilution to further water blanks of 90 ml, continue the sequence until the dilution of 10^{-7} is provided. On the basis of previous experience with the samples, it may be necessary to continue diluting until the 10^{-8} and 10^{-9} dilution is reached.

It should be mentioned here that the diluent selected is the investigator's choice. In most enumeration studies, water, either tap or distilled, may be used without affecting the relative values observed. However, in more specialized studies, a variety of standard diluents can be used satisfactorily (physiological saline, Ringer solution, peptone water and phosphate saline).

From each of five successive serial dilutions, usually 10^{-2}–10^{-6}, transfer 1 ml to each of five incubation vessels containing sterile enrichment solution. Incubate the vessels, preferably at 22°C, for 4 weeks in diffuse daylight, or under fluorescent or incandescent lighting. At weekly intervals during incubation, observe the inoculated containers for cyanobacteria (or other) growths, which are usually shown by surface rings or pellicles. Transfer a portion of the growth to a microscope slide or haemacytometer by means of a bacteriological transfer loop, and make a direct microscopic examination of the wet mount. Record the number of cultures that are positive for cyanobacteria at each dilution. Find the most probable number in the initial soil sample.

Calculation

For calculation see section on most probable number (MPN) technique.

Discussion

Errors in dilution counting may result in: a failure to obtain uniform dispersion of cyanobacteria that form gelatinous or cohesive cell masses; filaments that segment readily; or a large number of propagules as zoospores (Shield 1982).

Direct microscopy

(Allen 1957)

Principle of the method

It is a quantitative method based on direct microscopy of soil. Detection of cyanobacteria is facilitated by the fluorescence of their pigments. Individual species can be more readily identified by light transmission microscopes (Shield 1982).

Materials and apparatus

Microscope

The fluorescence microscope, the electron scanning technique, or the ordinary light transmission microscope can be used depending on the nature of investigation.

Counting chamber

Any haemacytometer chamber with cover glass can be used, providing a chamber depth of 0.1 mm is satisfactory.

Procedure

Place 20 g of soil into a sterile screw-capped bottle (225 or 450 ml) and add sterile water to bring the volume to 100 ml. Shake the bottle for 10 min on a mechanical shaker. Similarly prepare suspensions containing 10 and 5 g of soil, thus providing soil/water ratios of 1:5, 1:10 and 1:20. With fine-textured soils, these initial preparations are suitable for microscopic examination. Focus the condenser of the light source to produce a parallel beam. Insert the blue filter between the light source and the microscope. Add one drop of mineral oil between the top of the condenser and the bottom of the counting chamber. Reduce the light intensity and focus the microscope. Then increase the light to the optimum level.

Calculation

Using a ×10 objective lens and a ×5 eyepiece, count the cells in a total of 50 fields (filamentous cells appear as red lines against a blackground). Convert the average count per microscopic field (or haemacytometer square) to a number per gram of soil by dividing the average number of cells in a field or square by the grams of oven-dry soil represented by the volume of solution contained in one microscopic field or haemacytometer square.

Discussion

The direct microscopy provides a quantitative method for detecting cyanobacteria in a soil sample. The contrast which appears as red lines against the black background facilitates the identification of filaments, e.g. providing the soil particles are usually invisible, although they sometimes show a yellow or green inflorescence. Also individual species can be more readily identified by the light transmission microscope.

Although the fluorescence microscopy bypasses many of the objections raised against the dilution count and other cultural methods, the direct method itself is not without its disadvantages. The counting is tiresome and eye fatigue develops rapidly.

References

Abeliovich A (1987) Nitrifying bacteria in wastewater reservoirs. Appl Environ Microbiol 53: 754–760.

Alef K (1991) Methodenhandbuch Bodenmikrobiologie. Ecomed-Verlagsgesellschaft, Landsberg.

Alexander M (1982) Most probable number method for microbial populations. In: Methods of Soil Analysis. Part 2, Chemical and microbiological properties. Page AL, Miller RH, Keeney DR (eds). Agronomy Monograph 9, pp. 815–820.

Allen MM (1968) Simple conditions for growth of unicellular blue-green algae on plates. J Phycol 4: 1–4.

Allen ON (1957) Experiments in Soil Bacteriology, 3rd edn. Burgess Publishing Co., Minneapolis, MN.

Altemüller H-J, van Vliet-Lande B (1990) Soil thin section fluorescence microscopy. In: Soil Micromorphology. Douglas LA (ed.). Elsevier, Amsterdam, pp. 565–579.

Anderson JR, Slinger JM (1975a) Europium chelate and fluorescent brightener staining of soil propagules and their photomicrographic counting-I. Methods. Soil Biol Biochem 7: 205–209.

Anderson JR, Slinger JM (1975b) Europium chelate and fluorescent brightener staining of soil propagules and their photomicrographic counting-II. Efficiency. Soil Biol Biochem 7: 211–215.

Bååth E, Söderström B (1980) Comparisons of the agar-film and membrane-filter methods for the estimation of hyphal lengths in soil, with particular reference to the effect of magnification. Soil Biol Biochem 12: 385–387.

Babiuk LA, Paul EA (1970) The use of fluorescein isothiocyanate in the determination of the bacterial biomass of grassland soil. Can J Microbiol 16: 57–62.

Bakken LR (1985) Separation and purification of bacteria from soil. Appl Environ Microbiol 49: 1482–1487.

Bakken LR, Olsen RA (1983) Buoyant densities and dry-matter contents of microorganisms: conversion of a measured biovolume into biomass. Appl Environ Microbiol 45: 1188–1195.

Bamforth SS (1991) Enumeration of soil ciliate active forms and cysts by a direct count method. Agric Ecosyst Environ 34: 209–212.

Bjørnsen PK (1986) Automatic determination of bacterioplankton biomass by image analysis. Appl Environ Microbiol 51: 1199–1204.

Blakeslee D, Baines MG (1976) Immunofluorescence using dichlorotriazinylaminofluorescein (DTAF). I. Preparation and fractionation of labelled IgG. J Immunol Meth 13: 305–320.

Blandreau J (1983) Microbiology of the association. Can J Microbiol 29: 851–859.

Blandreau J (1986) Ecological factors and adaptive processes in N_2-fixing bacterial populations of the plant environment. Plant Soil 90: 73–92.

Bloem J, Ellenbroek FM, Bär-Gilissen MJB, Cappenberg ThE (1989) Protozoan grazing and bacterial production in stratified Lake Vechten estimated with fluorescently labeled bacteria and by thymidine incorporation. Appl Environ Microbiol 55: 1787–1795.

Bloem J, de Ruiter PC, Koopman GJ, Lebbink G, Brussaard L (1992a) Microbial numbers and activity in dried and rewetted arable soil under integrated and conventional management. Soil Biol Biochem 24: 655–665.

Bloem J, van Mullem DK, Bolhuis PR (1992b) Microscopic counting and calculation of species abundances and statistics in real time with an MS-DOS personal computer, applied to bacteria in soil smears. J Microbiol Meth 16: 203–213.

Bloem J, Veninga M, Shepherd J (1994a) Fully automatic determination of soil bacterial numbers, cell volumes and frequencies of dividing cells by confocal laser scanning microscopy and image analysis. Appl Env Microbiol (submitted).

Bloem J, Lebbink G, Zwart KB, Bouwman LA, Burgers SLGE, de Vos JA, de Ruiter PC (1994b) Dynamics of microorganisms, microbivores and nitrogen mineralization in winter wheat fields under conventional and integrated management. Agricult Ecosyst Environ 51: 129–143.

Bodenheimer FS, Reich K (1933) Studies on soil protozoa. Soil Sci 38: 259–265.

Bogdahn M (1985) Untersuchungen zum Metabolismus anorganischer Stickstoffverbindungen in einigen saccharolytischen Clostridia, dissertation, University of Bayreuth.

Bold HC, Alexopoulos C, Delevoryas T (1987) Morphology of Plants and Fungi. Harper and Row, New York.

Bollag JM, Orcutt ML, Bollag B (1970) Denitrification by isolated soil bacteria under various environmental conditions. Soil Sci Soc Am Proc 34: 875–879.

Bovee EC (1985) Rhizopoda. In: An Illustrated Guide to the Protozoa. Lee JJ, Hutner SH, Bovee EC (eds). Allen Press, Lawrence, KS, pp. 158–245.

Brakenhoff GJ, van Spronsen EA, van der Voort HTM, Nanninga N (1989) Three-dimensional confocal fluorescence microscopy. In Methods in Cell Biology, vol 30. Fluorescence microscopy of living cells in culture, Part B. Taylor DL, Wang YL (eds). Academic Press, San Diego, pp. 379–398.

Braun M, Schobert S, Gottschalk G (1979) Enumeration of bacteria forming acetate from H_2 and CO_2 in anaerobic habitats. Arch Microbiol 120: 201–204.

Brown VJ, Lowbury EJL (1965) Use of improved cetrimide agar medium and other culture methods for Pseudomonas aeruginosa. J Clin Path 18: 752–756

Campelo AB, Döbereiner J (1970) Ocorrência de Derxia sp. em solos de alguns estados brasileiros. Pesquisa Agropecuária Brasileira 5: 327–332.

Carr NG, Komaref M, Whitton BA (1973) Notes on isolation and laboratory culture. In: The Biology of Blue–green Algae. Carr CG, Whitton BA (eds). Blackwell, Oxford, pp. 525–530.

Cash J (1905, 1909) The British Freshwater Rhizopoda and Heliozoa, vols 1 and 2. The Royal Society, London.

Cash J, Wailes GH (1915, 1918) The British Freshwater Rhizopoda and Heliozoa, vols 3 and 4. The Royal Society, London.

Castenholz RW (1970) Laboratory culture of thermophilic cyanophytes. Schweiz Z Hydrol 32: 538–551.

Cavalcante VA, Döbereiner J (1988) A new acid tolerant nitrogen fixing bacterium associated with sugar cane. Plant Soil 108: 23–31.

Clarke FE (1965) Agar-plate method for total microbial count. In: Methods of Soil Analysis, Part 2: Chemical and Microbial Properties. Black CA, Evans DD, Ensminger LE, White JL, Clarke FE (eds). ASA-SSSA, Madison, WI, pp. 1460–1466.

Coelho RRR, Drozdowicz A (1978) The occurrence of actinomycetes in a cerrado soil in Brazil. Rev Ecol Biol Soil 15: 459–473.

Collins VG (1963) The distribution and ecology of bacteria in freshwater. Proc Soc Wat Treatm Examin 12: 40–73.

Couteaux MM (1967) Une technique d'observation des thecamoebiens du sol pour l'estimation de leur densite absolue. Revue Ecol Biol Sol 4: 593–596.

Cutler DW (1920) A method for estimating the number of active protozoa in the soil. J Agric Sci 10: 135–143.

Darbyshire JF, Wheatly RE, Greaves MP (1974) A rapid micromethod for estimating bacterial and protozoan populations in soil. Revue Ecol Biol Sol 11: 465–475.

Davidson RS, Hilchenbach MM (1990) The use of fluorescent probes in immunochemistry. Photochem Photobiol 52: 431–438.

Davies FL, Williams ST (1970) Studies on the ecology of actinomycetes in soil. I. The occurrence and distribution of actinomycetes in a pine forest soil. Soil Biol Biochem 2: 227–238.

De Man JC (1975) The probability of most probable numbers. J Appl Microbiol 1: 67–68.

De Man JC (1983) MPN tables, corrected. Eur J Appl Microbiol Biotechnol 17: 301–305.

Döbereiner J (1966) Azotobacter paspali sp. n. uma bactéria fixadora de nitrogênio na rizosfera de Paspalum. Pesquisa Agropecuária Brasileira 1: 357–365.

Döbereiner J (1992) The genera Azospirillum and Herbaspirillum. In: The Prokaryotes, 2nd edn. Ballows A, Trüper HG, Working M, Harder W, Schleifer KH (eds). Springer Verlag, Berling, pp. 2236–2253.

Drews G (1968) Mikrobiologisches Praktikum für Naturwissenschaftler. Springer Verlag, Berlin.

Edward JP, Ihnat J, Conroy M (1992) Nitrogen fixation by the benthic fresh water cyanobacterium *Lyngbya wollei.* Biologia 234: 59–64.

Faegri A, Torsvik VL, Goksöyr J (1977) Bacterial and fungal activities in soil: separation of bacteria and fungi by a rapid fractionated centrifugation technique. Soil Biol Biochem 9: 105–112.

Fernandes TA, Iyeer V, Apte SH (1993) Differential response of nitrogen-fixing cyanobacteria to salinity and osmotic stress. Appl Environ Microbiol Vol. 59, No. 3: 899–904.

Fisher RA, Yates F (1963) Statistical Tables in Biological, Agricultural, and Medical Research. Oliver and Boyd, Edinburgh.

Foissner W (1982) Okologie und Taxonomie der Hypotrichida (Protozoa: Ciliophora) einiger osterreicher Boden. Arch Protistenk 126: 19–143.

Foissner W (1985) Klassification und Phylogenie der Colpodea (Protozoa: Ciliophora). Arch Protistenk 129: 239–290.

Foissner W (1987) Soil Protozoa: Fundamental problems, ecological significance, adaptations in ciliates and testaceans, and guide to the literature. Prog Protistol 2: 69–212.

Frankland JC, Dighton J, Boddy L (1990) Methods for studying fungi in soil and forest litter. In: Methods in Microbiology, vol 22, Techniques in microbial ecology. Grigorova R, Norris JR (eds). Academic Press, London, pp. 343–404.

Fry JC (1990) Direct methods and biomass estimation. In: Methods in Microbiology, vol 22, Techniques in microbial ecology. Grigorova R, Norris JR (eds). Academic Press, London. pp. 41–85.

Gerhardt P, Murray RGE, Costilow RN, Nestor EG, Wood WA, Krieg NR, Phillips GB (1981) Manual of Methods for General Bacteriology. American Society for Microbiology (ASM), Washington, DC.

Gillis M, Kersters K, Hoste B, Janssens D, Kroppensted RM, Stephan MP, Teixeira KRS, Döbereiner J, De Ley J (1989) *Acetobacter diazotrophicus* sp. nov., a nitrogen-fixing acetic acid bacterium associated with sugar cane. Int J System Bacteriol 39: 361–364.

Gillis M, Döbereiner J, Pot B, Goor M, Falsen E, Hoste B, Reinhold B, Kersters K (1991) Taxonomic relationships between (Pseudomonas) rubrisubalbicans, some clinical isolates (E F group 1), *Herbaspirillum seropedicae* and *(Aquaspirillum) autotrophicum*. In: Nitrogen Fixation. Polsinelli M, Materassi R (eds). Kluwer Academic Publishers, Amsterdam, pp. 291–292.

Gray TRG (1990) Methods for studying the microbial ecology of soil. In: Methods in Microbiology, vol 22, Techniques in microbial ecology. Grigorova R, Norris JR (eds). Academic Press, London, pp. 309–342.

Griffiths BS, Ritz K (1988) A technique to extract, enumerate and measure protozoa from mineral soils. Soil Biol Biochem 20: 163–173.

Hamdi YA (1988) Blue–green algae. In application of nitrogen-fixing systems in soil improvements and management. FAO Soil Bull 49: 45–73.

Hammann R, Ottow JCG (1976) Isolation and characterization of iron-reducing, nitrogen-fixing saccharolytic clostridia from gley soils. Soil Biol Biochem 8: 357–364.

Harris RF, Keeney DR (1968) Towards standardization of soil extracts for microbial media. Can J Microbiol 14: 653–659.

Hassink J (1993) Relationship between the amount and the activity of the microbial biomass in Dutch grassland soils: Comparison of the fumigation–incubation method and the substrate-induced respiration method. Soil Biol Biochem 25: 533–538.

Hattori T (1976) Plate count of bacteria in soil on diluted nutrient broth as a culture medium. Rep Inst Agric Res Tohoku Univ 27.

Hattori T (1982) Analysis of plate count data of bacteria in natural environments. J Gen Appl Microbiol 28: 13–22.

Hattori R, Hattori T (1980) Sensitivity to salts and organic compounds of soil bacteria isolated on diluted media. J Gen Microbiol 26: 1–14.

Hesse W (1992) Walther and Angelina Hesse: early contributors to bacteriology. ASM News 58: 425–428.

Hirsch P, Rades-Rohkohl E (1983) Microbial diversity in a ground water aquifer in northern Germany. Dev Indust Microbiol 24: 183–200.

Hsu SC, Lockwood JL (1975) Powdered chitin agar as a selective medium for enumeration of actinomycetes in water and soil. Appl Microbiol 29: 422–426.

Hughes EO, Gorham PR, Zehnder A (1958) Toxicity of a unialgal culture of Microcystis aeruginosa. Can J Microbial 4: 225–236.

Hungate RE (1965) A roll tube method for cultivation of strict anaerobes. In: Methods in Microbiology, vol 3B. Norris JR, Ribbons DW (eds). Academic Press, New York, pp. 117–132.

Ingham ER, Klein DA (1982) Relationship between fluorescein diacetate-stained hyphae and oxygen utilization, glucose utilization, and biomass of submerged fungal batch cultures. Appl Environ Microbiol 44: 363–370.

Ingham ER, Klein DA (1984) Soil fungi: measurement of hyphal length. Soil Biol Biochem 16: 279–280.

Jacob HE (1970) Redox potential. In: Methods in Microbiology, vol 2. Norris JR, Ribbons DW (eds). Academic Press, New York, pp. 91–123.

Jager G, Bruins H (1975) Effect of repeated drying at different temperatures on soil organic matter decomposition and characteristics, and the soil microflora. Soil Biol Biochem 7: 153–159.

Jagnow G (1982) Growth and survival of *Azospirillum lipoferum* in soil and rhizospere as influenced by ecological stress conditions. Experientia 42: 100–107.

Jarvis B (1973) Comparison of an improved rose fungal-chlortetracycline agar with other media for selective isolation and enumeration of moulds and yeasts in foods. J Appl Bacteriol 36: 723–727.

Jenkinson DS, Ladd JN (1981) Microbial biomass in soil: measurement and turnover. In: Soil Biochemistry, vol 5. Paul EA, Ladd JN (eds). Marcel Dekker, New York, pp. 415–471.

Jones SE, Ditner SA, Freeman C, Whitaker CJ, Lock MA (1989) Comparison of a new inorganic membrane filter (Anopore) with a track-etched polycarbonate membrane filter (Nuclepore) for direct counting of bacteria. Appl Environ Microbiol 55: 529–530.

Kasper H, Tiedje JM (1982) Anaerobic bacteria and processes. In: Methods of Soil Analysis, Part 2, Chemical and microbiological properties. Page AL, Miller RH, Keeney (eds). 2nd edn. Agronomy Monograph No. 9.

Khammas KM, Ageron E, Grimont PAD, Kaiser P (1989) *Azospirillum irakense* sp. nov., a nitrogen-fixing bacterium associated with rice roots and rhizosphere soil. Res Microbiol 140: 679–693.

Koch AL (1981) Growth measurement. In: Manual of Methods for General Bacteriology. Gerhardt P, Murray RGE, Costilow RN, Nestor EG, Wood WA, Krieg NR, Phillips GB (eds). ASM, Washington, DC, pp. 179–207.

Kölbel-Boelke J, Tienken B, Nehrkorn A (1988) Microbial communities in the saturated groundwater environment. I: Methods of isolation and characterization of heterotrophic bacteria. Microb Ecol 16: 187–211.

Kooistra MJ, Lebbink G, Brussaard L (1989) The Dutch Programme on Soil Ecology of Arable Farming Systems. 2. Geogenesis, agricultural history, field site characteristics and present farming systems at the Lovinkhoeve experimental farm. Agric Ecosyst Environ 27: 361–387.

Korgonova GA, Geltser JG (1977) Stained smears for the study of soil Testacida (Protozoa: Rhizopoda). Pedobiologia 17: 222–225.

Krambeck C, Krambeck H-J, Overbeck J (1981) Microcomputer assisted biomass determination of plankton bacteria on scanning electron micrographs. Appl Environ Microbiol 42: 142–149.

Krieg NR (1981) Enrichment and isolation. In: Manual of Methods for General Bacteriology. Gerhardt P, Murray RGE, Costilow RN, Nestor EG, Wood WA, Krieg NR, Phillips GB (eds). ASM, Washington, DC, pp. 112–142.

Krieg NR, Döbereiner J (1984) Genus Azospirillum Tarrand, Krieg and Döbereiner 1970. In: Bergey's Mannual of Systematic Bacteriology, vol 1. Holt JG, Krieg NR (eds). Williams and Wilkins, Baltimore, pp. 94–104.

Kruger CL, Sheikh WW (1987) A new selective medium for isolating *Pseudomonas spp.* from water. Appl Environ Microbiol 53: 895–897.

Küster E, Williams ST (1964) Selection of media for the isolation of streptomycetes. Nature (London) 202: 928–929.

Ladha JK, Barraquio WL, Watanabe I (1983) Isolation and identification of nitrogen-fixing *Enterobacter cloacae* and *Klebsiella planticola* associated with rice plants. Can J Microbiol 29: 1301–1308.

Larkin JM (1972) Peptonized milk as an enumeration medium for soil bacteria. Appl Microbiol 23: 1031–1032.

Lee JJ, Hutner SH, Bovee ED (1985) An Illustrated Guide to the Protozoa. Allen Press, Lawrence, Ks.

Litchfield CD, Rake JB, Zinulis J, Watanabe RT, Stein DJ (1975) Optimization of procedures for the recovery of heterotrophic bacteria from marine sediments. Microb Ecol 1: 219–233.

Lodge DJ, Ingham ER (1991) A comparison of agar film techniques for estimating fungal biovolumes in litter and soil. Agric Ecosyst Environ 34: 131–144.

Lorch H-J, Lorenz S, Ottow JCG (1990) Populationsdichten verschiedener Bakteriengruppen in einer belüfteten Abwasserteichanlage. Forum Städte-Hyg 41: 133–138.

Lorch H-J, Ottow JCG, Gerhards K-H (1992) Nitrifikation und Denitrifikation in belüfteten Abwasserteichen mit zwischengeschalteter technischer Stufe. Korrespondenz Abwasser 39: 64–70.

Lousier JD, Bamforth SS (1990) Soil Protozoa. In: Soil Biology Guide. Dindal DL (ed.). John Wiley, New York, pp. 97–136.

Luftenegger G, Petz F, Foissner W, Adam H (1988) The efficiency of a direct counting method in estimating the numbers of microscopic soil organisms. Pedobiologia 31: 95–101.

Malek A, Hosny I, Eman NF (1974) Evaluation of media, used for enumeration of denitrifying bacteria. Zbl Bakt Abt II 130: 644–653.

Martin JK (1975) Comparison of agar media for counts of viable soil bacteria. Soil Biol Biochem 7: 401–402.

Martin JP (1950) The use of acid, rose bengal, and streptomycin in the plate method for estimating soil fungi. Soil Sci 69: 215–232.

Marxsen J (1988) Investigations into the number of respiring bacteria in the groundwater from sandy and gravelly deposits. Microb Ecol 16: 65–72.

Meynell GC, Meynell EW (1965) Theory and Practice in Experimental Bacteriology. Cambridge University Press, Cambridge.

Morgan JAW, Rhodes G, Pickup RW (1993) Survival of nonculturable *Aeromonas salmonicida* in lake water. Appl Environ Microbiol 59: 874–880.

Morgan P, Cooper CJ, Battersby NS, Lee SA, Lewis ST, Machin TM, Graham SC, Watkinson RJ (1991) Automated image analysis method to determine fungal biomass in soils and on solid matrices. Soil Biol Biochem 23: 609–616.

Mossel DAA (1977) Microbiological quality assurance of water in relation to food hygiene. Arch Lebensmittelhyg 28: 1–2.

Mossel DAA, Harrewijn GA, Nesselroy-Van Zadelihoff FM (1974) Standardization of the selective inhibitory effect of surface active compounds used in media for the detection of Enterobacteriaceae in foods and water. Health Lab Sci 11: 260–266.

Oberzill W (1970) Quantitative Erfassung sulfatreduzierender Bakterien im Boden und Grundwasser. Zbl Bakt Abt II 124: 91–96.

Ohata H, Hattori T (1983) Agromonas oligotrophic gen. nov., sp. nov., a nitrogen fixing oligotrophic bacterium. Antoni van Leeuwenhoek 49: 429–446.

Ohto H, Hattori T (1983) Oligotrophic bacteria on organic debris and plant roots in a paddy field soil. Soil Biol Biochem 15: 1–8.

Okon Y, Albrecht SL, Burris RH (1977) Methods for growing *Spirillum lipoferum* and for counting it in pure culture and in association with plants. Appl Environ Microbiol 33: 85–88.

Olivares FL, Baldani VLD, Baldani JI, Döbereiner J (1993) Ecology of *Herbaspirillum* spp. and ways of infection and colonization of cereals with these endophytic diazotrophs. Sixth International Symposium on Nitrogen Fixation with Non-legumes, Ismailia, Egypt.

Olson FCW (1950) Quantitative estimates of filamentous algae. Trans Am Microsc Soc 59: 272–279.

Ottow JCG (1969) The distribution and differentiation of iron-reducing bacteria in gley soils. Zbl Bakt Abt II 123: 600–615.

Ottow JCG (1972) Rose bengal as a selective aid in the isolation and enumeration of fungi and actinomycetes from natural sources. Mycologia 64: 304–315.

Ottow JCG, Glathe H (1968) Rose bengal–malt extract agar, a simple medium for the simultaneous isolation and enumeration of fungi and actinomycetes from soil. Appl Microbiol 16: 170–171.

Ottow JCG, Dörster W, Rüprecht W, Brietenbücher K (1984) Mikroorganismen der exotherm-aeroben Schlammstabilisierung. Landwirtsch Forsch 37: 268–276.

Page FC (1988) A New Key to Freshwater and Soil Gymnamoebae. Freshwater Biology Association, Ambleside.

Parkinson D, Gray TRC, Williams ST (1971) Methods for studying the ecology of soil microorganisms. International Biological Programme Handbook 19. Blackwell Scientific Publications, Oxford.

Patterson DJ, Larsen J (1991) The Biology of Free-living Heterotrophic Flagellates. Clarendon Press, Oxford.

Paul JH (1982) Use of Hoechst dyes 33258 and 33342 for enumeration of attached and planktonic bacteria. Appl Environ Microbiol 43: 939–944.

Pimentel JP, Olivares FL, Pitaed RM, Urquiaga S, Akiba F, Döbereiner J (1991) Dinitrogen fixation and infection of grass leaves by *Pseudomonas rubrisubalbicans* and *Herbaspirillum seropedicae*. Plant Soil 137: 61–65.

Ploem JS, Tanke HJ (1987) Introduction to Fluorescence Microscopy. Royal Microscopical Society, Microscopy Handbooks 10, Oxford University Press, Oxford.

Porter JN, Wilhelm JJ, Tresner HD (1960) Methods for the preferential isolation of actinomycetes from soils. Appl Microbiol 8: 174–178.

Postgate J (1981) Microbiology of free-living nitrogen fixing bacteria, excluding Cyanobacteria. In: Current Perspectives in Nitrogen Fixation. Gibson AH, Newton WE (eds). Australian Academy of Science, Canberra, pp. 217–228.

Postma J, Altemüller H-J (1990) Bacteria in thin soil sections stained with the fluorescent brightener calcofluor white M2R. Soil Biol Biochem 22: 89–96.

Postma J, van Elsas JD, Govaert JM, van Veen JA (1988) The dynamics of *Rhizobium leguminosarum* biovar trifolii introduced into soil as determined by immunofluorescence and selective plating techniques. FEMS Microbiol Ecol 53: 251–260.

Reinhold B, Hurek T, Fendrik I, Pot B, Gillis M, Kersters K, Thielemans D, De Ley J (1987) *Azospirillum halopraeferans* sp. nov. a nitrogen fixing organism associated with roots of Kallar grass (Leptochloa fusca (L) Kunth). Int J System Bacteriol 37: 43–51.

Reinhold-Hurek B, Hurek T, Gillis M, Hoste B, Vascanneyt M, Kersters K, De Ley J (1993) *Azoarcus* gen. nov., nitrogen fixing proteobacteria associated with roots of Kallar grass (*Leptochloa fusca* L. Kunth), and description of two species. *Azoarcus indigens* sp. nov. and *Azoarcus communis* sp. nov. Int J System Bacteriol 43: 574–584.

Robinson B (1992) Manual of Analytical Methods: Protozoology. State Water Laboratory, South Australia.

Rodriguez GG, Phipps D, Ishiguro K, Ridgway HF (1992) Use of a fluorescent redox probe for direct visualization of actively respiring bacteria. Appl Environ Microbiol 58: 1801–1808.

Roser DJ (1980) Ethidium bromide: a general purpose fluorescent stain for nucleic acid in bacteria and eucaryotes and its use in microbial ecology studies. Soil Biol Biochem 12: 329–336.

Russel TB, Tranvik L (1993) Impact of acidification and liming on the microbial ecology of lakes. AMBIO, vol 22 No. 5: 325–330.

Sato K, Antheunisse J, Mulder EG (1984) A possible effect of cellulose-decomposition on soil bacterial flora. J Gen Appl Microbiol 30: 1–14.

Schmider F, Ottow JCG (1981) Quantitative Differenzierung der denitrifizierenden Flora in unterschiedlich belasteten Biotopen (Gewässer, Böden und Abwasser). Landwirtsch Forsch Sonderheft 38: 667–677.

Schmidt EL, Paul EA (1982) Microscopic methods for soil microorganisms. In: Methods in Soil Analysis, Part 2. Page AL, Miller RH, Keeney DR (eds). American Society for Agronomy, Madison, WI, pp. 803–814.

Sherr BF, Sherr EB, Fallon RD (1987) Use of monodispersed, fluorescently labeled bacteria to estimate in situ protozoan bacterivory. Appl. Environ Microbiol 53: 958–965.

Shield LM (1982) Algae. In: Methods of Soil Analysis, Part 2. Chemical and microbial properties, 2nd edn. Agronomy Monograph No. 9, ASA-SSS0, Madison, WI, pp. 1093–1101.

Sieracki ME, Johnson PW, Sieburth JMcN (1985) Detection, enumeration, and sizing of planktonic bacteria by image-analyzed epifluorescence microscopy. Appl Environ Microbiol 49: 799–810.

Singh BN (1955) Culturing soil protozoa and estiminating their numbers in soil. In: Soil Zoology. Kevan, DKMcE (ed.). Butterworths, London, pp. 403–411.

Singh-Verma SB (1968) Zum Problem des quantitativen Nachweises der Mikroflora des Bodens mit der Methode Koch. III. Einfluss der verschiedenen Nährmedien auf die Gesamtkeimzahl. Zbl Bakt Abt II 122: 457–485.

Skinner FA (1971) The isolation of soil Clostridia. In: Isolation of Anaerobes. Shapton DA, Board RG (eds). Academic Press, New York, pp. 57–80.

Söderstrom BE (1977) Vital staining of fungi in pure cultures and in soil with fluorescein diacetate. Soil Biol Biochem 9: 59–63.

Söderström BE (1979) Some problems in assessing the fluorescein diacetate-active fungal biomass in the soil. Soil Biol Biochem 11: 147–148.

Stern KR (1988) Introductory Plant Biology. Wm C. Brown Publishers, Dubuqe, IA pp. 259–263.

Stout JD (1962) An estimation of microfaunal

populations in soils and forest litter. J. Soil Sci 13: 314–320.

Sugimoto EE, Hoitink HAJ, Tuovinen OH (1990) Enumeration of oligotrophic rhizosphere Pseudomonades with diluted selective media formulations. Biol Fertil Soils 9: 226–230.

Suwa Y, Hattori T (1984) Effects of nutrient concentration on the growth of soil bacteria. Soil Sci Plant Nutr 30: 397–403.

Tomlinson GA, Hochstein LI (1972) Isolation of carbohydrate-metabolizing, extremely halophilic bacteria. Can J Microbiol 18: 698–701.

Trolldenier (1973) The use of fluorescence microscopy for counting soil microorganisms. In: Modern Methods in the Study of Microbial Ecology. Rosswall T (ed.). Bulletins from the Ecological Research Committee, Stockholm, vol 17, pp. 53–59.

Turian G, Ottow JCG (1983) Einfluss biologisch-dynamischer Impfpräparate auf chemisch-physikalische und mikrobiologische Vorgänge bei der Nachkompostierung von Müll-Klärschlammkompost. Müll Abfall 5: 118–124.

Unger H (1958) 2,3,5-Triphenyltetrazoliumsalz als Hilfsmittel bei der Keimzählung nach dem Koch'schen Gussverfahren. Arch Mikrobiol 32: 20–24.

Van Veen JA, Paul EA (1979) Conversion of biovolume measurements of soil organisms grown under various moisture tensions, to biomass and their nutrient content. Appl Environ Microbiol 37: 686–692.

Warcup JH (1960) Methods for isolation and estimation of activity of fungi in soil. In: The Ecology of Soil Fungi. Parkinson D, Waid JS (eds). Liverpool University Press, Liverpool, pp. 3–21.

West AW (1988) Specimen preparation, stain type, and extraction and observation procedures as factors in the estimation of soil mycelial lengths and volumes by light microscopy. Biol Fertil Soils 7: 88–94.

Whang R, Hattori T (1988) Oligotrophic bacteria from a redzina forest soil. Antonie van Leeuwenhoek 54: 19–30.

Williams ST (1985) Oligotrophy in soil: fact or fiction? In: Bacteria in their Natural Environments. Fletcher MM, Floodgate GD (eds). Academic Press, London, pp. 81–110.

Williams ST, Davies FL (1965) Use of antibiotics for selective isolation and enumeration of actinomycetes in soil. J Gen Microbiol 38: 251–261.

Williams ST, Wellington EMH (1982) Principles and problems of selective isolation of microbes. In: Bioactive Microbial Products: Search and discovery. Bulock JD, Nisbet LJ, Winstanley DJ (eds). Academic Press, London, pp. 9–26.

Williams ST, Lanning S, Wellington EMH (1984) Ecology of actinomycetes. In: The Biology of the Actinomycetes. Goodfellow M, Mordarski M, Williams ST (eds). Academic Press, London, pp. 483–528.

Wolin EA, Wolfe RS, Wolin MJ (1964) Viologen dyaa inhibition of methane formation by *Methanobacillus omelianski.* J Bacteriol 124: 73–79.

Wollum AG II (1982) Cultural methods for soil microorganisms. In: Methods of Soil Analysis, Part 2, Chemical and microbiological properties, 2nd edn. Agronomy Monograph No. 9, ASA-SSSO, Madison, WI, pp. 781–802.

Wynn-Williams DD (1985) Photofading retardant for epifluorescence microscopy in soil microecological studies. Soil Biol Biochem 17: 739–746.

Yanni YG (1991) Efficiency of rice fertilization schedules including Cyanobacteria under soil application of phosphate and molybdate. World J Microbiol Technol 7: 415–418.

Zehnder JAB, Wuhrmann K (1977) Physiology of a *methanobacterium* strain AZ. Arch Microbiol 111: 199–205.

Estimation of microbial activities

5

The term **microbial activity** comprises all biochemical reactions catalysed by microorganisms in soil. Some reactions, such as respiration and heat output, can be conducted by most soil microorganisms whilst others, such as nitrification and nitrogen fixation, can only be conducted by a restricted number of microbial species. To avoid confusion, measurements of metabolic activities (aerobic and anaerobic) are performed under laboratory conditions on sieved soil samples devoid of visible flora and fauna so that only the contributions to the metabolic activities of living microorganisms inhabiting the soil are expressed. In contrast, the term **biological activity** implies the contribution to overall metabolic activity of all organisms inhabiting soil including microorganisms, fauna and flora (Nannipieri et al, 1990).

Measurements of microbial activity in soils are based on the presence of intact and active microbial cells; they reflect the physiological state of microbial cells. Estimation of microbial activity in the presence of an exogenous substrate, such as glucose, is frequently considered as **potential activity**, while the measurement performed in the absence of this substrate is termed as **actual activity** (Nannipieri et al, 1990). This distinction is not strictly correct because estimations carried out under laboratory conditions using sieved soils, optimal water content and optimal temperature should be considered as potential activities even if no substrate is added and no microbial growth is taking place. Actual activity implies only field measurements or estimations with undisturbed field soils carried out under natural conditions in the absence of added substrate and under fluctuating conditions of temperature and moisture. In this case it is very difficult to distinguish between the actual microbial and the actual biological activities (for a more detailed discussion on these problems see Chapter 1).

In this chapter, up-to-date methods to estimate microbial activities in soil are presented and discussed.

Methods in Applied Soil Microbiology and Biochemistry
ISBN 0–12–513840–7

Estimation of adenosine triphosphate in soils

Adenosine triphosphate (ATP) occurs in all living organisms, where it functions as an allosteric effector, as a group-carrying coenzyme and as a substrate. ATP is the most important and central coupling agent between exergonic and endergonic processes in all cells; in dead cells ATP is quickly degraded. Owing to its properties, ATP is proposed as a parameter for either estimating microbial activities or biomass in soil. It can be extracted from soil and estimated by the bioluminescence test system.

Bioluminescence is an enzyme-catalysed reaction that results in the emission of light. The enzyme luciferase catalyses the activation of D-luciferin by ATP and its subsequent oxidation to electronically excited oxyluciferin. The transition of oxyluciferin to its ground state results in light emission. In the assay of ATP, reaction conditions are arranged to provide light emission proportional to the ATP concentration. In this way the assay is very sensitive and allows the detection of ATP concentrations as low as 10^{-11} M.

In addition to the luciferin–luciferase system, ATP can be estimated by high pressure liquid chromatography (HPLC). The application of this technique is still very limited, presumably because of the relatively long analysis time and expensive equipment (Prevost et al 1991; Martens 1992).

An important advantage of the ATP estimation is the very high sensitivity of the luciferin–luciferase test system. A significant problem of this technique is the quantitative extraction of ATP from soil. The ATP extraction efficiency depends on the extractant used, the intensity and rapidity of the inactivation of ATPases and kinases, and the ATP adsorption by soil organic and mineral colloids (Sparling and Eiland 1983; Brookes et al 1987).

The inhibitory effect of different substances found in soil extracts (e.g. NO_3^-, Mg^{2+}, Ca^{2+}, Cl^-, Br^-, humic acids, etc.) on the luciferase activity represents a further problem (Denburg and McElroy 1970; Tobin et al 1978; Eiland and Nielson 1979). Furthermore, several elements in soil extracts, such as Fe^{3+} ions, form complexes with ATP and prevent reliable estimations (Eiland and Nielson 1979).

Another problem is that ATP can be extracted not only from microbial cells but also from plant roots and animal cells, which interferes with the estimation of microbial activities (Verstraete et al 1983; Sparling et al 1985; Martens et al 1988). The relatively high number of ATP extraction methods available is probably the result of the complexity of extraction and determination of ATP (Alef 1991, 1993).

In this chapter, the methods commonly used to estimate ATP in soils will be presented and discussed.

The trichloroacetic acid extraction method

K. Alef

(Jenkinson and Oades 1979; Tate and Jenkinson 1982)

Principle of the method

The ATP is determined, after its extraction from soil with a trichloroacetic acid (TCA)–phosphate–paraquat mixture, by the luciferin–luciferase test system.

Materials and apparatus

Luminometer and cuvettes
Ultrasonic and a 12.5 dai probe
Minicentrifuge (Eppendorf centrifuge and Eppendorf tubes)
Ice bath
Filter (Whatman no. 44)
pH meter
Finnpipette and sterile strips
Glass tubes
Polypropylene or glass centrifuge tubes (50 ml)

Chemicals and solutions

EDTA–magnesium arsenate buffer

Dissolve 31.2 g $Na_2HAsO_4{\cdot}7H_2O$ in 800 ml water, add 10 ml 0.2 M EDTA (tetrasodium salt) and then 2.46 g $MgSO_4{\cdot}7H_2O$ in 100 ml distilled water; adjust to pH 7.4 with 1 M H_2SO_4 and bring up to 1000 ml with distilled water to give a solution of 0.1 M with respect to arsenate, 10 mM with respect to Mg^{2+} and 2 mM to EDTA.

TCA–phosphate–paraquat extractant

Dissolve 81.6 g TCA and 89.6 g $Na_2HPO_4{\cdot}12H_2O$ in 600 ml distilled water. Dissolve 25.8 g paraquat dichloride (1,1-dimethyl-4,4-bipyridylium dichloride) in 100 ml distilled water and then add to the TCA–Na_2HPO_4 solution. Eventually the mixture is made up to 1000 ml with water to give a solution 0.5 M with respect to TCA, 0.25 M to phosphate and 0.1 M to paraquat. The solution has a pH of 1.6 and should be stored at −15°C.

Luciferin–luciferase mixture

A lyophilized mixture (Pico-Enzyme F) of purified luciferase and D-luciferin is dissolved in 2 ml distilled water per vial, according to the manufacturers' instructions (Packard Instrument Co.). Preparations obtained from other companies can also be used (follow the manufacturers' instructions).

ATP standard solution (10^{-3} M)

Dissolve 5.07 mg of acid-free ATP (or 6.23 mg disodium salt) in 7 ml of sterile extractant and bring up with extractant to 10 ml. Place each 0.5 ml of ATP solution into sterile Eppendorf tubes and store at −20°C.

Procedure

ATP extraction

Four portions of moist soil (< 2 mm sieve) each containing 2.5 g oven-dried soil are weighed into 50 ml polypropylene (or glass) centrifuge tubes and left to stand in ice; 25 ml cold TCA-phosphate-paraquat-extractant are added to all tubes. An ATP internal standard is added into two tubes. The soil is then dispersed for 1 min using a 12.5 dia probe driven by a Branson B12 150 W sonifier at full power. The probe tip should be about 1 cm under the surface of the liquid. After the ultrasonic treatment the tubes are cooled in ice for at least 5 min and filtered. The filtered extracts can be further used or frozen quickly and stored at −15°C. Instead of filtration, the treated soil extracts can be centrifuged at 4°C using cooled sterilized Eppendorf tubes.

Estimation of ATP

The soil filtrate or supernatant (50 μl) is mixed with 5 ml EDTA–Mg arsenate buffer and kept in ice. The luminometer should be put on at least 30 min before starting and adjusted at 30°C. The luciferin–luciferase mixture (50 μl) is pipetted in the cuvette and the relative light units measured should be less than 150 RLU/10 s (blank). 250 μl of the diluted filtrate (or supernatant) are then added to the cuvette, gently mixed and the relative light units (RLU)/10 s is measured again (three times at an interval of 1 min). A second blank with the diluted filtrate alone should also be performed (RLU/10 s less than 150).

Calibration curve

For calibration, dilute the ATP standard solution to obtain standard solutions with ATP concentrations ranging from 10^{-5} to 10^{-8} M and measure the RLU/10 s as described above (prepare a new calibration curve daily). The aim of the curve is to estimate the ATP concentration that should be added as an internal standard to the soil sample before ultrasonic treatment, and to prove the linearity of the ATP measurements.

Example

Assuming, that the RLU/10 s for soil extract is 3000, we can calculate from the calibration curve an equivalent ATP concentration, which should be added as ATP internal standard to the soil sample.

Calculation

ATP content in the soil can be calculated briefly

$$ATP\ (\mu g)/dwt\ (g) = \frac{A \times ATP_s \times 40}{B - A} \quad (5.1)$$

where A is the RLU/10 s measured for the soil extract, ATP_s is the added ATP internal standard in micrograms, B is the RLU/10 s measured for the soil extract containing the ATP internal standard, *dwt* is the dry weight of 1 g moist soil and 40 is the dilution factor.

Discussion

Paraquat can be omitted from the extractant because it only improves the extraction efficiency (Verstraeten et al 1983; Alef et al 1988; Inubushi et al 1989a). There are disadvantages to the use of paraquat; the compound is toxic and its pure commercial preparation is expensive.

The ATP internal standard was added to the soil sample before ultrasonic treatment. Several authors have omitted this step and have added the ATP standard to the soil extract in the cuvette.

The RLU/10 s should be measured several times (0–3 min). At the beginning a strong increase in the RLU can be observed, which then decreases slowly. The highest value should be taken for the calculation of results.

The cuvette can be cleaned as follows: wash the cuvette with distilled water and treat overnight with HCl (0.1 M), then wash several times with double distilled water and finally dry at 40°C.

This method is not suitable for ATP estimations in soils containing more than 10% $CaCO_3$ (Jenkinson and Oades 1979), because TCA is neutralized. With calcareous soils with a $CaCO_3$ content lower than 10%, it is necessary to interrupt the ultrasonic treatment after a few seconds to allow the bubbles to disappear.

TCA extracts ATP from microbial, plant and animal cells. Removing visible plant roots and animal tissue from soil samples is therefore recommended.

After every TCA extraction be sure that the pH of the soil suspension is less than 2.

Be sure that the pH of the luciferin–luciferase–soil filtrate mixture is 7.75. Otherwise the soil filtrate must be neutralized by using 0.1 M NaOH solution.

Strong dilution of the soil filtrate (1:100) is necessary to remove the quenching effect of TCA.

Ultrasonic treatment must be performed on ice to avoid a strong heat output.

Prepare a new ATP calibration curve daily.

The ATP concentrations extracted from soils depend very much on the ratio of soil weight to volume of extractant. This ratio must remain constant.

A thin probe can be used to disperse small soil samples (0.1 g ml^{-1} extractant).

TCA seems to extract ATP from microbial cells in the stationary phase (Verstraeten et al 1983).

The method can also be used for ATP estimation in anaerobic soils (Inubushi et al 1989b).

Special attention should be given to the stability of luciferin–luciferase preparation. Store luciferin in the dark and do not freeze the luciferase solution (see manufacturers' instructions).

Significant correlations were found between this method and the arginine ammonification, heat output, N-mineralization and glucose-induced respiration, but controversial correlations were found with the biomass estimated by the fumigation incubation method (Jenkinson et al 1979; Ross et al 1980; Sparling et al 1981; Verstraeten et al 1983; Alef et al 1988; Van de Werf 1989/1990).

The use of purified luciferase is proposed in this method. Lower ATP content in soil was found when crude luciferase preparations were used (Tate and Jenkinson 1982).

The TEA/NRB method

H. Van de Werf
G. Genouw
W. Verstraete

(Van de Werf and Verstraete 1979; Van de Werf 1989/1990)

Principle of the method

Van de Werf and Verstraete (1979) developed a method to extract selectively microbial ATP from soil. It was found that NRB (a quaternary detergent) was capable of releasing ATP selectively from unicellular organisms (bacteria, fungi, protozoa and algae).

Materials and apparatus

Kenwood Waring blender.
ATP photometer: the light production is followed by integrating the signal for 10 s using a Biocounter (Lumac BV), a Luminometer (LKB-Wallac) or a Luminescence Analyser (LKB-Wallac).

Chemicals and solutions

Tris-EDTA-NaN_3(TEA) buffer

Dissolve 1.2114 g of Tris (hydroxy methyl)-aminomethane, 0.336 g of EDTA (disodium salt) and 0.195 g of sodium azide (NaN_3) in 900 ml freshly distilled water. Adjust the pH to 7.75 with 0.1 N H_2SO_4 and bring up to 1000 ml with distilled water. The TEA buffer is stable for 2 days if stored at 4°C.

LumitR buffer

6.3 g of *N*-(2-hydroxyethyl)-piperazine-*N*-(3-propane-sulphonic acid) (HEPES), pH 7.75. The buffer is stable for 18 months when stored at 4°C.

LumitR PM

The purified luciferin–luciferase in a lyophilized form.

Preparation of the purified luciferase enzyme

Dissolve the lyophilized form in 7 ml LumitR buffer. This solution is stable for 4 weeks at −22°C.

NRB reagent

Nucleotide-releasing agent for bacteria (NRB) was obtained from Lumac BV, The Netherlands. The reagent is kept at room temperature.

ATP standard

As a freeze-dried form

The bottle contains 10 μg free adenosine 5′-triphosphate sodium salt (MW 605.2), which is stable for 1 year at −22°C.

Preparation of the ATP standards

A stock solution is prepared by dissolving the freeze-dried form in 100 ml of TEA buffer. The stock solution is further diluted to the desired concentrations. Usually the following concentrations are used: 0.25, 0.5 and 1 ng of ATP per 10 μl. These standards are stable for 4 weeks at −22°C.

All reagents should be brought to room temperature before use.

Procedure

Thirty grams of wet soil (3/4 FC) are suspended in 270 ml freshly prepared TEA buffer, homogenized for 1 min in a

Kenwood Waring blender, and allowed to equilibrate for 10 min (dilution 1:10). After 10 min, the suspension is carefully mixed and 10 ml of this suspension are then added to 90 ml of TEA buffer (dilution 1:100). Depending on the soil texture, further dilutions are performed, e.g. 1:200, 1:500, 1:1000 and 1:2000. The dilutions are allowed to equilibrate for 10 min but are mixed again before further use.

One hundred microlitres of the latter suspension are transferred to a 5.0 ml plastic vial, 100 µl of the ATP-releasing reagent NRB are added and the mixture is gently shaken for 10 s. The vial is inserted into the luminescence measuring apparatus, 100 µl of the luciferase enzyme preparation are injected into the vial and the light output is integrated over a period of 10 s. Three luminescence measurements are performed for each soil suspension. To convert the relative light units (RLUs) to ATP, a standard addition is practised for each soil sample analysed. Ten µl of the ATP standard, containing about the same amount of ATP present in the 100 µl soil suspension, are added to the vial just before NRB extraction and before the actual reading in the Biocounter.

Calculations

The ATP-TEA/NRB values are calculated and expressed on a dry weight basis, according to the following expression

ATP (µg)/dry soil (kg) =

$$\frac{(a - c).\ ATP_{st}\ .\ f}{(b - a)} \times \frac{(100 + z)}{10} \qquad (5.2)$$

where ATP_{st} is the ATP (in nanograms) added to 100 µl soil suspension, *a* is the average of the RLUs of the sample, *b* is the average of the RLUs of the internal standard, *c* is the average of the RLUs of the blank, *f* is the dilution factor and *z* is the moisture content of the soil by 3/4 FC (expressed in percentage water with regard to dry soil).

Discussion

The determination of the ATP content by this method is rapid and accurate. This depends on the standard addition and the use of high quality enzyme preparations (Verstraete et al 1983; Van de Werf 1989/90).

An important aim in biological studies in soil is to separate the soil biomass. By using selective extractants, the soil biomass can be separated in to microbial (ATP-TEA/NRB), faunal (ATP-TEA/NRB) and total ATP (ATP-TCA) (Van de Werf 1989/90).

The ATP-TEA/NRB values correlate very significantly with total soil microbial biomass as determined by the $CHCl_3$ fumigation–incubation method (Van de Werf and Verstraete 1984).

In soils with actively growing microbial biomass, the microbial ATP (TEA/NRB) and the total ATP (ATP-TCA) values are nearly equal (Verstraeten et al 1983). Soils for which the microbial ATP (TEA/NRB) is lower than the total ATP by a factor of about 2.4 can be considered normal and equilibrated (Verstraeten et al 1983). Soils for which ATP(TCA) surpasses ATP(TEA/NRB) by a factor of 4.0 or more, were considered as special and most probably rich in senescent plant material (Van de Werf and Verstraete 1984).

The ATP(TEA/NRB) extraction method can be used to study the effect of pollutants on soil microorganisms.

The sulphuric acid–phosphate extraction method

K. Alef

(Eiland 1983)

Principle of the method

The method is based on the extraction of ATP with sulphuric acid phosphate solution and estimation of ATP with the luciferin–luciferase test system in the presence of NRB.

Materials and apparatus

Luminometer and cuvettes
Ice bath
Plastic tubes (77 × 22 mm)
Further materials (see TCA method)

Chemicals and reagents

Buffer A

Dissolve 37.7 g Tris and 1.5 g EDTA in 700 ml distilled water, adjust the pH to 7.5 with diluted acetic acid and bring up with distilled water to 1000 ml to obtain a solution 250 mM with respect to Tris, and 4 mM to EDTA.

Buffer B

Dilute buffer A ten times with distilled water and sterilize by autoclaving for 20 min at 121°C.

Extractant A

Dissolve 27.3 ml of a 96% H_2SO_4 and 44.5 g $NaHPO_4{\cdot}2H_2O$ in 700 ml distilled water and bring up with distilled water to 1000 ml to obtain a solution of 0.5 M H_2SO_4 and 0.25 M Na_2HPO_4.

Extractant B

It may be either the quaternary detergent NRB (Lumac) or 10% Rodalon (Ferrosen, Copenhagen; containing a quaternary ammonium compound alkyldimethyl-benzylammonium chloride) dissolved in Tris–EDTA buffer B to give a 0.005% solution.

Luciferin–luciferase (Lumit)

The purified luciferin–luciferase enzyme, Lumit (Lumac), is dissolved in 5 ml HEPES buffer, containing 25 mM (*N*-2-hydroxyethylpiperazine-*N*-1,2-ethanesulphonic acid), 10 mM $MgSO_4$ and 0.02% NaN_3, pH 7.75 (Lumac), and stored in the dark for 2 h at 20°C before use.

ATP standard solutions

These are prepared by dissolving 10 μg crystalline disodium ATP (Lumac) in buffer B (25 mM Tris–0.4 mM EDTA, pH 7.5 at 25°C), giving an ATP range of from 0.2 to 4 μM. Standard ATP in these concentrations gives a reproducible linear response.

Procedure

Extraction of ATP

Soil samples (1 g) are placed into plastic tubes containing 10 ml of extractant A and kept on ice in plastic boxes. The ATP standards are performed by adding soil samples (1 g) to tubes containing 9.5 ml of extractant A and 0.5 ml of ATP standard solutions, similarly. The assay is routinely standardized by using 0.5 ml of a 4 μM ATP solution, which contained about the same amount of ATP as present in the soil suspension. To calculate the recovery of added ATP, 0.5 ml of a 4 μM ATP solution is added to tubes containing 10 ml of extractant A. All soil suspensions are shaken for 15 min on a reciprocating shaker at 0°C with 164 movements min^{-1}. Soil blank suspension for correcting ATP present in the enzyme solution is prepared by adding a sample (1 g) to 10 ml of extractant A. The suspension is then shaken for 15 min, sterilized by autoclaving for 20 min at 121°C and then cooled.

Measurement of ATP

Cooled soil blank suspension, samples and ATP standards are all measured using the following procedure. Shake the suspension for 5 s, which results in a homogeneous suspension, and pipette immediately 50 μl from the upper part of the suspension (1–2 cm layer) into 1.5 ml of buffer A, kept in an ice bath. After shaking for 3 s, pipette 50 μl of the latter mixture into a cuvette containing 50 μl NRB extractant and shake the mixture gently for 5 s. Place the cuvette in the luminometer for a further 5 s and then inject 100 μl of the luciferin–luciferase enzyme. The RLU is measured over a 0–10 s integration period. If the maximum light intensity is not reached during the first 10 s after injection of the enzyme, the ATP content is measured over a 0–30 s integration period. Two samples are taken, but from each acid suspension and measured. Three replicated soil samples are analysed.

Calibration curve and calculation of results

The internal ATP standard curve is prepared by adding different ATP concentrations to soil suspensions, which are then treated in the same way as the soil samples.

To obtain a linear ATP standard curve, the following measurements are made

(sample ATP + standard ATP) – (sample ATP); and, for the calibration of the samples: (sample ATP) – (soil blank suspension ATP).

The ATP content of soils is read from the standard curve and given in micrograms of ATP per gram of dry weight of soil corrected for recovery of added ATP, which can be calculated as follows

% recovery of ATP =

$$\frac{(A + B) - (A) \times 100}{C - D} \tag{5.3}$$

where A is the sample ATP, B is the standard ATP, C is the standard ATP measured in the extract and D is the reagent blank.

Discussion

An efficient ATP extraction is obtained when the pH of the soil suspension after adding the extractant A is lower than 2.

The effect of extractant to soil ratio on ATP extraction was studied only in a sandy soil (Eiland 1979). An extractant to soil ratio of 10:1 removed the same amount of ATP from the soil as the ratio of 25:1. However, it was difficult to obtain reproducible results with the latter ratio.

Significant correlations were found between this method and the biomass estimated by the fumigation–incubation method (Eiland 1979).

The ratio of ATP (this method)/biomass C is strongly affected by the storage conditions of soils (Eiland 1979).

The acidic phosphoric acid extraction method

C. Ciardi
P. Nannipieri

(Webster et al 1984; modified by Ciardi and Nannipieri 1990)

Principle of the method

According to Ciardi and Nannipieri (1990) the original extractant proposed by Webster et al (1984) was modified by omitting the use of Zwittergent to avoid foam production; thus the acidic extractant (PA) consists of H_3PO_4, urea, dimethylsulphoxide (DMSO), EDTA and adenosine. In addition, the buffering system was modified and the sonication time was increased to improve the efficiency of the ATP extraction.

Materials and apparatus

Luminometer and cuvettes
Beakers (50 ml)
Shaker
Ultrasonic apparatus with maximum frequency of 20 kHz
Ice bath

Chemicals and solutions

Extractant A

0.67 M H_3PO_4, 2 M urea, 20% (v/v) DMSO, 0.02% (w/w) adenosine, 20 mM EDTA.

Extractant B

Extractant A with the addition of ATP (to reach 5–10 ng in the cuvette during ATP measurement) as a spike.

Buffer

0.2 M Tris, 4 mM EDTA, 15 mM magnesium acetate brought to pH 10.2–10.8 with 1 M NaOH; the final pH of the buffer depends on the amount of soil extract to be used for ATP determination.

Luciferin–luciferase mixture (Monitor Reagent, BioOrbit, Tuku, Finland).

Internal standard (ATP standard, BioOrbit, Tuku, Finland).

Procedure

Soil samples are sieved under moist conditions and weighed (1 or 2 g) into beakers. Extractant A (1:10 soil:extractant ratio) is added and the mixture is homogenized and sonicated for 2.5 min at a 60% of the maximum frequency by using the medium size tip of a sonicor (New York) instrument. The beaker is cooled in ice. After capping the beaker with parafilm, the soil suspension is shaken on a water-shaking bath at 150 shakes min^{-1} for 30 min at 4°C. The mixture is then filtered by using a 42 Whatman filter. Soil extract (0.01–0.15 ml) is diluted to 0.8 ml with Tris–EDTA–Mg acetate buffer to reach a final pH of 7.3–7.8. A similar procedure is followed for the same soil sample by employing extractant B to quantify the efficiency of extraction.

Assay of ATP

An aliquot (0.2 ml) of monitor reagent is added to the fresh (or stored at −15°C) buffered extract (0.8 ml) in a polyethylene vial and then the mixture is shaken gently. The light output is then measured in a 1250 Luminometer (LKB-WALLAC, Finland) using an integration mode (10 s integration period). Three replicates are usually measured. After each determination, an internal ATP standard (50.7 ng) is added to correct for the influence of soil extract on the light output.

Calculation of results

The ATP content of soil is calculated by the internal standard procedure and by considering the ATP recovery obtained by the efficiency of extraction:

$$ATP = \frac{b - a}{c} \times d \qquad (5.4)$$

where a is the value of the blank (only the buffered extract), b is the value of sample (after the addition of monitor reagent), c is the value after addition of the internal standard and d is the amount of ATP standard added.

$$\%\ \text{Recovery of ATP} = \frac{(\text{ATP from extractant B}) - (\text{ATP from extractant A})}{\text{Added ATP in extractant B}} \times 100 \qquad (5.5)$$

Discussion

The acid extractant (PA) was found to be more effective in extracting ATP from soil than the TCA–phosphate–paraquat mixture used by Webster et al (1984) and Ciardi and Nannipieri (1990). The phosphoric acid and the adenosine of the PA mixture saturate, as do the NaH_2PO_4 and paraquat of the TCA–phosphate–paraquat extractant, the ATP and phosphate binding sites, respectively. Components such as EDTA, urea and DMSO present only in the PA mixture improve the ATP extraction because they separate cells from surfaces of soil colloid and lyse them; in addition, these compounds serve to chelate cations, to denature proteins and to prevent hydrogen bonding, respectively (Ciardi and Nannipieri 1990). When the two extractants were compared in organic forest soils, no differences in ATP extraction were found (Arnebrant and Baath 1991).

Estimation of the adenylate energy charge in soils

P.C. Brookes

The adenylate energy charge (AEC), defined by Atkinson (1977) as AEC = ([ATP] + 0.5 [ADP])/([ATP] + [ADP] + [AMP]) is a measure of the metabolic energy stored in the adenine nucleotide pool. Atkinson proposed that its measurement could provide an indication of the metabolic status of the organism or population under study. In theory, AEC could range from 0 (i.e. all AMP) to 1.0 (i.e all ATP). However, results mainly obtained from microorganisms growing in the chemostat show that organisms that are growing rapidly, indicated by such processes as active biosynthesis, have AECs that range from about 0.8 to 0.95, while values between about 0.5–0.75 indicate that the population is in a resting or stationary phase, incapable of biosynthesis. Values much below 0.4 are generally taken to indicate a moribund population, a situation that cannot usually be reversed. In contrast, viable spores may have an AEC below 0.1. *In vitro* both eukaryotic and prokaryotic microorganisms maintain similar AECs in the same growth phase (see Atkinson 1977).

Most AEC determinations and the empirical significance of AEC values have been obtained from studies on pure cultures of microorganisms or plant tissues. Fewer studies have been done on soil microorganisms *in vivo*, collectively known as the soil microbial biomass.

Evidence is now in the literature which shows that, far from becoming clearer with time, the interpretation of AEC values of the soil microbial biomass is, currently, not resolved.

All the proposed methods of measuring AEC in soil have several common features. These include disruption of the cells of the soil microorganisms, commonly with ultrasonics. During the process of cellular disruption the nucleotides are released into an appropriate extracting solution, which is then filtered to provide a soil extract. Concentrations of the three adenine nucleotides are then measured. ATP is measured first, usually using the firefly luciferin–luciferase system. ADP and AMP are then individually converted to ATP in separate portions of the extract and total ATP measured again by the luciferin–luciferase system. ADP and AMP are then calculated by difference. An approximate correction for incomplete extraction from soil is made by measuring percentage recovery of known amounts of adenine nucleotides added to soil extractant followed by the extraction of amended or unamended portions of the same soil. To understand the current situation with AEC determinations, and the problems involved, it may be helpful to review the background and history of AEC measurements in soil.

Brookes et al (1983) reported the first measurement of AEC in soil; an AEC of 0.85 was measured in an unamended grassland soil. On air-drying, AEC fell to 0.45 but recovered to 0.76 when the air-dried soil was rewetted for 1.5 h at 25°C.

In contrast, Martens (1985) reported much lower AEC values (0.3–0.4) in six agricultural soils. Additions of up to 10,000 μg C g^{-1} soil as glucose-C or plant-C increased the AEC, but only to a maximum of 0.67.

Thus, the data of Brookes et al (1983) suggested that the mainly dormant soil microbial biomass had an AEC similar to that of microorganisms growing exponentially in the chemostat. Martens (1985) measurements showed the precise opposite. They suggested a mainly dormant population, raising its AEC in response to inputs of fresh substrates, but not to the high level reported for microorganisms growing exponentially *in vitro*.

There were, however, significant differences between the nature of the extractants used by Brookes et al (1983) and Martens (1985). The extractant that Brookes et al used was that developed by Jenkinson and Oades (1979), a mixture of 0.5 M TCA, 0.5 M $Na_2H_2PO_4$ and 0.25 M paraquat with a pH of about 1.2. The phosphate and paraquat block positive and negative ATP absorption sites, respectively, on the soil surfaces, which otherwise markedly decrease the amounts of ATP extracted.

TCA is one of the most powerful inhibitors of phosphatases and ATPases known. Thus, Lundin and Thore (1975) found that TCA was the most effective of a number of reagents tested in inhibiting ATPase activity in microorganisms *in vitro*. Consequently, they reported that the measured AEC of five different microorganisms was generally highest with TCA (range 0.59–0.97) compared to other reagents, although AECs in extracts of perchloric or sulphuric acids were often similar to those with TCA. It was noticeable that AECs in KOH or Tris–EDTA extracts were consistently lower. For example, with Tris–EDTA, the AECs measured in the five organisms ranged from 0.54 to 0.77. Chloroform extraction gave the lowest AECs and the smallest measured ATP concentrations because it was particularly ineffective in deactivating dephosphorylating enzymes (Lundin and Thore 1975). Significant phosphatase activity was also reported by Brookes et al (1982) in soils held under chloroform.

Martens (1985) used alkaline 0.25 M $NaHCO_3$ containing 0.1 M KH_2PO_4 and 16 mM adenosine to extract ATP, ADP and AMP from soil. To 50 ml of the above solution was added 5 g soil and 15 ml $CHCl_3$, then the mixture was ultrasonicated. Brookes et al (1987) compared TCA and $NaHCO_3$ on a grassland soil, amended or not with 1100 μg ryegrass-C g^{-1} soil, at 0, 50 and 100 days of incubation at 50% water holding capacity (WHC) and 25°C. Some of their results are summarized in Table 5.1a. From this and other data in their paper they reported that:

1. AEC values in TCA soil extracts were high in all treatments (0.8–0.9) and independent of substrate addition or length of incubation.
2. AEC values in $NaHCO_3$ soil extracts were low (0.4) in fresh soil. AEC increased from 0.4 to 0.6 when the soil was incubated for 50 days whether or not substrate was added.
3. Substrate addition increased the total pool

Table 5.1a. Adenine nucleotides extracted by TCA or $NaHCO_3$ reagent* from a UK grassland soil, amended or not with ryegrass. Nucleotides were determined by the enzymic procedure.

Soil treatment	*Extractant*	*ATP†* *(nmol g^{-1} soil)*	*A_T†‡* *(nmol g^{-1} soil)*	*ATP as % of A_T*	*AEC*
Fresh soil	TCA	9.2	11.3	82	0.87
	$NaHCO_3$	4.5	14.3	32	0.39
Soil incubated for 50 days	TCA	11.7	15.4	76	0.88
	$NaHCO_3$	7.1	15.0	48	0.55
Soil incubated with ryegrass for 50 days	TCA	23.3	27.8	84	0.90
	$NaHCO_3$	12.1	22.1	55	0.60
LSD ($P = 0.05$)	TCA	1.99	3.14		0.12
	$NaHCO_3$	1.52	5.10		0.15

* For full analytical results see Brookes et al (1987).

† Corrected for incomplete recovery of added adenine nucleotides (see text).

‡A_T = ([ATP] + [ADP] + AMP]).

size of the three adenine nucleotides (A_T) as measured in both extractants. However, while A_T was very similar within treatments and within reagents, ATP as a percentage of A_T was very much lower with the $NaHCO_3$ reagent (32–55%) than with TCA (76–82%).

This finding of less ATP extracted from soil by the $NaHCO_3$ reagent than the TCA reagent is also in agreement with the results of Jenkinson et al (1979). They reported that $NaHCO_3$ only extracted about 30% of the ATP that was extracted by TCA from a range of arable, grassland and woodland soils. These soils covered a wide range of soil properties. For example, pH ranged from 3.9 to 8.3, soil organic C from about 0.8 to 3.9 and clay content from 21% to 43%.

Brookes et al (1987) therefore concluded that the main reason for the lower AECs observed with the $NaHCO_3$ reagent was that microbial ATPases were still active during extraction of soil and caused appreciable hydrolysis of microbial ATP to ADP and AMP. In contrast, they considered that their data showed, as reported by Lundin and Thore (1975), that TCA rapidly inactivated ATPases and was therefore preferable for extracting adenine nucleotides from soil.

The controversy over the AEC levels in the soil microbial biomass was therefore considered to be resolved. The evidence seemed to point clearly to the low AECs measured following extraction with $NaHCO_3$ being mainly due to enzymic hydrolysis of ATP to AMP and/or ADP during its extraction from soil. This also explains the previous finding of Jenkinson et al (1979) that soils extracted with TCA contained about three times as much measured ATP as soils extracted with $NaHCO_3$. The alternative explanation that ADP and AMP could be enzymically phosphorylated to ATP during extraction with 0.5 M TCA does not seem credible. The major processes that occur during or after cellular disruption are degradative.

All the above AEC analyses were done following enzymic conversion of ADP and AMP in soil extracts to ATP. Recently, however, Martens (1992) reported AEC measurements made on four soils following extraction with either a modified $NaHCO_3$ reagent (containing 6 mM EDTA and 20 mM inosine in place of adenosine) or with 0.5 M H_2SO_4 containing 0.25 M Na_2HPO_4 and 0.1 M paraquat. The extracted nucleotides were than analysed either by the enzymic procedure as described above or by a HPLC procedure as developed and described by Martens (1992). Agreement between nucleotide concentrations measured in the same extracts, either by enzymic proce-

Table 5.1b. Adenine nucleotides extracted from four unamended soils by modified H_2SO_4 or $NaHCO_3$ reagent and analysed by HPLC*.

Soil no.	*Soil clay content (%)*	*Extractant*	*ATP (nmol g^{-1} soil)*	*A_T (nmol g^{-1} soil)*	*ATP as % A_T*	*AEC*
1	49	H_2SO_4	5.1	8.1	63	0.76
		$NaHCO_3$	6.5	18.5	35	0.52
2	68	H_2SO_4	3.9	5.6	69	0.80
		$NaHCO_3$	4.5	11.7	39	0.53
3	23	H_2SO_4	2.3	3.4	68	0.80
		$NaHCO_3$	2.0	4.4	46	0.60
4	10	H_2SO_4	2.0	3.7	54	0.63
		$NaHCO_3$	1.7	4.1	49	0.58
LSD (P = 0.05)		H_2SO_4	0.78	0.90		0.16
		$NaHCO_3$	0.49	0.94		0.08

* For complete analytical results, see Martens (1992).

† Corrected for incomplete recovery of added adenine nucleotides (see text).

dures or by HPLC were generally close. However, these data are in disagreement with the findings of Brookes et al (1983) in many ways. Major discrepancies between the two sets of data are listed below:

1. Martens (1992) reported that in two out of the four soils, the amount of ATP extracted by the acidic reagent was significantly less, by 22% and 11%, respectively, than by the modified $NaHCO_3$ reagent. In the other two soils significantly more ATP was extracted with the acidic reagent (Table 5.1b).
2. Martens (1992) reported that the total amounts of adenine nucleotides (A_T) extracted from soils by the modified $NaHCO_3$ reagent were consistently higher than the amounts extracted by H_2SO_4. In the two soils with the highest clay content, A_T was about twice as great when they were extracted by $NaHCO_3$ than by H_2SO_4. In others, the difference was less (Table 5.1b). Martens (1992) considered the main reason for the differences between extractants was that $NaHCO_3$ consistently extracted more ADP and AMP than did H_2SO_4
3. The greater amounts of ADP and AMP extracted by the modified $NaHCO_3$ reagent caused lower calculated AECs with this reagent than with H_2SO_4. While Brookes et al (1987) found that ADP and AMP formed a larger proportion of A_T when extracted by $NaHCO_3$ than H_2SO_4, A_T was similar with both reagents. AECs with the modified $NaHCO_3$ reagent ranged from 0.52 to 0.60 while those with H_2SO_4 ranged from 0.75 to 0.80.

Martens (1992), considering that "High AEC values (0.8–0.9) conflict with the generally accepted hypothesis that the microbial biomass in soils is a largely dormant population with low metabolic and turnover rates" favoured the lower AEC values obtained following extraction with $NaHCO_3$. He considered that addition of 6 mM EDTA to the $NaHCO_3$ reagent would prevent degradation of ATP and ADP to AMP completely, although this has never been demonstrated in soil extracts. Indeed, he attributed the lower AEC values (0.3–0.40) reported in earlier work (Martens 1985) to partial dephosphorylation of ATP to ADP and AMP when soil was extracted with $NaHCO_3$ containing EDTA, although at a lower concentration.

All other reported AEC measurements in soil have been made using acidic reagents. Nannipieri et al (1990) reviewed the current literature and also considered that the main discrepancy was between results obtained when acid extractant mixtures were compared with alkaline $NaHCO_3$ reagents. Thus, in nearly all moist soils so far examined, which have first been given a conditioning incubation at between 15 and 20°C and then extracted by acidic reagents, AEC has been within the range 0.70 and 0.90. It can be altered, however, depending upon soil conditions. When soils were incubated at 5°C, AEC fell from 0.74 to 0.59 (Eiland 1985). Air-drying of a grassland soil also decreased AEC from 0.85 to 0.45 (Brookes et al 1983). AEC then increased to 0.76, nearly the level in fresh soil, when it was rewetted and incubated for 1.5 h at 25°C. Almost identical results for a sandy loam soil were also reported by Eiland (1985). Ciardi et al (1991) reported AEC values of 0.65 and 0.80 for a clay loam and organic clay soil sampled at 50% WHC and after storage at 4°C for 15 days. Another clay loam soil, stored moist for 3 years at 4°C before analysis, also had a high (0.70) AEC. These soils, all from Italy, were extracted by an acidic reagent based on 0.67 M H_3PO_4, 2 M urea and 20% dimethyl sulphoxide (Webster et al 1984). Upon air-drying and rewetting, the AECs of the three soils were 0.47, 0.62 and 0.32, respectively, rather lower than the values reported by Brookes et al (1983) or Eiland (1985). Whether this reflects true differences between the behaviour of the biomasses in northern and southern European soils or is related to differences in methodologies remains to be seen.

Ciardi et al (1993) showed that addition of NaN_3 to the same three Italian soils during rewetting decreased the net ATP produced following this rewetting by more than 50%. Following $CHCl_3$ fumigation, AEC fell to very low (0.06–0.1) levels, suggesting that nearly all the microbial cells had been killed. Surprisingly, A_T was increased by up to nearly 100% in the air-dried soils and in the air-dried–rewetted soils, with or without NaN_3 and up to 50% in the fumigated soils given the same treatments.

They considered that the larger pool of A_T following addition of NaN_3 was an artefact because more light per ATP unit was produced by the firefly luciferin–luciferase system in the presence of NaN_3. They also considered that the larger pools of A_T produced during air-drying–rewetting and $CHCl_3$ fumigation were due to hydrolysis of RNA during these processes. A similar argument was put forward by Rosacker and Kieft (1990) to account for transient decreases in AEC to 0.2–0.25 followed by an increase finally stabilizing at *c.* 0.4 when an air-dried grassland soil was rewetted. This further illustrates the difficulty in the correct measurement and interpretation of soil adenine nucleotide concentrations and AEC values. It can be seen, therefore, that soil AEC measurements remain controversial. The controversy mainly rests upon whether soil adenine nucleotides are most accurately extracted with acidic reagents or alkaline ones. The author considers that the case for recommending extraction with acidic reagents remains strong but cannot explain Martens (1992) finding that similar amounts of ATP were extracted by both types of reagent, while significantly more ADP and AMP were extracted by the modified $NaHCO_3$ reagent.

Adenylate energy charge as an indicator of microbiological activity in soils subjected to environmental stress

Residual heavy metals (e.g. Cu, Ni, Zn, Cd) from past field applications of metal-contaminated sewage sludge decreased the amounts of soil microbial biomass and microbial ATP, ADP and AMP by about half in an English sandy-loam soil. However, AEC was high and similar (0.85 and 0.89, respectively) in both the metal-contaminated and uncontaminated soils (Brookes and McGrath 1987). This suggests that AEC measurements may not be valid indicators of environmental stress in soils polluted with heavy metals. In the above case both soils mineralized similar amounts of soil organic C and N during laboratory incubations, which is consistent with them having comparable microbial activities, as measured by AEC.

ATP and AEC measurements have also been done in soils incubated anaerobically using the TCA reagent of Jenkinson and Oades (1979). Inubushi et al (1989b) reported little change in soil ATP when three well-drained soils from Rothamsted and three paddy soils from Konosu, Japan, were incubated aerobically. In contrast, ATP fell to less than 10% of its original amount in the Rothamsted soils after 80 days of anaerobic incubation. The decline in ATP in the Konosu paddy soils during anaerobic incubation was much more gradual. Final amounts after 80 days of anaerobic incubation were about 21–38% of the amounts in the original aerobically incubated soils.

AEC also decreased during anaerobic incubation, from 0.76 to 0.34 in the Rothamsted grassland soil and from 0.75 and 0.54 in one of the Konosu paddy soils. AEC similarly increased upon aeration but not to the level in fresh soil. These results suggest that the soil microbial biomass, when stressed by exposure to aerobic conditions, does not maintain ATP or AEC at the customary high level found in aerobic soil, as was also shown for the microbial biomass when stressed by the air-drying of soil. Upon aeration, however, restoration of both ATP and AEC occur rapidly, but not to the levels in fresh soil.

Enzymic procedure (TCA extraction)

(Brookes et al 1983)

Principle of the method

The method is based on the TCA extraction of ATP, ADP and AMP from the soil, followed by the determination of these adenylates by the luciferin–luciferase enzyme assay. All measurements are expressed on an oven-dried soil basis and in terms of molar quantities of the free acids ATP, ADP and AMP.

Soil treatment

Analyses may conveniently be carried out on moist soil, sieved < 2 mm, and incubated over water and soda lime at between 15 and 25°C for 7–10 days before analysis.

Materials and apparatus

Liquid scintillation counter
Sterile pipettes
Water bath, adjustable to 100°C
Water bath, adjustable to 30°C
Ice bath
Filter paper (Whatman No. 44)
Volumetric flasks (5 ml)

Chemicals and solutions

Reagents are all A.R. grade and distilled water is used throughout.

Stock 0.1 mM solutions of ATP, ADP and AMP are prepared in water and frozen at −15°C. Diluted standards were prepared immediately before use and kept over ice.

Phosphate buffer.

$Na_2HPO_4 \cdot 12H_2O$ (43.0 g) is dissolved in 600 ml water, adjusted to pH 7.4 with 2 M HCl and made up to 1 l.

Pyruvate kinase and phosphoenolpyruvate enzyme mixture

Pyruvate kinase (PK) and phosphoenolpyruvate (PEP) are dissolved in phosphate–magnesium buffer to give a solution containing 1.6 mM PEP, 24 mM Mg^{2+} and 88 μg PK 100 μl^{-1} and kept on ice. The phosphate–magnesium buffer is made by dissolving $MgCl_2 \cdot 6H_2O$ (121 mg) in 25 ml phosphate buffer immediately before use and cooling to 0°C.

Pyruvate kinase, phosphoenolpyruvate and myokinase enzyme mixture

The myokinase (MK), obtained as a suspension in 3.2 M $(NH_4)_2SO_4$ is dialysed in 1.5 l magnesium-free phosphate buffer to remove the $(NH_4)_2SO_4$. Phosphate–magnesium buffer (sufficiently enriched in Mg to give the required final Mg concentration) is then added. Final Mg, PEP and PK concentrations are as above, and MK is added to give 140 μg 100 μl^{-1} in the phosphate–magnesium buffer.

Luciferin–luciferase

Pico-Zyme F (a lyophilized mixture of purified D-luciferin and luciferase) is prepared according to the manufacturers' instructions. Sensitivity to ATP is increased by dissolving 500 μg luciferin ml^{-1} Pico-Zyme F solution. The preparation is kept on ice throughout.

Soil extractants

Replicate portions of each soil are extracted separately with four different sets of extractant, extractants A, B, C and D. Extractant A contains 0.5 M TCA, 0.25 M phosphate (as Na_2PO_4) and 0.1 M paraquat (Jenkinson and Oades 1979). Extractants B, C and D contain $0.12[25 + (w - 2.50)]$ml 0.1 mM AXP, where AXP = ATP, ADP or ADP, respectively, all made up to 250 ml with extractant A. The volumes of AXP all give 50 pmol AXP in the scintillation vial, assuming

100% recovery from soil (see later). The term w is the quantity of fresh soil containing 2.50 g oven-dried soil extracted by 25 ml extractant.

Procedure

Extraction of ATP, ADP and AMP from soil

To determine the AEC of a given soil, replicate portions of moist soil, each containing 2.50 g oven-dried soil, are weighed into 20 numbered centrifuge tubes and kept on ice. Aliquots (25 ml) of extractant A are added to tubes 1–5, extractant B to tubes 6–10, extractant C to tubes 11–15 and extractant D to tubes 16–20. After each addition of extractant, the soil is ultrasonicated (2 min) with a 20 kHz 140 W MSE sonifier with a 12.5 mm diameter probe at full power. After ultrasonic treatment, the tubes are cooled on ice (5 min), centrifuged and filtered (Whatman No. 44). The filtered extracts are stored at −15°C. Blank solutions of extractants A, B, C and D are treated similarly, but without the addition of soil.

Neutralization of soil extracts

The thawed extracts and appropriate blanks are each thawed and placed in 5.0 ml volumetric flasks. Sufficient EDTA (0.2 M) is added to the soil extracts to prevent precipitation of soil Ca and Mg, then both soil extracts and blanks are adjusted to pH 7.4 with NaOH (1 M).

Enzymic conversion of ADP and AMP to ATP

Neutralized soil extract (pH 7.4, 100 μl) and water or aqueous adenine nucleotide standards (1000 μl) are placed in 3 × 0.5 cm glass incubation vials (Table 5.2). For ATP measurements, phosphate–magnesium buffer (100 μl) is added; for ADP, phosphate–magnesium buffer (100 μl) containing PK and PEP; and for AMP, phosphate–magnesium buffer (100 μl) containing PK, PEP and MK. The vials are stoppered, mixed thoroughly and incubated in a water bath (30°C, 30 min). After incubation, all vials are heated on a boiling water bath (100°C, 3 min) to stop any further enzyme reactions. After remixing, the vials are cooled on ice for 30 min to allow the denatured enzyme to settle out. ATP is then measured by adding 1000 μl of the incubated solution to a scintillation vial containing 5 ml arsenate buffer, prepared as described by Jenkinson and Oades (1979). Pico-Zyme F (50 μl) is then added, the vial swirled and counted at 15 s and 1 h after the addition of enzyme on a Beckman S-250 liquid scintillation counter. (We currently use a Cambera Packard 2500 liquid scintillation counter.)

Measurement of soil adenine nucleotide concentration

The procedure for measuring and calculating the concentration of soil adenine nucleotides is given in Table 5.2. The ratio of counts obtained from incubated soil extractants B, C and D, less counts from incubated soil extractant A, compared to the counts from 50 pmol added ATP, ADP or AMP, again after incubation, give the fraction of added ATP, ADP or AMP recovered, respectively. Thus, soil ATP, ADP and AMP, corrected for recovery of added nucleotide, are calculated directly from the counts obtained from the appropriate incubated solutions. In this way, corrections are made for quenching due to the presence of soil extract or denatured enzyme in the counting vials during ATP analysis.

Table 5.2. Experimental plan for measurement of adenine nucleotides in soil by the enzymic procedure of Brookes et al (1983).

Nucleotide measured	Set no.*	Presence or absence of soil	Extractant used	Enzyme	Aqueous standard (pmol)† ATP	ADP	AMP	Counts† 0.1 min^{-1}	Soil adenine nucleotide content, corrected for recovery of added nucleotide (nmol g^{-1} soil)	% recovery of added adenine nucleotide
ATP	1	–	A	Buffer only	0	0	0	*a*		
	2	+	A	Buffer only	0	0	0	*b*		
	3	+	B	Buffer only	0	0	0	*c*		
	4	+	A	Buffer only	50	0	0	*d*	[ATP] = 0.48¶(*b*-*a*)[25 + (*w*‡-2.50)]/(*c*-*b*)	100(*c*-*b*)/(*d*-*b*)
	5	+	A§	Buffer only	100	0	0	*e*		
	6	–	B§	Buffer only	0	0	0	*f*		
ADP	7	–	A	PK + PEP	0	0	0	*g*		
	8	+	A	PK + PEP	0	0	0	*h*		
	9	+	A	PK + PEP	0	50	0	*i*		
	10	+	C	PK + PEP	0	0	0	*j*	[ADP] = 0.48(*h*-*g*)-(*b*-*a*)][25 + (*w*-2.50)]/(*j*-*h*)	100(*j*-*h*)/(*i*-*h*)
	11	+	A§	PK + PEP	50	0	0	*k*		
	12	–	C§	PK + PEP	0	0	0	*l*		
AMP	13	–	A	PK + PEP + MK	0	0	0	*m*		
	14	–	A	PK + PEP + MK	50	0	0	*n*		
	15	+	A	PK + PEP + MK	50	0	0	*o*		
	16	+	A	PK + PEP + MK	50	0	50	*p*	[AMP] = 0.48[*o*-*n*)-(*h*-*g*)][25 + (*w*-2.50)]/(*q*-*o*)	100(*q*-*o*)/(*p*-*o*)
	17	+	D	PK + PEP + MK	50	0	0	*q*		
	18	+	A§	PK + PEP + MK	100	0	0	*r*		
	19	–	D§	PK + PEP + MK	50	0	0	*s*		

* Each set contains five replicates, one for each replicate soil extraction.

† In final counting vial.

‡ *w* is the weight of moist soil containing 2.50 g oven-dried soil.

§ Although not used in calculations, set 5 is included to check that the relationships between counts and ATP concentration is linear, sets 11 and 18 to check that conversion of ADP and AMP to ATP is quantitative, and sets 6, 12 and 19 to check that extractants B, C and D are correctly prepared.

¶ The factor 0.48 is calculated from the volume of soil extract in the incubation vials (100 μl, diluted 1 in 2), the volume of incubated solution assayed for ATP (1000 μl of 1200 μl) and the weight of soil (2.50 g oven dry weight) extracted with 25 ml extractant.

HPLC procedure

(Martens 1992)

Principle of the method

The HPLC method of Martens (1992) for analysis of soil adenine nucleotides following extraction from soil by an alkaline $NaHCO_3$ reagent is given. The analytical details are virtually unchanged from those of Martens (1992).

Soil treatment

Analyses are performed on moist soils, previously sieved (< 2 mm) and stored at 4°C. Before analysis, the soils are adjusted to 45% WHC and kept at 22°C for 24 h.

Soil extractant

The extractant is based on the $NaHCO_3$–$CHCl_3$ reagent developed by Paul and Johnson (1977). The modified reagent contains inosine (20 mM) and EDTA (6 mM) in 0.5 M $NaHCO_3$. The final pH is 8.0.

Procedure

Soil extraction

To moist soil containing 2.0 g oven-dried soil is added extractant (20 ml) and $CHCl_3$ (8 ml). The mixture is sonicated (2 min) and filtered. Each extraction is replicated four times. Corrections are made for incomplete extraction of the adenine nucleotides by measuring the ^{14}C recovered in the extracts after separate additions of [^{14}C]ATP, [^{14}C]ADP and [^{14}C]AMP to the extractant, then extracting the soil as above.

HPLC analysis

Chemical derivatization of adenine nucleotides

For HPLC analysis, the adenine nucleotides in the soil extracts are first converted to fluorescent 1-N^6-ethanoadenosine derivatives of ATP, ADP and AMP. The soil extracts and standards (5 ml) are adjusted to pH 4.5 with concentrated acetic acid (200–300 μl). Then 1000 μl of the mixture is transferred to 30 ml conical flasks and 0.25 M sodium acetate buffer (500 μl, pH 4.5) and 55% chloracetaldehyde in water (100 μl) added. The flasks are incubated for 2 h at 60°C in a water bath, then about 90% of the reaction mixture is evaporated on a rotary evaporator to remove the excess chloracetaldehyde. The volume is readjusted to the initial with water and 200 μl aliquots used for the HPLC analysis.

HPLC system

The adenine nucleotides are analysed on a reverse-phase column KS-Nucleosil 120–3 C18 (3 μm) (Machery & Nagel, Germany) with a guard column (1 cm) packed with the same material. The mobile phase consists of two eluents: acetonitrile and 40 mM phosphate buffer with 5 mM tetrabutylammonium hydrogen phosphate as an ion-pair reagent. The elution of the adenine nucleotide derivatives is performed with a linear acetonitrile gradient with an initial composition of 7% acetonitrile and 93% phosphate buffer. Within 14 min the composition is changed to 20% acetonitrile and 80% phosphate buffer followed by an isocratic elution (4 min) with buffer of the initial composition. The flow rate is 1.0 ml min^{-1}. The separated adenine nucleotide derivatives are detected by a Shimadzu RF 530 Fluorescence Spectrometer with the wavelength settings at 270 nm for excitation and 410 nm for emission. The HPLC apparatus consisted of a ternary gradient former (Model 300 C) and an injection valve NZ 190 with a 200 μl sample loop, all purchased from

Gynkotek, Munchen, Germany. Signals from the fluorescence spectrometer are recorded and quantified via a Ramona radio chromatographic system (Raytest, Essen, Germany) equipped with a computer and a printer–plotter. This system includes a ^{14}C-detector, which can monitor the radioactivity of the separated compounds after their passage through the spectrometer.

Discussion

There is considerable evidence in the literature that TCA-based reagents extract more ATP from a wide range of soils than does alkaline $NaHCO_3$. The finding of contradictory results by Martens (1992) for two out of four soils must be considered with this in mind. Both findings cannot be correct.

The current available interpretations of AEC are contradictory. The high soil AECs (0.8–0.9) measured by many workers in both unamended and amended soils when extracted with acidic reagents cannot be reconciled with the significantly lower AECs (0.5–0.6) measured with alkaline $NaHCO_3$-based reagents.

All the evidence points to TCA being the most powerful inhibitor of ATPases currently used in soil ATP work. EDTA was included in Martens' (1992) $NaHCO_3$-based reagent to inhibit these enzymes but its effectiveness was not tested. If there is still significant enzyme activity in the $NaHCO_3$ reagent it could, in part, explain the discrepancy.

A further major cause of the discrepancy between AEC measurements by acidic reagents and $NaHCO_3$ (Martens 1992) was that the $NaHCO_3$ reagent extracted more ADP and AMP, but similar amounts of ATP to acidic reagents. One, unlikely, possibility is that Martens' (1992) $NaHCO_3$ reagent extracts a hitherto unknown pool of exocellular ADP and AMP. Another possibility is that there is endogenous hydrolysis of DNA or RNA to ADP and/or AMP during extraction with the $NaHCO_3$ reagent. This was put forward as a possible explanation by Rosacker and Kieft (1990) and Ciardi et al (1993) to account for increases in ADP and AMP, and corresponding decreases in AEC following rewetting of air-dried soils.

There is currently much interest in introducing either selected or genetically modified microorganisms to soil to perform specific functions, e.g. degradation of xenobiotics, N_2 fixation, P mineralization. In view of this it is important that soil microbial biomass ATP and adenylate energy charge relationships are properly understood. Their correct interpretation may hold the key to understanding the survival strategies adopted by native soil microorganisms. This, in turn, will help in controlling the survival of microorganisms introduced in to soil. Current discrepancies between methods of AEC must be resolved before firm conclusions can be drawn. The true significance of AEC values in the soil microbial biomass awaits this clarification.

Soil respiration

K. Alef

Active living cells need a constant supply of energy, which for heterotrophic microflora derives from the transformation of organic matter such as cellulose, proteins, nucleotides and humified compounds. Energy-supplying reactions in the cell are redox reactions based on the transfer of electrons from a donor to an acceptor. By respiration, that is the oxidation of organic matter by aerobic microorganisms, oxygen functions as the end acceptor of the electrons. The end products of the process are carbon dioxide and water. The metabolic activities of soil microorganisms can therefore be quantified by measuring the CO_2 production or O_2 consumption (Nannipieri et al 1990).

Soil respiration is one of the oldest and still the most frequently used parameter for quantifying microbial activities in soils (Kieft and Rosacker 1991).

The **basal respiration** is defined as the respiration without the addition of organic substrate to soil. **Substrate-induced respiration** (SIR) is the soil respiration measured in the presence of an added substrate such as glucose, amino acids, etc.

Respiration as an universal process is not only restricted to microorganisms but it is also carried out by other organisms inhabiting soils. Like other metabolic activities, it depends on the physiological state of the cells and is influenced by different factors in the soil.

Respiration is influenced by soil moisture, temperature, the availability of nutrients and soil structure. Air-drying reduces the soil respiration significantly. Remoistened soils, however, show very high initial activities, probably as a result of the release of concentrations of easily degradable organic compounds such as amino and organic acids caused by chemical and physical processes at the moistening of dry soils (Clark and Kemper 1967; Anderson 1975; Wilson and Griffin 1975a, 1975b; Kowalenko et al 1978; Kröckel and Stolp 1986; Kieft et al 1987).

The remoistening of air-dried soils containing carbonate also causes the release of abiotic CO_2. In this case it is recommended that the O_2 consumption is measured (Anderson 1982; Kieft et al 1987). Soil respiration decreases with the depth of soil and correlates significantly with soil organic matter (C_{org}) and most microbial parameters (Stotzky 1965; Thalmann 1968; Parkinson et al 1971; Alexander 1977; Gray and Williams 1977; Anderson and Domsch 1978a, 1978b; Domsch et al 1979; Sparling 1981a, 1981b; Sparling and Eiland 1983; Beck 1984a; Alef and Kleiner 1987; Alef et al 1988; Suttner and Alef 1988; Van de Werf 1989/1990; Alef 1990).

Soil respiration reacts differently to treatment and cultivation methods and has been used most frequently for the assessment of the side effects of chemicals such as pesticides and heavy metals, etc. (Jäggi 1976; Anderson 1984; Beck 1984b, 1985; Malkomes 1985; Carlisle and Trevores 1986; Domsch and Schröder 1986; Wilke 1986; Somerville and Graves 1987; Alef et al 1988; Schlosser 1988; Schuster 1988).

The basal soil respiration can be followed for long periods of time. However, the composition of aerobic microflora can occur during long-term incubation. In the case of the SIR method, a change in population is expected when the incubation period is longer than 4–6 h (Anderson and Domsch 1978b). The incubation temperature used varies between 20 and 30°C and the water-holding capacity between 50 and 70%. The pH value of the measurements is usually that of the soil in water.

Soil respiration can be determined by using simple techniques such as the incubation of soils in jars (Isermeyer 1952), closed Petri dishes (Pochon and Tardieux 1962) or in different types of flask (Jäggi 1976). CO_2 is usually adsorbed in NaOH and determined by HCl titration.

Other methods for determining CO_2 are based on changes in electrical conductivity of the NaOH solution (Anderson and Domsch 1978a; Cheng and Coleman 1989), or use gas chromatography (Brookes and Paul 1987) or infrared spectroscopy (Heinemeyer et al 1989). Labelled CO_2 ($^{14}CO_2$) is determined when the decomposition of specific organic compounds are monitored in soil (Naklas and Klein 1981). The consumption of O_2 can be estimated with the Warburg apparatus (Domsch 1962; Stotzky 1965), by the means of an electrorespirometer (Birch and Melville 1969; Kröckel and Stolp 1986; Alef et al 1988) or gas chromatography (Trevors 1985).

The respiratory quotient (RQ = volume of CO_2 evolved/volume of O_2 consumed) reflects the physiological state of the microbial biomass in soil and is assumed to have a value of 1. However, this is not fulfilled in soil due to several factors such as environmental conditions and the type of substrate.

Soil respiration can be determined directly in the field (Anderson 1982). Detailed information on field methods will be presented in Chapter 10. Laboratory methods are presented and discussed in this chapter.

Estimation of soil respiration in closed jars

(Isermeyer 1952)

Principle of the method

The estimation of CO_2 evolved during the incubation of soil in a closed system. CO_2 is trapped in an NaOH solution, which is then titrated with HCl.

Materials and apparatus

Jar (1 l) with a rubber ring and pegs or other suitable containers
Incubator adjustable at 25°C
Automatic titrator or simple equipment for titration
CO_2 absorption tubes

Chemicals and solutions

NaOH (0.05 M)
HCl (0.05 M)
Barium chloride solution (0.5 M)

Dissolve 122.14 g of $BaCl_2.12H_2O$ in water and make up with water to 1000 ml.

Indicator

Dissolve 0.1 g phenolphthalein in 80 ml ethanol (60% v/v) and bring up with ethanol to 100 ml.

CO_2-free water

Boil distilled water for about 2 min, cool and keep in a container equipped with CO_2 absorption tubes. CO_2-free water can also be obtained by cooling distilled water in a nitrogen atmosphere (see Chapter 6).

Procedure

Weigh 50 g sieved soil (55% WHC) in a beaker and place it in the bottom of a 1 l jar (Figure 5.1). Pipette 25 ml NaOH (0.05 M) in the jar and immediately make it airtight using a rubber ring and tow crossing pegs. Often 3–5 jars with NaOH (0.05 M) but without soil should be used as controls. Incubate all jars up to 3 days at 25°C (longer incubation periods could result in anaerobic conditions). This method can be used for determining the substrate-induced respiration. In this case mix the substrate (e.g. 0.5% glucose) with the soil sample and incubate for 4–6 h at 22°C.

Estimation of CO_2

Open the jars, take out the beakers and wash the external surface of the beaker with CO_2-free water to bring the NaOH solution completely into the jar. Add 5 ml barium chloride solution (0.5 M) and some drops of the indicator. Add HCl (0.05 M, dropwise) under continuous stirring until the colour changes from red to colourless.

Calculation of the results

The rate of the respiration is calculated by the following relationship:

$$CO_2(\text{mg})/SW/t = \frac{(V_o - V) \times 1.1}{dwt} \quad (5.6)$$

where SW is the amount of soil dry weight in grams, t is the incubation time in hours, V_o is the HCl used for titration (average value if more blanks are available) in millilitres, V is the HCl used for the soil sample (average value, in millilitres), dwt is the dry weight of 1 g moist soil and 1.1 is the conversion factor (1 ml 0.05 M NaOH equals 1.1 mg CO_2).

Discussion

Originally 100 g of soil were used (Isermeyer 1952). If jars with large surfaces are used, the evolved CO_2 is still linear to the amount of soil.

This method can be also used for measuring the glucose-induced respiration.

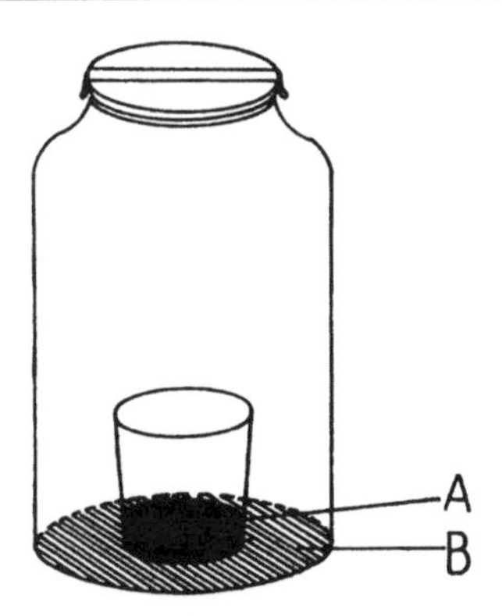

Figure 5.1. Estimation of soil respiration in closed jars (Alef 1991): (A) soil sample; (B) NaOH solution.

Estimation of soil respiration with closed bottles

(Jäggi 1976)

Principle of the method

See previous section on estimation of soil respiration in closed jars.

Materials and apparatus

Duran bottles (250 ml)
Glass or plastic tubes with small holes in the upper part.

Chemicals and reagents

See previous section on estimation of soil respiration in closed jars.

Procedure

Place the soil (20 g at 50–70% WHC) in a plastic or glass tube, then put it into a Duran bottle so that the tube is kept at the neck of the bottle to avoid any contact with the NaOH solution (Fig. 5.2). Three bottles with NaOH but without soil should be used as blanks. Close the bottles tightly and incubate for up to 3 days at 26°C. Other temperatures can be used according to the aim of the study. This method can also be used for measuring the glucose-induced respiration. In this case, mix ground glucose (0.5%) with the soil sample and incubate the closed bottles at 22°C for 4–6 h. However, with this technique it is not possible to determine the CO_2 evolution pattern during the incubation.

Estimation of CO_2

See previous section on estimation of soil respiration in closed jars.

Calculation of the method

See previous section on estimation of soil respiration in closed jars.

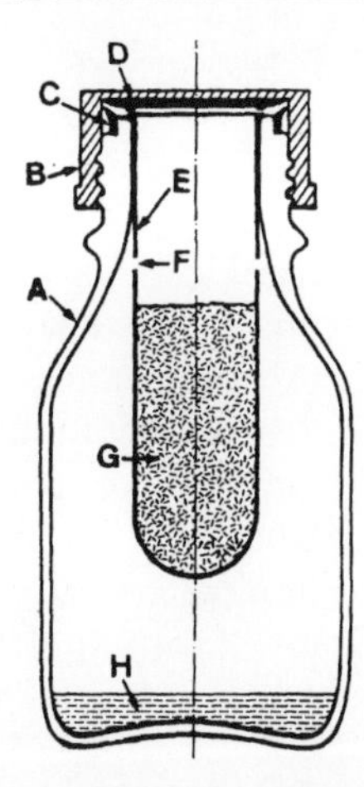

Figure 5.2. Estimation of soil respiration in closed bottles (Jäggi 1976; Alef 1991): (A) bottle (250 ml); (B) plastic cap; (C) and (D) polypropylene rings; (E) glass or polypropylene tube; (F) holes (2 mm); (G) soil sample; and (H) NaOH solution (20 ml, 0.05 M).

Discussion

To improve the gas exchange, the plastic or glass tubes present small holes. Care should be given on the location and size of the holes to avoid soil falling into the NaOH solution.

There are no data available concerning the relationship between CO_2 evolution and the amount of soil. However, Jäggi has used 10 g instead of 20 g soil (personal communication, 1989).

A relatively thick layer of soil in the tubes (method of Jäggi) or in the jar (method of Isermeyer) could limit the gas exchange between the soil sample and the air above it. In this case intensive respiration is reduced only in the upper part of soil.

The measurement can be significantly influenced by the abiotic CO_2 production with alkaline soils containing $CaCO_3$.

Another problem is the solubility of CO_2 in water, which depends on the pH value.

In such a closed system the composition of the gas phase changes during the incubation time so that constant conditions are unlikely (consumption of O_2, accumulation of CO_2). In this case it is recommended that jars are opened repeatedly (after 1 day) and titrated with the NaOH solution.

Estimation of soil respiration in the Sapromat

The estimation is based on the measurement of the O_2 uptake during the incubation of soil in a closed system. The O_2 is delivered in the system electrochemically. The apparatus, Sapromat type B6 or B12, consists of a regulated temperature water bath, containing the measuring units and an instrument for recording the results (Fig. 5.3). A measuring unit consists of a reaction vessel (A) with a CO_2 absorber (3), an O_2 producer (B) and a pressure meter (C). The vessels are connected with rubber hoses to form a closed unit, so that changes in the atmospheric pressure outside the units do not influence the measurements of O_2 uptake. The CO_2 produced will be adsorbed by a NaOH solution.

O_2 uptake during the soil respiration causes an underpressure. This is indicated by a pressure meter, which regulates the electrolytic oxygen prodution as well as the display and graphic recording of the results. The O_2 uptake is shown on the display in milligrams of O_2. A total of 50–100 g of soil should be used for the measurement. The O_2 uptake is not registered in the first 2 h, as this time is needed to balance the system. The measurements are usually performed at temperatures ranging from 20 to 30°C and at 55% of the WHC.

Discussion

1. The nitrogen to oxygen ratio of the gas phase remains constant for all the measuring period above the soil sample. In this way anaerobic conditions are avoided.
2. The measurements can be registered.

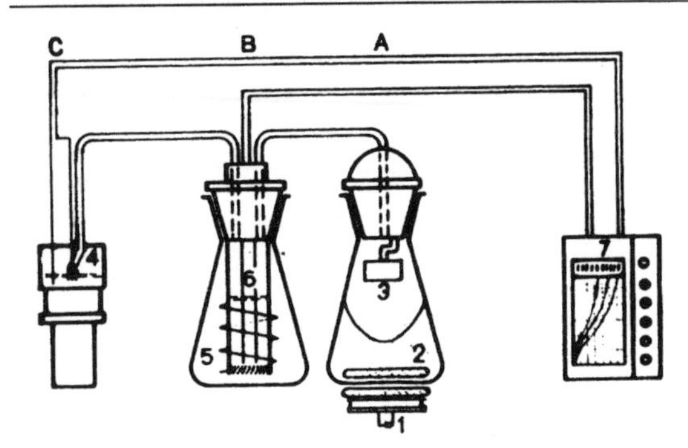

Figure 5.3. Estimation of soil respiration in a Sapromat (for key to numbers and letters, see text).

3. Twelve soil samples can be measured at the same time; this means measuring respiration of six different soils if duplicates are carried out.
4. Oxygen uptake is generally linear between 2 and 6 h and when the amount of soil used ranges from 50 to 100 g.
5. All the other problems discussed for measuring the respiration by the Isermeyer and Jäggi methods (gas diffusion, abiotic CO_2, production, etc.) are present in this method.

Estimation of soil respiration with the Wösthoff apparatus

This measurement is based on the estimation of CO_2 formation with the Wösthoff analyser Ultragas 3. The sample tubes are filled with moist sieved soil samples (100 g at 55% WHC) and incubated at 22°C (Anderson and Domsch 1978b). Every hour CO_2-free air is passed through the soil and CO_2 concentrations are measured for 10 min. The gas analyser Ultragas 3 measures changes in the electrolytic conductivity due to the reaction of CO_2 with NaOH solution. Such changes are proportional to the CO_2 concentration and the results can be automatically recorded (Fig. 5.4). The modern Ultragas US4-CO_2 is characterized by a high sensitivity and is thus suitable for measuring low concentrations of gas. CO_2 content lower than 1 ppm can easily be determined on a complete scale from 0 to 50 ppm.

Discussion

1. With this apparatus the diffusion problems are solved.
2. The respiration of four different soils can be measured at the same time.
3. The other problems discussed before such as the abiotic CO_2 production, the CO_2 solubility in water etc, are also present using this method.
4. Further information can be obtained from Wösthoff GmbH, Bochum, Germany.

The infrared gas analysis

(Heinemeyer et al 1989)

The method is based on the automatic estimation of CO_2 by the infrared gas analyser (IRGA). The soil samples are gassed with air, so that gas diffusion and CO_2 accumulation problems can be overcome.

The complete system is shown in Fig. 5.5. An air inlet system made of tubes (PVC, 32 mm i.d.) is installed 1 m above the roof top (12 m above ground) to ensure a disturbance-free supply of outside air. The set up consists of 24 identical sample lines, which are independent from each other. This allows different flow rates in different lines to be run in the range of 100–1000 ml min^{-1}. Each line contains a membrane gas pump, a 500 ml gas washing bottle filled with acidified deionized water as a humidifier and a sample tube (25 × 4 cm i.d.) made of acrylic glass. The soil samples are enclosed between porous polystyrene foam pads, and the complete tubes are connected to the system by rubber stoppers. Needle valves are adjusted individually to give the same total flow resistance for a sample line, regardless of whether the line is directed to the vent or the IRGA. This ensures a similar flow through the sample in both flow configurations. A thermal mass flow meter is used to measure the flow rate. An IRGA is used to measure CO_2 concentrations (for more details, see Heinemeyer et al 1989). The system is controlled by a computer. A sample tube without soil is used as a control. Soil samples of

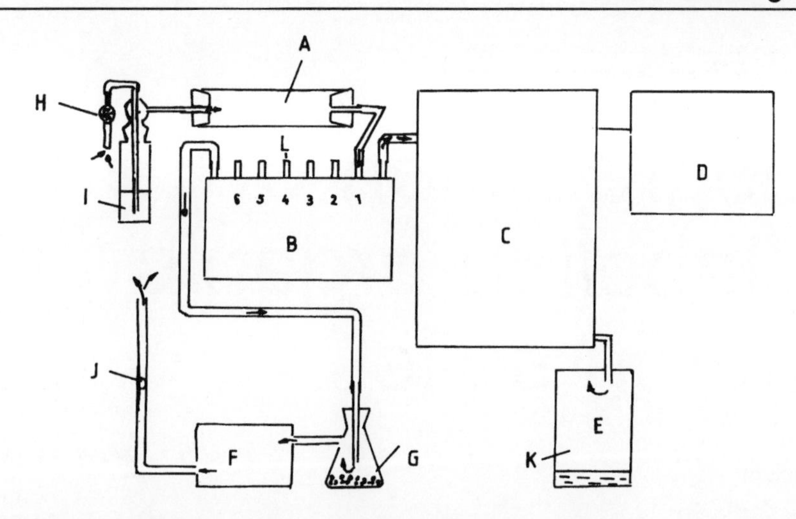

Figure 5.4. The Wösthoff apparatus for estimating soil respiration (Anderson 1982; Alef 1991): (A) soil sample; (B) sample changer; (C) CO_2-analyser; (D) recorder; (E) container; (F) air pump; (G) $CaCl_2$; (H) N_2CO_3; (I) water; (J) rotameter; (K) waste; and (L) sample inlets

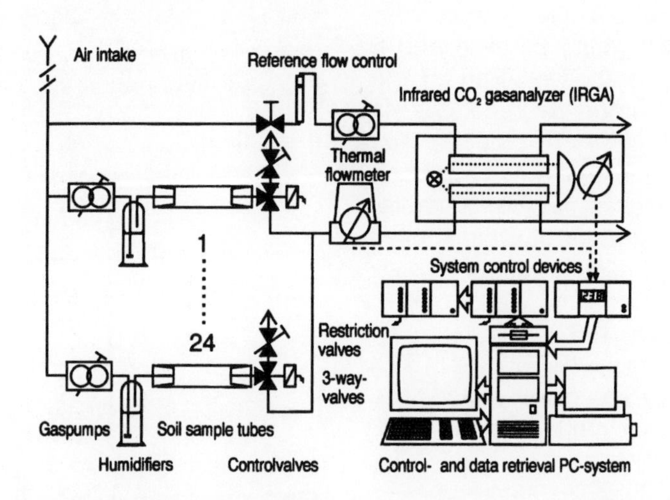

Figure 5.5. Estimation of soil respiration according to Heinemeyer et al (1989).

100 g (55% of the WHC) are normally used and the complete set up is kept at 22°C.

Discussion

1. The amount of soil can range from 5 to 100 g (Heinemeyer, personal communication, 1989) without affecting the linearity of CO_2 evolution. Other incubation temperatures can also be used.
2. This system is much more sensitive than the Sapromat and has good reproducibility (Beck 1990, personal communication).

Aerobic biodegradation of ^{14}C-labelled organic matter in soils

(Anderson 1990)

K. Alef

Biodegradation is the conversion of substrates into less complex intermediates or end products, catalysed by enzymes or intact microbial cells. It can be either a single- or multi-step conversion energy-generating process and it is collectively called catabolism. Part of the free energy derived from degradations is used to drive the processes of biosynthesis.

Several systems and instruments have been used for detecting and assaying the ^{14}C-radiation by studying the turnover of radiolabelled organic substances (C mineralization) in soils (Domsch et al 1973; Behera and Wagner 1974; Martens 1977; Anderson 1982).

The system of Andersen (1975, 1982, 1990) will be presented. It can be used to study the decomposition of organic matter in soils, compost, litter, etc.

Principle of the method

The method is based on the determination of evolved $^{14}CO_2$ during the biodegradation of radiolabelled substances or substrates added to soil, compost, etc. The moisture is adjusted at 50% of WHC and the incubation temperature is equal to 25°C.

Materials and apparatus

The chamber consists of an incubation flask (100–300 ml), into which a glass column (130 × 25 mm) is inserted. The column contains, from the bottom to the top, a cotton plug, about 10 g granulated soda lime, a second cotton plug and a second layer of 10 g of soda lime (Fig. 5.6).

A chamber for extracting and transferring

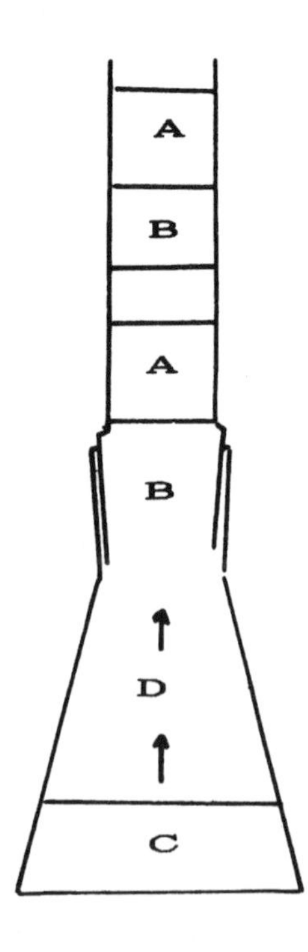

Figure 5.6. Incubation chamber for estimating the degradation of ^{14}C-labelled organic matter in soil (Anderson 1990): (A) soda lime; (B) cotton plug; (C) soil containing ^{14}C-organic matter; and (D) $^{14}CO_2$.

$^{14}CO_2$ from the soda lime into an inorganic solvent mixture for scintillation counting (Fig. 5.7). The chamber consists of a vacuum flask (300–500 ml) with an outlet, which is fitted with a thick-walled plastic tube (20 cm long), whose end contains a syringe needle (0.90 × 55 mm). The flask is sealed with a two-holed rubber stopper. A stainless steel cannula or a heavy-walled capillary glass is inserted in one hole. A plastic disposable syringe (50 ml) is used for acid delivery via the cannula into the flask. The second hole contains a glass tube connected to a one-way stopcock and needle valve for adjustment of N_2 flow rate (Fig. 5.7).

Water bath adjustable to 60°C

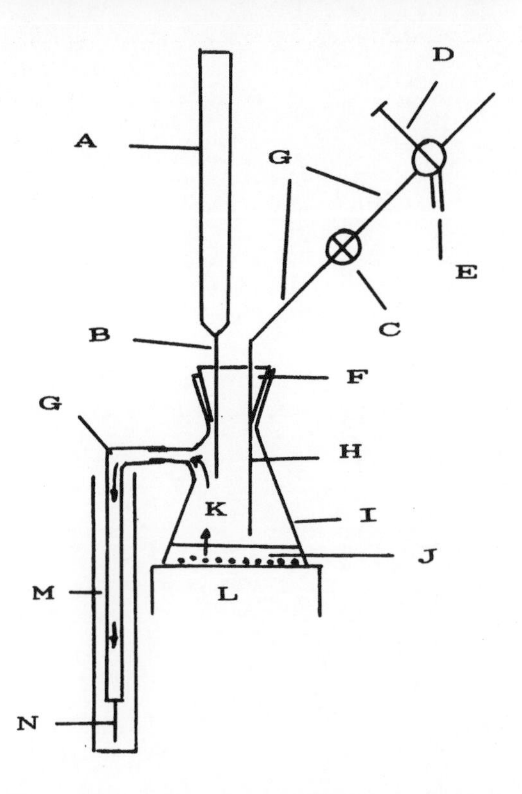

Figure 5.7. Chamber for transferring $^{14}CO_2$ to alkaline solution (Anderson 1990): (A) plastic syringe; (B) cannula; (C) one-way stopcock; (D) needle valve; (E) tube to N_2 flask; (F) rubber stopper; (G) plastic tube; (H) glass tube; (I) vacuum flask; (J) HCl (18% v/v) and dissolving granules of soda lime; (K) $^{14}CO_2$; (L) water bath (60°C); (M) 40 ml (NaOH, 1 M; and (N) syringe needle.

Tube (25 × 250 mm) containing the liquid CO_2-trapping agent during $^{14}CO_2$ transfer
Scintillation counter, vials and caps
Roll of cotton

Chemicals and solutions

HCl (18% v/v)
Soda lime (granulated, mesh size 1.5–2 mm)
Ethanolamine/methanol mixture (7:3, v/v)
Scintillation solution:
1000.0 ml Toluene
4.0 g 2,5-diphenyloxazole
0.3 g 2,2-*p*-phenylene-bis-(5-phenoxazole)

Procedure

Samples of 50 or 100 g of soil (treated with the labelled substrate) are placed in the vacuum flask (100 or 250 ml) of the incubation chamber (Fig. 5.6). At the beginning of the incubation, the weight of the soil plus the flask but minus that of the column is recorded. Every 7–14 days the water lost due to evaporation is replaced by spraying water on the soil surface. To avoid loss of $^{14}CO_2$ when opening the incubation chamber, this is placed under slight vacuum for about 30 s to allow complete trapping of $^{14}CO_2$. After the incubation the granules of soda lime are removed and transferred to the extraction chamber (Fig. 5.7), which is closed. After closing the one-way stopcock, the flask is placed in the water bath at 60°C. The syringe needle (plus part of the plastic tubing) attached to the flask outlet is inserted as far as possible into 25 ml of ethanolamine–methanol solution (Fig. 5.7). Fifty millilitres of HCl are then injected slowly (18%, v/v) into the flask to dissolve the soda lime and to release the absorbed $^{14}CO_2$. The acid (HCl) must be injected slowly so that the bubbling of ethanolamine–methanol solution is not too rapid. At the soda lime dissolution, the one-way valve is opened, and the N_2 is allowed to sweep the flask for about 20 min. The ethanolamine–methanol solution (25 ml) is transferred to a measuring cylinder (50 ml), and made up with the same solvent to 30 ml. Samples of 10 ml are removed and mixed with 10 ml of the scintillation solution and the counts per second are monitored for several minutes. The counting time depends on the scintillation counter and the quantity of radioactivity in the sample.

Calculation of results

Correct the data for the blank (background) and calculate briefly:

$$\text{Disintegrations per second} = \frac{\text{Counts per second} \times 3}{\text{Counter efficiency (\%)}} \times 100 \quad (5.7)$$

From these data, the percentage of ^{14}C-labelled material mineralized to $^{14}CO_2$ can be determined.

Discussion

The liquid scintillation counter is the most efficient instrument for detecting and assaying ^{14}C radiation.

The major technical difficulty of the incubation system is the aeration. However, the system described here gives satisfactory results for long- and short-term experiments (Anderson 1975; Anderson and Domsch 1978b; Marvel et al 1978).

Heat output

K. Alef

The energy of reaction products is different with respect to the energy of reactants, thus the process is accompanied by an absorption or liberation of energy in the form of heat. The difference in energy only depends on the initial and final states of the system; it is independent of the pathway of the reaction, no matter how complex the intermediate steps. Therefore, calorimetric measurements of living organisms represents an index of overall metabolic activity (Forrest 1972; Mortenson et al 1973) but not of single biochemical reactions.

Heat production is one of the most obvious changes occurring during the microbial degradation of organic matters. It is determined in a closed system by a microcalorimeter, where a very small temperature change (10^{-6}°C) can be recorded in microwatts. Heat output has often been used as a sensitive parameter to estimate microbial activity in soil in the presence or absence of one or several substrates (Ohms and Pollock 1966; Mortensen et al 1973; Konno 1976; Liungholm et al 1979a, 1979b; Sparling 1983; Alef et al 1988; Alef and Kleiner 1989).

Principle of the method

Determination of heat output of soil metabolic activity by using a microcalorimeter.

Materials and apparatus

Microcalorimeter
Ampoules (glass or stainless steel)
Incubator adjustable to 25°C

Chemicals and solutions

Glucose or other substrates

Procedure

The amount of soil to be used depends on the size of ampoule applied. Therefore, tests should be performed prior to the experiment to find the optimal amount of soil to be used.

Turn on the microcalorimeter at least 24 h before use and adjust the temperature to 22°C or 25°C.

Estimation of heat output without addition of substrate

1–10 g of moist soil are placed in the ampoule, which is then loosely capped and preincubated overnight at 25°C. The ampoules are then sealed and the heat output (µW) is measured against an empty ampoule. The calorimeter reaches a thermal equilibrium about 2 h after the introduction of the ampoule in the microcalorimeter.

Estimation of heat output in the presence of the substrate

1–10 g of moist soil are mixed with ground glucose (5–10 mg glucose g^{-1} soil) and placed in the ampoule, which is capped and preincubated for 3 h at 25°C. The ampoule is then sealed and the heat output (µW) is measured at 25°C.

Calculation of the results:

The measurements in microwatts obtained are proportional to the heat output; therefore the microbial activity can be expressed as µg W g^{-1}dwt.

Discussion

Zelles et al (1986), Alef et al (1988), and Alef and Kleiner (1989) have used 1–2 g of soil and 10 mg glucose g^{-1} soil. Sparling (1981a, 1981b) used 10 g soil and 5 mg glucose g^{-1} soil. An investigation on the best glucose/soil ratio for optimizing heat output is required.

Abiotic heat output, such as the heat of dissolving glucose in soil water, the hydration of CO_2, the ionization of H_2CO_3 and the evaporation of water, can interfere markedly with the measurements of microbial heat output. Accurate controls need to be performed.

Pollutants influence the heat output to different degrees (Zelles et al 1986; Hund et al 1988).

Significant correlations have been found between heat output, arginine ammonification, ATP, microbial biomass determined by the glucose-induced respiration, and protease activities in arable and grassland soils (Alef et al 1988; Alef and Kleiner 1989).

Sparling (1981a) found significant correlations between the heat output and the soil respiration, but not with ATP and microbial biomass estimated by means of the fumigation–incubation method.

Dimethyl sulphoxide reduction

(Alef and Kleiner 1989; Alef 1990)

K. Alef

Dimethyl sulphoxide (DMSO) is an intermediate in the global S cycle, derived from atmospheric photooxidation of dimethyl sulphide (DMS) (Andreae 1980). DMS arises chiefly from algal and microbial metabolism and has been indicated as the natural compound responsible for the transfer of sulphur from aquatic to land systems through the air (Andreae and Raemdonck 1983).

The DMSO reduction is a two electron ($2e^-$) transfer reaction

$$(CH_3)_2S{=}O + 2H + 2e^- \rightarrow (CH_3)_2S + H_2O \quad (5.8)$$

It seems to be catalysed by a variety of enzymes in the cell (Zinder and Brock 1978; Wood 1981; Barret 1985; Gibson and Large 1985; Suylen et al 1986). Results presented by Alef and Kleiner (1989), Zinder and Brock (1978) show that nearly all the microorganisms tested, including both prokaryotes and eukaryotes, aerobes and anaerobes, reduce DMSO to DMS.

Principle of the method

The method is based on the determination of produced DMS after the incubation of soil with DMSO solution for 3 h at 30°C.

Materials and apparatus

Gas chromatograph fitted with a flame ionization detector (FID) and a Porapack column (2 m), and an integrator. The temperature of the detector, injector and column are 220°, 200° and 160°C, respectively.
Reaction flasks (diameter 2.2 cm, height 4.4 cm, volume 8.5 ml) sealed tightly with rubber stoppers
Gas tight syringes (50 µl, 100 µl, 250 µl and 1000 µl)
Incubator or water bath adjustable to 30°C

Chemicals and solutions

DMSO solution

Dissolve 6.6 g DMSO in 100 ml distilled water and store at room temperature

DMS standard (99%, liquid)

store in sealed flask at 4°C

Procedure

Moist sieved soil (0.5 g) is placed in a small flask, closed with a rubber septum after the addition of 0.125 ml DMSO solution or water (control) and incubated for 3 h at 30°C, 0.1 to 1.0 ml of the headspace is analysed for DMS.

DMSO reduction can be estimated under anaerobic conditions by replacing the air in the reaction flask with N_2 by repeated evacuation and flushing using the apparatus described in Chapter 6 (see Fig. 6.5).

Estimation of DMSO reduction within a single aggregate

Using a minicone (length = 5 cm; diameter = 5 mm), a cylindrical sample is removed from a soil aggregate. The soil cylinder is divided into slices; 100–200 µl of 1.5% DMSO solution is added to 100–200 mg of each slice in the flask. DMS evolution is measured after several hours. The selected time depends on the intensity of DMS evolution and the reaction linearity. Because of the higher DMSO reduction rates, measurements under N_2 are recommended.

Estimation of DMSO reduction in compost samples

0.5 g of a well-homogenized compost is placed in reaction flask, and closed with a rubber septum after the addition of 0.5 ml of DMSO (1.5%) or water (blank). After incubation for 3 h at 30°C, 0.25 ml of the headspace is analysed for DMS.

Calibration curve

Transfer 10 µl (9.9 mg) cold (4°C) liquid DMS into a closed flask (20 ml) using a syringe and allow to evaporate and dissolve for at least 2 min at 30°C. For the preparation of the lowest DMS concentrations, further appropriate dilutions of this gas phase into other flasks are recommended. Use a clean syringe for each dilution step. To remove the DMS completely from the used syringe, incubate it at 80°C for about 30 min.

Calculation

Read the DMS concentrations from the calibration curve, correct for the blank and calculate as follows:

DMSO reduction rate *DMS* (ng)/*dwt* (g)/h =

$$\frac{DMS\,(\mathrm{ng})\,.\,V}{t.\,v.\,dwt} \qquad (5.9)$$

where t is the incubation time in hours, dwt is the dry weight of 1 g moist soil, V is the volume of the gas phase in the reaction flask (in the system described $V = 8$ ml) and v is the gas volume in millilitres injected in to the gas chromatograph.

Discussion

DMS is found quantitatively in the gas phase because it is insoluble in water (Zinder and Brock 1978). Thus no tedious and time-consuming extraction steps are needed.

In the controls no DMS production was detectable when DMSO was replaced by water. This significantly enhances the sensitivity of the method, and reduces the number of controls.

Under the conditions employed, the reaction remained linear for several hours and was strongly inhibited by HCN, NaN_3, toluene, $CHCl_3$ fumigation, autoclaving and air drying (sandy soils), but stimulated by glucose.

Analyses by gas chromatography of the headspace shows no other peaks behind the DMS peak. In anaerobic soils or waste water, other peaks could be detected. In this case a temperature programme should be run. DMS analysis can also be carried out by using gas chromatography/mass spectrometrical detection (GC/MSD) (Sklorz and Binert 1994).

A constant incubation temperature is very important for the estimation of the DMSO reduction.

Reaction flasks with different volumes or forms can be used. In this case special attention should be given to the reaction linearity. To obtain comparable results, flasks with the same volume and form should be used.

It is recommended that 0.25 ml of the headspace is used for DMS analysis. In the case of small samples (20–200 mg), 1.0 ml of the headspace should be analysed for DMS.

The sensitivity of the DMS determination can be increased significantly by using a flame photometric detector (FPD) with a sulphur filter. DMS concentrations as low as 10^{-12} mol can be estimated.

DMSO reduction has been applied successfully for assessments of microbial activities in a variety of soils (Alef and Kleiner 1989; Alef 1990; Bauer et al 1993; Sonnen and Bachmann 1993; Sparling and Searle 1993), food samples (Alef and Kleiner 1994), and monitoring of microbial efficiency in activated sludge (Rajbhandari et al 1991; Skorlz and Binert 1994).

Significant correlations between the DMSO reduction, biomass (glucose-induced respiration), arginine ammonification, heat output and organic carbon were found (Alef and Kleiner 1989; Alef 1990; Sparling and Searle 1993).

Dehydrogenase activity

K. Alef

The microbial oxidation of organic substances under aerobic conditions is linked to a membrane-bound electron transfer chain with O_2 as a final electron acceptor. The electron transport system is coupled with the synthesis of ATP, the so-called "oxidative phosphorylation".

NADH is the form in which electrons are collected from different substrates through the action of NAD-linked dehydrogenases. These electrons funnel into the chain via the flavoprotein NADH dehydrogenase. Other respiratory substrates are dehydrogenated by flavin-linked dehydrogenases, such as succinate dehydrogenase and acyl–CoA dehydrogenase, which funnel electrons into the chain via ubiquinone. NAD^+ and ubiquinone, thus serve to collect reducing equivalents from respiratory substrates oxidized by pyridine-linked and flavin-linked dehydrogenases, respectively. The electrons are further transferred to the cytochrome system, where they are oxidized by O_2 (Mahler and Cordes 1971; Lehninger 1975).

It became clear in the 1930s that the measurement of one or more enzyme activities in the respiratory chain could be used as an index for the total oxidative activities of the cell, therefore, dehydrogenase activity in soil has been used as a measure for overall microbial activity (Lenhard 1956, 1966; Casida et al 1964; Thalmann 1968; Klein et al 1971; Skujins 1973; Casida 1977; Chendrayan et al 1979).

One of the most frequently used methods to estimate dehydrogenase activity in soil is based on the use of triphenyltetrazolium chloride (TTC) as an artificial electron acceptor (Lenhard 1956). The TTC is water soluble, has a redox potential of about −0.08 V and functions as an electron acceptor for several dehydrogenases (Mattson et al 1947).

Many other flavin-containing enzymes catalysing specialized oxidation–reductions are not involved in the mainstream of electron transport, such as D-amino-acid oxidase, xanthine oxidase, oratate reductase and aldehyde oxidase, also reduce TTC to triphenyl formazan (TPF) (Jambor 1960).

Nearly all microorganisms reduce TTC to TPF, which can be colorimetrically estimated. It should be pointed out that plants as well as animals possess very active dehydrogenases.

Another artificial electron acceptor, the 2 (*p*-iodophenyl)-3-(*p* nitrophenyl)-5-phenyl tetrazolium chloride (INT) has been used as substrate for the dehydrogenase activity. Iodonitrotetrazolium chloride (INF) is the product of the reaction (Curl and Sandberg 1961; Benefield et al 1977; Trevors et al 1982; Trevors 1984, 1985; Von Mersi and Schinner 1991).

The TTC method

(Thalmann 1968)

Principle of the method

The method is based on the estimation of the TTC reduction rate to TPF in soils after incubation at 30°C for 24 h.

$$\text{TTC} + 2H^+ + 2e^- \rightleftharpoons \text{TPF} + HCl \quad (5.10)$$

TTC: $C_6H_5-C(=N-N-C_6H_5)(-N=N^+Cl^- -C_6H_5)$; TPF: $C_6H_5-C(=N-NH-C_6H_5)(-N=N-C_6H_5)$

Materials and apparatus

Test tubes (60 ml volume, 2.5 cm diameter) and rubber stoppers resistant to solvents. Incubator adjustable to 30°C

Chemicals and solutions

Tris–HCl buffer (100 mM)

Dissolve 12.1 g of Tris (hydroxy methyl)-aminomethane in 700 ml distilled water, adjust with HCl to pH 7.8 for acid soils with pH values less than 6, to pH 7.6 for neutral soils with pH values ranging from 6 to 7.5, and to pH 7.4 for alkaline soils with pH values higher than 7.5. Bring up with distilled water to 1000 ml.

TTC solution

Depending on the soil type, TTC quantities from 0.1 to 1.5 g are dissolved in 80 ml Tris buffer and made up with the same buffer to 100 ml:

0.1 g 100 ml^{-1} for sandy soils with low content of organic C
0.3–0.4 g 100 ml^{-1} for loamy sand, sandy loam and organic sand
0.6–0.8 g 100 ml^{-1} for loam, organic soil and sandy loam
0.8–1.0 g 100 ml^{-1} for clay loamy soil
1.0–1.5 g 100 ml^{-1} for clay soil and organic soil

Extractant

Acetone (analytical grade)

TPF standard solution

Dissolve 50 mg of TPF in 80 ml acetone (500 μg TPF ml^{-1}) and bring up with acetone to 100 ml.

Procedure

Because of the light sensitivity of TTC and TPF, all procedures should be performed under diffused light. Field-moist soil (5 g) is weighed into test tubes and mixed with 5 ml of TTC solution. The tubes are sealed with rubber stoppers and incubated for 24 h at 30°C. The control contains only 5 ml Tris buffer (without TTC). After the incubation, 40 ml acetone is added to each tube, and the tubes are shaken thoroughly and further incubated at room temperature for 2 h in the dark (shaking the tubes at intervals). The soil suspension (15 ml) is then filtered and the optical density of the clear supernatant is measured against the blank at 546 nm (red colour).

Calibration curve

Pipette 0, 0.5, 1.0, 2.0, 3.0 and 4.0 ml of TPF standard solution in a volumetric flask (50 ml), add 8.3 ml Tris buffer (pH 7.6) and bring up with acetone to 50 ml to obtain the following concentrations: 0, 5, 10, 20, 30 and 40 μg TPF ml^{-1}.

Calculation

Read the TPF concentrations (μg/ml) from the calibration curve, correct for the control value and calculate as follows:

Dehydrogenase activity *TPF* (μg)/*dwt*(g) =

$$\frac{TPF\ (\mu g)/ml \times 45}{dwt \times 5} \qquad (5.11)$$

where *dwt* is the dry weight of 1 g moist soil, 5 is the moist soil used (g) and 45 is the volume of solution added to the soil sample in the assay.

Discussion

The measured dehydrogenases are cell bound and not extracellular activities.

The procedure can be changed by using 10 g soil, 10 ml Tris buffer and an incubation at 27°C for 15 h.

TTC can be adsorbed by organic and inorganic colloids, therefore the TTC concentrations should be optimized.

TTC and TPF are light sensitive and it is

therefore recommended that the measurements are made in the dark.

It is important to use only clear filtrate to measure the optical density. If necessary, centrifuge the filtrate.

O_2 inhibits the TTC reduction, therefore it is important to have the same volume (same test tubes) when comparing the results from different tests (Thalmann 1968; Ross 1971; Benefield et al 1977).

In order to measure microbial activity by the TTC reduction, it is recommended that plant roots as well as visible animals should be removed from soil samples.

Acid soils (pH < 5) show very low dehydrogenase activities. Care should be paid in interpreting the results (Alef, unpublished results).

Besides O_2, NO_3^-, Fe^{3+} and NO_2^- inhibit the dehydrogenase activities. In contrast, inorganic P, Fe^{2+}, SO_4^-, Cl and Mnv stimulate the TTC reduction (Bremner and Tabatabai 1973).

A considerable discrepancy exists between the TTC reductions and those expected from respiration rates (Novak and Kubat 1972; Nannipieri et al 1990). This probably depends on the fact that TTC is reduced only when other acceptors in the soil are exhausted (Novak and Kubat 1972).

TPF is insoluble in water and remains on the reaction sites, therefore this reaction can be used to locate the site of reaction.

Pollutants influence the TTC reduction to different degrees (Doelman and Haanstra 1979; Brookes et al 1984; Malkomes and Wagner 1986; Wilke 1986, 1991).

In anaerobic soil, TTC can be chemically reduced (Thalmann 1968).

TTC reduction is not always linear with time (Alef, unpublished results).

TTC seems to be a very sensitive indicator for changes in soil microbial activity of the soil caused by different soil treatments and cultivation methods (Reddy and Faza 1989; Tiwari et al 1989).

This assay is not a valid parameter for estimating the microbial activity in copper-contaminated soils due to artifacts (Chander and Brookes 1991).

The INT method

(Von Mersi and Schinner 1991)

Principle of the method

The method is based on the incubation of soil with the substrate INT at 40°C for 2 h followed by colorimetric estimation of the reaction product INF.

Materials and apparatus

Autoclave
Incubator adjustable to 40°C
Shaker

Chemicals and solutions

HCl (3 M)
Tris buffer (1 M, pH 7.0)

Dissolve 30.28 g of Tris in 200 ml of distilled water, adjust the pH to 7.0 with HCl (3 M) and bring up to 250 ml with distilled water

INT solution (9.88 mM)

Dissolve 500 mg of INT (Serva 26840) into 2 ml *N,N*-dimethylformamide, then add 50 ml distilled water, sonicate the solution (ultrasonic bath) and bring up with distilled water to 100 ml. Store in the dark and use only freshly prepared solution.

Extractant

Mix 100 ml *N,N*-dimethylformamide with 100 ml ethanol.

INF standard solution (100 µg/ml)

Dissolve 10 mg INF (Sigma I-7375) in 80 ml extractant and bring with the same extractant to 100 ml.

Procedure

Field-moist soil (1 g) is weighed into test tubes and mixed with 1.5 ml Tris buffer and with 2 ml INT solution. The test tubes are sealed with rubber stoppers and incubated at 40°C in the dark for 2 h. The control is prepared with autoclaved soil (121°C/20 min) and treated like the samples. Each analysis is carried out in triplicate. After the incubation, every sample is mixed with 10 ml of the extraction solution and kept in the dark (the samples should be shaken vigorously at 20-min intervals). After filtration the developed INF is measured spectrophoto-metrically at 464 nm against the blank.

Calibration curve

Pipette 0, 1, 2 and 5 ml of INF solution into test tubes, add 13.5 ml extracting solution to each tube and mix thoroughly. The calibration concentrations are: 0, 100, 200 and 500 µg INF per test.

Calculation

The dehydrogenase activity is expresed as µg INF g^{-1}dwt 2 h^{-1} and calculated according the following relationship:

$$INF\ (\mu g\ g^{-1}\ dwt\ 2\ h^{-1}) = \frac{S_1 - S_o}{dwt} \qquad (5.12)$$

where S_1 is the INF (in micrograms) of the test, S_o is the INF (in micrograms) of the control and *dwt* is the dry weight of 1 g moist soil.

Discussion

INT is light sensitive, therefore the procedure must be performed in the dark.

INT is reduced chemically to INF by Fe^{2+}, high salt concentrations and ions, therefore, autoclaved controls are necessary.

In some soils, the measured INF concentrations in the autoclaved controls were higher than that of the test (Alef, unpublished results).

This method is standardized for arable soils. In the case of forest and grassland soils, the weight of soil samples, the INT concentrations and the incubation times should be optimized.

The INT seems to be a better electron acceptor than TTC (Benefield et al 1977; Trevors 1984). However, the observed O_2 uptake rates were still much higher than those calculated by the rates of formation of INF (Benefield et al 1977). These results also suggest that INT is not an optimal electron acceptor that can substitute O_2 (Nannipieri et al 1990).

Estimation of the hydrolysis of fluorescein diacetate

(Schnürer and Rosswall 1982)

K. Alef

Fluorogenic compounds are enzymatically transformed into fluorescent products that can be visualized by fluorescence microscopy. Fluorescein has the following formula and can be substituted with different groups (Rotman and Papermaster 1966).

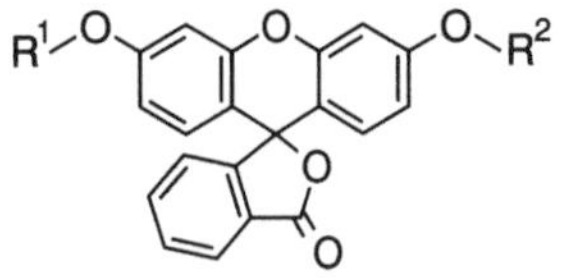

Fluorescein derivates can be hydrolysed by lipases, esterases and partially by proteases. The reaction can be catalysed by whole cells as well as extracellular enzymes. Microorganisms, algae, protozoa and animals tissues are able to catalyse this reaction (Rotman and Papermaster 1966; Medzon and Brady 1969; Ziegler et al 1975; Brunius 1980; Lundgren 1981; Barak and Chet 1986). Fluorescein esters are not polar and can be transported easily through the membrane of active cells. In contrast the fluorescent products are polar and remain inside the cell. Therefore, the technique can be used to stain active microbial cells in pure cultures and in soils (Brunius 1980; Lundgren 1981; Barak and Chet 1986; Bååth and Söderström 1988) but not spores or cells in the stationary growth phase (Söderström 1977; Lundgren 1981).

Schnürer and Rosswall (1982) used the hydrolysis of fluorescein diacetate (3,6-diacetyl-fluorescein) (FDA) to estimate microbial activities in straw and soil.

Principle of the method

The method is based on the estimation of fluorescein produced in soil treated with fluorescein diacetate solution and incubated at 24°C.

Materials and apparatus

Spectrophotometer
Shaker
Centrifuge and centrifuge tubes
Incubator adjustable to 24°C

Chemicals and solutions

Sodium phosphate buffer (60 mM, pH 7.6)

Adjust the pH to 7.6 with HCl

Acetone (analytical grade)

Fluorescein diacetate solution

2 mg/ml acetone, store at −20°C

Fluorescein standard solution

5 mg/ml acetone, store at −20°C

Procedure

Field-moist soil is sieved (2 mm mesh) and stored at 5°C before analysis. Soil (1–4 g dry weight) is placed in an Erlenmeyer flask

(250 ml) and treated with 25–100 ml phosphate buffer, then fluorescein diacetate (final concentration of 10 μg/ml) is added, and the mixture incubated on a rotary shaker at 24°C for up to 3 h. After the incubation the reaction is stopped by adding acetone to reach a final concentration of 50% v/v. The soil suspension is then centrifuged at 4000 rev min^{-1} for 10–15 min and the optical density of the clear supernatant is measured at 490 nm.

Calibration curve and calculation

The calibration curve should be prepared as described above, so that the relation between the concentration of fluorescein and the optical density is still linear; the results are expressed in micrograms of fluorescein per gram dry weight and per hour.

Discussion

The hydrolysis of FDA in soil is linear with time and soil up to 3 h and 4 g dry weight, respectively (Schnürer and Rosswall 1982). Zelles et al (1985) and Hund (1988) have used 0.5 and 1 g (fresh weight) soil and 20 ml phosphate buffer. The final concentration of FDA was 20 μg/ml. The relationship between activity and time was linear for up to 75 min at 23°C.

FDA hydrolysis has a pH optimum between 7 and 8. At pH 8.5 chemical hydrolysis of FDA occurs.

Acetone is necessary to stop the reaction and to extract fluorescein from the cells.

Some compounds extracted from soil can interfere with the colorimetric measurements.

Pollutants have different effects on the FDA hydrolysis (Zelles et al 1985, 1986; Hund 1988).

The FDA hydrolysis decreases with soil depth and is correlated with soil respiration (Schnürer and Rosswall 1982).

The FDA hydrolysis is completely inhibited after heat sterilization (Schnürer and Rosswall 1982).

Storage of FDA at −20°C is recommended. Chemical hydrolysis of FDA occurs at room temperature or at 8°C within a few days (Guilbault and Kramer 1964).

The FDA as well as fluorescein can be adsorbed on soil organic matter and clay minerals (Schnürer and Rosswall 1982).

Nitrogen mineralization in soils

K. Alef

All higher plants and microorganisms depend on combined nitrogen for their nutrition. Combined nitrogen in the form of ammonia, nitrate and organic compounds, often becomes the limiting factor for biological processes in soil (Haynes and Goh 1978). For this reason, the cyclic transformation of nitrogenous compounds, including the mineralization of nitrogenous organic matter, are of great importance in the total turnover of this element in soil.

The nitrogen mineralization consists of two different processes: the ammonification of organic compounds by a large number of heterotrophic microorganisms, and the oxidation of released ammonia to nitrite and nitrate (nitrification), mainly by autotrophic bacteria.

The mineralization of organic nitrogen depends mainly on temperature, moisture, aeration, type of organic N and pH. The inorganic N produced by mineralization is subject to N immobilization and fixation by clays. It can also be lost through denitrification and leaching.

Ammonification

Organic nitrogen (e.g. protein and nucleic acids) is converted by microbial decomposition to ammonia. The first step of protein hydrolysis, e.g. is the releasing of amino acids, which are then hydrolysed under aerobic or anaerobic conditions to ammonia (Ladd and Jackson 1982). The ammonification rates depend on the C/N ratio of the organic compound; high rates generally occur at low C/N ratios (Alef and Kleiner 1986a, 1986b; Nannipieri et al 1990; Speres 1981). Except for the hydrolysis of urea by extracellular urease, the ammonification is bound to the metabolism of active cells.

Ammonia can be released from organic N by different mechanisms:

1. Hydrolytical deamination:
   $$R\text{-}NH_2 + H_2O \rightarrow R\text{-}OH + NH_3 \quad (5.13)$$
2. Oxidative deamination:
   $$R\text{-}CHNH_2\text{-}COOH + H_2O \rightarrow R\text{-}CO\text{-}COOH + 2H^+ + NH_3 \quad (5.14)$$
3. Reductive deamination:
   $$R\text{-}CHNH_3\text{-}COOH + 2H^+ \rightarrow R\text{-}CH_2\text{-}COOH + NH_3 \quad (5.15)$$
4. Desaturative deamination:
   $$R\text{-}CH_2\text{-}CHNH_2\text{-}COOH \rightarrow R\text{-}CH{=}CH\text{-}COOH \quad (5.16)$$

The ammonification of amino acids has been proposed as a simple and rapid parameter to estimate microbial activity in soils (Alef et al 1988). Waring and Bremner (1964) have used the anaerobic ammonification as a technique to quantify the nitrogen availability in soils.

Nitrification

The conversion of ammonia to nitrate is brought about by two highly specialized groups of obligatory aerobic chemoautotrophic bacteria. Nitrification occurs in two steps: firstly, ammonia is oxidized to nitrite; and secondly, nitrite is oxidized to nitrate.

1. Ammonia oxidation:
   $$NH_4^+ + 1.5\ O_2 \rightarrow H_2O + 2H^+ \qquad G^o = -273.9\ \text{kJ/mol} \quad (5.17)$$
2. Nitrite oxidation:
   $$NO_2^- + 0.5\ O_2 \rightarrow NO_3^- \qquad G^o = -76.7\ \text{kJ/mol} \quad (5.18)$$

The energy derived from the oxidation of either NH_4^+ or NO_2^- can be used for the autotrophic CO_2 fixation. Generally, ammonia is rapidly oxidized to nitrate by heterotrophic bacteria and fungi. Both lithotrophic and heterotrophic nitrifying microorganisms can be cultivated in mineral mediums (Adams 1986a, 1986b; Rheinheimer et al 1988). Ammonia more than nitrate can be adsorbed on inorganic and organic colloids. Nitrate is easily leached and several compounds have been used to inhibit the soil nitrification (Azhar et al 1989; Guiraud et al 1989).

The nitrogen mineralization rates in laboratories can be estimated by incubation or leaching techniques, and in the field by using the lysometer bags (Eno 1960; Stanford and Smith 1972; Richter et al 1982; Rausch et al 1985; Harrington 1986; Heilmeier 1988).

The estimation of nitrogen mineralization rates can be used as a criterion to quantify microbial activity in soil; it has been used to study the influence of pollutants and environmental factors on soil microbial activity (Greaves et al 1980; Schlosser 1988; Schuster 1988; Somerville and Greaves 1987).

Estimation of nitrogen mineralization in the laboratory

(Keeney and Bremner 1966; modified by Beck 1983)

Principle of the method

The method is based on the incubation of water-saturated soil at 25°C followed by the determination of nitrate and ammonium.

Materials and apparatus

Spectrophotometer
Incubator adjustable to 25°C
Shaker
Plastic flasks (200 ml)
Centrifuge and centrifuge tubes

Chemicals and solutions

Reagents for ammonium determination

Dissolve 62.6 g of phenol (analytical grade) in ethanol, add 18.5 ml acetone (analytical grade) and bring up with ethanol to 100 ml (store at 4°C).

Dissolve 27 g of NaOH in distilled water and bring up with distilled water to 100 ml.

Shortly before use, mix 20 ml each of phenol and NaOH solution, and bring up to 100 ml with distilled water.

Sodium hypochlorite solution with an active Cl_2 content of 0.9%. At time intervals the content of active Cl_2 should be measured.

Potassium sulphate solution (1%)

Dissolve 10 g of potassium sulphate in 700 ml distilled water and bring up with distilled water to 1000 ml.

Sulphuric acid solution

Dilute 10 ml of sulphuric acid (con. d = 1.84) with distilled water and bring up with distilled water to 100 ml.

Copper sulphate solution (5%)

Dissolve 5 g of $CuSO_4 \cdot 5H_2O$ in 50 ml distilled water and bring up with distilled water to 100 ml.

Ammonium standard solution

Dissolve 3.821 g of ammonium chloride and 7.218 g of potassium nitrate in distilled water and bring up with distilled water to 1000 ml. Dilute 10 ml of this solution with 990 ml distilled water to obtain a solution of 10 µg NH_4-N/ml^{-1}.

Zinc granules (copperized)

Zinc granules (3–8 mm)

Copper sulphate (2.5%)

Dissolve 25 g of $CuSO_4.5H_2O$ in 700 ml distilled water and bring up with water to 1000 ml.

Preparation of the copperized granules

Zinc (250 g) is mixed with 150 ml distilled water and 15 ml sulphuric acid, and stirred until the surface of the zinc becomes clean. After removing the liquid phase, the zinc is washed with distilled water (3 × 150 ml). Distilled water (150 ml) is then added and, under continuous shaking, 25 ml of copper sulphate solution (dropwise) are pipetted until the zinc granules become completely black. Finally, the liquid phase is removed; the zinc granules are washed again, air-dried and stored in closed glass tubes.

Procedure

Soil (10 g) is placed in six Erlenmeyer flasks (100 ml) and each is moistened with 3 ml distilled water and loosely capped. Two are immediately stored at −20°C (control) and the other four flasks are incubated for 7–30 days at 25°C. After the incubation, all soil samples are suspended in 100 ml potassium sulphate solution (1%), transferred into plastic flasks (200 ml) and shaken for 1 h. Finally, the soil suspensions are centrifuged or filtered through an N-free filter.

Ammonium determination

The clear filtrate or supernatant (50 ml) is treated with 4 ml of the phenol/NaOH solution and 3 ml sodium hypochlorite solution in a volumetric flask (100 ml) and mixed thoroughly. After 90 min the optical density is measured at 630 nm.

Calibration curve

The calibration curve is prepared for concentrations between 0 and 100 µg NH_4-N 50 ml^{-1}.

Calculation: the concentrations can be calculated briefly:

$$NH_4\text{-N } (\mu g\ g^{-1}\ dwt) = \frac{NH_4\text{-N } (\mu g\ 50\ ml^{-1}) \text{ filtrate} \times 2}{dwt \times 10} \quad (5.19)$$

where *dwt* is the dry weight of 1 g moist soil, 10 is the weight of soil in the assay and 2 is the multiplication factor that takes into consideration the total soil suspension.

Nitrate determination

Depending on the nitrate concentrations, 2.5, 5 or 10 ml of the filtrate (or supernatant) are pipetted into a volumetric flask and made up to 100 ml with potassium sulphate. After the addition of 2 ml sulphuric acid, the volumetric flask is shaken. The filtrate (10 ml) of the control and 0.4 ml sulphuric acid are pipetted into glass tubes containing 1–2 zinc granules, mixed well and incubated for 4–5 h at room temperature (an overnight incubation is recommended). The optical density is measured against water at 210 nm by using quartz cuvettes.

Calibration plot

The calibration plot is prepared for concentrations between 0 and 100 µg NO_3-N/50 ml^{-1}.

Calculations

Nitrate concentrations can be calculated according to the following relationship:

$$NO_3\text{-N } (\mu g\ g^{-1} dwt) = \frac{NO_3\text{-N } (\mu g\ 50\ ml^{-1}) \text{ filtrate} \times 2}{dwt \times 10} \quad (5.20)$$

where *dwt* is the dry weight of 1 g moist soil, 10 is the weight of soil in the assay and 2 is the multiplication factor which keeps into consideration of the total soil suspension.

Calculation of the nitrogen mineralization rates

$$N_{min}\ (\mu g\ g^{-1}\ dwt\ day^{-1}) = \quad (5.21)$$

$$\frac{(NH_4\text{-}N_a + NO_3\text{-}N_a) - (NH_4\text{-}N_b + NO_3\text{-}N_b)}{t \times dwt}$$

where $NH_4\text{-}N_a$ is the NH_4-N concentration after the incubation, $NO_3\text{-}N_a$ is the NO_3-N concentration after the incubation, $NH_4\text{-}N_b$ is the NH_4-N concentration of the control, $NO_3\text{-}N_b$ is the NO_3-N concentration of the control, t is the incubation time and dwt is the dry weight of 1 g of moist soil.

Discussion

Ammonium and nitrate can be extracted from soils using 1 M K_2SO_4 solution (Wilke 1989), 1 M KCl (Kandeler and Gerber 1988) or 2 M KCl (Alef and Kleiner 1987). Ammonium can also be estimated according to various methods, including those of Alef and Kleiner (1987) and/or Kandeler and Gerber (1988). Nitrate concentrations can be also determined by the HPLC technique (Reynders and Vlassak 1981; Vlassak and Verstraeten 1986) or photometrically according to Norman and Stucki (1981).

The N mineralization, under the conditions used, has a temperature optimum at 50°C. At temperatures higher than 50°C, nitrification is inhibited, as is evident from the accumulation of ammonium (Beck 1983).

The N mineralization in soils is not always linear with time (up to 3 weeks). After remoistening dry soils, a lag phase has often been observed. Estimations of NH_4-N and NO_3-N during the soil incubation is recommended (Somerville and Greaves 1987; Katyal et al 1988).

The pH value of soil did not significantly influence the N-mineralization rate.

The assimilation of ammonium and nitrate by plants and soil microorganisms interferes with the results. Denitrification can also affect the results if the incubation is not carried out under well-aerated conditions (Woldendorf and Laabroek 1989).

Several authors suggest mixing soil with quartz sand to obtain an optimal gas diffusion (Keeney and Bremner 1966; Stanford and Smith 1972; Beck 1983; Katyal et al 1988).

Mineralization rates between 0.2 and 2, 1.5 and 6 µg N g^{-1}dwt day^{-} are found in agricultural and grassland soils, respectively (Beck 1983).

The soil N mineralization can be inhibited by pollutants (Domsch et al 1983; Minnich and McBride 1986; Somerville and Greaves 1987; Wilke 1989).

Significant correlations have been found between the N-mineralization rates, SIR, arginine ammonification, heat output and protease activity (Beck 1983; Alef et al 1988). No significant correlation has been found between the N-mineralization rate and the content of organic matter in tropic soils (Sahrawat 1981).

Estimation of the nitrogen mineralization potential of soils by the incubation–leaching method

(Stanford and Smith 1972)

Principle of the method

The method is based on the incubation of soil in a column at 35°C. At proper time intervals the mineral N is leached and determined.

Materials and apparatus

Vacuum pump
Leaching glass tubes
Glass wool pad (about ¼ inch)
Quartz sand (20 mesh)
Incubator adjustable to 35°C

Chemicals and solutions

Fine ground Devarda alloy (100 mesh)
$CaCl_2 \cdot 2H_2O$ (0.01 M)
Nutrient solution

0.002 M $CuSO_4 \cdot 2H_2O$, 0.0025 M $MgSO_4$, 0.005 M $Ca(H_2PO_4)_2 \cdot 2H_2O$ and 0.0025 M K_2SO_4

Reagents for the estimation of inorganic nitrogen (see Chapter 3)

Procedure

Duplicate 15 g samples of soil and 15 g quartz sand (20 mesh) are moistened with distilled water using a fine spray and mixed thoroughly. This procedure gives a homogeneous mixture and prevents particle-size segregation during transfer to leaching tubes. The soil is retained in the 50 ml leaching tube by means of a glass wool pad (about ¼ inch), which is placed over the soil to avoid the soil dispersion when solution is poured into the tube.

Mineral nitrogen initially present in the soil is leached with 100 ml $CaCl_2$ solution (in 5–10 ml samples), then 25 ml of a nutrient solution devoid of N are added to the soil. Excess water is removed under vacuum (60 cm Hg). The tubes are then stoppered and incubated at 35°C. The gaseous interchange through the open stem of the leaching tube is sufficient to maintain aerobic conditions during the incubation. After 2 weeks, mineral N is recovered by leaching with 0.01M $CaCl_2$ and "minus-N" solution, followed by applying suction as described above. Tubes are returned to the incubator for incubation for up to 30 weeks. Losses in soil moisture are negligible because of the intermittent leachings of mineral N and the successive restoration of optimal soil water contents.

Determination of inorganic N

Leachates are transferred to 800 ml Kjeldahl flasks and diluted with 300 ml of distilled water. Following addition of Devarda's alloy (0.5 g) and 2 ml of 10 M NaOH, mineral N (nitrate, nitrite and ammonium) is determined by acid titration (0.0025 M H_2SO_4) after distillation into boric acid (see urease estimation).

Calibration plot and calculation

See Chapter 3.

Discussion

Buchner funnels can be used instead of the leaching tubes (Lindemann et al 1989).

Minnich and McBride (1986) and Lindemann et al (1989) did not add the nutrient solution to the leached soil. Leaching causes a loss of soil nutrients. The estimation of such loss is very difficult.

Other methods than that described can be used to determine the mineral N (Lindemann et al 1989).

Arginine ammonification

(Alef and Kleiner 1987)

Amino acids released during the extracellular proteolysis are immediately taken up by soil microorganisms. Inside the cell the amino acids are deaminated (mainly oxidated deamination) and part of the amino group is excreted as ammonium. The excreted amount of ammonium depends on the C/N ratio of the amino acid, where high ammonium excretion occurs at a low ratio.

The ammonification of arginine seems to be a common process in microorganisms (Alef and Kleiner 1986; Speres 1981). A detailed description of the arginine metabolism has been reported by Cunin et al (1986) and Lehninger (1977).

Principle of the method

The method is based on the determination of ammonium after the incubation of soil with arginine for 3 h at 30°C.

Materials and apparatus

Centrifuge and centrifugation tubes (12 ml volume)
Rubber stoppers
Incubator adjustable to 30°C

Chemicals and solutions

Arginine solution

Dissolve 100 mg L-arginine in 30 ml distilled water and bring up with distilled water to 50 ml.

Potassium chloride (KCl 2 M)

Dissolve 149 g KCl in distilled water and make up with distilled water to 1000 ml.

Reagents for ammonium determination (all solutions are stable at 4°C for several weeks).

Phenolate (2%)

Dissolve 2 g of phenolate in distilled water and make up with distilled water to 100 ml.

Sodium nitroprusside solution

Dissolve 50 mg of sodium nitroprusside in 900 ml distilled water and bring up with distilled water to 1000 ml.

NaOH (5 g) and 25 ml NaOCl (15% active Cl_2) in distilled water and bring up with distilled water to 1000 ml.

Ammonium standard solutions.
Solution I

Dissolve 3.82 g of NH_4Cl in distilled water and bring up with distilled water to 1000 ml.

Solution II

Dilute 10 ml of solution I to 1000 ml with distilled water (10 µg NH_4-N ml^{-1}).

Procedure

Twenty-four hours before the assays the soil samples are incubated at room temperature. Routinely 2 g of soil are placed in six centrifugation tubes, loosely capped with rubber stoppers and incubated for 15 min at 30°C, then 0.5 ml of the arginine solution are added dropwise to three tubes. The tubes are incubated for a further 3 h at 30°C (loosely capped with stoppers). Distilled water (0.5 ml) is added to the other three tubes, which are immediately stored at −20°C (blank). Finally 8 ml KCl solution are added to the samples and the tubes closed with rubber stoppers and shaken for 30 min. The tubes are then centrifuged for 5–10 min. After centrifugation 0.5 ml of the supernatant are mixed with 1.5 ml KCl (2 M), 1 ml of the phenolate solution, 0.5 ml of the sodium nitroprusside solution and 0.5 ml of the NaOH/NaOCl solution. After incubation at 30°C for 30 min, the optical density is measured against a KCl blank at 630 nm.

Calibration plot

Pipette 0, 0.5, 1, 2.5 and 4 ml of solution II in volumetric flasks (10 ml), and bring up with 2 M KCl to 10 ml. The ammonium concentrations corresponding to 0, 0.5, 1, 2.5 and 4 µg NH_4-N ml^{-1} are estimated as described above.

Calculation

After correction for the control values (ammonification stopped at time zero), the values can be calculated from the following relationship:

$$NH_4\text{-N } (\mu g\ g^{-1}\ dwt\ h^{-1}) = \frac{NH_4\text{-N } (\mu g\ ml^{-1}) \times 4.25}{t \times dwt} \quad (5.22)$$

where *t* is the incubation time in hours, *dwt* in the dry weight of 1 g moist soil and 4.25 is the total volume of solution added to 1 g of soil in the assay.

Discussion

Under the conditions employed, the arginine ammonification is linear with time.

It is important to use only ammonium-free KCl.

For extinctions of about 1.0, it is recommended that less than 0.5 ml of the supernatant is used for the ammonium determination.

In forest and air-dried soils, high values for the control are found. This decreases the sensitivity of the method.

Arginine ammonification is not extracellular but is an intracellular process in the soil.

Significant correlations were found between the arginine ammonification, the substrate-induced respiration, the heat output, the DMSO reduction, the N-mineralization rate and the ATP content (Alef et al 1988; Suttner and Alef 1988; Alef and Kleiner 1989; Alef 1990).

The presence of easily available C-source in the soil sample (e.g. after the addition of organic matter like cellulose) can inhibit the deamination of arginine. This causes an underestimation of the arginine ammonification.

Like the determination of microbial biomass by the estimation of ATP and the substrate induced respiration, the equilibration of soil at constant moisture and temperature for several days prior to analysis is recommended. (Alef and Kleiner 1987, Nannipieri et al 1990, see chapter 8).

Pollutants and heavy metals influence the arginine ammonification to different degrees (Hund et al 1988; Wilke 1989, 1991).

Estimation of the nitrification in soil

(Beck 1976, 1979)

The nitrification activity of soil can be determined under short-term (a few hours or days) or long-term incubations. Usually the use of a long term assay is not preferred because changes in the composition of soil microflora can occur. Assays are also diversified on the basis of soil sampling, handling and storage. Superficial moist soils can be sieved (< 2 mm) before the assay; other assays are based on the use of soil cores, while in other methods, soils are air-dried, sieved (< 2 mm) and submerged in a solution to give a soil slurry.

Principle of the method

The method is based on the incubation of soil samples with ammonium solution up to 3 weeks at 25°C.

Materials and apparatus

See previous section on estimation of nitrogen mineralization in the laboratory.

Chemicals and solutions

Ammonium sulphate solution (1%)

Dissolve 1 g of ammonium sulphate in 80 ml distilled water and bring up with water to 100 ml.

Reagent for ammonium and nitrate determination

See previous section on estimation of nitrogen mineralization in the laboratory.

Procedure

Soil (10 g) is placed in four Erlenmeyer flasks (100 ml); 1 ml ammonium sulphate solution is added, and the soil WHC is adjusted to 50–60%. The flasks are then loosely capped. Two Erlenmeyer flasks are immediately stored at −20°C (controls) and the other two flasks incubated for up to 3 weeks at 25°C. After the incubation all samples are treated with 100 ml potassium sulphate solution (1%), transferred into plastic flasks (200 ml) and shaken for 1 h. Finally the soil suspension is centrifuged or filtered (N-free filter).

Calculation

The nitrification rates can be calculated from the nitrate produced or from the residual ammonium (see previous section on estimation of nitrogen mineralization in the laboratory).

Discussion

Nitrification is the oxidation of ammonium, therefore the soil aeration has an important role (Beck 1983; Lindemann et al 1989).

The nitrification has a temperature optimum at 25°C (Beck 1983).

Soil moisture has to be controlled during the incubation.

Low nitrification rates are observed in acid soils (Beck 1983; Adams 1986b).

Nitrification in soil is not always linear with the time course (up to 3 weeks). After remoistening dry soils, a lag mineralization phase can be observed.

Both ammonium and nitrate can be taken up by plants and soil microorganisms; this influences the results. Denitrification in well-aerated soils has no significant role (Woldendorf and Laabroek 1989).

Pollutants can significantly inhibit the nitrification in soil (Beck 1981; Rother et al 1982; Domsch et al 1983; Somerville and Greaves 1987; Wilke 1989).

Assay of nitrification (short-term estimations)

(Berg and Rosswall 1985)

Principle of the method

It is based on the determination of nitrite after the incubation of soil samples with $NaClO_3$ in the presence or absence of ammonium sulphate for 5 or 24 h at 25°C.

Materials and apparatus

Erlenmeyer flasks (100 ml)
Filter paper
Incubator adjustable to 25°C
Spectrophotometer
Shaker

Chemicals and solutions

Sodium chlorate solution.
Sodium chlorate solution I (1.5 M)

Dissolve 15.97 g of $NaClO_3$ in 70 ml distilled water and bring up with water to 100 ml.

Sodium chlorate solution II (75 mM)

Dilute 10 ml of the sodium chlorate solution I with distilled water to 200 ml.

Ammonium sulphate solution (1 mM)

Dissolve 0.13214 g of $(NH_4)SO_4$ in 800 ml distilled water and bring up with distilled water to 1000 ml.

KCl solution

Dissolve 149.12 g of KCl in 700 ml distilled water and bring up with distilled water to 1000 ml.

Buffer (0.19 M, pH 8.5)

Dissolve 10 g of NH_4Cl in distilled water, adjust the pH to 8.5 with NH_4OH and dilute with distilled water to 1000 ml.

Reagent for nitrite determination

Dissolve 2 g of sulphanilamide and 0.1 g naphthyl-diethylene-diammonium chloride in about 150 ml distilled water and add 20 ml phosphoric acid (conc.). After cooling dilute with distilled water to 200 ml.

Nitrite stock solution (1000 μg N ml^{-1})

Dissolve 4.9257 g of sodium nitrite in 700 ml distilled water and bring up with distilled water to 1000 ml. Store at 4°C.

Nitrite standard solution (10 μg N ml^{-1})

Dilute 5 ml of nitrite stock solution with distilled water to 500 ml.

Procedure

Estimation of nitrification in the absence of ammonium

Place 5 g of soil in each of three Erlenmeyer flasks, add 2.5 ml of sodium chlorate solution II, cap the flasks (Cap-o-test) and incubate two of them for 24 h at 25°C. To perform the control, store the third flask immediately at −20°C. After the incubation, add 5 ml distilled water, 10 ml of the KCl solution, mix thoroughly and filter immediately. Pipette 5 ml of the clear filtrate into glass test tubes, add 3 ml of buffer, 2 ml of the reagent for nitrite determination, shake well and allow to stand for 15 min at room temperature. Measure the colour intensity at 520 nm.

Estimation of nitrification in the presence of ammonium

Place 5 g of soil in each of three Erlenmeyer flasks, add 0.1 ml $NaClO_3$ I, 20 ml ammonium sulphate solution, cap the flasks (Cap-o-test) and incubate two of the flasks for 5 h at 25°C. To perform the control, store the third flask immediately at −20°C. After the incubation, add 5 ml of the KCl solution, mix thoroughly and filter the soil suspensions. Pipette 5 ml of the clear filtrate into glass tubes, add 3 ml buffer, 2 ml of the reagent for nitrite determination, shake well and allow to stand for 15 min at room temperature. Measure the colour intensity at 520 nm.

Calibration plot

Pipette 0, 2, 4, 6, 8 and 10 ml of the standard nitrite solution into volumetric flasks (100 ml), add 40 ml of the KCl solution (2 M) for estimation of nitrification in the absence of ammonium, or 20 ml for estimation of nitrification in the presence of ammonium, dilute with distilled water to 100 ml. Determine the nitrite as described above.

Calculation

Correct the measurements for the control and calculate as follows:

NO_2-N (μg g^{-1} dwt)/t =

$$\frac{NO_2\text{-N } (\mu g\ ml^{-1}) \text{ filtrate} \times v}{5 \times dwt} \quad (5.23)$$

where *dwt* is the dry weight of 1 g moist soil, 5 is the weight of the used moist soil, *t* is the incubation time in hours and *v* is the total volume of solutions added to soil sample in the assay (12.5 ml for the estimation of nitrification in the absence of ammonium, and 25.1 ml for estimation in presence of ammonium.

Discussion

The nitrification rates in soils with pH < 5 are very low, usually this method is used for soils with pH > 5 (e.g. agricultural soils).

$NaClO_3$ can be adsorbed by inorganic and organic colloids. For soils with an organic C content higher than 3.5%, higher chlorate concentrations should be used.

Filtrates can be stored overnight at 4°C.

The colour intensity should be measured within 4 h.

Estimation of nitrogenase activity of free-living bacteria in soils

(Hardy et al 1973)

Free-living nitrogen-fixing microorganisms are widely distributed and are found in nearly all soils. The ability to fix dinitrogen is widely distributed among the eubacteria and cyanobacteria; fixation by some streptomycetes also occurs (Postgate 1982). With the advent of the acetylene test for nitrogen fixation, the number of organisms capable of fixing N_2 has increased because new heterotrophs and phototrophs (aerobic, anaerobic, microaerobic, facultative anaerobic) have been discovered (Knowles 1982).

The nitrogenase enzyme system in the cells catalyses the dinitrogen reduction to ammonium, which is used for amino-acid synthesis. Nitrogenase catalyses also the reduction of acetylene, azide, cyanide, isocyanide, cyclopropene, diazirine and dinitrogen oxide. Nitrogenase theoretically consumes six electrons and six protons to reduce one molecule of dinitrogen to two of ammonia. There is evidence that one molecule of H_2 is released during the reaction which can be presented in this way:

$$8\ H^+ + N_2 + 8\ e^- \rightarrow 2\ NH_3 + H_2 \qquad (5.24)$$

This reaction requires about 16 mol ATP mol^{-1} N_2. This corresponds to about two molecules of ATP hydrolysed per electron transferred to the substrate. The synthesis and activity of nitrogenase in the cells is strongly inhibited by ammonium and markedly stimulated under nitrogen starvation (Alef et al 1981; Postgate 1982). The asymbiotic nitrogen fixation activity has been proposed as a sensitive assay for detecting pollution by heavy metals in soil (Skuzius et al 1986). However, this proposal has been criticized by Lorenz et al (1992) because: (1) this activity is almost absent in some soils; and (2) there is a significant inverse in the relationship between the concentration of metal and activity.

Principle of the method

The method is based on the estimation of ethylene formed during the incubation of soil with acetylene in an air-tight incubation system (flasks or jars) at temperatures ranging between 20 and 30°C, and soil at a holding capacity of 60%.

Materials and apparatus

Flasks and rubber septums (12 ml)
Jars with caps, fitted with an outlite to remove gas samples (500 ml)
2.3 Corer (6 × 12.5 cm or 8 × 7 cm)
Gas-tight syringes (1, 10, 50 and 100 ml)
Gas chromatograph fitted with flame ionization detector and Porapak R column (2 m)

Chemicals and solutions

Acetylene or calcium carbide (see discussion)
Ethylene (pure)
Glucose
N_2 (carrier gas, 5.0)
Air (5.0)
H_2 (5.0)

Procedure

Sieved (2 mm) moist soil (3–7 g, 60% WHC) or a moist core (about 500 g) is placed in the incubation system (flasks or jars). The incubation system is then closed tightly (with a rubber septum or a cap; if necessary use silicon fat). Replace 10% of the gas volume of the incubation system with acetylene and incubate the system at temperatures ranging from 20 to 30°C for up to 48 h. At the beginning of the experiment (after an equilibration time of about 5 min), a sample from the gas phase is removed and ethylene concentration is estimated. At time intervals (up to 10 h) 0.1–1 ml of the gas phase are then removed and analysed chromatogaphically for ethylene concentrations. Nitrogenase activity can also be measured in the presence of glucose; in this case, ground glucose is mixed with soil at a final concentration of 1% (Hund 1988).

Calibration plot

The calibration plot is prepared by injecting ethylene volumes ranging from 100 µl to 500 µl into the gas chromatograph.

Calculation

The concentration of the ethylene formed can be calculated as follows:

Ethylene (nmol g^{-1} dwt) t =

$$\frac{(C_1 - C_0) \times V \times P}{SW \times v \times R \times T \times dwt} \quad (5.25)$$

C_1 is the ethylene concentration at the defined incubation time in nanomoles, C_0 is the ethylene concentration at the beginning of the experiment in nanomoles, V is the volume of the gas phase in the incubation system in millilitres, v is the volume of the gas phase injected in the gas chromatograph in millilitres, P is the pressure in pascals, R is the gas constant (8.314 J mol^{-1} K^{-1}), T is the incubation temperature (295 K), t is the incubation time in hours and dwt is the weight of 1 g moist soil.

Discussion

The temperature of the detector, column and injector are 150, 50 and 70°C, respectively (Zechmeister-Boltenstern 1991). Hund (1988) used a column temperature of 80°C.

Hund (1988) used 10 g most soil in a 100 ml Erlenmeyer flask and used an acetylene final concentration of 4%. Zechmeister-Boltenstern (1991) used 7 g moist soil in a 12 ml flask.

Flasks or jars with different volumes can be used as incubation systems; it is important to have a linear relation between activity and soil weight and/or incubation time. The linearity of the reaction depends on the incubation time, and type and weight of soil used, therefore the calculation of nitrogenase activity for the total incubation time is recommended.

Generally the nitrogenase activity measured has a low reproducibility; therefore 8–10 replicates are recommended.

Soil storage at −20°C is not recommended (Zechmeister-Boltenstern 1991).

Acetylene can be prepared by dissolving CaC_2 in water in a special flask. The acetylene formed is then collected in the gas phase above the water.

Soils from humid climates with an organic matter content above 1.0% show a negligible activity, because of the ammonification rates of soils. Under laboratory conditions, the long incubation time and/or the addition of substrates (e.g. glucose) during the assay could result in the stimulation of nitrogenase activity in soils (because of depletion of ammonium). Therefore caution is needed in interpreting the results.

A control without C_2H_2 is necessary to permit the quantification of C_2H_4 by the reduction of C_2H_2; it permits the detection of any C_2H_4 released from rubber or other sources. Another control receiving a small quantity of C_2H_4 is recommended to detect any C_2H_4-metabolizing activity in the sample. A problem in this assay is that C_2H_2 inhibits the further metabolism of C_2H_4, so that the concentration of endogenous C_2H_4 might be greater in the presence than in the absence of C_2H_2 (Knowles 1982).

The acetylene reduction assay is very sensitive, cheap and simple. However, it presents several problems: (1) The C_2H_2 is not the physiological substrate of nitrogenase (David and Fay 1977). (2) It is almost impossible to match the *in situ* conditions (light intensity, O_2

concentration, nutrient and moisture availability) in a closed-system assay. In long-term incubations, changes in the composition of soil microflora may cause serious artifacts (Okon et al 1977), (3) Acetylene has side effects that do not occur with N_2; it inhibits the nitrogenase-dependent H_2 evolution, the uptake-hydrogenase activity, the conventional hydrogenase activity, the cell proliferation in *Clostridia*, the N_2O reduction by denitrifiers, the oxidation of ammonium to NH_2OH by *Nitrosomonas* and the methylotrophic bacteria (Gibson and Turner 1980; Knowles 1980).

The use of ^{15}N to study the nitrogen turnover in soils

R. Brumme
G. Aden

The use of the stable isotope ^{15}N provides reliable information on the complex nitrogen transformation processes in the soil–plant system. Studies using ^{15}N include: the effect of fertilizer in the agricultural practice, the pathway of nitrogen deposited in forest ecosystems, organic nitrogen turnover in mineralization and immobilization processes, nitrogen cycling in soils and in plant–soil systems, fixation–defixation of NH_4^+ by clay minerals, biological nitrogen fixation, gaseous loss of nitrogen in NH_3 volatilization and denitrification, and above-ground plant uptake of deposited nitrogen (Jansson 1971; IAEA 1991). The ^{15}N pool dilution method enables the estimation of the rates of nitrogen mineralization/immobilization, nitrification and NO_3 reduction in soils (Nishio et al 1985; Myrold and Tiedje 1986; Bjarnason 1988; Mulvaney 1991). A dynamic ^{15}N dilution method to determine nitrogen transformation rates in continuous flow systems (Brumme and Beese 1991) will be presented and discussed.

Determination of nitrogen transformation rates in continuous flow systems

Studies of the microbial transformation of nutrients in soils can be performed in sieved, non-aggregated soils where the mineralized nutrients accumulate. The disturbance of the soil structure alters the microbial processes in the soil (Hattori and Hattori 1976; Beese 1986; Parkin 1987). Soil microorganisms subjected to continuous water flow provide a more realistic simulation of the situation in nature (Starr et al 1974; Mochoge 1981; Mochoge and Beese 1983a, 1983b; Brumme and Beese 1991).

Principle of the method

Undisturbed soil columns are percolated at a constant flow rate with artificial rain (nutrient solution) containing ^{15}N-labelled ammonium. The movement of ^{15}N in the soil is used to calculate the transformation rate of soil nitrogen.

Materials and apparatus

PVC or Plexiglas tubes (30 cm long, 14 cm diameter)

Percolation system at the top of columns:

> peristaltic pump coupled with a timer (Khanna and Beese 1978); nozzle together with a vent and water under high pressure; other solutions (see Beese et al 1976).

Water removal system at the bottom of the column:

> suction plates/cups (KPM, Berlin) or membrane filters (0.2 µm) together with a constant partial vacuum of 100 hPa.

The columns are involved in an automated microcosm system, which also enables gaseous exchange of CO_2, N_2O and CH_4 to be measured (see Fig. 5.8).
Sand, washed in acid (0.2–0.6 mm diameter)
Chemicals for the percolation solution, depending on the chemical composition of the precipitation

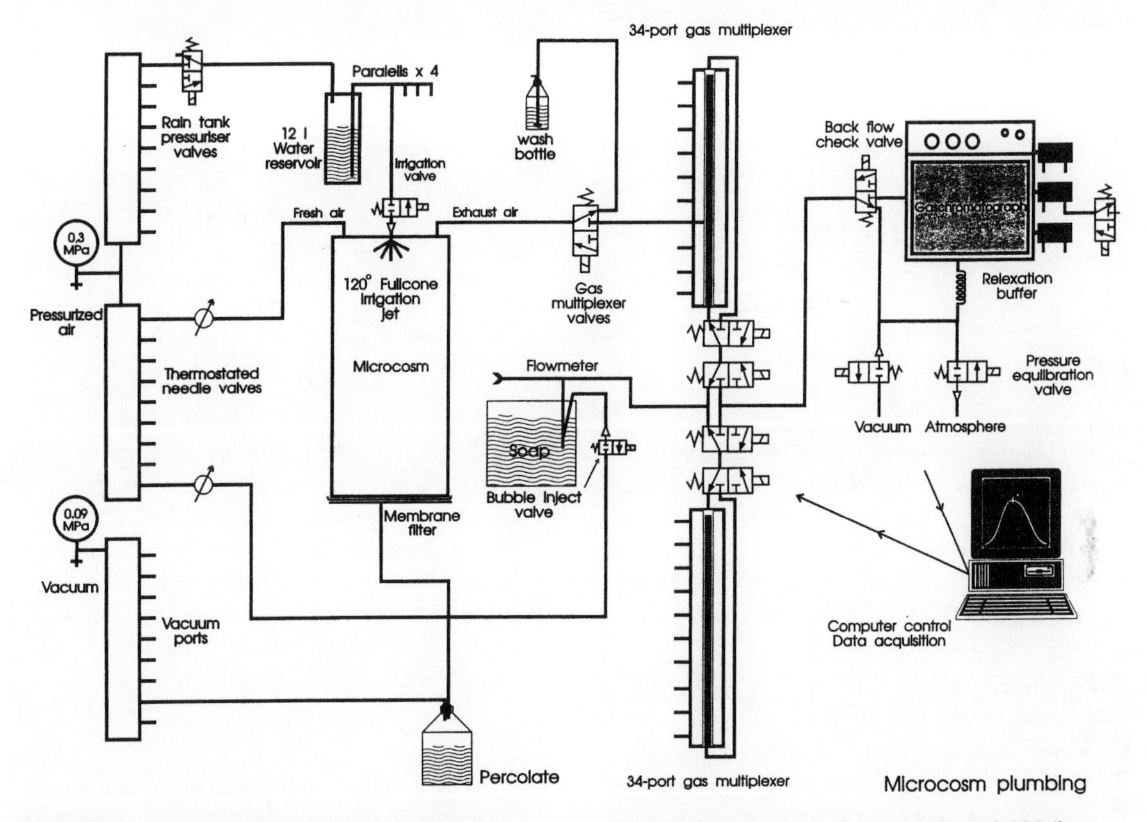

Figure 5.8. Automated microcosm system for studying ecological processes (Hantschel et al 1994).

Labelled $^{15}NH_4Cl$ (enrichment see discussion).
Refrigerator

Soil column sampling

Undisturbed soil columns are obtained after driving PVC or Plexiglas tubes into the soil. This can be done by a hydraulic system (Hantschel et al, 1994) or by hammering (Homeyer et al 1974). The surrounding soil must be removed beforehand (Fig. 5.9). About 5 cm should be left free at the top of the columns to enable gaseous exchange. March appears to be the best sampling time when investigating microbial activity and transformation rates.

Procedure

A constant flow of water through the soil columns is maintained by percolation of water at the top of the column and the removal of soil water at the bottom. A thin layer of acid-washed sand is placed between the bottom of the soil and the water removal system to achieve complete contact between both parts of the system (Fig. 5.8).

The columns are kept at a constant temperature within the range measured in

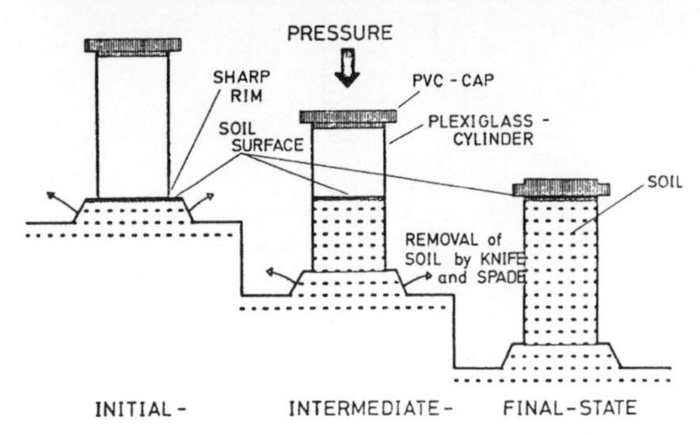

Figure 5.9. Sampling of undisturbed soil cores (Mochage 1981).

the field. A solution containing ammonium concentrations ranges between 2 mg Nl^{-1} (20–36 kg N ha^{-1} a^{-1}) and 4 mg N l^{-1} (40–72 kg N ha^{-1} a^{-1} s^{-1}) is percolated at a precipitation rate between 3 mm day^{-1} (1000 mm a^{-1}) and 5 mm day^{-1} (1800 mm a^{-1}).

Experimental conditions should be kept constant (daily percolation rate, percolation frequency, composition of the percolation solution, soil temperature, soil water potential) to avoid any change of microbial transformation or in the dilution intensity of the labelled nitrogen by soil-borne nitrogen. Before the addition of the ^{15}N-labelled ammonium to the percolation solution, a constant water flow through the columns (water input = water output) and a more or less constant microbial activity (e.g. soil respiration) should be established.

Usually the soil water in the columns (PV) is replaced once or twice to establish a continuous water flow. The concentration of the labelled nitrogen in the percolate should be also kept constant throughout the experiment. Once the labelled ammonium has been added, the soil water in the columns is replaced three or four times to obtain a nearly constant concentration of soil-borne and labelled nitrogen in the leached solution (Fig. 5.10). The soil solution is regularly analysed for nitrate and ammonium, and their enrichment with atom% ^{15}N.

At the end of the experiment, the ammonium, nitrate, organic nitrogen and microbial nitrogen content, as well as the enrichment of atom% ^{15}N is measured. Changes in exchangeable ammonium or in the labelled fraction within the columns during the experiment should be investigated by taking soil cores from separate soil columns just before a constant rate of leached nitrogen is reached and again at the end of the experiment.

Calculation

Simple soil nitrogen transformation model

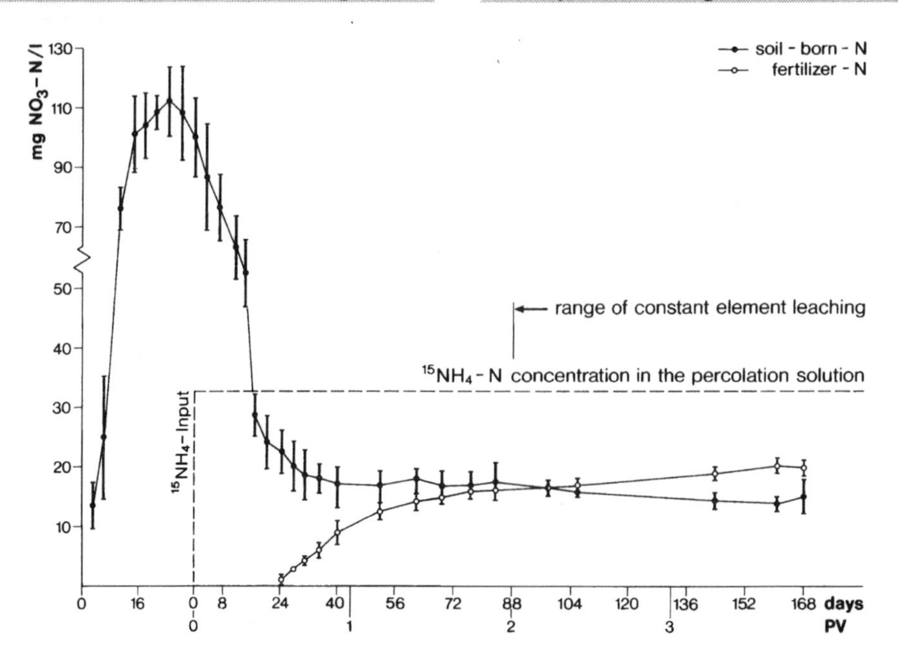

Figure 5.10. Soil-borne (●) and labelled (○) nitrate leached out of a microcosm (32 mg $^{15}NH_4$–N l^{-1} as $(NH_4)_2SO_4$ labelled with 97.7 atom% ^{15}N; percolation rate: 3 mm day^{-1}; 8°C; n=3; PV = pore water volume within the soil column) (Brumme and Beese 1991).

Nitrogen exists in solid form in soil organic matter (N_{org}), in soluble form in soil water (NO_3^-, NO_2^-, NH_4^+, N_{org}), in exchangeable form at the cation exchanger (NH_4^+ ex), in fixed form within clay minerals ($NH_4^+{}_{fix}$) and in gaseous form in soil air (N_2, N_2O, NO, NO_2, NH_3). Nitrogen fluxes exist between the different nitrogen pools as a result of microbial activity. Nitrogen transformation processes occur in soil, such as mineralization/immobilization, nitrification/nitrate reduction, fixation/defixation, sorption/desorption, and removal processes, e.g. gaseous losses and nitrogen leaching, in addition to nitrogen inputs from dry and wet deposition, addition of fertilizer, and N_2 fixation. The simple nitrogen model in Fig. 5.11 considers only the major processes that can be determined using labelled nitrogen.

Rate of net nitrification

The rate of net-nitrification (NN) (mg NO_3-N m^{-2} day^{-1}) is the amount of soil-borne nitrate (^{14}N) that is leached over a certain period of time at constant

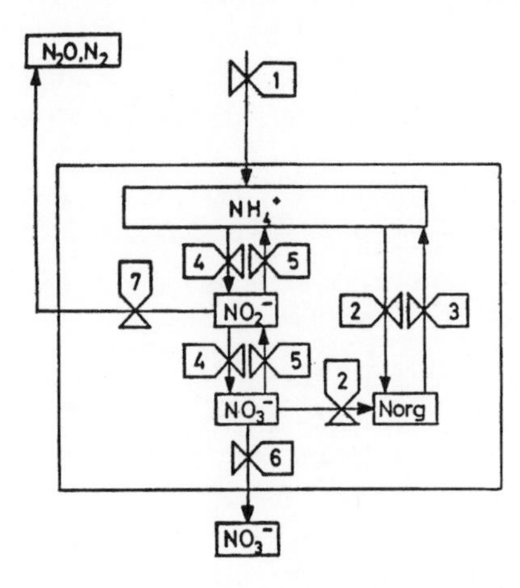

Figure 5.11. Simple soil nitrogen transformation model. Input variables: 1, $^{15}NH_4$ input. Transformation rates: 2, gross N immobilization; 3, ammonification; 4, gross nitrification; 5, gross nitrate reduction. Output variables: 6, leaching N losses (net nitrification); 7, gaseous net N losses.

seepage water nitrate concentrations and can be calculated by the sum of the nitrate leached during this period (see Fig. 10.5):

$$NN = {}^{14}NO_3\text{-N (mg/area/time)} \quad (5.26)$$

where area = the area of the column in square metres and time = the observed time in days.

Rate of net nitrogen mineralization

The rate of net nitrogen mineralization (NNM) (mg (NO_3 + NH_4)-N m^{-2} day^{-1}) is calculated from the amount of ammonium and nitrate that is leached and the change in the amount of exchangeable ammonium in the soil:

$$NNM = ({}^{14}NO_3 + {}^{14}NH_4)\text{-N (mg/area/time)} + {}^{14}NH_{4ex} \quad (5.27)$$

where $^{14}NH_4$ ex is the difference in the amount of exchangeable ammonium present in the soil as constant nitrate leaching commenced and the end of the experiment.

During the time of the constant nitrate leaching, the exchangeable ammonium should be constant and the leaching of ammonium negligible in nitrifying soils, the NNM becomes equal to NN:

$$NNM = NN \quad (5.28)$$

Rate constant of net nitrification

The rate constant of net nitrification (RNN) is calculated by dividing the output amount of labelled nitrate by the input amount of labelled ammonium within the period of constant leaching:

$$RNN = {}^{15}NO_3\text{-}N_{out}\text{ (mg)}/\ {}^{15}NH_4\text{-}N_{in}\text{ (mg)} \quad (5.29)$$

Assuming that the labelled ammonium underwent the same processes compared to the soil-borne ammonium, the RNN can also be calculated by the net nitrification divided by the gross nitrogen mineralization (GNM):

$$RNN = NN/GNM \quad (5.30)$$

Rate of net nitrogen immobilization

Assuming that the nitrogen that is immobilized by microorganisms passes through the ammonium pool, the rate of microbial net nitrogen immobilization (NNI) (mg N m^{-2}/day^{-1}) of soil-borne nitrogen can be calculated at the end of the experiment by analysing ^{15}N in the organic nitrogen pool (mg $^{15}N_{org}$), and the labelled and unlabelled amount of ammonium in the soil at the beginning and the end of the period of constant leaching (Hauck and Bremner 1976):

$$NNI = {}^{15}N_{org}\,(mg) \times [{}^{15}NH_4 + {}^{14}NH_4\,(mg)]/{}^{15}NH_4\,(mg/area/time) \quad (5.31)$$

where $^{15}NH_4$ (mg), $^{14}NH_4$ (mg) = the mean quantity of labelled and unlabelled ammonium in the soil present during the period of constant leaching (see notes and fourth discussion point).

Rate of net gaseous nitrogen loss (NNL)

A common way to calculate gaseous nitrogen loss of the applied labelled nitrogen is to use the ^{15}N balance:

$$^{15}N_{losses} = {}^{15}N_{applied} - {}^{15}N_{soil} - {}^{15}N_{leaching} \quad (5.32)$$

A direct measurement of all gasses, or of those presumed to be most significant in the soil studied, produces the best estimation of gaseous loss.

Rate of gross nitrogen mineralization

The difference between GNM and NNM is due to microbial NNI, gaseous NNL by denitrification, nitrification or NH_3 volatilization, and changes of exchangeable ammonium, assuming that no NH_4 fixation occurred:

$$GNM - NNM = NNL + NNI + NH_{4ex} \quad (5.33)$$

Assuming that these processes (NNI, NNL) occur after the formation of ammonium by ammonification, GNM can be calculated by using equations 5.29 and 5.30:

$$GNM = NN / RNN \quad (5.34)$$

Discussion

Our studies indicate a negligible influence of the percolation rate on the NN rates in the range 0.3–9 cm day^{-1} (Brumme 1986). Nevertheless, the percolation rate does influence the soil water content as well as the element concentration in the soil water solution. Depending on the percolation rate, we observed that the soil water content increased by about 2% and 20% when the rate of percolation increased from 0.3 to 3 cm day^{-1}, and from 3 to 9 cm day^{-1}, respectively. The nitrate concentration is inversely proportional to the percolation rate and can stimulate gaseous losses by denitrification.

For the enrichment of the ammonium tracer with ^{15}N, see ^{15}N analysis.

Gaseous NNL from soils is often estimated from the ^{15}N balance (Tiedje 1982). This method has the advantage that all gaseous forms of nitrogen (NH_3, N_2, N_2O, NO, NO_2) are considered. The disadvantage is that the accuracy is uncertain because the gaseous NNLs are obtained by subtraction (see equation 5.32). Further, no information about short-term changes is available. Direct estimation using gas chromatography provides more accurate rates of gaseous loss.

Soil microbiological parameters as well as NH_4 and $^{15}NH_4$ in the soil should be determined during the experiment for the use of equation 5.31, after sampling of small soil cores out of separate coil columns, which are not used for gas exchange measurements.

^{15}N analysis

The determination of ^{15}N in the different nitrogen pools of the soil demands different sample preparation prior to the isotope ratio mass spectrometry (Fig. 5.12) or the optical emission spectrometry analysis. Two sample preparation procedures exist for liquid samples: the steam distillation and Rittenberg procedure (Hauck 1982; Keeney and Nelson 1982) and the diffusion methods (O'Deen and Porter 1979; MacKown et al 1987; Brooks et al 1989; Burke et al 1990; Jensen 1991; Kelley et al 1991; Sorensen and Jensen 1991). A recent review on nitrogen isotope techniques has been published by Knowles and Blackburn (1992).

For the diffusion method, a combined system of an elemental analyser and an isotope mass spectrometer has to be established. For the distillation/Rittenberg procedure only a MS or ES is required.

^{15}N in solid samples such as soil material, and plant or organic matter are analysed directly by linking an elemental analyser to a MS or ES. Ground solid samples are placed in tin capsules, burned at more than 1000°C in an automated combustion instrument (elemental analyser) (see Chapter 3) and analysed in the linked MS or ES for atom% ^{15}N (Meier and Manersberg 1982; Preston and Owens 1983; Barrie and Workman 1984; Marshall and Whiteway 1985; Brooks et al 1989; Egsgaard et al 1989; Harris and Paul 1989; Schepers et al 1989; Burke et al 1990; Craswell and Esker 1991; Jensen 1991; Meier and Schmidt 1992; Mulraney 1992; Preston 1992; Reineking et al

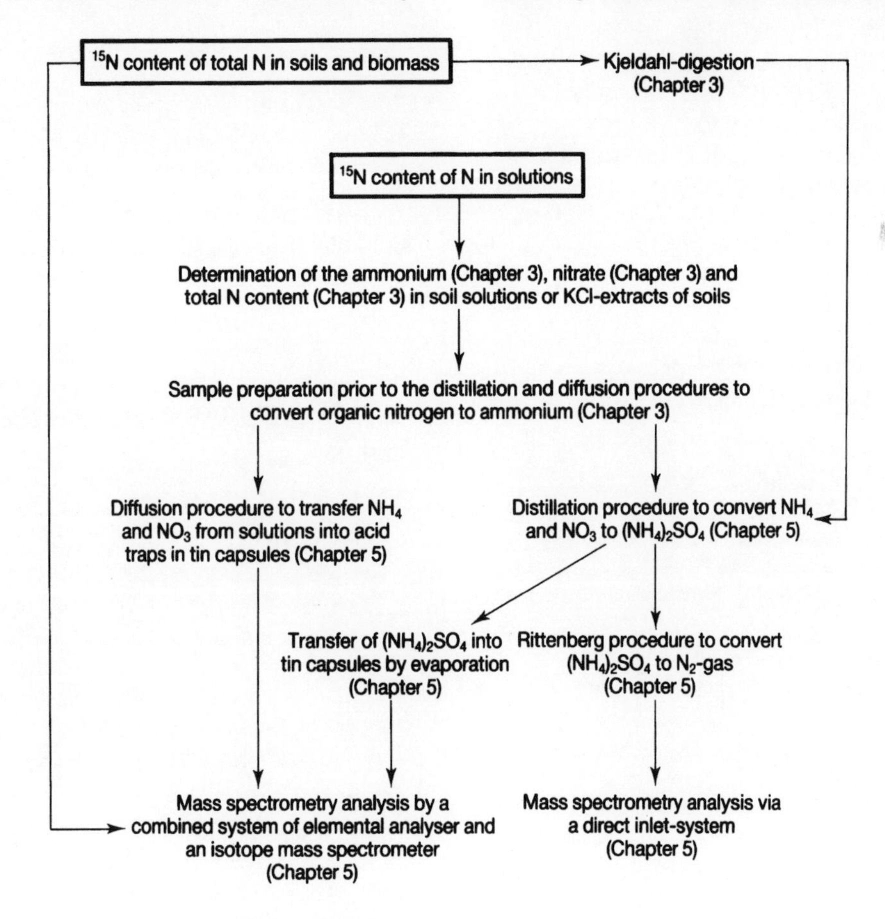

Figure 5.12. Scheme of an elemental analyser coupled to a mass spectrometer for the determination of the ^{15}N content of soil nitrogen (Reineking et al 1993).

1993). This innovative analysis has resulted in an increase in the application of ^{15}N methodology for research on nitrogen transformation in soils. If an EA/MS or EA/ES system is not available for solid sample analysis, a Kjeldahl digestion followed by distillation and the Rittenberg procedure can be performed prior to MS or ES analysis.

Diffusion method

(Jensen 1991, slightly modified)

Principle of the method

After a solution containing ammonium and nitrate is made alkaline, ammonium is transformed to free ammonia, which is then absorbed in an acid trap fixed above the solution. In the second step, nitrate is reduced to ammonium and the free ammonia is again caught in a second trap. The tin capsules containing the traps with ammonium sulphate are burned as for solid samples in an EA/MS or EA/ES system to determine the $^{15}N/^{14}N$ ratio.

Materials and apparatus

250 ml/500 ml polyethylene (PE) bottles with screw caps. A hooked piece of stainless steel wire is fixed with adhesive tape inside the cap.
GF/C glass fibre filter, 25 mm (Whatman), washed with acid (2.5 M $KHSO_4$), cleaned with distilled water and dried overnight at 550°C.
Tin capsules, 5 × 9 mm (Lüdi AG)
Glass beads
Tweezers
Polyethylene gloves
Desiccators
Microtest plates
Ultrasonic bath

Chemicals and solutions

2.5 M $KHSO_4$
Devarda's alloy
Magnesium oxide (heated overnight at 600°C)
Concentrated H_2SO_4
HCl (10%)

Procedure

To determine the ^{15}N content of a sample, the concentration, as well as the enrichment with ^{15}N (atom %^{15}N), has to be known. For the determination of the concentration of ammonium, nitrate and total N in soil solutions or salt extracts of soils, the procedures in Chapter 3 are recommended.

Sample preparation prior to the diffusion procedure:

Ammonium and nitrate in soil solutions are able to diffuse directly.

Exchangeable ammonium and nitrate in soils are able to diffuse directly out of salt extracts of soils (Chapter 3).

For the preparation of organic nitrogen from soil solutions or microbial biomass nitrogen samples (fumigation extraction method, Chapter 8) see the distillation method.

Diffusion

25–200 ml of the salt extract or soil solution containing 200–400 μg nitrogen are placed in 250 or 500 ml PE bottles with screw caps (diffusion containers). One glass bead is added to each bottle. Half of a 25 mm GF/C glass fibre filter is folded and placed into a tin capsule. A total of 30 μl of the absorbent (2.5 M $KHSO_4$) is pipetted on to the filter and the open tin capsule is then fixed to the hook. For the sequential diffusion of ammonium and nitrate, 0.2 g MgO is added to the solution, and ammonium is allowed to diffuse for 7 days. The tin capsule containing the glass fibre filter is changed and a Devarda's alloy (0.4 g) is added to the solution in order to reduce nitrate to ammonium. Ammonium is then allowed to diffuse for another 7 days.

The solutions are carefully rotated to mix the reagents after adding MgO and Devarda's alloy and allowed to stand at room temperature. After the diffusion procedure, the tin capsules containing the filters are dried for 1 or 2 days in a desiccator above concentrated H_2SO_4. For this purpose each tin capsule (more or less destroyed by $KHSO_4$) is placed into a new tin capsule and then placed on a microtest plate in the desiccator. After drying, both tin capsules are wrapped for MS- or ES-analysis. The PE bottles can be used several times if cleaned with 10% HCl and/or ultrasonic bath.

Mass or emission spectrometry analysis

The dried and closed capsules are analysed by using a elemental analyser in combination with a mass or emission spectrometer (Meier and Manersberger 1982) to determine the ^{15}N content.

Calculation

The amount of ^{15}N in a soil-N-fraction is calculated in consideration of the natural abundance of the sample:

$$^{15}N\ (mg) = [N\ (mg) \times (atom\%\ ^{15}N\ measured - atom\ \%\ ^{15}N\ in\ natural\ abundance)]/[100] \quad (5.35)$$

Where N(mg) is the measured N amount in a soil-N-fraction. The total amount of ^{15}N-labelled fertilizer in a soil-N-fraction is calculated using the ^{15}N enrichment of the fertilizer (Hauck and Bremner 1976):

$$N\ fertilizer\ (mg) = \frac{^{15}N\ (mg)}{(Atom\%\ ^{15}N\ in\ fertilizer - atom\%\ ^{15}N\ in\ natural\ abundance)} \times 100 \quad (5.36)$$

The amount of soil-borne nitrogen in a soil-N-fraction is then calculated by the difference between the fertilizer N and the measured N:

$$N_{soil\text{-}borne}(mg) = N(mg) - N_{fertilizer}(mg)$$

Discussion

To avoid cross-contamination with ^{15}N, the samples should be diffuse in order of increasing ^{15}N enrichment. Gloves should be used to wrap the samples.

Natural abundance of ^{15}N is generally about 0.3663 atom%. However, due to isotope fractionation, enrichments above and below this value are possible. Therefore the natural abundance should be determined for the different nitrogen fractions.

In case of high ^{15}N enrichment (>1–3 atom% ^{15}N) of the sample and/or high sensitivity of the mass spectrometer, isotope dilution with naturally labelled $(NH_4)_2SO_4$ may be necessary before mass spectrometry analysis. Use the following equation:

$$\frac{(a \times atom\%\ ^{15}N\ of\ a + b \times atom\%\ ^{15}N\ of\ b)}{c} = atom\%\ ^{15}N\ of\ c \quad (5.37)$$

where a is the amount of N in the sample, b is the amount of added N used for the dilution and $c = (a + b)$.

If an EA/ES system is used, no ^{15}N dilution is required. Emission spectrometers are less sensitive than isotope ratio mass spectrometers (Craswell and Eskew 1991), but enables measurement from 0.366 to 99 atom % ^{15}N.

Before using this method for routine measurements, it should be tested against standard solutions with known enrichments.

The best recovery rates are obtained when the nitrogen content is between 200 and 400 µg.

The coefficient of variation of atom% ^{15}N (standard solutions) for an EA/MS-system is lower when this modified diffusion method is used (0.2%).

The recovery of ammonium and nitrate in the traps ranges from about 91% to 100% (mean value 95%). To prevent isotope fractionation, all nitrogen must be converted to ammonia and diffused in the trap.

The distillation method

Principle of the method

The pH of a solution containing ammonium and nitrate is lowered, converting the ammonium to ammonia. The ammonia is distilled and collected in acidified water (Keeney and Nelson 1982) as ammonium sulphate. In a second step, nitrate is reduced to ammonium, and ammonia is distilled and collected in a second acidified solution.

For the analyses by mass or emission spectrometry, ammonium sulphate must be converted to N_2 gas, either by combustion in an elemental analyser or by the conversion of ammonium sulphate using hypobromite (Rittenberg procedure).

Materials and apparatus

Steam distillation apparatus
Round-bottomed flasks, 250 ml, with two necks
Beakers, 250 ml
Digital pH meter
Magnetic stirrer
Water bath
Test tubes
Sand bath
Tin capsules, 5 × 9 mm (Lüdi AG)
Rittenberg apparatus including a separatory funnel, a pumping system, a manometer and an oven (Fig. 5.13)
Dewar flasks
Molecular sieve (5 Å)
Vials

Chemicals and solutions

H_2SO_4 (0.05 M)
MgO
Devarda's alloy
HCl (10%)
LiOBr (10%)
Helium
Liquid N_2

Procedure

To determine the ^{15}N content of a sample, the concentration, as well as the enrichment with ^{15}N (atom % ^{15}N), has to be known.

For the determination of the concentration of ammonium, nitrate and total N in soil solutions or salt extracts of soils, see Chapter 3.

Sample preparation prior to the distillation procedure:

Ammonium and nitrate from soil solutions can be distilled directly without any sample preparation.

Exchangeable ammonium and nitrate from soils (Chapter 3) can also be distilled directly out of salt extracts of soils.

Organic nitrogen in soil solutions or microbial biomass N after fumigation–extraction method (Chapter 8) must first be converted to ammonium by digestion (Chapter 3). Generally a Kjeldahl procedure with concentrated H_2SO_4 containing selenium as a catalyst is used (Bremner and Mulvaney 1982, modified).

Distillation

The steam required for distillation is generated by heating distilled water in a 2 l flask. A maximum of 100 ml soil solution or digest is placed into the 250 ml round-bottomed flask, which is then fastened to the distillation apparatus with steel springs. A 250 ml beaker with 50 ml water (pH 4.5) is used to collect the distillate. The beaker is placed on a magnetic stirrer and the pH value is continuously controlled.

Ammonium

To distill ammonium, 0.2 g MgO (soil solution) or 50 ml 10 M NaOH (digest) is added to the solution through the second neck of the round-bottomed flask, which is then closed with a glass stopper. Under these conditions, ammonium is transformed to ammonia which is collected in the distillate. To avoid loss of ammonia, the pH of the distillate should immediately be adjusted to 4.5 using H_2SO_4. When 150 ml of the distillate has been collected, the beaker can be removed.

Nitrate

A new beaker with 50 ml distilled water (pH 4.5) is placed on the magnetic stirrer to collect the distillate. Before nitrate can be distilled, it must be reduced to ammonium. For the reduction, 0.4 g Devarda's alloy is added to the solution (already containing MgO or NaOH) through the second neck of the round-bottomed flask. Ammonium, derived from the nitrate, reacts to form ammonia. Repeat the distillation process.

Conversion to dinitrogen

Combustion procedure

The distillates containing more than 50 μg N (optimum 100–300 μg N) are acidified to pH 3–4 with H_2SO_4 and evaporated until dry. The residue is dissolved in 250–350 μl distilled water and the solution is then transferred to a tin capsule. The ammonium sulphate solution in the tin capsule is evaporated to dryness and then wrapped for the analyses of the $^{15}N/^{14}N$ ratio in an EA/MS or EA/ES system.

Rittenberg procedure (Ross and Martin 1970; Porter and O'Deen 1977; Reineking et al 1988)

The conversion of ammonium sulphate to N_2 gas is usually performed by the Rittenberg procedure as follows. A test tube containing solid ammonium sulphate (obtained after evaporating the distillate) is fixed to the apparatus below a separatory flask filled with LiOBr (Fig. 5.13). The LiOBr and complete vacuum system is flushed with He by opening three valves (+V3, +V2, +V1). After 30 s the system and the test tube are evacuated (−V1, −V3, +V4) until a pressure of −3 Pa is reached. After separation of the test tube from the apparatus (−V2, −V4) about 1 ml of the LiOBr solution is dropped on to the dry ammonium sulphate (+V1) until the solution in the test tube becomes yellow. The test tube is then frozen in liquid N_2 for 2–3 min to remove any water vapour, ammonia or NO_x compounds generated during the hypobromite reaction. During this time the small vial with a molecular sieve (5–7 balls, 5 Å), which was heated in an oven, is also dipped into the liquid N_2. The vacuum should be about −3 Pa. To transfer N_2 (formed from ammonium) into the small vial, valves V2 and V5 are opened. The pressure increases to more than 200 Pa. The N_2 gas passes a cool trap to remove any traces of water vapour and is then absorbed by the molecular sieve, which is located in the vial. When this procedure is completed, the vial is separated by melting; valve V5 is closed and the isotopic composition of the N_2 gas can be measured. After fixing a new vial to the apparatus and closing the valve V2 and opening V5, the system is cleaned momentarily of any LiOBr by evacuation (+V4). The vial is heated in the oven and the tip of the separatory flask is wiped to remove any LiOBr from the previous reaction. After closing valve V4 and attaching a new test tube, the procedure is repeated.

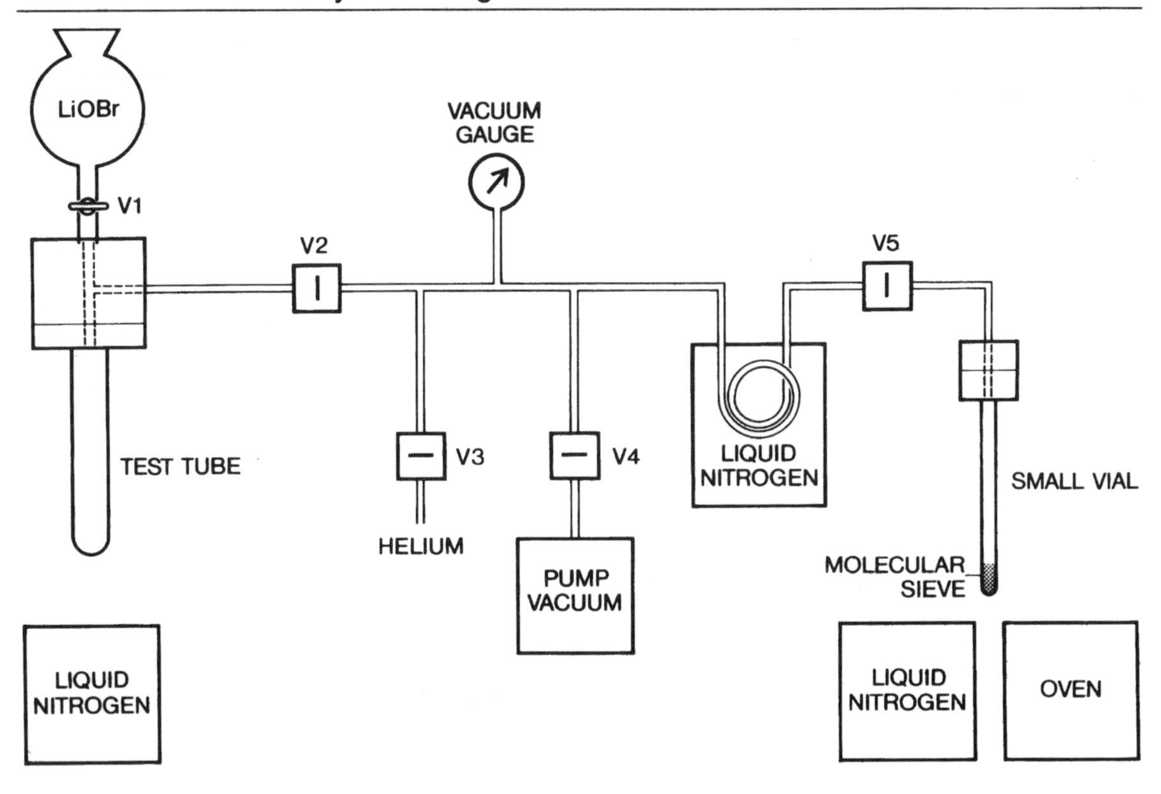

Figure 5.13. Rittenberg apparatus (Reineking et al 1988).

For mass spectrometry analysis of N_2-gas from the Rittenberg procedure, a volume of the soil solution or digest solution containing 1000 µg N is necessary. For emission spectrometry analysis of N_2-gas from the Rittenberg procedure, only 10 µg N_2 are necessary. Use vials recommended by the manufacturer (e.g. FAN, Leipzig) which can be used directly as discharge tubes.

Automated Rittenberg procedure with consecutive emission spectrometry

The conversion of ammonium sulphate to N_2 by Rittenberg, with consecutive emission spectrometry measurement of the $^{15}N/^{14}N$ of N_2 is enabled by the NO1–6PC by FAN, Leipzig. No more than 10µg N as ammonium sulphate is placed in small vials on an autosampler connected to the emission spectrometer. After the automated addition of a defined amount of NaOBr, the evolved N_2 is flushed to the emission spectrometer by helium to determine the $^{15}N/^{14}N$ ratio.

Mass or emission spectrometry analysis

The dried and closed tin capsules can be analysed by using a combination of an elemental analyser and an isotope mass spectrometer (Reineking et al 1993) or an emission spectrometer (Meier and Manersberger 1982). The ^{15}N content of the N_2 gas obtained by the Rittenberg procedure in the sealed tubes can be analysed directly in an emission spectrometer using the tubes as discharge tubes, or by sampling the tubes in the mass spectrometer via a direct inlet system.

Calculation

The concentration of nitrogen obtained from distillation is calculated from the consumption of 0.01 N H_2SO_4:

$$N\ (mg\ l^{-1}) = \frac{\text{consumption of } H_2SO_4\ (ml) \times 140 \times 1000/\text{volume of the sample (Ml)}}{1000} \quad (5.38)$$

For further calculations of the amount of ^{15}N and the total amount of N from fertilizer, see the calculation in the previous section on the diffusion method.

Discussion

To avoid cross-contamination errors, natural abundance samples and low ^{15}N enrichment samples should be distilled before the high ^{15}N enrichment samples (Reeder et al 1980). When the low ^{15}N enrichment samples are distilled after the high ^{15}N enrichment samples, the apparatus must be cleaned by distilling a flush of natural abundance $(NH_4)_2SO_4$.

If 1000 µg N per sample is not available or the ^{15}N enrichment is higher than 1–3 atom% ^{15}N, an isotope dilution is necessary before the mass spectrometry analysis.

A control, containing only water, must be distilled and considered in the calculation of the nitrogen concentrations.

A total of 1000 µg N per sample for mass spectrometer analyses is necessary to fill the volume of the gas inlet system of the mass spectrometer.

The tin capsules and test tubes containing the ammonium sulphate and the N_2 containing vials can be stored for a long time.

[^{3}H]Thymidine incorporation technique to determine soil bacterial growth rate

H. Christensen
S. Christensen

(Christensen 1991, 1993; Christensen et al 1992)

Determination of the total bacterial production of soil and rhizosphere samples may be obtained from the rate of DNA synthesis by incorporation of [^{3}H]thymidine (Tdr). This method provides an estimate of gross growth rate if degradation of [^{3}H]DNA produced from [^{3}H]Tdr is negligible during the labelling period. The [^{3}H]Tdr method has found use in quantifying bacterial production in absolute terms in aquatic systems (Moriarty 1986, 1990), in soil, and rhizosphere (Bååth 1990; Christensen et al 1992, 1993). To obtain absolute measures of bacterial growth rates, conversion factors between [^{3}H]Tdr incorporation and cell production have to be determined in the specific soil system for investigation. The [^{3}H]Tdr method has also been utilized to measure relative bacterial growth rates as influenced by factors such as mycorrhiza (Christensen and Jakobsen 1993) and heavy metals (Bååth 1992).

The present [^{3}H]Tdr labelling protocol is based on small samples labelled as slurries that provide optimal incorporation kinetics of [^{3}H]Tdr, and a quick one-tube format labelling and extraction procedure.

Principle of the method

Specific [^{3}H]Tdr labelling of soil bacteria is achieved by a 15 min pulse labelling of a soil slurry with 200 nM [^{3}H]Tdr. No labelling of soil fungi, plant roots or fauna, occurs under these conditions; these organisms require a higher concentration of Tdr or a longer labelling period to obtain significant [^{3}H]Tdr incorporation. Short-term soil manipulations before and during labelling will probably not affect the [^{3}H]Tdr incorporation rate, since the rate of DNA synthesis is regulated at the level of initiation of DNA replication. When DNA synthesis is initiated, replication will proceed at a constant rate for at least 45 min and, during this period, manipulations with the soil may be performed. The rate of bacterial production measured in the soil slurry therefore represents the *in situ* rate. ^{3}H-Labelled DNA is extracted from soil with warm base hydrolysis followed by centrifugation. DNA is precipitated by cold acid treatment of the supernatant. Labelled DNA is then collected by membrane filtration, and non-incorporated [^{3}H]Tdr is washed away from the filters. If absolute growth rates are required, a conversion factor between [^{3}H]Tdr incorporation and cell production needs to be determined by simultaneous measurement of [^{3}H]Tdr incorporation and change in bacterial biomass in incubation experiments, where bacterial death is minimized. A conversion factor derived for the specific soil system includes compensation for the recovery of DNA extraction, non-specific labelling of macromolecules besides DNA and isotope dilution.

Materials and apparatus

Centrifuge providing 10,000*g* with 5–10 ml tight-capped centrifuge tubes. Tubes are weighed with caps and autoclaved.
Filtration manifold for 25 mm diameter filters provided with stainless steel funnels, e.g. 24/25 mm 10-place filtration manifold (Sartorius, Göttingen), and cellulose acetate membrane filters, 0.45 µm, 25 mm diameter.
Reciprocal shaker run at 200 rev min^{-1}
Dry incubator at 60°C
Balance with detection limit of 0.1 mg
Liquid scintillation counter and glass scintillation vials, 20 ml
Pipettes for volumes of 2–1000 µl
Flasks, 100 ml, e.g. septum flasks

Chemicals and solutions

Stock solution of unlabelled Tdr is prepared by diluting 14.5 mg Tdr in 200 ml distilled water. The solution is autoclaved and stored at 4°C.

The [^{3}H]Tdr-labelling solution should contain 200 nM with ^{3}H specific activity of 30 Ci mmol^{-1}. The labelling solution is prepared under aseptic conditions from e.g. 301 µl [methyl$-^3$H]Tdr stock solution of specific activity 75 Ci mmol^{-1} (1 mCi ml^{-1}, 2% ethanol), 20 µl unlabelled Tdr solution and sterile distilled water up to 50 ml. This amount of labelling solution will suffice for 100 samples. Specific activity of the [^{3}H]Tdr stock solution may vary slightly and volumes of [^{3}H]Tdr and Tdr solutions need to be adjusted accordingly.

Aqueous 20% formalin solution (200 g l^{-1} formaldehyde).

For extraction of DNA, a 0.6 M NaOH solution, a 1 M HCl solution and a pH indicator are needed.

Trichloroacetic acid (TCA), 10 g l^{-1} solution, kept in a polyethylene spray bottle in ice (use reusable refrigerant, "cold-brick").

Tissue solubilizer (e.g. Solulyte, Baker Chemicals, Deventer) and scintillation cocktail for use with the tissue solubilizer (e.g. Lipofluor, Baker Chemicals).

Procedure for labelling soil with [^{3}H]Tdr

Soil labelling is initiated within 15 min of sampling and performed with sterilized equipment. Labelling may be performed in the field followed by extraction of label in the laboratory afterwards.

Aliquots of 0.5 ml [^{3}H]Tdr labelling solution (200 nM) are added to the centrifuge tubes and the temperature is allowed to adjust to the *in situ* soil conditions.

Soil samples of up to 10 mg of organic soil, 50 mg of mineral soil or 1–5 root segments of about 5 mm length with adhering soil are transferred to the centrifuge tubes, and mixed with the [^{3}H]Tdr labelling solution, taking care that all material is immersed. Six replicates of each treatment is recommended.

After 15 min, 0.5 ml of 20% formalin is added to stop [^{3}H]Tdr incorporation. This high concentration is required to stop the incorporation immediately. After approximately 30 min, 1 ml of water is added to avoid degradation of DNA during storage. Samples can then be stored frozen until extraction.

Blanks are prepared by adding 0.5 ml of the

20% formalin at least 10 min before addition of the [^{3}H]Tdr labelling solution to the soil sample. Blanks are then treated as samples incubated in the living condition.

Extraction of [^{3}H]Tdr-labelled soil samples

NaOH solution, 5 ml, is added to each centrifuge tube, and the contents of the tubes are mixed and incubated at 60°C for 1 h.

Centrifuge tubes are left on the shaker for 15 min at ambient temperature before being centrifuged at 10,000*g* for 20 min.

Supernatants are carefully transferred by pipette to 100 ml flasks, 50 ml distilled water is added and the flasks are placed on ice.

The centrifuge tubes with pellets are dried at 60°C, weighed and may be discarded.

Ice-cold supernatant extracts are acidified with a volume of ice-cold 1 M HCl to bring the pH below 2, which is checked by pH indicator.

Membrane filters are wetted in distilled water before being placed in the manifold. Stainless steel funnels are cooled on ice (use "cold-brick") thus providing a low temperature for the filtration. The acidified extracts are shaken by hand before being filtered. Filters are rinsed five times with 3 ml of ice-cold 1% TCA solution from the spray bottle, peeled off the filter manifold under suction and placed in scintillation vials.

Filters are digested with 0.5 ml tissue solubilizer in the scintillation vial, 10 ml of scintillation cocktail is added and samples are then stored overnight to reduce chemiluminiscence before counting the ^{3}H activity. To ensure that the chemiluminescence is reduced sufficiently so as not to give incorporation rates that are too high, ^{3}H spectra must be inspected for extreme distortion of peaks and, if distinct ^{3}H peaks are not observed, samples must be left for some hours further before being counted. All samples could eventually be recounted to confirm that the initial counts were not too high. Counting efficiency should be examined regularly by running internal ^{3}H standards.

Calculation of [^{3}H]Tdr incorporation and bacterial growth rate

[^{3}H] Tdr incorporation rate (measured as pmol Tdr g^{-1} dry wt soil h^{-1}) = (0.06 × dpm)/(mg dry wt), where dpm is counts in the ^{3}H channel corrected for counting efficiency, and mg dry wt is the weight of sample. This calculation is valid under conditions of 30 Ci $mmol^{-1}$ (= 6.66 × 1016 dpm mol^{-1}) specific activity of [^{3}H]Tdr and a labelling period of 15 min. For example, if 1000 dpm have been counted for a 50 mg dry wt soil sample, the [^{3}H]Tdr incorporation rate = (0.06 × 1000)/50 = 1.20 pmol Tdr g^{-1} dry wt soil h^{-1}.

The thymidine incorporation rate is corrected for blank incorporation. Data may now be used to compare relative rates of [^{3}H]Tdr incorporation between treatments.

Based on conversion factors from the literature, the limits of production of bacterial cells, and production of bacterial carbon may be obtained by multiplying [^{3}H]Tdr incorporation rate with 0.22–2.4 ×1018 cells mol^{-1} Tdr and 0.1–0.18 µg C pmol Tdr^{-1}, respectively.

To measure conversion factors specific for the soil system under investigation, the following procedure is proposed: the bacterial growth conditions are manipulated to obtain bacterial growth rate >> death rate, [^{3}H]Tdr incorporation is measured simultaneously with direct microscopic counts of bacteria and the conversion factor is then calculated as the increase in cell number for a given time interval divided by the integrated thymidine incorporation during that time interval.

To obtain a conversion factor between [^{3}H]Tdr incorporation and production of bacterial C, bacterial cell volumes are recorded simultaneously with direct microscopic counts. The bacterial biovolume is converted to bacterial carbon by the conversion factors of Simon and Azam (1989).

Natural soil with a low bacterial activity should be used; addition of finely ground plant material will then support moderate bacterial growth. Measurements are initiated simultaneously with the addition of nutrients and should only be continued for a few days to avoid development of a large population of bacterial predators. The population of protozoa could be checked by the MPN method (Darbyshire et al 1974). It is important to avoid vigorous bacterial growth on easily decomposable substrates (such as glucose), since that may cause low [^{3}H]Tdr incorporation efficiency. For further details on determination of conversion factors, see Christensen (1993).

Bacterial generation time may be calculated when the bacterial biomass or total cell number has been determined simultaneously with [^{3}H]Tdr labelling: generation time = (bacterial population size × ln2)/bacterial production.

Discussion

Rates of [^{3}H]Tdr incorporation for wheat seedlings grown in sand at low nutrient level were 1–3 pmol Tdr g^{-1} dry wt soil h^{-1} (Christensen 1993), and 10–35 pmol Tdr g^{-1} dry wt soil h^{-1} for barley grown in sandy loam soil under field conditions with rhizosphere effects of 2–3 (Christensen et al 1993). High variability of data between six replicates is often observed with coefficients of variation about 20% for unplanted soil, and 50–100% for rhizosphere samples. This is due to the spatial variability in microbial activity of soil and the consequence of the small sample size (Christensen et al 1992).

The method may not provide reliable estimates of bacterial growth with waterlogged soil, and in situations where a flush in bacterial growth occurs caused by extreme shifts in temperature or soil moisture content, or because soil has been partly sterilized. Low and variable [^{3}H]Tdr incorporation efficiency has been found under such conditions and measurements of [^{3}H]Tdr incorporation must then be accompanied by direct bacterial counts using microscopy.

Cell-specific conversion factors expressed as number of cells formed mol^{-1} Tdr incorporated has been reported as 0.54 × 1018 for soil bacterial isolates grown in continuous culture, as 2.4 × 1018 for soil bacteria in batch culture, as 0.22 × 1018 for mixed bacterial populations of rape seedlings, and as 2.1 × 1018 for bacteria on wheat seedlings (Christensen 1993; Michel and Bloem 1993). Carbon-specific conversion factors were 0.1–0.2 μg C $pmol^{-1}$ Tdr for soil bacteria in batch culture and bacteria on wheat seedlings (Christensen 1993). The high variability in the determination of these factors makes them very imprecise and these conversion factors must therefore only be used with caution to calculate a range of bacterial production. To estimate bacterial production more precisely, it is recommended that conversion factors for the given experimental setup are determined as described in the calculation section.

References

Adams JA (1986a) Identification of heterotrophic nitrification in strongly acid larch humus. Soil Biol Biochem 18: 339–341.

Adams JA (1986b) Nitrification and ammonification in acid forest litter and humus as affected by peptone and ammonium-N amendment. Soil Biol Biochem 18: 45–51.

Alef K (1990) Dimethylsulfoxid (DMSO)-Reduktion im Boden und Kompost. UWSF-Z Umweltchem Ökotox 2 (2): 76–78.

Alef K (1991) Methodenhandbuch Bodenmikrobiologie. Ecomed-Verlagsgesellschat, Landsberg.

Alef K (1993) Bestimmung mikrobieller Biomasse im Boden: Eine kritische Betrachtung. Z Pflanzernähr Bodenk 156: 109–114.

Alef K, Kleiner D (1986a) Arginine ammonification, a simple method to estimate microbial activity potentials in soils. Soil Biol Biochem 18: 233–235.

Alef K, Kleiner D (1986b) Arginine ammonification in soil samples. In: The application of enzymatic and microbiological methods in soil analysis. Veröff Landwirtsch-chem Bundesanstalt Linz/Donau 18: 163–168.

Alef K, Kleiner D (1987) Applicability of arginine ammonification as an indicator of microbial activity in different soils. Biol Fertil Soils 5: 148–151.

Alef K, Kleiner D (1989) Rapid and sensitive determination of microbial activity in soils and in soil aggregates by dimethylsulfoxide reduction. Biol Fertil Soils 8: 349–355.

Alef K, Kleiner D (1994) Rapid and sensitive determination of microbial activities in food samples by dimethylsulfoxide reduction. Chem Mikrobiol Technol Lebensm 16: 1–2.

Alef K, Arp D, Zumft WG (1981) Nitrogenase switch-off by ammonia in *Rhodopseudomonas palustris*: loss under nitrogen deficiency and independence from the adenylylation state of glutamine synthetase. Arch Microbiol 130: 138–142.

Alef K, Beck Th, Zelles L, Kleiner D (1988) A comparison of methods to estimate microbial biomass and N-mineralization in agricultural and grassland soils. Soil Biol Biochem 20: 561–565.

Alexander M (1977) Soil Microbiology. Wiley, New York.

Anderson JPE (1975) Einflüsse von Temperatur und Feuchte auf Verdampfung, Abbau und Festlegung von Diallat im Boden. Z Pflanzenkr Pflanzenschutz 7: 141–146.

Anderson JPE (1982) Soil respiration. In: Methods of Soil Analysis, Part 2, Chemical and microbiological properties. Page AL, Miller RH (eds). American Society of Agronomy, Madison, WI, pp. 831–871.

Anderson JPE (1984) Herbicide degradation in soil: influence of microbial biomass. Soil Biol Biochem 16: 483–489.

Anderson JPE (1990) Principles of and assay systems for biodegradation. Adv Appl Technol Ser 4: 129–145.

Anderson JPE, Domsch KH (1978a) Mineralization of bacteria and fungi in chloroform-fumigated soils. Soil Biol Biochem 10: 207–213.

Anderson JPE, Domsch KH (1978b) A physiological method for the quantitative measurement of microbial biomass in soils. Soil Biol Biochem 10: 215–221.

Andreae MO (1980) Dimethylsulphoxide in marine and freshwater. Limnol Oceano 25: 1054–1063.

Andreae MO and Raemdonck H (1983) Dimethylsulphide in surface ocean and the marine atmosphere: A global view. Science 221: 744–747.

Arnebrant K, Baath E (1991) Measurement of ATP in forest humus. Soil Biol Biochem 23: 501–506.

Atkinson DE (1977) Cellular Energy Metabolism and its Regulation. Academic Press, New York.

Azhar El Sayed, Van Cleemput O, Verstraete W (1989) The effect of sodium chlorate and nitrapyrin on the nitrification mediated nitrosation process in soils. Plant Soil 116: 133–139.

Bååth E (1990) Thymidine incorporation into soil bacteria. Soil Biol Biochem 22: 803–810.

Bååth E (1992) Measurement of heavy metal tolerance of soil bacteria using thymidine incorporation into bacteria extracted after homogenization–centrifugation. Soil Biol Biochem 24: 1167–1172.

Bååth E, Söderström B (1988) FDA-stained fungal mycelium and respiration rate in reinoculated sterilized soil. Soil Biol Biochem 20: 403–404.

Barak R, Chet I (1986) Determination by fluorescein diacetate staining of fungal viability during mycoparasitism. Soil Biol Biochem 18: 315–319.

Barrie A, Workman CT (1984) An automated analytical system for nutritional investigations using ^{15}N tracers. Spectros Int J 3: 439–447.

Barret EL (1985) Bacterial reduction of trimethylamine oxide. Ann Rev Microbiol 39: 131–149.

Bauer E, Pennerstrofer Ch, Kandeler E, Braun R (1993) Einfluβ von Kohlenwasserstoffen auf bodenbiologische Parameter. In: Bodensanierung, Bodenkontamination, Verhalten und ökotoxikologische Wirkung von Umweltchemikalien im Boden. Alef K, Fiedler H, Hutzinger O (eds). Ecoinforma 1992, Ecoinforma Press, Bayreuth, pp. 351–363.

Beck Th (1976) Verlauf und Steuerung der Nitrifikation im Bodenmodellversuch. Landwirtsch Forsch Sonderh 33/I: 85–94.

Beck Th (1979) Die Nitrifikation in Böden (sammelreferat). Z Pflanzenernaehr Bodenkd 142: 299–309.

Beck Th (1981) Untersuchungen über toxische Wirkung in der Siedlungsabfällen häufigen Schwermettalle auf die Bodenmikroflora. Z Pflanzenernähr Bodenk 144: 613–627.

Beck Th (1983) Die N-Mineralisierung von Böden im Laborbrutversuch. Z Pflanzenernähr Bodenk 146: 243–252.

Beck Th (1984a) Mikrobiologische und biochemische Charakterisierung landwirtschaflich genutzter Böden. II. Mitteilung. Beziehung zum Humusgehalt. Z Pflanzenernäh Bodenk 147: 467–475.

Beck Th (1984b) Der Einfluβ unterschiedlicher Bewirtschaftungsmaβnahmen auf bodenmikrobiologische Eigenschaften und die Stabilität der organischen Substanz in Böden. Kali-Briefe (Büntehof) 17: 331–340.

Beck Th (1985) Einfluβ der Bewirtschaftung auf das Bodenleben. VDLUFA-Schriftenreine 16: 31–46.

Beese F (1986) Parameter des Stickstoffumsatzes in Ökosystemen mit Böden unterschiedlicher Acitität. Göttinger Bodenk Ber 90: 1–344.

Beese F, Heuer C, Rodewald W (1976) Eine Einrichtung zur Beregnung von Bodensäulen. Z Pflanzenern Bodenkd 5: 581–587.

Behera B, Wagner HG (1974) Microbial growth rate in glucose-amended soil. Soil Sci Soc Am Proc 38: 591–594.

Benefield CB, Howard PJA, Howard DM (1977) The estimation of dehydrogenase activity in soil. Soil Biol Biochem 9: 67–70.

Berg P, Rosswall T (1985) Ammonium oxidizer numbers, potential and actual oxidation rates in two Swedish arable soils. Biol Fert Soils 1: 131–140.

Birch JW, Melville M (1969) An electrolytic respirometer for measuring oxygen uptake in soils. J Soil Sci 20: 101–110.

Bjarnason S (1988) Calculation of gross nitrogen immobilization and mineralization in soil. J Soil Sci 39: 393–406.

Bremner JM, Mulvaney CS (1982) Nitrogen–total. In: Methods of Soil Analysis, Part 2, Chemical and microbiological properties. Page AL, Miller RH (eds). American Society of Agronomy, Madison, WI, pp. 595–624.

Bremner JM, Tabatabai MA (1973) Effects of some inorganic substances on TTC assay of dehydrogenase activity in soils. Soil Biol Biochem 5: 385–396.

Brookes PC, McGrath SP (1987) Adenylate energy charge in metal-contaminated soil. Soil Biol Biochem 19: 219–220.

Brookes PD, Paul EA (1987) A new automated technique for measuring respiration in soil samples. Plant Soil 101: 183–187.

Brookes PC, Powlson DS, Jenkinson DS (1982) Measurement of microbial biomass phosphorus in soil. Soil Biol Biochem 14: 319–329.

Brookes PC, Tate KR, Jenkinson DS (1983) The adenylate energy charge of the soil microbial biomass. Soil Biol Biochem 15: 9–16.

Brookes PC, McGrath SP, Klein DA, Elliott ET (1984) Effects of heavy metals on microbial activity and biomass in field soils treated with sludge. In: Environmental Contamination. CEP Ltd, Edinburgh, pp. 574–583.

Brookes PC, Landman A, Pruden G, Jenkinson DS (1985) Chloroform fumigation and the release of soil nitrogen: a rapid direct extraction method to measure microbial biomass nitrogen in soil. Soil Biol Biochem 17: 837–842.

Brookes PC, Newcombe A, Jenkinson DS (1987) Adenylate energy charge measurements in soil. Soil Biol Biochem 19: 211–217.

Brooks PD, Stark JM, McInteer BB, Preston T (1989) Diffusion method to prepare soil extracts for automated nitrogen-15 analysis. Soil Sci Soc Am J 53: 1707–1711.

Brumme R (1986) Modelluntersuchungen zum Stofftransport und Stoffumsatz in einer Terra fusca-Rendzina auf Muschelkalk. Ber Forschungsz Waldökosysteme/Waldsterben Univ Göttingen A 24: 1–206.

Brumme R, Beese F (1991) Simultaneous determination of nitrogen transformation rates in soil columns using ^{15}-N: N-model of a terra fusca–Rendzina. Z Pflanzenernähr Bodenk 154: 205–210.

Brunius G (1980) Technical aspects of the use of 3′–6′-diacetyl fluorescein for vital fluorescent staining of bacteria. Curr Microbiol 4: 321–323.

Burke IC, Mosier AR, Porter LK, O'Deen LA (1990) Diffusion of soil extracts for nitrogen and nitrogen-15 analyses by automated combustion/mass spectrometry. Soil Sci Soc Am J 54: 1190–1192.

Carlisle SM, Trevores JT (1986) Effect of the herbicide glyphosphate on respiration and hydrogen consumption in soil. Water Air Soil Poll 27: 391–401.

Casida LE Jr (1977) Microbial metabolic activity in soil as measured by dehydrogenase determinations. Appl Environ Microbiol 34: 630–636.

Casida LE Jr, Klein DA, Santoro T (1964) Soil dehydrogenase activity. Soil Sci 98: 371–376.

Chander K, Brookes PC (1991) Is the dehydrogenase assay invalid as a method to estimate microbial activity in copper-contaminated soils? Soil Biol Biochem 23: 909–915.

Chendrayan K, Adhya TK, Sethunathan N (1979) Dehydrogenase and invertase activities of flooded soils. Soil Biol Biochem 12: 217–273.

Cheng W, Coleman DC (1989) A simple method for measuring CO_2 in a continuous air-flow system: modifications to the substrate-induced respiration technique. Soil Biol Biochem 21: 385–388.

Christensen H (1991) Growth rate of rhizosphere bacteria as measured by the thymidine method. In: The Rhizosphere and Plant Growth. DL Keister and PB Cregan (eds). Kluwer Academic, Dordrecht, p. 105.

Christensen H (1993) Conversion factors for the thymidine incorporation technique estimated with bacteria in pure culture and on seedling roots. Soil Biol Biochem 25: 1085–1096.

Christensen H and Jakobsen I (1993) Reduction of bacterial growth by a vesicular-arbuscular mycorrhizal fungus in the rhizosphere of cucumber (Cucumis sativus L). Biol Fertil Soils 15: 253–258.

Christensen H, Griffiths B and Christensen S (1992) Bacterial incorporation of tritiated thymidine and populations of bacteriophagous fauna in the rhizosphere of wheat. Soil Biol Biochem 24: 703–709.

Christensen H, Rønn R, Ekelund F and Christensen S (1993) Bacterial production determined by 3H-thymidine incorporation in field rhizospheres as evaluated by comparison to rhizodeposition. Soil Biol Biochem (submitted).

Ciardi C, Nannipieri P (1990) A comparison of methods for measuring ATP in soil. Soil Biol Biochem 22: 725–727.

Ciardi C, Ceccanti B, Nannipieri P (1991) Method to determine the adenylate energy charge of soil. Soil Biol Biochem 23: 1099–1101.

Ciardi C, Ceccanti B, Nannipieri P, Casella S, Toffanin A (1993) Effects of various treatments on contents of adenine nucleotides and RNA of Mediterranean soils. Soil Biol Biochem 25: 739–746.

Clark FE, Kemper WD (1967) Microbial activity in relation to soil water and soil aeration. Agronomy 11: 472–480.

Craswell ET, Eskew DL (1991) Nitrogen and Nitrogen-15 analysis using automated mass and emission spectrometers. Soil Sci Soc Am J 55: 750–756.

Cunin K, Glansdorff N, Pierrard A, and Stalon V (1986) Biosynthesis and metabolism of arginine in bacteria. Microbiol Rev 50: 314–352.

Curl H Jr, Sandberg J (1961) The measurement of dehydrogenase activity in marine organisms. J Mar Res 19: 123–138.

Darbyshire JF, Wheatley RE, Greaves MP, Inkson RHE (1974) A rapid micromethod for estimating bacterial and protozoan populations in soil. Rev Ecol Biol Sol 11: 465–475.

David KAV, Fay P (1977) Effects of long-term treatment with acetylene on nitrogen-fixing microorganisms. Appl Environ Microbiol 34: 640–646.

Denburg JL, McElroy (1970) Anion inhibition of firefly luciferase. Arch Biochem Biophys 141: 668–675.

Doelman P, Haanstra L (1979) Effect of lead on soil respiration and dehydrogenase activity. Soil Biol Biochem 11: 475–479.

Domsch KH (1962) Bodenatmung, Sammelbericht über Methoden und Ergebnisse. Tentr Bakteriol Parasitenk Abt II, 116: 33–78.

Domsch KH, Schröder M (1986) Einfluβ einiger Herbizide auf den mikrobiellen Biomasse-Kohlenstoff und den Mineralstickstoffgehalt des Bodens. In: DFG-Forschungsbericht Herbizide II. VCH Verlagsgesellschaft mbH, Weinheim, pp. 225–233.

Domsch KH, Anderson JPE, Ahlers R (1973) Method for simultaneous measurement of radioactive and inactive CO_2 evolved from soil samples during incubation with labeled substrates. Appl Microbiol 25: 819–824.

Domsch KH, Beck Th, Anderson JPE, Söderström B, Parkinson D, Trolldenier G (1979) A comparison of methods for soil microbial population and biomass studies. Z Pflanzenernäh Bodenk 142: 520–533.

Domsch KH, Jagnow G, Anderson TH (1983) An ecological concept for the assessment of side-effects of agrochemicals on soil microorganisms. Residue Rev 86: 66–105.

Egsgaard H, Larsen E, Jensen ES (1989) Evaluation of automated determination of nitrogen-15 by on-line combustion. Anal Chim Acta 226: 345–349.

Eiland F (1979) An improved method for determination of adenosine triphosphate (ATP) in soil. Soil Biol Biochem 11: 31–35.

Eiland F (1983) A simple method for quantitative determination of ATP in soil. Soil Biol Biochem 15: 665–670.

Eiland F (1985) Determination of adenosine triphosphate (ATP) and adenylate energy charge (AEC) in soil and use of adenine nucleotides as measures of soil microbial biomass and microbial activity. Danish J Plant Soil Sci Rep No 1777: 1–93.

Eiland F, Nielsen BS (1979) Influence of cation content on adenosine triphosphate determinations in soil. Microbial Ecol 5: 129–137.

Eno F (1960) Nitrate production in the field by incubating the soil in polyethylene bags. Soil Sci Soc Am Proc 24: 277–279.

Forrest WW (1972) Microcalorimetry. In: Method in Microbiology, Band 6B. Norris JR, Ribbons DW (eds). Academic Press, London, pp. 285–318.

Gibson RM, Large PJ (1985) The methionine sulphoxide reductase activity of the yeast dimethylsulphoxide reductase system. FEMS Microbiol Lett 26: 95–99.

Gibson AM, Turner GL (1980) Measurement of nitrogen fixation by indirect assaies. In: Methods for Evaluating Biological Nitrogen-fixation. Bergersen FS (ed.). Wiley, New York.

Gray TRG, Williams ST (1977) Soil Microorganisms. Longman, London.

Greaves MP, Poole NJ, Domsch KH, Jagnow G, Verstraete W (1980) Recommended tests for assessing the side-effect of pesticides on soil mikroflora. Tech Rep Agric Res Council, Weed Res Organisation No. 59: 1–15.

Guilbault GG, Kramer DN (1964) Fluorometric determination of lipase, acylase, alpha- and gamma-chemotrypsin and inhibitors of these enzymes. Anal Chem 36: 409–412.

Guiraud G, Marol C, Thabaud MC (1989) Mineralization of nitrogen in the presence of nitrification inhibitor. Soil Biol Biochem 21: 29–34.

Hantschel RE, Flessa H, Beese F (1994) An automated microcosm system for studying soil ecological processes. Soil Sci Soc Am J 58: 401–404.

Hardy RW, Burns RC, Holsten RD (1973) Application of acetylene–ethylene assay for measurement of nitrogen fixation. Soil Biol Biochem 5: 47–81.

Harrington RA (1986) Aeration status and nitrogen mineralization rates of soils incubated in polyethylene bags. IV. International Congress of Ecology, Syracuse, New York, 10–16 August.

Harris D, Paul EA (1989) Automated analysis of ^{15}N and ^{14}C in biological samples. Commun Soil Sci Plant Anal 20: 935–947.

Hattori T, Hattori R (1976) The physical environment in soil microbiology: an attempt to extend principles of microbiology to soil microorganisms. CRC Crit Rev Microbiol 4: 423–459.

Hauck RD (1982) Nitrogen–isotope-ratio analysis. In: Methods of Soil Analysis, Part 2, Chemical and microbiological properties. Page AL, Miller RH (eds). American Society of Agronomy, Madison, WI, pp. 735–779.

Hauck RD, Bremner JM (1976) Use of tracers for soil and fertilizer nitrogen research. In: Advances in Agronomy, vol 28, NC Brady (ed.). pp. 219–266. American Society of Agronomy, Madison Wis.

Haynes RJ, Goh KM (1978) Ammonium and nitrate nutrition of plants. Biol Rev 53: 465–510.

Heilmeier H (1988) Das Wachstum einer biennen Pflanze unter besonderer Berücksichtigung der Bedeutung von Stickstoff am Beispiel von *Arctium*, Compositae. Dissertation, University of Bayreuth.

Heinemeyer O, Insam H, Kaiser EA, Walenzik (1989) Soil microbial biomass and respiration measurements: An automated technique based on infrared gas analysis. Plant Soil 116: 191–195.

Homeyer KL, Labenski KO, Meyer B, Thermann A (1974) Herstellung von Lysimetern mit Böden in natürlicher Lagerung (Monolith-Lysimeter) als Durchlauf-, Unterdruck- oder Grundwasserlysimeter. Z Pflanzenern Bodenkd 136: 242–245.

Hund K (1988) Bewertung von allgemeinen und den N-Kreislauf betreffenden Methoden zur Bestimmung der mikrobiellen Aktivität des Bodens nach Zugabe von Umweltchemikalien. Dissertation, Institut für Chemie, Lehrstuhl für ökologische Chemie der technischen Universität München, Weihenstephan-Freising.

Hund K, Zelles L, Scheunert I and Korte F (1988) A critical estimation of methods for measuring side-effects of chemicals on microorganisms in soils. Chemosphere 17: 1183–1188.

IAEA (1991) Proceedings of a Symposium on Stable Isotopes in Plant Nutrition, Soil Fertility and Environmental Studies. IAEA, Vienna, 1–5 October 1990.

Inubushi K, Brookes PC, Jenkison DS (1989a) Influence of paraquat on the extraction of adenosine triphosphate from soil by trichloracetic acid. Soil Biol Biochem 21: 741–742.

Inubushi K, Brookes PC, Jenkinson DS (1989b) Adenosine-5-triphosphate and adenylate energy charge in waterlogged soil. Soil Biol Biochem 21: 733–739.

Isermeyer H (1952) Eine einfache Methode zur Bestimmung der Bodenatmung und der Karbonate im Boden. Z Pflanzenernäh Bodenk 56: 26–38.

Jäggi W (1976) Die Bestimmung der CO_2-Bildung als Maβ der bodenbiologischen Aktivität. Schweiz Landwirtschaft Forschung Band 15, Heft 314: 317–380.

Jambor B (1960) Tetrazoliumsalze in der Biologie. Universät Jena, Jena.

Jansson SL (1971) Use of ^{15}N in studies of soil nitrogen. In: Soil Biochemistry, vol 2, McLaren AD, Skujins J (eds). Marcel Dekker, New York, pp. 129–166.

Jenkinson DS, Oades JM (1979) A method for measuring adenosine triphosphate in soil. Soil Biol Biochem 11: 193–199.

Jenkinson DS, Davidson SA, Powlson DS (1979) Adenosine triphosphate and microbial biomass in soil. Soil Biol Biochem 11: 521–527.

Jensen ES (1991) Evaluation of automated analysis of ^{15}N and total N in plant material and soil. Plant Soil 133: 83–92.

Kandeler E, Gerber H (1988) Short-term assay of soil urease activity using colorimetric determination of ammonium. Biol Fertil Soils 6: 68–72.

Katyal JC, Carter MF, Vlek PLG (1988) Nitrification activity in submerged soils and its relation to denitrification loss. Biol Fertil Soils 7: 16–22.

Keeney DR, Bremner JM (1966) Comparison and evaluation of laboratory methods of obtaining an index of soil nitrogen availability. Agron J 58: 498–503.

Keeney DR, Nelson DW (1982) Nitrogen – inorganic forms. In: Methods of Soil Analysis, Part 2, Chemical and microbiological properties. Page AL, Miller RH (eds). American Society of Agronomy, Madison, WI, pp. 643–698.

Kelley KR, Ditsch DC, Alley MM (1991) Diffusion and automated nitrogen-15 analysis of low-mass ammonium samples. Soil Sci Soc Am J 55: 1016–1020.

Khanna PK, Beese F (1978) The behaviour of sulfate

on salt input in podzolic brown earth. Soil Sci 125: 16–22.

Kieft TL, Rosacker LL (1991) Application of respiration- and adenylate-based soil microbiological assay to deep subsurface terrestrial sediments. Soil Biol Biochem 23: 563–568.

Kieft TL, Soroker E, Firestone (1987) Microbial biomass response to a rapid increase in water potential when dry soil is wetted. Soil Biol Biochem 12: 119–126.

Klein DA, Loh TC, Goulding RL (1971) A rapid procedure to evaluate the dehydrogenase activity of soils low in organic matter. Soil Biol Biochem 3: 385–387.

Knowles R (1980) Nitrogen fixation in natural plant communities and soils. In: Methods for Evaluating Biological Nitrogen-fixation. Bergerson FS (ed.). Wiley, New York, pp. 557–582.

Knowles R (1982) Free-living dinitrogen fixing bacteria. In: Methods of Soil Analysis, Part 2, Chemical and microbiological properties. Page AL, Miller RH (eds). American Society of Agronomy, Madison, WI, pp. 1071–1092.

Knowles R and Blackburn TH (eds) (1992) Nitrogen Isotope Techniques. Academic Press, London.

Konno T (1976) Calorimetric application on measurements of activity of soil microorganisms. Netsu 3: 148–151.

Kowalenko CG, Ivarson KC, Cameron DR (1978) Effect of moisture content, temperature and nitrogen fertilization on carbon dioxide evolution from field soils. Soil Biol Biochem 10: 417–423.

Kröckel L, Stolp H (1986) Influence of the water regime on denitrification and aerobic respiration in soil. Biol Fertil Soils 2: 15–21.

Ladd JN, Jackson RB (1982) Biochemistry of ammonification. In: Nitrogen in Agricultural Soils. Stevenson FE (ed.), American Society of Agronomy, Madison, WI, pp 173–228.

Lehninger AL (1975) Biochemistry. Worth Publishers, Inc., New York.

Lehninger AL (1977) Biochemie. Verlag Chemie, Weinheim, New York.

Lenhard G (1956) The dehydrogenase activity in soil as a measure of the activity of soil microorganisms. Z Pflanzenernäh Düng Bodenkd 73: 1–11.

Lenhard G (1966) The dehydrogenase activity for the study of soils and river deposits. Soil Sci 101: 400–402.

Lindemann WC, Fresquez PR, Cadenas M (1989) Nitrogen mineralization in coal mine spoil and topsoil. Biol Fertil Soils 7: 318–324.

Liungholm K, Noren B, Scöld R, Wadsö I (1979a) Use of microcalorimetry for the characterization of microbial activity in soil. Oikos 33: 15–23.

Liungholm K, Noren B, Wadsö I (1979b) Microcalorimetric observations of microbial activity in normal and acidified soils. Oikos 33: 24–30.

Lorenz SE, McGroth SP, Giller KE (1992) Assessment of free-living nitrogen fixation activity as biological indication of heavy metal toxicity in soil. Soil Biol Biochem 24: 601–606.

Lundgren B (1981) Fluorescein diacetate as a stain of metabolically active bacteria in soil. Oikos 36: 17–22.

Lundin A, Thore A (1975) Comparison of methods for determination of bacterial adenine nucleotides determined by firefly assay. Appl Microbiol 30: 713–721.

MacKown CT, Brookes PD, Smith MS (1987) Diffusion of nitrogen-15 Kjeldahl digests for isotope analysis. Soil Sci Soc Am J 51: 87–90.

Mahler HR and Cordes EH (1971) Biological Chemistry. Harper and Row, New York.

Malkomes HP (1985) Einfluβ des Herbizids Dinoseb-acetat und dessen Kombination mit einem Phospholipid auf bodenmikrobiologische Aktivitäten unter Labor- und Gewächshausbedingungen. Z Pflanzenkrankh Pflschutz 92: 489–501.

Malkomes HP und Wagner K (1986) Einfluβ von Clopyralid (Lontrel 100) auf mikrobielle Aktivitäten unter Laborbedingungen. Zentralbl Mikrobiol 141: 603–614.

Marshall RB, Whiteway JN (1985) Automation of an interface between a nitrogen analyser and an isotope ratio mass spectrometer. Analyst 110: 867–871.

Martens R (1977) Degradation of endosulfan-8, $9-{}^{14}C$ in soil under different conditions. Bull Environ Contam Toxicol 17: 438–446.

Martens R (1985) Estimation of the adenylate energy charge in unamended and amended agricultural soils. Soil Biol Biochem 17: 765–772.

Martens R (1992) A comparison of soil adenine nucleotide measurements by HPLC and enzymatic analysis. Soil Biol Biochem 24: 639–645.

Martens R, Bunte D, Borkott H (1988) Carbon–ATP ratios of active soil animals and their possible influence on total biomass-C–ATP ratios of soils. Soil Biol Biochem 20: 965–967.

Marvel JT, Brightwell BB, Malik JM, Sutherland ML, Rueppel ML (1978) A simple apparatus and quantitative method for determining the persistence of pesticides in soil. J Agric Food Chem 26: 1116–1120.

Mattson AM, Jensen CO, Dutcher AR (1947) Triphenyltetrazolium-chloride as a dye for vital tissue. Science 106: 294–295.

Medzon E, Brady ML (1969) Direct measurement of acetylesterase in protist cells. J Bacteriology 97: 402–415.

Meier G and Manersberger (1982) NO1-6-Ein never emission spectrometrischer Analysator. Isotopenpraxis 18: 164–170.

Meier G and Schmidt G (1992) Ertahrungen bei der simultanen Bestimmungen von besamtstickstoff und ${}^{15}N$ durch kepplung von "DUMAS"-Stickstoffbestimmungsgeraten mit den NO1–6. Isotopenpraxis, Isotope in Environmental and Health Studies 28: 85–100.

Michel PH and Bloem J (1993) Conversion factors for estimation of cell production rates of soil bacteria from [^{3}H]thymidine and [^{3}H]leucine incorporation. Soil Biol Biochem 25: 943–950.

Minnich MM, McBride MB (1986) Effect of copper activity on carbon and nitrogen mineralization in field-aged copper-enriched soils. Plant Soil 91: 231–240.

Mochoge B (1981) The behaviour of nitrogen fertilizers in neutral and acid loess soils. Göttinger Bodenkundl Ber 69: 1–173.

Mochoge B, Beese F (1983a) The behaviour of nitrogen fertilizer in neutral and acid loess soils. I. Transport and transformation of nitrogen. Z Pflanzenernaeh Bodenk 146: 89–100.

Mochoge B, Beese FD (1983b) The behaviour of nitrogen fertilizer in neutral and acid loess soils. II. Distribution and balances of ^{15}N-tagged nitrogen. Z Pflanzenernaehr Bodenk 146: 504–515.

Moriarty DJW (1986) Measurement of bacterial growth rates in aquatic systems from rates of nucleic acid synthesis. Adv Microb Ecol 9: 245–292.

Moriarty DJW (1990) Techniques for estimating bacterial growth rates and production of biomass in aquatic environments. In: Methods in Microbiology, vol 22. Academic Press, London, pp. 211–234.

Mortensen U, Noren B, Wadsö I (1973) Microcalorimetry in the study of the activity of microorganisms. In: Modern Methods in the Study of Microbial Ecology. Rosswall R (ed.). Bulletin of Ecological Research Communication, vol 17. Swedish Natural Science Research Council, Stockholm, pp. 189–197.

Mulvaney RL (1991) Some recent advances in the use of 15N for research on nitrogen transformations in soil. Proceedings of a Symposium on Stable Isotopes in Plant Nutrition, Soil Fertility and Environmental Studies. IAEA, Vienna, 1–5 October 1990, pp. 283–296.

Mulvaney RL (1992) Mass spectrometry. In: Nitrogen Isotope Techniques. Knowles R and Blackburn TH (eds.) Academic Press, London, pp. 11–58.

Myrold DD, Tiedje JM (1986) Simultaneous estimation of several nitrogen cycle rates using 15N: theory and application. Soil Biol Biochem 18: 559–568.

Naklas JP, Klein DA (1981) Use of amino acid mixture to estimate the mineralization capacity of grassland soils. Soil Biol Biochem 13: 427–428.

Nannipieri P, Grego S, Ceccanti B (1990) Ecological significance of the biological activity in soil. In: Soil Biochemistry, vol 6. Bollag JM, Stotzky G (eds). Marcel Dekker, New York, pp. 293–355.

Nishio T, Kanamori T, Fujimoto T (1985) Nitrogen transformations in an aerobic soil as determined by a $^{15}NH_4^+$ dilution technique. Soil Biol Biochem 17: 149–154.

Norman RJ, Stucki JW (1981) The determination of nitrate and nitrite in soil extracts by ultraviolet spectrophotometry. Soil Sci Soc Am J 45: 347–353.

Novak B, Kubat J (1972) On the relation between dehydrogenase activity and CO_2 evolution of soil. Zentrabl Bakteriol Parasitenkd 12: 246–252.

O'Deen WA, Porter LK (1979) Digest tube diffusion and collection of ammonia for nitrogen-15 and total nitrogen determination. Anal Chem 51: 586–589.

Ohms J, Pollock GE (1966) Calorimetry. In: Biology and the Exploration of Mars. Pittendrigh CS, Vishniac W, Pearman JPT (eds). Publication no. 1296. National Canadian Society of Natural Research.

Okon Y, Albrecht SL, Buvis RM (1977) Methods for growing *Spirillum lipoforum* and for counting it in pure culture and in association with plant. Appl Environ Microbiol 33: 85–88.

Parkin TB (1987) Soil microsites as a source of denitrification variability. Soil Sci Soc Am J 51: 1194–1199.

Parkinson D, Gray TRG, Williams ST (1971) Methods for Studying the Ecology of Soil Microorganisms. IBP Handbook 19. Blackwell Scientific Publications, Oxford.

Paul EA, Johnson RL (1977) Microscopic counting and adenosine 5′-triphosphate measurement in determining microbial growth in soils. Appl Environ Microbiol 34: 263–269.

Pochon J, Tardieux P (1962) Techniques d' Analyses en Microbiologie du Sol. Edition de la Tourelle, St Mandé.

Porter LK, O'Deen WA (1977) Apparatus for preparing nitrogen from ammonium chloride for nitrogen-15 determinations. Anal Chem 49: 514–516.

Postgate JR (1982) The Fundamentals of Nitrogen Fixation. Cambridge University Press, Cambridge.

Preston M (1992) Optical Emission Spectrometry. In: Nitrogen Isotope Techniques. Knowles R and Blackburn TH (eds.) Academic Press, London, pp. 59–88.

Preston T, Owens NJP (1983) Interfacing an automatic elemental analyser with an isotope ratio mass spectrometer: the potential for fully automated total nitrogen and nitrogen-15 analysis. Analyst 108: 971–977.

Prevost D, Angers DA, Nadean P (1991) Determination of ATP in soils by high performance liquid chromatography. Soil Biol Biochem 23: 1143–1146.

Rajbhandari KK, Lorch HJ, Ottow JCG (1991) Charakterisierung der Biomasseaktivität im Schlamm verschiedener Reinigungstufen einer belüfteten

Abwasserteichanlage mit Hilfe der Dimethylsulphoxidreduktase- und Dehydrogenase-Aktivität. VDLUFA-Schrifter 33: 642.

Rausch H, Luttich M, Freytag HE (1985) Quantifizierung der Stickstoffmineralisierung aus der organischen Bodensubstanz mit Hilfeder Stanford-Methode. Arch Acker- Pflanzenb Bodenk 29: 77–83.

Reddy GB, Faza A (1989) Dehydrogenase activity in sludge amended soil. Soil Biol Biochem 21: 327.

Reeder JD, O'Deen WA, Porter LK, Lober RW (1980) A comparison of cross-contamination in distillation units used in total nitrogen and nitrogen-15 analyses. Soil Sci Soc Am J 44: 1262–1267.

Reineking A, Schulte HD, Langel R (1988) Bedienungsanleitung zur Rittenberg-Apparartur. Isotopenlabor für biologische und medizinische forschung der Georg-August-Universität, Göttingen.

Reineking A, Langel R, Schikowski J (1993) ^{15}N, ^{13}C-online measurements with an elemental analyser (Carlo Erba, NA 1500), a modified trapping box and a gas isotope mass spectrometer (Finnigan, MAT 251). Istopenpraxis, Environ. Health Stud. 29: 169–174.

Reynders L, Vlassak K (1981) A rapid and sensitive determination method for soil nitrate status. Z Pflanzenernähr Bodenk 144: 628–636.

Rheinheimer G, Hegemann W, Raff J, Sekonlov I (1988) Stickstoffkreislauf im Wasser. Oldenbourg-Verlag, Munich, Vienna.

Richter J, Nuske A, Habenicht W, Bauer J (1982) Optimized N-mineralization parameters of Loess-soil from incubation experiments. Plant Soil 68: 379–388.

Rosacker LL, Kieft TL (1990) Biomass and adenylate energy charge of a grassland soil during drying. Soil Biol Biochem 22: 1121–1127.

Ross DJ (1971) Some factors influencing the estimates of dehydrogenase activities of some soils under pasture. Soil Biol Biochem 3: 97–110.

Ross DJ, Tate KR, Cairns A, Pansier EA (1980) Microbial biomass estimations in soils from tussock grasslands by three biochemical procedures. Soil Biol Biochem 12: 375–383.

Ross PJ, Martin AE (1970) A rapid procedure for preparing gas samples for nitrogen-15 determination. Analyst 95: 817–822.

Rother JA, Millbank JW, Thornton I (1982) Effects of heavy-metal additions on ammonification and nitrification in soils contaminated with cadmium, lead and zinc. Plant Soil 69: 239–258.

Rotman R, Papermaster BW (1966) Membrane properties of living mammalian cells as studied by enzymatic hydrolysis of fluorogenic esters. Proceedings NAS 55: 134–141.

Sahrawat KL (1981) Ammonification in air-dried tropical lowland histosols. Soil Biol Biochem 13: 323–324.

Schepers JS, Francis DD, Thompson MT (1989) Simultaneous determination of total C, total N, and ^{15}N on soil and plant material. Commun Soil Sci Plant Anal 20: 949–959.

Schlosser HJ (1988) Auswertung ökotoxikologischer Forschungen zur Belastung von Ökosystemen durch Chemikalien. Dissertation, Biologische Bundesanstalt für Land- und Forstwirtschaft, Berlin, pp. 32–41.

Schnürer J, Rosswall T (1982) Fluorescein diacetate hydrolysis as a measure of total microbial activity in soil and litter. Appl Environ Microbiol 43: 1256–1261.

Schuster E (1988) Einfluβ von Pflanzenschutzmittel-Spritzfolgen und -Kombinationen auf die mikrobiologische Aktivität des Bodens. Freiland- und Laborversuche. Dissertationsarbeit, Fach Bodenkunde Universität Trier.

Simon M, Azam F (1989) Protein content and protein synthesis rates of planktonic marine bacteria. Mar Ecol Prog Ser 51: 201–213.

Sklorz M, Binert J (1994) Determination of microbial activity in activated sewage sludge by dimethyl sulphoxide reduction: evaluation of method and application. Environmental Science and Pollution Research 3: 140–145.

Skujins J (1973) Dehydrogenase: an indicator of biological activities in soils. Bulletin of Ecology Research Communications. NFR Statens Naturvetensk. Forskningsrad 17: 235–241.

Skuzius J, Novestedt HO, Oden S (1986) Development of sensitive biological methods for the determination of low level toxic contamination in soils. Swedish J Agric Sci 16: 113–118.

Somerville L, Greaves MP (1987) Pesticide effects on soil microflora. Somerville L, Greaves MP (eds). Taylor & Francis, London New York. pp. 115–132.

Sonnen H, Bachmann F (1993) Die quantifizierte Bestimmung und Bewertung biologischer Bodenaktivität: Fundamintale Grundlage für biologische Sanierungsmaβnahmen. In: Bodensanierung, Bodenkontamination, Verhalten und ökotoxikologische Wirkung von Umweltchemikalien im Boden. Alef K, Fiedler H, Hutzinger O (eds). Ecoinforma 1992, Ecoinforma Press, Bayreuth, pp. 125–138.

Söderström BE (1977) Vital staining of fungi in pure cultures and in soil with fluorescein diacetate. Soil Biol Biochem 9: 59–63.

Sorensen P, Jensen ES (1991) Sequential diffusion of ammonium and nitrate from soil extracts to a polytetrafluoroethylene trap for ^{15}N determination. Anal Chim Acta 252: 201–203.

Sparling GP (1981a) Microcalorimetry and other methods to assess biomass and activity in soil. Soil Biol Biochem 13: 93–98.

Sparling GP (1981b) Heat output of the soil biomass. Soil Biol Biochem 13: 373–376.

Sparling GP (1983) Estimation of microbial biomass and activity in soil using microcalorimetry. J Soil Sci 34: 381–390.

Sparling GP, Eiland F (1983) A comparison of methods for measuring ATP and microbial biomass in soils. Soil Biol Biochem 15: 227–229.

Sparling GP, Searle PL (1993) Dimethyl sulfoxide reduction as sensitive indicator of microbial activity in soil: the relationship with microbial biomass and mineralization of nitrogen and sulphur. Soil Biol Biochem 25: 251–256.

Sparling GP, Ord BG, Vaughan D (1981) Microbial biomass and activity in soils amended with glucose. Soil Biol Biochem 13: 99–104.

Sparling GP, West AW, Whale KN (1985) Interference from plant roots in the estimation of soil microbial ATP, C, N and P. Soil Biol Biochem 17: 275–278.

Speres AB (1981) Diversity of ammonifying bacteria. Hydrobiologia 83: 343–350.

Stanford G, Smith SJ (1972) Nitrogen mineralization potentials of soils. Soil Sci Soc Am Proc 36: 465–472.

Starr JL, Broadbent FE, Nielsen DR (1974) Nitrogen transformations during continuous leaching. Soil Sci Soc Am 38: 283–289.

Stotzky G (1965) Microbial respiration. In: Methods of Soil Analysis, Part 2. Black CA, Evans DD, Ensminger LE, White JL, Clark FC (eds). Agronomy, Inc, Madison, WI, pp. 1550–1572.

Suttner T, Alef K (1988) Correlation between the arginine ammonification, enzyme activities, microbial biomass, physical and chemical properties of different soils. Zentralbl Mikrobiol 143: 569–573.

Suylen GM, Stefess GC and Kuenen (1986) Chemolithotrophic potential of a Hyphomicrobium species, capable of growth on methylated sulphur compounds. Arch Microbiol 146: 192–198.

Tate KR, Jenkinson DS (1982) Adenosine triphosphate measurement in soil: an improved method. Soil Biol Biochem 14: 331–335.

Thalmann A (1968) Zur Methodik der Bestimmung der Dehydrogenaseaktivität im Boden mittels Triphenyltetrazoliumchlorid (TTC). Landwirtsch Forsch 21: 249–258.

Tiedje JM (1982) Denitrification. In: Methods of Soil Analysis, Part 2, Chemical and microbiological properties. Page AL, Miller RH (eds). American Society of Agronomy, Madison, WI, pp. 1011–1026.

Tiwari SC, Tiwari BK, Mishra RR (1989) Microbial populations, enzyme activities and nitrogen–phosphorus–potassium enrichment in earthworm casts and in the surrounding soil of a pineapple plantation. Biol Fertil Soils 8: 178–182.

Tobin RS, Ryan JF, Afghan BK (1978) An improved method for the determination of adenosine triphosphate in environmental samples. Water Res 12: 783–792.

Trevors JT (1984) Dehydrogenase activity in soil: a comparison between the INT and TTC assay. Soil Biol Biochem 16: 673–674.

Trevors JT (1985) Oxygen consumption in soil: Effect of assay volume. Soil Biol Biochem 17: 385–386.

Trevors JT, Mayfield CI, Inniss WE (1982) Measurement of electron transport system (ETS) activity in soil. Microbial Ecol 8: 163–168.

Van de Werf H (1989/1990) A respiration-simulation method for estimating active soil microbial biomass. Ph.D. thesis, Faculty of Agricultural Sciences, Gent.

Van de Werf H, Verstraete W (1979) Direct measurement of microbial ATP in soils. Proceedings of an International Symposium on Analytical Applications of Bioluminescence and Chemiluminescence. Schram E, Stanley P (eds). State Printing & Publishing, Inc., Westlake Village, CA, pp. 333–338.

Van de Werf H, Verstraete W (1984) ATP measurement by bioluminescene: environmental applications. In: Analytical Applications of Bioluminescence and Chemiluminescence. Kricka LJ, Stanley PE, Thorpe GHG, Whitehead TP (eds). Academic Press, London, pp. 33–48.

Verstraete W, Van de Werf H, Kucnerowicz F, Iliawi M, Verstraeten LMJ, Vlassak K (1983) Specific measurement of soil microbial ATP. Soil Biol Biochem 15: 391–396.

Verstraeten LMJ, De Coninck K, Vlassak K, Verstraete W, Van de Werf H, Ilaiwi M (1983) ATP content of soils estimated by two contrasting extraction methods. Soil Biol Biochem 15: 397–402.

Vlassak K, Verstraeten LMJ (1986) Use of nitrate reductase as a simple and sensitive nitrate determination assay. Veröff. Landwirtsch.-chem. Bundesanstalt, Linz/Donau.

Von Mersi W, Schinner F (1991) An improved and accurate method for determining the dehydrogenase activity of soils with iodonitrotetrazolium chloride. Biol Fertil Soils 11: 216–220.

Waring SA, Bremner JM (1964) Ammonium production in soil under waterlogged conditions as an index of nitrogen availability. Nature 201: 951–952.

Webster J, Hampton G, Leach F (1984) ATP in soil: a new extractant and extraction procedure. Soil Biol Biochem 16: 335–342.

Wilke BM (1986) Einfluβ verschiedener potentieller anorganischer Schadstoffe auf die mikrobielle Aktivität von Waldhumusformen unterschiedlicher Pufferkapazität. Bayreuther Geow. Arbeiten.

Wilke BM (1989) Long-term effects of different inorganic pollutants on nitrogen transformations in a sandy cambisol. Biol Fertil Soils 7: 254–258.

Wilke BM (1991) Effects of single and successive additions of cadmium, nickel and zinc on carbon dioxide evolution and dehydrogenase activity in a sandy luvisol. Biol Fertil Soils 11: 34–37.

Wilson JM, Griffin DM (1975a) Water potential and respiration of microorganisms in the soil. Soil Biol Biochem 7: 199–204.

Wilson JM, Griffin DM (1975b) Respiration and radial growth of soil fungi at two osmotic potentials. Soil Biol Biochem 7: 269–274.

Woldendorf JW, Laanbroek HJ (1989) Activity of nitrifiers in relation to nitrogen nutrition of plants in natural ecosystems. Plant Soil 115: 217–228.

Wood PM (1981) The redox potential for dimethylsulphoxide reduction to dimethylsulphide. FEBS Lett 124: 11–14.

Zechmeister-Boltenstern S (1991) Bodenbiologische Arbeitsmethoden. Schinner F, Öhlinger R, Kanndeler E (eds). Springer Verlag, Berlin, pp. 120–125.

Zelles L, Scheunert I, Korte F (1985) Side effects of some pesticides on non-target soil microorganisms. J Environ Sci Health 20: 457–488.

Zelles L, Scheunert I, Korte F (1986) Comparison of methods to test chemicals for side effect on soil microorganisms. Ecotoxicol Environ Safety 12: 53–69.

Ziegler GB, Ziegler E, Witzenhausen R (1975) Nachweis der Stoffwechselaktivität von Mikroorganismen durch vital-Fluorochromierung mit 3′–6′-Diacetylfluorescein. Zentralblatt für Bakteriol Parasit Infektionskrankheiten Hygiene Abt 1, Orig Reihe A 230: 252–264.

Zinder SH, Brock TD (1978) Dimethylsulphoxide reduction by microorganisms. J Gen Microbiol 105: 335–342.

Anaerobic microbial activities in soil

6

Most soils, for example, agricultural soils, can be considered as being dominated by aerobic conditions. However, at any site in a soil, anaerobic and aerobic conditions are created as soon as the oxygen demand exceeds the oxygen diffusion to this site.

Generally, soil anaerobiosis is caused by high soil moisture resulting from a high water table or heavy rains. Reducing conditions in the soil are also associated with the solubilization of Fe, Mn, Al and other cations. In flooded soils, the diffusion of oxygen from the air into the soil is markedly decreased by the water layer, the amount of dissolved O_2 in the soil decreases rapidly at an early stage of waterlogging because of its consumption by aerobes. The rate of O_2 consumption depends primarily on the amount of available carbon for respiration, which is affected by water availability and temperature. In addition to the water content, the O_2 supply rate depends on the physical properties of the soil, especially porosity, which is influenced by structure and texture (Skinner 1975). Anaerobiosis results in great changes in the organic matter and nutrient dynamics. Processes such as loss of nitrogen by denitrification, production of organic acids, nitrogen fixation by free-living heterotrophic bacteria, rapid mineralization of pollutants, etc. are markedly affected by anaerobiosis.

In aerobic sites of soils, microbial activity is dominated by the activities of obligate aerobic bacteria, facultative anaerobic bacteria carrying out aerobic metabolism and fungal activities. Anaerobic conditions result in dramatic changes in the community structure. After the oxygen has been consumed by the facultative anaerobes, the growth of obligate anaerobes occurs. The absence of O_2, however, is not the sole factor determining the growth of anaerobic species in the soil; growth will greatly depend on the reducing intensity of the environment, on the availability of nutrients, and on the physical factors such as temperature and water content. Anaerobic microbial processes occurring in soil are listed in Table 6.1. The microorganisms that carry out the

Table 6.1. Anaerobic microbial processes in soil (Tiedje et al 1984).

Process	*Reaction*
Fe^{3+}, Mn^{4+} reduction	$OM + Fe^{3+}, Mn^{4+} \longrightarrow Fe^{3+}, Mn^{2+}$
Denitrification	$OM + NO_3^- \longrightarrow N_2O, N_2$
Fermentation	$OM \longrightarrow$ organic acids, principally acetate and butyrate
Nitrate respiration	$OM + NO_3^- \longrightarrow NO_2^-$
Dissimilatory NO_3^- ammonification	$OM + NO_3^- \longrightarrow NH_4^+$
Sulphate reduction	OM or $H_2 + SO_4^{2-} \longrightarrow S^{2-}$
Carbon dioxide reduction	$H_2 + CO_2 \longrightarrow CH_4$, acetate
Acetate splitting	Acetate $\longrightarrow CO_2 + CH_4$
Proton reduction	Fatty acids and alcohols $+ H^+ \longrightarrow H_2 +$ acetate $+ CO_2$

OM = organic matter.

Methods in Applied Soil Microbiology and Biochemistry
ISBN 0–12–513840–7

Fe and Mn reduction, dissimilatory nitrate reduction and denitrification are all facultative anaerobes, and thus easily make the shift from aerobic to anaerobic growth. The last four processes are carried out by obligate anaerobes that are expected to be significant in flooded soils or intensively anaerobic conditions.

The methane can be formed from CO_2 and H_2 or by splitting acetate into CH_4 and CO_2 by **methanogenic** bacteria; the proton reduction is carried out by **acetogens**, which oxidize butyrate and propionate to acetate anaerobically. Only small amounts of methane are likely to be formed from natural or cultivated soils. The dissimilatory **sulphate-reducing bacteria** are obligate anaerobes, which occur widely in nature and flourish in wet or waterlogged soils, and permanent muds, provided that sulphate and organic matter are available (Tiedje et al 1984).

Most obligate anaerobes of soil are spore-forming bacteria of the genus *Clostridium.* They are capable of degrading cellulose, protein and polysaccharides, and producing ammonium from nitrate. Therefore, the isolation of obligate anaerobes from soils can not be used as an indicator for anaerobic microbial activities or for anaerobic conditions, since these microorganisms can be found in the form of spores even in dry soils. In this chapter several techniques will be presented for creating anaerobic conditions and quantifying anaerobic microbial activities in soils.

Sampling, transport and storage of anaerobic soils

K. Alef

Quantification of microbial processes in anaerobic soil requires the complete removal of oxygen during the estimation. Sampling and transport of anaerobic soils, therefore, must be performed in the absence of oxygen (Alef 1991). Anaerobic soil can be transported in e.g. jars (Fig. 6.1) that are previously gassed with pure nitrogen or helium. Another way to minimize the effect of oxygen on soil samples is to transport them on dry ice. If needed, soils can then be sieved in the laboratory in an anaerobic system (box or gloves). Before sieving, all equipment (jars containing the soil samples, sieve, scale, funnel, reaction flasks, etc.) should first be placed in the anaerobic system (Fig. 6.2). Anaerobic conditions in the system should be continuously controlled; the use of methylene blue strips is recommended.

Figure 6.1. Jars for storing soil under anaerobic conditions.

Figure 6.2. Anaerobic sack.

In the absence of oxygen, the strips remain colourless and become blue in the presence of traces of oxygen.

After sieving, anaerobic estimations can be performed immediately, or the soils can again be placed in anaerobic jars and stored at 4°C or at the recommended temperature. It should be pointed out that there is no need for such procedures when anaerobic estimations are performed with soils sampled from aerobic sites.

Redox potential measurement

J. Zausig

In waterlogged soil or under anoxic conditions, redox potential is the best measure of soil status and therefore an important measure to be considered in determining anaerobic microbial activity.

The redox potential (E_h) can be calculated according to the Nernst formula:

$$E_h = E_o + (RT)(n\ F)^{-1} \ln (a_{ox}\ a^{-1}{}_{red}) \qquad (6.1)$$

where E_o is the reduction potential under standard conditions, R is the molar gas constant (1.99 cal/degree), T is 298 K, n is the number of electrons transferred per mole, F is the Faraday constant (23.063 cal/v-equiv.), a_{ox} is the oxidized form and a_{red} is the reduced form.

Principle of the method

Measurement of the potential of bright platinum or gold electrodes in soil or other media with respect to the potential of a calomel electrode.

Materials and apparatus

Bright platinum (Pt) or gold electrode of the desired size and shape
Calomel electrode or silver/silver chloride electrode as reference electrode
Voltmeter with high input resistance (10^{10}–10^{13} Ω; this requirement is fulfilled by all modern pH meters)

Chemicals and solutions

HNO_3 (conc.) for washing of platinum electrodes.

Calibration solution

Solution A or B can be used.

Solution A

Saturated $K_2Cr_2O_7$ solution, with H_2SO_4 to pH 2 (E_h +590 mV versus calomel electrode; E_h + 636 mV versus Ag/AgCl electrode; rH 33).

Solution B

Saturated chinhydron solution, with buffer solution to pH 4 (E_h +220 mV versus calomel electrode; E_h + 266 mV versus Ag/AgCl electrode; rH 24).

Calibration

Before the calibration, the Pt or gold electrodes should be conditioned in concentrated HNO_3, washed under running water and finally rinsed in deionized water for at least 1 h. For calibration, the Pt or gold electrode is placed in a calibration vessel containing the desired solution together with the reference electrode. The electrodes are connected to the voltmeter or pH meter. The calibration solution should be stirred, and the temperature should be constant during calibration and measurement. After reaching constant values, the millivolt reading can be corrected according to the standard potential (see above).

Procedure

The measuring electrode (Pt or gold) and the reference electrode (calomel or silver/silver chloride) are both placed in the sample (bulk soil, sieved soil or soil suspension) and connected to the voltmeter. While systems with low buffering capacity show redox potential drift under presence of oxygen, measurements under

Figure 6.3. Set up for redox potential measurements with soil aggregates: (A) soil aggregate; (B) redox electrode with 0.1 mm diameter Pt wire melted in a glass capillary with the Pt tip exposed; (C) reference silver chloride electrode; (D) micromanipulator.

nitrogen atmosphere give constant values after 20–50 min. Before and after redox measurements the pH value should be checked to correct the measured redox potential values to redox potential at pH 7 (E_h7). Figure 6.3 shows the setup for redox potential measurements in soil aggregates.

Calculation

The redox potential values are referred to a standard hydrogen electrode by adding the potential of the reference electrode (247 mV for a calomel and 293 mV for an Ag/AgCl reference electrode at 20°C). By adding the correction $E = (\text{pH} - 7) \times 59$ mV, the values are referred to pH 7 (E_h 7 redox potential).

Discussion

Soil samples should be fresh to avoid changes in the redox potential. The use of an inner part of soil monolithes is possible if transport or storage of some hours to days is unavoidable.

Air-dried soils or soils with low moisture content do not allow good electric contact between the electrode and soil particles. In this case inaccurate potentials are obtained.

Measuring the redox potential in soils at 50–60% of water-holding capacity is recommended.

Electrodes have to be moved carefully in the soil suspension so as to avoid sedimentation processes and related drifts in the redox potential.

In most natural systems redox potential measurements can only be interpreted qualitatively. A quantitative interpretation is only possible if the electrode kinetic parameters and the nature of sensor effective redox couples is known (Stolzy et al 1981; Peiffer et al 1992).

Anaerobic conditions and testing

K. Alef

Preparation of oxygen-free gases

A satisfactory and economical way to remove the oxygen from gases is the use of a vertical column of heated copper. The column is heated electrically to about 350°C. Such systems are commercially available. Details concerning the use and regeneration of the column are usually supplied by the manufacturer.

Preparation of anaerobic solutions

(Hungate 1969; Alef 1991)

Principle of the method

The method is based on the removal of dissolved oxygen from heat-stable solutions by boiling under nitrogen atmosphere.

Materials and apparatus

Round-bottomed flasks (250, 500, 1000 ml)
Ice bath
Column containing reduced copper
Gassing line with syringes and gassing needles
Glass tubes or flasks with butyl rubber stoppers
Pipettes with PVC tubes

Chemicals and reagents

Oxygen-free nitrogen (pure)
Oxygen-free helium (pure)

Procedure

Place about 100 ml of the required heat-stable solution (substrate, buffer, distilled water) into a 250 ml round-bottomed flask, put the gassing needle into the solution and flush it with small stream of O_2-free nitrogen or helium. Boil the solution vigorously under continuous gassing with the O_2-free gas for about 1 min. The solution can then be kept anaerobic by replacing the vapour with O_2-free nitrogen or helium as the container cools in ice. After cooling, the anaerobic solution can then be transferred into the vessel (flask or tube) using a pipette. An O_2-free gas stream must flow into the dispensing and the receiving vessel at a rate sufficient to exclude air.

If the solution is heat labile, it can be freed of oxygen by bubbling O_2-free gas through it but the rate of removal is slow unless the gas bubbles are very abundant. Oxygen escapes by diffusion through the gas–liquid interface, making the rate of removal a function of the interface area. About 1 h is usually required to rid a solution of most of its dissolved oxygen but more extensive bubbling is necessary to equal the effect of boiling. Closing the vessel is achieved by inserting the stopper with the thumb and finger of one hand, holding the tube with the other three fingers, with the gassing needle still in place. The stopper should be loose enough to allow the gassing needle to be withdrawn with the other hand without moving the stopper. The removal of the needle should be rapid, as its end comes by the stopper. Butyl rubber (impermeable to oxygen) stoppers are used to close the vessels (Fig. 6.4). If necessary,

Figure 6.4. Preparation of anaerobic solution.

the closed glass tubes or bottles can then be autoclaved (see Chapter 4).

Discussion

The gassing line should be reconstructed by using a black pipe made of polypropylene because air can pass through the PVC tubes.

If sterile solutions are required, the glass syringe should be filled with a cotton plug and sterilized (by autoclaving at 121°C for 30 min) to prevent any microbial contamination.

If required, a reducing agent can be added to the solution to achieve a negative redox potential (see Chapter 4).

Achieving anaerobic conditions and anaerobic testing

(Alef 1991)

Basic techniques will be presented that can be modified to obtain optimal anaerobic conditions.

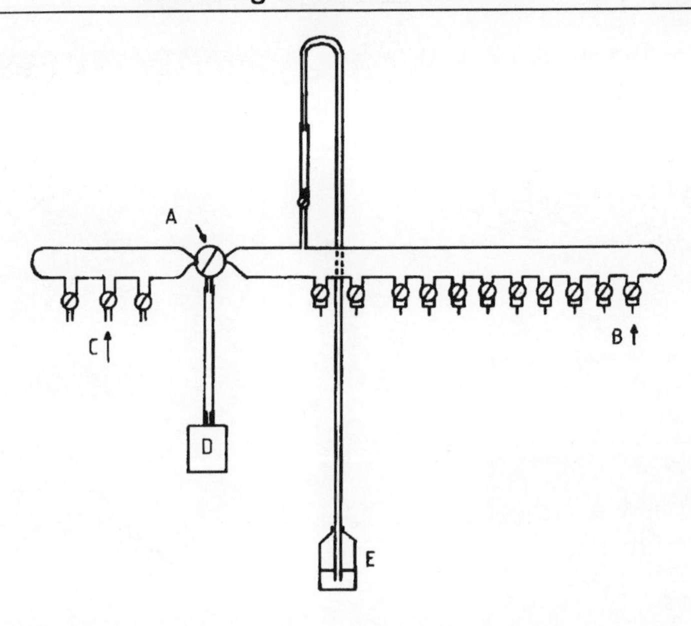

Figure 6.5. Apparatus used for replacing the air phase with another gas (Alef and Kleiner 1989): (A) tap; (B) needle; (C) gas (N_2, He, Ar, etc.); (D) vacuum pump; (E) mercury.

Principle of the method

In the first method, soil is placed in the reaction flask and treated with the required solution under anaerobic conditions. The second method is based on the estimation of microbial activity in waterlogged soil.

Materials and apparatus

Gassing line with syringe and needles
Reaction flasks with rubber stoppers
Gas-tight syringes (1, 5, 10, 20 ml)
Vacuum line
Anaerobic system (box or gloves)
Jars containing anaerobic soil

Chemicals and solutions

Methylene blue strips
O_2-free nitrogen (pure)
O_2-free helium (pure)
Anaerobic solutions (substrate, buffer, etc.)

Procedure

Measurements in anaerobic soil

To carry out an anaerobic test, the reaction flasks must be free of oxygen. This can be achieved by replacing the air with oxygen-free N_2, He or Ar by repeated evacuation and flushing using an apparatus consisting of a glass manifold connected to a manometer, a mechanical pump, an N_2 bomb and 10 outlets equipped with syringe needles for attaching the rubber stoppers of the reaction flasks (Fig. 6.5: Alef and Kleiner 1989).

The anaerobic reaction flasks as well as the jar(s) containing anaerobic soil (Fig. 6.1) should then be placed in the anaerobic box or gloves (Fig. 6.2). After controlling the anaerobic conditions in the system by using methylene blue strips, the flasks and the jar are opened. The required weight of soil is placed in the reaction flasks, and the flasks and jars are sealed tightly. Finally, the flasks are removed from the anaerobic box.

For estimating anaerobic microbial activities such as dentrification, ammonification, etc., the anaerobic

A

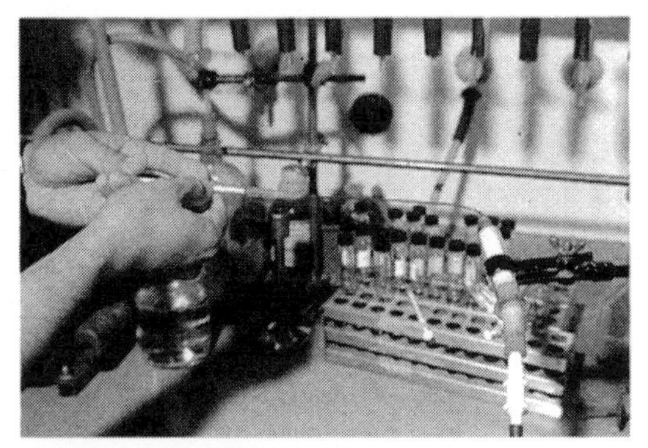

B

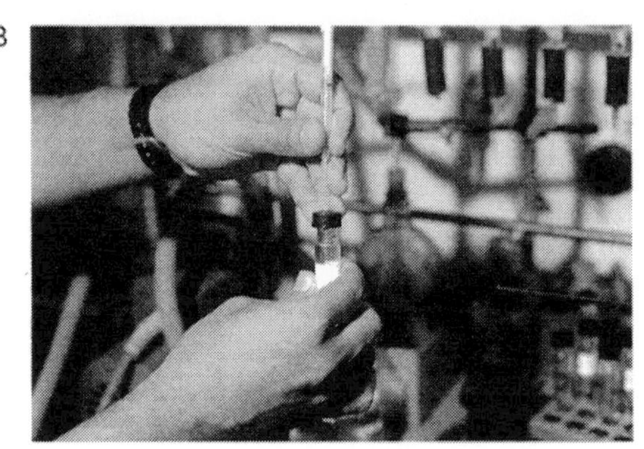

C

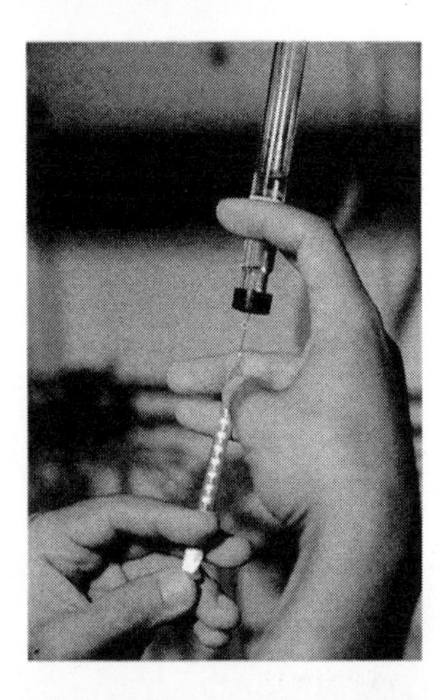

Figure 6.6. Transfer of anaerobic solution.

solutions (anaerobic substrate or buffer) should be added to the anaerobic soil. This can be performed as follows: a gas-tight syringe is flushed several times with O_2-free gas. A volume of gas (equal to the volume of solution to be added) is removed from the gas phase of the reaction flask and injected into the tubes containing the anaerobic solution (e.g. substrate, buffer, etc.). An equal volume of the anaerobic solution is removed and injected into the reaction flask. In cases where different anaerobic solutions are added to the soil, the procedure is repeated (Fig. 6.6). After the anaerobic solutions have been added to the soil, the reaction flasks (containing soil, substrate, buffer, etc.) are incubated at the required temperature for the required time.

Measurements in waterlogged soil:

The quantification of anaerobic microbial processes can also be carried out in waterlogged soils as follows: a moist or dry soil sample (5–10 g) is placed in a glass or centrifugation tube and flooded with aerobic solution (distilled water, substrate, buffer, etc.), then the tube is sealed tightly and incubated under the desired conditions (Welp and Brümmer 1985; Inubushi et al 1989). In this system the oxygen present will only be removed by the activity of the soil microorganisms. The basic procedures presented can be modified to achieve optimal anaerobic microbial activities in soil samples.

Assay of the anaerobic ammonification

(Waring and Bremner 1964; Keeney 1982)

K. Alef

Principle of the method

The method is based on the incubation of waterlogged soil for 7 days at 40°C followed by the estimation of the accumulated ammonium.

Materials and apparatus

Spectrophotometer
Incubator adjustable to 40°C
Shaker
Filter paper (N-free)
Glass tubes (16 × 150 cm) with rubber stopper and screw caps

Chemicals and solutions

KCl solution (4 M)

Dissolve 298 g of KCl in about 800 ml distilled water, bring up with distilled water to 1000 ml.

Reagents for ammonium determination

See Chapters 3 and 5.

Procedure

Place 5 g of moist soil in glass tubes (three tubes), add 12.5 ml distilled water, and incubate two tubes under continuous shaking at 40°C for 7 days. To perform the control, store the third tube immediately at −20°C. After the incubation, filter all soil suspensions and use the clear filtrate for ammonium estimation (Tabatabai and Bremner 1972; Alef and Kleiner 1987; Kandeler and Gerber 1988).

Calculation

Correct the results to the blank and calculate as follows:

$$NH_4\text{–N } (\mu g)/dwt \text{ soil}/ 7 \text{ days} = \frac{N\text{–}NH_4 \ (\mu g) \ ml^{-1} \text{ filtrate} \times v}{dwt} \quad (6.2)$$

where v is the volume of the filtrate in millilitres and dwt is the dry weight of 1 g moist soil.

Discussion

It takes some time for the oxygen to be consumed completely by soil microbial activity. The time needed depends on the soils used.

Waring and Bremner (1964) found a significant correlation between the anaerobic ammonification and the nitrogen mineralization in different soils ($n = 39$, $r = 0.96$).

The estimation can be performed with moist or dry soils. A linear time course of the ammonification within 7 days is very unlikely.

By using soil samples from anaerobic sites, high ammonium concentrations in the blanks can be expected.

Anaerobic nitrate reduction

Many facultative anaerobic bacteria contain **dissimilatory nitrate reductases** catalysing the reaction

$$NO_3^- + 2e^- + 2H^+ \rightarrow NO_2^- + H_2O \quad (6.3)$$

The ability to reduce nitrate to nitrite, however, does not permit normal growth under anaerobic conditions. A few facultative anaerobic bacteria can use nitrate as a terminal electron acceptor by reducing it to NO_2^-, N_2O and N_2. The ability to form N_2 from nitrate is termed **denitrification**. This process is linked to the normal aerobic respiratory electron transport chain, which enables electrons to be transferred to nitrate and its partial reduction products. It is therefore an alternative mode of respiratory energy-yielding metabolism used by denitrifying bacteria to support growth in the absence of oxygen.

In the presence of air, even when nitrate is present, respiration proceeds entirely through the aerobic electron transport chain, since the enzymes responsible for denitrification are not synthesized in the presence of oxygen. Because of the importance of nitrogen turnover in nature, methods to estimate nitrate reductase, nitrate ammonification and denitrification in anaerobic soil and sediment will be presented and discussed.

Assay of dissimilatory nitrate reductase activity

(Abdelmagid and Tabatabai 1987, modified by Schinner et al 1991)

K. Alef

Principle of the method

Determination of nitrite produced after incubation of soil samples with nitrate solution under waterlogged conditions for 24 h at 25°C.

Materials and apparatus

Spectrophotometer
Test tubes (180 × 18 cm)
Volumetric flasks (100 ml)
Filter paper

Chemicals and solutions

2, 4-Dinitrophenol (DNP) solution (0.9 mM)

Dissolve 166.6 mg of DNP in about 800 ml distilled water (used hot water) and bring up with distilled water to 1000 ml.

Nitrate solution (25 mM)

Dissolve 2.53 g of KNO_3 in about 800 ml distilled water and bring up with distilled water to 1000 ml.

KCl solution (2 M)

Dissolve 298.24 g of KCl in about 800 ml distilled water and bring up with distilled water to 1000 ml.

Buffer (0.19 M, pH 8.5)

Dissolve 10 g NH_4Cl in distilled water, adjust the pH to 8.5 with NH_4OH and dilute with distilled water to 1000 ml.

Reagent for nitrite determination

Dissolve 2 g of sulphanilamide and 0.1 g naphthyl-diethylene-diammonium chloride in about 150 ml boiling distilled water. After cooling dilute with distilled water to 200 ml.

Nitrite stock solution (1000 μg N ml^{-1})

Dissolve 4.9257 g of sodium nitrite in about 800 ml distilled water and bring up with distilled water to 1000 ml. Store at 4°C.

Nitrite standard solution (10 μg N ml^{-1})

Dilute 5 ml of nitrite stock solution with distilled water to 500 ml.

Procedure

Place 5 g of soil in test tubes (three tubes), add 4 ml of DNP solution, 1 ml of nitrate solution and 5 ml distilled water. Mix well and incubate two tubes for 24 h at 25°C. To perform the blank, store the third tube immediately at −20°C. After the incubation, add 10 ml of KCl solution, mix thoroughly and filter immediately. Pipette 5 ml of the clear filtrate into glass test tubes, add 3 ml buffer, 2 ml reagent for nitrite determination, shake well and allow to stand for 15 min at room temperature. Measure the colour intensity at 520 nm.

Calibration curve

Pipette 0, 2, 4, 6, 8 and 10 ml of nitrite standard solution into a volumetric flask

(100 ml), add 50 ml of KCl solution and dilute with distilled water to 100 ml. Perform the nitrite determination as described above.

Calculation

Correct the measurements to the blank and calculate as follows:

$$NO_2{-}N\ (\mu g\ g^{-1}\ dwt\ 24\ h^{-1}) = \frac{\text{Nitrite}{-}N\ (\mu g\ ml^{-1})\ \text{filtrate} \times 20}{5 \times dwt} \quad (6.4)$$

where *dwt* is the dry weight of 1 g moist soil, 20 is the volume of the filtrate and 5 is the used weight of moist soil in grams.

Discussion

Abdelmagid and Tabatabai (1987) found a lag phase of 10 h due to inhibition with O_2.

To obtain anaerobic conditions, the procedure described in the previous section on anaerobic soils is recommended.

Nitrate reductase activity was strongly inhibited at 40°C (Abdelmagid and Tabatabai 1987).

Fu and Tabatabai (1989) used this test to estimate the side effects of chemicals on soil microorganisms.

Changes in the community structure can be expected, therefore such estimation does not support information on the actual state of microbial populations.

Assay of denitrification

(Ryden et al 1979à, 1979b, modified by Schinner et al 1991)

K. Alef

Principle of the method

The method is based on the determination of N_2O formed after the incubation of soil samples in the presence of acetylene for 48 h at 25°C.

Materials and apparatus

Gas chromatograph with electrical conductivity detector (ECD), Porapak Q column (80–100 mesh, 3 m)
Gas-tight syringes (1, 10 ml)
Erlenmeyer flasks with screw caps and silicon stoppers
Anaerobic box or sack

Chemicals and solutions

Carrier gas (helium, purity grade 5.0)
N_2O (standard)
Acetylene

Procedure

Place 30 g of moist soil in an Erlenmeyer flask (100 ml) and seal with a screw cap and silicon stopper. Replace 10 ml of the air phase with 10 ml of acetylene and incubate for 48 h at 25°C. To estimate the denitrification under anaerobic conditions, replace the gas phase with helium as described in the previous section on achieving anaerobic conditions and anaerobic testing. Replace 10 ml of helium with 10 ml of acetylene and incubate at 25°C for up to 200 h. The addition of a carbon source (glucose 180 µg C g^{-1} dwt) and nitrate (200 µg N g^{-1} dwt) for studying the effect of pollutants on soil microorganisms is recommended. After the incubation, gas samples (0.5 or 1.0 ml) were removed and analysed on N_2O. The injector, column, and detector (HCD or ECD) temperature were 100, 40 and 80°C, respectively. For measuring the volume of the gas phase in the Erlenmeyer flask used in the assay, the flask was filled with water at the end of the incubation time (ml added water = ml gas phase).

Calibration curve

The calibration curve was prepared by analysing different volumes of pure N_2O (e.g. 50 µl).

Calculation

From the calibration curve read the (N_2O (µg ml^{-1}) gas phase and calculate as follows:

$$N_2O\ (\mu g\ 30\ g^{-1}\ dwt)\ t = \frac{N_2O{-}N(\mu g) \times v}{dwt} \qquad (6.5)$$

where t is the incubation time in hours, v is the volume of the gas phase of the Erlenmeyer flask in millilitres and dwt is the dry weight of 1 g moist soil.

Discussion

The electron capture detector is very sensitive for N_2O.

Estimation of denitrification can also be measured in a soil column or in the field (Ryden et al 1979b) (see the next section on nitrate reduction in sediments and waterlogged soil measured by ^{15}N techniques and Chapter 10).

Oxygen strongly inhibits the denitrification; N_2O production occurs only at low O_2 potential.

This procedure can be useful for studying the nitrogen turnover in the soil or the effect of pollutants on soil microorganisms under anaerobic conditions.

Nitrate reduction in sediments and waterlogged soil measured by ^{15}N techniques

N. Risgaard-Petersen
S. Rysgaard

The use of $^{15}NO_3^-$ in experimental set ups with undisturbed sediment cores or waterlogged soil, provides detailed information on the complex microbial nitrogen cycle. Goering and Pamatmat (1970) were among the first to use the reduction of $^{15}NO_3^-$ to $^{15}N_2$ gas as an assay for denitrification in sediments; following this pioneering work, the method has been improved by many modifications (Nishio et al 1983; Enoksson and Samulsson 1987; Binnerup et al 1992; Rysgaard et al 1993a, 1994; Risgaard-Petersen et al 1994). Recently, a new method has been developed for simultaneous determination of coupled nitrification/denitrification, and denitrification of NO_3^- supplied from the water column (Nielsen 1992). The methods presented have mainly been developed and used for estimating nitrate reduction in sediment. After slight modifications, these methods can also be applied to waterlogged soil.

Principle of the method

^{15}N-labelled NO_3^- is added to the water overlying the sediment or soil, and diffuses down through the oxic sediment or soil strata. Unlabelled NO_3^- ($^{14}NO_3^-$) originating both from the overlying water and from the nitrification process mixes with the added $^{15}NO_3^-$, and within 10–15 min a stable profile concentration of $^{14}NO_3^-$ and $^{15}NO_3^-$ is established. Just below the oxic/anoxic interface, dissimilatory NO_3^- reducers will reduce this isotopic mixture of NO_3^- to either N_2 (denitrification) or NH_4^+ (dissimilatory nitrate reduction to ammonia).

Denitrification

Assuming random association of the homogeneously distributed ^{15}N and ^{14}N molecules in the denitrification zone, it is possible to calculate the rate of $^{14}N^{14}N$ production from the content of $^{14}N^{15}N$ and $^{15}N^{15}N$ isotopes by simple probability mathematics (Fig. 6.7). The production of ^{14}N gas represents the *in situ* denitrification. When the ^{15}N atom% of NO_3^- in the water column is known, the rate of denitrification of NO_3^- supplied from the water column (D_w) can be

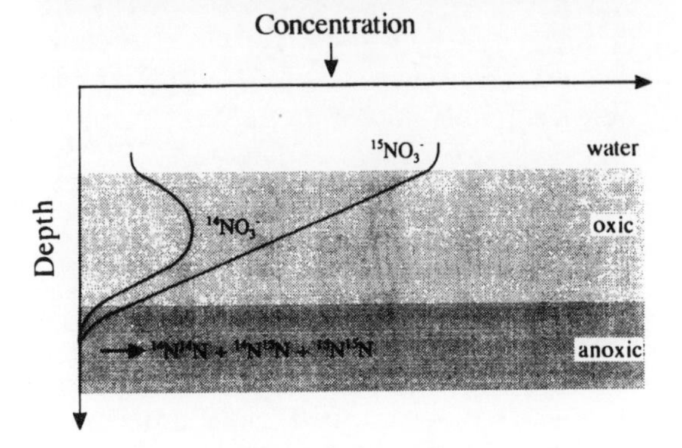

Figure 6.7. Concentration profiles of $^{15}NO_3^-$ and $^{14}NO_3^-$ within the sediment.

quantified. The rate of denitrification coupled to nitrification can then be estimated by subtracting D_w from the rate of *in situ* denitrification.

Dissimilatory nitrate reduction to ammonia

Assuming that dissimilatory nitrate reduction to ammonia (DNRA) is the only significant $^{15}NH_4^+$-producing process, the rate of dissimilatory reduction of $^{15}NO_3^-$ to NH_4^+ is quantified from the production of $^{15}NH_4^+$. When the ^{15}N atom% of NO_3^- in the water column or waterlogged soil is known, the *in situ* rate of DNRA of NO_3^- supplied from the water column ($DNRA_w$) or waterlogged soil can be quantified. Assuming that denitrification and DNRA take place in the same sediment or waterlogged soil strata, the ^{15}N atom% of NO_3^- reduced to NH_4^+ will be the same as the ^{15}N atom% of NO_3^- reduced to N_2. The total *in situ* rate of dissimilatory $^{14}NO_3^-$ reduction to NH_4^+ (DNRA total) can then be estimated from the production of $^{15}NH_4^+$ and the ^{15}N atom% of NO_3^- in the anaerobic NO_3^- reduction zone. The latter can be estimated from the $^{14}N^{14}N$, $^{15}N^{14}N$ and $^{15}N^{15}N$ production. DNRA coupled to nitrification is hereafter estimated by subtracting $DNRA_w$ from $DNRA_{total}$.

Sediment core sampling

Undisturbed sediment cores are collected with a PVC, Plexiglass or glass tube, which is driven directly into the sediment. After closing the tube with a rubber stopper, the core can be withdrawn from the surrounding sediment. If cores are transported from the sampling site to the laboratory, a 20 cm high water column should be left above the sediment core in order to obtain a sufficient oxygen and nutrient reservoir. It is recommended that the sediment cores are sampled by scuba diving whenever possible to minimize the disturbance of the sediment.

Incubation of sediment or waterlogged soil

Intact sediment or waterlogged soil cores can either be incubated with $^{15}NO_3^-$ in a batch mode (Nielsen 1992) or in a continuous flowthrough system (Nishio et al 1983; Binnerup et al 1992; Rysgaard et al 1993a and 1994; Risgaard-Petersen et al 1994).

Batch incubation of intact sediment cores

Materials and apparatus

PVC, Plexiglass or glass tubes (30 cm long, 5 cm i.d.)
Rubber stoppers
Open reservoir (50 l)
Rotating external magnet with motor supply
Teflon-coated magnet (3 cm long)
Temperature controller (e.g. Hetero Heater, Birkerod, Denmark)
Metal, Teflon or glass stick
Glass syringe (30 ml), supplied with a 10 cm long gas-tight tube (e.g. Isoversenic, Vernet, France)
Gas-tight glass vial (6–12 ml), with screw cap and a rubber septum (e.g. Exetainer[R], Labco, UK)
Plastic vials (20 ml and 50 ml)

Chemicals and solutions

$ZnCl_2$ solution (7 M)
$K^{15}NO_3$ or $Na^{15}NO_3$ (99 atom% ^{15}N)
Winkler reagents or an oxygen electrode

Procedure

The sediments in all tubes are adjusted to an equal height (about 10 cm) and the small Teflon-coated magnet is placed approximately 5 cm above the sediment surface. The tubes are then placed around the large rotating magnet in an open reservoir with *in situ* water at the *in situ* temperature, oxygen and light conditions (Fig. 6.8). The external magnet drives the small magnets in each tube, which provides a homogeneous

Batch incubation

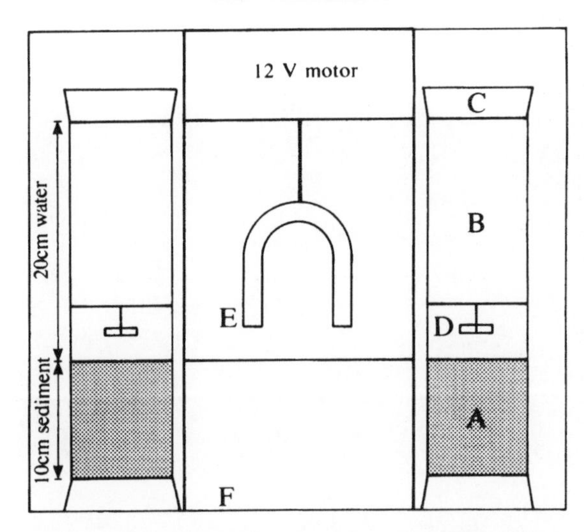

Figure 6.8 The batch incubation system. Two plexiglass cores are situated around a central rotating magnet. (A) intact sediment; (B) water column; (C) rubber stoppers; (D) Teflon-coated magnet; (E) external magnet; (F) open reservoir.

stirring of the water above the sediment, whereby it is possible to simulate the *in situ* diffusive boundary layer.

The oxygen flux across the sediment–water interface is then estimated. This is accomplished by incubating the rubber stoppered tubes for 1–2 h, and measuring the resulting change of the oxygen concentration in the water above the sediment or soil. Oxygen concentrations are measured by using an oxygen electrode. The maximum incubation time for the ^{15}N experiment is calculated from the oxygen data. As a general rule, the duration of the incubation should not result in more than a 20% alteration of the initial oxygen concentration in the water overlying the sediment or soil. Within this period all concentration changes are approximately linear (this can be verified by sampling cores at different time intervals and then plotting concentration against time).

The ^{15}N experiment is initiated by adding $^{15}NO_3^-$ to the reservoir at a concentration of at least 30 μM ^{15}N. The water in the tubes is then replaced with water from the reservoir using a syringe and allowed to equilibrate with the sediment porewater for 10–15 min. After this period the tubes are closed with rubber stoppers and the incubation begins. ^{15}N is estimated at the beginning and the end of the incubation.

Samples to be analysed for NO_3^- concentration and ^{15}N enrichment are collected from the water column in the tubes with a syringe and frozen in 20 ml vials. Samples to be analysed for ^{15}N, N_2 and NH_4^+ are collected from both the water column and the porewater with the syringe. The latter is accomplished by carefully mixing the sediment and water column with a metal stick after the addition of 250 μl $ZnCl_2$ solution to the sediment surface. Water samples and sediment water suspension samples for N_2 isotope analysis are preserved in gas-tight glass vials with 2% (v/v) $ZnCl_2$ solution, while samples for NH_4^+ concentrations and isotope analysis of the thoroughly mixed sediment–water suspensions are frozen in 50 ml vials. Samples for the determination of initial concentrations of ^{15}N isotopes are collected from the reservoir.

Calculation

Production rates of ^{15}N isotopes are calculated as follows:

$$p^{15}N_x = \frac{(C_{sed}-C_{ini})(\phi V_{sed}+V_2)+(C_{water}-C_{ini})(V_1-V_2)}{t} \times \frac{1}{A} \quad (6.6)$$

where $p^{15}N_x$ is the production rate of the relevant isotope ($^{15}NH_4^+$, $^{14}N^{15}N$ or $^{15}N^{15}N$), C_{sed} and C_{water} are the concentrations of the isotope in the sediment–water (or soil–water) suspension and in the water column, respectively, C_{ini} is the initial concentration of the isotope, V_{sed} is the volume of the sediment core (or soil), ϕ is the sediment (soil) porosity, V_1 and V_2 are the volumes of the water column before and after sample collection, respectively, t is the incubation time and A is the surface area.

Incubation of intact sediment cores in a continuous flow through system

Materials and apparatus

Glass chamber with gas-tight lid
Gas-tight tubing (e.g. Isoversenic, Vernet, France)
Peristaltic pump (e.g. Watson Marlow, Cornwall, UK)
Glass bottle (20 l)
Open reservoir (50 l)
Teflon-coated magnets
Large external rotation magnet
Temperature controller (e.g. Hetero Heater, Birkerod, Denmark)
Gas-tight glass vial (6–12 ml) with a screw cap and a rubber septum (e.g. Exetainer[R], Labco, UK)
Plastic vials (20 ml and 50 ml).

Chemicals and solutions

$ZnCl_2$ solution (7 M)
$K^{15}NO_3$ or $Na^{15}NO_3$ (99 atom% ^{15}N)

Procedure

The sediments (or soil) in all chambers are adjusted to an equal height and the small Teflon-coated magnet is placed approximately 2 cm above the sediment (or soil) surface. The chambers are then placed around the large rotating magnet in an open reservoir that contains water held at *in situ* temperature. After closing the chambers with glass lids, they are connected to a 20 l glass bottle containing *in situ* water to which $^{15}NO_3^-$ is added. Water is then pumped continuously from the glass bottle over the intact sediment (or soil) in the glass chamber (Fig. 6.9) with a peristaltic pump.

Continuous flow through incubation

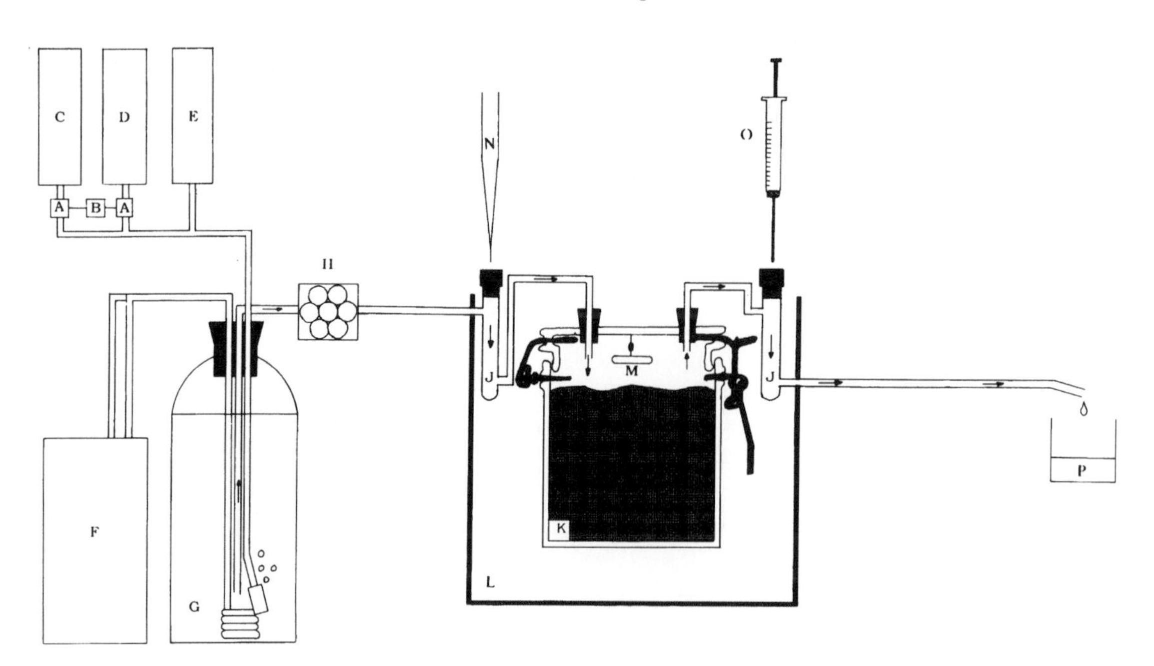

Figure 6.9. The continuous flow-through system. (A–E) gas mixing system; (F) temperature controller; (G) reservoir; (H) peristaltic pump; (J) inflow and outflow sampling chambers; (K) sediment chamber; (L) open thermostated reservoir; (M) Teflon-coated magnet; (N) oxygen sensor; (O) glass syringe; (P) outlet.

Water samples to be analysed for N_2 concentration and isotopic composition are collected from the inlet and outlet of the incubation chamber in gas-tight glass vials containing 2% (v/v) $ZnCl_2$ solution. Water samples to be analysed for concentrations and isotopic composition of NO_3^- and NO_2^-, and NH_4^+ are also collected from the outlet and inlet, and then frozen in 20 ml plastic vials. In a continuous flow-through system, the concentration change between inlet and outlet reflects the production or consumption rates within the sediment system in a steady state. The concentration gradients of ^{15}N-labelled N_2 and NH_4^+ formed by NO_3^- reduction processes within the sediment need time (from hours to days) to reach this steady state. In addition there is a delay between the actual exchange of isotopes across the sediment–water (or soil–water) interface and its monitoring, due to the turnover time of the water phase in the chamber. A temporary change in the production rates of ^{15}N isotopes will therefore appear as a gradual temporal approach to a new equilibrium. It is, therefore, recommended to monitor the concentration changes of ^{15}N-labelled components during the initial incubation in order to ensure steady-state conditions.

Calculation

The flux of isotopes ($p^{15}N_x$) across the sediment–water (or soil–water) interface is estimated from the general flux equation:

$$p^{15}N_x = \frac{(C_{in} - C_{out}) \times V}{A} \tag{6.7}$$

where C_{in} and C_{out} are the concentrations of the relevant isotope in the inflowing and outflowing water, respectively, A is the surface area of the sediment and V is the flow rate. As mentioned above, fluxes represent the production rate of the relevant isotope when the sediment–water system is in a steady state.

Calculation of nitrate reduction rates

The rates of dissimilatory NO_3^- reduction can be estimated from the ^{15}N production rates according to Nielsen (1992):

$$D_{15} = p\,(^{14}N^{15}N) + 2p\,(^{15}N^{15}N) \tag{6.8}$$

$$D_{14} = \frac{p\,(^{15}N^{14}N)}{2p\,(^{15}N^{15}N)} \times D_{15} \tag{6.9}$$

where D_{15} and D_{14} are the rates of denitrification of $^{15}NO_3^-$ and $^{14}NO_3^-$, respectively, $p\,(^{15}N^{14}N)$ and $p\,(^{15}N^{15}N)$ are the production rates of the two ^{15}N-labelled N_2 species ($^{15}N^{14}N$ and $^{15}N^{15}N$), respectively.

The *in situ* rate of denitrification of NO_3^- supplied from the water column (D_w) is calculated from D_{15} and the ^{15}N atom% of NO_3^- in the water column:

$$D_w = \left(\frac{100}{^{15}N \text{ atom\% } NO_3^-} \times D_{15}\right) - D_{15} \tag{6.10}$$

Coupled nitrification denitrification (D_n) is then calculated by the difference:

$$D_n = D_{14} - D_w \tag{6.11}$$

In situ rates of dissimilatory nitrate reduction to ammonia of NO_3^- supplied from the water column are estimated from the difference:

$$DNRA_w = \left(\frac{100}{^{15}N \text{ atom\% } NO_3^-} \times p^{15}NH_4^+\right) - p^{15}NH_4^+ \tag{6.12}$$

where $p^{15}NH_4^+$ is the production rate of $^{15}NH_4^+$, and ^{15}N atom% NO_3^- is the ^{15}N atom% of NO_3^- in the water column. Assuming that DNRA in the same sediment layer is equal to the denitrification rate the ^{15}N atom% of NO_3 reduced to ammonia equals the ^{15}N atom% of NO_3^- reduced to N_2. Total DNRA of $^{14}NO_3^-$ can then be calculated as:

$$DNRA_{total} = p^{15}NH_4^+ \times (D_{14}/D_{15}) \quad (6.13)$$

The rate of DNRA coupled to nitrification ($DNRA_n$) is then estimated from the difference:

$$DNRA_n = DNRA_{total} - DNRA_w \quad (6.14)$$

Discussion

Denitrification

A fundamental assumption in the method is that a uniform mixing of $^{15}NO_3^-$ and $^{14}NO_3^-$ takes place in the denitrifying microenvironment. Heterogeneous topography, bioturbation, etc. can lead to local variation of the nitrification activity and of the vertical transport of NO_3^- to the denitrification zone. Such heterogeneity can lead to the existence of subpools with different $^{14}N/^{15}N$ ratios in the denitrifying microenvironment. It can be shown that this situation leads to an underestimation of D_{14} because $^{14}N^{15}N$ is produced in lower amounts than those predicted from the homogeneity assumption (Boast et al 1988). However, as illustrated in Fig. 6.10 the underestimation depends on the NO_3^- level applied. At increasing $^{15}NO_3^-$ concentrations, increasing amounts of $^{14}NO_3^-$ will be trapped as $^{14}N^{15}N$ and the estimated D_{14} will converge toward the correct value. This is, however, only true when the nitrifying and denitrifying microenvironments are situated close to the sediment (or soil) surface, and the rate of denitrification of NO_3^- supplied from the water column is proportional to NO_3^- concentrations. It is obvious that the method can not be used in macrophyte-covered sediments with two or more discrete zones of nitrification and denitrification.

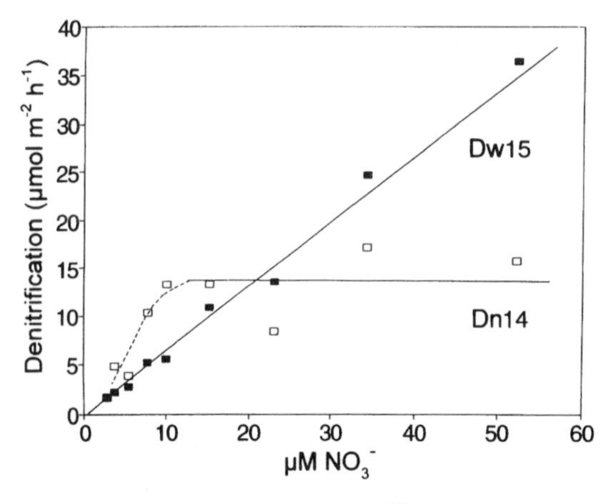

Figure 6.10. Denitrification of $^{15}NO_3^-$ (Dw15) and $^{14}NO_3^-$ (Dn14) as a function of the $^{15}NO_3^-$ concentration in the water column. Nine parallel sediment cores were incubated each with a different $^{15}NO_3^-$ concentration in the overlying water. Then Dw15 was measured and Dn14 calculated (Nielsen and Glud, unpublished).

Dissimilatory nitrate reduction to ammonia

A fundamental assumption of the method is that no alternative $^{15}NH_4^+$-producing processes occur within the sediment or waterlogged soil. Assimilatory NO_3^- reduction to NH_4^+ may interfere with the dissimilatory NH_4^+ production. The assimilation of NO_3^- can be highly significant in sediments (or waterlogged soil) colonized by benthic microphytes (Risgaard-Petersen et al 1994). If *in situ* NO_3^- concentrations are low (0–5 µM), the microphytes will reduce a part of the added $^{15}NO_3^-$, to $^{15}NH_4^+$ and partly release it into the environment (Risgaard-Petersen, unpublished). Therefore, if the goal of the experiment is to quantify DNRA, it is recommended that experiments are performed in non-phototrophic systems or after a period of at least 3 days of preincubation in the dark, in order to avoid interference from microphytobenthic NO_3^- assimilation (Rysgaard et al 1993a and 1994).

Analysis of ^{15}N

Gas analysis

Materials and apparatus

Gas chromatograph (e.g. Roboprep-G+, Europa Scientific, UK) in line with a triple

collector isotope ratio mass spectrometer (e.g. TracerMass, Europa Scientific)
Hypodermic needles
Gas-tight tubing (e.g. Isoversenic, Vernet, France)
Helium

Procedure

$^{15}N^{15}N$ and $^{14}N^{15}N$ are extracted from the water in the glass vials by introducing a He headspace. This is accomplished by intersecting a hypodermic needle in line with a He flask through the rubber septum of the vial. With a glass syringe 4 ml water are removed and subsequently replaced by an equivalent volume of He (Fig. 6.11). The vial is then shaken vigorously for 5 min and thereafter more than 98% of the N_2 will be found in the headspace (Weiss 1970).

The isotopic composition of N_2 in the headspace is then determined by a mass spectrometer. The vials are placed in an autosampler in line with a gas chromatograph and a mass spectrometer. The entire headspace (4 ml) of the vial is then transferred to a carrier flow of He (99.9995% purity) using a double, concentric needle (Fig. 6.12). The sample is then passed through a drying tube (10 mm × 200 mm) packed with $Mg(ClO_4)_2$ to remove water vapour and Carbosorb (10–20 mesh) to remove CO_2. After passing a gas chromatography column (3 mm × 45 mm, packed with Carbosieve G and held at 50°C), the sample is diverted through a reduction column (15 mm × 300 mm) packed with Cu wires at 650°C to remove O_2 coeluting with N_2. The latter is performed to minimize formation of NO in the ion source of the mass spectrometer, which will interfere at m/z 30. After the removal of O_2 on the sample, the N_2 is directed to a triple-collector mass spectrometer to obtain the isotopic composition of N_2.

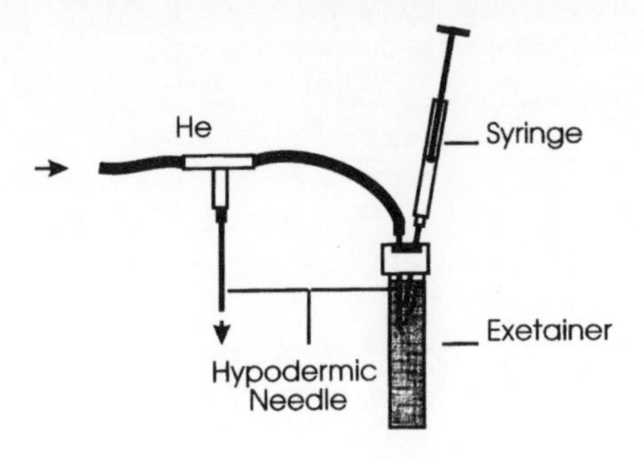

Figure 6.11. Introduction of an He headspace into a gas-tight Exetainer using needles and a syringe.

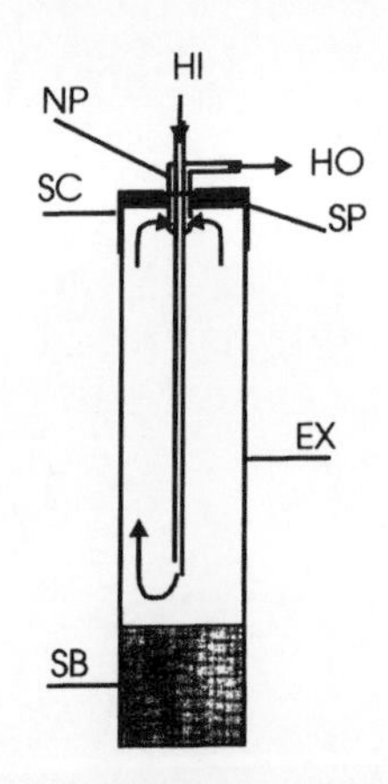

Figure 6.12. Gas sampling from the headspace with a double concentric needle. (HI) He inflow; (HO) He outflow; (SC) screw cap; (SP) rubber septum; (NP) double concentric needle; (EX) exetainer; (SB) water sample.

Nitrate analysis

Several methods for determining the ^{15}N isotopic composition of NO_3^- have been described (O'Dean and Porter 1980; Crumpton et al 1987; Prosser et al 1993). In general, these methods require more than 2 µmol nitrogen for a reliable analysis. Recently, Risgaard-Petersen et al (1993) demonstrated that the ^{15}N atom% of NO_3^- could be determined precisely in samples containing less than 5 nmol of NO_3^-. The method, based on biological reduction of NO_3^- to N_2, followed by mass spectrometer analysis, will be presented below.

Materials and apparatus

Gas-tight glass vials (5 ml) with screw caps (e.g. Exetainer, Labco, UK)
Needles and syringes
Pure culture of denitrifiers (e.g. *Pseudomonas nautica*, or *Pseudomonas aeroginosa*, DSM, Germany)
NO_3^- free media for the specific stain
Thiourea

Procedure

A sample volume of 3.6 ml is transferred to the 5 ml glass vial together with 1 ml of NO_3^--free media, 0.2 ml 3 mM thiourea, and 0.2 ml of the bacteria culture. The vial is then closed with the screw cap, and incubated for 12–24 h at 30°C in order to reduce all NO_3 to N_2 by denitrification. The isotopic composition of the N_2 produced is then analysed as described above.

Calculation

Nitrogen in the vial derives from two different sources in the vial: (1) the atmospheric N_2 dissolved in the sample; and (2) the N_2 produced during denitrification of NO_3^- having an unknown ^{15}N enrichment, *X*. As demonstrated by Risgaard-Petersen et al (1993), the association of ^{14}N and ^{15}N will be random during the formation of N_2. The relative amounts of the two N_2 isotopes $^{14}N^{15}N$ and $^{15}N^{15}N$, originating from the reduction of NO_3^-, can thus be described by simple probability mathematics. Complete mixing of the gases from the two sources (atmospheric N_2 and reduced NO_3^-) will, therefore, result in the following relative amounts of $^{14}N^{15}N$ and $^{15}N^{15}N$ in the headspace:

$$f_{29}H = (1 - q)\, f_{29}A + q\,[2X\,(1-X)] \quad (6.15)$$

$$f_{30}H = (1-q)\, f_{30}A + qX2 \quad (6.16)$$

where q is the concentration of NO_3^- reduced to N_2 relative to the total N_2 concentration, $f_{29}H$ and $f_{30}H$ are the relative amount of $^{14}N^{15}N$ and $^{15}N^{15}N$ in the headspace, $f_{29}A$ and $f_{30}A$ are the relative amount of $^{14}N^{15}N$ and $^{15}N^{15}N$ in atmospheric N_2 and X is the unknown ^{15}N enrichment of NO_3^-. It can be shown that equations 6.15 and 6.16 can be solved with respect to X, resulting in an expression where q is excluded (Risgaard-Petersen et al 1993):

$$X = \frac{1 + [1 - (R + 2)\,(f_{29}\,A - Rf_{30}A)]^{\frac{1}{2}}}{R + 2} \quad (6.17)$$

where $R = (f_{29}\,H - f_{29}\,A)/(f_{30}\,H - f_{30}\,A)$. The ^{15}N atom% of NO_3^- can thereafter be calculated as:

$$^{15}NO_3^- \text{ atom\%} = X \times 100 \quad (6.18)$$

Discussion

It has been demonstrated that the assay makes it possible to determine precisely and accurately the ^{15}N atom% of NO_3^- (0.1 SD) in samples containing less than 2 nmol of NO_3^- (Rysgaard and Risgaard-Petersen, unpublished).

Ammonium analysis – Exchangeable ammonium (modified after Hatton and Pickering 1990; Laima 1992):

Materials and apparatus

Centrifuge
KCl (1 M)
Polypropylene centrifuge tubes
Chloroform

Procedure

Homogenized sediment (5 g wet weight) is added to a 50 ml (65 ml) polypropylene centrifuge tube together with 50 ml 1 M KCl (sediment/KCl 1:10 w/v). A few drops of chloroform are added to stop bacterial activity. The slurry is then shaken for 30 min and centrifuged (3000 rev min^{-1} for 10 min). The supernatant is analysed for $^{15}NH_4^+$ content and the remaining sediment is used for a second extraction. The procedure is repeated until the concentration of NH_4^+ in the supernatant is undetectable. Two to three extractions ae usually sufficient to extract all the exchangeable NH_4^+.

$^{15}NH_4^+$ analysis

Several methods are available for ^{15}N analysis of NH_4^+ (Blackburn 1979; O'Dean and Porter 1980; Paaske and Kristensen 1982; Rönner et al 1983; Dudek et al 1986; Selmer and Sörenssen 1986; MacKown et al 1987; Brooks et al 1989). We present a modification of one method, which requires 1 μmol N for a reliable analysis.

Materials and apparatus

Elemental analyser (e.g. RoboPrep-C/N, Europa Scientific, UK)
Mass spectrometer (e.g. TracerMass, Europa Scientific, UK)
Gas-tight glass containers (10 ml) with screw caps (e.g. Exetainerr[R], Labco UK)
Hypodermic needles with the tip hammered flat
Tin capsule (4.0 × 6.0 mm, Micro Kemi AB, Sweden)
N-free Al_2O_3
$H_2 SO_4$ (5 M)
NaOH (10 M)

Procedure

2 ml of a sample are transferred to the gas-tight glass vial together with the hypodermic needle. The tin capsule containing 50 mg purified Al_2O_3 is placed in the needle cap (Fig. 6.13). The purified Al_2O_3 is prepared by pipetting 10 ml of 5 N H_2SO_4 g to the N-free Al_2O_3, which is then dried at 80°C for 24 h. The diffusion assay is initiated by increasing the pH of the sample with the addition of 200 μl 10 M NaOH and closing the vial with a screw cap. Ammonium is then removed from the solution by diffusion of NH_3, which is then concentrated on the acidified Al_2O_3 in the tin capsule. Our experience is that > 95% of the NH_4^+ contained in a 2 ml water sample will be effectively trapped in the capsule at room temperature within 48 h of incubation. The tin capsules containing the diffused NH_4^+ are then carefully packed, and subsequently combusted in an elemental analyser in line with an isotope ratio mass spectrometer in order to determine the isotopic composition and concentration of NH_4^+. It is recommended that a standard sample is run for every five samples containing a known amount of NH_4^+ with a known ^{15}N enrichment.

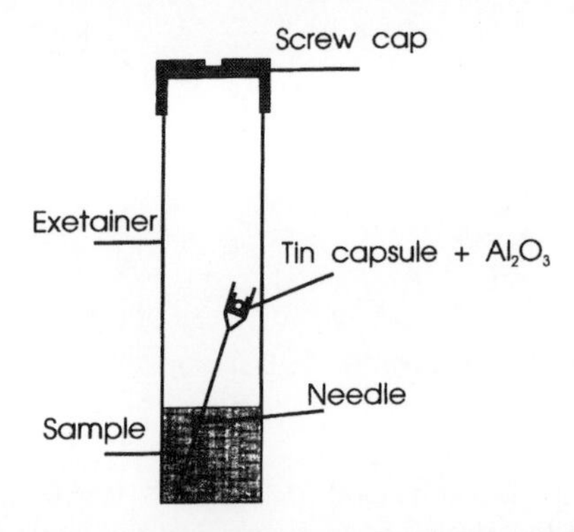

Figure 6.13. The $^{15}NH_4^+$ microdiffusion assay.

Calculation

The ^{15}N atom% of NH_4^+ is calculated as follows:

$$^{15}\text{N atom\%} = \frac{R_1 + 2R_2}{2\,(1 + R_1 + R_2)} \times 100 \quad (6.19)$$

where R_1 and R_2 equals m/z $^{14}N^{15}N/^{14}N^{14}N$ and $^{15}N^{15}N/^{14}N^{14}N$, respectively. In addition to the $^{15}NH_4^+$, the concentration of NH_4^+ can be obtained by using the gas chromatographic device fitted on the mass spectrometer.

Iron(III) reduction test

G. Welp
G. Brümmer

(Welp and Brümmer 1985)

If soils become saturated with water due to impeded drainage, a rising ground water table or flooding, a characteristic sequence of redox processes starts. Initially, the oxygen is reduced by aerobic and facultative anaerobic microorganisms oxidizing organic material (first stage). After the consumption of oxygen, the number of anaerobic bacteria increases (second stage). Their activity mainly depends on the amount of organic substances that can be mineralized and on the amount of other nutrients. Instead of gaseous and dissolved oxygen in the soil then the oxygen of nitrate, manganese oxides, iron oxides and sulphate is used for microbial respiration (Ottow 1969; Brümmer 1974; Gotoh and Patrick 1974). This is also a thermodynamically determined order of reduction. Nitrate and nitrite as well as manganese oxides always serve as electron acceptors before iron oxides and sulphates are reduced (Ottow 1969; Sorensen 1982).

Fe(III) can be reduced by an enzyme-mediated transfer of electrons to Fe(III) oxides and/or via microbial assimilation products that are formed under anaerobic conditions (Ponnamperuma 1972; Brümmer 1974; Munch and Ottow 1977). The transfer of electrons possibly is induced by an enzyme that is also engaged in nitrate reduction (dissimilatory nitrate reductase). Because this process is coupled with a phosphorylation, the reduction of Fe(III) oxides could be understood as an "anaerobic respiration" (Alef 1991). Fe(III) oxides are sparingly soluble in water. Therefore, they have to be solubilized before they can penetrate into cells of microorganisms (or higher plants). This can be done by complexing agents (siderophores) that are produced, especially in the rhizosphere, by some microbes and plant roots (Neilands 1981; Marschner 1986). Although the specific importance of the different mechanisms of Fe(III) reduction in soils still are discussed, it is generally accepted as an anaerobic bacterial process (Hund et al 1990). This view is confirmed by experiments showing that the formation of Fe^{2+} ions under waterlogged conditions is retarded or completely suppressed after sterilizing soils or adding toxic substances like heavy metals and pesticides (Premi 1971; Brümmer 1974; Pal et al 1979; Reddy and Gambrell 1985). Under standardized conditions the degree of the bacterial reduction of insoluble Fe(III) oxides to soluble Fe^{2+} ions therefore can be used as a measure of the bacterial activity of soils.

The test was designed to determine the effects of environmental chemicals on the bacterial activity in soils (Welp and Brümmer 1985). The procedure cannot be used directly to calculate the bacterial state of a soil because the Fe(III) reduction capacity also depends on the amount and quality of Fe(III) oxides present in the soil samples. With the method described, good results can be obtained e.g. when different amounts of a toxicant are added to different samples of the same soil material in order to establish dose–response curves (Welp et al 1991; Welp and Brümmer 1992). With such dose–response curves the effects of different chemicals on soil bacteria can be measured for soils of different composition. In addition to being an easy and reproducible method, the Fe(III) reduction test covers both the activity of aerobic (first stage of the test procedure) and anaerobic bacteria (second stage).

Principle of the method

Bacterial reduction of insoluble Fe(III) oxides to soluble Fe^{2+} ions or soluble organic Fe(II) complexes after addition of easily decomposable organic matter (yeast extract) and water saturation.
Determination of soluble Fe after a 5-day incubation period.

Sphere of application

All soils with pH ($CaCl_2$) above 3.5 to 4 and sufficient amounts of easily reducible Fe(III) oxides.

Materials and apparatus

Polyethylene tubes, 50 ml, with screw caps
End-over-end shaker
Büchner funnels (or other vacuum filtration systems)
Filter paper circles (e.g. Schleicher and Schuell, 5893, blue ribbon)
Vacuum pump
Suitable pipettes and glassware
Incubator or constant temperature room (20°C)
Atomic absorption spectrometer (AAS) (or other equipment to measure Fe)

Chemicals and solutions

KCl solution (1 M)
HNO_3 (conc.)
Yeast extract
Soil suspension with active bacteria

Supernatant of a mixture of 10 g fresh humic soil material and 20 ml distilled water.

Procedure

Samples of 20 g air-dried soil and 20 ml distilled water (or distilled water spiked with different amounts of a toxicant) are filled into 50 ml polyethylene tubes (three replicates). Glass tubes are more suitable for toxicity tests with volatile chemicals or hydrophobic substances that tend to be adsorbed by plastic materials. The tubes are shaken for 48 h in an end-over-end shaker at 20°C. After opening the tubes, 0.3 g of yeast extract and 50 µl of a soil extract with active bacteria are added. Then the closed tubes with the soil suspensions are incubated for 5 days at 20°C. On the second, third and fourth day of incubation, the samples are shaken for half a minute thoroughly by hand. At the end of the incubation period 20 ml 1 M KCl solution are added. The tubes are closed again, shaken thoroughly by hand for a short time and after a 15-min settling period the solution is sucked through filter paper via a Büchner funnel into 50 ml flasks. The filtration has to be done quickly to avoid Fe^{2+} oxidation (therefore, it is recommended to carry out vacuum filtration). The solution is stabilized with 0.25 ml nitric acid and analysed for Fe with the AAS.

Discussion

Besides using the older colorimetric methods, nowadays a measurement with a flame AAS or ICP-OES is recommended. To avoid matrix problems (high salt concentration, high content of dissolved organic matter) the filtrates should be diluted 20 times. Calibration standards should be made in a range of 1–20 mg Fe l^{-1}.

The influence of different drying and storage conditions on the microbial toxicity of chemicals cannot be clearly assessed. According to Wingfield et al (1977), microbes in fresh soil samples are less sensitive to the herbicide dalapon as compared to those of air-dried samples. Welp and Brümmer (1985) did not find clear differences for the toxicity of the anionic detergent LAS and the herbicide 6-hydroxy-2, 4-dichlorophenoxyacetic acid (2,4-D) when comparing fresh, frozen (−18°C) and air-dried soil samples. In the interest of easy handling, the use of air-dried soil samples is recommended. The microbial, physical and chemical changes in soils that are induced by drying are partly compensated for by shaking the soil suspension for 48 h and by adding active bacteria to the soil extract. It is also possible to run the test with fresh soil material. In this case it should be guaranteed that the free iron in the sample is still in an oxidized form. A good mixing of the soil sample is necessary in order to obtain good reproducibility for the results.

The addition of a carbon and energy source is necessary to obtain a sufficient Fe(III) reduction within a relatively short period. Because of its versatile composition, yeast extract is suitable to meet the requirements of aerobic and anaerobic microorganisms. Other media can be used (see Welp 1987), e.g. glucose, which was proposed by Zelles et al (1986). Using glucose instead of yeast extract often induces a more sensitive reaction of the microflora towards a toxicant but in some soils it does not lead to an intensive Fe(III) reduction (Welp and Brümmer, unpublished data).

The differences in the intensity of Fe(III) reduction between control and contaminated samples are influenced by the duration of the incubation period (Welp 1987). After a prolonged incubation of more than 5 days these differences often decrease due to the adaptation of bacteria, growth of less sensitive microbes, fixation or decomposition of the chemical, or a changing chemical environment. The incubation time and the dose of yeast extract were optimized in such a way that: (1) an intensive Fe(III) reduction is obtained within a fairly short time; and (2) distinct toxic effects of chemicals can be recognized in relation to the applied doses. In an alternative procedure Zelles et al (1986) proposed a 7-day incubation. This longer period could be the main reason for the relatively low sensitivity towards toxic chemicals observed in their experiments.

A comparison of the Fe(III) reduction test with common procedures to calculate microbial activity (respiration, substrate induced respiration, dehydrogenase) revealed comparable results for low doses of toxicants. Higher doses of chemicals in most cases decrease Fe(III) reduction more strongly than other parameters of microbial activities (Welp 1987). The higher sensitivity of the Fe(III) reduction test is related to the fact that this assay measures the influence of toxic chemicals on bacterial activities. Chemicals with a high toxic effect on fungi, such as fungicides, show no effect on Fe(III) reduction (Welp and Brümmer, unpublished data).

Besides a microbial reduction, Fe(III) oxides can be dissolved by acids. Therefore, in soils with pH ($CaCl_2$) below 3.5–4.0, the biotic reduction and abiotic dissolution of Fe(III) occurs simultaneously. Therefore measuring Fe(III) reduction in acid soils is not recommended. Short-term reduction of Fe(III) oxides by bacteria in soils is mainly restricted to poor crystalline forms like ferrihydrite. Thus, a small amount of these easily reducible Fe(III) oxides (in some podzols and ferralsols) can be the reason for a very low level of Fe^{2+} that prevents a reasonable evaluation of the test results.

Bacterial sulphate reduction in waterlogged soils measured by the $^{35}SO_4^{2-}$ radiotracer technique

K. Ingvorsen

Sulphate-reducing bacteria (SRB) are anaerobic prokaryotes and are physiologically characterized by their ability to obtain energy from oxidation of organic material (or H_2) using sulphate as a terminal electron acceptor. This energy-yielding process is referred to as dissimilatory (or respiratory) sulphate reduction.

A general equation of anaerobic degradation of organic matter in sediment or soil by dissimilatory sulfate reduction is:

$$2(CH_2O) + SO_4^{2-} = H_2S + 2HCO_3^- \qquad (6.20)$$

Part of the sulphide (H_2S) produced is subsequently converted into iron sulphides (FeS), pyrite (FeS_2), elemental sulphur (S^o) and organic sulphur. The relative distribution of biogenic sulphide into these major sulphur pools varies greatly, and depends on the chemical and biological characteristics of the sediment or soil.

The term TRIS (Total Reduced Inorganic Sulphur) is sometimes used in connection with sulphate reduction measurements in sediments. TRIS includes both Acid-Volatile Sulphide (AVS: HS^- + FeS) and Chromium-Reducible Sulphur (CRS: S^o + FeS_2).

Numerous studies have demonstrated the importance of dissimilatory sulphate reduction in anaerobic carbon mineralization in sulphate-rich ecosystems, e.g. in shallow-water coastal sediments and salt marshes (Jørgensen and Fenchel 1974; Skyring and Chambers 1976; Jørgensen and Cohen 1977; Jørgensen 1977; Howarth and Giblin 1983; Howarth 1984; Howarth and Jørgensen 1984; Skyring 1987). The concentration of sulphate in freshwater is usually 100-fold lower than in seawater. Nevertheless, it has been shown that bacterial sulphate reduction at least in some freshwater environments may catalyse a considerable amount of the total carbon mineralization (Ingvorsen et al 1981; Ingvorsen and Brock 1982; Hordijk et al 1985; Bak and Pfennig 1991). In freshwater sediments, however, the zone of active sulphate reduction is confined to the uppermost centimetres of the sediment due to a limited diffusional supply of sulphate into the sediment from the overlaying water (Hordijk et al 1987; Bak and Pfennig 1991).

SRB are also widespread in soils (Furusaka 1968; Wakao and Furusaka 1973, 1976; Durbin and Watanabe 1980; Postgate 1984) and bacterial sulphate reduction has been demonstrated in anoxic waterlogged soils, e.g. rice paddies (Engler and Patrick 1973; Skyring et al 1979; Postgate 1984; Furusaka 1968). Our present knowledge of the importance of SRB in organic matter degradation in soil ecosystems is very limited. However, water-saturated (or flooded soils) can operationally at least be regarded as a "sediment" and existing ^{35}S radiotracer techniques may be readily adapted to estimate sulphate reduction rates in flooded soils after they become anoxic.

Aerated, well-drained soils are complex ecosystems usually consisting of an heterogeneous assemblage of oxic and anoxic microenvironments with large spatial variations in chemical, physical and microbial parameters. Use of e.g. the $^{35}SO_4^{2-}$ radiotracer technique in soils which contain oxygen will result in an underestimation of the sulphate-reducing activity due to reoxidation of ^{35}S-sulphide. The reoxidation of biogenic ^{35}S-sulphide occurs both biologically and abiologically (Jørgensen 1988). A further complication with well-drained soils is that it may not be possible to determine the actual concentration of sulphate available to the SRB in the experimental sample.

In aerated, well-drained soils, therefore, only potential rates of dissimilatory sulphate reduction can be obtained. This can be done by saturating samples with anoxic water (with or without amendment of non-labelled sulphate) and proceeding experimentally as for the waterlogged soils. Such experiments will not yield realistic estimates of *in situ* sulphate reduction rates but give an estimate of the potential activity of SRB in the sample.

A number of different methods have been developed for estimating the sulphate reduction activity in sediments. These include:

1. $^{35}SO_4^{2-}$ radiotracer techniques for measuring the turnover rate of the interstitial sulphate pool in intact sediment cores or sediment slurries (Jørgensen 1978a, 1978c; Ingvorsen et al 1981; Hordijk et al 1985; Bak and Pfennig 1991).
2. Batch-type incubation techniques for measuring the depletion rate of the non-radioactive sulphate pool ($^{32}SO_4^{2-}$) in sediment slurries (Hordijk et al 1985; Bak et al 1991).
3. Mathematical modelling (Berner 1964; Jørgensen 1978b).

The modelling approach (method 3), an indirect method based on measured depth profiles of sulphate in the sediment, requires steady-state conditions that are unlikely to exist in soils and, for this reason, it will not be considered further.

Direct $^{35}SO_4^{2-}$ radiotracer techniques have been used for the measurement of sulphate reduction rates in a wide variety of sediments (see references above) using intact sediment cores or homogenized sediment slurries. The core injection technique has been thoroughly evaluated by Jørgensen (1978a, 1978c), Howarth and Jørgensen (1984) and Fossing and Jørgensen (1989). This technique causes minimal disturbance of the chemical and microbial gradients, and was found to give reliable estimates of bacterial sulphate reduction rates.

A procedure for measuring bacterial sulphate reduction in a sediment (or waterlogged soil) using radiolabelled sulphate ($^{35}SO_4^{2-}$) is described below. The reader is referred to the literature for various modifications of the [^{35}S] sulphate radiotracer technique (Sorokin 1962; Skyring et al 1979; Hordijk et al 1985; Skyring 1988).

Principle of the method
(core injection version)

Trace amounts of $^{35}SO_4^{2-}$ are injected into a sediment core at different depths. At the end of the incubation and after stopping the metabolic activity, the sulphate and sulphide radioactivities are separated by a distillation procedure, and quantified by liquid scintillation counting. By determining the concentration of non-radioactive sulphate ($^{32}SO_4^{2-}$) in a separate core, the sulphate reduction rate can be calculated (see calculation).

Materials and apparatus

Acrylic sampling tubes with silicone injection seals
Piston
Microlitre syringe
Metal spatulas
Analytical balance
Centrifuge tubes with sealing caps (50 ml)
Centrifuge

Disposable plastic syringes
Distillation apparatus (Fig. 6.14)
Pressure cylinder with oxygen-free N_2
Scintillation vials
Liquid scintillation spectrometer
Pressure filtration apparatus
Membrane filters (0.2 μm)
HPLC system (ion chromatograph)

Chemicals and solutions

Carrier-free $^{35}SO_4^{2-}$ solution
Zinc acetate (20% w/v)
Zinc acetate (2% w/v) in 0.1% acetic acid
Ethanol
Antifoam (polyethylene glycol)
Scintillation liquid
HCl (12 M)
$CrCl_3$ ($6H_2O$)
Mossy zinc
Acid Cr^{2+} solution is prepared according to Fossing and Jørgensen (1989).

Procedure

Sampling

Undisturbed sediment cores may be obtained by hand coring using transparent acrylic sampling tubes fitted with silicone injection seals at 1.0 cm depth intervals. Detailed descriptions of sampling procedures are given by Jørgensen (1978a). Acrylic tubes with sediment cores (stoppered at the bottom) are kept in the dark at *in situ* temperature until injection with [^{35}S]sulphate.

Addition of $^{35}SO_4^{2-}$ – Incubation

Before injecting the isotope solution, the supernatant water (if any) is carefully removed with a syringe and the position of the sediment core within the sampling tube is adjusted such as to position the first injection seal 0.5 cm below the sediment surface. To obtain a "10-segment" depth profile of sulphate reduction, 5 μl of carrier-free $^{35}SO_4^{2-}$ solution is injected horizontally at 1 cm intervals into each core segment with a 50 μl microsyringe, down to a depth of 9.5 cm. The isotope solution is deposited in a line through the core segment by slowly withdrawing the syringe while injecting. After injection the core is incubated *in situ* or in the dark at the *in situ* temperature.

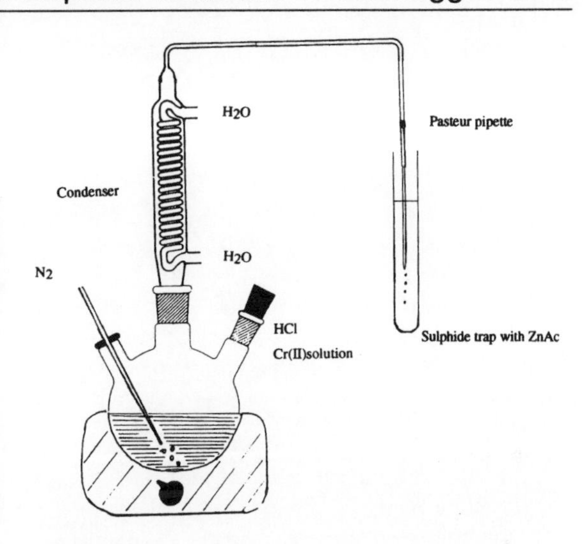

Figure 6.14. Distillation unit for separating [^{35}S]-sulphate and [^{35}S] sulphide radioactivities. Several distillation units may be connected in line for simultaneous processing of samples.

Note

After incubation the sediment core is pushed out of the acrylic tube from below using a piston and each core segment (1 cm) is sliced off and transferred to 10 ml of 20% zinc acetate (ZnAc) in a 50 ml plastic centrifuge tube and mixed. This treatment stops the metabolic activity in the sample and simultaneously fixes the sulphide as zinc sulphide (ZnS). If the fixed sediment samples are not processed within a few days, they should be stored at −25°C.

Sulphate reduction rates in different sediment environments vary greatly (Jørgensen 1983). Major factors governing sulphate reducing rates in aquatic sediments are: sulphate concentration, availability of organic substrates, numbers of bacteria, pH and temperature. The optimal incubation time and amount of isotope injected must, therefore, be determined in preliminary experiments, e.g. by making time-course experiments. The incubation time should be kept to a minimum. Ideally, the decrease of the [^{35}S]sulphate pool should not exceed a few per cent during incubation.

As a guideline, we routinely inject 5 µl of isotope solution containing 60 kBq into each core segment (diameter 30 mm, height 10 mm). Marine sediments are usually incubated for 2–6 h. Owing to their low sulphate content (and usually a high turnover rate of the sulphate pool), freshwater sediments are incubated for a much shorter time, e.g. 0.25–0.5 h, in order to avoid large changes in the *in situ* sulphate concentrations during incubation.

The use of carrier-free [^{35}S]sulphate solutions allows measurement of low sulphate reduction rates using short-term incubations. When measuring rates less than 1 nmol SO_4^{2-} cm^{-3} day^{-1}, the amount of label added to the sample must be increased (e.g. to 600 or even 6000 kBq) in order to obtain a realistic incubation time. When using high amounts of radioactive sulphate, an extra "$^{35}SO_4^{2-}$ washing procedure" should be applied prior to distillation in order to minimize physical carryover of $^{35}SO_4^{2-}$ into the ZnAc traps (Fossing and Jørgensen 1989).

Recovery of [^{35}S]sulphate and [^{35}S]sulphide radioactivities using a single-step distillation procedure

The sediment/ZnAc slurries containing both [^{35}S]sulphate and [^{35}S]sulphide are homogenized and centrifuged at 4000 rev min^{-1} for 15 min. For measuring the $^{35}SO_4^{2-}$ radioactivity, a 100 µl portion of the supernatant is transferred to a scintillation vial containing 2 ml H_2O and 10 ml Ecoscint A (National Diagnostics, GA, USA) scintillation fluid. The rest of the supernatant is discarded. The remaining sediment pellet in the centrifuge tube is homogenized and a subsample (1–2 g) is placed in a reaction flask (Fig. 6.14) containing 5 ml of ethanol.

The reaction flask is connected via a condenser to a H_2S trap containing 10 ml 2% (w/v) ZnAc in 0.1% acetic acid and a few drops of antifoam (Fig. 6.14). The distillation apparatus is closed and gassed with oxygen-free N_2. After 15 min of gassing, 16 ml of reduced chromium solution (1 M Cr^{2+} in 0.5 N HCl) and 8 ml of 12 N HCl are added by syringe to the reaction flask through the side port. The sediment slurry is then refluxed at gentle boiling for 25 min. During this period, AVS and CRS are converted to H_2S and carried to the ZnAc trap by the N_2 stream (Fossing and Jørgensen 1989). After distillation, the ZnS suspension in the ZnAc trap is resuspended using a vortex mixer, and a 2 ml subsample is removed and mixed with 10 ml of scintillation fluid (Ecoscint A) and counted for [^{35}S]sulphide radioactivity. The counting efficiency is determined by standard procedures used for liquid scintillation counting.

Determination of sulphate concentration

The sulphate concentration in core segments is determined on a separate core. Pore water is obtained by pressure filtration under N_2 or by centrifugation and 0.2 µm filtration (Jørgensen 1977; Hordijk et al, 1985; Bak et al, 1991). Chemical analysis of sulphate in pore water is performed using ion-chromatography methods (Hordijk et al 1984; Hordijk and Cappenberg 1985; Bak et al 1991) or gravimetrically by barium precipitation. The gravimetric method is not very sensitive but works well for marine sediments (Jørgensen and Cohen 1977).

For soil and freshwater samples the turbidimetric sulphate assay of Tabatabai (1974) may also be applied. For reasons that are unknown, this assay sometimes gives spurious results when applied to pore water from freshwater sediments (Ingvorsen, unpublished results).

Sulphate reduction rates are normally expressed as nanomoles of sulphate reduced per cubic centimetre of sediment per day (i.e. nmol SO_4^{2-} cm^{-3} day^{-1}). The following formula is used to calculate the porosity of the sediment and thus to convert the pore water sulphate concentration from millimolar to nanomoles per cubic centimetre:

$$\phi = \sigma \times H_2O \qquad (6.21)$$

where ϕ is the sediment porosity (ml cm^{-3}), σ is the sediment density (g cm^{-3}) and H_2O is the water content (ml g^{-1}). The water content and density of sediment segments are determined on separate cores as described by Jørgensen and Cohen (1977).

Calculation

The rate of sulphate reduction (SRR) is calculated (in nmol SO_4^{2-} cm^{-3} day^{-1}) for each segment according to the following formula:

$$SRR = \frac{a}{(A + a)t} \times [SO_4^{2-}] \times 1.06 \qquad (6.22)$$

where a is the total radioactivity of AVS + CRS (as ZnS), A is the total radioactivity of the sulphate pool after incubation, t is the incubation time (in days), $[SO_4^{2-}]$ is the sulphate concentration in sediment (nmol cm^{-3}) and 1.06 is the correction factor for microbial isotope fractionation between ^{32}S and ^{35}S.

Discussion

Slurry incubation techniques may be used to estimate sulphate reduction rates in sediments, which can not be sampled as cores. The turnover of the sulphate pool in sediment slurries may be determined with [^{35}S]sulphate (Jørgensen 1978a; Ingvorsen et al 1981; Hordijk et al 1985) or by chemical measurement of [^{32}S]sulphate depletion. The latter approach has been applied in freshwater sediments by taking advantage of newly developed and highly sensitive ion chromatography techniques for analysis of sulphate in freshwater (Hordijk et al 1985; Bak and Pfennig 1991; Bak et al 1991).

Recent investigations in freshwater and marine sediments have provided strong evidence that sulphide is oxidized to sulphate under anoxic conditions (Aller and Rude 1988; Jørgensen 1990a, 1990b; Elsgaard and Jørgensen 1992). The reaction pathways of anoxic sulphide oxidation in sediments are not known, and both abiological and biological processes have been proposed.

Relatively little is known about the sulphur cycle in soils compared to the cycle in marine sediments. It is reasonable to assume that anoxic oxidation of sulphide also occurs in waterlogged soils. Furthermore, several investigators have shown that a signficant amount of the sulphate in soils is bound in organic sulphate esters and other organic compounds (Fitzgerald 1976; Saggar et al 1981; Luther et al 1986). Some of the ^{35}S-labelled sulphur added to soils may be converted into sulphur compounds, which escape detection by the acid chromium distillation procedure described previously. It is, therefore, important to perform a number of control experiments in order to evaluate the ^{35}S-radiotracer technique in soil systems. For example, tests for recovery of [^{35}S] sulphate and [^{35}S]sulphide under different incubation conditions should be made. The possibility of [^{35}S]sulphate adsorption should be checked using zero-time and killed (or inhibited) controls.

Anaerobic heat output

B.P. Albers
A. Hartmann
F. Beese

Principle of the method

The technique of microcalorimetry (Chapter 5) is usually carried out under aerobic conditions in a closed ampoule mode (Sparling 1981, 1983; Kimura and Takahashi 1985; Heilmann and Beese 1992). This has the disadvantage of O_2 depletion and CO_2 enrichment for long-term measurements (Ljungholm et al 1979). A flow microcalorimeter enables measurements of heat output in a controlled aerobic or anaerobic environment. The simultaneous detection of heat output and CO_2 emission allows the calculations of the calorimetric/CO_2 (Cal/CO_2) ratio (Albers et al 1994) or calorimetric/respirometric (C/R) ratio (Gnaiger and Kemp 1991).

The determination of heat output in a flow microcalorimeter in combination with respirometric CO_2 measurements allows the characterization of the actual degree or the potential of anaerobic metabolism in soils.

Materials and apparatus

Climatic chamber (22°C)
Microcalorimeter (4-channel) with 20 ml flow ampoules (TAM 2277, Thermometric, Järvalla, Sweden)
Pulse-free microlitre pump or exact valve
Gas chromatograph with integrator
Gas-tight tubes

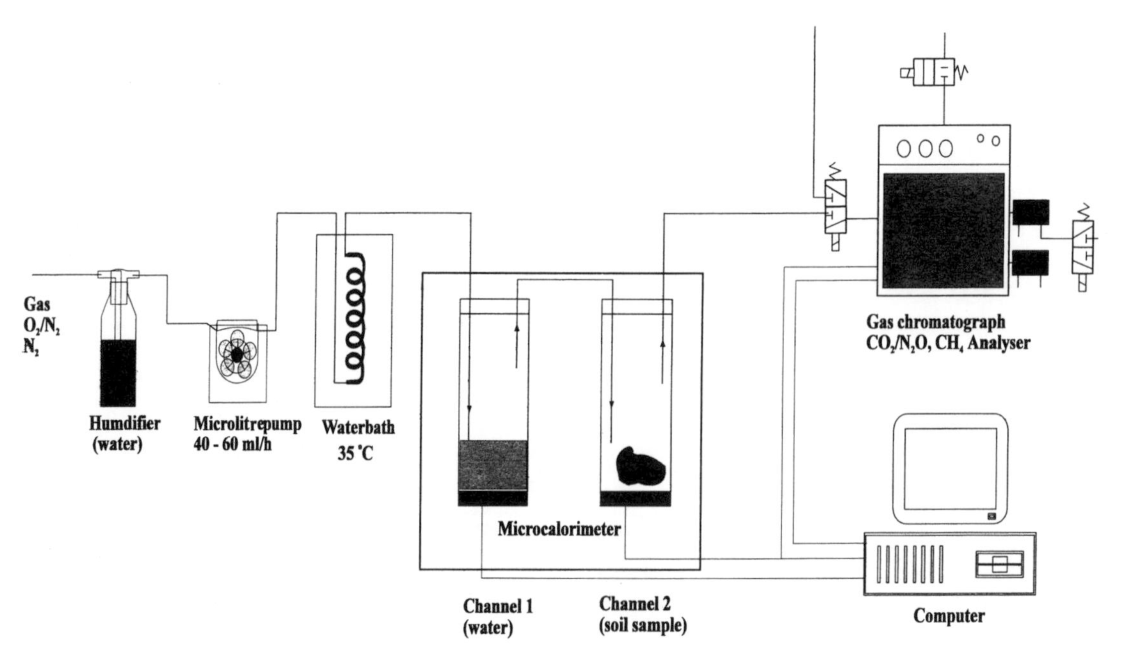

Figure 6.15. Flow calorimeter in combination with a gas chromatograph.

Water bath (35°C)
Humidifier

Chemicals and solutions

Glucose
Gas cylinder with a defined atmosphere (air, N_2, N_2/CO_2)

Procedure

Description of flow microcalorimeter system

The flow of a defined gas atmosphere (O_2/N_2 20/80, N_2 from a gas bottle) through the system described in Fig. 6.15 is regulated by a microlitre pump or an exact valve to a flow rate of 40–60 ml h^{-1}. The gas is moistened in a humidifier with distilled water and is preheated to 35°C in a water bath. The gas passes an ampoule filled with 10 ml distilled water to control the correct temperature of the gas inlet. The ampoule with the soil sample is connected with the water ampoule by a short and isolated tube. The outlet gas is analysed in a gas chromatograph.

Soil samples

Two channels of a 4-channel microcalorimeter are equipped with 20 ml perfusion vessels. One is filled with water. The heat output of this vessel should be stable and near zero during an experiment. The other contains 5–10 g (dry weight) of the soil, which is amended with glucose or other substrates. For comparison, a third channel operates in a closed ampoule system, which is filled with the same amount of a similarly treated soil sample. The remaining channel contains an empty vessel to monitor the baseline stability during an experiment. All experiments are carried out in a climatic chamber at 22°C. Both ampoules loaded with the soil samples are held for 1 h out of the measuring position for temperature equilibration. The values of the heat output can initially be used after 2 h in the measuring position. The outlet gas is automatically measured in hourly intervals in a stop flow mode in a gas chromatograph.

Calculation of the Cal/CO_2 ratio

The Cal/CO_2 ratio can be calculated from the total calorimetric heat flux and the CO_2 production of soils:

$$Cal/CO_2 = \frac{Cal[(-)\ \mu W \times g^{-1}]}{CO_2\ (nmol \times s^{-1} \times g^{-1})}$$

$$= \frac{Cal\ [(-)kJ]}{CO_2\ (mol)} \qquad (6.23)$$

where μW is the heat output in micro Watt and s = seconds. The Cal/CO_2 ratio for an exothermal reaction is negative.

Discussion

The metabolic conservation of glucose to CO_2 and H_2O results in a theoretical enthalpy change of −477 kJ/mol CO_2 (Wieser 1986). For six different sieved soils, moistened to 40–60% of their water holding capacity, a ratio of 435 (± 11.5) kJ/mol CO_2 was found (Albers et al 1994).

There is no significant difference in the heat output between the closed ampoules and the perfusion vessels under aerobic conditions during the first 15 h of an experiment.

Albers et al (1993) found a decrease in the Cal/CO_2 ratio due to anaerobic metabolism under an O_2/N_2 atmosphere depending on aggregate sizes and water potentials. A Cal/CO_2 ratio of −141 (± 10.8) kJ/mol was found for soil aggregates of different sizes amended with glucose and incubated under a N_2 atmosphere.

The quantification of the anabolic reaction is an uncertainty in the microcalorimetric measurements. The heat output of a microcalorimeter can be divided into a catabolic (exergonic) and an anabolic (endergonic) reaction. Gustafsson (1991) calculated that 1.5% of the total enthalpy change corresponds to anabolic processes (aerobic growth of yeasts with glucose), and Belaich (1980) found up to 8% of anabolism for fermentative processes. However, for anabolic reactions with a large degree of reduction between the substrate and the biomass, the influence of the anabolic reaction could be substantial (Gustafsson 1991).

Evaporation or condensation processes cause an abiotic heat change. For this reason the upper limit of the flow rate is 60 ml h^{-1}.

In the case of glucose amendment (1%) and high water content of the soil, anaerobicity is reached, depending on the soil type, after 24–48 h incubation at 22°C.

The flow microcalorimetry combined with a gas chromatograph can be used to measure the anaerobic potential activity after the addition of an exogenous substrate such as glucose. This system can also be used to investigate the effect of chemicals (fertilizers, pollutants or different substrates) on soil microbial activity under anaerobic conditions.

The flow microcalorimeter in combination with a gas chromatograph optionally allows the simultaneous measurement of the heat flux (microbial activity) in relation to the anaerobic production of other trace gases such as CH_4, N_2O, etc.

References

Abdelmagid HM, Tabatabai MA (1987) Nitrate reductase activity of soils. Soil Biol Biochem 19: 412–427.

Albers BP, Hartmann A, Beese F (1993) Kalorimetrische und respiratorische Messungen mit einem Durchflußmikrokalorimeter und GC zur Bestimmung von anaeroben Umsetzungsprozessen in Böden. Mittl Dtsch Bodenk Gesell 71: 305–308.

Albers BP, Beese F, Hartmann A (1994) Flow-microcalorimetry as a tool for aerobic and anaerobic microbial activity measurements. Biol Fertil Soils (in press).

Alef K (1991) Methodenhandbuch Bodenmikrobiologie. Ecomed, Landsberg/Lech.

Alef K, Kleiner D (1987) Applicability of arginine ammonification as an indicator of microbial activity in different soil. Biol Fertil Soils 5: 148–151.

Alef K, Kleiner D (1989) Rapid and sensitive estimation of microbial activity in soils and soil aggregates by dimethylsulfoxide reduction. Biol Fertil Soils 8: 349–355.

Aller RC, Rude PD (1988) Complete oxidation of solid phase sulfides by manganese and bacteria in anoxic marine sediments. Geochim Cosmochim Acta 52: 751–765.

Bak F, Pfennig N (1991) Microbial sulfate reduction in littoral sediment of Lake Constance. FEMS Microbiol Ecol 85: 31–42.

Bak F, Scheff G, Jansen K-H (1991) A rapid and sensitive ion chromatographic technique for the determination of sulfate and sulfate reduction rates in freshwater lake sediments. FEMS Microbiol Ecol 85: 23–30.

Belaich JP (1980) Growth and metabolism in bacteria. In: Biological Microcalorimetry. Beezer AE (ed.). Academic Press, London, New York.

Berner RA (1964) An idealized model of dissolved sulfate distribution in recent sediments. Geochim Cosmochim Acta 28: 1497–1503.

Binnerup SJ, Jensen K, Revsbech NP, Jensen, MH, Sorensen J (1992) Denitrification, dissimilative reduction of nitrate to ammonium, and nitrification in a bioturbated estuarine sediment as measured with ^{15}N and microsensor techniques. Appl Environ Microbiol. 58: 303–313.

Blackburn TH (1979) Method for measuring rates of NH_4^+ turnover in anoxic marine sediments, using a $^{15}N-NH_4^+$ dilution technique. Appl Environ Microbiol 37: 760–765.

Boast CW, Mulvaney RL, Baveye P (1988) Evaluation of Nitrogen-15 tracer techniques for direct measurement of denitrification in soil: 1. Theory. Soil Sci Soc Am J 52: 1317–1322.

Brooks PD, Stark JM, McInteer BB, Preston T (1989) Diffusion method to prepare soil extracts for automated nitrogen-15 analysis. Soil Sci Soc Am J 53: 1707–1711.

Brümmer G (1974) Redoxpotentiale und Redoxprozesse von Mangan-, Eisen- und Schwefelverbindungen in hydromorphen Böden und Sedimenten. Geoderma 12: 207–222.

Crumpton WG, Isenhart TM, Hersh CM (1987) Determination of nitrate in water using ammonia probes and reduction by titanium (111). Wastewater Anal 59: 905–908.

Dudek N, Brezinski MA, Wheeler PA (1986) Recovery of ammonium nitrogen by solvent extraction for the determination of the relative ^{15}N abundance in regeneration experiments. Mar Chem 18: 59–70.

Durbin KJ, Watanabe I (1980) Sulfate-reducing bacteria and nitrogen fixation in flooded rice soils. Soil Biol Biochem 12: 11–14.

Elsgaard L, Jørgensen BB (1992) Anoxic transformations of radiolabeled hydrogen sulfide in marine and freshwater sediments. Geochim Cosmochim Acta 56: 2425–2435.

Engler RM, Patrick WH Jr (1973) Sulfate reduction and sulfide oxidation in flooded soils as affected by chemical oxidants. Soil Sci Soc Amer Proc 37: 685–688.

Enoksson V, Samulsson MO (1987) Nitrification and dissimilatory ammonium production and their effects on nitrogen flux over the sediment–water interface in biotubated coastal sediments. Mar Biol Pro Ser 36: 181–189.

Fitzgerald JW (1976) Sulfate ester formation and hydrolysis: a potentially important yet often ignored aspect of the sulfur cycle of aerobic soils. Bacteriol Rev 40: 698–721.

Fossing H, Jorgensen BB (1989) Measurement of bacterial sulfate reduction in sediments: evaluation of a single-step chromium reduction method. Biogeochem 8: 205–222.

Fu MH, Tabatabai MA (1989) Nitrate reductase activity in soils: effects of trace elements. Soil Biol Biochem 21: 943–946.

Furusaka C (1968) Studies on the activity of sulfate reducers in paddy soil. Bull Agric Res Inst Tohoku Univ 19: 101–184.

Gnaiger E, Kemp RB (1991) Anaerobic metabolism in aerobic mammalian cells: information from the ratio of calorimetric heat flux and respirometric oxygen flux. Biochim Biophys Acta 1016: 328–332.

Goering GM, Patmatmat MM (1970) Denitrification in sediments of sea off Peru. Invest Pesq 35: 233–242.

Gotoh S, Patrick WH Jr (1974) Transformation of iron in a waterlogged soil as influenced by redox potential and pH. Soil Sci Soc Am Proc 38: 66–71.

Gustafsson L (1991) Microbiological calorimetry. Thermochim Acta 193: 145–171.

Hatton D, Pickering WF (1990) Modified procedure for determination of exchangeable ammonium ions in lake sediments. Chem Spec Bioavail 2: 139–147.

Heilmann B, Beese F (1992) Miniaturized method to measure carbon dioxide production and biomass of soil microorganisms. Soil Sci Soc J 56: 596–598.

Hordijk CA, Cappenberg TE (1985) Sulfate analysis in pore water by radio-ion chromatography employing 5-sulfoisophtalic acid as a novel eluent. J Microbiol Meth 3: 205–214.

Hordijk CA, Hagenaars CPMM, Cappenberg TE (1984) Analysis of sulfate at the mud–water interface of freshwater lake sediments using indirect photometric chromatography. J Microbiol Meth 2: 49–56.

Hordijk C, Hagenaars CPMM, Capenberg TE (1985) Kinetic studies of bacterial sulfate reduction in freshwater sediments by high-pressure liquid chromatography and microdistillation. Appl Environ Microbiol 49: 434–440.

Hordijk CA, Snieder M, van Engelen JJM, Cappenberg TE (1987) Estimation of bacterial nitrate reduction rates at in situ concentrations in freshwater sediments. Appl Environ Microbiol 53: 217–223.

Howarth RW (1984) The ecological significance of sulphur in the energy dynamics of salt marsh and coastal marine sediments. Biogeochem 1: 5–27.

Howarth RW, Giblin A (1983) Sulfate reduction in the salt marshes at Sapelo Island, Georgia. Limnol Oceanogr 28: 70–82.

Howarth RW, Jørgensen BB (1984) Formation of ^{35}S-labelled elemental sulfur and pyrite in coastal marine sediments (Limfjorden and Kysing Fjord, Denmark) during short-term $^{35}SO_4^{2-}$ reduction measurements. Geochim Cosmochim Acta 48: 1807–1818.

Hund K, Fabig W, Zelles L (1990) Vergleichende Untersuchung von Methoden zur Bestimmung der mikrobiellen Aktivität im Boden. Agrobiol Res 43: 131–138.

Hungate RE (1969) A roll tube method for cultivation of strict anaerobes. In: Methods in Microbiology, 3B. Norris JR, Ribbons DW (eds). Academic Press, New York pp. 117–132.

Ingvorsen K, Brock TD (1982) Electron flow via sulfate reduction and methanogenesis in the anaerobic hypolimnion of Lake Mendota. Limnol Oceanogr 27: 559–564.

Ingvorsen K, Zeikus JG, Brock TD (1981) Dynamics of bacterial sulfate reduction in a eutrophic lake. Appl Environ Microbiol 42: 1029–1036.

Inubushi K, Brookes PC, Jenkinson DC (1989) Adenosin-5-triphosphate and adenylate energy charge in waterlogged soil. Soil Biol Biochem 21: 733–739.

Jørgensen BB (1977) The sulfur cycle of a coastal marine sediment (Limfjorden, Denmark). Limnol Oceanogr 22: 814–832.

Jørgensen BB (1978a) A comparison of methods for the quantification of bacterial sulfate reduction in coastal marine sediments. I. Measurement with radiotracer techniques. Geomicrobiol J 1: 11–27.

Jørgensen BB (1978b) A comparison of methods for the quantification of bacterial sulphate reduction in coastal marine sediments. II. Calculation from mathematical models. Geomicrobiol J 1: 29–47.

Jørgensen BB (1978c) A comparison of methods for the quantification of bacterial sulfate reduction in coastal marine sediments. III. Estimation from chemical and bacteriological field data. Geomicrobiol J 1: 49–64.

Jørgensen BB (1983) The microbial sulfur cycle. In: Microbial Geochemistry. Krumbein WE (ed.). Blackwell Scientific Publications, Oxford, pp. 91–118.

Jørgensen BB (1988) Ecology of the sulphur cycle: oxidative pathways in the sediment. In: The Nitrogen and Sulphur Cycles. Cole JA, Ferguson S (eds). Cambridge University Press, Cambridge, pp. 31–63.

Jørgensen BB (1990a) The sulfur cycle of freshwater sediments: role of thiosulfate. Limnol Oceanogr 35: 1329–1342.

Jørgensen BB (1990b) A thiosulfate shunt in the sulfur cycle of marine sediments. Science 249: 152–154.

Jørgensen BB, Cohen Y (1977) Solar Lake (Sinai). 5. The sulfur cycle of the benthic cyanobacterial mats. Limnol Oceanogr 22: 657–666.

Jørgensen BB, Fenchel T (1974) The sulfur cycle of a marine sediment model system. Marine Biol 24: 189–201.

Kandeler E, Gerber H (1988) Short-term assay of soil urease activity using colorimetric determination of ammonium. Biol Fertil Soil 6: 68–72.

Keeney DR (1982) Nitrogen-availability indices. In: Methods of Soil Analysis, Part 2. Page AL, Miller RH, Keeney DR (eds). American Society of Agronomy, Inc., Madison WI, p. 711–773.

Kimura T, Takahashi K (1985) Calorimetric studies of soil microbes: quantitative relation between heat evolution during microbial degradation of glucose and changes in microbial activity in soil. J Gen Microbiol 131: 3083–3089.

Laima MC (1992) Extraction and seasonal variation of NH_4^+ pools in different types of coastal marine sediments. Mar Ecol Prog Ser 82: 75–84.

Ljungholm K, Noren B, Sköld R, Wadsö I (1979) Use of microcalorimetry for the characterization of microbial activity in soil. Oikos 33: 15–23.

Luther GW III, Church TM, Scudlark JR, Cosman M (1986) Inorganic and organic sulfur cycling in salt-marsh pore waters. Science 232: 746–749.

MacKown CT, Brooks PD, Smith MS (1987) Diffusion of nitrogen-15 Kjeldahl digest for isotope analysis. Soil Sci Soc Am J 51: 89–90.

Marschner H (1986) Mineral Nutrition of Higher Plants. Academic Press, London.

Munch JC, Ottow JCG (1977) Modelluntersuchungen zum Mechanismus der bakteriellen Eisenreduktion in hydromorphen Böden. Z Pflanzenernähr Bodenkde 140: 549–562.

Neilands JB (1981) Microbial iron compounds. Ann Rev Biochem 50: 715–731.

Nielsen LP (1992) Denitrification measured at in situ conditions in stream sediment. FEMS Microbiol Ecol 86: 357–362.

Nishio T, Koike I, Hattori A (1983) Estimates of denitrification and nitrification in coastal and estuarine sediments. Appl Environ Microbiol 45: 444–450.

O'Dean WA, Porter LK (1980) Devarda's alloy reduction of nitrate and tube diffusion of the reduced nitrogen for indophenol ammonium and nitrogen-15 determinations. Anal Chem 52: 1164–1166.

Ottow JCG (1969) Der Einfluβ von Nitrat, Chlorat, Sulfat, Eisenoxidform und Wachstumsbedingungen auf das Ausmaβ der bakteriellen Eisenreduktion. Z Pflanzenernähr Bodenkde 124: 238–253.

Paaske E, Kristensen S (1982) Ammonium regeneration by microzooplancton in the Oslofjord. Mar Biol 69: 55–63.

Pal SS, Sudhakar-Barik N, Sethunathan N (1979) Effects of benomyl on iron and manganese reduction and redox potential in flooded soil. J Soil Sci 30: 155–159.

Peiffer S, Klemm O, Pecher K, Hollerung R (1992) Redox measurements in aqueous solutions-theoretical approach to data interpretation, based on electrode kinetics. J Contamin Hydrol 10: 1–18.

Ponnamperuma FN (1972) The chemistry of submerged soils. Adv Agron 24: 29–96

Postgate JR (1984) The Sulphate-reducing Bacteria, 2nd edn. Cambridge University Press, Cambridge.

Premi PR (1971) Effect of addition of compounds of copper, manganese, zinc, chromium, and calcium on extractable iron from soil incubated under aerobic and anaerobic conditions. Agrochim Pisa 15: 202–211.

Prosser SJ, Gojon A, Barrie A (1993) Fast, automated analysis of nitrogen-15 nitrate from plant and soil extracts. Soil Sci Soc Am J 57: 410–414.

Reddy KS, Gambrell RP (1985) The rate of soil reduction as affected by levels of methyl parathion and 2, 4-D. J Environ Sci Health 20: 275–298.

Risgaard-Petersen N, Rysgaard S, Revsbech NP (1993) A sensitive assay for determination of $^{14}N/^{15}N$ isotope distribution in NO_3^-. J Microbiol Meth 17: 155–164.

Risgaard-Petersen N, Rysgaard S, Nielsen LP, Revsbech NP (1994) Diurnal variation in denitrification of nitrate originating from the water as well as from nitrification, in sediments colonized by benthic microalgae. Limnol Oceanogr (in press).

Ryden JC, Lund LJ, Focht DD (1979a) Direct measurement of dentrification loss from soils: I. Laboratory evaluation of acetylene inhibition of nitrous oxide reduction. Soil Sci Soc Am J 43: 104–110.

Ryden JC, Lund LJ, Focht DD (1979b) Direct measurement of denitrification loss from soils: II. Development and application of field methods. Soil Sci Soc Am J 43: 110–118.

Rysgaard S, Risgaard-Petersen N, Nielsen LP, Revsbech NP (1993a) Nitrification and denitrification in lake and estuarine sediments measured by the ^{15}N dilution technique and isotope pairing. Appl Environ Microbiol 59: 2093–2098.

Rysgaard S, Risgaard-Petersen N, Sloth NP, Jensen K, Nielsen LP (1993b) Oxygen regulation of nitrification and denitrification in freshwater sediments (submitted).

Saggar S, Bettany JR, Stewart JWB (1981) Sulfur transformations in relation to carbon and nitrogen in incubated soils. Soil Biol Biochem 13: 499–511.

Schinner F, Öhlinger R, Kandeler E (1991) Bodenbiologische Arbeitsmethoden. Spreniger Verlag, Berlin, New York.

Selmer JS, Sörensson F (1986) New procedure for extraction of ammonium from natural waters for ^{15}N isotopic ratio determinations. Appl Environ Microbiol 52: 577–579.

Skinner FA (1975) Anaerobic bacteria and their activities in soil. In: Soil Microbiology: A Critical Review. Walker N (ed.). Halsted Press, New York, pp. 1–19.

Skyring GW (1987) Sulfate reduction in coastal ecosystems. Geomicrobiol J 5: 295–374.

Skyring GW (1988) Acetate as the main energy substrate for the sulfate-reducing bacteria in Lake Eliza (South Australia) hypersaline sediments. FEMS Microbiol Ecol 53: 87–94.

Skyring GW, Chambers LA (1976) Biological sulphate reduction in carbonate sediments of a coral reef. Aust J Mar Freshwater Res 27: 595–602.

Skyring GW, Oshrain RL, Wiebe WJ (1979) Sulfate reduction rates in Georgia marshland soils. Geomicrobiol J 1: 389–400.

Sorensen J (1982) Reduction of ferric iron in anaerobic, marine sediment and interaction with reduction of nitrate and sulfate. Appl Environ Microbiol 46: 319–324.

Sorokin YI (1962) Experimental investigation of bacterial sulfate reduction in the Black Sea using ^{35}S. Microbiology (Engl. Transl.) 31: 329–335.

Sparling GP (1981) Heat output of the soil biomass. Soil Biol Biochem 13: 373–376.

Sparling GP (1983) Estimation of microbial biomass and activity in soil using microcalorimetry. J Soil Sci 34: 381–390.

Stolzy LH, Focht DD, Flühler H (1981) Indicators of soil aeration status. Flora 171: 236–265.

Tabatabai MA (1974) Determination of sulfate in water samples. Sulphur Inst J 10: 11–13.

Tabatabai MA, Bremner JM (1972) Assay of urease activity in soil. Soil Biol Biochem 4: 479–478.

Tiedje JM, Sexstone AJ, Parkin TB, Revsbech NP (1984) Anaerobic processes in Soil. Plant Soil 76: 197–212.

Wakao N, Furusaka C (1973) Distribution of sulphate-reducing bacteria in paddy-field soil. Soil Sci Plant Nutr 19: 47–52.

Wakao N, Furusaka C (1976) Presence of microaggregates containing sulfate-reducing bacteria in paddy-field soil. Soil Biol Biochem 8: 157–159.

Waring SA, Bremner JM (1964) Ammonium production in soil under waterlogged conditions as an index of nitrogen availability. Nature 201: 951–952.

Weiss RF (1970) The solubility of nitrogen, oxygen and argon in water and seawater. Deep Sea Res 17: 721–735.

Welp G (1987) Einfluß des Stoffbestandes von Böden auf die mikrobielle Toxizität von Umweltchemikalien. Ph.D. thesis, Kiel.

Welp G, Brümmer G (1985) Der Fe(III)-Reduktionstest-ein einfaches Verfahren zur Abschätzung der Wirkung von Umweltchemikalien auf die mikrobielle Aktivität in Böden. Z Pflanzenernaehr Bodenk 148: 10–23.

Welp G, Brümmer GW (1992) Toxicity of organic pollutants to soil microorganisms. In: Effects of organic contaminants in sewage sludge on soil fertility, plants and animals. Hall JEM, Sauerbeck DR, L'Hermite P (eds). Proc. of a seminar. Office for Official Publ. of the European Commun., Luxembourg, 161–168, ISBN 92-826-3878-2.

Welp G, Brümmer GW, Rave G (1991) Dosis-Wirkungs-Beziehungen zur Erfassung von Chemikalienwirkungen auf die mikrobielle Aktivität von Böden: I. Kurvenverläufe und Auswertungsmöglichkeiten. Z Pflanzenernähr Bodenkde 154: 159–168.

Wieser W (1968) Bioenergetik. Thieme, Stuttgart, New York.

Wingfield GJ, Davies HA, Greaves MP (1977) The effect of soil treatment on the response of the soil microflora to the herbicide dalapon. J Appl Bacteriol 43: 39–46.

Zelles L, Scheunert I, Korte F (1986) Comparison of methods to test chemicals for side effects on soil microorganisms. Ecotoxicol Environ Safety 12: 53–69.

7 Enzyme activities

Enzymes are among the most remarkable biomolecules because they show extraordinary specificity in catalysing biological reactions. They have usually been named by adding the suffix "-ase" to the name of the substrate, the molecule on which the enzyme exerts its catalytic action. For example, urease catalyses the hydrolysis of urea to ammonia and CO_2, phosphatase catalyses the hydrolysis of phosphate esters to orthophosphate and the corresponding ester. However, this nomenclature has not always been practical and therefore a systematic classification of enzymes has been adopted on the recommendation of an International Enzyme Commission. The new system divides enzymes into six major classes, which are subdivided further into subclasses, according to the type of reaction catalysed. A recommended name, a systematic name and a classification number is given to each enzyme. For example, the systematic name for phosphodiesterase is phosphoric diester hydrolase, while its classification number is EC 3.1.4.

Research into soil enzymes has increased steadily over the last 30 years; new theoretical approaches and methods have been introduced, and a wealth of information on various enzyme reactions in soil has been collected.

Various activities associated with biotic and abiotic components contribute to the overall activity of soil enzymes. According to Burns (1982), an enzyme may be associated physically with proliferating animal, microbial and plant cells, and it may be located in the cytoplasm, in the periplasm of gram-negative bacteria or attached to the outer surface of cells. The enzyme may be present in non-proliferating cells (microbial spores or protozoan cysts), in entirely dead cells or in cell debris. The enzyme may also be present as an extracellular soluble molecule, temporarily associated in enzyme–substrate complexes, adsorbed to clay minerals or associated with humic colloids. Some of these categories may represent various stages in the life of an enzyme; an intracellular enzyme in a viable cell may still function after the cell dies and thus it becomes associated with cell debris; it may be released in the aqueous phase as cell membranes are broken and eventually be adsorbed by soil colloids that are still active. Usually enzyme–clay and enzyme–organic polymer complexes show a remarkable resistance to proteolytic and thermal denaturation (Sarkar et al 1980, 1989; Burns 1982, 1986; Trasar-Cepeda and Gil-Sotres 1987, 1988; Nannipieri et al 1988).

Using current methods, it is impossible to decide which combination of activities in the categories above has been determined experimentally (Burns 1982; Nannipieri et al 1990; Nannipieri 1994). Both chemical and physical bacteriostatic agents have been used to inhibit enzyme production and assimiliation of reaction products by growing populations of microorganisms. Toluene, one of the most frequently used bacteriostatic agents, can serve as a carbon source for fungi and bacteria; in addition, it may have a plasmolytic action and cause the release of intracellular enzymes. The effect of toluene depends on soil type, assay conditions and the enzyme assayed (Ladd 1978). Sterilization of soils by high-energy electron beams or gamma irradiation has frequently been used to inactivate soil microorganisms in order to determine the level of enzyme activity that is not associated with living and active microbial cells (Beck and Poschenrider 1963; Powlson and Jenkinson 1976; Skujins 1978). However, this approach suffers from some of the objections regarding

Methods in Applied Soil Microbiology and Biochemistry
ISBN 0–12–513840–7

the use of toluene (Ladd 1978; Nannipieri 1994).

Measurements of soil enzyme activities have to be interpreted with caution. These measurements represent the maximum potential rather than the actual enzyme activity because the incubation conditions of enzyme assays are chosen to ensure optimum rates of catalysis. The concentration of substrate is in excess and optimal value of pH and temperature are selected so as to permit the highest rate of enzyme activity, and the volume of the reaction mixture is such that it allows free diffusion of substrate.

Problems arising from the interpretation of measured soil enzyme activity have often led to the conclusion that soil enzyme assays have no meaning in ecological and agricultural terms (Nannipieri 1994). It is noteworthy that other experimental methods in soil microbiology (for example, the measurement of actual nitrification *in situ*) are based on the use of unrealistic conditions. In addition, enzyme measurements answer qualitative questions about specific metabolic processes and, in combination with other measurements (ATP, AEC, CO_2 evolution, etc.), may increase our understanding of the effect of agrochemicals, cultivation practices, and environmental and climatic factors on the microbiological activity of soil (Skujins 1978; Nannipieri 1994).

Current methods for the determination of enzyme activities of soil are presented in this chapter.

Protease activity

K. Alef
P. Nannipieri

(Ladd and Butler 1972)

Proteins are the most abundant organic molecules in cells, constituting 50% or more of their dry weight (Bremner 1967; Alexander 1977; Burns 1978; Warman and Isnor 1989). Protease activities, detected in microorganisms, plants and animals, catalyse the hydrolysis of proteins to polypeptides, and oligopeptides to amino acids. Because of the high molecular weight of proteins, the first enzymatic step in protein degradation occurs outside microbial cells. Compounds with low molecular weight such as amino acids can then be transported into the cells by specific transport systems and deaminated there (Burns 1978; Ladd 1978; Ladd and Jackson 1982; Warman and Bishop 1987; Lähdesmäki and Piispanen 1989).

Nearly all microorganisms in soils are capable of protein degradation, which is generally linked to ammonium release (Alexander 1977; Alef and Kleiner 1986; Morra and Freeborn 1989). In soil, proteases are present in living and active cells, in dead cells, as free enzymes, and adsorbed to organic, inorganic or organomineral particles (Sarkar et al 1980, 1989; Loll and Bollag 1983). The protease activity of soil has a pH optimum of eight and a temperature optimum of about 55°C. At temperatures above 60°C, the enzyme is denaturated (Hoffmann and Teicher 1957; Ladd 1972; Ladd and Butler 1972; Mayaudon et al 1975).

Air drying or storage of different soils at −8, 4 or 22°C significantly decreases the protease activity (Ladd 1972; Speir and Ross 1975; Speir et al 1981; Beck 1986). In the field, protease activity varied with the season; it is not correlated with changes in microbial populations (Ladd 1978). Nevertheless, significant correlations are mostly found between the protease activity and several microbial parameters estimated under laboratory conditions like the arginine ammonification, the substrate-induced respiration, heat output, nitrogen mineralization and ATP (Beck 1984a, 1984b; Holz 1986a, 1986b; Schulz-Berendt 1986; Alef et al 1988; Suttner and Alef 1988). It decreased with the depth and increased after soil treatment with organic compounds such as straw (Holz 1986a; Kandeler 1986; Loll and Bollag 1983). Protease can be extracted from soils and the activity estimated in crude extracts (Burns 1978; Ladd 1978; Nannipieri et al 1980, 1982; Loll and Bollag 1983). Casein-hydrolysing enzymes, as any other extracellular enzymes acting on substrates with high molecular weight, are supposed to be short lived in soil (Ladd 1978; Nannipieri et al 1980); indeed any protection of the enzyme against proteolytic degradation by soil colloids may render it inaccessible to the high molecular weight substrate.

Several assays differing in the type of substrate and procedures used for determining products and incubation conditions are available to estimate protease activity in soils. Casein, azocasein, gelatin, peptides and albumins have been used as substrates, while both short and long incubation times (between 2 and 16 h) have been employed (Hoffman and Teicher 1957; Ladd and Butler 1972; Beck 1973; Ross et al 1975).

The method of Ladd and Butler (1972) is presented and discussed.

Principle of the method

The determination of amino acids released after incubation of soil with sodium caseinate for 2 h at 50°C using Folin-Ciocalteu reagent.

Materials and apparatus

Spectrophotometer
Shaking water bath (adjustable to 50°C)
Centrifuge and centrifuge tubes (25 ml)
Folded filter paper
Glass tubes with screw caps

Chemicals and solutions

Tris buffer (50 mM, pH 8.1)

Dissolve 6.05 g Tris (hydroxy methyl) amino methane in 700 ml distilled water, adjust the pH to 8.1 with HCl and dilute to 1000 ml with distilled water.

Sodium caseinate (2%)

Suspend 10 g sodium caseinate in warm distilled water (50°C) and bring up with distilled water to 500 ml (use a stirrer). If only casein is available, dissolve 10 g in Tris buffer, adjust the pH to 8.1 with NaOH solution and bring up to 500 ml with Tris buffer (50 mM, pH 8.1)

Trichloroacetic acid (15%)

Dissolve 75 g trichloroacetic acid (TCA) in about 300 ml distilled water and dilute to 500 ml with distilled water.

Alkaline reagent

Dilute 60 ml of NaOH (1 M) with distilled water before dissolving 50 g Na_2CO_3 (water free) in the solution and bring up to 1000 ml with distilled water.

Dissolve 0.5 g $CuSO_4.5H_2O$ in distilled water and dilute to 100 ml with distilled water.

Dissolve 1 g potassium sodium tartrate ($C_4H_4KNaO_6.4H_2O$) in distilled water and dilute to 100 ml with distilled water.

Mix 1000 ml of NaOH/Na_2CO_3 solution with 20 ml of $CuSO_4$ solution and 20 ml of potassium sodium tartrate solution.

Folin-Ciocalteu reagent (33%)

Dilute 167 ml of Folin reagent to 500 ml with distilled water.

Tyrosine standard solution (500 μg ml^{-1})

Dissolve 50 mg tyrosine in Tris buffer and dilute to 100 ml with Tris buffer.

Procedure

Place 1 g of moist, sieved soil (2 mm) in a centrifuge tube, add 5 ml Tris buffer and 5 ml sodium caseinate solution. Stopper the tubes, mix the contents and incubate for 2 h at 50°C on a shaking water bath. At the end of incubation, add 5 ml of TCA solution and mix the contents thoroughly. To perform the controls, add 5 ml of Na caseinate solution at the end of the incubation and immediately before adding the TCA solution. Centrifuge the resulting soil suspensions (10,000–12,000 rev min^{-1}, 10 min). Pipette 5 ml of the clear supernatant into tubes, mix with 7.5 ml of the alkaline reagent, and incubate for 15 min at room temperature. After adding 5 ml of the Folin reagent, filter the mixtures through paper filter into glass tubes and measure the absorbance after exactly 1 h at 700 nm (measure the absorbance several times until the measured value becomes constant).

Calibration curve

Pipette 0, 1, 2, 3, 4 and 5 ml of the tyrosine solution into glass tubes, add 5 ml Na caseinate and bring up to 10 ml with Tris buffer. Then add 5 ml TCA solution and perform the measurements as described above.

Calculation

Correct the measured absorbance for the controls and calculate as follows:

Protease activity (μg tyrosine g^{-1} dwt 2 h^{-1}) =

$$\frac{C \times 15}{dwt} \qquad (7.1)$$

where *dwt* is the dry weight of 1 g of moist soil, 15 is the final volume of solutions added to the soil in the assay and *C* is the measured tyrosine concentration (μg ml^{-1} supernatant or filtrate).

Discussion

The method is based on the determination of TCA-soluble tyrosine derivatives by the Folin reagent. The diluted Folin-Ciocalteu reagent is unstable and should be prepared immediately before use. The diluted Folin-Ciocalteu reagent should be stored at 4°C.

The pH and temperature conditions are those reported by Ladd and Butler (1972). It is recommended that the activity measurements should be carried out at different pH and temperature values in order to find the optimal conditions for the soil to be analysed. Different soils may present diverse results from those tested by Ladd and Butler (1972).

In the control, casein should be added immediately before TCA at the end of incubation time because the precipitation is not complete if the substrate is added after the acidic solution.

Calibration curves prepared from tyrosine standards may differ from day to day and therefore it is recommended that tyrosine standard should be included in each analysis.

Supernatants can be stored at 4°C for 5 h but not longer before being analysed for their tyrosine content.

Chloroform fumigation did not influence the casein-hydrolysing activity of different soils (Ladd 1978).

An automated technique to determine the protease activity of soil was developed by Holz (1980, 1986a, 1986b, 1986c).

Urease activity

K. Alef
P. Nannipieri

The enzyme urease catalyses the hydrolysis of urea to CO_2 and NH_3 with a reaction mechanism based on the formation of carbamate as an intermediate (Tabatabai 1982).

$$H_2NCONH_2 + H_2O \rightarrow 2NH_3 + CO_2 \quad (7.2)$$

This enzyme is very widely distributed in nature, being present in microbial, plant and animal cells. It also catalyses the hydrolysis of hydroxyurea, dihydroxyurea and semicarbazid; it contains nickel and its molecular weight may range from 151,000 to 480,000 Da (Bremner and Mulvaney 1978; Blakeley and Zerner 1984).

A variety of methods has been used to assay the urease activity of soil (Bremner and Mulvaney 1978; Gosewinkel and Broadbent 1984; Simpson et al 1984, 1985; Kandeler and Gerber 1988; McCarty et al 1989). Most of these methods involve determination of ammonia liberated on the incubation of toluene-treated soil with buffered urea solution. Other methods involve the estimation of the rate of urea hydrolysis in soils by determining the residual urea or the $^{14}CO_2$ liberated after incubation (Skujins and McLaren 1969; Douglas and Bremner 1970, 1971; Bremner and Mulvaney 1978; Mulvaney and Bremner 1979). Other commonly adopted methods have not involved the use of buffer to control pH or addition of toluene to inhibit microbial proliferation (Hoffman and Schmidt 1953; Galstyan 1965; Tabatabai and Bremner 1972; Zantua and Bremner 1975a, 1975b; Frankenberger and Johanson 1986; Kandeler and Gerber 1988).

There are controversial reports regarding the pH optimum of urease activity in soil: both neutral, pH 6–7 (Hoffman and Schmidt 1953), and alkaline, pH 8.8–10 (Tabatabai and Bremner 1972; May and Douglas 1976; Kandeler and Gerber 1988; Perez-Mateos and Gonzalez-Carcedo 1988) values have been observed.

Urease in soils is tightly bound to soil organic matter and soil minerals and K_m values have been found to range between 1.3 to 213 mM (Gosewinkel and Broadbent 1984; Bonmati et al, 1985; Kandeler and Gerber, 1988; Gianfreda et al 1992; Lai and Tabatabai 1992).

A temperature optimum as high as 60°C has been observed and the urease is usually denaturated at 70°C. The incubation temperature of assays ranged from 15 to 42°C. However, soils are generally incubated at 30°C (Sumner 1951; Zantua and Bremner 1977; Bremner and Mulvaney 1978; Kissel and Cabrera 1988; Moyo et al 1989).

Ureases extracted from soil have been found to be resistant to thermal and proteolytic denaturation (Burns et al 1972; Nannipieri et al 1974). Enzyme–organomineral complexes with high molecular weights have been found to be more resistant than complexes with lower molecular weights (Nannipieri et al 1978a). According to Burns et al (1972) urease–organic complexes of high molecular weight are likely to possess molecular arrangements that permit the movement of substrates and products toward the enzyme, but not that of large molecules, such as proteases.

The urease activity in soils is very stable and rarely influenced by air drying, irradiation or storage at temperatures between −60 and 22°C (McLaren 1969; Pancholy and Rice 1973; Zantua and Bremner 1975b; Kandeler and Gerber 1988; Fenn et al 1992).

Urease activity was not significantly correlated with microbial biomass and was affected differently by heavy metals, oxygen concentrations and nitrogen availability in different types of soils (Tabatabai 1977; Nor 1982; Doelman and Haanstra 1986; Cochran et al 1989;

McCarty and Bremner 1991; McCarty et al 1992).

Urease inhibitors have been used frequently to prevent a rapid hydrolysis of urea in agricultural soils (Fillery et al 1984, 1986; Simpson et al 1984, 1985; Haare and White 1985; Cai et al 1989; McCarty et al 1989; Zhengping et al 1991; Xiaoyan et al 1992).

The methods involving the estimation of ammonia released have been evaluated thoroughly and are therefore presented here.

Estimation of urease activity

(Tabatabai and Bremner 1972)

Principle of the method

The method is based on determination of ammonia released after the incubation of soil samples with urea solution for 2 h at 37°C.

Material and apparatus

Steam distillation apparatus
Incubator adjustable to 37°C
pH meter
Volumetric flasks (50, 100, 1000, 2000 ml)
Automated titration
Erlenmeyer flask (100 ml)

Chemicals and solutions

Toluene

Tris (hydroxy methyl) amino methane buffer (50 mM, pH 9)

Dissolve 6.1 g Tris in 700 ml distilled water, adjust the pH to 9.0 with H_2SO_4 (0.2 M) and bring up to 1000 ml with distilled water.

Urea solution (200 mM)

Dissolve 1.2 g of urea in about 80 ml of Tris buffer and bring up to 100 ml with the same buffer. Prepare a fresh solution daily and store at 4°C.

Potassium chloride (2.5 M)–silver sulphate (100 mg l^{-1}) solution

Dissolve 100 mg of reagent grade Ag_2SO_4 in 700 ml of distilled water, dissolve 188 g of reagent grade KCl in the solution and dilute to 1000 ml with distilled water.

Reagents for determination of NH_4-N

H_2SO_4 (0.005 M)

Indicator solution

Dissolve 0.66 g bromocresol green and 0.33 g of methyl red in ethanol (95%) and bring up to 1000 ml with ethanol.

Boric acid indicator solution

Pipette 40 ml of indicator solution and 400 ml of ethanol (95%) into a volumetric flask (2 l). Mix 40 g of boric acid with 1400 ml warm distilled water. After cooling, place the solution into the volumetric flask (2 l). Pipette NaOH (0.05 M) in the volumetric flask until the colour of 1 ml of the solution changes from pink to light green by the addition of 1 ml distilled water. Finally bring up the volume to 2000 ml with distilled water.

MgO

Treat a heat stable MgO at 600–700°C for 2 h. Cool and store over NaOH or silica gel (in a desiccator).

Ammonia standard solution

Dissolve 0.234 g ammonium sulphate in distilled water and dilute to 1000 ml with distilled water (50 µg NH_4-N ml^{-1}).

Procedure

Place 5 g of moist soil in a volumetric flask (50 ml), add 0.2 ml of toluene and 9 ml Tris buffer, mix the contents, add 1 ml of urea solution and mix again for a few seconds. Then stopper the flasks and incubate for 2 h at 37°C. After the incubation, add approximately 35 ml of KCl–Ag_2SO_4 solution, swirl the flask for a few seconds, and allow the flask to stand until the contents have cooled to room temperature (about 5 min). Bring up the contents to 50 ml by addition of KCl–Ag_2SO_4 solution and mix the contents thoroughly. To perform controls, follow the procedure described for assay of urease activity, but make the addition of 1 ml of 0.2 M urea solution after the addition of 35 ml of KCl–Ag_2SO_4 solution. It is recommended that at least three replications are carried out. The reaction shows a linear time course up to 5 h.

Estimation of released ammonia

Pipette 5 ml of boric acid indicator solution into an Erlenmeyer flask and put it in its special place (see Chapter 3, Bremner and Edwards 1965). Pipette 20 ml of the resulting soil suspension into a distillation flask (100 ml), add 0.2 g MgO and steam distillate the contents until 30 ml of distillate are collected in the Erlenmeyer flask. Titrate the distillate with 0.005 M H_2SO_4 (Bremner and Keeney 1966); 1 ml of H_2SO_4 (0.005 M) is equivalent to 70 µg NH_4-N.

Calculation

Correct the results for the control, obtain the µg NH_4-N ml^{-1} soil suspension, and calculate briefly:

Urease activity (µg NH_4-N g^{-1} dwt $2h^{-1}$) =

$$\frac{C \times 50}{dwt \times 5} \tag{7.3}$$

where C is the measured NH_4-N concentration (µg NH_4-N ml^{-1} soil suspension), dwt is the dry weight of 1 g moist soil, 5 is the weight of used soil in the test and 50 is the total volume of the soil suspension.

Estimation of urease activity

(Kandeler and Gerber 1988)

Principle of the method

Colorimetric determination of released ammonia after the incubation of soil samples with urea solution for 2 h at 37°C.

Materials and apparatus

Incubator adjustable to 37°C
Shaker
Filter paper
Spectrophotometer
Volumetric flasks (100, 500, 1000, 2000 ml)
Erlenmeyer flasks (50, 100 ml)

Chemicals and solutions

Urea solution

Dissolve 2.4 g urea in 400 ml distilled water and make up with distilled water to 500 ml (prepare daily).

KCl solution

Dissolve 74.6 g KCl in distilled water, add 10 ml of 1 M HCl (32% HCl is equal to 10 M) and bring up with distilled water to 1000 ml.

NaOH (0.3 M)

Dissolve 12 g NaOH in distilled water and bring up with distilled water to 1000 ml.

Sodium salicylate solution

Dissolve 17 g Na-salicylate and 120 mg sodium nitroprusside in distilled water and bring up with distilled water to 100 ml.

Na-salicylate/NaOH solution

Mix equal volumes of the NaOH and sodium salicylate solutions, and distilled water (to be prepared daily).

Sodium dichloroisocyanide solution (0.1%)

Dissolve 0.1 g sodium dichloroisocyanide in 100 ml distilled water (prepare the solution shortly before using).

Borate buffer (pH 10)

Dissolve 56.85 g disodium tetraborate, or 30 g disodium tetraborate (water free) in 1500 ml warm distilled water. After cooling, adjust to pH 10 with NaOH solution (20%) and bring up to 2000 ml with distilled water.

Ammonium standard solution

Solution I

Dissolve 3.82 g ammonium chloride in distilled water and bring up with distilled water to 1000 ml (1000 μg NH_4-N ml^{-1}). The solution is stable for several weeks at 4°C.

Solution II

Pipette 0.0, 1.0, 1.5, 2.0, 2.5 ml of solution I into volumetric flasks (100 ml) and bring up to 100 ml with KCl solution.

Procedure

Procedure for the non-buffered method

Place 5 g of moist soil in an Erlenmeyer flask (100 ml) and add 2.5 ml urea solution. Then stopper the flasks and incubate for 2 h at 37°C. After the incubation, add 50 ml of KCl solution and shake the flask for 30 min. After filtering the resulting suspension the filtrates are analysed for the ammonium content. Perform the blanks as described above but with 2.5 ml distilled water, and add the urea solution at the end of the incubation and immediately before KCl addition. It is recommended that at least three replications are carried out.

Ammonium determination

Pipette 1 ml of the clear filtrate into an Erlenmeyer flask (50 ml), then add 9 ml of distilled water, 5 ml of the Na salicylate/NaOH solution and 2 ml of the sodium dichloroisocyanide solution and allow to stand at room temperature for 30 min prior to measuring the optical density at 690 nm.

Procedure for the buffered method

Place 5 g of moist soil samples in Erlenmeyer flask (100 ml), add 2.5 ml urea solution and 20 ml borate buffer. Further steps are performed as described for the non-buffered assay, with the addition of 30 ml KCl solution at the end of incubation.

Calibration curve

Pipette 1 ml of the ammonium standard solution II in glass tubes, dilute with 9 ml distilled water, and determine the ammonium concentrations (0, 1, 1.5, 2, 2.5 μg NH_4-N ml^{-1}).

Calculation

Correct the results to the blanks and calculate as follows

Urease activity (μg NH_4-N g^{-1} dwt 2 h^{-1}) =

$$\frac{\mu g\ NH_4\text{-}N\ ml^{-1} \times V \times 10}{dwt \times 5} \qquad (7.4)$$

where *dwt* is the dry weight of 1 g moist soil, *V* is the total volume of the extract (52.5 ml), 10 is the dilution factor and 5 is the weight of the soil used in the assay.

Discussion

According to Tabatabai (1982), the Tris buffer has to be used in the assay involving the ammonium determination because it prevents NH_4^+ fixation by soil. The urease activity is higher in buffered than in unbuffered soil suspensions.

The buffer of the assay proposed by Tabatabai and Bremner (1972) has to be prepared by using H_2SO_4 instead of HCl.

The latter has a stimulating effect on the reaction catalysed by jack bean urease.

The KCl–Ag_2SO_4 solution must be prepared by dissolving KCl into the Ag_2SO_4 solution because Ag_2SO_4 does not dissolve in KCl.

The addition of KCl–Ag_2SO_4 stops the enzymatic urea hydrolysis and the suspension can stand for 2 h before the determination of ammonium, because no additional ammonium is released.

The inhibitor is not used in the method of Kandeler and Gerber (1988). The sample dilution by KCl may considerably slow down the enzymatic activity rates; however, times for ammonium extraction and filtration should not differ among different measurements.

Phenyl mercuric borate has been used as an inhibitor for urease activity in soil (Perez-Mateos and Gonzalez-Carcedo 1988).

The filtration of soil suspensions should be carried out with nitrogen-free filter paper to avoid nitrogen contamination of the filtrate.

The colour developed in the ammonium analysis (method of Kandeler and Gerber 1988) is stable for at least 8 h at room temperature.

The method based on colorimetric determination of urea is not proposed because it has certain deficiencies such as the instability of some reagents or the chromogen compounds, low reproducibility, lack of sensitivity and lack of linearity at low urea concentrations; in addition, the procedure is time consuming (Tabatabai 1982).

The leaching technique can also be used to estimate the urease activity in soils (Perez-Mateos and Gonzalez-Carcedo 1988).

Amidase activity in soils

W.T. Frankenberger, Jr
M.A. Tabatabai

(Frankenberger and Tabatabai 1980a)

Amidase (acylamide amidohydrolase, EC 3.5.1.4) is the enzyme that catalyses the hydrolysis of amides, and produces ammonia and the corresponding carboxylic acid

$$R{:}CONH_2 + H_2O \rightarrow NH_3 + R{:}COOH \quad (7.5)$$

Amidase acts on C–N bonds other than peptide bonds in linear amides. It is specific for aliphatic amides, and aryl amides cannot act as substrates (Kelly and Clarke 1962; Florkin and Stotz 1964). This enzyme is widely distributed in nature. It has been detected in animals and microorganisms (Bray et al 1949a; Clarke 1970). Amidase is also present in the leaves of corn (*Zea mays* L.), sorghum (*Sorghum bicolor* L. Moench), alfalfa (*Medicago sativa* L.) and soybeans (*Glycine max* L.) (Frankenberger and Tabatabai 1982). Microorganisms shown to possess amidase activity include bacteria (Clarke 1970; Frankenberger and Tabatabai 1985), yeast (Joshi and Handler 1962) and fungi (Hynes 1970, 1975).

Information concerning the distribution, specificity and kinetic properties of amidase in soils is needed because its substrates (amides) are potential nitrogen fertilizers (Frankenberger and Tabatabai 1981c). Among the various nitrogen compounds that can serve as substrates for amidase, the sparingly soluble oxamide and formamide have been tested as nitrogen fertilizers (Beaton et al 1967).

Air-drying field-moist samples results in a decrease of amidase activity ranging from 14 to 33% (average 21%) (Frankenberger and Tabatabai 1981a). Freezing of field-moist samples at −20°C for 3 months resulted in activity increases ranging from 3 to 16% (average 9%). Heating of field-moist and air-dried samples for 2 h before assaying amidase activity showed that this enzyme is inactivated at temperatures above 50°C.

Studies of the distribution of amidase in soils showed that it is concentrated in surface soils and decreases with soil depth (Frankenberger and Tabatabai 1981a). Statistical analysis indicated that the activity of this enzyme is significantly correlated with organic carbon in surface soils (r = 0.74***) and in soil profiles (r = 0.89**). Amidase activity also was significantly correlated with total nitrogen (r = 0.74***), percentage of clay (r = 0.69***) and urease activity (r = 0.73***) in the 21 surface soil samples studied. There was no significant relationship between amidase activity and soil pH, or percentage sand. Amidase activity and microbial counts obtained with acetamide or propionamide as a substrate in the absence of toluene indicated that these substrates induced production of this enzyme by soil microorganisms.

Principle of the method

This method involves determination of the NH_4-N released by amidase activity when soil is incubated with buffered (0.1 M Tris (hydroxy methyl) amino methane (THAM), pH 8.5) amide solution and toluene at 37°C. The ammonium released is determined by a rapid procedure involving treatment of the incubated soil sample with 2.5 M KCl containing an amidase inhibitor (uranyl

** $P < 0.01$; *** $P < 0.001$.

acetate) and steam distillation of an aliquot of the resulting suspension with MgO for 3.3 min. The procedure gives quantitative recovery of NH_4-N added to soils and does not cause chemical hydrolysis of the substrate.

Materials and apparatus

Volumetric flasks (50 ml)
Incubator (37°C) or temperature-controlled waterbath
Distillation flask (100 ml)
Steam distillation apparatus (see Keeney and Nelson, 1982)

Chemicals and solutions

Toluene: Fisher-certified reagent (Fisher Scientific Co., Chicago, IL)
Tris–sulphuric acid buffer (0.1 M, pH 8.5)

Dissolve 12.2 g of Tris (hydroxymethyl) amino methane (Fisher-certified reagent) in about 800 ml of water, adjust the pH to 8.5 by titration with about 0.1 M H_2SO_4 and dilute the solution with water to 1 l.

Amide solutions (0.5 M)

Add 2.0 ml, 2.95 g, or 3.65 g of formamide (Aldrich certified), acetamide (Sigma certified), or propionamide (Aldrich certified), respectively, into a 100 ml volumetric flask. Make up the volume by adding Tris buffer, and mix the contents. Store the solution in a refrigerator.

Potassium chloride (2.5 M)–uranyl acetate (0.005 M) solution

Dissolve 2.12 g of reagent-grade $UO_2(C_2H_3O_2)_2{:}2H_2O$ in about 700 ml of water, dissolve 188 g of reagent grade KCl in this solution, dilute the solution to 1 l with water and mix thoroughly. Prepare this solution immediately before use.

Reagents for the determination of ammonium (magnesium oxide, boric acid indicator solution, 0.0025 M H_2SO_4)

Prepare as described by Keeney and Nelson (1982).

Procedure

Place 5 g of soil (< 2mm) in a 50 ml volumetric flask, add 0.2 ml of toluene and 9 ml of Tris buffer, swirl the flask for a few seconds to mix the contents, add 1 ml of 0.5 M amide solution and swirl the flask again for a few seconds. Then stopper the flask and place it in an incubator at 37°C.

The incubator time varies with the specific substrate added. Hydrolysis of formamide is relatively rapid and can be measured quantitatively after 2 h of incubation. The rates of hydrolysis of acetamide and propionamide are somewhat slower than the rate of hydrolysis of formamide, and they can be determined quantitatively after 24 h of incubation.

After incubation, remove the stopper, add approximately 35 ml of KCl–$UO_2(C_2H_3O_2)_2{:}2H_2O$ solution, swirl the flask for a few seconds, and allow the flask to stand until the contents have cooled to room temperature (about 5 min). Then dilute the volume to 50 ml by the addition of KCl-$UO_2(C_2H_3O_2)_2{:}2H_2O$ solution, stopper the flask and invert it several times to mix the contents.

To determine NH_4-N in the resulting soil suspension, invert the flask several times and pipette a 20 ml aliquot of the suspension into a 100 ml distillation flask and determine the NH_4-N released by steam distillation of this aliquot with 0.2 g of MgO for 3.3 min as described by Bremner (1965a).

Controls should be performed in each series of analyses to account for the NH_4-N that is not derived from the amide through amidase activity. To perform controls, follow the procedure described for assay of amidase activity, but make the addition of 1 ml of 0.5 M amide solution after the addition of KCl-$UO_2(C_2H_3O_2)_2{:}2H_2O$

reagent. The KCl–$UO_2(C_2H_3O_2)_2.2H_2O$ reagent should be stirred continuously before use; if this solution is allowed to stand without stirring, KCl will precipitate slowly out of the solution with time. The purpose of adding the KCl–uranyl acetate solution to the soil sample after incubation is to inactivate amidase and to allow quantitative determination of the NH_4-N released (Jakoby and Fredericks 1964). The soil suspension analysed for NH_4-N must be mixed thoroughly immediately before sampling for NH_4-N analysis.

Discussion

The optimum pH observed for amidase activity (pH 8.5) (Frankenberger and Tabatabai 1980a) is somewhat higher than that reported for amidase activity of *Pseudomonas fluorescens* isolated from a garden soil (pH 6.5) (Jakoby and Fredericks 1964). This deviation is the pH optimum for amidase activity in solution from that in soil is expected because it is well known that the pH optima of enzymes in solutions are about 2 pH units lower than the same enzymes in soils (McLaren and Estermann 1957). The choice of buffer in the method described is based on the finding that the NH_4-N released by amidase activity is considerably higher in the presence of Tris than in the presence of phosphate, citrate or modified universal buffer. The other reason for choosing this buffer is that no ammonium fixation by soils is observed in the presence of Tris buffer.

For a valid assay of enzyme activity it is necessary to ensure that the substrate concentration is not limiting the reaction rate in the assay procedure. A study on the effect of varying the substrate concentration in the method described indicated 0.05 M is satisfactory for the assay of amidase activity in soils (Frankenberger and Tabatabai 1980a). At this concentration, the enzyme is saturated with the substrate and the reaction rate essentially follows zero-order kinetics (Frankenberger and Tabatabai 1980b). Enzyme-catalysed reactions usually show linear relationships between the amount of product formed and time of incubation. Formation of NH_4^+ with formamide as a substrate follows zero-order kinetics for at least 4 h (Frankenberger and Tabatabai 1980a). When acetamide and propionamide are used as substrates, the observed relationship existed up to 60 and 36 h, respectively. Hydrolysis of acetamide and propionamide is considerably slower than the hydrolysis of formamide. The former two substrates have to be incubated for 24 h to measure differences in amidase activity quantitatively in soils. Interference by microbial activity does not seem to occur during the 24 h incubation time as evident by the linear relationship between the amount of NH_4-N released and the time of incubation in the presence of toluene. Toluene should be added to soil samples before adding 9 ml of buffer because tests indicated that it is less effective in preventing amidase induction when added after the Tris buffer treatment described.

Enzyme-catalysed reactions proceed at a faster rate with increasing temperature of incubation within a certain temperature range as long as the enzyme is stable and retains its full activity. At a temperature above this range, the activity begins to decrease because of enzyme inactivation. Studies of the effect of temperature on amidase activity in soils showed that inactivation of amidase occurs at about 65°C (Frankenberger and Tabatabai 1980a). Steam sterilization inactivates amidase and addition of formaldehyde and toluene decrease the activity. Amidase is induced in the presence of the substrates without toluene. Among various chemical salts, NaF and $NaAsO_2$ have an inhibitory effect on amidase activity. The inhibition by F^- suggests that amidase requires Mg^{2+} and/or Ca^{2+}. Brown et al (1973) found that neither hydroxyl nor thiol groups are directly involved in the catalytic active sites of amidase even though thiol groups seem necessary for stabilization of the active enzyme. The effects of several salts selected to provide a variety of cations and anions showed that soil amidase activity is not affected when the Tris buffer contains 5 mM with respect to the following ions: K^+, Na^+, NH_4^+, NO_3^-, NO_2^-, N_3^-, Cl^-, CN^-, S_2^- and SO_4^{2-} (Frankenberger and Tabatabai 1980a). Enzyme kinetics reveal that As(III) is a competitive inhibitor of amidase, whereas As(I), Hg(II), and Se(IV) are non-competitive inhibitors (Frankenberger and Tabatabai 1981b).

L-Asparaginase activity of soils

W.T. Frankenberger, Jr
M.A. Tabatabai

(Frankenberger and Tabatabai 1991a, 1991b)

The enzyme, L-asparaginase (L-asparagine amidohydrolase, EC 3.5.1.1) has an important role in nitrogen mineralization of soils. The chemical nature of nitrogen in soils is such that a large proportion (15–25%) of the total soil nitrogen is often released as NH_4^+ by acid hydrolysis (6 M HCl). Some evidence suggests that a portion of the released NH_4^+ comes from the hydrolysis of amide (asparagine and glutamine) residues in soil organic matter (Sowden 1958). Bremner (1955) reported that after acid hydrolysis of humic preparations, 7.3–12.6% of the total nitrogen was in the form of amide-N. L-Asparaginase activity was first detected in soils by Drobni'k (1956). This enzyme catalyses the hydrolysis of L-asparagine, producing L-aspartic acid and ammonia as shown below:

$$H-\underset{\underset{\underset{\underset{NH_2}{|}}{CO}}{\underset{|}{CH_2}}}{\overset{\overset{NH_2}{|}}{\underset{|}{C}}}-COOH + H_2O \xrightarrow{\text{L-asparaginase}} H-\underset{\underset{COOH}{\underset{|}{CH_2}}}{\overset{\overset{NH_2}{|}}{\underset{|}{C}}}-COOH + NH_3 \qquad (7.6)$$

The enzyme is widely distributed in nature. It has been detected in both plants and microorganisms (Wriston 1971). Plants may contribute substantial amounts of L-asparagine to soils and by the action of soil L-asparaginase, release NH_4^+ to the inorganic nitrogen pool. Soils have been tested for L-asparaginase activity by Beck and Poschenrieder (1963), simply by adding L-asparagine to soils and monitoring the NH_4^+ released, but the actual assay was not tested thoroughly using systematic studies of factors affecting the release of NH_4^+. Frankenberger and Tabatabai (1991a) developed a simple and sensitive method for the assay of L-asparaginase activity in soils and determined various parameters that affected the observed activity. The method developed involves determination of the NH_4^+-N released by L-asparaginase activity when soil is incubated with buffered (0.1 M Tris, pH 10) L-asparagine solution and toluene at 37°C. Studies on the distribution of L-asparaginase in soil profile samples revealed that its activity generally decreases with sample depth and is accompanied by a decrease in organic C content (Frankenberger and Tabatabai 1991b). Statistical analyses indicated that L-asparaginase activity was significantly correlated (**$p < 0.01$) with organic C ($r = 0.86$**) and total N ($r = 0.78$**) in 26 surface soil samples examined. There was no significant relationship between L-asparaginase activity and the percentage of clay or sand. There was, however, a significant correlation between L-asparaginase activity and amidase ($r = 0.82$**) and urease ($r = 0.79$**) activities in the surface samples studied (Frankenberger and Tabatabai 1991b).

Principle of the method

Frankenberger and Tabatabai (1991a) developed a simple, precise and sensitive method to assay L-asparaginase

** $P < 0.01$

activity in soils. This method uses steam distillation to determine the NH_4^+ produced by L-asparaginase activity when soil is incubated with buffered (0.2 M Tris, pH 10) L-asparagine solution and toluene at 37°C for 2 h. The procedure developed gives quantitative recovery of NH_4-N added to soils and does not cause chemical hydrolysis of L-asparagine.

Materials and apparatus

Volumetric flasks (50 ml)
Incubator (37°C) or temperature-controlled waterbath
Distillation flask (100 ml)
Steam distillation apparatus (see Keeney and Nelson 1982)

Chemicals and solutions

Toluene

Fisher-certified reagent (Fisher Scientific Co., Chicago, IL).

Tris buffer (0.1 M, pH 10) is prepared by dissolving 12.2 g of Tris (hydroxy methyl) amino methane (Fisher-certified reagent) in about 800 ml of water, adjusting the pH to 10 by titration with 0.1 M NaOH and diluting the solution with water to 1 l.

L-Asparagine solution (0.5 M) is prepared by dissolving 1.65 g L-asparagine (Sigma Chemical Co., St Louis, MO) into a 25 ml volumetric flask. The volume is adjusted by adding Tris buffer. The contents are mixed while running hot tap water over the flask.

Potassium chloride (2.5 M)–silver sulphate (100 parts 10^{-6}) solution is prepared by dissolving 100 mg reagent grade Ag_2SO_4 in about 700 ml water, dissolving 188 g reagent grade KCl in this solution and diluting the solution to 1 l.

Reagents for the determination of ammonium (magnesium oxide, boric acid-indicator solution, 0.005 N sulphuric acid) is prepared as described by Keeney and Nelson (1982).

Procedure

A 5 g soil sample (< 2 mm) in a 50 ml volumetric flask is treated with 0.2 ml toluene and 9 ml Tris buffer. The flask is swirled for a few seconds to mix the contents; 1 ml 0.5 M L-asparagine solution is then added and the flask is swirled again for a few seconds. The flask is stoppered and placed in an incubator at 37°C for 2 h.

After incubation, the stopper is removed and approximately 35 ml KCl–Ag_2SO_4 solution is added. The flask is swirled for a few seconds and allowed to stand until the contents have cooled to room temperature (about 5 min). Then the volume is diluted to 50 ml by the addition of KCl–Ag_2SO_4 solution, and the flask is stoppered and inverted several times to mix the contents.

To determine NH_4-N in the resulting soil suspension, the flask is inverted several times and a 20 ml aliquot of the suspension is pipetted into a 100 ml distillation flask and the NH_4-N released is determined by steam distillation of this aliquot with 0.2 g MgO for 3.3 min, as described by Keeney and Nelson (1982).

In the controls, the procedure described for the assay of L-asparaginase activity is followed but the 1 ml 0.5 M L-asparagine solution is added after the addition of the KCl–Ag_2SO_4 reagent. The purpose of adding the KCl–Ag_2SO_4 solution to the soil sample after incubation is to inactivate L-asparaginase and allow a quantitative determination of the NH_4-N released.

Discussion

The rate of NH_4^+ released by L-asparaginase activity exhibits two pH optima, 8.5 and 10.0 (Frankenberger and Tabatabai 1991a). Although there is slightly more activity at pH 10, the maximum activity is somewhat stable from pH 8 to 12.

Varying the substrate (L-asparagine) concentration in the method described showed that the concentration adopted (50 mM) is satisfactory for the assay of L-asparaginase activity in soils (Frankenberger and Tabatabai 1991a). At this concentration, L-asparaginase is saturated with L-asparagine and the reaction rate essentially follows zero-order kinetics. The apparent K_m constant averaged 6.1 mM in nine soils (Frankenberger and Tabatabai 1991a). The D-isomer of asparagine is also hydrolysed in soils but the activity is only about 16% of the activity of the L-isomer at a saturating substrate concentration.

L-Asparaginase activity can be determined using 1 g soil. However, a linear relationship between the amount of soil and the amount of NH_4^+ released showed that 5 g of soil was ideal for assaying L-asparaginase activity (Frankenberger and Tabatabai 1991c). The amount of NH_4-N released becomes non-linear when greater amounts of soil are used, indicating either that the substrate concentration is becoming a limiting factor or that one of the products (NH_4^+ or L-aspartic acid) released can inhibit L-asparaginase activity.

There is a linear relationship between L-asparaginase activity in soils and time of incubation. Release of NH_4^+ (derived from L-asparagine) in soils showed zero-order reactions for at least 3 h (Frankenberger and Tabatabai 1991a). However, with a longer incubation time, the reaction rate becomes non-linear. Therefore, the incubation time chosen for the assay of L-asparaginase activity in soils is 2 h, which allows ample time for the accumulation of NH_4^+. The linear relationship within 2 h indicates that there is little or no assimilation of the enzymatic products by microorganisms during incubation.

The optimal temperature for the L-asparaginase reaction in soils occurred at 60°C (Frankenberger and Tabatabai 1991a). Denaturation occurs at about 65°C, which is similar to that reported for amidase (Frankenberger and Tabatabai 1980a). The enzyme L-asparaginase appears to be stable to freezing and thawing, and to storage in the frozen state for periods of up to 6 months or longer (Law and Wriston 1971). The energy of activation of the reaction catalysed by L-asparaginase in nine soils (expressed in kJ mol^{-1}) ranged from 20.2 to 34.1 (average 26.6 kJ mol^{-1}) (Frankenberger and Tabatabai 1991a). The temperature coefficients for L-asparaginase activity in eight soils, for temperatures between 10 and 50°C, ranged from 1.12 to 1.70, with an average of 1.39 (Frankenberger and Tabatabai 1991a).

Steam sterilization is known to inactivate soil enzymes, but Frankenberger and Tabatabai (1991a) found trace amounts of L-asparaginase activity after this treatment. Toluene-treated soils exhibit greater L-asparaginase activity than untreated soils. This increase in activity in the presence of toluene may be due to a change in the permeability of the microbial cell membrane to substrates and enzyme reaction products, as suggested by Skujins (1967). Treatment of soils with formaldehyde severely inhibits L-asparaginase activity.

The effects of $HgCl_2$, *p*-(chloromercuri)benzoic acid and iodoacetic acid on L-asparaginase activity suggest that a free sulphydryl moiety is necessary to maintain the active enzyme. The effects of several salts selected to provide a variety of cations and anions showed that soil L-asparaginase activity is not affected when the Tris buffer used is prepared to contain 5 mM with respect to the following ions: K^+, Na^+, Mn^{2+}, Cl^-, NO_3^-, PO_4^{3-} and SO_4^{2-} (Frankenberger and Tabatabai 1991a). Treatment of soils with NaF inhibits L-asparaginase activity. This inhibition seems to be related to the binding of Mg^{2+} required for activation of the enzyme. Both Ca^{2+} and Mg^{2+} activate L-asparaginase activity in soil.

L-Glutaminase activity of soils

W.T. Frankenberger, Jr
M.A. Tabatabai

(Frankenberger and Tabatabai 1991c, 1991d)

L-Glutaminase is among the amidohydrolases that supplies available nitrogen to plants. This hydrolase is specific and acts on C–N bonds other than peptides. L-Glutaminase (L-glutamine amidohydrolase, EC 3.5.1.2) activity in soils was first detected by Galstyan and Saakyan (1973). The reaction catalysed by this enzyme involves the hydrolysis of L-glutamine yielding L-glutamic acid and NH_3:

$$NH_2-CH(COOH)-CH_2-CH_2-CO-NH_2 + H_2O \xrightarrow{\text{L-glutaminase}} NH_2-CH(COOH)-CH_2-CH_2-COOH + NH_3 \quad (7.7)$$

L-Glutaminase is widely distributed in nature and has been detected in several animals (Sayre and Roberts 1958), plants (Bidwell 1974) and microorganisms (Imada et al 1973). Microorganisms that have shown to contain L-glutaminase activity include bacteria, yeasts and fungi. Plants and microorganisms are probable sources of L-glutaminase activity in soils, however, the main source is believed to be microbial in nature. Among the bacteria, very high levels of L-glutaminase activity have been reported in Achromobacteraceae soil isolates (Roberts et al 1972). Fungal species that are known to produce L-glutaminase include *Tilachlidium humicola, Verticillium malthousei* and *Penicillium urticae* (Imada et al 1973).

L-Glutaminase activity in soil profile samples generally decrease with sample depth (Frankenberger and Tabatabai 1991d). The relationship between L-glutaminase activity and organic C, using pooled data (33 samples) from five soil profiles, was highly correlated (r = 0.92**). The soil properties that related to L-glutaminase activities in 25 surface soils included organic C (r = 0.79**) and total N (r = 0.76**). There was no significant relationship between L-glutaminase activity and pH, percentage of clay or sand. There was, however, a significant correlation between L-glutaminase activity and amidase (r = 0.82**), urease (r = 0.78**) and L-asparaginase (r = 0.92**) activities in surface soil samples studied. In the presence of trace elements (5 μmol^{-1} g^{-1} soil), the average inhibition of L-glutaminase in three soils showed that Ag(I), Hg(II), Sn(II), Cr(III), Ti(IV) and W(VI) were the most effective inhibitors (average >25%) (Frankenberger and Tabatabai 1991b).

Principle of the method

L-Glutaminase catalyses the hydrolysis of L-glutamine to produce ammonia and L-glutamic acid. A simple, precise, rapid and sensitive method to assay its activity was developed by Frankenberger and Tabatabai (1991c), which involves the determination of NH_4^+ released by L-glutaminase activity

** $P < 0.01$

when soil is exposed to L-glutamine, THAM buffer and toluene at 37°C for 2 h. The NH_4-N released is determined by treatment of the soil sample with 2.5 M KCl containing a L-glutaminase inhibitor (Ag_2SO_4) and steam distillation of an aliquot of the resulting soil suspension.

Materials and apparatus

Volumetric flasks (50 ml)
Incubator (37°C) or temperature-controlled water bath
Distillation flask (100 ml)
Steam distillation apparatus (see Keeney and Nelson 1982)

Chemicals and solutions

Toluene

Fisher-certified reagent (Fisher Scientific Co., Chicago, IL).

THAM buffer (0.1 M, pH 10)

Prepared by dissolving 12.2 g of Tris (THAM, Fisher certified reagent) in about 800 ml of water, adjusting the pH to 10 by titration with 0.1 M NaOH, and diluting with water to 1 litre.

L-Glutaminase solution (0.5 M)

Prepared by dissolving 1.82 g of L-glutamine (Sigma Chemical Co., St Louis, MO) in a 25 ml volumetric flask. The volume is adjusted by adding THAM buffer (0.1 M, pH 10), and the contents are mixed while running hot tap water over the flask.

Potassium chloride (2.5 M)–silver sulphate (100 mg ml^{-1}) solution

Dissolve 100 mg of reagent grade Ag_2SO_4 in about 700 ml of water, dissolve 188 g of reagent grade KCl in this solution and dilute the solution to 1 l with water.

Reagents for determination of ammonium (magnesium oxide, boric acid indicator solution, 0.005 M sulphuric acid)

Prepared as described by Keeney and Nelson (1982).

Procedure

A 5 g soil sample (<2 mm) in a 50 ml volumetric flask is treated with 0.2 ml toluene and 9 ml Tris buffer. The flask is swirled for a few seconds to mix the contents; 1 ml 0.5 M L-glutaminase solution is added and the flask is swirled again for a few seconds. The flask is stoppered and kept at 37°C for 2 h.

After incubation, the stopper is removed and approximately 35 ml KCl–Ag_2SO_4 solution is added. Other steps involved in this procedure are described by Frankenberger and Tabatabai (1991a). The NH_4-N released is determined as described by Keeney and Nelson (1982). To perform controls, the procedure for the assay of L-glutaminase is followed, except the 1 ml of 0.5 M L-glutamine solution is added after the addition of the KCl–Ag_2SO_4 reagent. A second control is included, but without soil to account for partial chemical hydrolysis of the substrate during incubation.

Discussion

L-Glutamine (50 mM) decomposes upon heating in the presence of Tris buffer (0.1 M, pH 10). The rate of chemical hydrolysis follows an exponential curve with increasing temperature and ranges from 0.5 to 52% at 30° and 85°C, respectively (Frankenberger and Tabatabai 1991c). Decomposition of L-glutamine is also concentration dependent in buffer solutions. L-Glutamine does not decompose in water, however, in the presence of Tris, its decomposition increases with an increase in concentration and time. Approximately 2% of L-glutamine (50 mM) decomposes in 2 h in the presence of Tris buffer, whereas 18% is decomposed in 24 h at 37°C. Similar results were reported by Bray et al (1949b), who found that 28% of L-glutamine (20 mM) was decomposed in 20 h in the presence of phosphate buffer but only 6% in a phosphate-free (veronal) buffer. Frankenberger and Tabatabai (1991c) observed *ca.* 2% decomposition of L-glutamine during the 2 h assay and temperature of 37°C, but this was accounted for by using the proper controls.

L-Glutaminase activity in soils exhibits high activity within a remarkably broad range of pH. The optimum pH of L-glutaminase activity in soil is 10 (Frankenberger and Tabatabai 1991c). The apparent pH optimum for clay-adsorbed enzymes generally is displaced one or two pH units to more alkaline values. This shift in pH optimum to higher values occurs because acidity at the clay surface is significantly greater than in bulk solution (Boyd and Mortland 1990).

Frankenberger and Tabatabai (1991c) showed that 50 mM of L-glutamine is satisfactory for the assay of L-glutaminase activity in soils. At this concentration, L-glutaminase becomes saturated with L-glutamine and the rate essentially follows zero-order kinetics. The D-isomer of glutamine is also hydrolysed in soils but at only 7% of the activity of the L-isomer at saturating concentrations of the substrate (50 mM). The activity of L-glutaminase is proportional to the amount of soil used in the assay. A linear relationship has been established between the enzyme activity and the amount of soil used, up to 5 g of soil (Frankenberger and Tabatabai 1991c). The amount of NH_4-N released deviates from linearity when greater amounts of soil (>5 g) are used, indicating that either the substrate concentration is limiting the reaction rate or one of the products (NH_4 or L-glutamic acid) is inhibiting L-glutaminase activity. The observed deviation from linearity of the amount of NH_4^+ produced with increasing the amount of soil appears to be due to a substrate limitation (Frankenberger and Tabatabai 1991c). Although the formation of NH_4^+ in soils shows zero-order reactions for at least 4 h, the reaction rates deviate from linearity with prolonged incubation. Maximal L-glutaminase activity in soil occurs at 50°C under the conditions of assay. The activation energy of the reaction catalysed by L-glutaminase of nine soils ranged from 20.3 to 39.9 kJ mol^{-1} (average 32.4) (Frankenberger and Tabatabai 1991c). The substrate concentration activity curves of L-glutaminase in soils obey Michaelis–Menten kinetics plots. The Michaelis constants (K_m) ranged from 8.2 to 38.6 mM (Frankenberger and Tabatabai 1991a). The Q_{10} values of the reaction catalysed by L-glutaminase in nine soils exposed to temperatures of 10–50°C ranged from 1.19 to 1.85 (average 1.49) (Frankenberger and Tabatabai 1991c). Steam sterilization (121°C, 1 h) inactivates L-glutaminase but there is some residual activity remaining after this treatment. L-Glutaminase activity is greater in toluene-treated soils than in the absence of toluene. L-Glutaminase in soils is strongly inhibited by several sulphydryl group reagents. The effects of $HgCl_2$, *p*-(chloromercuri)benzoic acid, and iodoacetic acid suggests that a free sulphydryl moiety is necessary to maintain the active enzyme. This is further supported by the finding that L-glutaminase is not only inhibited by Hg^{2+} and *p*-mercuribenzoate, but also by Ag^+, Pb^{2+} and Cu^{2+} at somewhat higher concentrations (Hartman 1971). Among the chemical salts, only NaF has an inhibitory effect on soil L-glutaminase activity. Inhibition by F^- suggests that L-glutaminase requires Mg^{2+} or Ca^{2+} to be active. At 5 mM, both Ca^{2+} and Mg^{2+} activated L-glutaminase activity by an average of 4 and 12%, respectively (Frankenberger and Tabatabai 1991c). Also, when soils are treated with 5 mM Na_2-EDTA, L-glutaminase activity in soils increases. L-Glutaminase activity is not affected when the Tris buffer used contains 5 mM with respect to the following ions: K^+, Na^+, Mn^{2+}, Cl^-, NO_3^-, PO_4^{3-} or SO_4^{2-}.

L-Histidine ammonia lyase activity

W.T. Frankenberger, Jr
M.A. Tabatabai

(Frankenberger and Johanson 1982)

Histidine ammonia-lyase (EC 4.3.1.3) is the enzyme that catalyses the irreversible non-oxidative deamination of L-histidine to urocanate and ammonia:

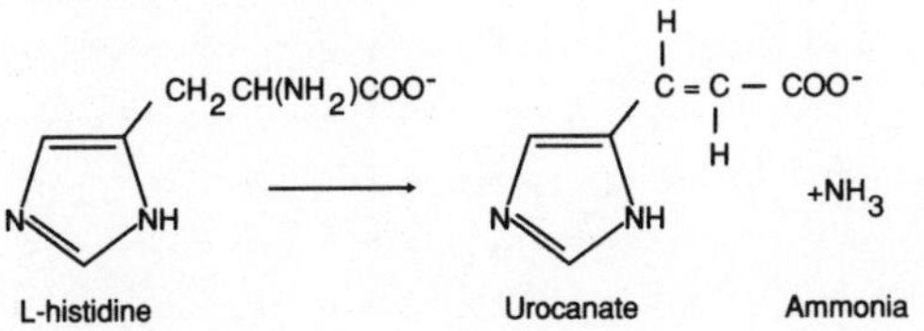

L-histidine → Urocanate + Ammonia ($+NH_3$) (7.8)

The enzyme, histidase (L-histidine NH_3-lyase) was first discovered in 1926 by Gyorgy and Rothler (1926) and Edlbacher (1926) who observed an increase in NH_4^+ when L-histidine was added to aqueous liver extracts. In 1930, Edlbacher et al (1930) later observed that glutamic acid, NH_3 and a reduced acid (formic acid) were the reaction products:

$$\text{L-Histidine} + 4H_2O \rightarrow \text{Glutamic acid} + NH_3 \text{ formic acid} \quad (7.9)$$

Since then, the enzyme that catalyses the deamination of L-histidine to urocanate has been studied in both mammalian and microbial systems and was referred to as "histidase", "histidinase", and "histidine-α-deaminase". The International Enzyme Commission has now named this enzyme "histidine NH_3^- lyase" (Florkin and Stotz 1964).

The amino acid, L-histidine, was first detected in soils by Schreiner and Shorey (1910). It comprises approximately 10% of the total basic amino acids found in soil hydrolysates (6 M HCl) (Bremner 1965a). Histidine is unique in that it bears a weakly basic imidazolium group, and it is the only amino acid whose R group has a pK near 7.0. The imidazole group of L-histidine can function as a nucleophilic group in the binding of substrates in various enzymatic reactions.

L-Histidine is widely distributed in nature. In plants, animals and microorganisms it can account for <1 to 8% of the amino acids found in proteins (Tabor 1954). Among the microorganisms, L-histidine NH_3-lyase preparations have been purified from *Aerobacter, Bacillus, Klebsiella, Pseudomonas* and *Salmonella* species (Lessie and Neidhardt 1967).

The distribution of L-histidine ammonia lyase activity in soils is correlated with organic carbon ($r = 0.70^{***}$) and total N ($r = 0.55^{*}$) in the topsoil (Frankenberger and Johanson 1983a). There was no significant correlation between L-histidine NH_3-lyase activity and soil pH, cation exchange capacity, percentage of clay and percentage of sand. The activity of this enzyme is concentrated in surface soils and decreases with profile depth. Air-drying of field-moist soil samples increased L-histidine NH_3-lyase activity by an average of 18%.

Principle of the method

This method involves determination of the NH_4-N released by L-histidine NH_3-lyase

* *P* =
*** *P* =

activity when soil is incubated with buffered (0.1 M Tris, pH 9.0) L-histidine solution and toluene at 37°C. The NH_4^+ released is determined by a procedure involving treatment of the incubated soil sample with 2.5 M KCl containing a L-histidine NH_3 lyase inhibitor (uranyl acetate) and steam distillation of an aliquot of the resulting suspension with MgO. This procedure gives quantitative recovery of NH_4-N released in soils and does not cause chemical hydrolysis of the substrate.

Materials and apparatus

Volumetric flasks (50 ml)
Incubator (37°C) or temperature-controlled water bath
Distillation flasks (100 ml)
Steam distillation apparatus (see Keeney and Nelson, 1982)

Chemicals and solutions

Toluene

Fisher-certified reagent (Fisher Scientific Co., Chicago, IL).

Tris–sulphuric acid buffer (0.1 M, pH 9.0)

Dissolve 12.2 g of Tris (hydroxy methyl) amino methane (Fisher-certified reagent) in about 800 ml of water, adjust the pH to 9.0 by titration with 0.1 M H_2SO_4 and dilute the solution with water to 1 l.

L-Histidine solution (0.5 M)

Add 5.22 g of L-(+)-histidine hydrochloride monohydrate (Aldrich-certified) into a 50 ml volumetric flask. Make up the volume by adding Tris buffer and mix the contents. Store the solution in a refrigerator.

Potassium chloride (2.5 M)–uranyl acetate (0.005 M) solution

Dissolve 2.12 g of reagent grade UO_2-$(C_2H_3O_2)_2$:$2H_2O$ in about 700 ml of water, dissolve 188 g of reagent grade KCl in this solution, dilute the solution to 1 l with water and mix thoroughly. Prepare this solution immediately before use.

Reagents for the determination of ammonium (magnesium oxide, boric acid indicator solution, 0.0025 M sulphuric acid)

Prepare as described by Keeney and Nelson (1982).

Procedure

Place 5 g of soil (< 2 mm) in a 50 ml volumetric flask, add 0.2 ml of toluene and 9 ml of Tris buffer, swirl the flask for a few seconds to mix the contents, add 1 ml of 0.5 M L-histidine solution and swirl the flask again for a few seconds. Then stopper the flask and place it in an incubator at 37°C for 48 h.

After incubation, remove the stopper, add approximately 35 ml of the KCl–$UO_2(C_2H_3O_2)_2$ solution, swirl the flask for a few seconds and allow the flask to stand until the contents have cooled to room temperature (about 5 min). Then dilute the volume to 50 ml by the addition of KCl–$UO_2(C_2H_3O_2)_2$ solution, stopper the flask and invert it several times to mix the contents.

To determine NH_4-N in the resulting soil suspension, invert the flask several times and pipette a 20 ml aliquot of the suspension into a 100 ml distillation flask and determine the NH_4-N released by steam distillation of this aliquot with 0.2 g of MgO for 3.3 min as described by Keeney and Nelson (1982).

Controls should be performed in each series of analyses to account for the NH_4-N not derived from L-histidine through L-histidine NH_3-lyase activity. To perform controls, follow the procedure described for the assay of L-histidine NH_3-lyase activity, but make the addition of 1 ml of 0.5 M L-histidine solution after the addition of KCl–$UO_2(C_2H_3O_2)_2$ reagent. The purpose of adding the KCl–uranyl acetate solution to the soil sample after incubation is to inactivate the enzyme, L-histidine NH_3-lyase and to allow quantitative determination of the NH_4-N released. The soil suspension analysed for NH_4-N must be mixed thoroughly immediately before sampling for NH_4-N analysis.

Discussion

Under the conditions of the assay procedure, optimal activity of L-histidine NH_3-lyase occurs at pH 9.0 with Tris/H_2SO_4 buffer (Frankenberger and Johanson 1982). The pH optimum of soil L-histidine NH_3-lyase is in good agreement with the value (pH 9.0) detected in *Pseudomonas* spp. preparations as reported by Rechler (1969). The choice of buffer in the method described was based on the findings that the NH_4-N released by L-histidine NH_3-lyase activity is considerably higher in the presence of Tris/H_2SO_4 than in the presence of phosphate, citrate and modified universal buffer. Another reason for choosing this buffer is that no ammonium fixation by soils occurs in the presence of Tris/H_2SO_4 buffer (Frankenberger and Tabatabai 1980a).

Toluene was added to the assay because urocanase (involved in the breakdown of urocanate to NH_3 and L-4-imidiazolone-5-propionate) is inhibited in the presence of toluene.

L-Histidine NH_3-lyase in soils does not yield hyperbolic velocity curves as displayed by Michaelis–Menten kinetics. Instead, a sigmoidal response occurs when the initial reaction rate is plotted against the substrate concentration (Frankenberger 1983). The sigmoidal curves of soil L-histidine NH_3-lyase activity accelerate exponentially at 1/2 V_{max}. For a valid assay of enzymatic activity it is necessary to ensure that the enzyme substrate concentration is not limiting the reaction rate in the assay procedure. The concentration adopted (50 mM) is satisfactory for the assay of L-histidine-NH_3 lyase activity in soils. At this concentration, the reaction rates approach zero-order kinetics (Frankenberger and Johanson 1982).

A linear relationship between the amount of soil and NH_4-N released indicated that 5 g of soil was satisfactory for assaying L-histidine NH_3 lyase activity (Frankenberger and Johanson 1982). However, the amount of NH_4-N released deviates from linearity when greater amounts of soil are used, indicating either that the substrate concentration is a limiting factor or one of the products released (NH_4^+ or urocanate) inhibits the L-histidine NH_3-lyase reaction.

Under the standard conditions described, the accumulation of NH_4-N derived from L-histidine NH_3-lyase activity is linear for at least 168 h (Frankenberger and Johanson 1982). Interference by microbial activity does not seem to occur during the 48 h incubation period as evident by a linear relationship established between the amount of NH_4-N released and time of incubation in the presence of toluene.

A study of L-histidine NH_3-lyase activity in soils as a function of temperature showed maximal activity of 80°C under the conditions of assay (Frankenberger and Johanson 1982). Denaturation occurred at 85°C. The assay of L-histidine NH_3 lyase activity in soils is performed at 37°C because tests showed that temperatures below 37°C make it difficult to detect quantitatively low activities exhibited by some soils.

The addition of sulphydryl compounds such as 2-mercaptoethanol and L-cysteine to soils prior to the assay activates L-histidine NH_3-lyase activity (Frankenberger and Johanson 1982). Treatment of soils with formaldehyde severely inhibits this enzyme. Inhibition of L-histidine NH_3-lyase activity by phenyl mercuric acetate suggests that a free sulphydryl moiety is necessary to maintain the active enzyme. When soils are treated with Na-EDTA, L-histidine NH_3-lyase activity is decreased. The effects of several salts selected to provide a variety of cations and anions showed that L-histidine NH_3-lyase activities are not affected when the Tris buffer used is made to contain 5 mM with respect to the following ions: K^+, Na^+, Cl^-, NO_2^-, PO_4^{3-} and SO_4^{2-} (Frankenberger and Johanson 1982). Treatment of soils with NaF and $NaAsO_2$ inhibit L-histidine NH_3-lyase activity. The inhibition by F^- suggests that L-histidine NH_3-lyase requires Mg^{2+} and/or Ca^{2+} to be active. Both Ca^{2+} and Mg^{2+} activate L-histidine NH_3 lyase activity.

Phosphatase activity

K. Alef
P. Nannipieri
C. Trazar-Cepeda

Phosphatases catalyse the hydrolysis of phosphate esters and are enzymes with relatively broad specificity, capable of acting on a number of different structurally related substrates, but at widely different rates. They have received trivial names according to their substrates; phytase, nucleotidases, sugar phosphatases and glycerophosphatase belong to the group of phosphoric monoester hydrolases (EC 3.1.3), whose mechanism of reaction is reported in

$$R-O-\overset{\overset{O}{\|}}{\underset{\underset{O^{\ominus}}{|}}{P}}-O^{\ominus} + H_2O \longrightarrow HO-\overset{\overset{O}{\|}}{\underset{\underset{O^{\ominus}}{|}}{P}}-O^{\ominus} + R-OH \qquad (7.10)$$

Nucleases, which catalyse hydrolysis of ribo- and deoxyribonucleic acids to their individual nucleotides, and phospholipases, which catalyse the hydrolysis of phospholipids, belong to the group of phosphoric diester hydrolases (EC 3.1.4), whose mechanism of reaction is reported in

$$R^1O-\overset{\overset{OH}{|}}{\underset{\underset{R^2O}{|}}{P}}=O + H_2O \longrightarrow HO-\overset{\overset{OH}{|}}{\underset{\underset{R^2O}{|}}{P}}=O + R^1OH \qquad (7.11)$$

where R^1 and R^2 are alcohol, phenol or nucleotide groups.

The other three groups of enzymes, phosphoric triester hydrolases (EC 3.1.5), enzymes acting on phosphoryl-containing anhydrides (EC 3.6.1) and enzymes acting on P–N bonds (EC 3.9), such as phosphoamidases (EC 3.9.1.1), are also called phosphatases (Florkin and Stotz 1964). According to Malcolm (1983), the term phosphatases also includes enzymes hydrolysing pyrophosphates, metaphosphates and inorganic polyphosphates.

Phosphomonoesterases have been extensively studied in soil because they catalyse the hydrolysis of organic phosphomonoester to inorganic phosphorus which can be taken up by plants. According to their optimum pH, phosphomonoesterases are classified as acid, neutral and alkaline phosphatases; the first two enzymes have been detected in animal, microbial and plant cells. On the other hand alkaline enzymes have been found only in microorganisms and animals (Beck 1973; Burns 1978; Chonkar and Tarafdar 1981; Dick and Tabatabai 1983). Both acid and alkaline phosphomonoesterases are supposed to play an important role in plant nutrition because their activity in the ectorhizosphere is higher than in bulk soil (Tarafdar and Jungk 1987), while the content of organic phosphorus shows an opposite trend.

Both acid (4–6.5 as pH optimum) and alkaline phosphatase (9–10 as pH optimum) have been found in soil (Speir and Ross 1978). According to Eivazi and Tabatabai (1977), acid phosphatase is predominant in acid soils, while alkaline phosphatase prevails in alkaline soils. Usually assays are carried out near neutral pH (6.5–7.0) (Speir and Ross 1978; Nannipieri et al 1988). The optimum temperature has been found to range from 40 to 60°C and assays are usually carried out at 37°C (Hoffmann 1967; Tabatabai and Bremner 1969; Nannipieri et al 1988; Doelman and Haanstra 1989).

Since the early 1960s, artificial substrates have replaced natural ones in assays for determining phosphatase activity of soil (Speir and Ross 1978). Artificial substrates such as phenyl phosphate or *p*-nitrophenyl phosphate are low molecular weight esters that are more rapidly hydrolysed than natural substrates, and they contain an organic moiety which is easily determined; methods using natural substrates are based on the determination of inorganic P and present the problem that phosphate can not be recovered quantitatively from soil (Speir and Ross 1978).

Michaelis constants of phosphomonoesterases have been determined in soil (Dick and Tabatabai 1984; Trasar-Cepeda and Gil-Sortes 1987, 1988; Juma and Tabatabai 1988). According to Cervelli et al (1973), Michaelis–Menten kinetics cannot be applied without a correction factor that takes into consideration the adsorption of the substrate *p*-nitrophenyl phosphate by soil. Kinetic parameters of phosphomonoesterases extracted from soil have also been determined (Nannipieri et al 1988; Kandeler 1990); humus–phosphatase complexes extracted by sodium pyrophosphate and fractionated by ultrafiltration and gel chromatography present at least two enzymes (or two forms of the same enzyme) catalysing the same reaction and characterized by markedly different K_m and V_{max} values (Nannipieri et al 1988). Humus–phosphomonoesterases complexes with higher molecular weight are more resistant to proteolytic and thermal denaturation than complexes with lower molecular weight (Nannipieri et al 1988).

The activity of phosphomonoesterases is strongly influenced not only by pH and temperature values, but also by the organic matter content, soil moisture and anaerobiosis. Due to these effects, phosphomonoesterase activity in soil varies with the season (Speir and Ross 1978; Beck 1984a; Sparling et al 1986; Pulford and Tabatabai 1988; Rastin et al 1988).

Air-drying decreases the phosphomonoesterase activity of soil (Speir and Ross 1978; Sparling et al 1986); enzyme activity increases after the remoistening of air-dried soil. The phosphomonoesterase decreases with soil depth (Speir and Ross 1978; Beck 1984a; Cochran et al 1989) and does not correlate to bacterial number (Nannipieri et al 1978b; Speir and Ross 1978). The influence of pollutants, soil treatment and cultivation on enzyme activity and phosphate mineralization has been studied extensively in soil (Juma and Tabatabai 1977; Mathur and Sanderson 1978; Speir and Ross 1978; Mathe and Kovacs 1980; Beck 1984a; Gadkari 1984; Nakas et al 1987; Rastin et al, 1988; Wilke 1988; Cochran et al 1989; Doelman and Haanstra 1989). Usually orthophosphate inhibits phosphatase activity in soil (Kiss et al 1974; Nannipieri et al 1978b; Speir and Ross 1978; Spiers and McGill 1979; Appiah et al 1985; Lopez-Hernandez et al 1989).

Phosphomonoesterase activity

Assay of phosphomonoesterase activity

(Hoffman 1967, modified by Beck 1984a)

Principle of the method

Determination of released phenol after the incubation of soil samples with phenyl phosphate solution for 3 h at 37°C.

Materials and apparatus

Photometer
Incubator adjustable to 37°C
Shaker
Filter paper (Whatman 2v)
Volumetric flasks (100 ml)

Chemicals and solutions

Di-sodium phenyl phosphate solution (27 g/l).

Buffers

Borate buffer (pH 10)

Dissolve 12.404 g boric acid in 700 ml distilled water, adjust the pH to 10 with NaOH and dilute to 1000 ml with distilled water.

Citrate buffer (pH 7)

Dissolve 300 g of potassium citrate in 700 ml, adjust the pH to 7.0 with HCl and dilute to 1000 ml with distilled water.

Acetate buffer (pH 5)

Dissolve 136 g sodium acetate ($NaCH_3COOH.3H_2O$) in 700 ml distilled water, adjust the pH to 5 with acetic acid and dilute to 1000 ml with distilled water.

2,6-Dibromo-quinone-chlorimide solution

200 mg 100 ml^{-1} ethanol.

Toluene

Standard solution

10 µg phenol ml^{-1}.

Procedure

Place 10 g of moist, sieved soil (2 mm) in a volumetric flask (100 ml), add 1.5 ml toluene (5 ml for peat soils), and allow to stand for 15 min at room temperature. Then add 10 ml of di-sodium phenyl phosphate solution, 20 ml buffer (borate buffer for alkaline phosphatase; citrate buffer for neutral phosphatase; or acetate buffer for acid phosphatase). Stopper the flask, mix the contents, and incubate for 3 h at 37°C. After the incubation, dilute the soil suspension with warm distilled water (38°C) to 100 ml, mix thoroughly, and filter immediately. To perform the blanks, add 10 ml of distilled water instead of the di-sodium phenyl phosphate substrate solution. All measurements are carried out in duplicate with one blank.

Determination of phenol

Pipette 5 ml of buffer solution and 1–8 ml of the filtrate into a volumetric flask (100 ml), swirl the flask and dilute with distilled water to 25 ml. Then add 1 ml of the 2,6-dibromo-quinone-chlorimide solution and incubate for 20–30 min at room temperature. Finally dilute with distilled water to 100 ml and measure the optical density at 600 nm.

Calibration curve

Prepare a calibration curve (0–200 µg phenol ml^{-1}) by using 0–20 ml standard phenol solution as described above.

Calculation

Correct the results for the blanks and calculate according to the following relationship:

$$\text{Phenol } (\mu g\ g^{-1}\ dwt\ h^{-1}) = \frac{C \times 100}{dwt \times t \times 10} \qquad (7.12)$$

where C is the measured phenol concentration (µg phenol ml^{-1} filtrate), dwt is the dry weight of 1 g moist soil, t is the incubation time in hours, 100 is the total volume of the soil suspension in millilitres and 10 is the weight of the soil used in the test

Discussion

Hoffman used a substrate concentration of 6.75 g l^{-1} but Beck (1984a) recommended the use of higher concentration (27 g l^{-1}) to saturate the enzyme with soils having high activity.

The reaction is linear for 6 h.

In the original paper, air-dried soils were used (Hoffmann 1967); since air-drying significantly decreases phosphatase activity (Speir and Ross 1978; Beck 1984a; Sparling et al 1986), the use of moist soils is recommended.

An immediate filtration or centrifugation of the soil suspension at 4°C is necessary to stop the reaction. To get a rapid filtration, the use of only a part of the soil suspension is recommended.

With this method it is not possible to estimate the phosphatase activity without the addition of buffer.

Phosphomonoesterase activity

(Tabatabai and Bremner 1969; Eivazi and Tabatabai 1977)

Principle of the method

The method is based on the determination of *p*-nitrophenol released after the incubation of soil with *p*-nitrophenyl phosphate for 1 h at 37°C.

Materials and apparatus

Photometer
Incubator adjustable to 37°C
Filter paper (Whatman no. 2v)
Erlenmeyer flasks
Volumetric flasks (100 ml)

Chemicals and solutions

Toluene

Modified universal buffer (MUB) stock solution

Dissolve 12.1 g of Tris, 11.6 g of maleic acid, 14 g of citric acid and 6.3 g of boric acid (H_3BO_3) in about 500 ml of NaOH (1 M), and dilute the solution to 1000 ml with distilled water. Store at 4°C.

Modified universal buffer, pH 6.5 and 11

Titrate 200 ml of MUB stock solution to pH 6.5 under continuous stirring with HCl (0.1 M) and dilute to 1000 ml with distilled water. Titrate another 200 ml of the MUB stock solution to pH 11 by using NaOH (0.1 M) and dilute to 1000 ml with distilled water.

p-Nitrophenyl phosphate solution (PNP, 15 mM)

Dissolve 2.927 g or disodium *p*-nitrophenyl phosphate tetrahydrate in about 40 ml MUB (pH 6.5 or 11) and bring up to 50 ml with the buffer of the same pH. Store at 4°C.

$CaCl_2$ (0.5 M) solution

Dissolve 73.5 g of $CaCl_2.H_2O$ in distilled water and dilute with distilled water to 1000 ml.

NaOH (0.5 M) solution

Dissolve 20 g of NaOH in distilled water and bring up with distilled water to 1000 ml.

NaOH (0.1 M) solution

Dissolve 4 g of NaOH in distilled water and dilute with distilled water to 1000 ml.

Standard *p*-nitrophenol solution

Dissolve 1 g *p*-nitrophenol in about 70 ml of distilled water and dilute the solution to 1000 ml with distilled water. Store at 4°C.

Procedure

Soil (1 g) is placed in an Erlenmeyer flask (50 ml) and treated with 0.25 ml of toluene, 4 ml of MUB (pH 6.5 for the assay of acid phosphatase, pH 11 for the assay of alkaline phosphatase), and 1 ml of *p*-nitrophenyl phosphate solution made in the same buffer. After stoppering the flasks, the contents are mixed and incubated for 1 h at 37°C. After the incubation, add 1 ml of $CaCl_2$ (0.5 M) and 4 ml of NaOH (0.5 M). Mix the contents and filter the soil suspension through a Whatman no. 2v folded filter paper. Measure the absorbance at 400 nm. To perform the controls, add 1 ml of PNP solution after the additions of $CaCl_2$ (0.5 M) and 4 ml of NaOH (0.5 M) and immediately before filtration of the soil suspension. All measurements are performed in triplicate.

Calibration curve

Dilute 1 ml of standard *p*-nitrophenol solution to 100 ml with distilled water in a volumetric flask. Then pipette 0, 1, 2, 3, 4 and 5 ml aliquots of this diluted standard solution into Erlenmeyer flasks (50 ml), adjust the volume to 5 ml by addition of distilled water, and proceed as described for *p*-nitrophenol analysis of the incubated soil sample.

Calculation

Correct the results for the control and calculate the *p*-nitrophenol per millilitre of the filtrate by reference to the calibration curve.

p-Nitrophenol ($\mu g\ g^{-1}$ dwt h^{-1}) =

$$\frac{C \times v}{dwt \times SW \times t} \qquad (7.13)$$

where C is the measured concentration of *p*-nitrophenol ($\mu g\ ml^{-1}$ filtrate), dwt is the dry weight of 1 g moist soil, v is the total volume of the soil suspension in millilitres, SW is the weight of soil sample used (1 g) and t is the incubation time in hours.

Discussion

Criticism about the use of phenolic esters as substrates was made by Cosgrove (1977) because they might not be suitable for assaying the hydrolysis of natural esters, such as inositol phosphates and nucleotides. Caution is required in determining phosphomonoesterase activity of acid soils because under these conditions the *p*-nitrophenyl phosphate is rapidly hydrolysed to *p*-nitrophenol.

The enzymatic reaction is linear for up to 12 h.

Using short time assays (1–2 h), toluene can be omitted (Nannipieri et al 1978b; Gadkari 1984; Doelman and Haanstra 1989).

A rapid centrifugation can be applied instead of the filtration.

The enzymatic reaction can be stopped by placing soil suspensions in ice after the end of the incubation time (Nannipieri et al 1978b; Gadkari 1984).

The addition of $CaCl_2$ prevents dispersion of clays during the subsequent treatment with NaOH, which is usually used to extract *p*-nitrophenol from soil.

Care is required in choosing a buffer, because soil phosphatase activity is inhibited by acetate and citrate–phosphate buffers, while nuclease is inhibited by both borate and phosphate buffers (Speir and Ross 1978). The inhibition by citrate-phosphate buffer is probably related to the effect of phosphate (Malcolm 1983).

Incomplete extraction of *p*-nitrophenol was observed in soils with high organic matter content and extractable Fe and Al. A quantitative recovery of phenol released from phenylphosphate was observed in these soils (Harrison 1979).

The method of Tabatabai and Bremner (1969) uses a concentration of 23 mM substrate in the soil suspension, while lower concentrations (from 1 to 7.5 mM) have been used by others (Malcolm 1983). Concentrations as low as 1 mM of *p*-nitrophenyl phosphates have to be avoided because K_m values of 1.37–5.69 have been reported for soil phosphomonoesterases (Malcolm 1983).

The colour due to *p*-nitrophenol is stable for 24 h.

Phosphodiesterase activity

(Browman and Tabatabai 1978)

Phosphodiesterase has been detected in microorganisms, plants and animals (Browman and Tabatabai 1978). Usually the activity of phosphodiesterase is much lower than the phosphomonoesterase in soils (Eivazi and Tabatabai 1977). Air-drying did not affect phosphodiesterase activity, while steam sterilization for 20 h at 120°C inactivated it (Eivazi and Tabatabai 1977). On the other hand

Sparling et al (1986) observed an inactivation of the enzyme activity after air-drying. Orthophosphate inhibited the enzyme activity in a competitive way (Browman and Tabatabai 1978). The enzyme activity is significantly correlated with the organic C content of soils (Browman and Tabatabai 1978). Phosphodiesterase has been extracted from a forest soil and chromatographically fractionated into seven fractions (Hayano 1987); 2,4-cyclic-nucleotide-2-phosphodiesterase (EC 3.1.4.1.6) or 2,3-cyclic-nucleotide-3-phosphodiesterase (EC 3.1.4.3.7) seem to be enzymes of one of these fractions.

Principle of the method

The method is based on the determination of the released *p*-nitrophenol after the incubation of soil samples with bis-*p*-nitrophenyl phosphate solution for 1 h at 37°C.

Materials and apparatus

See estimation of phosphomonoesterase.

Chemicals and solutions

Toluene

Tris buffer (50 mM, pH 8)

Dissolve 6.1 g of Tris (hydroxy methyl) amino methane in 800 ml distilled water, adjust the pH to 8 with (0.1 M) H_2SO_4 and bring up to 1000 ml with distilled water.

Sodium bis-*p*-nitrophenyl phosphate (5 mM)

Dissolve 0.1811 g of sodium bis-*p*-nitrophenyl phosphate in 80 ml of Tris buffer and dilute to 100 ml with the same buffer. Store at 4°C.

$CaCl_2$ (0.5 M)

See estimation of phosphomonoesterase.

Tris–NaOH extractant solution (0.1 M, pH 12)

Dissolve 12.2 g of Tris (hydroxy methyl) amino methane in 800 ml of distilled water, adjust the pH to 12 with NaOH (0.5 M) and bring up to 1000 ml with distilled water.

Tris solution (0.1 M, pH 10)

Dissolve 12.2 g of Tris (hydroxy methyl) amino methane in 800 ml distilled water and dilute to 1000 ml with distilled water.

Standard *p*-nitrophenol solution

See estimation of phosphomonoesterase.

Procedure

Place 1 g of sieved (2 mm), moist soil in an Erlenmeyer flask (50 ml), add 0.2 ml toluene, 4 ml Tris buffer and 1 ml sodium bis-*p*-nitrophenyl phosphate substrate. Mix the contents, stopper the flasks and incubate for 1 h at 37°C. After the incubation, add 1 ml $CaCl_2$ and 4 ml Tris–NaOH extractant solution. Mix the contents thoroughly and filter the suspension through a Whatman no. 2v folded filter paper. Measure the absorbance at 400 nm. To perform the blanks, make the addition of substrate after the addition of $CaCl_2$ and the extractant solution and immediately before filtration of the soil suspension. At high colour intensity, dilute the filtrate with Tris solution until the reading falls within the limits of the calibration curve.

Calibration curve

See estimation of phosphomonoesterase.

Calculation

See estimation of phosphomonoesterase.

Discussion

The assay should be performed only for a short incubation time because the product, *p*-nitrophenyl phosphate, can be further hydrolysed by phosphomonoesterase (Browman and Tabatabai 1978).

Anaerobic incubation of soil samples caused an increase or a decrease in the activity (Pulford and Tabatabai 1988).

The choice of buffer is important because orthophosphate, EDTA and citrate (5 mM) significantly inhibits the phosphodiesterase activity of soil (Browman and Tabatabai 1978).

K_m values between 1.3 and 2 mM for the substrate di-*p*-nitrophenyl phosphate have been found (Browman and Tabatabai 1978).

Phosphotriesterase activity

(Eivazi and Tabatabai 1977)

Paraoxanase, the enzyme responsible for catalysing the cleavage of diethyl-*p*-nitrophenyl phosphate (paraoxon) into *p*-nitrophenol and diethyl phosphate is indicated as a phosphotriesterase. However, according to Schmidt and Laskowski (1961), enzymes like paraoxanase should be classified as a phosphoric anhydride hydrolase rather than phosphatase because of acidic properties of nitro-substituted phenols. Since the method for assaying the activity of these enzymes in soil is based on the use of Tris-*p*-nitrophenyl phosphate, the term "phosphotriesterase" may not be a suitable name for identifying these enzymes.

Phosphotriesterase activity in soil shows an optimum pH around 10; it is increased by air-drying and inactivated by steam sterilization (Eivazi and Tabatabai 1977).

Principle of the method

The method is based on the colorimetric determination of the *p*-nitrophenol released after the incubation of soil with buffered Tris-*p*-nitrophenylphosphate.

Materials and apparatus

The same materials and apparatus described in the earlier section on phosphomonoesterase activity are required as well as 100-mesh glass beads.

Chemicals and solutions

Toluene

Modified universal buffer stock solution

Prepare as described in the section on phosphomonoesterase activity.

MUB pH 10.0

Titrate 200 ml of MUB stock solution to pH 10.0 under continuous stirring with 0.1 M NaOH and then dilute to 1000 ml with distilled water.

Sodium Tris-*p*-nitrophenylphosphate in solid form.

$CaCl_2$ 0.5 M solution

Prepare as described in the section on phosphomonoesterase activity.

NaOH 0.5 M solution

See estimation of phosphomonoesterase.

NaOH 0.1 M solution

See estimation of phosphomonoesterase.

Standard *p*-nitrophenol solution

Prepare as indicated in the section on phosphomonoesterase activity.

Procedure

Place 23 mg of sodium Tris-*p*-nitrophenylphosphate (insoluble in water) in a 50 ml Erlenmeyer flask, mixed with 1 g of moist sieved (2 mm) soil and covered with 2 g of 100-mesh glass beads. Then add 0.25 ml of toluene (added dropwise to the surface of the glass beads) and 5 ml of MUB, pH 10.0. Stopper the flasks, mix the contents and incubate for 1 h at 37°C. After the incubation, add 1 ml of 0.5 M $CaCl_2$ and 4 ml of 0.5 M NaOH, mix the contents, and filter the soil solution through a Whatman no. 2v folded filter paper. Measure the optical density of the filtrate at 400 nm. As for the assay of the other two phosphatases, controls should be performed with each soil analysed; in these controls, the substrate (23 mg of Tris-*p*-nitrophenylphosphate) should be added after 0.5 M $CaCl_2$ and 0.5 M NaOH, but immediately before filtering the soil suspension.

Calibration curve

Proceed as indicated for phosphomonoesterase estimation.

Discussion

It has been found that addition of toluene greatly increased the phosphotriesterase activity (Eivazi and Tabatabai 1977), probably due to a better dissolution of the substrate (insoluble in water) and/or to better soil–substrate contact.

Because the Tris-*p*-nitrophenylphosphate is insoluble in water, the K_m and V_{max} Michaelis–Menten constants cannot be determined.

Pyrophosphatase activity

(Dick and Tabatabai 1978)

The enzyme catalysing the hydrolysis of pyrophosphate to orthophosphate is the inorganic pyrophosphatase (pyrophosphate phosphohydrolase, EC 3.6.1.1) and has been extracted from bacteria, animals, and plants (Dick and Tabatabai 1978). Interest in this enzyme activity in soils derives from the fact that polyphosphates are used as fertilizers.

It is difficult to determine pyrophosphatase activity in soil not only due to problems concerning the extraction and determination of orthophophate, but also because the substrate (pyrophosphate) may continue to be hydrolysed abiotically (for example, at low pH) and the presence of pyrophosphate may inhibit the measurement of orthophosphate.

A pH optimum of 8 has been observed in soil (Dick and Tabatabai 1978) while heating of moist soils over 60°C inactivates enzyme activity (Tabatabai and Dick 1979). Toluene and EDTA have no effect on the pyrophosphatase activity, while inhibition is observed when soil is steam-sterilized or treated with formaldehyde, fluoride, oxalate and carbonate (Dick and Tabatabai 1978). The activity decreased with depth, and it is positively correlated with organic C and clay content. The pyrophosphatase activity of soil is negatively correlated with the $CaCO_3$ content of soil (Tabatabai and Dick 1979). The best method for preserving enzyme activity is to store the field-moist soil for 2 months at 5°C (Tabatabai and Dick 1979).

Principle of the method

The assay of inorganic pyrophosphatase is based on the determination of the orthophosphate released when the soil is incubated with buffered pyrophosphate solution; the orthophosphate is extracted by 1 M H_2SO_4, which gives the quantitative recovery of orthophosphate added to soils. The colorimetric method used for determining the orthophosphate extracted in the presence of pyrophosphate is specific for orthophosphate (Pi) and has been developed to determine Pi in aqueous solutions containing labile organic and inorganic P compounds (Dick and Tabatabai 1977). This method involves a rapid formation of heteropoly blue in the presence of ascorbic acid–trichloroacetic acid reagent and formation of complexes of the excess molybdate ions by a citrate–arsenite reagent to prevent further formation of blue colour due to the Pi liberated by hydrolysis of pyrophosphate.

Materials and apparatus

Centrifuge tubes (50 ml) with stoppers
Incubator or temperature-controlled water bath, adjustable to 37°C
Reciprocal (end-over-end) shaker
High-speed centrifuge (12,000 rev min^{-1})
Spectrometer
Volumetric flasks (25, 100, 500, 1000 ml)

Chemicals and solutions

Modified universal buffer stock solution

Prepare as described in the section on phosphomonoesterase activity.

Pyrophosphate solution, 50 mM

Dissolve 2.23 g of sodium pyrophosphate decahydrate in approximately 20 ml of MUB stock solution, titrate the solution to pH 8.0 with 0.1 м HCl and dilute to 100 ml with distilled water. This solution must be prepared daily.

Modified universal buffer pH 8.0

Titrate 200 ml of MUB stock solution to pH 8.0 under continuous stirring with 0.1 м HCl and then dilute to 1000 ml with distilled water.

Sulphuric acid (0.1 м)

Add 28 ml of concentrated H_2SO_4 to approximately 700 ml of water and dilute to 1000 ml with distilled water.

Ascorbic acid (0.1 м)-trichloroacetic acid (0.5 м) (reagent A)

Dissolve 8.8 g of ascorbic acid and 41 g of trichloroacetic acid in approximately 400 ml distilled water and adjust the volume to 500 ml with distilled water. This reagent must be prepared daily.

Ammonium molybdate (0.015 м) reagent (reagent B)

Dissolve 9.3 g of ammonium molybdate tetrahydrate in about 450 ml of distilled water and adjust the volume to 500 ml with distilled water.

Sodium citrate (0.15 м)–sodium arsenate (0.3 м)–acetic acid (7.5%) reagent (reagent C)

Dissolve 44.1 g of sodium citrate and 39 g of sodium arsenite in about 800 ml of distilled water and adjust the volume to 1000 ml with distilled water.

Standard phosphate stock solution

Dissolve 0.4390 g of potassium dihydrogen phosphate in about 700 ml of distilled water and dilute to 1000 ml with water. This solution contains 10 µg of orthophosphate-P ml^{-1}.

Procedure

Place 1 g of moist soil in a 50 ml centrifuge tube, add 3 ml of pyrophosphate solution and mix the contents. Stopper the tube and incubate at 37°C for 5 h. Then remove the stopper and immediately add 3 ml of MUB, pH 8.0 and 25 ml of 1 м H_2SO_4. Stopper the tube and shake for 3 min in a end-over-end shaker. Centrifuge the soil suspension for 30 s at 12,000 rev min^{-1} and take 1 ml of the supernatant for orthophosphate (Pi) analysis. For determining orthophosphate place 10 ml of reagent A in a 25 ml volumetric flask and add the 1 ml aliquot of the supernatant, and immediately add 2 ml of reagent B and 5 ml of reagent C. Swirl the flask after the addition of the supernatant and each of the reagents B and C, and make up the volume with distilled water. After 15 min measure the optical density at 700 nm. Controls should be performed in each series of analyses to allow for Pi not derived from pyrophosphatase activity. To perform controls, add 3 ml of MUB pH 8.0 to 1 g of soil and incubate for 5 h. Then add 3 ml of 50 mм pyrophosphate solution and immediately add 25 ml 1 м H_2SO_4, and then continue as described above.

Calibration curve

Calculate the Pi content of the aliquot analysed by reference to a calibration graph. To prepare this graph, pipette aliquots of 5, 10, 15, 20 and 25 ml of the standard orthophosphate stock solution into 100 ml volumetric flasks, adjust the volumes with distilled water and mix thoroughly; in this way, you will have standards containing 0, 5, 10, 15, 20 and 25 µg of orthophosphate-Pi. Analyse 1 ml aliquots of the dilute standards by the procedure described for samples.

Discussion

The blue colour developed during the determination of orthophosphate is stable in laboratory light for at least 4 h.

The soil suspension is centrifuged for 30 s to minimize the hydrolysis of pyrophosphate during extraction. Filtration takes time and may lead to hydrolysis of pyrophosphate present in the soil suspension.

Pyrophosphate is chemically hydrolysed at temperatures above 40°C, so special controls should be included, if higher temperatures have to be used.

Dick and Tabatabai (1978) found that at a pyrophosphate concentration above 60 mM the enzyme reaction decreased. This result can be due to two factors: (1) inhibition of the pyrophosphatase activity at high concentrations of the substrate; (2) the concentration of Mg^{2+} required for activation of pyrophosphate, and binding the substrate to the soil enzyme may not be sufficient at high pyrophosphate concentrations (Butler 1971).

It is important that the extraction and the analysis of the orthophosphate liberated from pyrophosphatase be carried out immediately, because pyrophosphate hydrolyses slowly with time in the presence of the extractant.

The reagent should be added immediately after addition of reagent B to remove the excess molybdate ions and avoid hydrolysis of acid-labile phosphate.

Cellulase activity

K. Alef
P. Nannipieri

Cellulose is the most abundant structural polysaccharide of plant cell walls. Microbial degradation of cellulose is therefore an important process in the degradation of plant debris (Rai and Srivastava 1983; Sinsabaugh and Linkins 1988). Cellulose is a linear polymer of D-glucose with β(1–4) glucosidic linkages. The minimum molecular weight of cellulose from different sources has been estimated to vary from about 50,000 to 250,000 in different species, equivalent to 300–15,000 glucose moieties. Although cellulose has a high affinity for water, it is completely insoluble in it.

Cellulase catalyses the hydrolysis of cellulose to D-glucose and consists of at least three enzymes: endo-β-1,4-glucanases (EC 3.2.1.4); exo-β-1,4-glucanases (EC 3.2.1.91); and β-glucosidases (EC 3.2.1.22) (Lee and Fan 1980; Eriksson and Wood 1985; Sinsabaugh and Linkins 1989). Exocellulases bind to crystalline cellulose and cleave celluloligosaccharides from the non-reducing ends of cellulose molecules while endocellulases randomly cleave glucosidic linkages along non-crystalline parts of cellulose; β-glucosidases release glucose from celluloligosaccharides and aryl-β-glucosides.

The most important properties affecting the susceptibility of the cellulose to enzymatic hydrolysis are the degree of crystallinity, the nature of the associated substances and the surface area (Marsden and Gray 1986).

The degradation of cellulose in soils is a slow process and depends on the concentration, location, and mobility of cellulases (Hayano 1986). Type of litter, substrate concentrations, pH, temperature and water content significantly affect cellulose degradation (Hunt 1977; Schröder and Gewehr 1977; Schröder and Urban 1985; Sinsabaugh and Linkins 1988; Tateno 1988; Kshattriya et al 1992). A differential adsorption of the various cellulases on the lignocellulosic material has been hypothesized. According to Sinsabaugh and Linkins (1988), exocellulase is largely associated with the cellulose component, while endocellulase is adsorbed by the lignin component. Cellulose is the first component that is hydrolysed and the ratio of endocellulase to exocellulase activity increases during the decomposition of deciduous leaf litters.

The cellulase activity of soil generally has optimum activity at pH 5–6 and at a temperature ranging from 30 to 50°C (Benefield 1971; Pancholy and Rice 1973; Hope and Burns 1987). Air-drying strongly inactivates cellulase activity in soil (Speir and Ross 1981). An increase in cellulase activity occurs in rhizosphere soil, as compared to the activity of non-rhizosphere soil. The effect of vegetation, season and type of agricultural managements on cellulase activity of soil has been studied extensivelly (Kiss et al 1978).

Soil cellulases are mainly produced by fungi (Clark and Stone 1965; Yamana et al 1970; Hayano 1986; Rhee et al 1987). Only a few species of *Actinomyces* are able to grow on cellulose as their only carbon source (Stutzenberger 1972). *Clostridium* seems to be the main cellulolytic microorganism under anaerobic conditions (Gottschalk et al 1981; Joliff et al 1989).

Several methods are available to estimate cellulase activity of soil. They are based on the determination of either released reducing sugars or evolved $^{14}CO_2$. Cotton strips, radioisotope-labelled cellulose and carboxy methyl cellulose (CMC) are used as the substrate for the enzyme (Benefield 1971; Pancholy and Rice 1973; Latter and Howson 1977; Ibister et al 1980; Sato 1981; Schinner and von Mersi 1990).

Assay of cellulase activity

(Schinner and Von Mersi 1990)

Principle of the method

The method is based on the determination of released reducing sugars after the incubation of soil samples with CMC for 24 h at 50°C. The activities of endoglucanase and β-glucosidase can only be estimated.

Materials and apparatus

Spectrophotometer
Incubator adjustable to 50°C
Erlenmeyer flasks (100 ml)
Filter paper (Schleicher & Schüll)
Glass funnel
Stirrer
Water bath adjustable to 100°C

Chemicals and solutions

Acetate buffer (2 M, pH 5.5)

Dissolve 164.08 g of sodium acetate (water-free) in 700 ml distilled water, adjust the pH to 5.5 with acetic acid and dilute to 1000 ml with distilled water.

Carboxymethyl cellulose sodium salt solution (0.7% w/v)

Dissolve 7 g CMC in 1000 ml of acetate buffer and stir at 45°C for 2 h. This solution can be stored for 1 week at 4°C.

Reagents for the determination of reducing sugars

Reagent A

Dissolve 16 g of anhydrous sodium carbonate (water-free) and 0.9 g potassium cyanide in distilled water, and dilute to 1000 ml with distilled water.

Reagent B

Dissolve 0.5 g of potassium ferric hexacyanide in distilled water and dilute to 1000 ml with distilled water. Store in brown bottles.

Reagent C

Dissolve 1.5 g of ferric ammonium sulphate, 1 g of sodium dodecyl sulphate and 4.2 ml of concentrated H_2SO_4 in distilled water (50°C), and bring up to 1000 ml with distilled water.

Glucose monohydrate solution

28 mg 1000 ml^{-1} distilled water.

Procedure

Place 10 g (arable) or 5 g (forest) of moist sieved (2 mm) soil in an Erlenmeyer flask, add 15 ml of acetate buffer and 15 ml of CMC solution, cap the flask, and incubate for 24 h at 50°C. After the incubation, filter the resulting soil suspension. The control is prepared by adding 15 ml CMC solution after the incubation but immediately before filtration. Dilute 1 ml of the filtrate with distilled water to 20 ml (agricultural soils) or 30 ml (forest soils). Pipette 1 ml of the diluted filtrate into glass tubes, add 1 ml reagent A and 1 ml reagent B. Close the tubes, mix well and boil in a water bath (100°C) for exactly 15 min. After cooling in a water bath at 20°C for 5 min, add 5 ml reagent C, mix well and allow to stand at 20°C for 60 min for colour development. Measure the optical density within the following 30 min at 690 nm against the blank.

Calibration curve

Pipette in test tubes 0, 0.1, 0.2, 0.3, 0.4, 0.5, 0.6, 0.7, 0.8, 0.9 and 1.0 ml of glucose monohydrate solution. Dilute with distilled water to 1ml and determine the reducing sugars as described above.

Calculation

Correct the measurements for the control and calculate briefly:

Glucose equivalent ($\mu g\ g^{-1}$ dwt 24 h^{-1}) =

$$\frac{C \times v \times f}{sw \times dwt} \quad (7.14)$$

where C is the measured glucose concentration (μg glucose ml^{-1} filtrate), v is the volume of the test suspension (in this system 30 ml), f is the dilution factor (20 for agricultural and 40 for forest soils), sw is the weight of the moist soil used (10 or 5 g) and dwt is the weight of 1 g moist soil.

Discussion

By means of this method, the endoglucanase and β-glucosidase but not the exoglucanase activity can be estimated.

This method gives low values in arable soils but it is recommended for forest soils. The authors also recommended that the activity of moist soil is determined because air-drying decreases the activity (Schinner and von Mersi 1990).

The reaction is not linear with time (Schinner, personal communication).

High concentrations of salts, ammonium ions, silver, manganese oxalate, oxalic acid, hydroxide and fluoride disturb the glucose determination. For these reasons, it is recommended to dilute the filtrate.

No significant differences in the cellulase activity rates were found when the measurements were performed in the absence or presence of toluene (Schinner and Von Mersi 1990).

Strict control of the pH during the determination of reducing sugars is necessary. After the addition of reagents A and B the pH must be > 10.5; it must be lower than 2 after the addition of reagent C.

The determination of reducing sugars is a modification of the methods of Hagedorn and Jensen (1923), and Park and Johnson (1949).

There are no significant correlations between the cellulase activities and other microbial parameters of soils (Kiss et al 1978).

Disposal of the cyanide-containing solutions can be performed by oxidation with H_2O_2 at an alkaline pH value.

Assay of cellulase activity

(Hope and Burns 1987)

Principle of the method

The method is based on the determination of liberated reducing sugars after the incubation of soil samples with Avicel for 16 h at 40°C. By this method, the total cellulase (endoglucanase, exoglucanase and β-glucosidase) activity can be estimated.

Materials and apparatus

Spectrophotometer
Shaking water bath adjustable at 40 and 100°C
Erlenmeyer flasks (25 ml)
Centrifuge with centrifuge tubes (50 ml)

Chemicals and solutions

Acetate buffer (0.1 M, pH 5.5 containing 0.2% azide)

Dissolve 13.6 g sodium acetate tetrahydrate in about 700 ml distilled water and adjust the pH to 5.5 with dilute acetic acid. Then dissolve 2 g of sodium azide and dilute to 1000 ml with distilled water.

Washed Avicel (obtained from Honeywill and Stein, Willington, UK).

Reagents for the determination of reducing sugars:

Copper reagent:

Solution I

Dissolve 15 g Na–K-tartarate and 30 g Na_2CO_3 in about 300 ml distilled water. Then dissolve 20 g $NaHCO_3$ in the solution.

Solution II

Dissolve 180 g Na_2SO_4 in 500 ml distilled water, boil the solution to remove dissolved air. Allow the solution to stand at room temperature for cooling.

Solution III

Mix solutions I and II, and dilute to 1000 ml with distilled water.

Solution IV

Dissolve 5 g $CuSO_4.5H_2O$ and 45 g Na_2SO_4 in distilled water and bring up to 250 ml with distilled water.

Solution V

Shortly before use, mix one volume of solution III and one volume of solution IV.

Arsenate–molybdate solution

Dissolve 25 g ammonium–molybdate in 450 ml distilled water, add 21 ml of concentrated H_2SO_4 under continuous stirring. Finally add 25 ml of $Na_2HAsO_4.7H_2O$ (3 g 25 ml^{-1} water) and mix thoroughly. Allow the solution to stand for 2 days at 37°C and fill in brown bottles. Shortly before use, dilute one volume of this solution with two volumes of H_2SO_4 (0.75 M).

Glucose-monohydrate standard solution 79 mg 1000 ml^{-1} distilled water.

Procedure

Place 1 g of moist soil in an Erlenmeyer flask (25 ml), add 5 ml of acetate buffer and 0.5 g Avicel. Cap the flasks and incubate for 16 h at 40°C in a shaking water bath. Finally, stop the reaction by centrifugation (2500*g*, 10 min). To prepare the control, make the addition of Avicel after the incubation period and immediately before centrifugation. All measurements should be carried out at least in duplicate. Determine the reducing sugars in the supernatants according to the method of Nelson and Somogyi (in Spiro 1966); pipette 1 ml of the supernatant into test tubes, add 1 ml solution V, cap the tubes with glass balls and boil in a water bath for 20 min. After cooling, add 1 ml of the diluted arsenate–molybdate solution and mix well. Finally dilute the mixture with 3 ml of distilled water and measure the colour optical density at 520 nm.

Calibration curve

See assay of cellulase activity.

Calculation

Correct the measurements for the blank and calculate briefly:

Glucose ($\mu g\ g^{-1}$ dwt 16 h^{-1}) =

$$\frac{C \times v}{dwt} \qquad (7.15)$$

where C is the measured glucose concentration ($\mu g\ ml^{-1}$ supernatant), v is the volume of the soil suspension (in this system, 5.5 ml) and *dwt* is the dry weight of 1 g moist soil.

Discussion

The enzymatic reaction is linear for 16 h (Hope and Burns 1987).

Incubation of soil suspension with azide prevented the assimilation of glucose by soil microorganisms. For these reasons it is suggested that inhibitors are used (Hope and Burns 1987). Since azide does not always inhibit microbial mineralization of available carbon, the efficiency of the inhibitor has to be verified for each soil.

This method does not permit discrimination between the various cellulases and β-glucosidase activity.

The presence of high glucose oxidase activity in soils can falsify the estimation of cellulase activity (Ross 1974).

β-Glucosidase activity

K. Alef
P. Nannipieri

β-Glucosidase (β-D-glucoside glucohydrolase, EC 3.2.1.21) is the rate limiting enzyme in the microbial degradation of cellulose to glucose. The enzyme catalyses the hydrolysis of glucosides according to the following reaction:

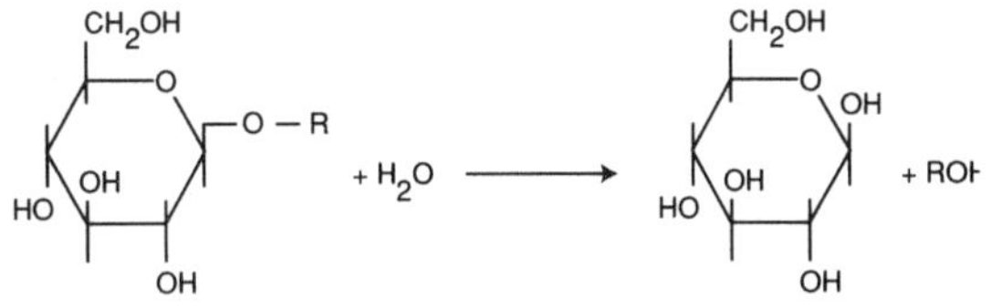

(7.16)

It has been detected in microorganisms, plants and animals (Bahl and Agrawal 1972; Dey and Pridham 1972; Wallenfels and Weil 1972). This enzyme is included in the category of glucosidases that hydrolyse disaccharides. α-Glucosidase (called maltase EC 3.2.1.20), which catalyses the hydrolysis of α-D-glucopyranoside, is also included among glucosidases. Other glucosidases are α-galactosidase (EC 3.2.1.2) and β-galactosidase (called lactase EC 3.2.1). β-Glucosidase is more prominent in soil than α-glucosidase and α and β galactosidases (Tabatabai 1982). *p*-Nitrophenyl-β-D-glucoside has been used as a substrate to estimate β-glucosidase activity in soils. The K_m values ranged from 1.3 to 2.4 mM for this substrate (Eivazi and Tabatabai 1988). The β-glucosidase was inactivated at 70°C and shows a significant correlation with the organic matter of soil (Eivazi and Tabatabai 1988).

Assay of the β-glucosidase activity

(Tabatabai 1982; Eivazi and Tabatabai 1988)

Principle of the method

The method is based on the determination of the released *p*-nitrophenol after the incubation of soil with *p*-nitrophenyl glucoside solution for 1 h at 37°C:

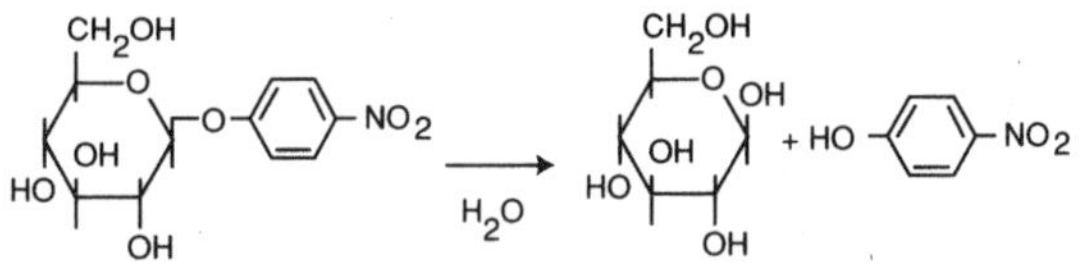

(7.17)

Materials and apparatus

Spectrophotometer
Erlenmeyer flasks (50 ml)
Water shaking bath adjustable to 37°C
Filter paper (Whatman 2v)

Chemicals and solutions

Toluene
Modified universal buffer (MUB), pH 6.0

See estimation of phosphomonoesterase

$CaCl_2$ (0.5 M)

See estimation of phosphomonoesterase

Tris buffer (0.1 м, pH 12 and 10)

See estimation of phosphomonoesterase

p-Nitrophenol standard solution

See estimation of phosphomonoesterase

p-Nitrophenyl-β-D-glucoside (PNG) solution (25 mM)

Dissolve 0.377 g of PNG in 40 ml of MUB buffer and dilute to 50 ml with the same buffer. Store at 4°C.

Procedure

Place 1 g of moist, sieved (2 mm) soil in an Erlenmeyer flask (50 ml), add 0.25 ml of toluene, 4 ml of MUB solution, 1 ml PNG solution, stopper the flasks, and mix the contents thoroughly and incubate for 1 h at 37°C. After the incubation, add 1 ml of $CaCl_2$ solution, 4 ml of Tris buffer, pH 12, swirl the flasks and filter the soil suspensions immediately (Whatman filter 2v). Measure the colour intensity at 400 nm. If the optical intensity is too high, dilute the filtrate with Tris buffer pH 10. To prepare the blanks, make the addition of the substrate PNG at the incubation before adding the $CaCl_2$ and Tris buffer. All measurements are carried out in triplicate with one blank.

Calibration curve

See estimation of phosphomonoesterase.

Calculation of results

See estimation of phosphomonoesterase.

Discussion

The enzyme activity had an optimum at pH 5–6.8 (Agrawal and Bahl 1968; Tabatabai 1982). Toluene stimulates the β-glucosidase activity (Eivazi and Tabatabai 1988). The enzyme activity had an optimum at a temperature of 60°C; nevertheless, a temperature of 37°C is used for the assay (Eivazi and Tabatabai 1988).

The PNG solution was stable for several days at 4°C.

The addition of $CaCl_2$ and Tris buffer pH 12 was necessary for an efficient extraction of *p*-nitrophenol from soils.

The reaction was linear up to 2 g soil (Eivazi and Tabatabai 1988).

Hayano and Tubakil (1985) assumed that the β-glucosidase in soils originates from fungi.

Assay of β-glucosidase activity

(Hoffmann and Dedeken 1965)

Principle of the method

The method is based on the determination of released salignin (2-oxymethyl-phenol) after the incubation of soil with salicin 2-(hydroxymethyl)phenyl β-D-glucopyranoside: β-glucoside selignin for 3 h at 37°C.

Materials and apparatus

See estimation of phosphomonoesterase.

Chemicals and solutions

Acetate buffer (2 м, pH 6.2)

Dissolve 164 g of sodium acetate (water-free) in 700 ml of distilled water, adjust the pH to 6.2 with dilute acetic acid and bring up to 1000 ml with distilled water.

Borate buffer (pH 9.6)

Dissolve 56.85 g of sodium borate.10 H_2O or 30 of sodium tetraborate in 1.5 l of warm distilled water, adjust the pH to

9.6 with NaOH (20%) and dilute with distilled water to 2000 ml.

Dibromo quinone chlorimide solution

200 mg 100 ml^{-1} ethanol.

Substrate solution (2.306 g β-glucoside-salignin 100 ml^{-1})

Dissolve 2.306 g β-glucoside-salignin in about 85 ml of distilled water and bring up to 100 ml with distilled water.

Salignin standard solution

0.1 mg ml^{-1} distilled water.

Toluene

Procedure

Place 5 g of moist, sieved (2 mm) soil into a volumetric flask (50 ml), add 1 ml toluene and allow to stand for 15 min at room temperature. Then add 10 ml of substrate solution, 20 ml of acetate buffer and incubate for 3 h at 37°C. After the incubation dilute the contents to 50 ml with distilled water, mix thoroughly and filter immediately. Pipette 1–3 ml of the filtrate into a volumetric flask (50 ml), add 20 ml acetate buffer, 2 ml borate buffer and 0.5 ml dibromo quinone chlorimide solution. Dilute with distilled water to 50 ml and after 90 min measure the optical density at 578 nm. To perform the control, add 10 ml of distilled water instead of the substrate solution. Measurements are carried out in triplicate with one control.

Calibration curve

Pipette 0, 1, 3, 5 and 10 ml of the standard solution into volumetric flasks (50 ml), add 20 ml acetate buffer, 2 ml borate buffer, 0.5 of dibromo quinone chlorimide solution, and dilute to 50 ml with distilled water. After 90 min, measure the optical density at 578 nm.

Calculation

Correct for the control and calculate briefly:

Salicin (μg g^{-1} dwt 3 h^{-1}) =

$$\frac{C \times 50}{dwt \times 5} \quad (7.18)$$

where C is the measured salicin concentration (μg ml^{-1} filtrate), dwt is the dry weight of 1 g moist soil, 50 is the total volume of the reaction mixture and 5 is the weight of soil used.

Discussion

With the buffer used in this assay, the pH value of the reaction mixture remains stable during the incubation time.

Because of the short incubation time, toluene can be omitted from the assay.

Saccharase activity

K. Alef
P. Nannipieri

Maltose, lactose and sucrose are the most common water-soluble disaccharides, which consist of two monosaccharides joined by a glycosidic linkage. One of the most abundant sugars in plants is sucrose, a disaccharide of glucose and fructose [O-β-D-fructofuranosyl-(2,1)-α-D-glucopyranoside]. Unlike most disaccharides and oligosaccharides, sucrose contains no free anomeric carbon atoms. For this reason, sucrose does not undergo mutarotation and does not act as a reducing sugar. It is much more readily hydrolysed than other disaccharides.

Saccharase or invertase (β-D-fructofuranoside fructohydrolase EC 3.2.1.26) catalyses the hydrolysis of sucrose to D-glucose and D-fructose, and is widely distributed in microorganisms, animals and plants. In soil it has optimum activity at pH 5.0–5.6 and temperature of 50°C (Roberge 1978; Frankenberger and Johanson 1983c). The K_m values of saccharase in soils have been found to range from 16.3 to 42.1 mM (Frankenberger and Johanson 1983c).

Air-drying decreases saccharase activity of soil (Hoffmann 1959; Ross 1965, 1968; Frankenberger and Johanson 1983b), while activity is very stable at 4°C and decreased with soil depth (Hoffman and Elias-Azar 1965; Dutzler-Franz 1977; Holz 1986a, 1986b). Toluene and sterilization with gamma rays had no influence on the enzyme activity (Voets et al 1965; Roberge 1978). Hoffmann and Pfitscher (1982) found significant correlations between saccharase activity and the organic C in soil. In contrast, Dutzler-Franz (1977) found no significant correlations ($n = 54$).

Saccharase activity in soil was the only carbohydratase activity significantly correlated with the carbohydrate:lignin ratio of added plant material (Rice and Mollik 1977). In a climosequence of soils in tussock grassland, the ratio saccharase to amylase activity was greatest in the less developed soils (Ross 1975).

Assay of saccharase activity

(Schinner and Von Mersi 1990)

Principle of the method

The method is based on the determination of released reducing sugars after the incubation of soil with sucrose for 3 h at 50°C.

Materials and apparatus

See estimation of cellulase activity.

Chemicals and solutions

Acetate buffer (2 M, pH 5.5)

See estimation of cellulase activity.

Sucrose solution (1.2%)

Dissolve 12 g of sucrose in acetate buffer, stir at 45°C for 2 h and dilute to 1000 ml with acetate buffer. The solution can be stored for 1 week at 4°C.

Reagents for the determination of reducing sugars:

Reagent A

See estimation of cellulase activity.

Reagent B

See estimation of cellulase activity.

Reagent C

See estimation of cellulase activity.

Glucose monohydrate solution

See estimation of cellulase activity.

Procedure

Place 5 g of moist sieved soil (2 mm) in an Erlenmeyer flask (100 ml), add 15 ml of sucrose solution and 15 ml of acetate buffer, stopper the flask and incubate for 3 h at 50°C. After the incubation filter the resulting soil suspension. In the control, the substrate (15 ml of sucrose solution) is added at the end of the incubation and immediately before filtration. Dilute 1 ml of the filtrate to 40 ml with distilled water and pipette 1 ml of the diluted filtrate into glass tubes, add 1 ml reagent A, and 1 ml reagent B. Close the tubes, mix well and boil in a water bath (100°C) for exactly 15 min. After cooling in a water bath at 20°C for 5 min, add 5 ml of reagent C, mix well and allow to stand at 20°C for 60 min for colour development. Measure the optical density within the following 30 min at 690 nm against the control. It is recommended that all measurements are carried out in triplicate.

Calibration curve

See estimation of cellulase activity.

Calculation of results

Correct the measurements for the control and calculate according to the following relationship:

Glucose (μg g^{-1} dwt 24 h^{-1}) =

$$\frac{C \times v \times f}{5 \times dwt} \qquad (7.19)$$

where C is the glucose concentration of the solution in micrograms per millilitre, v is the volume of the suspension (in this case 30 ml), f is the dilution factor (here 40), 5 is the weight of the moist soil used in grams and dwt is the dry weight of 1 g of moist soil.

Discussion

We recommend the method by Schinner and Von Mersi (1990) for the determination of CMC-cellulase activity in soil, because it is based on a shorter incubation time than the method proposed by Frankenberger and Johanson (1983a).

Assay of saccharase activity

(Hoffmann and Pallauf 1965)

Principle of the method

The method is based on the determination of released reducing sugars after the incubation of soil samples with sucrose solution for 3 h at 37°C.

Materials and apparatus

Spectrophotometer
Shaking bath adjustable to 37°C
Volumetric flasks (100 ml)
Filter paper (Schleicher and Schüll 512, 1/2)
Water bath adjustable to 100°C

Chemicals and solutions

Acetate buffer (2 M, pH 5.5)

See estimation of cellulase activity.

Sucrose solution

200 g/1000 ml distilled water.

Copper reagent

150 g $CvSO_4.5H_2O$ 500 ml^{-1} distilled water (solution 1).

Dissolve 25 N_2CO_3 (water-free), 25 g potassium ferric hexacyanide, 20 g

$NaHCO_3$ (water-free) and 200 g Na_2SO_4 (water-free) in about 700 ml distilled water, respectively, warm the mixture and dilute to 1000 ml with distilled water. Finally, add a few drops of toluene and store at 37°C (solution 2).

Mix one volume of solution 1 and 25 volumes of solution 2

Na_2HPO_4 solution

Dissolve 17.9 g of $Na_2HPO_4.12H_2O$ in 700 ml distilled water and bring up with distilled water to 1000 ml.

Molybdate solution

Ammonium heptamolybdate (5%)

Dissolve 5 g of ammonium heptamolybdate in 70 ml distilled water and bring up with distilled water to 100 ml.

Dilute 200 ml of concentrated H_2SO_4 with 800 ml distilled water.

Mix one volume of ammonium heptamolybdate solution and one volume of H_2SO_4 solution (solution 3).

Toluene

Standard solution

Dissolve 100 mg glucose and 100 mg fructose in 160 ml distilled water and bring up with distilled water to 200 ml. After the addition of few drops of toluene, store at 4°C.

Procedure

Place 10 g moist, sieved (2 mm) soil in a volumetric flask (100 ml), add 2 ml of toluene and allow to stand for 15 min. Then add 10 ml of sucrose solution, 10 ml of acetate buffer, stopper the flask and incubate for 3 h at 37°C. After the incubation, dilute the contents to 100 ml with distilled water (38–40°C) and filter immediately. In the control, make the addition of sucrose solution after the incubation period and immediately before the filtration. A second control without soil should be carried out. It is recommended that all measurements are carried out in triplicate. Pipette 5 ml filtrate into a volumetric flask (100 ml), add 4 ml of copper reagent and boil in a water bath for 25 min. After cooling, add 2 ml of Na_2HPO_4 solution and 5 ml of molybdate solution 3, mix thoroughly, and allow to stand for 60 min. Finally, dilute to 100 ml with distilled water and after 15 min measure the optical density at 578 nm.

Calibration curve

Pipette 0, 5, 10, 15 and 20 ml of the standard solution into volumetric flasks (100 ml), add 10 ml of acetate buffer and dilute to 100 ml with distilled water. Determine the reducing sugar concentrations as described above. The final sugar concentrations are 0, 50, 100, 150, 200 $\mu g\ ml^{-1}$.

Calculation

Correct the measurements for the control and calculate briefly:

Sugar ($\mu g\ 10\ g^{-1}$ dwt $3\ h^{-1}$) =

$$\frac{C \times 100 \times F}{dwt} \tag{7.20}$$

where *dwt* is the dry weight of 1 g of moist soil, 100 is the final volume of the diluted filtrate, *C* is the measured concentration ($\mu g\ ml^{-1}$ diluted filtrate or supernatant) and *F* = 20 (dilution factor).

Discussion

The resulting colour complex is stable for up to 12 h.

When using soils with a high clay content (>25%), filtration of a portion of the soil suspension is recommended to avoid a long filtration time.

Xylanase activity

K. Alef
P. Nannipieri

(Schinner and Von Mersi 1990)

Xylans, plant homopolysaccharides, consisting of xylose, arabinose, glucose, mannose and galactose, can be degraded and used as carbon sources by soil microorganisms. Xylanases (3.2.1.8 xylan 4-xylanhydrolase) include endoxylanases, which catalyse the hydrolysis of xylans to oligosaccharides, and exoxylanases, which catalyse the hydrolysis of oligosaccharides to reduced monomers. Like the cellulases, the estimation of xylanase activity in soils is based on the determination of reducing sugars (Hashimoto et al 1971; Schinner and Von Mersi 1990). The pH optimum of xylanase activity has been shown to range from 5 to 7, while its temperature optimum is generally around 50°C (Schinner and von Mersi 1990).

Principle of the method

The method is based on the determination of released reducing sugars after the incubation of soil samples with xylan for 3 h at 50°C.

Materials and apparatus

See estimation of cellulase activity.

Chemicals and solutions

Acetate buffer (2 M, pH 5.5)

See estimation of cellulase activity.

Xylan suspension

Dissolve 12 g xylan (Serva 3800) in acetate buffer, stir at 45°C for 2 h and dilute to 1000 ml with acetate buffer.

Reagents for the determination of reducing sugars

Reagent A

See estimation of cellulase activity.

Reagent B

See estimation of cellulase activity.

Reagent C

See estimation of cellulase activity.

Glucose monohydrate solution

See estimation of cellulase activity.

Procedure

Place 5 g of moist sieved (2 mm) soil in an Erlenmeyer flask (100 ml), add 15 ml of xylan suspension, 15 ml of acetate buffer, cap the flask and incubate for 24 h at 50°C. After the incubation, filter the soil suspension. In the control, the addition of 15 ml xylan suspension is carried out after the incubation and immediately before filtration. Dilute 1 ml of the filtrate to 40 ml with distilled water (forest soils). Pipette 1 ml of the diluted filtrate into glass tubes, add 1 ml reagent A, and 1 ml reagent B. Close the tubes, mix well and boil in a water bath (100°C) for exactly 15 min. After cooling in a water bath at 20°C for 5 min, add 5 ml reagent C, mix well, and allow to stand at 20°C for 60 min for colour development. Measure the optical density within the following 30 min at 690 nm against the blank. All measurements should be carried out in triplicate.

Calibration curve

See estimation of cellulase activity.

Calculation of results

Correct the measurements for the control and calculate as follows:

Glucose-equivalent (µg g^{-1} dwt 24 h^{-1}) =

$$\frac{C \times v \times f}{sw \times dwt} \quad (7.21)$$

where *C* is the measured glucose concentration in micrograms per millilitre, *v* is the volume of the soil suspension (in this system, 30 ml), *f* is the dilution factor (in this system 40), *sw* is the weight of the used moist soil in grams and *dwt* is the weight of 1 g moist soil in grams.

Discussion

Strict control of the pH during the determination of reducing sugars is necessary. The pH of the solution has to be higher than 10.5 after adding reagents A and B, and lower than 2 after adding reagent C.

Like the carboxy methyl cellulase, xylanase showed a negative correlation with soil respiration, dehydrogenase, protease and urease activity.

High concentrations of salts, ammonium ions, silver, manganese, oxalate and fluoride disturb the reducing sugars determination. For these reasons, it is recommended that the filtrate is diluted.

The disposal of cyanide-containing solutions can be performed by the oxidation with H_2O_2 at an alkaline pH value.

Lipase activity

K. Alef
P. Nannipieri

(Cooper and Morgan 1981, modified by Schinner et al 1991)

Lipids are a heterogeneous group of compounds, having in common the property of insolubility in water but solubility in non-polar solvents such as chloroform, hydrocarbons or alcohols. They rarely exist in an organism in the "free" state but are typically combined with proteins or carbohydrates as lipoproteins or lipopolysaccharides (see Chapter 9). A significant proportion of lipids enter soil in the form of triacylglycerols, the primary storage fat in plant and animal tissue. Therefore the initial degradation step involves the enzyme lipase (triacylglycerol acylhydrolase, EC 3.1.1.3), with the liberation of fatty acids and glycerol. The reaction can also stop at the monoglyceride stage (Desnuelle 1972).

Fatty acids are long chain carboxylic acids. The parent molecules are the long, straight-chain saturated acids, but there may be many modifications or substitutions in the chain that produce branched, unsaturated, hydroxy-, keto-, epoxy- or cyclic acids. The most abundant natural fatty acids are the *cis*-monounsaturated or polyunsaturated derivatives. Under aerobic conditions, fatty acids are degraded by a stepwise cleavage of 2-carbon fragments (β-oxidation) with final production of CO_2 and H_2O; acetyl-CoA produced by this type of degradation is further metabolized to yield ATP (Mahler and Cordes 1971). Other oxidation mechanisms (α-oxidation, σ-oxidation) do not completely break down the fatty acid molecule, but result in the formation of oxygenated or dicarboxylic acids. Under anaerobic conditions the fatty acids are often accumulated in the environment. Several methods are available to estimate the lipase activity in soils (Pokorna 1964; Pancholy and Lynd 1972; Cooper and Morgan 1981). Because of the short incubation time, the method of Cooper and Morgan (1981), modified by Schinner et al (1991), is presented and discussed.

Principle of the method

The estimation of lipase activity is based on the measurement of the fluorescent 4-methyl umbelliferone (4-MU) released after the incubation of soil with 4-methyl umbelliferone heptanoate (4-MUH) as a substrate for 10 min at 30°C.

Materials and apparatus

Fluorometer
Water shaking bath adjustable to 30°C
Centrifuge with tubes (10 ml)
Erlenmeyer flasks (25 ml)

Chemicals and solutions

Ethylene glycol monomethyl ether

4-Methyl umbelliferone heptanoate (10 mM)

Dissolve 58 mg in 10 ml ethylene glycol monomethyl ether and bring up with the same solvent to 20 ml (store at −20°C). Prepare the solution daily.

Tris buffer (0.1 M, pH 7.5)

Dissolve 12.11 g of Tris in about 700 ml distilled water, adjust the pH to 7.5 with HCl (0.1 M) and bring up to 1000 ml with distilled water.

4-Methyl umbelliferone standard solution (10 mM)

Dissolve 176 mg MU in 70 ml ethylene glycol monomethyl ether and bring up with the same solvent to 100 ml. The solution is stable at 4°C for several days.

Procedure

Place 0.1 g field-moist soil in an Erlenmeyer flask (25 ml), add 4.5 ml Tris buffer, cap the flasks and incubate at 30°C in a shaking water bath for 10 min. Then add 0.5 ml of the 4-methyl-umbelliferone heptanoate substrate solution, mix the contents well and incubate the capped Erlenmeyer flasks in the shaking water bath (30°C) for an additional 10 min. At the end of the incubation, cool the flasks in an ice bath. To prepare the control, add the substrate solution immediately after placing the soil mixture on ice. Finally centrifuge soil mixtures at 2°C and 4000*g*. Measure the released MU in the supernatant spectrofluorometrically at an excitation wavelength of 340 nm and emission at 450 nm.

Calibration curve

Dilute the MU standard solution with distilled water (1:100). Place 0.1 g soil in an Erlenmeyer flask, add 4.5 ml of Tris buffer, and 0, 0.1, 0.2, 0.3, 0.4 and 0.5 ml of the diluted MU standard solution. Add 0.5, 0.4, 0.3, 0.2, 0.1 and 0.0 ml distilled water to reach a final volume of 5 ml. Proceed as described in the procedure. The concentrations of the standards are 0, 10, 20, 30, 40 and 50 nM. The calibration curve must be prepared for each soil.

Calculation

Correct the measurements for the control, read the concentrations of the MU formed (nM) and calculate according to the following relationship:

MU (nM g^{-1} dwt^{-1} 10 min^{-1}) =

$$\frac{MU}{0.1 \times dwt} \qquad (7.22)$$

where *MU* is the measured 4-methyl umbelliferone concentration in nanomoles per litre, *dwt* is the dry weight of 1 g moist soil and 0.1 is the dry weight of the soil sample used.

Discussion

The method proposed by Cooper and Morgan (1981) measured lipase activity in soil extracts. Obviously, the enzyme activity of the soil extracts does not reflect the activity in bulk soil, because a complete extraction of the enzyme is almost impossible. In addition, the state of the enzyme in the extract is different from that of the soil enzyme.

The problem with the method is the incomplete extraction of 4 MU from soil, due to its adsorbance by organic and inorganic colloids.

The lipase activity of soil can also be estimated by using acylesters of MU substituted with C8, C9 and C12 chains.

Chitinase activity

K. Alef
P. Nannipieri

(Rodriguez-Kabana et al 1983, modified by Rössner 1991)

Chitin, a homopolymer of *N*-acetyl-glucosamine in β(1,4) linkages is the major organic element in the exoskeleton of insects, crustaceans, and many species of fungi and other organisms. Enzymatic hydrolysis of chitin to acetyl glucosamine is mediated by two hydrolases, chitinase (EC 3.2.1.14) and chitobiase. Chitinases are common in nature and are produced by bacteria, fungi, plants and the digestive glands of animals that consume chitin-containing materials. In plants this enzyme is induced and accumulated in response to microbial infections (Boiler 1985). Chitinase released by rhizobacteria is believed to play an active role in the biological control of plant pathogenic fungi. Soil chitinase activity decreases with increasing soil depth. Significant correlation between chitinase activity and nitrogen content has also been found (Veon et al 1990).

Principle of the method

The method is based on the incubation of soil with chitin suspension for 16 h at 37°C. The liberated *N*-acetyl glucosamine is extracted with KCl solution and determined spectrophotometrically.

Materials and apparatus

Centrifuge and centrifugation tubes
Cellulose dialysis tube
Spectrophotometer
Polypropylene textile (mesh about 70 μm)
Water bath adjustable to 100°C
Erlenmeyer flasks (25, 100, 1000 ml)
Volumetric flasks (50 ml)
Paper filter
Shaker

Chemicals and solutions

Chitin suspension (5% w/v)

15 g of chitin are mixed with 200 ml of concentrated HCl and stirred for 3 h at room temperature. The suspended chitin is then filtered through the polypropylene textile under vacuum. After dialysing the particle-free solution overnight against water flow, the chitin suspension is centrifuged. The supernatant is then removed and the pellet (chitin) resuspended in distilled water. This procedure (dialysis) must be repeated until the chitin suspension shows a pH value of 5.5–6.0. After determining the dry weight of the chitin, a 5% suspension in water is prepared. Finally, the chitin suspension is homogenized by sonification and 0.2 g NaN_3 1000 ml^{-1} suspension is added. Store at 4°C.

Phosphate buffer (0.12 M, pH 6.0)

Dissolve 16.13 g of KH_2PO_4, 0.2 g of $Na_2HPO_4.2H_2O$ and 0.2 g NaN_3 in about 700 ml of distilled water, adjust the pH to 6.0 with NaOH or HCl and bring up with distilled water to 1000 ml.

Borate buffer (0.8 M, pH 9.1)

Dissolve 4.95 g H_3BO_3 in 40 ml of 0.8 M KOH, add about 20 ml of distilled water, adjust the pH to 9.1 with KOH (0.8 M) and bring up with distilled water to 100 ml.

KCl solution (2 M)

Dissolve 149.12 g of KCl in about 700 ml distilled water and dilute with distilled water to 1000 ml.

4-(Dimethyl amino)benzo aldehyde stock solution (DMBA)

Dissolve 10 g DMBA in a mixture of 87.5 ml of concentrated acetic acid and 12.5 ml concentrated HCl and store at 4°C.

DMBA diluted solution

Mix one volume of DMBA stock solution with four volumes of concentrated acetic acid. Prepare a fresh solution daily.

N-Acetyl glucosamine (GlcNAc) standard solution (45 mM)

Dissolve 0.5 g of *N*-acetyl glucosamine in 30 ml of distilled water and bring up with distilled water to 50 ml. Prepare a fresh solution daily.

Procedure

Place 1.0 g of moist soil in an Erlenmeyer flask and add 5 ml of phosphate buffer. After the addition of 5 ml of the chitin substrate solution, the flask is closed with a rubber cap and incubated for 16 h at 37°C. In the control, the substrate solution is added immediately at the end of the incubation period. After the incubation, the flask received 10 ml of KCl solution, before being shaken at room temperature for 30 min. After filtration of the soil suspension, 0.5 ml of the filtrate is mixed with 1.5 ml distilled water, 0.4 ml borate buffer and boiled for 3 min in a water bath. After cooling to room temperature, 5 ml of diluted DMBA solution are added, the mixture mixed thoroughly and incubated at 35°C for 30 min. The optical density is measured at 585 nm within 20 min. All estimations must be carried out at least in duplicate.

Calibration curve

Pipette 0, 0.5, 1.0 and 2.5 ml of the *N*-acetyl glucosamine solution into volumetric flasks (50 ml), add 12.5 ml of phosphate buffer, 25 ml of KCl solution and bring up with distilled water to 50 ml. Pipette 0.5 ml of each solution into glass tubes and perform the determination of *N*-acetyl glucosamine as described above. The *N*-acetyl glucosamine concentrations in the calibration curve are 0, 50, 100, 150, 200 and 250 µg ml^{-1}.

Calculation

Correct the data for the blank and calculate as follows:

N-acetyl glucosamine (µg g dwt^{-1} 16 h^{-1}) =

$$\frac{C \times v}{dwt} \qquad (7.23)$$

where C is the measured *N*-acetyl glucosamine concentration in micrograms per millilitre, v is the final volume of solutions added to the soil (20 ml) and *dwt* is the dry weight of 1 g moist soil.

Discussion

The reaction is linear between 4 and 24 h (Rössner 1991)

A higher GlcNAc yield was achieved by using KCl as an extractant.

The colorimetric determination method of GlcNAc is a modification of that described by Reissig et al (1955).

Rössner (1991) used NaN_3 instead of toluene to inhibit microbial activity.

Catalase activity

K. Alef
P. Nannipieri

(Beck 1971)

The enzyme catalase (hydrogen peroxidase oxidoreductase, EC 1.11.1.6) has a detoxifying function in cells by catalysing the following reaction:

$$H_2O_2 \rightarrow H_2O + \tfrac{1}{2}O_2 \qquad (7.24)$$

Induced hydrogen peroxide is poisonous for cells, oxidizing the SH-groups of proteins. Partial reduction products of oxygen like the superoxide anion $(O_2)^-$ and hydrogen peroxide may be formed during electron transport to molecular oxygen via the aerobic respiration, as well as in various hydroxylation and oxygenation reactions. These products are extremely reactive and capable of irreversibly damaging various biomolecules.

Aerobic cells generally contain the enzyme superoxide dismutase, which converts superoxide into hydrogen peroxide and molecular oxygen:

$$2O_2^- + 2H^+ \rightarrow H_2O_2 + O_2 \qquad (7.25)$$

Enzyme systems like the L- and D-amino-acid oxidases contain tightly bound FMN or FAD as prosthetic groups. These enzymes catalyse the oxidative deamination of amino acids:

$$\text{L-Amino acid} + H_2O + \text{E-FMN} \rightarrow \alpha\text{-Keto acid} + NH_3 + \text{E-FMNH}_2 \qquad (7.26)$$

$$\text{D-Amino acid} + H_2O + \text{E-FAD} \rightarrow \alpha\text{-Keto acid} + NH_3 + \text{E-FADH}_2 \qquad (7.27)$$

The reduced forms of the L- and D-amino acid oxidases can be reoxidized directly with molecular oxygen to form hydrogen peroxide:

$$\text{E-FMNH}_2 + O_2 \rightarrow \text{E-FMN} + H_2O_2 \qquad (7.28)$$

$$\text{E-FADH}_2 + O_2 \rightarrow \text{E-FAD} + H_2O_2 \qquad (7.29)$$

All aerobic and most of the facultative anaerobic bacteria contain catalase activity. This enzyme has not been detected in obligate anaerobic bacteria. Catalase was the first enzyme studied in soil (König et al 1906, 1907; May and Gile 1909). Its estimation is based on the determination of released O_2 (Weetall et al 1965; Beck 1971; Kuprevich and Shcherbakova 1971a; Trevors 1984; Holz 1986b). Catalase activity is detected in plant and animal cells. This interferes with the estimation of microbial catalase activity in soil (Beck 1971; Ladd 1978). In addition, Mn and Fe, as well as organic substances in soils catalyse the liberation of O_2 from H_2O_2 (Baeyens and Livens 1936; Beck 1971; Kuprevich and Shcherbakova 1971b). Catalase activity is very stable in soil. It shows significant correlation with the content of organic carbon and decreases with the soil depth (Beck 1971; Ladd 1978). No relation has been found between the catalase activity and soil fertility (Skujins 1978).

Principle of the method

The method is based on the volumetric determination of oxygen liberated after incubation of soil with hydrogen peroxide for 3 min at room temperature.

Materials and apparatus

Erlenmeyer flasks (200 ml)
Magnetic stirrer
Scheibler apparatus (see Reuter 1976; Stefanic et al 1984). If Scheibler apparatus is not available, a U-glass tube filled with Brodie's solution can be used for the volumetric measurement.

Chemicals and solutions

Phosphate buffer (0.2 M, pH 6.8)

Dissolve 35.63 g of disodium hydrogen phosphate in 700 ml distilled water, adjust to pH 6.8 with HCl and dilute to 1000 ml with distilled water.

Sodium carbonate solution (10%)

Dissolve 100 g Na_2CO_3 in about 800 ml distilled water and bring up with distilled water to 1000 ml.

H_2O_2 (3%)

Prepare immediately before use.

Sodium azide solution (6.5%)

6.5 g 100 ml^{-1} distilled water.

Brodie's solution (obtained from Braun, Melsungen, Germany).
MnO_2.

Procedure

Place 5–10 g moist or dry soil into an Erlenmeyer flask (200 ml) and add 20 ml of phosphate buffer. To prepare the controls, add 2 ml azide solution and 20 ml phosphate buffer. Swirl the flasks and allow to stand for 30 min.

Pipette 10 ml of H_2O_2 (3%) in the small tank of the Scheibler apparatus. After closing the flask, mix the H_2O_2 with the soil sample and incubate under continuous stirring for 3 min at room temperature (20°C). Then read the change of the gas volume in the Scheibler apparatus or in the U-glass tube.

For the determination of oxygen formed, place 0.5 g of MnO_2 and 20 ml of sodium carbonate solution (10%) instead of soil in an Erlenmeyer flask, add 10 ml of H_2O_2 (3%) and read the change in the gas volume in the Scheibler apparatus (or U-glass tube). The oxygen formed is considered as the 100%.

Calculation

$$\text{Catalase index} = \frac{D_o - D_A}{D_{Mn} \times dwt} \times 100 \quad (7.30)$$

where D_o is the O_2 developed in the absence of azide in millilitres, D_A is the O_2 developed in the presence of azide in millilitres, D_{Mn} is the O_2 developed in the presence of Mn in millilitres and *dwt* is the soil dry weight (%).

Discussion

Azide has been used to calculate the abiotic H_2O_2 hydrolysis (Beck 1971; Ladd 1978).

Storage of moist or air-dried soils at room temperature for 4 months had no effect on catalase activity (Beck 1971).

Pesticides can either inhibit or activate the catalase activity of soil, depending on the compound (Ladd 1978).

Good correlations have been found between catalase and dehydrogenase activity, but only weak or no correlations between catalase and amylase activity or bacterial number (Beck 1971).

Arylsulphatase activity

K. Alef
P. Nannipieri

(Tabatabai and Bremner 1970a)

Sulphatases catalyse the hydrolysis of organic sulphate esters and have been classified according to the type of the ester in arylsulphatases: alkylsulphatases, steroid sulphatases, glucosulphatases, chondrosulphatases and myrosulphatases (Tabatabai 1982).

Arylsulphatase (EC 3.1.6.1) catalyses the irreversible reaction:

$$R.OSO_3^- + H_2O \rightarrow R.OH + H^+ + (SO_4)^{2-} \quad (7.31)$$

and has been detected in microorganisms, plants and animals (Nicholls and Roy 1971). The enzyme also catalyses the hydrolysis of *p*-nitrophenyl sulphate, potassium phenyl sulphate, potassium nitrocatechol sulphate and potassium phenolphthalein sulphate.

The K_m values were found to range from 0.2 to 0.95 mM, when *p*-nitrophenyl sulphate was used as the substrate, and were dependent on the soil type (Perucci and Scarponi 1984). Arylsulphatase activity decreased with soil depth and showed significant correlations with the content of soil organic carbon, total nitrogen and cation exchange capacity (Tabatabai and Bremner 1970b; King and Klug 1980; Appiah and Ahenkorah 1989). Inorganic sulphur [$(SO_4)^{2-}$, S(IV) and S(VI)] inhibit arylsuphatase activity of soil (Dodgson and Rose 1976; Fitzgerald 1976; Jarvis et al 1987). Strong inhibition has also been detected with $(PO_4)^{3-}$, $(AsO_4)^{3-}$, $(MoO_4)^{2-}$ and $(WO_4)^{2-}$ but not with NO_3^-, NO_2^- and Cl^- (Al-Khafaji and Tabatabai 1979; Tabatabai and Bremner 1970a). The role of arylsulphatase in sulphur mineralization in soils is not clear. The importance of this enzyme activity in the sulphur mineralization is derived from the finding that most of the total S found in surface soils is present in the form of organic sulphates (Tabatabai 1982). However, Jarvis et al (1987) found no correlation between arylsulphatase activity and sulphur mineralization. Another problem may develop because *p*-nitrophenyl sulphate, which was probably used as the substrate, does not occur in nature.

Fitzgerald et al (1985) expressed doubts about the importance of the *p*-nitrophenyl sulphate dependent-activities in soils and suggested the use of [^{35}S]tyrosine sulphate as a substrate.

Principle of the method

The method is based on the determination of *p*-nitrophenol released after the incubation of soil with *p*-nitrophenyl sulphate for 1 h at 37°C.

Materials and apparatus

See estimation of phosphomonoesterase.

Chemicals and solutions

Toluene.

Acetate buffer (0.5 M, pH 5.8)

Dissolve 68 g of sodium acetate trihydrate in 700 ml of distilled water, adjust the pH with concentrated acetic acid to 5.8 and bring up to 1000 ml with distilled water.

p-Nitrophenyl sulphate solution (25 mM)

Dissolve 0.312 g of potassium *p*-nitrophenyl sulphate in about 40 ml acetate buffer and dilute the solution to 50 ml with buffer. Store at 4°C.

$CaCl_2$ (0.5 M)

See estimation of phosphomonoesterase.

NaOH (0.5 M)

p-Nitrophenol standard solution

Dissolve 1.0 g *p*-nitrophenol in about 70 ml distilled water and dilute to 1000 ml with distilled water. Store at 4°C.

Procedure

Place 1 g of moist sieved (2 mm) soil in Erlenmeyer flask (50 ml), add 0.25 ml toluene, 4 ml of acetate buffer, 1 ml of *p*-nitrophenyl sulphate solution and mix the contents. Cap the flasks and incubate for 1 h at 37°C. After the incubation add 1 ml of $CaCl_2$ (0.5 M) and 4 ml of NaOH (0.5 M), mix the contents, filter the soil suspension and measure the optical density at 400 nm. To prepare the control, make the addition of 1 ml of *p*-nitrophenyl sulphate after the addition of $CaCl_2$ and NaOH, i.e. immediately before filtration of the soil suspension.

Calibration curve

See estimate of phosphomonoesterase activity.

Calculation

See estimation of phosphomonoesterase activity.

Discussion

Air-drying usually increased the arylsulphatase activity (Tabatabai and Bremner 1970a).

It has been observed that acetate buffer gives higher activities than other types of buffers (Tabatabai 1982).

The substrate concentration in the assay is equal to 5 mM; however, in soils with higher enzyme activities, higher *p*-nitrophenyl sulphate concentrations should be used (Tabatabai and Bremner 1970a; Sarathchandra and Perrott 1981).

The addition of toluene (0.1–1.0 ml) increased the enzyme activity.

To decrease the intensity of the control, Mathur and Rayment (1977) and Sarathchandra and Perrott (1981) used diethyl ether to extract *p*-nitrophenol from soils with high organic matter content.

Pollutants in soils have been found to affect arylsulphatase activity. (Press et al 1985; Wilke 1986; Haanstra and Doelman 1991).

References

Agrawal KML, Bahl OP (1968) Glucosidase of *Phaseolus vulgaris*. J Biol Chem 234: 103–111.

Alef K, Kleiner D (1986) Arginine ammonification in soil samples. Veröff Landwirtsch-Chem Bundesanstalt Linz/Donau 18: 163–168.

Alef K, Beck Th, Zelles L, Kleiner D (1988) A comparison of methods to estimate microbial biomass and N-mineralization in agricultural and grassland soil. Soil Biol Biochem 20: 561–565.

Alexander M (1977) Soil Microbiology. Wiley, New York.

Al-Khafaji AA, Tabatabai MA (1979) Effects of trace elements on arylsulfatase activity in soils. Soil Sci 127: 129–133.

Appiah MR, Ahenkorah Y (1989) Arylsulphatase activity of different latosol soils of Ghana cropped to cocoa (*Theobroma cacao*) and coffee (*Coffea conephora* var. robusta). Biol Fertil Soils 7: 186–190.

Appiah MR, Halm BJ, Ahenkorah Y (1985) Phosphatase activity of soil as affected by cocoa pod ash. Soil Biol Biochem 17: 823–826.

Baeyens J, Livens J (1936) Catalytic power of a soil and fertility. Agricoltura 30: 145–155.

Bahl OP, Agrawal KML (1972) Alfa-Galactosidase, and β-glucosidase, and β-*N*-acetylglucosaminidase from *Aspergillus niger*. In: Methods in Enzymology, vol 28. Ginsburg V (ed.). Academic Press, New York, pp. 728–734.

Beaton JD, Hubbard WA, Speer RC (1967) Coated urea, thiourea, urea, urea-formaldehyde, hexamine, oxamide, glycoluril, and oxidized nitrogen-enriched coal as slowly available sources of nitrogen for orchardgrass. Agron J 59: 127–133.

Beck Th (1971) Die Messung der Katalaseaktivität von Böden. Z Pflanzenernähr Bodenkd 130: 68–81.

Beck Th (1973) Über die Eignung von Modellversuchen bei der Messung der biologischen Aktivität von Böden. Bayer Landw Jb 50: 270–288.

Beck Th (1984a) Mikrobiologische und biochemische Charakterisierung landwirtschaflich genutzter Böden. I. Mitteilung: Die Ermittlung einer bodenmikrobiologischen Kennzahl. Z Pflanzenernähr Bodenkd 147: 456–466.

Beck Th (1984b) Mikrobiologische und biochemische Charakterisierung landwirtschaflich genutzter Böden. II. Mitteilung: Beziehung zum Humusgehalt. Z Pflanzenernähr Bodenkd 147: 467–475.

Beck Th (1986) Aussagekraft und Bedeutung enzymatischer und mikrobiologischer Methoden bei der Charakterisierung des Bodenlebens von landwirtschaflichen Böden. Veröff Landwirtsch-chem Bundesanstalt Linz/Donau 18: 75–100.

Beck Th, Poschenrieder H (1963) Experiments on the effect of toluene on the soil microflora. Plant Soil 18: 346–357.

Benefield CB (1971) A rapid method for measuring cellulase activity in soils. Soil Biol Biochem 3: 325–329.

Bidwell RGS (1974) Plant Physiology. Macmillan, New York, pp. 173–206.

Blakeley RL, Zerner B (1984) Jack bean urease: the first nickel enzyme. J Mol Catal 23: 263–292.

Boiler T (1985) Induction of hydrolases as a defense reaction against pathogens. IN: Cellular and Molecular Biology of Plant Stress. Key JL, Kosuge T (eds). Liss, New York, pp. 247–262.

Bonmati M, Pujola M, Sana J, Soliva M, Felipo MT, Garau M, Ceccanti B, Nannipieri P (1985) Chemical properties, populations of nitrite oxidizers, urease and phosphatase activities in sewage sludge amended soils. Plant Soil 84: 79–91.

Boyd SA, Mortland MM (1990) Enzyme interactions with clays and clay–organic matter complexes. In: Soil Biochemistry, vol 6. Bollag J-M, Stotzky G (eds). Dekker, New York, pp. 1–28.

Bray HG, James SP, Raffan IM, Ryman BE, Thorpe WV (1949a) The fate of certain organic acids and amides in the rabbit. Biochem J 44: 618–625.

Bray HG, James SP, Raffan IM, Thorpe WV (1949b) The enzymic hydrolysis of glutamine and its spontaneous decomposition in buffer solutions. Biochem J 44: 625–627.

Bremner JM (1955) Studies on soil humic acids: I. The chemical nature of humic nitrogen. J Agric Sci 46: 247–256.

Bremner JM (1965a) Inorganic forms of nitrogen. In: Methods of Soil Analysis, Part 2. Black CA et al (eds). Agronomy 9. American Society of Agronomy, Madison, WI, pp. 1179–1237.

Bremner JM (1965b) Organic nitrogen in soils. In: Soil Nitrogen. Bartholomew WV, Black CA, Evans DD, Ersminger LE, White JL, Clark FE (eds). Agronomy 10. American Society of Agronomy, Madison, WI, pp. 93–132.

Bremner JM (1967) Nitrogenous compounds. In: Soil Bochemistry. McLaren AD, Peterson GH (eds). Dekker, New York, pp. 19–66.

Bremner JM, Edwards AP (1965) Determination and isotope-analysis of different forms of nitrogen in soils: I. Apparatus and procedure for distillation and determination of ammonium. Soil Sci Soc Am Proc 29: 504–507.

Bremner JM, Keeney DR (1966) Determination and isotope-ratio analysis of different forms of nitrogen in soils. 3. Exchangable ammonium, nitrate and nitrite by extraction–distillation methods. Soil Sci Soc Am Proc 30: 577–582.

Bremner JM, Mulvaney RL (1978) Urease activity in soils. In: Soil Enzymes. Burns RG (ed.). Academic Press, New York, pp. 149–196.

Browman MG, Tabatabai MA (1978) Phosphodiesterase activity of soils. Soil Sci Soc Am J 42: 284–290.

Brown PR, Smyth MJ, Clarke PH, Rosemeyer MA (1973) The subunit structure of the aliphatic amidase from *Pseudomonas aeruginosa*. Eur J Biochem 34: 177–187.

Burns RG (1978) Enzyme activity in soil, some theoretical and practical consideration. In: Soil Enzymes. Burns RG (ed.). Academic Press, New York, London, pp. 73–75.

Burns RG (1982) Enzyme activity in soil: Location and possible role in microbial ecology. Soil Biol Biochem 14: 423–427.

Burns RG (1986) Interaction of enzymes with soil minerals and organic colloids. In: Interactions of Soil Minerals with Natural Organics and Microbes. Huang PM, Schnitzer M (eds). Soil Science Society of America, Special Publication Number 17, Madison, WI, pp. 429–451.

Burns RG, Pukite AH, McLaren AD (1972) Concerning the location and persistence of soil urease. Soil Sci Soc Am Proc 36: 308–315.

Butler LG (1971) Yeast and other inorganic pyrophosphatases. In: The Enzymes, vol 4. Boyer PD (ed.). Academic Press, New York, pp. 529–541.

Cai GX, Freney JR, Muirhead WA, Simpson JR, Chen DL, Trevitt ACF (1989) The evaluation of urease inhibitors to improve the efficiency of urea as N-source for flooded rice. Soil Biol Biochem 21: 137–145.

Cervelli S, Nannipieri P, Ceccanti B, Sequi P (1973) Michaeles constant of soil acid phosphatase. Soil Biol Biochem 5: 841–845.

Chonkar PK, Tarafdar JC (1981) Characteristics and location of phosphatases in soil–plant systems. J Indian Soc Soil Sci 29: 215–219.

Clark AE, Stone BA (1965) Properties of a β-1,4-glucan hydrolase from *Aspergillus niger*. Biochem J 96: 802–807.

Clarke PH (1970) The aliphatic amidases of *Pseudomonas aeruginosa*. Adv Microb Physiol 4: 179–222.

Cochran VL, Elliott LF, Lewis CE (1989) Soil microbiol biomass and enzyme activity in subarctic agricultural and forest soils. Biol Fertil Soils 7: 283–288.

Cooper AB, Morgan HW (1981) Improved fluorometric method to assay for soil lipase activity. Soil Biol Biochem 13: 307–311.

Cosgrove DJ (1977) Microbial transformations in the phosphorous cycle. Adv Microb Ecol 1: 95–134.

Desnuelle P (1972) The lipase. In: The Enzymes. Bayer PO (ed.). Academic Press, London, pp. 575–615.

Dey PM, Pridham JB (1972) Biochemistry of alfa-galactosidases. In: Advances in Enzymology, vol 36. Meister A (ed.). Wiley, New York, pp. 91–130.

Dick WA, Tabatai MA (1977) Determination of orthophosphate in aqueous solutions containing labile organic and inorganic phosphorus compounds. J. Environ Qual 6: 82–85.

Dick WA, Tabatabai MA (1978) Inorganic pyrophosphatase activity of soils. Soil Biol Biochem 10: 59–65.

Dick WA, Tabatabai MA (1983) Effects of soils on acid phosphatases and inorganic pyrophosphatase of corn roots. Soil Sci 136: 19–25.

Dick WA, Tabatabai MA (1984) Kinetic parameters of phosphatases in soils and organic waste materials. Soil Sci 137: 7–15.

Dodgson KS, Rose FA (1976) Sulfohydrolases. In: Metabolism of Sulfur Compounds, vol 7, Metabolic Pathways. Greenberg DM (ed.). Academic Press, New York, pp. 359–431.

Doelman P, Haanstra L (1986) Short- and long-term effects of heavy metals on urease activity in soils. Biol Fertil Soils 2: 213–218.

Doelman P, Haanstra L (1989) Short- and long-term effects of heavy metals on phosphatase activity in soils: an ecological dose–response model approach. Biol Fertil Soils 8: 235–241.

Douglas LA, Bremner JM (1970) Extraction and colorimetric determination of urea. Soil Sci Soc Am Proc 34: 859–868.

Douglas LA, Bremner JM (1971) A rapid method of evaluating different compounds as inhibitors of urease in soils. Soil Biol Biochem 3: 309–315.

Drobni'k J (1956) Degradation of asparagine by the soil enzyme complex. Cesk Mikrobiol 1: 47.

Dutzler-Franz G (1977) Einfluβ einiger chemischer und physikalischer Bodenmerkmale auf die Enzymaktivität verschiedener Bodentypen. Z Pflanzenernähr Bodenkd 140: 329–350.

Edlbacher S (1926) A communication on intermediates of histidine metabolism. Part I. Z Physiol Chem 157: 106–114.

Edlbacher S, Kraus J, Scheurich N (1930) A communication on intermediates of histidine metabolism. Part II. Z Physiol Chem 191: 225–242.

Eivazi F, Tabatabai MA (1977) Phosphatases in soils. Soil Biol Biochem 9: 167–172.

Eivazi F, Tabatabai MA (1988) Glucosidases and galactosidases in soils. Soil Biol Biochem 20: 601–606.

Eriksson KE, Wood TM (1985) Biodegradation of cellulose. In: Biosynthesis and Biodegradation of Wood Components. Higuchi T (ed.). Academic Press, London, pp. 469–503.

Fenn LB, Tipton JL, Tatum G (1992) Urease activity in two cultivated and non-cultivated arid soils. Biol Fertil Soils 13: 152–154.

Fillery IRP, Simpson JR, De Datta SK (1984) Influence of field environment and fertilizer management on ammonia loss from flooded rice. Soil Sci Soc Am J 48: 914–920.

Fillery IRP, De Datta SK, Craswell ET (1986) Effect of phenyl phosphorodiamidate on the fate of urea applied to wetland rice fields. Fertilizer Res 9: 251–263.

Fitzgerald JW (1976) Sulfate ester formation and hydrolysis: a potentially important yet often ignored aspect of the sulfur cycle of aerobic soils. Bacteriol Rev 40: 698–721.

Fitzgerald JW, Watwood ME, Rose FA (1985) Forest floor and soil arylsulphatase: hydrolysis of tyrosin sulphate, an environmental relevant substrate for the enzyme. Soil Biol Biochem 17: 885–887.

Florkin M, Stotz EH (1964) Comprehensive Biochemistry, 13. Elsevier, Amsterdam, pp. 126–134.

Frankenberger WT Jr (1983) Kinetic properties of L-histidine ammonia-lyase activity in soils. Soil Sci Soc Am J 47: 71–74.

Frankenberger WT Jr, Johanson JB (1982) L-Histidine ammonia-lyase activity in soils. Soil Sci Soc Am J 46: 943–948.

Frankenberger WT Jr, Johanson JB (1983a) Distribution of L-histidine ammonia-lyase activity in soils. Soil Sci Soc Am J 136: 347–353.

Frankenberger WT, Johanson JB (1983b) Method of measuring invertase activity of soils. Plant Soil 74: 301–311.

Frankenberger WT, Johanson JB (1983c) Factors affecting invertase activity in soils. Plant Soil 74: 313–323.

Frankenberger JR, Johanson JB (1986) Use of plasmolytic agents and antiseptics in soil enzyme assays. Soil Biol Biochem 18: 209–213.

Frankenberger WT Jr, Tabatabai MA (1980a) Amidase activity in soils: I. Method of assay. Soil Sci Soc Am J 44: 282–287.

Frankenberger WT Jr, Tabatabai MA (1980b) Amidase activity in soils: II. Kinetic parameters. Soil Sci Soc Am J 44: 532–536.

Frankenberger WT Jr, Tabatabai MA (1981a) Amidase activity in soils: III. Stability and distribution. Soil Sci Soc Am J 45: 333–338.

Frankenberger WT Jr, Tabatabai MA (1981b) Amidase activity in soils: IV. Effects of trace elements and pesticides. Soil Sci Soc Am J 45: 1120–1124.

Frankenberger WT Jr, Tabatabai MA (1981c) Fate of amide nitrogen added to soils. Agricul Food Chem 29: 152–155.

Frankenberger WT Jr, Tabatabai MA (1982) Amidase and urease activity in plants. Plant Soil 64: 153–166.

Frankenberger WT Jr, Tabatabai MA (1985) Characteristics of an amidase isolated from a soil bacteria. Soil Biol Biochem 17: 303–308.

Frankenberger WT Jr, Tabatabai MA (1991a) L-Asparaginase activity of soils. Biol Fertil Soils 11: 6–12.

Frankenberger WT Jr, Tabatabai MA (1991b) Factors affecting L-asparaginase activity in soils. Biol Fertil Soils 11: 1–5.

Frankenberger WT Jr, Tabatabai MA (1991c) L-Glutaminase activity of soils. Soil Biol Biochem 23: 869–874.

Frankenberger WT Jr, Tabatabai MA (1991d) Factors affecting L-Glutaminase activity of soils. Soil Biol Biochem 23: 875–879.

Gadkari D (1984) Influence of the herbicide goltix on extracellular urease and phosphatase in suspended soil. Zbl Mikrobiol 139: 415–424.

Galstyan ASH (1965) A method of determining the activity of hydrolytic enzymes in soil. Soviet Soil Sci 2: 170–175.

Galstyan ASH, Saakyan EG (1973) Determination of soil glutaminase activity. Doklady Acad Nauk SSSR 209: 1201–1202.

Gianfreda L, Rao MA, Violante A (1992) Adsorption, activity and kinetic properties of urease on montmorillonite, aluminium hydroxide, and Al $(OH)_x$-montmorillonite complexes. Soil Biol Biochem 24: 51–58.

Gosewinkel U, Broadbent FE (1984) Conductometric determination of soil urease activity. Commun Soil Sci Plant Anal 15: 1377–1389.

Gottschalk G, Andressen JR, Hippe H (1981) The Prokaryotes, vol II. Starr MP, Stolp H, Trüper HG, Balows A, Schlegel HG (eds). Springer Verlag, Berlin, pp. 1767–1803.

Gyorgy P, Rothler H (1926) Conditions for the autolytic formation of ammonia in nature. Series II. Determination of ammonia derived from amino acids and nitrogen containing substances. Biochem Z 173: 334–347.

Haanstra L, Doelman P (1991) An ecological dose–response model approach to short- and long-term effects of heavy metals on arylsulfatase activity in soil. Biol Fertil Soils 11: 18–23.

Haare EA, White WC (1985) Fertilizer market profile. In: Fertilizer Technology and Use. Engelstad OP (ed.). American Society of Agronomy, Madison, WI, pp. 1–24.

Hagedorn HC, Jensen BN (1923) Zur Mikrobestimmung des Blutzuckers mittels Ferrizyanid. Biochem Z 135, 46–58.

Harrison AF (1979) Variation of four phosphorus properties in woodland soils. Soil Biol Biochem 11: 393–403.

Hartman SC (1971) Glutaminase and γ-glutamyltransferases. In: The Enzymes, vol 4. Boyer PD (ed.) Academic Press, New York, pp. 79–100.

Hashimoto S, Muramatsu T, Fumatsu M (1971) Studies on xylanase from *Trichoderma viride*. Agr Biol Chem 35: 501–508.

Hayano K (1986) Cellulase complex in tomato field

soil: induction, localization and some properties. Soil Biol Biochem 18: 215–219.

Hayano K (1987) Characterization of a phosphodiesterase component in a forest soil extract. Biol Fertil Soils 3: 159–164.

Hayano K, Tubakil (1985) Origin and properties of β-glucosidase activity of tomato-field soil. Soil Biol Biochem 17: 553–557.

Hoffmann E, Schmidt W (1953) Über das Enzymsystem unserer Kulturböden. II. Urease. Biochem Z 323: 125–127.

Hoffmann G (1959) Distribution and origin of some enzymes in soil. Z Pflanzenernähr Düng Bodenkd 77: 243–251.

Hoffmann G (1967) Eine photometrische Methode zur Bestimmung der Phosphatase-Aktivität in Böden. Z Pflanzenernaehr Bodenkd 118: 161–172.

Hoffmann G, Dedken M (1965) Eine Methode zur Kolorimetrischen Bestimmung der β-Glucosidase-Aktivität im Boden. Z Pflanzenernach Budenk 108: 193–198.

Hoffmann G, Elias-Azar K (1965) Verschiedene Faktoren der Bodenfruchtbarkeit nordiranischer Böden und ihre Beziehung zur Aktivität hydrolytischer Enzyme. Z Pflanzenernähr Düng Bodenkd 108: 199–217.

Hoffmann G, Pallauf J (1965) Eine kilorimetrische Methode zur Bestimmung der Saccharaseaktivität in Böden. Z Pflanzenernähr Düng Bodenkd 110: 193–201.

Hoffmann G, Pfitscher A (1982) Korrelationen von Enzymaktivitäten im Boden. Z Pflanzenernähr Bodenkd 145: 36–41.

Hoffmann G, Teicher K (1957) Das Enzymsystem unserer Kulturböden VII. Protease II. Z Pflanzenernähr Düng Bodenkd 3: 243–251.

Holz F (1980) Automatisierte, enzymatische-photometrische Lysinbestimmung und ihre Anwendung als Screeningsmethode in Züchtungsprogrammen. Landwirtsch Forschung 33: 272–289.

Holz F (1986a) Automatische photometrische Bestimmung der Aktivität von Bodenenzymen durch Anwendung (enzymatisch-)oxydativer Kupplungsreaktionen im Durchfluβ. II Mitteilung: Die Bestimmung der Saccharaseaktivität. Landwirtsch Forschung 39: 245–259.

Holz F (1986b) Automatisierte, photometrische Bestimmung der Aktivität von Bodenenzymen durch Anwendung (enzymatisch-)oxydativer Kupplungsreaktionen im Durchfluβ. I. Mitteilung: Die Bestimmung der Katalaseaktivität. Landwirtsch Forschung 39: 139–153.

Hope CFA, Burns RG (1987) Activity, origins and location of cellulase in a silt loam soil. Biol Fertil Soils 5: 164–170.

Hunt H (1977) A simulation model for decomposition in grasslands. Ecology 58: 469–484.

Hynes MJ (1970) Induction and repression of amidase enzymes in *Aspergillus nidulans*. J Bacteriol 103: 482–487.

Hynes MJ (1975) Amide utilization in *Aspergillus nidulans*: Evidence for a third amidase enzyme. J Gen Microbiol 91: 99–109.

Ibister JD, Shippen RS, Caplan J (1980) A new method for measuring cellulose and starch degradation in soils. Bull Environ Contam Toxicol 24: 570–574.

Imada A, Igarasi S, Nakahama K, Isono M (1973) Asparaginase and glutaminase activities of microorganisms. J Gen Microbiol 76: 85–89.

Jakoby WB, Fredericks J (1964) Reactions catalyzed by amidases. Acetamidase. J Biol Chem 239: 1978–1982.

Jarvis BW, Lang GE, Wieder RK (1987) Arylsulphatase activity in peat exposed to acid precipitation. Soil Biol Biochem 19: 107–109.

Joliff G, Edelman A, Klier A, Rapoport G (1989) Inducible secretion of a cellulase from *Clostridium thermocellum* in *Bacillus subtilis*. Appl Environ Microbiol 55: 2739–2744.

Joshi JG, Handler P (1962) Purification and properties of nicotinamidase from *Torula cremoris*. J Biol Chem 237: 929–935.

Juma NG, Tabatai MA (1977) Effects of trace elements on phosphatase activity in soils. Soil Sci Soc Am J 41: 343–346.

Juma NG, Tabatabai MA (1988) Comparison of kinetic and thermodynamic parameters of phosphomonoesterases of soils and of corn and soybean. Soil Biol Biochem 20: 533–539.

Kandeler E (1986) Aktivität von Proteasen in Böden und ihre Bestimmungsmöglichkeiten. VDLUFA-Schriftenreihe 20: 829–847.

Kandeler E (1990) Characterization of free and adsorbed phosphatases in soils. Biol Fertil Soils 8: 199–202.

Kandeler E, Gerber H (1988) Short-term assay of soil urease activity using colorimetric determination of ammonium. Biol Fertil Soils 6: 68–72.

Keeney DR, Nelson DW (1982) Nitrogen Inorganic forms. In: Methods of Soil Analysis, Part 2, 2nd edn. Page AL, Miller RH, Keeney DR (eds). American Society of Agronomy, Madison, WI, pp. 643–698.

Kelly M, Clarke PH (1962) An inducible amidase produced by a strain of *Pseudomonas aeruginosa*. J Gen Microbiol 27: 305–316.

King GM, Klug MJ (1980) Sulfohydrolase activity in sediment of Wistergreen Lake, Kalamazoo Country, Michigan. Appl Environ Microbiol 29: 950–956.

Kiss S, Stefanic G, Dragan-Bularda M (1974) Soil enzymology in Romania (Part I). Contrib Bot Univ Babes-Bolyia Cluj pp. 207–209.

Kiss S, Dragan-Bularda M, Radulescu D (1978) Soil polysaccharidases: activity and agricultural importance. In: Soil Enzymes. Burns RG (ed.). Academic Press, New York, pp. 117–147.

Kissel DE, Cabrera ML (1988) Factors affecting urease activity. In: Ammonia Volatilization from Urea Fertilizer. Bock BR, Kissel DE (eds). TVA, National Fertilizer Development Center, Muscle Shoals, pp. 53–66.

König J, Hasenbäumer J, Coppenrath E (1906) Several new properties of cultivated soils. Landw Versuchs-Stationen 63: 471–478.

König J, Hasenbäumer J, Coppenrath E (1907) Relationships between the properties of soil and the nutrient uptake by plants. Landw Versuchs-Stationen 66: 401–461.

Kshattriya S, Sharma GD, Mishra RR (1992) Enzyme activities related to litter decomposition in the forests of different age and altitude in North East India. Soil Biol Biochem 24: 265–270.

Kuprevich VF, Shcherbakova TA (1971a) Soil Enzymes. US Department of Commerce, National Technical Information Service, Springfield, Va. (Pochvennaya Enzimologiya Nauka tekh. Minsk, 1966; translated from Russian.)

Kuprevich VF, Shcherbakova TA (1971b) Comparative enzymatic activity in diverse types of soil. In: Soil Biochemistry, vol 2. McLaren AD, Skujins JJ (eds). Marcel Dekker, New York, pp. 167–201.

Ladd JN (1972) Properties of proteolytic enzymes extracted from soils. Soil Biol Biochem 4: 227–237.

Ladd JN (1978) Origin and range of enzymes in soil. In: Soil Enzymes. Burns RG (ed.). Academic Press, New York, pp. 51–96.

Ladd JN, Butler JHA (1972) Short-term assays of soil proteolytic enzyme activities using proteins and dipeptide derivatives as substrates. Soil Biol Biochem 4: 19–30.

Ladd JN, Jackson RB (1982) Biochemistry of ammonification. In: Nitrogen in Agricultural Soils. Stevenson FJ (ed.). Agronomy Monograph 22, American Society of Agronomy, Madison, WI, pp. 173–228.

Lähdesmäki P, Piispanen R (1989) Changes in concentrations of free amino acids during humification of spruce and aspen leaf litter. Soil Biol Biochem 21: 975–978.

Lai CM, Tabatai MA (1992) Kinetic Parameters of immobilized urease. Soil Biol Biochem 24: 225–228.

Latter PM, Howson G (1977) The use of cotton strips to indicate cellulose decomposition in the field. Pedobiologia 17: 145–155.

Law AS, Wriston JC Jr (1971) Purification and properties of *Bacillus coagulans* L-asparaginase. Arch Biochem Biophys 147: 744–752.

Lee YH, Fan LT (1980) Properties and mode of action of cellulase. Adv Biochem Eng 17: 101–129.

Lessie TG, Neidhardt FC (1967) Formation and operation of the histidine-degrading pathway in *Pseudomonas aeruginosa*. J Bacteriol 93: 1800–1810.

Loll MJ, Bollag JM (1983) Protein transformation in soil. Adv Agron 36: 351–382.

Lopez-Hernandez D, Nino M, Nannipieri P, Fardeau JC (1989) Phosphatase activity in *Nasutitermes ephratae* termite nests. Biol Fertil Soils 7: 134–137.

Mahler RH, Cordes EH (1971) Biological Chemistry. Harper & Row, New York, London.

Malcolm RE (1983) Assessment of phosphatase activity in soils. Soil Biol Biochem 15: 403–408.

Marsden WL, Gray PP (1986) Enzymatic hydrolysis of cellulose in lignocellulose materials. CRC Crit Rev Microbiol 3: 235–276.

Mathe P, Kovacs G (1980) Effect of Mn and Zn on the activity of phosphate in soil: I. Phosphatase activity of a calcareous chernozen soil under maize. Agrokem Talajtan 29: 441–446.

Mathur SP, Sanderson RB (1978) Relationship between copper contents, rates of soil respiration and phosphatase activities of some histozols in an area of Southwestern Quebec in the summer and the fall. Con J Soil Sci 58: 125–134.

Mathur SP, Rayment AF (1977) Influence of trace element fertilization on the decomposition rate and phosphatase activity of a mesic fibrisol. Can J Soil Sci 57: 397–408.

May DW, Gile PL (1909) The catalase of soils. Puerto Rico Agr Exp Sta Circular no. 9: 3–13.

May PB, Douglas LA (1976) Assay for soil urease activity. Plant Soil 45: 301–305.

Mayaudon J, Batistic L, Sarkar JM (1975) Properties of proteolytically active extracts from fresh soils. Soil Biol Biochem 7: 281–286.

McCarty GW, Bremner JM (1991) Production of urease by microbial activity in soils under aerobic and anaerobic conditions. Biol Fertil Soils 11: 228–230.

McCarty GW, Bremner JM, Chac HS (1989) Effects of *N*-(*n*-butyl)thiophosphorictriamide on hydrolysis of urea by plant, microbial, and soil urease. Biol Fertil Soils 8: 123–127.

McCarty GW, Shogern DR, Bremner JM (1992) Regulation of urease production in soil by microbial assimilation of nitrogen. Biol Fertil Soils 12: 261–264.

McLaren AD (1969) Radiation as a technique in soil biology and biochemistry. Soil Biol Biochem 1: 63–73.

McLaren AD, Estermann EF (1957) Influence of pH on the activity of chymotrypsin at the solid–liquid interface. Arch Biochem Biophys 68: 157–160.

Morra MJ, Freeborn LL (1989) Catalysis of amino acid deamination in soils by pyridoxal-5-phosphate. Soil Biol Biochem 21: 645–650.

Moyo CC, Kissel DE, Cabrera ML (1989) Temperature effects on soil urease activity. Soil Biol Biochem 21: 935–938.

Mulvaney RL, Bremner JM (1979) A modified diacetyl monoxime method for colorimetric determination of urea in soil extracts. Commun Soil Sci Plant Anal 10: 1163–1170.

Nakas JP, Gould WD, Klein DA (1987) Origin and expression of phosphatase activity in a semiarid grassland soil. Soil Biol Biochem 19: 13–18.

Nannipieri P (1994) Productivity, Sustainability and Pollution. In : Soil Biota. Management in Sustainable Farming Systems (Parkhurst CE, Doube BM, Gupta VV, Grace PR, eds), CSIRO, Adelaide, Australia, pp. 238–244.

Nannipieri P, Ceccanti B, Cervelli S, Sequi P (1974) Use of 0.1 M pyrophosphate to extract urease from a podzol. Soil Biol Biochem 6: 359–362.

Nannipieri P, Ceccanti B, Cervelli S, Sequi P (1978a) Stability and kinetic properties of humus–urease complexes. Soil Biol Biochem 10: 143–147.

Nannipieri P, Johnson RL, Paul EA (1978b) Criteria for measurement of microbial growth and activity in soil. Soil Biol Biochem 10: 223–229.

Nannipieri P, Ceccanti B, Cervelli S, Matarese E (1980) Extraction of phosphatase, urease, protease, organic carbon and nitrogen from soil. Soil Sci Soc Am J 44: 1011–1016.

Nannipieri P, Cecconti B, Conti C, Bianchi D (1982) Hydrolases extracted from soil: their properties and activities. Soil Biol Biochem 14: 257–263.

Nannipieri P, Ceccanti B, Bianchi B (1988) Characterization of humus–phosphate complexes extracted from soil. Soil Biol Biochem 20: 683–691.

Nannipieri P, Grego S, Ceccanti B (1990) Ecological significance of the biological activity in soil. In: Soil Biochemistry, vol 6. Bollag J-M, Stotzky G (eds). Marcel Dekker, New York, Basel, pp. 293–355.

Nicholls RG, Roy AR (1971) Arylsulfatase. In: The Enzymes, vol 5, 3rd edn. Boyer PD (ed.). Academic Press, New York, pp. 21–41.

Nor YM (1982) Soil urease activity and kinetics. Soil Biol Biochem 14: 63–65.

Pancholy SK, Lynd JQ (1972) Quantitative fluorescence analysis of soil lipase activity. Soil Biol Biochem 4: 257–259.

Pancholy SK, Rice EL (1973) Soil enzymes in relation to old field succession: amylase, cellulase, invertase, dehydrogenase and urease. Soil Sci Soc Am Proc 37: 47–50.

Park JT, Johnson MJ (1949) A submicrodetermination of glucose. J Biol Chem 181: 149–151.

Peréz-Mateos M, Gonzalez-Carcedo S (1988) Assay of urease activity in soil columns. Soil Biol Biochem 20: 567–572.

Perucci P, Scarponi L (1984) Arylsulphatase activity in soils amended with crop residues: kinetics and thermodynamic parameters. Soil Biol Biochem 16: 605–608.

Pokorna V (1964) Method of determining the lypolytic activity of upland and lowland peats and mucks. Pochvovedenie 106: 85–87.

Powlson DS, Jenkinson DS (1976) The effects of biocidal treatments on metabolism in soil. II. Gamma irradiation, autoclaving, air-drying and fumigation. Soil Biol Biochem 8: 179–188.

Press MC, Henderson J, Lee JA (1985) Arylsulphatase activity in peat in relation to acidic deposition. Soil Biol Biochem 17: 99–103.

Pulford ID, Tabatabai MA (1988) Effect of waterlogging on enzyme activities in soils. Soil Biol Biochem 20: 215–219.

Rai B, Srivastava AK (1983) Decomposition and competitive colonization of leaf litter by fungi. Soil Biol Biochem 15: 115–117.

Rastin N, Rosenplänter K, Hüttermann A (1988) Seasonal variation of enzyme activity and their dependence on certain soil factors in beech forest soil. Soil Biol Biochem 20: 637–642.

Rechler MM (1969) The purification and characterization of L-histidine ammonia-lyase (*Pseudomonas*). J Biol Chem 244: 551–559.

Reissig JI, Strominger JL, Leloir LF (1955) A modified method for the estimation of *N*-acetylamine sugars. J Biol Chem 217: 959–966.

Reuter G (1976) Gelände- und Laborpraktikum der Bodenkunde. VEB Verlag, Berlin.

Rhee YH, Hah YC, Hong SW (1987) Relative contributions of fungi and bacteria to soil carboxymethylcellulase activity. Soil Biol Biochem 19: 479–481.

Rice L, Mollik MAB (1977) Causes of decreases in residual carbohydrase activity in soil during oldfield succession. Ecology 58: 1297–1309.

Roberge MR (1978) Methodology of soil enzyme measurement and extraction. In: Soil Enzymes. Burns RG (ed.) Academic Press, New York, pp. 341–370.

Roberts J, Holcenberg JS, Dolowy WC (1972) Isolation, crystallization, and properties of Achromobacteraceae glutaminase–asparaginase with antitumor activity. J Biol Chem 247: 84–90.

Rodriguez-Kabana R, Godoy G, Morgan-Jones G, Shelby RA (1983) The determination of soil chitinase activity; conditions for assay and ecological studies. Plant Soil 75: 95–106.

Ross DJ (1965) Effects of air-dry, refrigerated, and frozen storage on activities of enzymes hydrolysing sucrose and starch. J Soil Sci 16: 86–94.

Ross DJ (1968) Some observations on the oxidation of glucose by enzymes in soil in the presence of toluene. Plant Soil 28: 1–11.

Ross DJ (1974) Glucose oxidase activity in soil and its possible interference in assay of cellulase activity. Soil Biol Biochem 6: 303–306.

Ross DJ (1975) Studies on a climosequence of soils in tussock grasslands. Invertase and amylase activities of topsoils and their relationship with other properties. N Zealand J Soil Sci 18: 511–518.

Ross DJ, Speir TW, Giltrap DJ, McNeilly BA, Molloy LF (1975) A principal components analysis of some biochemical activities in climosequence of soils. Soil Biol Biochem 7: 349–355.

Rössner H (1991) Bestimmung der Chitinase-Aktivität. In: Bodenbiologische Arbeitsmethoden. Schinner F, Öhlinger R, Kandeler E (eds). Springer Verlag, Berlin, pp. 66–70.

Sarathchandra SU, Perrott KW (1981) Determination of phosphatase and arylsulphatase activities in soils. Soil Biol Biochem 13: 543–545.

Sarkar JM, Batistic L, Mayaudon J (1980) Les hydrolases du sol et leur association avec les hydrates de carbone. Soil Biol Biochem 12: 325–328.

Sarkar JM, Leonowicz A, Bolag JM (1989) Immobilization of enzymes on clays and soils. Soil Biol Biochem 21: 223–230.

Sato K (1981) Relations between soil microflora and CO_2 evolution upon decomposition of cellulose. Plant Soil 61: 251–258.

Sayre FW, Roberts E (1958) Preparation and some properties of a phosphate-activated glutaminase from kidneys. J Biol Chem 233: 1128–1134.

Schinner F, Von Mersi W (1990) Xylanase-, CM-cellulase- and invertase activity in soil: an improved method. Soil Biol Biochem 22: 511–515.

Schinner F, Öhlinger R, Kandeler E (1991) Bodenbiologische Arbeitesmethoden. Springer Verlag, Berlin, New York.

Schmidt G, Laskowsky M Sr. (1961) Phosphate ester cleavage (survey). In: The Enzymes, 2nd edn. Boyer PD, Lardy H, Myrbach K (eds). Academic Press, New York, pp. 3–35.

Schreiner O, Shorey EC (1910) The presence of arginine and histidine in soils. J Biol Chem 8: 381–384.

Schröder D, Gewehr H (1977) Stroh- und Zelluloseabbau in verschiedenen Bodentypen. Z Pflanzenernähr Bodenkd 140: 273–284.

Schröder D, Urban B (1985) Bodenatmung, Celluloseabbau und Dehydrogenaseaktivität in verschiedenen Böden und ihre Beziehungen zur organischen Substanz sowie Bodeneigenschaften. Landwirtsch Forschung 38: 166–172.

Schulz-Berendt V (1986) Der Stickstoff-Haushalt eines Ruderalstandortes als Grundlage der Beurteilung von Ökosystem-Veränderungen. Dissertation, University of Bremen.

Simpson JR, Freney JR, Westelaar R, Muirhead WA, Leuning R, Denmead OT (1984) Transformations and losses of urea nitrogen after application to flooded rice. Aust J Agric Res 35: 189–200.

Simpson LR, Freney JR, Muirhead WA, Leuning R (1985) Effects of phenylphosphodiamidate and dicyandiamide on nitrogen loss from flooded rice. Soil Sci Soc Am J 49: 1426–1431.

Sinsabaugh RL, Linkins AE (1988) Adsorption of cellulase components by leaf litter. Soil Biol Biochem 20: 927–931.

Sinsabaugh RL, Linkins AE (1989) Cellulase mobility in decomposing leaf litter. Soil Biol Biochem 21: 205–209.

Skujins JJ (1967) Enzymes in soil. In: Soil Biochemistry, vol 1. McLaren AD, Peterson GH (eds). Marcel Dekker, New York, pp. 371–414.

Skujins J (1978) History of abiotic soil enzyme research. In: Soil Enzymes. Burns RG (ed.). Academic Press, London, pp. 1–49.

Skujins JJ, McLaren AD (1969) Assay of urease activity ^{14}C-Urea in stored, geogically preserved, and in irradiated soils. Soil Biol Biochem 1: 89–99.

Sowden FJ (1958) The forms of nitrogen in the organic matter of different horizons of soil profiles. Can J Soil Sci 38: 147–154.

Sparling GP, Speir TW, Whale KN (1986) Changes in microbial biomass C, ATP content, soil phosphomonoesterase and phospho-diesterase activity following air-drying of soils. Soil Biol Biochem 18: 363–370.

Speir R, Lee R, Pansier EA, Cairns A (1981) A comparison of sulphatase, urease and protease activities in planted and in fallow soils. Soil Biol Biochem 12: 281–291.

Speir TW, Ross DJ (1975) Effects of storage on the activities of protease, urease, phosphatase, and sulphatase in three soils under pasture. N Zealand Sci 18: 231–237.

Speir TW, Ross DJ (1978) Soil phosphatase and sulphatase. In: Soil Enzymes. Burns RG (ed.). Academic Press, London, pp. 197–250.

Speir TW, Ross DJ (1981) A comparison of the effects of air-drying and acetone dehydration on soil enzyme activities. Soil Biol Biochem 13: 225–229.

Spiers GA, McGill WB (1979) Effects of phosphorus addition and energy supply on phosphatase production and activity in soils. Soil Biol Biochem 11: 3–8.

Spiro RG (1966) Analysis of sugars found in glycoproteins. In: Methods in Enzymology, vol 8. Neufeld EF, Ginsberg V (eds). Academic Press, London, pp. 326.

Stefanic G, Beck Th, Schwimmer J, Hartmann F, Varbanciu A (1984) Apparatus for measuring the soil catalase activity. Fifth Symposium on Soil Biology, Romania IASI, pp. 47–50.

Stutzenberger FG (1972) Cellulolytic activity of *Thermomonospora curbata*: optimal assay conditions, partial purification and product of the cellulase. Appl Microbiol 24: 83–90.

Sumner JB (1951) Urease. In: The Enzymes, vol 1. Sumner JB (ed.). Academic Press, New York, pp. 873–892.

Suttner T, Alef K (1988) Correlation between arginine ammonification, enzyme activities, microbial biomass, physical and chemical properties of different soils. Zentralbl Mikrobiol 143: 569–573.

Tabatabai MA (1977) Effects of trace elements on urease activity in soils. Soil Biol Biochem 9: 9–13.

Tabatabai MA (1982) Soil enzymes. In: Methods of Soil Analysis, Part 2, Chemical and Microbiological Properties. Page AL, Miller EM, Keeney DR (eds). American Society of Agronomy, Madison, WI, pp. 903–947.

Tabatabai MA, Bremner JM (1969) Use of *p*-nitrophenyl phosphate for assay of soil phosphatase activity. Soil Biol Biochem 1: 301–307.

Tabatabai MA, Bremner JM (1970a) Arylsulphatase activity of soils. Soil Sci Soc Am Proc 34: 225–229.

Tabatabai MA, Bremner JM (1970b) Factors affecting soil arylsulfatase activity. Soil Sci Soc Am Proc 34: 427–429.

Tabatabai MA, Bremner JM (1972) Assay of urease activity in soil. Soil Biol Biochem 4: 479–487.

Tabatabai MA, Dick WA (1979) Distribution and stability of pyrophosphatase in soil. Soil Biol Biochem 11: 655–659.

Tabor H (1954) Metabolic studies on histidine, histamine, and related imidazoles. Pharmacol Rev 6: 229–343.

Tarafdar JC, Jungk A (1987) Phosphatase activity in the rhizosphere and its relation to the depletion of soil organic phosphorous. Biol Fertil Soils 3: 199–204.

Tateno M (1988) Limitation of available substrates for the expression of cellulase and protease activities in soil. Soil Biol Biochem 20: 117–118.

Trasar-Cepeda MC, Gil-Sotres F (1987) Phosphatase activity in acid high organic matter soils in Galicia (NW Spain). Soil Biol Biochem 19: 281–287.

Trasar-Cepeda MC, Gil-Sotres F (1988) Kinetics of acid phosphatase activity in various soils of Galicia (NW Spain). Soil Biol Biochem 20: 275–280.

Trevors JT (1984) Rapid gas chromatographic method to measure H_2O_2 oxidoreductase (catalase) activity in soil. Soil Biol Biochem 16: 525–526.

Veon M, Mikyeshite K, Sawado Y, Obe Y (1990) Assay of chitinase and *N*-acetylglucosamidase activity in forest soils with 4-methylumbelliferyl derivatives. Z Pflanzenernähr Bodenkd 154: 171–175.

Voets JP, Dedeken M, Besseme E (1965) The behaviour of some amino acids in gamma irradiated soils. Naturwissenschaften 52: 476.

Wallenfels K, Weil R (1972) β-Galactosidase. In: The Enzymes, vol 7, 3rd edn. Boyer PD (ed.). Academic Press, New York, pp. 617–663.

Warman PR, Bishop C (1987) Amino-N compounds found in soil organic matter hydrolysates of a loamy sand using an immobilized protease reactor column. Biol Fertil Soils 5: 219–224.

Warman PR, Isnor RA (1989) Evidence of peptides in low-molecular-weight fractions of soil organic matter. Biol Fertil Soils 8: 25–28.

Weetall HH, Weliky N, Vango SP (1965) Detection of microorganisms in soil by their catalytic activity. Nature 205: 1019–1021.

Wilke B-M (1986) Einfluβ verschiedener potentieller anorganischer Schadstoffe auf die mikrobielle Aktivität von Waldhumusformen unterschiedlicher Pufferkapazität. Bayreuther Geowissenschaftliche Arbeiten, Band 8, University of Bayreuth.

Wilke BM (1988) Langzeitwirkungen potentieller anorganischer Shadstoffe auf die mikrobielle Aktivität einer sandigen Braunerde. Z Pflanzenernähr Bodenkd 151: 131–136.

Wriston JC Jr (1971) L-Asparaginase. In: The Enzymes, vol 4. Boyer PD (ed.). Academic Press, New York, pp. 101–121.

Xiaoyan Z, Likai Z, Guanyun Wu (1992) Urea hydrolysis in a brown soil: effect of hydroquinone. Soil Biol Biochem 24: 165–170

Yamana K, Suzukki H, Nishizawa K (1970) Purification and properties of extracellular and cellbound cellulase components of *Pseudomonas fluorescens var. cellulosa*. J Biochem 67: 19–35.

Zantua MI, Bremner JM (1975a) Comparison of methods of assaying urease activity in soils. Soil Biol Biochem 7: 291–295.

Zantua MI, Bremner JM (1975b) Preservation of soil samples for assay of urease activity. Soil Biol Biochem 7: 297–299.

Zantua MI, Bremner (1977) Stability of urease in soils. Soil Biol Biochem 9: 135–140.

Zhengping W, Van Cleemput O, Demeyer P (1991) Effect of urease inhibitors on urea hydrolysis and ammonia volatilization. Biol Fertil Soils 11: 43–47.

Microbial biomass

8

Microbial biomass has been defined as the part of the organic matter in soil that constitutes living microorganisms smaller than 5–10 μm^3. It is generally expressed in milligrams of carbon per kilogram soil or micrograms of carbon per gram dry weight, typically biomass carbon ranges from 1 to 5% of soil organic matter (Jenkinson and Ladd 1981; Sparling 1985; Smith and Paul 1990). The interest in estimating soil microbial biomass is related to its function as a pool for subsequent delivery of nutrients, and its role in structure formation and stabilization of soil and as an ecological marker (Smith and Paul 1990). Estimations of the microbial biomass have usually involved treatment of the biomass as a single component, although it is known that a diversity of populations with different biochemical characteristics are present.

Several methods have been used to estimate microbial biomass in soil. They are based on staining and counting of microbial cells (Babiuk and Paul 1970; Trolldenier 1973; Anderson and Slinger 1975; Paul and Johanson 1977; Söderström 1977; Torsvik and Goksoyr 1978; Lundgren 1981), on the use of physiological parameters such as ATP, respiration and heat output (Anderson and Domsch 1978; Sparling et al 1981; Van de Werf 1989/1990; Sparling and West 1990), or on the application of the fumigation technique (Shen et al 1984; Jenkinson and Powlsen 1976a, 1976b, 1976c; Brookes et al 1985; Vance et al 1987a, 1987b, 1987c; Joergensen et al 1990).

To avoid confusion, it should be recognized that the estimation of biomass in soils comprises two different aspects (Jenkinson and Ladd 1981; Alef 1993):

1. Measurement of a suitable indicator of microbial biomass: a quantitative indicator of microbial biomass in soil should only be found in living microbial cells and rapidly degraded once released into the soil environment. In addition, its concentration should be constant in the cells and the compound should be extracted quantitatively from soils. Reliable methods for estimating this indicator should be available.
2. The possibility of calibrating the methods used and calculation of data into biomass should be available.

These aspects are interdependent, because a highly sensitive and very reliable technique to estimate the biomass-indicator in soil would be useless, if there were serious objections to and criticisms of the calibration method.

In this chapter methods for estimating microbial biomass, based on different principles, are presented and discussed. Their validity, the application of different calculation factors, the principal and technical difficulties of these methods are also discussed.

Methods in Applied Soil Microbiology and Biochemistry
ISBN 0–12–513840–7

The fumigation incubation method

R.G. Joergensen

Fumigation with $CHCl_3$ caused a flush of mineralized CO_2 and NH_4^+ once the fumigant is removed and the soil incubated. Störmer (1908) suggested that this flush was due to the decomposition of the soil organisms killed. This hypothesis was verified by studying the effects of partial sterilization of soil on the decomposition of ^{14}C-labelled ryegrass (Jenkinson 1966). It was postulated that the size of the soil microbial biomass can be estimated by the CO_2 flush and incubating fumigated soils.

In a series of five papers, Jenkinson and coworkers investigated various aspects of $CHCl_3$ fumigation (Jenkinson 1976; Jenkinson and Powlson 1976a, 1976b; Jenkinson et al 1976; Powlson and Jenkinson 1976) and devised the fumigation incubation method for determining the size of microbial biomass.

The use of this method is based on the following assumptions (Jenkinson and Ladd 1981):

1. The soil fumigation kills the microbial biomass and it does not affect the non-living organic matter; therefore the flush exclusively derives from the microbial biomass.
2. The number of organisms killed in the unfumigated soil is negligible compared with that in fumigated soil.
3. The fraction of dead microbial biomass carbon mineralized over a given time period does not differ in different soils.

Principle of the method

Moist soil is exposed to ethanol-free chloroform for 24 h, the fumigant is removed by repeated evacuation, the soil is inoculated and then incubated at 25°C for 10 days at 50% of its water-holding capacity.

Materials and apparatus

Incubator adjustable to 25°C
Vacuum desiccators
Vacuum line (water pump or electric pump)
Burette or automatic titrator

Chemicals and solutions

Ethanol-free $CHCl_3$ (LiChrosolv, Merck no. 2444)
Soda lime (p.a.)
1 M NaOH (p.a.)
1 M HCl (Titrisol)
0.1 M HCl (Titrisol)
1.5 M $BaCl_2 \times 2H_2O$ (p.a.; 366.42 g l^{-1})
1% phenolphthalein solution
0.1% methyl orange solution
2 M KCl (p.a.; 149.12 g l^{-1})

Procedure

Fumigation and incubation of soil (Jenkinson and Powlson 1976b)

After sieving, homogenization and adjustment to 40% of the water-holding capacity (WHC), moist soils (100 g dry weight) are split into two samples; one is fumigated while the other is not treated and used as a control, each of which (50 g dry weight) are placed in a 100 ml glass vial. The vials are placed in two separate desiccators lined with wet paper and containing a 50 ml vial with soda lime; they are pre-incubated in the dark for 7–10 days at 25°C to allow the effects of sampling to subside.

For the fumigation, a beaker containing 25 ml ethanol-free $CHCl_3$ and some boiling chips are placed in a desiccator, which is evacuated until the $CHCl_3$ has boiled vigorously for 2 min and then incubated in the dark at 25°C for 24 h. After fumigation, $CHCl_3$ is removed by repeated (six-fold) evacuation. The unfumigated soil is also incubated in the dark at 25°C for 24 h. All soils (fumigated and unfumigated) are transferred at the same time to 1 l bottles, each containing 20 ml of 1 M NaOH in a 50 ml beaker, plus 20 ml water in the bottom of the bottle.

The inoculum is prepared by shaking 10 g of soil with 100 ml water. The water content of the soils is adjusted to 50% WHC with this thin suspension of soil colloids as a microbial inoculum for the fumigated soils. Fumigated samples in the closed 1 l bottles and three closed bottles without soil (blanks) are incubated for 10 days at 25°C.

Extraction of microbial biomass nitrogen (Jenkinson and Powlson 1976a; Powlson and Jenkinson 1976)

At the end of the 10-day incubation period, the whole amount (50 g dry weight) of each fumigated and unfumigated soil is transferred from the 1 l bottles into 250 ml flasks and then immediately extracted with 200 ml 2 M KCl [the ratio extractant:soil (dry weight) is 4:1 v/w] for 30 min under oscillated shaking at 200 rev min^{-1}; then the soil suspension is filtered through a paper filter (Whatman 42) and ammonium and nitrate concentrations in the soil extract are measured. Different methods for measuring ammonium and nitrate are available; the distillation method has been described in the section of urease activity in Chapter 7 (Keeney and Nelson 1982).

CO_2 measurements:

Titration to pH 8.3 in the presence of $BaCl_2$ (Anderson 1982)

An aliquot (5 ml) of the 1 M NaOH is placed in a titration vessel. Evolved CO_2 is absorbed by NaOH producing Na_2CO_3. One millilitre of 1.5 M $BaCl_2$ solution is added to precipitate the carbonate as insoluble $BaCO_3$. A few drops of phenolphthalein are added as an indicator. Then, unreacted NaOH is brought to pH 8.3 by slowly adding 0.1 M HCl under magnetic stirring (disappearance of the colour). The acid must be added slowly to avoid any possible dissolution of the precipitated $BaCO_3$.

Titration from pH 8.3 to pH 3.7 (Jenkinson and Powlson 1976b)

In order to determine the evolved CO_2 which forms Na_2CO_3 after absorbtion in NaOH, an aliquot (5 ml) of the 1 M NaOH is placed in a titration vessel. A few drops of phenolphthalein and methyl orange are added as indicators. The pH of the NaOH solution is brought to about 10 by slow addition of 1 M HCl and then to pH 8.3 by slow addition of 0.1 M HCl. During the acid addition the solution is magnetically stirred. At pH 8.3, all Na_2CO_3 is changed to $NaHCO_3$ (colour change: pink to colourless). The solution is then titrated with 0.1 M HCl to pH 3.7 to remove all CO_2 from the solution (colour change: light red to orange-yellow).

Calculation

Calculation of evolved CO_2 (titration to pH 8.3 in the presence of $BaCl_2$):

$$CO_2\text{-C } (\mu g\ g^{-1} \text{ soil}) = (B - S) \times M \times E \times A : DW \quad (8.1)$$

where B is the amount of acid needed to titrate the NaOH in the blank bottles (without soil) to the end point (pH 8.3) in microlitres, S is the amount of acid needed to titrate the NaOH in the bottles containing the soil to the endpoint (pH 8.3) in microlitres, M is the molarity of the HCl, E = 6 (equivalent weight to express the data as carbon), A is the ratio total volume of the

NaOH to volume of the NaOH aliquot and *DW* sample is the dry weight of the soil in gramm.

Calculation of evolved CO_2:

$$CO_2\text{–C } (\mu g\ g^{-1} \text{ soil}) = (S - B) \times M \times E \times A : DW \quad (8.2)$$

where *S* is the amount of acid needed to titrate the NaOH in the bottles containing the soil samples from pH 8.3 to pH 3.7 in microlitres, *B* is the amount of acid needed to titrate the NaOH in the blank bottles (without soil) from pH 8.3 to pH 3.7 in microlitres, *M* is the molarity of the HCl, $E = 12$ (equivalent weight to express the data in terms of carbon), *A* is the ratio total volume of the NaOH to volume of the NaOH aliquot and *DW* sample is the dry weight of the soil in gramm.

Calculation of microbial biomass carbon:

$$\text{Biomass C} = F_C{:}k_C \quad (8.3)$$

where F_C = (CO_2–C evolved from fumigated soil in the 0–10 day incubation period) - (CO_2–C evolved from unfumigated soil samples in the 0–10 day incubation period) and $k_C = 0.45$, the fraction of the killed biomass mineralized to CO_2 over the 10-day incubation period (Jenkinson 1988). The calibration was performed by adding a known quantity of microorganisms to a soil, fumigating it and measuring the proportion of the added microbial carbon mineralized to CO_2. Calibration factors obtained from microorganisms grown *in vitro* are assumed to be applicable to the endogeneous soil microflora (Jenkinson 1988).

Calculation of microbial biomass nitrogen:

$$\text{Biomass N} = F_N/k_N \quad (8.4)$$

where F = [NH_4–N + NO_3–N mineralized in fumigated soil during the 0–10 day incubation period) – (NH_4–N + NO_3–N; mineralized in unfumigated soil during the 0–10 day incubation period) and $k_N = 0.57$, the fraction of the killed biomass mineralized to NH_4–N over the 10-day incubation period (Jenkinson 1988). The calculation of the k_N factor by simply adding microorganisms grown *in vitro* to soil, fumigating them and measuring how much of microbial nitrogen is mineralized to NH_4 under standardized conditions presents more problems than the calculation of k_C factor. The nitrogen content depends on the type of microorganism and environmental conditions (Jenkinson 1988). In addition, nitrogen mineralization can occur during the incubation following the fumigation.

However, the C:N ratio of native microbial biomass seems to be relatively constant in soils that do not contain large quantities of freshly added plant material of wide C:N ratio. On this assumption, the k_N factor can be calculated as follows (Jenkinson 1988):

$$k_N = \beta{:}\alpha\ k_C \quad (8.5)$$

where β = biomass C:N = 6.7 (Anderson and Domsch 1980 ; Shen et al 1984), $\alpha = F_C/F_N = 5.31$ (Jenkinson 1988) and $k_N = 6.7{:}5.31 \times 0.45 = 0.57$. The ratio $F_C{:}F_N$ was taken from a large group of soils (104) sampled under different environmental conditions (Jenkinson 1988).

Discussion

Two other ways of calculating microbial biomass carbon exist, each based on the use of a different control:

(A) Biomass = [(CO_2 (Y–C evolved from fumigated soil during the 0–10 day incubation period) – (CO_2–C evolved from unfumigated soil during the 10–20 day incubation period)]:k_C (8.6)

as proposed by Jenkinson and Powlson (1976b) for fresh soils.

(B) Biomass C = [(CO_2–C evolved from fumigated soil during the 0–10 day incubation period) – (CO_2–C evolved from fumigated soil during the 10–20 day incubation period)]:k_C (8.7)

as proposed by Chaussod and Nicolardot (1982). None of the three controls are entirely satisfactory because a large variety of microorganisms are living in the unfumigated soil while a much smaller and specialized population is present in the fumigated soil. The calculation of biomass carbon is based on the assumption that both fumigated and unfumigated soils respire at the same rate, which usually occurs after the first few days of the incubation (Jenkinson and Powlson 1976a), However, all three ways of calculation give similar values of biomass in pre-incubated soils (Shen et al 1987).

Another determination of microbial biomass carbon does not involve a control (Voroney and Paul 1984). The CO_2-C evolved from non-living organic matter, under the standard incubation conditions, causes an overestimation of biomass carbon (Shen et al, 1987). For this reason Voroney and Paul's procedure is not recommended.

An incubation period of 7 days at a higher temperature (28°C) has been used (Chaussod and Nicolardot 1982). A lower incubation temperature (22°C) has been proposed by Anderson and Domsch (1978). A change in incubation conditions (temperature, incubation period) affects the respiration rate and the total amount of CO_2 mineralized.

The fumigation incubation method is not recommended for acidic (pH $H_2O < 4.5$) soils because: (1) soil inoculation is difficult under these conditions (Chapman 1987; Vance et al, 1987a); (2) the k_C declines sharply below pH(H_2O) 5 (Jenkinson 1988); (3) the respiration rate is higher in unfumigated than in the fumigated soil once the flush is over (Powlson and Jenkinson 1981; Sparling and Williams 1986). The modified fumigation incubation method for acid soils (Vance et al, 1987b) does not involve the use of a control and it presents the same limitations already mentioned for the method of Voroney and Paul (1984).

The fumigation incubation method is unsuitable for soils recently treated with organic matter, because large microflora of the unfumigated soil decomposes the substrate more effectively than the smaller microflora of the fumigated soil; this makes the biomass values too small or even negative (Martens 1985).

In waterlogged soils, both CH_4 and CO_2 are produced from the decomposition of microbial debris. Thus, the modified fumigation incubation procedure for waterlogged soils can only be used for determining for microbial biomass nitrogen (Inubushi et al, 1984).

With calcareous soils that are low in organic matter, errors can occur due to the decomposition of bicarbonate. This error can be reduced by placing beakers with soda lime in desiccators holding fumigated and unfumigated soils (Jenkinson and Powlson 1976b).

The fumigation incubation method should only be used after pre-incubating well-drained soils above pH (H_2O) 4.5 as originally devised (Jenkinson and Powlson 1976b). In all other cases the fumigation extraction method is recommended.

Direct microscopy cannot be used to prove the effects of fumigation. Large amounts of dead microbial cells, stained with phenol-aniline, remained after incubating fumigated soils for more than 50 days (Jenkinson 1976).

The desiccator must be kept under vacuum for 24 h to ensure the presence of a $CHCl_3$ atmosphere, which kills virtually all soil microorganisms. After fumigation, the number of bacterial and fungal plate counts decreases by 93.2–99.9% of the initial value (Shields et al 1974; Lynch and Panting 1980). The soil immediately after fumigation contains about 2×10^5 viable bacteria g^{-1} soil (Shields et al 1974).

The fumigation incubation method cannot be used when fresh roots are present in soil because cell membranes of young living roots are affected by $CHCl_3$ fumigation (Mueller et al 1992). Thus, additional cell material is mineralized during the incubation following $CHCl_3$ fumigation (Martin and Foster 1985; Sparling et al 1985).

Ethanol-free $CHCl_3$ must be used; ethanol cannot be completely removed from soil after fumigation (Jenkinson 1988) and it is used as a substrate, being mineralized to CO_2, and thus is incorrectly measured as biomass C.

The samples are fumigated for 24 h at 25°C with $CHCl_3$. After death, the autolysis of microbial cells depends on time and temperature. Low fumigation temperatures and short fumigation times may affect the factors (k_C and k_N) and thus biomass estimation (Joergensen and Brookes 1991).

Other fumigants like methyl bromide (Powlson and Jenkinson 1976) and carbon disulphide (Kudeyarov and Jenkinson 1976) can be used.

Carbonic anhydrase, used by Jenkinson and Powlson (1976b) decreases the drift and facilitates titration (Underwood 1961).

The CO_2-C can be determined in different ways. After its adsorption in the alkaline solution it can be titrated (Jenkinson and Powlson 1976b) or determined by colourimetric method (Chaussod et al 1986). It can also be determined by gas chromatography (Martens 1985) after its accumulation in the head space (Anderson and Domsch 1978; Sparling 1981; Chaussod and Nicolardot 1982). This determination is not recommended in neutral and alkaline soil, where evolved CO_2 can remain as HCO_3^- in the soil solution. This leads to an underestimation of the respiration rate (Martens 1987).

The microbial biomass C determined by the fumigation incubation method is significantly correlated to data obtained by direct microscopy (Jenkinson et al 1976; Vance et al 1987b), fumigation extraction (Vance et al 1987c; Kaiser et al 1992), ATP content (Jenkinson 1988), the heat output (Sparling 1981), the substrate-induced respiration method (Anderson and Domsch 1978; Kaiser et al 1992) and to biomass estimates by mathematical analysis of respiration curves (Van de Werf and Verstraete 1987a). The fumigation incubation method is a basic calibration procedure to convert measured data into biomass or biomass carbon for most of these methods. However, the correlation between biomass C estimates by fumigation incubation and those by direct counts was sometimes poor (Schnürer et al 1985). Some doubts still remain about the calibration procedures due to statistical problems (Wardle and Parkinson 1991).

Another way of calculating biomass nitrogen is based on the use of a different control to that on p. 372.

$$F_N = [(NH_4\text{-N mineralized in fumigated samples during 0–7 days}) - (NH_4\text{-N mineralized in fumigated samples during 7–14 days})] : k_N \qquad (8.8)$$

according to Nicolardot and Chaussod (1986).

It is unnecessary to measure changes in NO_3-N concentrations because fumigated soils do not nitrify (Jenkinson and Powlson 1976a; Harden et al 1993b).

The use of a control soil is less important for determining biomass nitrogen than biomass carbon, because the ratio N mineralized by fumigated soil to N mineralized by unfumigated soil is much greater than the ratio C mineralized by fumigated soil to C mineralized by unfumigated soil (Ayanaba et al 1976; Voroney and Paul 1984). Inorganic-N flush values were not influenced by inoculation size (Ross 1990).

Nitrogen immobilization during the incubation may not be a serious problem

(Jenkinson 1988) because: (1) the recolonizing population is much smaller than the killed population (Jenkinson and Powlson 1976a; Martens 1985); (2) the C:N ratio of the recolonizing population is not very different from that of the killed population (Harden et al 1993b).

The use of the fumigation incubation method has provided an estimate of the size of the soil microbial biomass. This has been useful in studies based on the use of isotopes (Jenkinson and Powlson 1976a; Amato and Ladd 1980; Kassim et al 1981; Carter and Rennie 1984; Bottner 1985; Merckx et al 1985; Nannipieri et al 1985; Nicolardot et al 1986; Schnürer and Rosswall 1987; Vance et al 1987a). Thus it has been possible to study and model nutrients cycling in soil (Jenkinson and Rayner 1977; Jenkinson and Ladd 1981; Nannipieri 1984; Van Veen et al 1984). The formation of microbial biomass also depends on the turnover of root-derived material (Merckx et al 1985). Changes in microbial biomass can be detected long before changes in total organic matter carbon or nitrogen (Powlson and Jenkinson 1981; Powlson et al 1987). Thus the determination of microbial biomass has been useful for detecting the effects of tillage (Carter 1986), crop rotation and residue management (Collins et al 1992).

Biomass measuresments serve as a sensitive indicator for soil pollution by heavy metals (Brookes and McGrath 1984) or pesticides (Soulas et al 1984; Duah-Yentumi and Johnson 1986). It has been also used to measure the effects of water potential increase (Kieft et al 1987).

The fumigation extraction method

(Vance et al 1987)

R. Joergensen

Fumigation and extraction

(Brookes et al 1985; Vance et al 1987c)

Principle of the method

Chloroform fumigation of soil kills and lyses microbial cells with the release of cytoplasm into the soil environment; thus the cell material can be extracted from soil (Powlson and Jenkinson 1976). Organic carbon (Vance et al 1987), total nitrogen and NH_4-N (Brookes et al 1985), inorganic P (Brookes et al 1982), ninhydrin-reactive nitrogen (Joergensen and Brookes 1990; Badalucco et al 1992), carbohydrate carbon (Joergensen et al 1990, 1994), phenol-reactive carbon, orcinol*-reactive carbon, deoxyribose-containing compounds, Folin-Ciocolteu's reagent-reactive carbon and nitrogen compounds (Badalucco et al 1992) are extracted by 0.5 M K_2SO_4.

With an appropriate calibration, some of these compounds have been used to measure microbial biomass in soil. Biomass phosphorus, which is extracted by 0.5 M $NaHCO_3$ (Brookes et al 1982; Hedley and Stewart 1982), is less closely related to biomass carbon (Brookes et al 1984; Paul and Clark 1988).

* anthrone-reactive carbon (Badalucco et al, 1990, 1992)

Materials and apparatus

Room or incubator adjustable to 25°C
Vacuum desiccators
Vacuum line (water pump or electric pump)
Horizontal or overhead shaker
Freezer
Paper filter (Whatman 42)

Chemicals and solutions

Ethanol-free $CHCl_3$ (LiChrosolv, Merck no. 2444)
Soda lime (p.a.)
0.5 M K_2SO_4 (87.135 g l^{-1})

Procedure

Moist soils (50 g dry weight) are split into two samples (each 25 g dry weight). The unfumigated control is placed in a 250 ml bottle and then immediately extracted with 100 ml 0.5 M K_2SO_4 (ratio extractant: soil (dry weight) is 4:1 v/w) for 30 min in an oscillating shaker at 200 rev min^{-1} (or 45 min for an overhead shaker at 40 rev min^{-1}) and then filtered through a paper filter (Whatman 42). The fumigation is carried out using a 50 ml glass vial which contains the moist soil; the vial is placed in a desiccator lined with wet tissue paper and a vial with soda lime. A beaker containing 25 ml ethanol-free $CHCl_3$ and a few boiling chips is added, and the desiccator evacuated until the $CHCl_3$ has boiled vigorously for 2 min. The desiccator is then incubated in the dark at 25°C for 24 h. After fumigation, $CHCl_3$ is removed by repeated (six-fold) evacuations and the soil is transferred to 250 ml bottles for extraction with 0.5 M K_2SO_4 as mentioned for the unfumigated sample. All extracts are stored at −15°C prior to analysis.

Discussion

The soil samples must be sieved at approximately 40% WHC only if homogeneous samples are required (Ocio and Brookes 1990).

Soil weights can range from 200 mg (Daniel and Anderson 1992) to 200 g (Ocio and Brookes 1990a).

The water content of soil must be higher than 30% of the WHC (Joergensen, Unpublished results). In dry soils, microorganisms are apparently less affected by $CHCl_3$ (Sparling and West 1989; Sparling et al 1990). Also the rate of enzyme activity and, thus, autolysis is slower at moisture content of less than 30% WHC.

The soil water content can fluctuate widely. The microbial biomass C and N of soils at 40–50% WHC were similar to those in saturated soils (Widmer et al, 1989; Mueller et al 1992) provided they were not incubated anaerobically (Inubushi et al 1991).

Reliable biomass measurements can be obtained in waterlogged soils (e.g. paddy soils) by fumigation extraction (Inubushi et al 1991).

Problems arise for fumigation and extraction in very compressed soils which cannot be dispersed (Ross 1988; Joergensen, unpublished results).

Soda lime absorbs CO_2 evolved during the pre-incubation and fumigation period. It keeps the level of inorganic CO_2 low, which might have toxic effects or interfere with the organic carbon measurements.

The samples are fumigated for 24 h at 25°C with $CHCl_3$. Fumigation depends on the incubation time and temperature; lower fumigation temperatures and shorter fumigation times may affect biomass estimates (Joergensen and Brookes 1991).

No significant differences were obtained by filtering extracts with membrane and glass fibre filters: Whatman 42, Whatman GF/A, Sartorious 13430 (Mueller 1992), Schleicher & Schuell 595 1/2 (Scholle et al 1992). Other filter material should be used only after testing.

Soil microbial biomass carbon is extracted by 0.5 M K_2SO_4. The high potassium concentration flocculates the soil and prevents the adsorption of NH_4^+ released by fumigation. The relatively high salt concentration also inhibits microbial decomposition of decomposable microbial material extracted after fumigation. However, if the extracts have to be stored for a long period, they must be frozen at −15°C (Joergensen and Brookes 1991).

A white $CaSO_4$ precipitate may be formed during the storage of the 0.5 M K_2SO_4 soil extract especially if the sample is frozen. It is very difficult and in any case unnecessary to dissolve this excess $CaSO_4$ because it does not interfere with any of the analytical procedures described in this chapter. Sodium hexametaphosphate (Wu et al 1990; Joergensen et al 1990) or citric acid buffer (Joergensen and Brookes 1990; Badalucco et al 1992) are added to the extract to keep the excess $CaSO_4$ in solution.

Young living root cells are also affected by $CHCl_3$ fumigation (Martin and Foster 1985; Sparling et al 1985). Thus, root-cell material released after fumigation can be extracted by K_2SO_4 (Mueller et al 1992). In soils containing large amounts of living roots, pre-extraction procedure must be carried out.

In soils containing more than 20% organic matter, the ratio soil:extractant of 1:4 must be increased to 1:20 when the litter layers contain as much as 94% organic matter (Scholle et al 1992).

The fumigation extraction method can be used in soils recently amended with organic substrates such as glucose or straw (Ocio and Brookes 1990b).

It is possible to use the fumigation extraction method in decomposition studies of ^{14}C- and ^{15}N-labelled organic substrates. Examples of the use of ^{14}C have been reported by Sparling et al (1990), Chander and Brookes (1991a), Harden et al (1993a), Wu et al (1993) and ^{15}N are Ocio et al (1991a, 1991b).

There is only very little evidence that non-living soil organic matter is made extractable by K_2SO_4 after $CHCl_3$ fumigation (Jenkinson and Powlson 1976; Joergensen et al 1990; Badalucco et al 1990, 1992).

The fumigation extraction method can be used to estimate microbial biomass carbon and nitrogen in acid soils (Vance et al 1987; Sparling and West 1988; Martikainen and Palojaervi 1990).

Fumigation and extraction after pre-extraction

(Mueller et al 1992)

Additional materials and apparatus

Centrifuge

Additional chemicals and solutions

0.05 M K_2SO_4 (8.714 g l^{-1})

Procedure

Moist soils (50 g dry weight) are split into two samples; each sample (25 g dry weight) is placed in a 250 ml glass bottle, pre-extracted with 100 ml 0.05 M K_2SO_4 for 20 min in an oscillating shaker at 200 rev min^{-1} and passed through a sieve (arable soils: 2 mm mesh; grassland soils: 3 mm mesh). Roots (and small stones) on the sieve are carefully washed to remove soil particles with an additional 75 ml of 0.05 M K_2SO_4. The soil suspension is centrifuged in the original glass bottle for 15 min at approximately 500*g*. The supernatant is then separated from the soil and measured; three drops of liquid $CHCl_3$ are added to the soil to be fumigated. The bottles containing these soils are placed in a desiccator and exposed to $CHCl_3$ vapour, as described in the previous procedure. Fumigated and unfumigated soils are then extracted with 0.5 M K_2SO_4, again as in the previous procedure.

Discussion

The procedure with the pre-extraction, sieving and centrifuging must be used when the soil contains living roots. In addition, it increases the accuracy of microbial biomass nitrogen by decreasing any background of inorganic nitrogen (Widmer et al 1989). It also diminishes the problems of measuring microbial biomass in dry soils (Sparling and West 1989); therefore it is recommended for measuring fluctuations in soil microbial biomass C and N during the year (Mueller 1992; Mueller et al 1992; Joergensen et al 1994).

The pre-extraction was not tested with clay soils. Problems may arise if the soil cannot be dispersed.

Microbial biomass carbon by dichromate oxidation

(Kalembasa and Jenkinson 1973; Vance et al 1987c)

Principle of the method

In the presence of a strong acid, organic matter is oxidized and Cr(+VI) is reduced to Cr(+III). The amount of dichromate left is back titrated.

Additional materials and apparatus

Liebig condenser
Round-bottomed flasks, 250 ml
Burette

Additional chemicals and solutions

66.7 mM $K_2Cr_2O_7$ (p.a.; 19.6125 g l^{-1})
Concentrated H_3PO_4 (85% p.a.)
Concentrated H_2SO_4 (98% p.a.)
40.0 mM ferrous ammonium sulphate $[(NH_4)_2Fe(SO_4)_2 \times 6H_2O]$ (p.a.)
25 mM 1,10-phenanthroline-ferrous sulphate complex solution

H_2SO_4/H_3PO_4 mixture

Two parts concentrated H_2SO_4 are mixed with one part concentrated H_3PO_4 (v/v).

Titration solution

Ferrous ammonium sulphate (15.69 g l^{-1}) is dissolved in distilled water, acidified with 20 ml concentrated H_2SO_4 and made up to 1000 ml with distilled water.

Procedure

Two millilitres of 66.7 mM (0.4 N) $K_2Cr_2O_7$ and 15 ml of the H_2SO_4/H_3PO_4 mixture are added to 8 ml of the filtered extract in a 250 ml round-bottomed flask. The whole mixture is gentle refluxed for 30 min, cooled and diluted with 20–25 ml water, which is added through the condenser as a rinse. The residual dichromate is measured by back titration with 40.0 mM ferrous ammonium sulphate solution using 25 mM 1,10-phenanthroline–ferrous sulphate complex solution as an indicator.

Calculations

Calculation of the extracted organic C:

$$C\ (\mu g\ ml^{-1}) = (H-S) : C \times M \times D : A \times E \times 1000 \quad (8.8)$$

where S is the titration solution consumed by the sample in millilitres, H is the titration solution consumed by the hot (refluxed) blank in millilitres, C is the titration solution consumed by the cold (unrefluxed) blank in millilitres, M is the normality (N) of the $K_2Cr_2O_7$ solution, D is the volume of the $K_2Cr_2O_7$ solution added to the reaction mixture, A is the aliquot of the extract and E = 3 (conversion of Cr (+VI) to (Cr +III)

$$C\ (\mu g\ g^{-1}\ soil) = C\ (\mu g\ ml^{-1}) \times (K{:}DW + W) \quad (8.9)$$

where K is the amount of the extractant, DW is the dry weight of the sample in g and W is the soil water (% dry weight:100).

Calculation of microbial biomass C:

$$\text{Biomass C} = E_C : k_{EC} \quad (8.10)$$

where E_C = (organic C extracted from the fumigated soil) − (organic C extracted from the unfumigated soils) and k_{EC} = 0.38 (Vance et al 1987c). The calibration factor k_{EC} was calculated by relating the results of fumigation incubation and fumigation extraction (12 soils).

Discussion

HgO can be omitted after the chloroform has been removed (Ross 1989).

The method is based on the following asumptions: (1) the average oxidation level of extracted soil organic C is (±0); and (2) dichromate is only consumed by the extracted organic C.

Soil microbial biomass C measured by the fumigation extraction method is significantly correlated to microbial biomass obtained by the fumigation incubation method (Vance et al 1987c; Tate et al 1988; Sparling et al 1990; Jordan and Beare 1991; Kaiser et al 1992), substrate-induced respiration (Sparling and West 1988a, 1988b; Sparling et al 1990; Joergensen et al 1991; Kaiser et al 1992; Harden et al 1993b) and direct microscopy (Martikainen and Palojaervi 1990). It is also significantly correlated to the adenosine triphosphate content of soil (Joergensen et al 1989; Chander and Brookes 1991a), the basal respiration rate (Powlson and Jenkinson 1976; Kaiser et al 1992), the arginine ammonification rate (Kaiser et al 1992), and soil organic C and N contents (Joergensen et al 1989; Wolters and Joergensen 1991; Badalucco et al 1992; Kaiser et al 1992).

The conversion factor k_{EC} is derived from the relationship between microbial biomass determined by the fumigation extraction and the fumigation incubation method (Vance et al 1987c). They used a few soils covering a wide range of properties. A similar k_{EC} value was obtained by Kaiser et al (1992) and Sparling et al (1990) using a different approach. Data from direct calibration (*in situ* ^{14}C-labelling of the microbial population) are rare (Sparling et al 1990). Caution is required in using the k_{EC} factor, in particular, with soils that have recently received organic matter, e.g. 1–5 days after addition of an actively decomposing substrate (Harden et al 1993a), in waterlogged soils (Inubushi et al 1991) and in organic layers of forest soils (Scholle et al 1992). Under these conditions the indirect calibration of the fumigation extraction with fumigation incubation method may not be valid. In addition, doubts still remain about the calibration procedures (Wardle and Parkinson 1991).

No significant relationships could be detected between the k_{EC} factor and soil properties such as organic matter, clay content (Kaiser et al 1992) and pH (Vance et al 1987c).

Measurements of biomass C by the fumigation extraction method are a sensitive indicator of soil pollution, due to heavy metals (Chander and Brookes 1991a, 1991b, 1991c, 1991d, 1992), pesticides (Fournier et al 1992; Jones et al 1992; Harden et al 1993a, 1993b), environmental stress on soil microflora due to change in moisture content (Ross 1989), high salt concentration (Schimel et al 1989) or increased acidification (Wolters and Joergensen 1991, 1992). The time-course of the soil microbial biomass was measured under fallow (Ocio et al 1991b) and crop (Mueller 1992) to study immobilization and mineralization (Ocio et al 1991; Mueller 1992; Joergensen et al 1994) and the decomposition of straw (Ocio et al 1991b). The estimation of microbial biomass C by the fumigation extraction method has been used to study the interaction between the fauna and microflora in soil (Daniel and Anderson 1992; Scholle et al 1992; Wolters and Joergensen 1992).

Biomass carbon by ultraviolet persulphate oxidation

(Wu et al 1990)

Principle of the method

In the presence of $K_2S_2O_8$, the extracted organic carbon is oxidized by ultraviolet (UV) light to CO_2, which can be measured by using infrared (IR) or photospectrometry.

Additional materials and apparatus

Automatic carbon analyser with an IR detector (e.g. Dohrman DC 80, Maihak

Tocor 4) or continuous-flow systems with colorimetric detection [Skalar, Perstorp]

Additional chemicals and solutions

$K_2S_2O_8$ (p.a.)
Concentrated H_3PO_4 (85% p.a., Merck no. 573)
Sodium hexametaphosphate $[(NaPO_4)_6)n]$ (extra pure)

Twenty grams of $K_2S_2O_8$ are dissolved in 900 ml of distilled water, acidified to pH 2 with concentrated H_3PO_4 and made up to 1000 ml.

Fifty grams of sodium hexametaphosphate are dissolved in 900 ml distilled water, acidified to pH 2 with concentrated H_3PO_4 and made up to 1000 ml.

Procedure

For the automated UV persulphate oxidation method, 5 ml K_2SO_4 soil extract are mixed with 5 ml sodium hexametaphosphate solution. Any precipitate of $CaSO_4$ in the soil extract is dissolved by this procedure. The $K_2S_2O_8$ is automatically fed into the UV oxidation chamber, where the oxidation to CO_2 is activated by UV light.

Calculations

Calculation of the extracted organic C:

$$C\ (\mu g\ g^{-1}\ \text{soil}) = [(S \times D_s) - (B \times D_B)] \times (K{:}DW + W) \qquad (8.11)$$

where S is the carbon of the soil in micrograms per millilitre, B is the carbon of the blank in micrograms per millilitre, D_S is the dilution of sample with sodium hexametaphosphate, D_B is the dilution of the blank with sodium hexametaphosphate, K is the amount of the extractant, DW is the dry weight of the sample in g and W is the soil moisture (% dry weight:100).

Calculation of microbial biomass C:

$$\text{Biomass C} = E_C : k_{EC} \qquad (8.12)$$

where E_C = (organic C extracted from the fumigated soil) − (organic C extracted from the unfumigated soil) and k_{EC} = 0.45 (Wu et al 1990). The conversion factor was calculated by relating the values found using the fumigation incubation method to those using the fumigation extraction method (23 soils).

Discussion

Samples containing large amounts of soluble organic substances (e.g. compost), should be diluted with H_2SO_4 (Joergensen, unpublished results).

The UV persulphate oxidation gives on average 19% larger results for extracted organic C than the dichromate oxidation (Wu et al 1990; Joergensen and Brookes 1991).

Identical C concentrations were measured after either the UV persulphate oxidation or oven combustion at 800°C of a range of K_2SO_4 extracts from soil covering a wide range of organic C concentrations (Joergensen and Brookes 1991).

It is impossible to measure organic C with $K_2Cr_2O_7$ in the presence of high concentrations of chloride. In this case, it is also impossible to use the UV persulphate oxidation method because chloride absorbs a large amount of energy in the UV range.

The fumigation extraction method for microbial biomass nitrogen

(Brookes et al 1985)

R.G. Joergenson

Fumigation and extraction

See earlier sections in this chapter entitled "Fumigation and extraction" and "Fumigation and extraction after pre-extraction" under the fumigation extraction method.

Determination of total nitrogen

(Pruden et al 1985)

Principle of the method

Total nitrogen is determined after reduction of nitrate to ammonium under strong acidic conditions by the Kjeldahl digestion.

Additional materials and apparatus

Digestion block
Steam distillation apparatus
Burette or autotitrator

Additional chemicals and solutions

Zinc powder (p.a.)
Concentrated H_2SO_4 (98% p.a.)
10 μM HCl (Titrisol)
0.19 M $CuSO_4 \times 5H_2O$ (p.a.) (47.441 g l^{-1})
$KCr(SO_4)_2 \times 12H_2O$ (p.a.)
10 M NaOH (p.a.; 400 g l^{-1})
2% H_3BO_3 (extra pure; 20 g l^{-1})

The reducing agent is prepared by dissolving 50 g $KCr(SO_4)_2$ in approximately 700 ml deionized water, adding 200 ml of concentrated H_2SO_4, cooling and diluting to 1000 ml.

Procedure

Ten millilitres of the reducing reagent and approximately 300 mg zinc powder are added to 30 ml of the K_2SO_4 soil extract and left for at least 2 h at room temperature, before the beginning of the Kjeldahl digestion, which is performed by adding 0.6 ml of 0.19 M $CuSO_4$ and 8 ml of concentrated H_2SO_4. The mixture is heated gently for 2 h until all the water has disappeared, then for 3 h at the maximum temperature. The digest is allowed to cool before distillation with 40 ml 10 M NaOH; the evolved NH_3 is adsorbed in 2% H_3BO_3 and the resulting solution is titrated with 10 μM HCl to pH 4.8.

Calculation

Calculation of the extracted total nitrogen:

$$N\ (\mu g\ g^{-1}\ soil) = (S - B){:}A \times M \times N \times 1000\ (K{:}DW + W) \quad (8.13)$$

where S is the HCl consumed by the sample in millilitres, B is the HCl consumed by the control in millilitres, A is the sample aliquot in millilitres, M is the molarity of HCl, $N = 14$ (the molecular weight of nitrogen), K is the amount of the extractant, DW is the dry weight of the sample in g and W is the soil moisture (% dry weight:100).

Calculation of microbial biomass nitrogen (Brookes et al 1985):

$$\text{Biomass N} = E_N{:}k_{EN} \quad (8.14)$$

where E_N = (total N extracted from the fumigated soil) − (total N extracted from the unfumigated soil) and $k_{EN} = 0.45$ (Jenkinson 1988). The conversion factor k_{EN} was empirically determined by comparing values obtained from 37 soils with the fumigation incubation method and fumigation extraction method, respectively (Brookes et al 1985; Jenkinson 1988). It was obtained by multiplying k_N for 0.79 (Brookes et al 1985), therefore it is equal to:

$$k_{EN} = 0.79 \times 0.57 = 0.45 \quad (8.15)$$

(Jenkinson 1988).

Discussion

The Kjeldahl digestion can be modified as reported by Bremner and Mulvaney (1982).

A method is available in which the extracted nitrogen is oxidized to nitrate, which is then determined colorimetrically (Valderrama 1981; Sparling and West 1988b; Cabrera and Beave 1993). This method can be used when automated continuous flow analysing systems are available (Houba et al 1987).

Ammonia may be volatilized from a soil during the vacuum incubation and fixed by another. This may lead to ammonia exchange among different soils (Vittori Antisari et al 1990).

If losses of nitrate occur during the fumigation period (and if the necessary NO_3-N measurements were available), they can be corrected by considering the difference between the NO_3-N extracted initially and the NO_3-N extracted at the end of the fumigation period (Brookes et al 1985).

The proportionality constants of biomass nitrogen and carbon are identical. This is presumably incorrect. Nicolardot (1986) found that the cytoplasmic fraction had a smaller C:N ratio than the intact cells. The extracted organic carbon and total nitrogen after soil fumigation mainly derive from the cytoplasmic fraction (Jenkinson 1988).

The microbial biomass nitrogen can be used as a parameter for modelling the turnover of soil organic nitrogen (Jenkinson and Parry 1989). The role of microbial biomass in the nitrogen dynamics in soil has been studied in laboratory (Ocio et al 1991a) and field experiments (Ocio et al 1991b; Mueller 1992; Joergensen et al 1994).

Determination of inorganic and organic nitrogen

(Joergensen and Meyer 1990)

Principle of the method

Inorganic NH_4^+, NO_3^- and organic nitrogen are separately measured by pH titration after their separation by steam distillation.

Additional materials and apparatus

Digestion block or digestion stand
Steam distillation apparatus
Autotitrator

Additional chemicals and solutions

MgO (extra pure)
Devarda alloy (p.a.)
Concentrated H_2SO_4 (98% p.a.)
10 μM HCl (Titrisol)
Selenium catalyst mixture (p.a.)
10 M NaOH (p.a.; 400 g l^{-1})
2% H_3BO_3 (extra pure; 20 g l^{-1})

Procedure

Both NH_4^+ and NO_3^- contents are analysed by steam distillation (Bremner and Keeney 1966; Keeney and Nelson 1982). A sample of the soil extract (75 ml) is pipetted into a Kjeldahl flask and 200 mg of MgO are added to volatize NH_3 under alkaline conditions immediately before the beginning of the distillation. When the distillate reaches the 30 ml mark on the receiver flask (a 50 ml Erlenmeyer flask containing 5 ml of 2% H_3BO_3), the first distillation is stopped; then 200 mg of Devarda alloy are rapidly added to the distillation flask to reduce nitrate and nitrite to ammonia, which is volatilized under the alkaline conditions of the distillation flask. When the distillate again reaches the 30 ml mark of the receiver flask, the second distillation is stopped. Then 10 ml of concentrated H_2SO_4 and 500 mg selenium catalyst mixture are added and the flask is heated gently until all the water has been eliminated. Thereafter, the mixture is boiled for 3 h until the digest becomes clear. The sample is cooled and, after the addition of 40 ml of 10 M NaOH, the NH_4^+ formed is steam distilled into the H_3BO_3 solution. In each of the three distillates, NH_4-N is determined with 10 μM HCl by titration to pH 4.8.

Calculation

Calculation of the extracted amounts of NH_4-N, NO_3-N and organic N:

$$\text{N } (\mu g\ g^{-1} \text{ soil}) = (S - B){:}A \times M \times N \times 1000\ (K{:}DW + W) \qquad (8.16)$$

where S is the HCl consumed by the soil in millilitres, B is the HCl consumed by the control in millilitres, A is the sample aliquot in millilitres, M is the molarity of the HCl, N = 14 (the molecular weight of nitrogen), K is the amount of the extractant, DW is the dry weight of the sample in g and W is the soil moisture (% dry weight/100).

Calculation of the microbial biomass nitrogen (Brookes et al 1985):

$$\text{Biomass N} = E_N{:}\ k_{EN} \qquad (8.17)$$

E_N = (NH_4-N + organic N extracted from the fumigated soil) − (NH_4-N + organic N extracted from the unfumigated soil) and k_{EN} = 0.45 (Jenkinson 1988).

Discussion

Steam distillation methods are applicable to coloured extracts and permit isotope-ratio analysis in tracer studies concerning the fate of ^{15}N-enriched compounds in soil (e.g. Ocio et al 1991a, 1991b).

The fractionated distillation makes it possible to measure microbial biomass nitrogen in soils containing large amounts of nitrate without the pre-extraction step.

Because of the high variations of the control, the estimation of NH_4^+ by the steam distillation is less sensitive than that by the automatic colourimetric analysis.

Only a few data have been published on microbial biomass nitrogen (Harden et al 1993b).

The use of the ninhydrin nitrogen reaction for estimating microbial biomass

(Amato and Ladd 1988; Joergensen and Brookes 1990)

R.G. Joergensen

Fumigation and extraction

See earlier sections in this chapter entitled "Fumigation and extraction" and "Fumigation and extraction after pre-extraction" under the fumigation extraction method.

Determination of ninhydrin-reactive nitrogen

(Joergensen and Brookes 1990)

Principle of the method

Ninhydrin forms a purple complex with molecules containing α-amino nitrogen and with ammonium and other compounds with free α-amino groups such as amino acids, peptides and proteins (Moore and Stein 1948). The presence of reduced ninhydrin (hydrindantin) is essential to obtain quantitative colour development with ammonium (Lamonthe and McCormick 1973). According to Amato and Ladd (1988), the amounts of ninhydrin-reactive compounds, released from the microbial biomass during the $CHCl_3$ fumigation and extracted by 2 M KCl, is strongly related to the initial soil microbial biomass carbon content.

Additional materials and apparatus

Boiling water bath
Spectrophotometer

Additional chemicals and solutions

Ninhydrin
Hydrindantin
Dimethyl sulphoxide (p.a.)
Lithium acetate dihydrate (p.a.)
Acetic acid (96% p.a.)
Citric acid (p.a.)
NaOH (p.a.)
Ethanol (95%)
L-Leucine
$(NH_4)_2SO_4$ (p.a.)

Ninydrin reagent

Ninhydrin (2 g) and hydrindantin (0.3 g) are dissolved in dimethyl sulphoxide (75 ml), 25 ml of 4 M lithium acetate buffer at pH 5.2 are then added (Moore 1968) and the mixture is flushed for 30 min with N_2-free O_2 .

Lithium acetate buffer

Lithium acetate (408,04 g) is dissolved in water (400 ml), adjusted to pH 5.2 with acetic acid and finally made up to 1 l with water. A lithium acetate buffer ready to use is commercially available (Pierce No. 27203)

Citric acid buffer

Citric acid (42 g) and NaOH (16 g) are dissolved in water (900 ml), adjusted to pH 5 with 10 M NaOH, if required, then finally made up to 1 l with water.

Dilution solution

Ethanol (95%) and distilled water are mixed in the ratio 1:1.

Standard solutions

A 10 mM L-Leucine (1.312 g l^{-1}) and a 10 mM ammonium-N [0.661 g $(NH_4)_2SO_4$ l^{-1}] solution are separately prepared in 0.5 M K_2SO_4 and diluted within the range 0–1000 μM N.

Procedure

The standard nitrogen solutions, K_2SO_4 soil extracts or blank (0.6 ml) and the citric acid buffer (1.4 ml) are added to 20 ml test tubes. The ninhydrin reagent (1 ml) is then added slowly, mixed thoroughly and closed with loose aluminium lids. The test tubes are then heated for 25 min in a vigorously boiling water bath. Any precipitate formed during the addition of the reagents then dissolves. After heating, the ethanol:water mixture (4 ml) is added, the solutions are thoroughly mixed again and the absorbance read at 570 nm (1 cm path length).

Calculation

Calculation of extracted ninhydrin-reactive N (N_{nin}):

$$N_{nin}\ (\mu g\ g^{-1}\ \text{soil}) = (S - B){:}L \times N \times (K{:}DW + W) \times 1000 \qquad (8.18)$$

where S is the absorbance of the sample, B is the absorbance of the blank, L is the molar absorbance coefficient of leucine, N = 14 (the molecular weight of nitrogen), K is the amount of the extractant, DW is the dry weight of the sample in g and W is the soil moisture (% dry weight:100).

Calculation of ninhydrin-reactive N in the microbial biomass:

$$B_{nin} = (N_{nin}\ \text{extracted from the fumigated soil}) - (N_{nin}\ \text{extracted from the unfumigated soil}) \qquad (8.19)$$

Calculation of microbial biomass carbon

$$\text{Biomass C} = B_{nin} \times 20.6 \qquad (8.20)$$

(Joergensen and Brookes 1990). The conversion factor was calculated by relating the microbial biomass C and B_{nin} values of the same extracts obtained in 12 soils by the fumigation extraction method.

Discussion

At 100°C the reaction with free amino groups of proteins and amino acids is essentially complete within 15 min (e.g. leucine reaches the maximum optical density after approximately 5 min). However, the reaction of hydrindantin with ammonium requires 25 min (Joergensen and Brookes 1990).

The low solubility of $CaSO_4$ and K_2SO_4 in the presence of organic solvents causes serious analytical problems. The use of citric acid overcomes this problem because it complexes Na, K and Ca ions. It was found that the optimum ratio [ninhydrin reagent to (citric acid buffer + sample)] was 1:2. If a greater ratio was used, the colour intensity (especially for ammonium) decreased (Joergensen and Brookes 1990).

The ratio between the volume of the sample and that of the citric acid should not be closer than 0.75:1.75 to avoid the formation of a precipitate after the addition of the ninhydrin reagent (Joergensen and Brookes 1990).

2-Methoxyethanol is the most common solvent used in the ninhydrin method

(Amato and Ladd 1988; Carter 1991; Badalucco et al 1992). However, because it is an ether it tends to form peroxides that destroy ninhydrin and hydrindantin. Dimethyl sulphoxide (boiling point of 189°C) is peroxide free, it is a better solvent for hydrindantin than 2-methoxyethanol and has lower toxicity. The ninhydrin reagent prepared in dimethyl sulphoxide gives a more stable colour than in 2-methoxyethanol (Joergensen and Brookes 1990).

The ninhydrin method proposed by Amato and Ladd (1988) for 2 M KCl extracts does not require the use of citric acid buffer. The optimum reagent to sample ratio is 1:2.

Different relationships have been used to calculate microbial biomass C from B_{nin}. Carter (1991) proposed the relationship: biomass C = 24.3 B_{nin}, by using 33 soils with different properties. Soils were extracted with 2 M KCl and microbial biomass C values ranged from 2 to 20 µg g^{-1}. Ocio et al (1991b) obtained the relationship: biomass C = 21.4 B_{nin}, by studying the fluctuation of microbial biomass in a soil throughout the year. Badalucco et al (1992) found the relationship: biomass C = 28.2 B_{nin}, by using K_2SO_4 as an extractant.

Measurements of microbial biomass by the ninhydrin-reactive nitrogen method are a sensitive indicator of soil pollution due to heavy metals (Chander and Brookes 1991b). They are also used to measure microbial biomass in waterlogged soils (Inubushi et al 1991), to compare the effect of different tillage systems (Carter 1991), and to monitor the effects of straw incorporation in microbial biomass under laboratory (Ocio and Brookes 1990b) and field conditions (Ocio et al 1991b).

The fumigation extraction method for microbial biomass phosphorus

(Brookes et al 1982)

R.G. Joergensen

Fumigation

See earlier section in this chapter entitled "Fumigation and extraction", under the fumigation extraction method.

Principle of the method

The difference between the amount of phosphate extracted by 0.5 M $NaHCO_3$ (pH 8.5) from soil fumigated with chloroform and the amount from unfumigated soil is used to calculate the phosphorus content of microbial biomass.

Some of the phosphate released by the fumigation is adsorbed by soil colloids. An approximate correction of this fraction is made by incorporating a known quantity of phosphorus during extraction and correcting for recovery (Brookes et al 1982).

Additional materials and apparatus

Spectrophotometer

Additional chemicals and solutions

0.5 M $NaHCO_3$
KH_2PO_4 (p.a.) (250 mg P l^{-1})
Concentrated H_2SO_4 (98% p.a.)
L-Ascorbic acid (p.a.)
Ammonium heptamolybdate tetrahydrate
$[(NH_4)_6MO_7O_4 \times 4H_2O]$ (p.a.)
Potassium antimony tartrate
$[K(SbO)C_4H_4O_6 \times 0.5H_2O]$ (extra pure)
NaOH (p.a.)

Extractant

$NaHCO_3$ (42 g) is dissolved in water (900 ml), adjusted to pH 8.5 with 10 M NaOH, then finally made up to 1 l with water (Olsen and Sommers 1982).

Reagent A

Ammonium heptamolybdate tetrahydrate (12.5 g) is dissolved in 125 ml of bidistilled water; 0.5 g potassium antimony tartrate is dissolved in 20 ml bidistilled water. These two solutions are combined in a 500 ml bottle, mixed thoroughly, filled up with 4.5 M H_2SO_4 (258 ml of concentrated H_2SO_4 diluted to 1 l) to 500 ml and stored in a brown glass bottle at 4°C.

Reagent B

Ascorbic acid (50 g) is dissolved in about 300 ml of bidistilled water, then 250 ml of 4.5 M H_2SO_4 are added, filled up with bidistilled water to 1 l and stored in a brown glass bottle at 4°C.

Procedure

Fumigation and extraction

Moist soils (30 g dry weight) are split into three samples of 10 g dry weight. The first sample (unfumigated control) is placed in a 250 ml centrifuge bottle and then extracted with 200 ml of 0.5 M $NaHCO_3$ for 30 min by oscillating shaking at 150 rev min^{-1}. The ratio extractant:soil (dry weight) is 20:1 (v/w). The resulting suspension is centrifuged (2000 *g*) and then filtered through a paper filter (Schleicher & Schuell 595 1/2). The second sample (the unfumigated recovery control) is also placed in a 250 ml centrifuge bottle and then extracted as described above with 200 ml of 0.5 M $NaHCO_3$ and additionally 1 ml of KH_2PO_4 solution containing 250 µg phosphorus is added to give a spike of 25 µg g^{-1} soil. For the fumigation treatment, the third sample is placed in a desiccator containing wet tissue paper and a vial of soda lime. A beaker containing 25 ml of ethanol-free $CHCl_3$ is added and the desiccator evacuated until the $CHCl_3$ has boiled for 2 min vigorously. The desiccator is then incubated in the dark at 25°C for 24 h. After the removal of $CHCl_3$ by repeated evacuations, the extraction of the fumigated soil is performed as reported for the unfumigated control.

Measurement of phosphate

An aliquot of the soil extract (75 ml) in a 250 ml Erlenmeyer flask is acidified with 5 ml of 4.5 M H_2SO_4 and the mixture is allowed to stand for 24 h at 4°C; the pH value of the mixture is determined with pH paper. If the pH value exceeds 1.5, additional 4.5 M H_2SO_4 is added.

The extract is filtered (Schleicher & Schuell, 595 1/2) immediately before the determination of inorganic P is analysed by a modified ammonium molybdate–ascorbic acid method (Olsen and Sommers 1982). Aliquots (1.0–23.8 ml, usually 2.5–5 ml) of the filtered extract containing 1–20 µg of P are pipetted into 25 ml volumetric flasks. Then 0.4 ml of reagent A and 0.8 ml of reagent B are added, and the mixture is made up to 25 ml with distilled water and thoroughly mixed. The colour is stable for 24 h and the maximum intensity develops in 10 min. The absorbance of the blue colour formed by the reaction is read at 882 nm. The method is calibrated using a standard P solution (Olsen and Sommers 1982).

Calculation

Calculation of extracted phosphate:

$$P\ (\mu g\ g^{-1}\ soil) = (S - B):P \times D_H \times D_A \times (K:DW + W) \quad (8.20)$$

where *S* is the absorbance of the sample, *B* is the absorbance of the blank, *P* is the absorbance coefficient of phosphate (KH_2PO_4), D_H is the dilution by the 4.5 M H_2SO_4 and D_A is the dilution by the ammonium molybdate-ascorbic acid method, *K* is the amount of the extractant, *DW* is the dry weight of the sample in g and *W* is the soil moisture (% dry weight:100).

Calculation of microbial biomass P (Brookes et al 1982):

$$\text{Biomass P} = E_P:k_{EP} \quad (8.21)$$

$$E_P = (F - U):(Z - U) \times 25 \quad (8.22)$$

where *F* is the PO_4-P extracted from the fumigated soil, *U* is the PO_4-P extracted from the unfumigated soil, *Z* is the PO_4-P extracted from the unfumigated soil with the spike of 25 µg g^{-1} soil and $k_{EP} = 0.40$ (Brookes et al 1982). The calibration was performed by adding a known quantity of microorganisms to a soil, fumigating and measuring the proportion of the added microbial P extracted (Brookes et al 1982).

Discussion

The method proposed by Hedley and Stewart (1982) was based on the difference between total P extracted by $NaHCO_3$ from fumigated and unfumigated soil samples. However, the method of Brookes et al (1982) is recommended because the measurement of phosphate is more accurate than that of total P. The oxidation method of Valderrama (1981) could improve the precision of total P analysis in soil extracts.

Soda lime absorbs CO_2 evolved during the pre-incubation and fumigation period. It keeps low levels of inorganic CO_2. High concentrations of CO_2 might have toxic effects on soil microflora or they may change the solubility of inorganic P due to the formation of carbonate.

The recovery of P is linear in the range of P concentrations released in soil after chloroform fumigation (Meyer et al 1993).

According to McLaughlin et al (1986) the pre-incubation period is not necessary in the case of measurements of microbial biomass P. Significantly lower values of microbial biomass P were obtained when soils were pre-incubated prior to measurements.

The chloroform treatment does not significantly affect the recovery of added phosphate (Brookes et al 1992; Meyer et al 1993).

In the case of interference in the P determination, this can be overcome simply by diluting the soil extract. The interference is evident if the P concentration of the diluted sample is proportionately greater or lower than that calculated by considering the dilution factor (Olsen and Sommers 1982).

It is possible to use the fumigation extraction method for determining microbial biomass P in studies concerning the decomposition of ^{32}P-labelled substrates (McLaughlin and Alston 1985; McLaughlin et al 1986).

Microbial biomass P was related to microbial biomass C in soil with a contrasting agricultural history (Brookes et al 1984), or measured to assess the influence of plant material applied to field soils (McLaughlin and Alston 1985; McLaughlin et al 1986). They were also used to measure the effects of alternate land use on nutrient flux in dry tropical forest soils (Srivastava and Singh, 1991). The effect of environmental conditions on soil microbial P has been studied in temperate and humid European beech forest soils (Joergensen et al 1994).

The substrate-induced respiration method

G.P. Sparling

The substrate-induced respiration (SIR) method utilizes the physiological respiration response of soil organisms to substrate amendment to provide an estimate of soil microbial biomass carbon. Anderson and Domsch (1974, 1975) observed that bacterial and fungal respiration was stimulated within minutes into maximal activity in short-term (1–6 h) experiments by the addition of saturating quantities of a readily-available substrate such as glucose. In each case it was necessary to determine the saturating amount of glucose response by preliminary experiments.

The enhanced rate of respiration stayed reasonably stable for up to 6–8 h, after which it began to increase. The further increase after 6–8 h was attributed to cell division and population growth. Anderson and Domsch (1973, 1975, 1978) suggested that the maximal rate of respiration during the "plateau" phase following substrate addition, was a characteristic of the initial microbial population. Because the method depends on the stimulation of respiration by the addition of a substrate, it is now commonly known as the substrate-induced respiration method.

The development of the fumigation incubation method (Jenkinson and Powlson 1976) to measure soil microbial carbon, allowed Anderson and Domsch (1978) to calibrate their SIR technique and hence to relate the SIR response to soil microbial biomass. The SIR response of a range of soils from Germany was measured and plotted against microbial biomass carbon estimated by the fumigation incubation method. The resulting plot was a straight line, suggesting that the SIR response of the soil biomass was a reasonably constant characteristic and that the biomass from different soils behaved in a generally similar manner. Hence, after calibration, the biomass carbon could be estimated from the SIR response.

Assumptions implicit in the method are: (1) the SIR response of different organisms is reasonably constant; (2) the majority of the soil microbiota will respond during the period of measurement; (3) glucose is a suitable substrate to induce the maximal response; (4) the contribution to microbial carbon from non-glucose-metabolizing organisms is insignificant or consistently low. Subsequent work by others using German, British and New Zealand soils (West and Sparling 1986; Martens 1987; Sparling and West 1988a, 1988b; Ocio and Brookes 1990) has tended to support these assumptions, although alternative ways to measure SIR have been described, revised calibrations have been suggested, and there are indications that the relationship between SIR response and the microbial biomass carbon is not completely linear (see later discussion).

Principle of the method

The method is based on the estimation of soil respiration by detecting the O_2 uptake or CO_2 evolution immediately after the amendment with saturating quantities of glucose. The SIR is calibrated against the biomass carbon measured by one of the fumigation methods.

Soil preparation and storage

Soils are usually prepared for SIR analyses by sieving < 2 mm which is necessary to remove roots and macrofauna and to disperse the soil to obtain good distribution of glucose substrate and release of evolved CO_2. Some authors have used soils sieved to < 6.3 mm rather than < 2 mm and obtained similar results (Sparling and Searle 1993), but there are no records of the technique being attempted on undisturbed soil cores. The author's personal opinion is that this would be difficult to apply because of the contribution from living plant material to respiration, and uneven distribution of substrate and efflux of CO_2.

Air-drying of soil should be avoided as this kills part of the microbial biomass, but if soils are naturally dry it is possible to apply the SIR method to soils at low moisture contents (West and Sparling 1986; Sparling et al 1986). Rewetting and pre-incubation of dry soils is not necessary (see below), if the user wishes to determine the biomass carbon of the dry soil.

Storage of soils, if absolutely necessary, is recommended at 5°C in the field-moist state, in containers that allow gaseous exchange such as loosely sealed polyethylene bags. Freezing of the soil causes substantial microbial death and a decline in the SIR response, and is not recommended as a storage method. During storage at 5°C over several weeks, there is a gradual decline in the SIR response which may be more rapid at higher temperatures (Anderson and Domsch 1985). Some of the decline is attributable to microbial death (Anderson and Domsch 1985), but in other cases a decline in the SIR response during long-term storage at 5°C was not matched by a decline in microbial carbon as determined by the fumigation methods (Ross 1988, 1989). The SIR technique relies on a rapid physiological response from the soil organisms following substrate amendment and it is not unreasonable that this response declines gradually during storage. Consequently, it is recommended that soils are analysed promptly after collection and storage should be for a maximum of 3–4 weeks at 5°C. The soils should be allowed to return to room temperature before commencing the SIR test.

Materials and apparatus

A variety of apparatus can be used to measure soil respiration (see procedure below and also the section on soil respiration in Chapter 5)

Ultragas 3 (Wosthoff) CO_2 analyser
Perspex "microcosm" chambers (Ineson and Anderson 1982)
Gas chromatograph
Flow-through systems (Cheng and Coleman 1989)
Differential respirometer (Sapromat, Warburg flasks)
Infrared gas analyser

Chemicals and solutions

Glucose

For dry addition, the glucose is finely ground in a mortar and pestle, and mixed with talcum power (about 1:3 ratio) to assist even mixing with the soil (about 0.5 g talcum powder per 100 g soil). Optimum rates of glucose to achieve maximal respiration response need to be determined from preliminary experiments, levels have ranged from 1 to 8 mg g^{-1}. For addition as solution, the glucose is dissolved in water immediately prior to use. Recommended solution concentrations have varied from 8 to 30 mg ml^{-1}.

Talcum

Further reagents may be required for the measurement of respiration depending upon the method used.

Procedure

Measurement of the SIR response

Many different methods to measure the SIR response have been described. A selection of the methods with technical detail and comment is given below.

Ultragas 3 (Wosthoff) CO_2 analyser

Anderson and Domsch (1978) used an Ultragas 3 CO_2 analyser (Wosthoff Company, Bochum, Germany) to determine the respiration of amended soils. This is a purpose-built instrument in which the CO_2 is flushed out of soil in a stream of air and collected in alkali traps containing 0.04 N NaOH. The electrical conductivity of the traps changes depending on the amount of CO_2 in solution and is automatically recorded. The apparatus used by Anderson and Domsch (1978) was housed in a temperature-controlled room (22°C). Automatic switching between samples and aeration trains allows for minimal operator attendance and for a number of samples to be measured simultaneously.

The equipment requires large (*c.* 100 g) soil samples. Anderson and Domsch (1975, 1978, 1985) added the glucose as a finely ground powder, mixed with 0.5 g talcum powder (about 1:3 glucose:talc ratio) to assist even distribution. The glucose/talc mixture was blended with the soil using a food mixer Esge Co. Type M 301 fitted with a single blade and operated at *c.* 1600 rev min^{-1} for 25–30 s. The soil was held in a 1 l plastic beaker during mixing. The soil samples were poured into 25 × 4 cm plastic tubes, stoppered with polyurethane plugs and connected to the analyser. Moisture content of the soils was not specified but can be assumed to be moist to allow the glucose powder to dissolve. The level of glucose addition necessary to achieve substrate saturation and maximum rate of respiration was determined from preliminary experiments. Levels of substrate required were typically 500–8000 ppm.

Incubations were carried out at 22°C and an intermittent flushing system was employed. Once each hour, CO_2-free air was drawn through the soils for 20 min and the rate of CO_2 production measured for 10 min. Measurements were continued for several hours until the pattern of CO_2 respiration became apparent. This system, being fully automated, is convenient, has reasonable throughput, but is bulky, complicated and requires high capital outlay.

Martens (1987) compared the intermittent flushing system of the Ultragas 3 with a self-assembled continously flushing aeration train. He found that when neutral and alkaline soils were analysed in the Ultragas apparatus, some CO_2 remained in solution and was incompletely removed during the flushing phase, and gave rise to overestimation of the rate of respiration. Martens (1987) recommended the continuous flushing system, which gave a more reliable measure of the rate of respiration. Heinemeyer et al (1989) reached a similar conclusion. In the continuous system, soil was held in glass tubes 13 × 3 cm i.d. and were flushed continuously with a current of water-saturated CO_2-free air at 15 ml min^{-1}. CO_2 was collected into 4 ml of 2 M NaOH and analysed hourly using a gas chromatography method (see below).

Conductimetric methods

Ineson and Anderson (1982) measured respiration conductimetrically with a much simpler system than the Ultragas 3 apparatus. Perspex "microcosm" chambers containing soil or leaf litter were incubated statically. Each chamber contained a conductivity cell with KOH absorbant to provide a measurement of respiration. The conductivity cell was claimed to have advantage over titrimetric methods in that it reduced volumetric errors and permitted continuous monitoring. A solution of 75 mM KOH was found to be adequate to measure the rate of respiration of several grams of leaf litter over 6 h. For 50 mM

KOH the maximum absorption rate is around 0.34 mg ml^{-1}. During measurements the apparatus was sealed and connected to a MEL Conductivity Bridge, Pye Unicam, Cambridge, UK. The system is compact and simple but the conductimetric cells require individual calibration and all conductivity cells need to be compensated for any temperature variations.

Gas chromatography

Chromatography is a rapid and sensitive method to measure soil gases. Sparling (1981) applied the method to estimate both basal and SIR respiration. West and Sparling (1986) modified the SIR methods to allow them to estimate the microbial biomass of soils under moisture stress. Both substrate and moisture were supplied as an aqueous glucose solution. To ensure even distribution of substrate and water through the soil they used a slurry of one part soil to two parts solution. This approach is only possible with small soil samples (1 g) because the static method of incubation depends on rapid equilibration of respired CO_2 with the headspace gases. West and Sparling (1986) advised the method was not suitable for soils of pH > 6.5 because of CO_2 remaining dissolved in the soil solution.

The soils were incubated in MacCartney bottles sealed with Vacutainer stoppers to allow sampling of the headspace gas. Soil (1 g equivalent dry weight) was added to the MacCartney bottle (volume 28.5 ml), and glucose solution added equivalent to 2 ml final solution volume (allowing for soil water) and a final solution concentration of 30 mg ml^{-1}. Bottles were sealed with a Vacutainer stopper (Becton Dickinson, MA, USA), incubated at 25°C and shaken vigorously on a vortex mixer for 5 s immediately prior to taking headspace gas samples. These were obtained with a hyperdermic needle and syringe, and taken 30 min and 150 min after amendment of the soil with glucose solution. The gas samples (1 ml) were analysed by Carle 8000 GC, fitted with a 2.2 m stainless steel column (1 mm i.d.) stationary phase of 80–100 mesh Poropak T, and a Katharometer detector. Helium was used as the carrier gas (30–40 ml min^{-1}), and oven and column temperatures were 105°C (Orchard and Cook 1983). Standards of known CO_2 concentration are used to calibrate the system. The method is simple and rapid with throughput of *c.* 40 samples per hour. The major cost incurred is that of the gas chromatograph and standard gases.

Ritz and Wheatley (1989) also used GC to analyse CO_2; they used a Hewlett Packard 5890 gas chromatograph with thermal conductivity detector, 1.8 m column 2 mm i.d. with Poropak Q as the stationary phase, column temperature of 70°C, and He as carrier. A range of soil weights and solution volumes were used by Ritz and Wheatley (1989) who found that with 5 g soil and 10 ml solution in a 28.5 ml MacCartney bottle, there was (predictably) very low recovery of CO_2 in the headspace gas. The use of such extreme solution:headspace ratios should be avoided with the "static" GC method, which depends on diffusion and rapid equilibration between the solution and gaseous phases.

Martens (1987) also used GC to analyse the CO_2 content of NaOH solutions. The solutions (50 ml in sealed medical flats) were acidified to pH 1 with 5 N H_2SO_4. After vigorous shaking, headspace gas samples were taken via syringe and injected on to a GC system with a TCD detector and Poropak T on a 1.5 m column (further details not supplied). Recovery of CO_2 from the NaOH was estimated to be 88.3%.

Simple flow-through systems

A simple low-cost flow-through apparatus was described by Cheng and Coleman (1989). Soil (30 g) was contained in a 125 ml conical flask connected to an input stream of CO_2-free air. The inlet air was obtained from an aquarium air pump and bubbled through an "air-stone" (Aqua-Mist Cylinder 11 mm, Penn-Plax Inc., USA) in a flask containing 1 or 2 M KOH. The outlet gas stream from each soil flask was passed through 50 ml of 10–20 mM NaOH, using a flow rate of 50–80 ml min^{-1} controlled using needle valves. These concentrations and flow rates were found to give equilibration of CO_2 output within 30 min and good recovery of respired CO_2 as checked against infrared gas analysis (IRGA). Use of NaOH of < 8 mM to trap CO_2, and flow-rates of < 10 ml min^{-1}, were not recommended because of poor trapping efficiency and the long time (as much as 2 h) needed to reach a steady rate of respiration.

Cheng and Coleman (1989) added glucose as a solution equivalent to 120% of field water-holding capacity, and found solution concentrations of 8 mg glucose g^{-1} soil were adequate for maximal initial respiration response. The CO_2 in the NaOH trap was estimated by titration against HCl of known molarity after first precipitating out carbonates with an excess of aqueous $BaCl_2$ and titration to pH 8.6 using phenolphthalein (Sparling 1990, and else where in this volume).

The system has the advantage of being very simple and not requiring sophisticated equipment, but it does require a high level of operator input and, compared to the gas chromatography (GC) and IRGA methods, has low sensitivity and a low rate of throughput.

Differential respirometer

The SIR response of soils has been measured using respirometers such as the Gilson apparatus or Warburg flasks (Ross 1980; Sparling and West 1990). Respirometers were considered more reliable than the static GC method for use with alkaline soils because the CO_2 is removed from solution into the alkali trap, and both O_2 uptake and CO_2 efflux can be measured (Sparling and West 1990). Experimental conditions used on the Gilson Differential Respirometer by Sparling and West (1988a, 1988b) were: flask volume 19 ml, 0.5 g soil, 0.3 ml of 5 M KOH as the CO_2 trap in the centre well, 1 ml glucose solution concentration 30 mg ml^{-1}, shaken at 100 strokes min^{-1}, respiration was measured every 30 min via direct reading of the digital micromanometers. Ross (1980) used a similar set-up but with 2 g soil and 2 ml of glucose solution.

Infrared gas analysis

IRGA is an extremely sensitive method to analyse for CO_2 and has been applied to measure respiration of soils in a variety of situations including the SIR technique (Wardle and Parkinson 1990a, 1990b, 1990c; Brooks and Paul 1987; Heinemeyer et al 1989). Details of the method are provided elsewhere in this volume.

Generalised method

Because there are so many different ways to perform the SIR test, it is only possible to present a generalized procedure. Greater technical detail has been supplied under the headings for each of the individual methods, and further information on the measuring of soil respiration can be found elsewhere in this volume:

1. Prior to any SIR test the soil should have been sieved and any adjustment to the water content completed.

2. Add the required amount of glucose to the soil either as solution and directly

in the incubation vessel, or as a powder (mixed with talcum powder), mixed and transferred into the incubation vessel.

3. Incubate the soil (usually at 25°C) and measure the maximal initial respiration rate of the soil over the next 0.5–6 h, using one of the methods detailed above.

4. Calculate the hourly rate of respiration when the maximal response occurred. Convert the rate of respiration to an estimate of the microbial carbon by reference to a previously established calibration curve between the SIR response and microbial C, or use one of the published factors.

Calibration

Anderson and Domsch (1978) calibrated the SIR response against the microbial biomass carbon estimated by fumigation incubation and obtained the linear relationship: biomass C (mg C 100 g^{-1} soil) = 40.04 (ml CO_2 100 g^{-1} soil h^{-1}) + 0.37; this relationship being valid for incubations at 22°C. The regression was highly significant ($r = 0.96$) but the presence of outlier datum points greatly strengthen (bias) this relationship. Anderson and Domsch (1978) noted that the rate of evolution of CO_2 from their samples was not always uniform during the 6 h following amendment and selected what they thought was the most appropriate period to obtain a calibration. This was the first hour after amendment if the rate showed a steady increase (type III soils); the beginning of the plateau phase if the soil showed a later increase (type II soils); or at the minimum point if the rate of respiration declined before increasing (type I soils).

Other workers have found their soils to give generally linear rates of respiration for up to 6 h after amendment (West and Sparling 1986; Cheng and Coleman 1989). West and Sparling (1986) used the rate of respiration 0.5–2.5 h after amendment with glucose solution and initially calibrated this against biomass carbon estimated from biovolume measurements. They obtained the relationship: biomass C (μg g^{-1} soil) = 433 × log(μl CO_2 g^{-1} h^{-1}) + 59.2, R2 = 71%, for incubations at 25°C. Later West et al (1986) used the fumigation methods for calibration and obtained the relationship: biomass C (μg g^{-1} soil) = 40.92 × (μl CO_2 g^{-1} h^{-1}) + 12.9, which was very similar to that originally proposed by Anderson and Domsch (1978).

Subsequently, Sparling et al (1990) recalibrated their SIR data with a wider range of soils using the fumigation methods and ^{14}C-labelling, and obtained the revised relationship: biomass C (μg g^{-1} soil) = 50.4 × (μl CO_2 g^{-1} h^{-1}), r = 0.88. Martens (1987) also recalibrated the SIR method against the fumigation method to allow for his improved continuous flushing technique and obtained the relationship: biomass C (mg 100 g^{-1} soil) = 49.5 × (ml CO_2 100 g^{-1} soil h^{-1}) − 16.7, r = 0.98; in reasonable agreement with Anderson and Domsch (1978) and very similar to Sparling et al (1990).

It is recommended that workers should perform their own calibrations for the analytical method of their choice. If this is not practical, then a calibration should be chosen that was obtained using similar analytical methods to measure the SIR response.

Discussion

Glucose is the usual choice of substrate. Anderson and Domsch (1978) tested a range of alternative substrates including complex types and casamino acids but maximum response was obtained with glucose. The glucose is added to soil either by mixing the powder with moist soil (Anderson and Domsch 1978), or by adding glucose solution to soil (Ross 1980; West and Sparling 1986; Cheng and Coleman 1989).

Analytical methods

The "static" method of incubation has major limitations when applied to soils above pH 6.0, particularly when there is a low soil:headspace ratio because substantial amounts of CO_2 remain dissolved in the slurry or soil solution resulting in underestimation of the respiration (Ritz and Wheatley 1989; West and Sparling 1986, 1990). This is a particular problem with alkaline or neutral soils with both the Ultragas 3 (Martens 1987) or GC methods whenever there is static incubation, and the CO_2 is detected by analyses of headspace gases (Cheng and Coleman 1989; Sparling and West 1990). For this reason, West and Sparling recommended that the GC static incubation method should not be applied to soils of pH 6.5 or above, while Ritz and Wheatley recommended acidification of the soil solution to displace any dissolved CO_2, prior to GC analysis of the headspace gases. Strong alkali traps or the continuous flushing systems appear to be the most reliable techniques, particularly for neutral and alkaline soils.

Agreement with other estimates of microbial carbon

Wardle and Parkinson (1990b) examined the relationships between various methods (including SIR) to estimate microbial carbon. They concluded that none of the statistical relationships were reliable and for most of their work used the SIR response without conversion to biomass carbon (Wardle and Parkinson 1990a, 1990b, 1900c). Ineson and Anderson (1982) found poor agreement between SIR estimates of microbial carbon of leaf litter and that estimated by biovolume, but considered that the biovolume estimates were also subject to error. In contrast, Sparling and Williams (1986) obtained reasonable agreement between SIR and biovolume estimates of microbial biomass in forest humus samples, and recommended the SIR method for use on acid forest soils where the fumigation incubation method was unreliable. With a series of mineral soils, Sparling and West (1988b) obtained good agreement between SIR and the fumigation extraction methods, and also with the fumigation incubation method except on a single acidic sample. Sparling et al (1986) and West et al (1988a, 1988b) recommended their modified SIR method as being more reliable than the fumigation incubation method to estimate the biomass carbon of soils under moisture stress. Dumontet and Mathur (1989), working with metal-contaminated soils, reported that the SIR biomass, calculated using the Anderson and Domsch (1978) factors, correlated poorly with the biomass measured by fumigation incubation. They derived a new relationship from the SIR response in the first hour after amendment and used the kinetic model of Chaussod and Nicolardot (1982) to estimate microbial carbon. The relationship obtained was: biomass C (mg kg^{-1}) = 40.0 × (ml CO_2 kg^{-1} h^{-1}) + 48.6. This seems a very minor change to the original proposed by Anderson and Domsch (1978) and would not be expected to greatly improve the agreement between the SIR and fumigation methods. It may have been that, for contaminated soils, the fumigation incubation technique was showing greater variability than the SIR method.

Advantages of the SIR method

The main advantage of the SIR method is the rapidity and flexibility of the assay. With the various forms of equipment and analyses available, SIR can be undertaken using very simple cheap equipment (e.g. Cheng and Coleman 1989) or with increasing complexity, more sensitivity, automation and high levels of throughput (e.g. Brooks and Paul 1987). The method has advantages over the fumigation incubation method in that it can be applied to soils of low pH, and to leaf litter and forest floor materials (Hart et al 1986). Very small soil samples can be analysed (West and Sparling 1986) and the relative contributions by bacteria and fungi to rhizosphere populations have been estimated (Vancura and Kunc 1977).

Limitations and disadvantages of SIR

Caution is needed if the SIR method is used on soils that have recently received substrate. Anderson and Domsch (1978) examined a range of organisms in laboratory culture and showed that, expressed per unit of cell C, the SIR response of organisms in exponential growth was several times greater than when in stationary phase. Consequently, if organisms in soil are actively growing, then the SIR method could be expected to overestimate the microbial biomass carbon. In most soils periods of active growth are short, and because available substrates are normally in low supply, the majority of organisms are in resting or stationary phase of growth (Jenkinson and Ladd 1981). On such soils the SIR method could be expected to give a reliable estimate of microbial C.

The limitations of some versions of the SIR method on alkaline soils are given above.

The SIR method cannot be used to estimate the turnover of carbon isotopes through the microbial biomass. Because microbial carbon is estimated indirectly (usually by calibration against a fumigation technique) there is no means to recover the carbon and measure the specific activity. Roots and non-microbial organisms can contribute to the respiration, interfere with the assay and cause overestimation of the SIR response. This is also true of other methods to estimate microbial biomass. Consequently soils are nearly always sieved prior to analyses to remove roots and macrofauna.

Martens (1987) noted that the relationship between SIR and microbial carbon, although highly significant, was not strictly linear, and soils with low biomass contents gave a higher than expected SIR response while the response from soils with a high biomass content was lower than expected. Ocio and Brookes (1990) detected a similar trend in straw-amended soils, indicating that the SIR respiration response per unit of microbial carbon may not be constant over the whole biomass range. However, the errors are not great, and the relationship between SIR response and microbial carbon is usually sufficiently consistent for the majority of applications. It is also notable that the majority of regressions do not pass through zero, suggesting that there is a proportion of the biomass measured by the fumigation methods that does not respond within the duration of an SIR assay.

The determination of active microbial biomass by the respiration simulation method

H. Van der Werf
G. Genouw
L. Van Vooren
W. Verstraete

Van der Werf and Verstraete (1987a) proposed a method for estimating the active soil biomass by measuring the O_2 consumption, which is directly related to the physiology of soil microorganisms. In this method, biokinetic parameters were estimated by a mathematical model based on the kinetics of oxygen consumption, substrate utilization and biomass growth.

Principle of the method

The method is based on the continuous monitoring of oxygen uptake by soil supplied with a readily degradable organic substrate in a respirometer. From the oxygen consumption curves derived, only the lower (lower limit) and the upper (upper limit) curves are considered. These curves are mathematically analysed according to conventional microbial growth kinetics (Van de Werf and Verstraete 1987a, 1987b). In this way it is possible to derive not only the quantity of active microbial biomass (X_o), but also the maximum specific growth rate (μ_{max}), the substrate affinity constant (K_s), the maximum cell yield factor (Y_{max}) and the cell maintenance coefficient (m) of microbial biomass.

According to the mathematical model the length of the lag phase is directly related to the initial amount of the active microbial biomass. The measurement with the method presented can be carried out within 10 h. In this manner, the procedure of determination of active microbial biomass is simplified considerably.

Materials and apparatus

Sapromat apparatus (Voith, Germany).
Personal computer with soil biomass respiration simulation program (SB-SIMULA) or the soil biomass respiration parameter optimization program (SB-OPTIMA). With the SB-SIMULA program it is possible to simulate the cumulative oxygen consumption for 120 h given a set of parameters: the initial substrate concentration (S_o), X_o, μ_{max}, K_s, Y_{max} and m. The program calculates the squared distances between the simulated curve and the observed lower and upper oxygen uptake curves; it checks if the simulated oxygen uptake curve fits between the lower and upper limits. By successive manual iterations of parameters, the optimum parameter combination can be found. However, the manual adjustment of the different parameters is time consuming, and many simulations are needed to find the optimal combination. The SB-OPTIMA program, which is an extension of the SB-SIMULA program, calculates automatically the optimum parameter combination (X_o,

μ_{max}, K_s, Y_{max} and m) for each given lower and upper data set and the S_o. Both programs can be obtained from the Laboratory of Microbial Ecology, University of Gent, Coupure L. 653, B-9000 Gent, Belgium.

Chemicals and solutions

Glucose monohydrate
Yeast extract
NH_4Cl
$MgSO_4.7H_2O$
KH_2PO_4
Granulated soda lime

Procedure

Preparation of the soil sample

The soil sample is homogenized by sieving through a 4 mm sieve and adjusted to 3/4 FC (field capacity). Then it is incubated in the dark in a plastic bag covered with polyethylene film at 20°C. After 2 weeks, the soil is subjected to the respiration test.

The respiration test:

The extended method (120 h)

After the pre-incubation, 100 g of soil (wet weight), are treated with 120 mg of glucose monohydrate, 30 mg of yeast extract, 45 mg of NH_4Cl, 12 mg of $MgSO_4.7H_2O$ and 10 mg of KH_2PO_4. This amendment corresponds to a 1436 mg chemical oxygen demand (COD) kg^{-1} soil wet weight by bichromate oxidation. The nutrient-supplemented soil is well mixed and then placed in a Sapromat flask. The flask is then closed by a stopper, equipped with granulated soda lime. The Sapromat Erlenmeyer flask is incubated in the respirometer at 20°C at a constant oxygen partial pressure of 200 HPa. The amount of oxygen consumed is registered continuously over a period of 120 h (5 days). The experiment is performed in triplicate.

The short method (10 h)

This procedure is identical to the extended method but the incubation time is now only 10 h.

Calculation

The extended method

The mathematical model used to determine the active microbial biomass present in soil is based on Monod kinetics and three differential equations:

$$\mu = \mu_{max} \frac{S}{K_s+S} \quad (8.23)$$

$$-\frac{dS}{dt} = \frac{1}{Y_{max}} \times \frac{dX}{dt} + mX \quad (8.24)$$

$$\frac{dX}{dt} = \mu X - mY_{max}X \quad (8.25)$$

$$\frac{dy}{dt} = \frac{(1 - fY_{max})}{Y_{max}} \mu X + mX \quad (8.26)$$

where f is a conversion factor (for biomass containing 50% carbon on a dry weight basis, $f = 1.33$), K_s is the substrate affinity constant (mg COD kg^{-1} soil wet weight), m is the cell maintenance coefficient (mg COD utilized mg^{-1} biomass dry weight h^{-1}), μ is the specific growth rate (h^{-1}), μ_{max} is the maximum specific growth rate (h^{-1}), S is the substrate concentration, expressed as COD (mg COD kg^{-1} soil wet weight), S_o is the substrate concentration at the onset of the respiration test, expressed as COD (mg COD kg^{-1} soil wet weight), X is the microbial biomass (mg biomass dry weight kg^{-1} soil wet weight), X_o is the active soil microbial biomass, capable of starting to metabolize the substrate added directly (mg biomass dry weight kg^{-1} soil wet weight), y is the oxygen concentration (mg oxygen kg^{-1} soil wet weight) and Y_{max} is the maximum yield factor (mg biomass dry weight formed mg^{-1} COD utilized).

The equations are integrated by a fourth order Runga–Kutta algorithm. The values μ_{max}, K_s, Y_{max}, m and X_o are successively iterated by trial and error parameter estimation (SB-SIMULA) or by an optimization routine (SB-OPTIMA) until the simulated curve fits between the upper and lower limit oxygen uptake curves. The simulated and experimental curves are finally visualized on the screen.

The short method

If only the determination of the active microbial biomass (X_o) of soil is required, the short method can be used. The amount of biomass (X_o) is calculated by means of an empirical relationship based on oxygen consumption during a 10-h incubation period in a Sapromat apparatus:

$$X_o = 0.769\ \bar{y}_{O_2,10} \qquad (8.27)$$

where X_o is the active soil microbial biomass, capable of directly metabolizing the substrate added (mg biomass dry weight kg^{-1} soil wet weight) and $\bar{y}_{O_2,10}$ is the average (mean of duplo) oxygen consumption during a 10-h incubation period (mg O_2 kg^{-1} soil wet weight). The empirical equation is derived from 220 experiments and has a correlation coefficient of 0.997. The equation is only valid for an incubation temperature of 20 ± 0.5°C.

Discussion

The respiration simulation method is based on the oxygen consumption and simulation by a mathematical model based on growth kinetics and microbial physiology. All the other methods available are validated using approximative and indirect measurements (Anderson and Domsch 1978).

The respiration simulation optimization method only relates to the microbial soil biomass capable of metabolizing immediately after glucose addition. Other fractions of the soil biomass are not taken into account.

The extended respiration simulation method is time consuming (120 h) with respect to the registration of the respiration curves. The short method (10 h), though less informative, is much faster.

In the extended method, K_s and μ_{max} were validated by comparing them with the Lineweaver–Burk transformation method. Both methods gave the same results. For further details, the reader is referred to Van de Werf and Verstraete (1987a, 1987b).

The respiration simulation method has been used to estimate active microbial biomass in a range of soils subjected to different agricultural and management practices (Van de Werf and Verstraete 1987d; Wardle and Parkinson 1990; Kaiser et al 1992; Scheu 1992).

Example

The extended method presented here has been used to measure active microbial biomass of Nazareth soil, which is a sandy soil with a pH of 4.0, organic carbon

Table 8.1. Experimental cumulative oxygen consumption as a function of time for the Nazareth soil.

Time (h)	*Cumulative oxygen consumption (mg O_2 kg^{-1} soil wet weight)* Lower limit	Upper limit
0	0	0
10	16	29
20	80	113
30	244	304
40	407	445
50	469	500
60	507	533
70	537	560
80	552	580
90	562	595
100	567	605
110	572	613
120	577	620

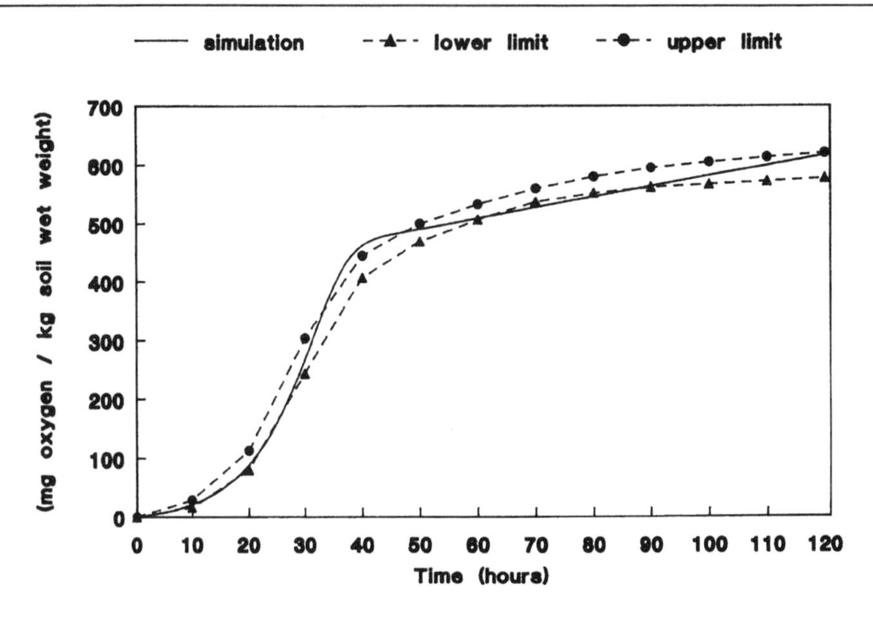

Figure 8.1. Experimental and simulated (————) cumulative oxygen consumption curves for nutrient-amended Nazareth soil. Upper (-●-) and lower (-▲-) limits of the experimental measurements are shown.

content of 3.12% and a total nitrogen content of 0.114%.

Determination of the cumulative oxygen consumption in intervals of 10 h from the experimental curves. The results are arranged in a table, which is characterized by a lower and an upper limit data set (Table 8.1).

With the data set of Table 8.1 and with S_o = 1436 mg COD kg^{-1} soil wet weight, the optimized parameters estimated with the SB-OPTIMA program are: X_o = 14.5 mg kg^{-1} wet soil, μ_{max} = 0.16 h^{-1}, K_s = 496 mg COD kg^{-1} wet soil, Y_{max} = 0.51 mg biomass dry weight mg^{-1} COD and m = 0.0027 mg COD mg^{-1} biomass dry weight h^{-1}. Figure 8.1 shows a graph of the measured and simulated cumulative oxygen consumption as a function of the incubation time.

Microscopic methods

K. Alef

These methods are based on the calculation of bacterial total count as well as the total length of fungi mycelium found in the soil biomass. Staining and direct counting techniques have been presented in Chapter 4. Using these techniques, microbial biomass can be calculated.

The bacterial biomass is calculated as follows:

$$B\ (\text{g g}^{-1}\ \text{dwt}) = V \times D \times BN \qquad (8.28)$$

where B is the fresh bacterial biomass, V is the average volume of the bacterial cell in cubic centimetres, D is the average cell density, BN is the bacterial count measured in 1 g dry soil and dwt is the dry weight of 1 g moist soil. An average volume of a bacterial cell is found to be 1 μm^3, while the average cell density ranged from 1.3 to 1.5.

The fungi biomass in soil can be calculated as follows:

$$\text{Fresh weight}\ (\text{g g}^{-1}\text{dwt}) = \text{I}\ r^2 \times \text{I} \times D \qquad (8.29)$$

where r is the average radius of the fungi mycelium in centimetres, l is the total length of fungi mycelium measured in 1 g dry soil in centimetres, dwt is the dry weight of 1 g moist soil and D is the average cell density (1.3 g cm^{-3}). For the calculation of dry microbial biomass, the biomass fresh weight is multiplied by the factor 0.25 (assuming that the dry weight is equal to 25% of fresh biomass) (Van Veen and Paul 1979; Jenkinson and Ladd 1981; Schmidt and Paul 1982).

Discussion

The distinction between live and dead cells is often difficult, and sometimes there is an uncertainty that all organisms actually present have been counted or measured.

Cell volume and density have been calculated using laboratory cultures. Properties of cells grown on artificial media could significantly differ from those that live in soils.

ATP content as a parameter to estimate microbial biomass

K. Alef

ATP occurs in all organisms and has been used as a measure of microbial activity (Nannipieri 1990). This compound is present in living microorganisms and is absent in dead cells and in other non-living forms of soil organic matter. It can be extracted from soils and determined by very sensitive test systems (e.g. luciferin–luciferase). Several methods to estimate microbial ATP in soils have been presented and discussed in Chapter 5.

ATP content has also been used as a parameter of microbial biomass in soils. As calibration methods, microscopic, physiological and fumigation techniques have been used. However, caution must be taken in the calculation of extracted ATP in the biomass for the following reasons:

1. ATP concentrations in the cells are not constant and dependent on the cell physiological state (Setlow and Kornberg 1970; Lee et al 1971; Ausmus 1973; Sparrow and Doxtader 1973; Coleman et al 1977; Scott and Ellar 1978; Karl 1980; Sparling et al 1981; Ahmed et al 1982; Kaczmarker et al 1982; Verstaete et al 1983; Fairbanks et al 1984).
2. A serious obstacle to interpreting and comparing ATP levels in different soils results from the various methods used to extract ATP from soil. The completeness of the extraction can not be checked practically. A complete desorption of ATP from clay soils could not be achieved.
3. The ATP/biomass carbon ratios obtained are not constant and dependent on the extraction methods used, the methods applied to estimate the biomass and the storage conditions of the soils. The ATP/biomass C ratios found ranged from 69 to 858 (Oades and Jenkinson 1979; Ross et al 1980; Sparling 1981; Van de Werf 1989/1990).

In conclusion, ATP measurements reflect the status of microbial activity and not the biomass size in soil. The principal and technical difficulties bound to ATP estimation make a reliable estimation of ATP biomass unlikely.

References

Ahmed M, Oades JM, Ladd JN (1982) Determination of ATP in soils: effect of soil treatments. Soil Biol Biochem 14: 273–279.

Alef (1993) Bestimmung mikrobieller Biomasse in Boden: Eine kritische Betrachtung. Z Pflanzenernaehr Bodenk 156: 109–114.

Amato M, Ladd JN (1980) Studies of nitrogen immobilization and mineralization in calcareous soils. V. Formation and distribution of isotope-labelled biomass during decomposition of ^{14}C- and ^{15}N-labelled plant material. Soil Biol Biochem 12: 405–411.

Amato M, Ladd JN (1988) Assay for microbial biomass based on ninhydrin-reactive nitrogen in extracts of fumigated soil. Soil Biol Biochem 20: 107–114.

Anderson JM, Ineson P (1982) A soil microcosm system and its application to measurements of respiration and nutrient leaching. Soil Biol Biochem 14: 415–416.

Anderson JPE (1982) Soil respiration. In: Methods of Soil Analysis, Part 2: Chemical and microbiological properties, 2nd edn. Page AL, Miller RH, Keeney DR (eds). Agronomy 9/2, American Society of Agronomy, Madison, WI, pp. 831–871.

Anderson JPE, Domsch KH (1973) Selective inhibition as a method for estimation of the relative activities of microbial populations in soils. In: Modern Methods in the Study of Microbial Ecology. Rosswall T (ed.). Bulletins from the Ecological Research Committee (Stockholm), vol 17, pp. 281–282.

Anderson JPE, Domsch KH (1974) Use of selective inhibitors in the study of respiratory activities and shifts in bacterial and fungal populations in soil. Ann Microbiol 24: 189–194.

Anderson JPE, Domsch KH (1975) Measurement of bacterial and fungal contributions to respiration of selected agricultural and forest soils. Can J Microbiol 21: 314–322.

Anderson JPE, Domsch KH (1978) A physiological method for the quantitative measurement of microbial biomass in soils. Soil Biol Biochem 10: 215–221.

Anderson JPE, Domsch KH (1980) Quantities of plant nutrients in the microbial biomass of selected soils. Soil Sci 130: 211–216.

Anderson JR, Slinger JM (1975) Europium chelate and fluorescent brightner staining of soil propagules and their photomicrographic counting: I. Methods. Soil Biol Biochem 7: 205–209.

Anderson T-H, Domsch KH (1985) Determination of ecophysiological maintenance carbon requirements of soil microorganisms in a dormant state. Biol Fertil Soils 1: 81–89.

Ausmus BS (1973) The use of the ATP assay in terrestrial decomposition studies. In: Modern Methods in the Study of Microbial Ecology. Rosswall T (ed.). Bulletins from the Ecological Research Committee (Stockholm), vol 17, pp. 223–234.

Ayanaba A, Tuckwell SB, Jenkinson DS (1976) The effects of clearing and cropping on the organic reserves and biomass of tropical forest soils. Soil Biol Biochem 8: 519–525.

Babiuk LA, Paul EA (1970) The use of fluorescein isothiocyanate in the determination of the bacterial biomass of grassland soil. Can J Microbiol 16: 57–62.

Badalucco L, Nannipieri P, Grego S, Ciardi C (1990) Microbial biomass and anthrone-reactive carbon in soils with different organic matter contents. Soil Biol Biochem 22: 899–904.

Badalucco L, Gelsomino A, Dell'Orco S, Grego S, Nannipieri P (1992) Biochemical characterization of soil organic compounds extracted by 0.5 M K_2SO_4 before and after chloroform fumigation. Soil Biol Biochem 24: 569–578.

Bottner P (1985) Response of microbial biomass to alternate moist and dry conditions in a soil incubated with ^{14}C- and ^{15}N-labelled plant material. Soil Biol Biochem 17: 329–337.

Bremner JM, Keeney DR (1966) Determination and isotope-ratio analysis of different forms of nitrogen in soil: 3. Exchangeable ammonium, nitrate by extraction–distillation methods. Soil Sci Soc Am Proc 30: 577–583.

Bremner JM, Mulvaney CS (1982) Nitrogen – total. In: Methods of Soil Analysis, Part 2: Chemical and Microbiological Properties, 2nd edn. Page LA, Miller RH, Keeney DR (eds). Agronomy 9/2, American Society of Agronomy, Madison, WI, pp. 595–624.

Brookes PC, McGrath SP (1984) Effects of metal toxicity on the size of the soil microbial biomass. J Soil Sci 35: 341–346.

Brookes PC, Powlson DS, Jenkinson DS (1982) Measurement of microbial biomass phosphorus in soil. Soil Biol Biochem 14: 319–329.

Brookes PC, Powlson DS, Jenkinson DS (1984) Phosphorus in the soil microbial biomass. Soil Biol Biochem 16: 169–175.

Brookes PC, Landman A, Puden G, Jenkinson DS (1985) Chloroform fumigation and the release of soil nitrogen: a rapid direct extraction method to measure microbial biomass nitrogen in soil. Soil Biol Biochem 17: 837–842.

Brooks PD, Paul EA (1987) A new automated technique for measuring respiration in soil samples. Plant Soil 101: 183–187.

Cabrera ML, Beare MH (1993) Alkaline persulfate oxidation for determining total nitrogen in microbial biomass extracts. Soil Sci Soc Am J 57: 1007–1012.

Carter MR (1986) Microbial biomass as an index for tillage-induced changes in soil biological properties. Soil Tillage Res 7: 29–40.

Carter MR (1991) Ninhydrin-reactive N released by the fumigation–extraction method as a measure of microbial biomass under field conditions. Soil Biol Biochem 23: 139–143.

Carter MR, Rennie DA (1984) Dynamics of soil microbial biomass N under zero and shallow tillage for spring wheat, using ^{15}N urea. Plant Soil 76: 157–164.

Chander K, Brookes PC (1991a) Microbial biomass dynamics during the decomposition of glucose and maize in metal-contaminated and non-contaminated soils. Soil Biol Biochem 23: 917–925.

Chander K, Brookes PC (1991b) Effects of heavy metals from past applications of sewage sludge on microbial biomass and organic matter accumulation in a sandy loam and silty loam UK soil. Soil Biol Biochem 23: 927–932.

Chander K, Brookes PC (1991c) Is the dehydrogenase assay invalid as a method to estimate microbial activity in copper-contaminated soils? Soil Biol Biochem 23: 909–915.

Chander K, Brookes PC (1991d) Plant inputs of carbon to metal-contaminated soil and effects on the soil microbial biomass. Soil Biol Biochem 23: 1169–1177.

Chander K, Brookes PC (1992) Synthesis of microbial biomass from added glucose in metal-contaminated and non-contaminated soils following repeated fumigation. Soil Biol Biochem 24: 613–614.

Chapman SJ (1987) Inoculum in the fumigation method for soil biomass determination. Soil Biol Biochem 19: 83–87.

Chaussod R, Nicolardot B (1982) Mesure de la biomasse microbienne dans les sols cultivés. I. Approche cinétique et estimation simplifiée du carbonne facilement minéralisable. Rev Écol Biol Sol 19: 501–512.

Chaussod R, Nicolardot B, Catroux G (1986) Mesure en routine de la biomasse microbienne des sols par la methode de fumigation au chloroform. Science Du Sol, pp. 201–211.

Cheng W, Coleman DC (1989) A simple method for measuring CO_2 in a continuous air-flow system: modifications to the substrate-induced respiration technique. Soil Biol Biochem 21: 385–388.

Coleman DC, Cole CV, Anderson RV, Blaha M, Capion MK, Clarholm M, Elliott ET, Hunt HW, Schaefer B, Sinclair J (1977) An analysis of rhizosphere–saprophage interactions in terrestrial ecosystems. Eco Bull (Stockholm) 25: 299–309.

Collins HP, Rasmussen PE, Douglas CL jr (1992) Crop rotation and residue management effects on soil carbon and microbial dynamics. Soil Sci Soc Am J 56: 783–788.

Daniel O, Anderson JM (1992) Microbial biomass and activity in contrasting soil materials after passage through the gut of the earthworm *Lumbricus rubellus* Hoffmeister. Soil Biol Biochem 24: 465–470.

Duah-Yentumi S, Johnson DB (1986) Changes in soil microflora in response to repeated applications of some pesticides. Soil Biol Biochem 18: 629–635.

Dumontet S, Mathur SP (1989) Evaluation of respiration-based methods for measuring microbial biomass on metal-contaminated acidic mineral and organic soils. Soil Biol Biochem 21: 431–436.

Faegri A, Torsuik VG, Goksoyr J (1979) Bacterial and fungal activities in soil: separation of bacteria and fungi by a rapid fractionated centrifugation method. Soil Biol Biochem 9: 105–112.

Fairbanks BC, Woods LE, Bryant RJ, Elliott ET, Cole CV, Coleman DC (1984) Limitation of ATP estimates of microbial biomass. Soil Biol Biochem 16: 549–558.

Frankland JC (1974) Importance of phase-contrast microscopy for estimation of total fungal biomass by the agar-film technique. Soil Biol Biochem 6: 409–410.

Fournier FC, Froncek B, Gamouh A, Collu T (1992) Comparison of three methods to test the side effects of pesticides on soil microbial biomass. In: Proceedings of the International Symposium on Environmental Aspects of Pesticide Microbiology. Anderson JPE, Arnold DJ, Lewis F, Torstensson L (eds). Swedish University of Agricultural Sciences, Uppsala, pp. 100–104.

Hamilton RD, Holm-Hansen O (1967) Adenosine triphosphate content of marine bacteria. Limnol Oceano 12: 319–324.

Harden T, Joergensen RG, Meyer B, Wolters V (1993a) Mineralization and formation of soil microbial biomass in a soil treated with simazine and dinoterb. Soil Biol Biochem 25: 1273–1276.

Harden T, Joergensen RG, Meyer B, Wolters V (1993b) Soil microbial biomass estimated by fumigation–extraction and substrate-induced respiration in two pesticide treated soils. Soil Biol Biochem 25: 679–683.

Hart PBS, Sparling GP, Kings JA (1986) Relationship between mineralisable nitrogen and microbial biomass in a range of plant litters, peats, and soils of moderate to low pH. N Zealand J Agric Res 29: 681–686.

Hedley MJ, Stewart JWB (1982) Method to measure

microbial phosphate in soils. Soil Biol Biochem 14: 377–385.

Heinemeyer O, Insam H, Kaiser EA, Walenzik G (1989) Soil microbial biomass and respiration measurements: an automated technique based on infra-red gas analysis. Plant Soil 116: 191–195.

Houba VJG, Novozamsky I, Uittenbogaard J, Lee JJ (1987) Automatic determination of "total soluble nitrogen" in soil extracts. Landwirtschaftl Forsch 40: 295–302.

Ineson P, Anderson JM (1982) Microbial biomass determinations in deciduous leaf litter. Soil Biol Biochem 14: 607–608.

Inubushi K, Wada R, Takai Y (1984) Determination of microbial biomass nitrogen in submerged soil. Soil Sci Plant Nut 30: 455–459.

Inubushi K, Brookes PC, Jenkinson DS (1991) Soil microbial biomass C, N and ninhydrin-N in aerobic and anaerobic soils measured by the fumigation–extraction method. Soil Biol Biochem 24: 737–741.

Jenkinson DS (1966) Studies on the decomposition of plant material in soil. II. Partial sterilization of soil and the soil biomass. J Soil Sci 17: 280–302.

Jenkinson DS (1976) The effects of biocidal treatments on metabolism in soil. IV. The decomposition of fumigated organisms in soil. Soil Biol Biochem 8: 203–208.

Jenkinson DS (1988) The determination of microbial biomass carbon and nitrogen in soil. In: Advances in Nitrogen Cycling in Agricultural Ecosystems. Wilson JR (ed.) CAB International, Wallingford, pp. 368–386

Jenkinson DS, Ladd JM (1981) Microbial biomass in soil: measurement and turnover. In: Soil Biochemistry, vol 5. Paul EA, Ladd JN (eds). Dekker, New York, pp. 415–471.

Jenkinson DS, Parry LC (1989) The nitrogen cycle in the broadbalk wheat experiment: a model for the turnover of nitrogen through the soil microbial biomass. Soil Biol Biochem 21: 535–541.

Jenkinson DS, Rayner JH (1977) The turnover of soil organic matter in some of the Rothamsted classical experiments. Soil Sci 123: 298–305.

Jenkinson DS, Powlson DS (1976a) The effect of biocidal treatments on metabolism in soil. I. Fumigation with chloroform. Soil Biol Biochem 8: 167–177.

Jenkinson DS, Powlson DS (1976b) The effect of biocidal treatments of metabolism in soil. V. A method for measuring soil biomass. Soil Biol Biochem 8: 209–213.

Jenkinson DS, Powlson DS (1976c) The effects of biocidal treatments on metabolism in soil. III. The relationship between soil biovolume, measured by optical microscopy, and flush of decomposition caused by fumigation. Soil Biol Biochem 8: 189–202.

Jenkinson DS, Powlson DS, Wedderburn RWM (1976) The effects of biocidal treatments on metabolism in soil. III. The relationship between soil biovolume measured by optical microscopy, and the flush of decomposition caused by fumigation. Soil Biol Biochem 8: 189–202.

Joergensen RG, Brookes PC (1990) Ninhydrin-reactive nitrogen measurements of microbial biomass in 0.5 M K_2SO_4 soil extracts. Soil Biol Biochem 22: 1023–1027.

Joergensen RG, Brooks PC (1991) Soil microbial biomass estimations by fumigation extraction. Mitt Deut Bodenk Gesell 66: 511–514.

Joergensen RG, Meyer B (1990) Nutrient changes in decomposing beech leaf litter assessed using a solution flux approach. J Soil Sci 41: 279–293.

Joergensen RG, Brookes PC, Jenkinson DS (1989) Some relationships between microbial ATP and soil microbial biomass, measured by the fumigation–extraction procedure, and soil organic matter. Mitt Deut Bodenk Gesell 59: 585–588.

Joergensen RG, Brookes PC, Jenkinson DS (1990) Survival of the soil microbial biomass at elevated temperatures. Soil Biol Biochem 22: 1129–1136.

Joergensen RG, Harden T, Meyer B, Wolters V (1991) Einfluss von bioziden Substanzen auf die Bodenmikroflora nach Zugabe von ^{14}C-markiertem Stroh. Mitt Deut Bodenk Gesell 66: 515–518.

Joergensen RG, Kübler H, Meyer B, Wolters V (1994) Microbial biomass phosphorus in soils of temperate forests. Soil Biol Biochem (in press)..

Joergensen RG, Meyer B, Mueller T (1994) Time-course of the soil microbial biomass under wheat. – A one year field study. Soil Biol Biochem 26: 987–994.

Jones SE, Jones AL, Johnson DB (1992) Effects of differential pesticide inputs on the size and composition of soil microbial biomass: results from the Boxworth and SCARAB projects. In: Proceedings of the International Symposium on Environmental Aspects of Pesticide Microbiology. Anderson JPE, Arnold DJ, Lewis F, Torstensson L (eds). Swedish University of Agricultural Sciences, Uppsala, pp. 30–36.

Jordan D, Beare MH (1991) A comparison of methods for estimating soil microbial biomass carbon. Agric Ecol Environ 34: 35–41.

Kaczmarek W, Kaszubiak H, Pedziwilk Z (1976) The ATP content in soil microorganisms. Ekologia Polska 24: 399–406.

Kaiser EA, Mueller T, Joergensen RG, Insam H, Heinemeyer O (1992) Evaluation of methods to estimate the soil microbial biomass and the relationship with soil texture and organic matter. Soil Biol Biochem 24: 675–683.

Kalembasa SJ, Jenkinson DS (1973) A comparative study of titrimetric and gravimetric methods for the determination of organic carbon in soil. J Sci Food Agric 24: 1085–1090.

Karl DM (1980) Cellular nucleotide measurements

and applications in microbial ecology. Microbial Rev 44: 739–796.

Kassim G, Martin JP, Haider K (1981) Incorporation of a wide variety of organic substrate carbons into soil biomass as estimated by the fumigation procedure. Soil Sci Soc Am J 45: 1106–1112.

Keeney DR, Nelson DW (1982) Nitrogen – inorganic forms. In: Methods of Soil Analysis, Part 2: Chemical and Microbiological Properties, 2nd edn. Page AL, Miller RH, Keeney DR (eds). Agronomy 9/2, American Society of Agronomy, Madison, WI, pp. 643–698.

Kieft TL, Soroker E, Firestone MK (1987) Microbial biomass response to a rapid increase in water potential when dry soil is wetted. Soil Biol Biochem 19: 119–126.

Kudeyarov VN, Jenkinson DS (1976) The effects of biocidal treatments on metabolism in soil VI. Fumigation with carbon disulphide. Soil Biol Biochem 8: 375–378.

Lamonthe PJ, McCormick PG (1973) Role of hydrindantin in the determination of amino acids using ninhydrin. Anal Chem 243: 1906–1911.

Lee CC, Harris RF, Williams JDH, Armstrong DE (1971) Adenosine triphosphate in lake sediment: I. Determination. Soil Sci Soc Am Proc 35: 82–86.

Lundgren B (1981) Fluorescein diacetate as a stain of metabolically active bacteria in soil. Oikos 36: 17–22.

Lynch JM, Panting LM (1980) Cultivation and soil biomass. Soil Biol Biochem 12: 29–33.

Martens R (1985) Limitations in the application of the fumigation technique for biomass estimations in amended soils. Soil Biol Biochem 17: 57–63.

Martens R (1987) Estimation of microbial biomass in soil by respiration method: importance of soil pH and flushing methods for the measurement of respired CO_2. Soil Biol Biochem 19: 77–81.

Martikainen PJ, Palojaervi A (1990) Evaluation of the fumigation–extraction method for the determination of microbial C and N in a range of forest soils. Soil Biol Biochem 22: 797–802.

Martin JK, Foster RC (1985) A model system for studying the biochemistry and biology of the root–soil interface. Soil Biol Biochem 17: 271–274.

McLaughlin MJ, Alston AM (1985) Measurement of phosphorus in the soil microbial biomass: Influence of plant material. Soil Biol Biochem 17: 271–274.

McLaughlin MJ, Alston AM, Martin JK (1986) Measurement of phosphorus in the soil microbial biomass: a modified procedure for field soils. Soil Biol Biochem 18: 437–443.

Merckx R, Den Hartog A, van Veen JA (1985) Turnover root-derived material and related microbial biomass formation in soils of different texture. Soil Biol Biochem 17: 565–569.

Meyer B, Kübler M, Wolters, V, Joergensen RG (1993) Die Messung von mikrobiell gebundenem Phosphor in Laubwald-Böden. Mitt Deut Bodenk Gesell 71: 365–366.

Moore S (1968) Amino acid analysis: aqueous dimethyl sulfoxide as solvent for the ninhydrin reaction. J Biol Chem 243: 6281–6283.

Moore S, Stein WH (1948) Photometric ninhydrin method for use in the chromatography of amino acids. J Biol Chem 243: 6281–6283.

Mueller T, Joergensen RG, Meyer B (1992) Estimation of soil microbial biomass C in the presence of fresh roots by fumigation–extraction. Soil Biol Biochem 24: 179–181.

Mueller T (1992) Zeitgang der mikrobiellen Biomasse in der Ackerkrume einer mitteleuropäischen Löß-Parabraunerde. Eine Ursache für die Mineralisation und Immobilisation des Bodenstickstoffs? Ph.D. thesis, University of Göttingen.

Nannipieri P (1984) Microbial biomass and activity measurements in soil: ecological significance. In: Current Perspectives in Microbial Ecology. Klug MJ, Reddy CA (eds). American Society for Microbiology, Washington, DC, pp. 515–521.

Nannipieri P, Ciardi C, Palazzi T (1985) Plant uptake, microbial immobilization, and residual soil fertilizer of urea–nitrogen in a grass–legume association. Soil Sci Soc Am J 49: 405–411.

Nannipieri P, Ciardi C, Badalucco L, Casella S (1986) A method to determine soil DNA and RNA. Soil Biol Biochem 18: 275–281.

Nicolardot B (1986) Étude comparée de la minéralisation de différentes fractions cellulaires d'*Aspergillus flavus* dans un sol fumié ou non au chloroforme. CR Acad Sci Paris 303 (III): 489–494.

Nicolardot B, Chaussod R (1986) Mesure de la biomasse microbienne dans les sols cultivés. III. Approche cinétique et estimation simplifiée de l'azote facilement minéralisable. Rev Écol Biol Sol 23: 233–247.

Nicolardot B, Guiraud G, Chaussod R, Catroux G (1986) Minéralisation dans le sol de matériaux microbiens marqués au carbonne 14 et a l'azote 15: quantification de l'azote de la biomasse microbienne. Soil Biol Biochem 18: 263–273.

Oades JM, Jenkinson DS (1979) A method for measuring adenosine triphosphate in soil. Soil Biol Biochem 11: 193–199.

Ocio JA, Brookes PC (1990a) Soil microbial biomass measurements in sieved and unsieved soils. Soil Biol Biochem 22: 999–1000.

Ocio JA, Brookes PC (1990b) An evaluation of methods for measuring the microbial biomass in soils following recent additions of wheat straw and the characterization of the biomass that develops. Soil Biol Biochem 22: 685–694.

Ocio JA, Martinez J, Brookes PC (1991a) Contribution of straw-derived N to total microbial biomass following incorporation of cereal straw to soil. Soil Biol Biochem 23: 655–659.

Ocio JA, Brookes PC, Jenkinson DS (1991b) Field incorporation of straw and its effects on soil microbial biomass and soil inorganic N. Soil Biol Biochem 23: 171–176.

Olsen SR, Sommers LE (1982) Phosphorus. In: Methods of Soil Analysis, Part 2: Chemical and Microbiological Properties, 2nd edn. Page AL, Miller RH, Keeney DR (eds). Agronomy 9/2, American Society of Agronomy, Madison, WI, pp. 403–430.

Orchard VA, Cook FJ (1983) Relationship between soil respiration and soil moisture. Soil Biochem 15: 447–453.

Paul EA, Clark FE (1988) Soil Microbiology and Biochemistry. Academic Press, San Diego.

Paul EA, Johnson RL (1977) Microscopic counting and adenosine 5′-triphosphate measurement in determining microbial growth in soils. Appl Environ Microbiol 34: 263–269.

Powlson DS, Jenkinson DS (1976) The effects of biocidal treatments on metabolism in soil. II. Gamma irradiation, autoclaving, air-drying and fumigation with chloroform or methyl bromide. Soil Biol Biochem 19: 179–188.

Powlson DS, Jenkinson DS (1981) A comparison of the organic matter, biomass adenosine triphosphate and mineralizable nitrogen contents of ploughed and direct-drilled soils. J Agr Sci 97: 713–721.

Powlson DS, Brookes PC, Christensen BT (1987) Measurement of soil microbial biomass provides an early indication of changes in total soil organic matter due to straw incorporation. Soil Biol Biochem 19: 159–164.

Pruden G, Kalembasa SJ, Jenkinson DS (1985) Reduction of nitrate prior to Kjeldahl digestion. J Sci Food Agric 36: 71–73.

Ritz K, Wheatly RE (1989) Effects of water amendment on basal and substrate-induced respiration rates of mineral soils. Biol Fertil Soils 8: 242–246.

Ritz K, Griffiths BS, Wheatley RE (1992) Soil microbial biomass and activity under a potato crop fertilised with N and without C. Biol Fertil Soil 12: 265–271.

Ross DJ (1980) Evaluation of a physiological method for measuring microbial biomass in soils from grasslands and maize fields. N Zealand J Sci 23: 229–236.

Ross DJ (1988) Modifications to the fumigation procedure to measure microbial biomass C in wet soils under pasture: influence on estimates of seasonal fluctuations in the soil biomass. Soil Biol Biochem 20: 377–383.

Ross DJ (1989) Estimation of soil microbial C by a fumigation–extraction procedure: influence of soil moisture content. Soil Biol Biochem 21: 767–772.

Ross DJ (1990) Measurement of microbial biomass C and N in grassland soils by fumigation–incubation procedures: influence of inoculum size and the control. Soil Biol Biochem 22: 289–294.

Ross DJ, Tate KR, Cairns A, Pansier EA (1980) Microbial biomass estimations in soils from tussock grasslands by three biochemical procedures. Soil Biol Biochem 12: 375–383.

Schimel J, Scott W, Killham K (1989) Changes in cytoplasmic carbon and nitrogen pools in a soil bacterium and a fungus in response to salt stress. Appl Environ Microbiol 55: 1635–1637.

Scheu S (1992) Automated measurement of the respiratory response of soil microcompartments: active microbial biomass in earthworm faeces. Soil Biol Biochem 24: 1113–1118.

Schmidt EL, Paul EA (1982) Microscopic methods for soil microorganisms. In: Methods in Soil Analysis, Part 2. Page AL, Miller RH, Keeney DR (eds). American Society for Agronomy, Madison, WI, pp. 803–814.

Schnürer J, Rosswall T (1987) Mineralization of nitrogen from ^{15}N labelled fungi, soil microbial biomass and roots and its uptake by barley plants. Plant Soil 102: 71–78.

Schnürer J, Clarholm M, Rosswall T (1985) Microbial biomass and activity in an agricultural soil with different organic matter contents. Soil Biol Biochem 17: 611–618.

Scholle G, Wolters V, Joergensen RG (1992) Effects of mesofauna exclusion on the microbial biomass in two moder profiles. Biol Fertil Soils 12: 253–260.

Scott IR, Ellar DJ (1978) Metabolism and the triggering of germination of *Bacillus megaterium*. Concentrations of amino acids, organic acids, adenine nucleotides and nicotinamide nucleotides during germination. Biochem J 174: 627–634.

Setlow P, Kornberg A (1970) Biochemical studies of bacterial sporulation and germination. XXII. Energy metabolism in early stages. J Biol Chem 245: 3637–3644.

Shen SM, Pruden G, Jenkinson DS (1984) Mineralization and immobilization of nitrogen in fumigated soil and the measurement of microbial biomass nitrogen. Soil Biol Biochem 16: 437–444.

Shen SM, Brookes PC, Jenkinson DS (1987) Soil respiration and the measurement of microbial biomass C by the fumigation technique in fresh and air-dried soil. Soil Biol Biochem 19: 153–158.

Shields JA, Paul EW, Lowe WE (1974) Factors influencing the stability of labelled microbial biomass nitrogen. Soil Biol Biochem 16: 437–444.

Smith LJ, Paul EA (1990) The significance of soil microbial biomass estimations. Soil Biochemistry, vol 6. Bollag JM, Stotzky G (eds). Marcel Dekker, New York, pp. 357–396.

Söderström BE (1977) Vital staining of fungi in pure cultures and in soil with fluorescein diacetate. Soil Biol Biochem 9: 51–63.

Soulas G, Chaussod R, Verguet A (1984) Chloroform

fumigation technique as a means of determining the size of specialized soil microbial populations: application to pesticide-degrading microorganisms. Soil Biol Biochem 16: 497–501.

Sparling GP (1981) Microcalorimetry and other methods to assess biomass and activity in soils. Soil Biol Biochem 13: 93–98.

Sparling GP (1985) The soil biomass. In: Soil Organic Matter and Biological Activity. Vaughan D, Malcolm RE (eds). Martinus Nijhoff/Dr W Junk, Dordrecht, Boston, Lancaster, p. 223.

Sparling GP (1990) Soil biomass evaluation. In: PACIFICLAND Workshop on the Establishment of Soil Management Experiments on Sloping Lands. IBSRAM Technical Notes no. 4, International Board for Soil Research and Management, Bangkok 1990, pp. 163–184.

Sparling GP, Searle PL (1993) Dimethyl sulphoxide reduction as a sensitive indicator of microbial activity in soil: the relationship with microbial biomass and mineralization of nitrogen and sulphur. Soil Biol Biochem 25: 251–256.

Sparling GP, West AW (1988a) A direct extraction method to estimate soil microbial C: calibration in situ using microbial respiration and ^{14}C labelled cells. Soil Biol Biochem 20: 337–343.

Sparling GP, West AW (1988b) Modifications to the fumigation–extraction technique to permit simultaneous extraction and estimation of soil microbial C and N. Commun Soil Sci Plant Anal 19: 327–334.

Sparling GP, West AW (1989) Importance of soil water content when estimating soil microbial C, N and P by the fumigation–extraction method. Soil Biol Biochem 21: 245–253.

Sparling GP, West AW (1990) A comparison of gas chromatography and differential respirometer methods to measure soil respiration and to estimate the soil microbial biomass. Pedobiologia 34: 103–112.

Sparling GP, Williams BL (1986) Microbial biomass in organic soils: estimation of biomass C and effect of glucose or cellulose amendments on the amounts of N and P released by fumigation. Soil Biol Biochem 18: 507–513.

Sparling GP, Ord BG, Vaughan D (1981) Microbial biomass and activity in soils amended with glucose. Soil Biol Biochem 13: 99–104.

Sparling GP, West AW, Whale KN (1985) Interference from plant roots in the estimation of soil microbial ATP, C, N, and P. Soil Biol Biochem 17: 275–278.

Sparling GP, Speir TW, Whale KN (1986) Changes in microbial biomass C, ATP content, soil phosphomonoesterase and phosphodiesterase activity following air-drying of soils. Soil Biol Biochem 18: 363–370.

Sparling GP, Feltham CW, Reynolds J, West AW, Singleton P (1990) Estimation of soil microbial C by a fumigation–extraction method: use on soils of high organic matter content, and a reassessment of the k_{EC}-factor. Soil Biol Biochem 22: 301–307.

Sparrow EB, Doxtader KG (1973) Adenosine triphosphate (ATP) in grassland soil: its relationship to microbial biomass and activity. Grassland Biome U.S.J.B.P. Technical Report no. 224.

Srivastava SC, Singh JS (1991) Microbial C, N and P in dry tropical forest soils: effects of alternate land-uses and nutrient flux. Soil Biol Biochem 23: 117–124.

Störmer (1908) Über die Wirkung des Schwefelkohlenstoffs und ähnlicher Stoffe auf den Boden. Zentral Bakteriol II 20: 282–286.

Tate KR, Ross DJ, Feltham CW (1988) A direct extraction method to estimate soil microbial C: effects of experimental variables and some different calibration procedures. Soil Biol Biochem 20: 329–335.

Torsvik VL, Goksoyr J (1978) Determination of bacterial DNA in soil. Soil Biol Biochem 10: 7–12.

Trolldenier G (1973) The use of fluorescence microscopy for counting soil microorganisms. In: Modern Methods in the Study of Microbial Ecology. Roswall T (ed). Bulletins from the Ecological Research Committee (Stockholm) vol 17, pp. 53–59.

Underwood AL (1961) Carbonic anhydrase in the titration of carbon dioxide solutions. Anal Chem 33: 955–966.

Valderrama JC (1981) The simultaneous analysis of total nitrogen and total phosphorus in natural waters. Marine Chem 10: 109–122.

Vance ED, Brookes PC, Jenkinson DS (1987a) Microbial biomass measurements in forest soils: determination of k_c values and tests of hypotheses to explain the failure of the chloroform fumigation–incubation method in acid soils. Soil Biol Biochem 19: 689–696.

Vance ED, Brookes PC, Jenkinson DS (1987b) Microbial biomass measurements in forest soils: the use of the chloroform fumigation–incubation method in strongly acid soils. Soil Biol Biochem 19: 697–702.

Vance ED, Brookes PC, Jenkinson DS (1987c) An extraction method for measuring soil microbial biomass C. Soil Biol Biochem 19: 703–707.

Vancura V, Kunc F (1977) The effect of streptomycin and actidione on respiration in the rhizosphere and nonrhizosphere soil. Zbl Bakteriol Parasitenk Infektionskrankh Hygiene 132: 472–478.

van de Werf H (1989/1990) A respiration–simulation method for estimating active soil microbial biomass. Ph.D thesis, University of Gent, Belgium.

van de Werf H, Verstraete W (1987a) Estimation of active soil microbial biomass by mathematical analysis of respiration curves: development and verification of the model. Soil Biol Biochem 19: 253–260.

van de Werf H, Verstraete W (1987b) Estimation of active soil microbial biomass by mathematical analysis of respiration curves: calibration of the test procedure. Soil Biol Biochem 19: 261–265.

van de Werf H, Verstraete W (1987c) Estimation of active soil microbial biomass by mathematical analysis of respiration curves: relation to conventional estimation of total biomass. Soil Biol Biochem 19: 267–271.

van de Werf H, Verstraete W (1987d) Estimation of microbial biomass by stimulation of respiration. In: Pesticide Effects on Soil Microflora. Sommerville L, Greaves MP (eds). Taylor and Francis Publishing, Basingstoke, pp. 147–170.

van Veen JA, Paul EA (1979) Conversion of biovolume measurements of soil organisms, grown under various moisture tensions, to biomass and their nutrient content. Appl Environ Microbiol 37: 686–692.

van Veen JA, Ladd JN, Frissel MJ (1984) Modelling C and N turnover through the microbial biomass in soil. Plant and Soil 76: 257–274.

Verstraete W, Van de Werf H, Kucnerowicz F, Imas Ilaiwi M, Verstraeten L, Vlassak K (1983) Specific measurement of soil microbial ATP. Soil Biol Biochem 15: 391–396.

Vittori Antisari L, Ciavatta C, Sequi P (1990) Volatilization of ammonia during the chloroform fumigation of soil for measuring microbial biomass N. Soil Biol Biochem 22: 225–228.

Voroney RP, Paul EA (1984) Determination of K_c and K_n in situ for calibration of the chloroform fumigation–incubation method. Soil Biol Biochem 16: 9–14.

Wardle DA, Parkinson D (1990a) Effects of three herbicides on microbial biomass and activity. Plant Soil 122: 21–28.

Wardle DA, Parkinson D (1990b) Comparison of physiological techniques for estimating the response of the soil microbial biomass to soil moisture. Soil Biol Biochem 22: 817–823.

Wardle DA, Parkinson D (1990c) Response of the soil microbial biomass to glucose, and selective inhibitors, across a soil moisture gradient. Soil Biol Biochem 22: 825–834.

Wardle DA, Parkinson D (1991) A statistical evaluation of equations for predicting total microbial biomass carbon using physiological and biochemical methods. Agric Ecol Environ 34: 75–86.

West AW, Sparling GP (1986) Modification to the substrate-induced respiration method to permit measurement of microbial biomass in soils of differing water contents. J Microbiol Meth 5: 117–189.

West AW, Sparling GP, Grant WD (1986) Correlation between four methods to estimate total microbial biomass in stored, air-dried and glucose-amended soils. Soil Biol Biochem 18: 569–576.

West AW, Sparling GP, Speir TW, Wood JM (1988a) Comparison of microbial biomass C, N-flush and ATP and certain enzyme activities of different textured soils subjected to gradual drying. Aust J Soil Res 26: 217–219.

West AW, Sparling GP, Speir TW, Wood JM (1988b) Dynamics of microbial C, N-flush and ATP and enzyme activities of gradually dried soils from a climosequence. Aust J Soil Res 26: 519–530.

Widmer P, Brookes PC, Parry LC (1989) Microbial biomass nitrogen measurements in soils containing large amounts of inorganic nitrogen. Soil Biol Biochem 21: 865–867.

Wolters V, Joergensen RG (1991) Microbial carbon turnover in beech forest soils at different stages of acidification. Soil Biol Biochem 23: 897–902.

Wolters V, Joergensen RG (1992) Effects of *Aporrectodea caliginosa* (Savigny) on microbial carbon turnover in beech forest soils at different stages of acidification. Soil Biol Biochem 24: 171–177.

Wu J, Brookes PC, Jenkinson DS (1993) Formation and destruction of microbial biomass during the decomposition of glucose and rye grass in soil. Soil Biol Biochem 25: 1435–1441.

Wu J, Joergensen RG, Pommerening B, Chaussod R, Brookes PC (1990) Measurement of soil microbial biomass C – an automated procedure. Soil Biol Biochem 22: 1167–1169.

Community structure

9

The quantification of the microbial community in soil is complicated because microorganisms are commonly attached to soil minerals and organic matter, and they occur in consortia usually containing different physiological and morphological types (Ellwood et al 1979; Bitton and Marshall 1980). Morphologically diverse microbial microcolonies can be detected and characterized *in situ* by using scanning electron microscopy and fluorescent antibody staining. Nevertheless these techniques do not give information on the microbial community structure or the physiological state of the different populations (Sieburth 1975; Ward and Frea 1979). Techniques for isolation and enumeration of soil microorganisms underestimate the microbial community and do not give information on its ecological importance.

Valuable information on community structure can be obtained by measuring the "biomarkers", which are biochemical components of the microbial cells and their extracellular products. Methods of quantifying biomarkers require neither growth with its attendant problems of microbial selection nor removal of cells from their natural environments (White 1983; Smith et al 1986; McKinley et al 1988). Ergosterol can be used as a biomarker for fungal biomass, while muramic and diaminopimelic acids as biomarkers for prokaryotic cells (King and White 1977; Fazio et al 1979; West et al, 1987; Zelles et al 1987a, 1987b; Zelles 1988). Lipopolysaccharide lipid A fatty acids and teichonic acids are used as biomarkers for gram-negative and gram-positive bacteria, respectively (Saddler and Wardlaw 1980; Parker et al 1982; Gehorn et al, 1984; Zelles et al 1992; Zelles and Bai 1993). Hydroxy fatty acids as well as plasmalogens and sphingolipids are biomarkers for anaerobic bacterial populations (Rizza et al 1970; Thompson 1972; Lechevalier 1977; Lovley and Klug 1982). Information on the nutritional status of microbial community can be obtained by monitoring the properties of specific endogenous storage compounds like triglyceride glycerol relative to the microbial biomass. The accumulation of poly β-hydroxyalkanoate (PHA) in bacterial cells indicates that the cells have insufficient total nutrients to allow growth with division (Costerton et al 1981; Fazio et al 1982; Findlay and White 1983; Findlay et al 1985; Odham et al 1986).

Recently DNA has been isolated and identified from soil to study the composition of soil microflora (DeLong et al 1989; Hahn et al 1989). Compounds like streptomycin or acitdion can be used selectively to quantify bacterial and fungal biomass in soil (Parkinson 1982).

Methods in Applied Soil Microbiology and Biochemistry
ISBN 0-12-513840-7

Differentiation by selective inhibition techniques

K. Alef
G.P. Sparling

(Anderson and Domsch 1975)

One way to study microbial community structure in soil is to inhibit selectively the metabolic activity of different microbial groups. This determination of the contribution of individual components of the population to the total metabolism is of interest, especially in connection with actual and chronic effects of environmental factors such as soil contaminations, the application of pesticides, and soil treatment and cultivation on soil microbial community. The assays are based on the stimulation of the overall metabolism of the active microbial population in soil by adding a readily degradable substrate. Simultaneously, subsamples of soil are also amended with inhibitors, which are specific for the different microbial groups. Specific inhibitors are used to repress the respiration of the bacterial or fungal components of soil. It is important that the inhibitors repress only their target population and that they function in soil. Anderson and Domsch (1974, 1975) recommended streptomycin sulphate to inhibit bacteria and cycloheximide (actidione) to inhibit fungi. Wardle and Parkinson (1990) and West (1986) also used these inhibitors. Both compounds block protein synthesis and are therefore most effective against the more actively synthesizing organisms. It is rare that complete supression of the soil respiration is obtained, even when the inhibitors are added together to the soil at high concentrations. Concentrations required for maximum suppression differ from soil to soil.

Principle of the method

The method is based on the estimation of glucose dependent soil respiration in the presence and absence of streptomycin and/or cycloheximide.

Materials and apparatus

See Chapter 4.

Chemicals and solutions

Glucose
Streptomycin
Cycloheximide
Talcum powder

Procedure

Sieved moist soil (100 g dry weight) is mixed with glucose, streptomycin or cycloheximide (the substances are mixed first with 0.5 g of talcum powder). The soil respiration (CO_2 formation) is then measured at 22°C for up to 6 h (see Chapter 5). A series of experiments should be performed to obtain the concentrations of antibiotics, causing maximal inhibition of soil respiration.

Calculation of results

Calculate the results as follows:

The respiration of fungi (%) =
$$100\ (A - C)/(A - D) \quad (9.1)$$

The respiration of bacteria (%) =
$$100\ (A - B)/A - D) \quad (9.2)$$

where A is the soil respiration in the absence of antibiotics, B is the soil respiration in the presence of streptomycin, C is the soil respiration in the presence of cycloheximide, and D is the soil respiration in the presence of streptomycin and cycloheximide. This calculation is only valid as long as $A-[(A-B)+(A-C)] = D \pm 5$ (Anderson and Domsch 1975).

Discussion

Depending on the soil, the glucose concentrations necessary to achieve maximum respiration response ranged from 500 to 8000 ppm; concentrations of antibiotics to achieve maximum repression of the response have ranged from 500 to 3000 ppm of streptomycin and 500 to 4000 ppm of cycloheximide (Anderson and Domsch 1975). However, it is unusual for addition of both antibiotics to result in 100% reduction in respiration (Anderson and Domsch 1975). West (1986) sought to improve the efficiency and contact with target organisms by adding antibiotics to sieved soil in a solution of glucose (2 ml solution g^{-1} soil) 3.5 h before the start of any measurements.

Several factors combine to reduce the effectiveness of the antibiotics. They may be sorbed to soil organic matter and minerals, and they are rapidly degraded in soil. Distribution of the antibiotics through the soil is difficult, meaning that some organisms may not come into contact with the antibiotic or may encounter only low concentrations. The diversity of soil organisms is such that organisms may show resistance, and soil factors such as pH affect the solubility, rate of diffusion and adsorption of antibiotics. Consequently, there is rarely any absolute specificity of the antibiotic, this is usually soil and concentration dependent.

Consequently, it is important that the optimum antibiotic concentration be determined for each soil, and that the measurements be only short-term. Using this approach, and considering only the repressed populations, Anderson and Domsch (1975, 1978) demonstrated the dominance of fungi over bacteria in forest soils, and Vancura and Kunc (1977) showed the dominance of bacteria in rhizosphere soils.

Biomarkers

L. Zelles
K. Alef

Ergosterol

Estimation of ergosterol

(Zelles et al 1987a, 1987b)

Fungi play an important role in the decomposition of organic matter such as cellulose and lignin in soil. The quantification of fungal biomass is mainly restricted to staining and microscopic methods (Bååth et al 1981; Parkinson 1982). Ergosterol (ergosta-5,7,22, trien-3β-ol) content has been shown to be a sensitive and reliable indicator of fungal growth (Seitz et al 1977, 1979; Lee et al 1980; Matcham et al 1985; Johnson and McGill 1990). It is generally accepted that ergosterol is the main sterol of most of Ascomycetes, Basidiomycetes and fungi imperfecti. There are also fungi like Mucorales, which possess ergosterol but not as a principal sterol. Phycomycetes, Saprolegniales, Leptomytales and generally lower fungi produce sterols other than ergosterol (Kato 1986). Sterols represent about 0.7–1.0% of the fungal dry weight. They are mainly localized in the membrane (Weete 1989). Low amounts of ergosterol have been also detected in some green microalgae (Newell et al 1987). The interest in fungal sterols is increased with the discovery of anti-fungal substances blocking ergosterol biosynthesis. The determination of ergosterol in soil has been reported by West et al (1987), Zelles et al (1987a,b, 1990, 1991) and Davis and Lamar (1992). West et al (1987) reported a linear correlation between the ergosterol content and the fungal surface area. West et al (1987) measured 0.2–0.31 μg sterol g^{-1} in air-dried and substrate-amended arable and grassland soils. Depending on the soil, ergosterol concentrations ranging from 0.1 to 96.2 μg g^{-1} soil could be detected (Grant and West 1986; Zelles et al 1990, 1991; Davis and Lamar 1992). Matcham et al (1985) reported detection limits of 25–200 μg g^{-1} fungal biomass. The ergosterol content of fungal cells varies depending on species and environmental conditions (Tunlid and White 1992). Yields of ergosterol from hyphae and from fungal-colonized soil were greater in alkaline than in neutral extractants (Zelles et al 1987a, 1987b; Davis and Lamar 1992). The recovery of ergosterol was quantitative in two soils amended with fungal tissue, however, only *c.* 66% were recovered from a clay subsoil (Davis and Lamar 1992).

Ergosterol concentrations in soils decreased with the depth and air drying, while substrate amendment increased the ergosterol concentration (West et al 1987).

Principle of the method

The method is based on the alkaline extraction of ergosterol from soil, followed by quantitative determination by HPLC. The limit of detection ranged from 8 to 15 μg biomass g^{-1} soil (Davis and Lamar 1992).

Materials and apparatus

HPLC equipment, precision pump, solvent delivery system, UV detector operating at 282 nm and automatic injector.
C-18 reverse-phase analytical column, 125 ×4.6 mm (for instance, hypersil 5 μm, Gynkothek Munich, Germany).
Rotary evaporator.
Reflux apparatus.
Round-bottomed flasks (100 ml), glassware, pipettes.

Chemicals and solutions

Mobile phase for HPLC separation (methanol:water, 95:5)
Petroleum ether (b.p. 50–70°C)
KOH pellets
Ergosterol standards (Sigma Chemicals)
Methanol
Ethanol

Procedure

Fresh soil (equivalent to 1 g dry weight) is placed into 100 ml round-bottomed flasks, and treated with 20 ml methanol, 5 ml ethanol and 2 g of KOH. The saponification is carried out at 70°C for 90 min in a reflux apparatus. After cooling, 5 ml of distilled water is added to the solution. In a separatory funnel the unpolar substances are extracted in two steps by adding first 30 ml and then 20 ml of petroleum ether. The collected petroleum ether is evaporated at 40°C by a rotary evaporator to dryness and the unpolar fraction is dissolved in 2 ml methanol. The final extract (20 μl) is usually injected into the HPLC. When the ergosterol level of the sample is high, the final extract is diluted so that the amount of ergosterol injected would be within the calibrated range. The separation is carried out with methanol:water 95:5 as the mobile phase, at a flow rate of 1.5 ml min^{-1}. The isolated peak is identified as ergosterol based on its co-chromatography and identical absorption spectrum with pure standard. Peak areas are measured by a computing integrator (Hewlett-Packard 3396A). Six replications are usually measured.

Calibration curve

Pure ergosterol from Sigma is recrystallized twice from pure ethanol and dried under vacuum in the dark at room temperature. With the standard solution a calibration curve is prepared. The relationship between ergosterol concentration and peak area is linear in the range 0–1000 ng.

Calculation

The amount of ergosterol is obtained using the calibration curve. The results are expressed as nanograms, or micrograms per gram dry weight of soil. The dilutions of ergosterol and the used amounts of soil have to be considered.

Discussion

The quantitative recovery of ergosterol depends on soil type. A high clay content reduces the recovery (Davis and Lamar 1992).

Rapid freezing treatment before lyophilization prevents losses of ergosterol (Davis and Lamar 1992).

Storage of soils causes losses in ergosterol (West et al 1987; Zelles et al 1991), this might be interpreted as ergosterol degradation occurring rapidly in soil after fungal death (Davis and Lamar 1992).

Recoveries from soil treated with pure ergosterol are not equal to those obtained with fungal tissue added to soil and containing the same amounts of ergosterol. Therefore, the determinations of extraction efficiencies should be based upon recoveries from fungal tissue added to soil (Davis and Lamar 1992).

Green algae and plant material also contain ergosterol in small amounts. It is recommended that plant material is removed from the soils.

Little is known about the mineralization of ergosterol in soils. Its extraction from dead material is possible.

Since the ergosterol content of fungal cells varies depending on species and environmental conditions (Tunlid and White 1992), care is required in converting ergosterol concentrations in biomass values.

Estimation of ergosterol

(Grant and West 1986)

Principle of the method

The method is based on hexane extraction of ergosterol from soil and its quantification by the means of HPLC analysis. Methanol/ H_2O has been used as a mobile phase.

Materials and apparatus

HPLC unit with a UV detector
HPLC column (Waters μBondpak C-18)
Ultrasonic with microtip probe
Rotary evaporator
Reflux
Filter paper (Whatman GFC)
Separatory funnels
Ice bath
Teflon filter (0.2 μm, Gelman Acrodisc CR)

Chemicals and solutions

Mobile phase: methanol/H_2O (95:5 v/v)
KOH pellets
Methanol
Ethanol
Redistilled hexane
Ergosterol standard

Procedure

Soil samples (40–60 g fresh weight) and methanol (150 ml, 0°C) are sonicated (3 min at maximum power, microtip probe) then kept at 0°C for 0.5 h. The suspension is then filtered and the residue is washed with 50 ml methanol. The combined filtrate is rotary evaporated (30–35°C) to 125 ml, then ethanol (25 ml) and KOH (12 g) are added, the mixture is saponified under reflux for 0.5 h, and then cooled to room temperature. After adding 30 ml water, ergosterol is extracted with 3 × 90 ml of redistilled hexane. The combined extracts are rotary evaporated (25–30°C) to dryness and stored at 4°C under nitrogen.

Immediately prior to assay, ergosterol is redissolved in methanol at 60°C, cooled, made up to 2.0 ml with methanol and filtered through 0.2 μm Teflon filters. Assays are performed by HPLC with C-18 column using a methanol/H_2O (95:5 v/v) mobile phase, a flow rate of 2 ml min^{-1} and UV detection at 282 nm. The retention time of ergosterol is 7–10 min. For each assay, three replicate samples of each soil are extracted and chromatographed.

Calibration curve

The ergosterol assay has a sensitivity of about 10 ng g^{-1} soil. The calibration curve should be performed at concentrations between 0 and 1000 ng (see Zelles et al 1987a, 1987b).

Discussion

Soils are extracted in the dark to avoid the photodecomposition of ergosterol.

Chromatography of authentic ergosterol added to soil extracts showed no splitting or distortion of the ergosterol peak.

Physical losses of ergosterol were very low (8%).

Muramic acid

Zelles (1988)

Muramic acid (2-amino-3-*o*-(1-carboxyethyl)-2-deoxy-D-glucose) is only found in prokaryotes, therefore it has been used as a measure of bacterial and cyanophyte biomass. This amino sugar is a component of the peptidoglycan of bacterial cell wall. The peptidoglycan consists of glycan chains of alternating units of *N*-acetylglucosamine and *N*-acetylmuramic acid. The individual chains are interconnected by short peptide bridges containing specific amino acids including *m*-diaminopimelic acid. The muramyl peptide in gram-positive bacteria accounts for 10–50% of the dry weight of the cell wall, while in gram-negative bacteria it is about 10% (Tunlid and White 1992). Millar and Cassida (1970) obtained muramic acid values ranging from 0 to 150 $\mu g\ g^{-1}$ dry weight in 33 different soils. Durska and Kaszubiak (1983) concluded that muramic acid in soil mainly occurs in humic acids, after incorporation of dead bacterial cells during the humification. More recent studies (Balkwill et al 1988) in subsurface soils have demonstrated an equivalence in microbial biomass estimations between measurements based on the muramic acid content, other chemical measurements and microscopic counts. Similar results have been obtained when muramic acid measurements have been compared with other biochemical determinations in different stored soils (Zelles et al 1990, 1991).

Principle of the method

The method is based on the acetic extraction of muramic acid from soil, derivatization with ophthalaldehyde, separation and quantitative determination by HPLC.

Materials and apparatus

HPLC equipment, precision pump, solvent delivery system, injector valve, fluorescence monitor with the excitation wavelength set at 340 nm and the emission set at 445 nm.
C-18 column (OSD Hypersil 5 µm, 125 × 4.6 mm).
Rotary evaporator, reflux apparatus.
Centrifuge tubes (about 30 ml volume), microcentrifuge with tubes (about 2 ml volume).
Round-bottomed flasks (100 ml), pear-shaped flasks (10 ml), glasswear, pipettes.

Chemicals and solutions

Mobile phase for HPLC separation

(0.05 M sodium citrate and 0.05 M sodium acetate adjusted at pH 5.3 in a 1:1 ratio):methanol:tetrahydrofuran 90:8.5:1.5.

Borate buffer

Boric acid solution (0.4 M) adjusted to pH 9.5 with 1 M NaOH.

o-Phthalaldehyde-2-mercaptoethanol reagent for derivatization of muramic acid

270 mg of *o*-phthalaldehyde (Fluka, Switzerland) is dissolved in methanol (chromatography grade) and 200 µl of 2-mercaptoethanol (Merck, Germany) are added. This solution is adjusted to 50 ml with borate buffer. The reagent mixture is allowed to "age" for at least 12 h before use. Its efficacy is maintained by adding 20 µl 2-mercaptoethanol every 2 days. This solution is stable at 4°C in dark for a week.

Reference compound muramic acid.

Concentrated HCl.

Procedure

Fresh soil (1 g dry weight) is placed into a 100 ml round-bottomed flask, and then 10 ml 6 M HCl are added; the weight of the mixture is determined and hydrolysed for 3 h at 100°C by reflux. After cooling the hydrolysate is adjusted to the original weight. From the supernatant (after centrifugation of the solution at 3000 rev min^{-1}, 10 min), 300 µl are removed and dried in a pear-shaped flask at about 50°C in a rotary evaporator. This dry hydrolysate is then redissolved with 300–1000 µl of *o*-phthalaldehyde reagent. This solution is then placed into microcentrifuge tubes and centrifuged for about 1 min, prior to its injection (20 µl) into the HPLC apparatus.

Calibration curve

One, 10, 100, 1000 and 10,000 pg muramic acid are derivatized and a linear function of the relative fluorescence yield is obtained (1.0–1000 relative units).

Calculation

The amount of muramic acid is obtained from the calibration curve. The dilutions, the amounts of soils, the amount of derivatization reagents used and the injection volumes have to be considered.

Discussion

In order to achieve the maximum release of muramic acids, the extraction parameters (HCl concentration, relation between the volume of HCl and the amount of soil, and time of hydrolysis) have to be found for each soil (Zelles 1988).

The derivatization of the hydrolysate with the *o*-phthalaldehyde reagent must be carried out under alkaline conditions. To achieve this, three different possibilities are available: (1) the hydrolysate residue is redissolved in distilled water and then evaporated (this procedure can be repeated several times); (2) a higher concentration of *o*-phthalaldehyde reagent is used; and (3) lower amounts of hydrolysate can be used for derivatization (Zelles 1988).

The derivatization time should be short and constant (no longer than 3 min) (Zelles 1988).

Gas chromatography has been used to determine muramic acid but the length of this procedure renders the use of this technique impractical for routine analyses (Mimura and Romano 1985).

The analysis takes about 30 min (Zelles 1988).

A number of investigations (Millar and Cassida 1970; Jenkinson and Ladd 1981) suggest that soil muramic acid is present mainly in dead material, therefore the content of this compound is not suitable for determining microbial biomass.

Despite the criticisms, comparative investigations with other biochemical methods suggest that muramic acid gives useful information (Hicks and Newell 1983; Balkwill et al 1988; Zelles et al 1990, 1991).

Estimation of teichoic acid components

(Gehorn et al 1984)

Teichoic acids, which make up from 20 to 40% of the dry weight of the cell walls of gram-positive bacteria, are polymeric chains of glycerol or ribitol molecules linked to each other by phosphodiester bridges. Both teichoic acids and the polysaccharides of bacterial cell walls are antigenic. This property is useful in the taxonomic classification of bacteria (Tonn and Gander 1979; Ward 1981).

Gehorn et al (1984) used the teichoic acids as a biomarker to estimate the biomass of gram-positive bacteria in soil and sediment.

Principle of the method

The method is based on the specific hydrolysis of teichoic acid components (polyglycerol, polyribitol) with concentrated acid (Glaser and Burger 1964; Lang et al 1982). The quantification is performed by gas chromatography (GC) or by gas chromatography/mass spectrometry (MS).

Materials and apparatus

Gas chromatograph (FID); vitreous silica capillary column (25 m, 0.2 mm internal diameter, coated with the polar 25% phenyl–25% cyanopropyl silicon chemically bonded BP-15 phase). The gas chromatograph is operated in the splitless mode.
GS/MSD
Rotary evaporator
Separatory funnels
Centrifuge and polypropylene tubes
Magnetic stirrer

Freezer (−70°C)
Desiccator with vacuum pump
Erlenmeyer flasks with glass stopper (250 ml)
Hydrolysis tubes
Nitrogen
Filter paper (Whatmann 2V)
Water bath adjustable at 60°C
Shaker
Dry freezer

Chemicals and solutions

Chloroform
Methanol
Chloroform/methanol (1:2)
Hydrofluoric acid (concentrated)
KOH pellets
Phosphorus pentoxide (P_2O_5)
HCl (6 M)
KOH (1 M)
KOH (6 M)
1,9-Nonandiol
Acetic anhydride
Pyridine
Acetic anhydride/pyridine (1:1)
Tartaric acid (20%)

Procedure

Soil samples are cooled on ice immediately after sampling, transported to the laboratory and extracted.

Lipid extraction

Soils are extracted with 8–10 times the volume of the one-phase chloroform–methanol (1:2 v/v) extractant. After at least 2 h, one volume of chloroform and one volume of water are added, and the suspension is shaken vigorously. The suspension is centrifuged at 5000*g* for 20 min and the supernatant poured into a separatory funnel. After partitioning, the lower chloroform phase is recovered by filtration through a Whatman 2V filter into a round-bottomed flask and the solvent removed *in vacuo* on a rotary evaporator. The extracted residues are recovered and lyophilized.

Hydrofluoric acid (HF) hydrolysis

Concentrated HF in 50 ml polypropylene centrifuge tubes is placed in the −70°C freezer. After the tubes are thoroughly cooled, the lyophilized lipid-extracted residue is added to the tubes with a final ratio of 2 ml HF g^{-1} of residue. The tubes are capped and the mixture stirred for 72 h at 4°C with a Teflon-covered magnetic stirring bar. The sample is then centrifuged at 36,000*g* for 15 min at 4°C and the supernatant, containing the teichoic acid components, is decanted quantitatively into new tubes. The acid-treated residue is washed once with one volume of water and water is added to the supernatant. The HF is removed in a glass desiccator *in vacuo* over P_2O_5 and KOH pellets at room temperature. The glass vessel etches superficially but maintains its strength. The strong vacuum that can be maintained in the desiccator speeds the removal of the HF. Generally the samples are dry after 24–48 h in the desiccator. The vacuum pump is protected by a 5 × 34 cm polyvinylchloride pipe containing sodium fluoride in series with a glass trap cooled with dry ice acetone. After over 100 h of pumping, the glass trap showed no evidence of etching.

Hydrochloric acid hydrolysis

Hydrochloric acid (6 M) is added to dried teichoic acid hydrolysate (0.5 ml g^{-1}) and heated to 60°C for 1 h. The hydrolysate is neutralized with 6 M KOH with a final adjustment to pH 7.8–8.0 with 1 M KOH with the formation of a precipitate. The mixture is centrifuged at 36,000*g* for 15 min at 4°C, and the supernatant containing the hydrolysate is quantitatively decanted into a glass test tube and reduced to dryness *in vacuo* at 50°C with the Haak–Buchler rotary evapo-mix solvent remover.

Peracetylation

The internal standard of 0.2 μmol 1,9-nonanediol dissolved in methanol is added to the test tubes and the solvent removed in a stream of nitrogen at 30°C. One millilitre of acetic anhydride:pyridine (1:1 v/v) is added, the solution mixed with a Vortex mixer for 5 min and then heated for 2 h at 60°C. After cooling, 2 ml of chloroform is added and the solution partitioned against 2 ml of 20% tartaric acid with vortex mixing followed by centrifugation at 100*g*. The aqueous tartaric acid is removed and the chloroform washed twice more with tartaric acid. The chloroform is transferred to a clean test tube and the solvent removed in a stream of nitrogen. The derivatized material is then dissolved in a small volume of chloroform for gas–liquid chromatography (GLC).

Gas–liquid chromatography

One microlitre of a sample is injected into the gas chromatograph. The temperature programme is initiated at 80°C and increased to 140°C at 5°C min^{-1}, which is followed by a 2°C min^{-1} rise to 220°C and a 10 min isothermal period. The hydrogen carrier gas is at a flow rate of 1.5 ml min^{-1} at 0.92 kg cm^{-2}. The detection is by hydrogen flame. Under these conditions the molar response ratio of the peractylated glycerol and ribitol to the 1,9-nonanediol is 2.0 and 1.5, respectively.

Mass spectrometry

Capillary gas chromatography/mass spectral fragmentography is performed using the same column and chromatographic conditions as utilized in GLC. The mass spectrometer is autotuned with decafluorotriphenylphosphine and utilized at 70 meV fragmentation energy. Spectra are recorded in the peak finder scan mode at a scan speed of 380 AMU s^{-1} (4 samples/0.1 AMU) between 50 and 350 AMU. The threshold of detectability is set at 100 linear counts at an electron multiplier voltage of 1800 V.

Identification

The putative triacetylglycerol is co-eluted with an authentic standard and showed major ions at m/z 103 (100%), 145 (76%), 116 (45%), 115 (32%), 86 (19%) and 73 (15%) of the authentic compound in mass spectral fragmentography. Pentaacetylribitol is co-eluted with authentic standards and showed major ions at m/z 60 (100%), 85 (83%), 98 (52%), 103 (47%), 115 (88%), 145 (45%), 187 (19%) and 217 (10%).

Discussion

The assay has a sensitivity of about 10 pmol.

The recovery of teichoic acid glycerol and ribitol after HF and HCl hydrolysis from the mixture is 114 ± 11% and 123 ± 47%.

This method has been mainly used to estimate the biomass of gram-positive bacteria in sediments (Gehorn et al 1984), while further research is needed in soil.

Little is known about the decomposition of teichoic acids in soils (Moriaty 1975; King and White 1977).

Teichoic acids increase with the depth of soil due to the increasing number of gram-positive bacteria present at deeper levels, while the microbial biomass decreases.

The ratio of biomass of gram-positive bacteria to microbial biomass or to other microbial parameters can be a useful indicator for microbial processes in soil.

Lipopolysaccharide (lipid A) fatty acids

(Parker et al 1982; Zelles and Bai 1993)

The outer membrane of gram-negative bacterial cells is mainly composed of lipopolysaccharide (LPS) polymers consisting of a lipid (lipid A), a core polysaccharide and an O-specific side chain (Wilkinson 1988).

Generally, LPS fatty acids seem to constitute about 15–20% of total cellular fatty acids of gram-negative bacteria (Jantzen and Bryn 1985). Saddler and Wardlaw (1980) demonstrated in sediments a rapid degradation of LPS of dead bacteria. The fatty acids of LPS have been used for biomass determination of gram-negative bacteria. The analysis of β-hydroxy myristic acid in soil has been used to determine the LPS content (Smith et al 1986). There is also a potential present in soil to analyse the community composition based on the variance of the hydroxy acids in LPS, which has been already demonstrated in sediments (Parker et al 1982; Goosens et al 1985).

Principle of the method

The soil residue that has been extracted by the Bligh and Dyer procedure to obtain phospholipid fatty acids is acidic hydrolysed. The released fatty acids are then transesterified, separated on a solid-phase column, and the hydroxy fatty acid fraction is determined in GC/MS.

Materials and apparatus

Refluxing apparatus
Gas chromatograph/mass spectrometer
Capillary column (50 m × 0.2 mm i.d., 0.33 mm film thickness) coated with cross-linked 5% phenyl methyl silicone gum phase
Rotary evaporator
Sample concentrator, with nitrogen supply
Centrifuge, centrifuge tubes (about 12 ml volumes)
Horizontal shaker for 1 l round-bottomed flasks
Mechanical shaker for centrifuge tubes
Water bath
Filter funnel
Separation funnel
Round-bottomed flasks
Funnels

Chemicals and solutions

Phosphate buffer

Dissolve 8.7 g of K_2HPO_4 in about 800 ml distilled water, neutralize with 1 M HCl to pH 7.4 and bring up with distilled water to 1000 ml

Reagents for acidic methylation:

Methanol:chloroform:HCl (37%) (10:1:1 by volume), 2% NaCl, hexane:toluene (1:1 v/v)

Reagents for TMSi derivatization

0.5 ml pyridine:*N,O*-bis-(trimethylsilyl) trifluoroacetamide:hexamethyldisilazane: trimethylchlorosilane (0.2:1:2:1 by volume)

Solid-phase extraction column (SPE-NH_2), chloroform, hexane, dichloromethane, ethylacetate
Celite 545
Anhydrous Na_2SO_4

Procedure (see also Fig. 9.1.)

Lipid extraction

After collecting the residues on the Celite funnel, the soil is quantitatively transferred to a 500-ml round-bottomed flask, dried *in vacuo* on a rotary evaporator and suspended with 120 ml 4 M HCl. After refluxing at 100°C for 5 h and cooling, the slurry is filtered on a pre-prepared filter funnel with *c.* 2 cm layer of Celite 545. The supernatant is transferred to a 250 ml separatory funnel to which a small amount of water is added, then extracted with about 60, 40, 40 ml of chloroform. The combined chloroform fractions containing free fatty acids is dried over Na_2SO_4 and the solvent is reduced under vacuum.

Esterification of free fatty acids

The lipid-fraction is transferred to a 4 ml vial and the solvent is removed in a stream of nitrogen at 40°C. The sample then is esterified by adding 2 ml of methanol:chloroform:concentrated HCl (10:1:1 v/v) and stored at 60°C overnight. The contents are transferred to a centrifuge tube and 2 ml of 2% NaCl and 4 ml hexane:toluene (1:1 v/v) are added and the mixture is then centrifuged to separate the solution in two phases. This procedure is repeated three times, then the fraction is dried by passing through a pre-prepared funnel containing a layer of 1 cm anhydrous sodium sulphate. The volume of chloroform is reduced by evaporation and stored at −20°C.

Separation of hydroxy fatty acids

The separation of hydroxy-substituted fatty acids (hydroxy fatty acid methyl esters, OH-FAMEs) from unsubstituted fatty acids is carried out on a solid-phase extraction (SPE-NH_2 column, size 1, or 2 gr) (Zelles and Bai 1993). The column is conditioned with one volume of dichloromethane and one volume of hexane. The sample is dissolved in a small volume of hexane:dichloromethane (3:1 v/v) and applied on the column. The total phospholipid fatty acid methyl esters (PL-FAMEs) are recovered with two column-volumes of hexane:dichloromethane (3:1 v/v), and the OHFAMEs with two column-volumes of dichloromethane:ethylacetate (9:1 v/v). The unsubstituted FAMEs are discarded.

Silane derivative formation

To increase the sensitivity of OHFAMEs they are trimethyl silylated (Parker et al 1982) by adding 0.5 ml of freshly prepared pyridine:N,O-bis-(trimethylsilyl) trifluoroacetamide: hexamethyldisilazane: trimethylchlorosilane (TMSi), 0.2:1:2:1

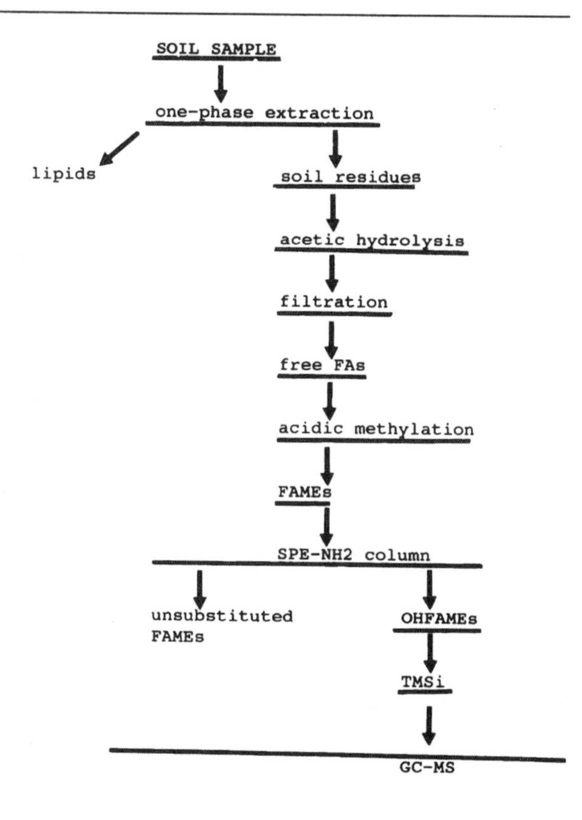

Figure 9.1. Flow diagram of the analytical steps for the extraction and separation of the fatty acids from lipopolysaccharides.

v/v prepared by adding each reagent in the order given with mixing) and then heating for 15 min at 60°C; the solvent is then removed under a stream of nitrogen.

GC/MS determination

The different fatty acid methyl esters are redissolved in 0.1 ml hexane containing methyl nonadecanoate (19:0) as an internal standard. One microlitre of the final solution is injected into the GC/MS for qualitative and quantitative measurements. A selection must be taken between the presented mass spectrometric methods SCAN and SIM. A more detailed description is presented later in the section on determination of phospholipid fatty acid.

GC/MS operation

Samples are injected into a capillary column (50 m × 0.2 mm i.d., 0.33 mm film thickness) coated with cross-linked 5% phenylmethyl silicone gum phase in the splitless mode with a 0.75 min venting time by an automatic sampler. Helium is used as the carrier gas at a flow rate of 0.25 ml min^{-1}. The temperature programme of the oven is initiated at 70°C (kept for 2 min) and increased to 160°C at a rate of 40°C min^{-1}, then to 280°C at a rate of 3°C min^{-1} (injector temperature 290°C). The electron multiplier voltage is between 1800 V and 2000 V. The transfer line is kept at 300°C. The GC/MS is periodically autotuned with perfluorotributylamine (PFTBA) with an ionization energy of 70 eV.

Calculation

A representative soil extract sample containing all the presented FAMEs are taken as a relative standard. This sample is injected for the SCAN and for the SIM mode. The amounts of each OHFAME in this soil extract are estimated by the SCAN chromatogram using the following relationship:

$$C = \frac{R_f\, k\, m}{R_i\, A\, MW} \times 1000 \qquad (9.3)$$

where C is the amount of the compound of interest in the soil (nM g^{-1}), R_f and R_i are the responses of the compound of interest and the internal standard, respectively, k is the factor for the compound of interest against the internal standard in terms of the response, m is the amount of internal standard injected (ng), A is the amount of soil extract injected (mg) and MW is the molecular weight of compound of interest.

Discussion

Different methods have been used to release fatty acids from LPS. They are based on either acidic or alkaline hydrolysis; a combination of both methods has been also used (Wollenweber and Rietschel 1990).

The method described above is about five times more sensitive than the classical phenol–water or trichloroacetic acid method, when applied to marine sediments (Parker et al 1982).

The ester-bound fatty acids are released by the extraction procedure of Bligh and Dyer (1959); these phospholipids are usually extracted before the extraction of LPS fatty acids.

Amide-linked hydroxy fatty acids are encountered exclusively in sphingolipids, lipopolysaccharides and ornithine lipids (Goosens et al 1985). All these types of hydroxy fatty acids (OH-FAs) can be released by the acetic hydrolysis.

Significant correlations have been found in the subsurface between microbial biomass based on the content of lipid A OH-FAs and that of phospholipid and muramic acid. Conversion factors were those derived from studies on bacterial monocultures (Tunlid and White 1992).

The minimum detection limit of the present method is approximately 1 ng g^{-1}.

Phospholipid fatty acids

(Zelles and Bai 1993)

Phospholipids are found in the membranes of all living cells but not in the storage products of microorganisms (Kates 1964). In the modern concept of membrane structure of the fluid mosaic model of Singer and Nicholson (1972), the phospholipids are the main components of the membranes and the composition of their fatty acids play an important role in the physiological conditions of the microorganisms. Phospholipids are actively metabolized during the growth of bacterial monocultures (White and Tucker 1969). A study of the degradation of labelled phosphatidylcholine in soils suggested a rapid turnover of microbial phospholipids (Tollefson and McKercher 1983).

Lipids are the signature components most often used for determining the community composition of microorganisms in ecological studies (Guckert and White 1986; Vestal and White 1989; Tunlid and White 1992).

Bacterial groups can be characterized on the basis of their lipid composition:

1. Archaebacteria, whose fatty residues are ether-linked to glycerol.
2. Anaerobic bacteria containing sphingolipids and/or plasmalogens, which are largely absent from aerobes.
3. Bacteria with saturated or monounsaturated fatty acids ester-linked to glycerol.
4. Cyanobacteria (and also eukaryotes) with lipids containing polyunsaturated fatty acids.
5. The lipids of gram-negative bacteria containing more hydroxylated fatty acids.
6. The gram-positive bacteria, containing more branched fatty acids (Ratledge and Wilkinson 1988).

By using the differences listed above, populations in a microbial community can be identified by specific "signature" phospholipid fatty acids.

Principle of the method

Phospholipids are extracted and purified from soils using the Bligh and Dyer (1959) extraction procedure, followed by transesterification to methyl ester. After the structurally different fatty acids are separated on solid-phase extraction (SPE) columns they are injected into GC/MS for identification and quantification.

Materials and apparatus

Gas chromatograph and mass spectrometer
Capillary column (50 m × 0.2 mm i.d., 0.33 mm film thickness) coated with cross-linked 5% phenylmethyl silicone gum phase (e.g. Ultra 2, part No. 19091B-105, Hewlett Packard)
Rotary evaporator
Sample concentrator, with nitrogen supply
Centrifuge with tubes (about 12 ml volume)
Horizontal shaker for 1 l round-bottomed flasks
Mechanical shaker for centrifuge tubes
Water bath
Filter funnel
Separation funnel
Round-bottomed flasks
Funnels

Chemicals and solutions

Phosphate buffer

Dissolve 8.7 g K_2HPO_4 in about 800 ml distilled water, neutralize with 1 M HCl to pH 7.4 and bring up with distilled water to 1000 ml.

Reagents for mild alkaline methanolysis

Methanol:toluene (1:1 v/v), 0.2 M KOH in chloroform (freshly prepared), 1 mol acetic acid.

Reagents for acidic methylation of the unsaponifiable fraction

Methanol:chloroform:HCl (37%) (10:1:1 by volume), 2% NaCl, hexane:toluene (1:1 v/v)

Reagents for TMSi derivatization 0.5 ml of pyridine: *N,O*-bis- (trimethylsilyl) trifluoroacetamide: hexamethyldisilazane: trimethylchlorosilane (0.2:1:2:1 by volume)

Dimethyl disulphide (gold label, Aldrich Chemicals Co.), sodium thiosulphate, iodine solution (6% w/v in diethyl ether)
Separation on SPE columns

SI column; chloroform, methanol
NH_2 column; chloroform, hexane, dichloromethane, ethylacetate
SCX column; dichloromethane, acetonitrile, acetone, silver nitrate

Celite 545
Anhydrous Na_2SO_4
KOH
Different fatty acid methyl esters standards

Procedure (see also Fig. 9.2)

Lipid extraction

The modified one-phase extraction procedure of Bligh and Dyer (White et al 1979a) is used to extract the lipids from soil. Duplicate samples (100 g equivalent to dry weight) of soil are placed into a 1 l round-bottomed flask and suspended in 200 ml of 0.05 M phosphate buffer (pH 7.4). The water content of the used soil is subtracted from the buffer, and 500 ml anhydrous methanol and 250 ml chloroform are then added. The suspension is shaken vigorously for 2 h, then 250 ml of water and 250 ml chloroform are additionally mixed with the suspension and allowed to separate for 24 h. The chloroform and the soil slurry are decanted through a filter funnel containing a Celite 545 layer about 2 cm thick. The organic phase is transferred to a separation funnel and separated, then dried by passing through a prepared funnel containing a layer of about 1 cm anhydrous sodium sulphate over a plug of cotton. The chloroform is removed by evaporation and finally adjusted to 10 ml; the fraction is stored at −20°C.

Separation of phospholipids

The lipid extracts are separated by liquid chromatography on a bonded-phase column (SPE SI, size 2 g, Analytical Chem International, USA). After conditioning the column using one column-volume of chloroform, neutral, glyco- and phospholipids are separated by using one volume of chloroform, one volume of acetone and four volumes of methanol as eluent. The final fraction is regarded as the phospholipid fraction and is subjected to the mild alkaline methanolysis.

Mild alkaline methanolysis (White et al 1979a)

After removal of chloroform the residue of phospholipids is resuspended in 1 ml of methanol/toluene (1:1 v/v). To the resulting suspension 5 ml of freshly prepared 0.2 M KOH in chloroform are added. The mixture is incubated for 15 min at 37°C, neutralized to pH 6.0 with 1 M acetic acid. Then 10 ml of chloroform and 10 ml water are added to the suspension, which is then mixed vigorously for 5 min on a mechanical mixer. The suspension is centrifuged, the fatty acid methyl esters and non-saponifiable phospholipids are recovered in the organic phase, and the glycerol phosphate esters derived from the diacyl phospholipids are in the water phase. This step is repeated twice, the chloroform fraction is dried over Na_2SO_4 and reduced to a small volume.

Separation of the functionally different FAMEs on solid-phase extraction columns

Separation of unsubstituted fatty acids from OH-FAMEs and from non-saponifiable ones (Zelles and Bai 1993)

The separation is carried out on a SPE NH_2 column (size 1 or 2 gr). The column is conditioned with one volume of dichloromethane and one volume of hexane. The sample is dissolved in a small volume of hexane:dichloromethane (3:1 v/v) and added to the column. The unsubstituted PL-FAMEs are recovered with two volumes of hexane:dichloromethane (3:1 v/v), the OH-FAMEs with two volumes dichloromethane:ethylacetate (9:1 v/v) and the non-saponifiable ones with two volumes of 2% acetic acid in methanol.

Silver ion chromatography

A solid-phase column (SCX, column size 0.5 or 1 gr) impregnated with silver nitrate is employed to separate the unsubstituted FAMEs from saturated fatty acids (SATFAs), monounsaturated fatty acids (MUFAs) and polyunsaturated fatty acids (PUFAs) exactly as described by Christie (1989). A solution of 20 mg silver nitrate in 0.25 ml acetonitrile:water (10:1 v/v) is allowed to percolate through a solid-phase extraction column, wrapped in aluminium foil to the top of the adsorbant bed. The silver nitrate-treated column is flushed with acetonitrile (2 volumes), acetone (2 volumes), and dichloromethane (4 volumes). The column is then ready for use. The FAMEs are dissolved in a small volume of dichloromethane:hexane (7:3 v/v) and are applied to the column. The SATFA fraction is eluted with two volumes of the same solvent. Two volumes of dichloromethane:acetone (9:1 v/v) are used for eluting the MUFAs. PUFAs are eluted by using four volumes of acetone:acetonitrile (9:1 v/v). Solvent mixtures are allowed to flow under gravity (approximately 0.5 ml min^{-1} and fractions are collected manually.

Treatment of the unsaponifiable fraction

After evaporation to dryness, the fraction is dissolved in 2 ml methanol:chloroform:HCl (37%) (10:1:1 by volume) and kept at 60°C in a sealed vial overnight. After the addition of 2 ml of 2% NaCl and 4 ml hexane:toluene (1:1 v/v), it is centrifuged. This last step is repeated three times. The organic phase contains the FAMEs of the unsaponifiable fraction. This can be separated again on a SPE NH_2 column.

Preparation of the samples for GC/MS injection

The SATFA fraction is ready to be injected into the GC/MS but the other fractions need to be treated before they are injected. To increase the sensitivity, the OH-FAs are trimethylsilylated (Parker et al 1982). To determine the position of unsaturations in MUFAs, the fraction is treated with dimethyl disulphide (Nichols et al 1986).

Formation of silane derivative

The FAMEs are trimethylsilylated by adding 0.5 ml of freshly prepared pyridine:*N*,*O*-bis-(trimethylsilyl) trifluoroacetamide: hexamethyldisilazane; trimethylchlorosylane (TMSi), 0.2:1:2:1 (v/v) (prepared by adding each reagent in the order given with mixing), heated for 15 min at 60°C. The solvents are removed under a stream of nitrogen.

Dimethyl disulphide (DMDS) derivatization

The residues of the monounsaturated FAMEs are dissolved in 0.05 ml of hexane. DMDS (0.1 ml) and 1.2 drops of I_2 (6% in ether, w/v) are added and the mixture is kept at 60°C for 72 h. The excess I_2 is removed by the addition of 5% sodium thiosulphate. The adduct is extracted by hexane (repeated three times). The hexane phase is combined and evaporated nearly to dryness.

GC/MS determination

The different FAMEs are redissolved in 0.1 ml of hexane containing methyl nonadecanoate (19:0) as an internal standard. One microlitre of the final solution is injected into the GC/MS for qualitative and quantitative measurements. A selection must be taken between the presented mass spectrometric methods SCAN and SIM.

GC/MS operation

Samples are injected into a capillary column (50 m × 0.2 mm i.d., 0.33 mm film thickness) coated with cross-linked 5% phenylmethyl silicone gum phase in the splitless mode with a 0.75 min venting time by an automatic sampler. Helium is used as the carrier gas at a flow rate of 0.25 ml min^{-1}. The temperature programme of the oven is begun at 70°C (for 2 min) and increased to 160°C at 40°C min^{-1}, followed to 280°C at 3°C min^{-1} (injector temperature 290°C). The second programme is started at 100°C and increased to 210°C at 50°C min^{-1}, followed by 300°C at 3°C min^{-1} (injector temperature 300°C). The latter operational variables are only used

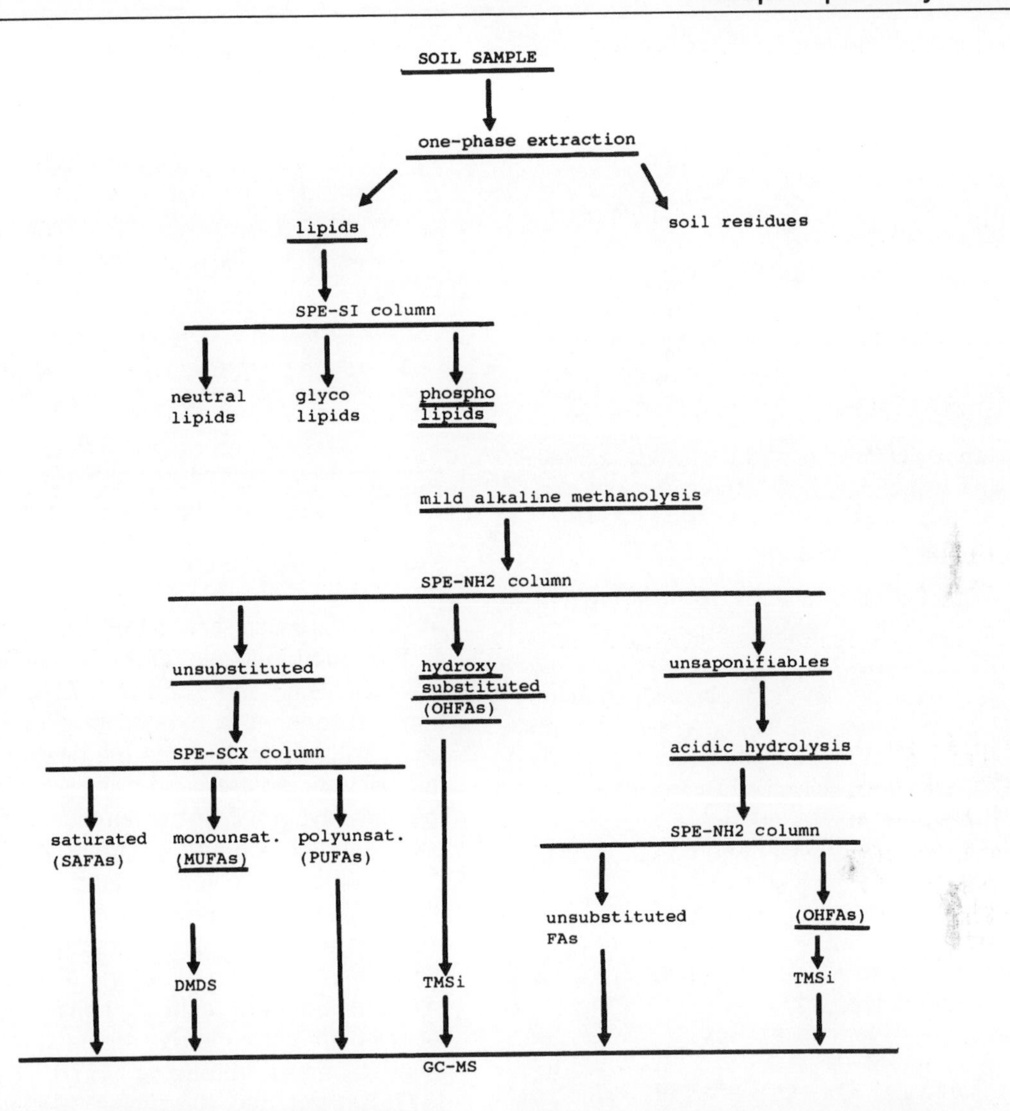

Figure 9.2. Flow diagram of the analytical steps for the separation of lipids and fatty acid methyl esters.

to measure DMDS derivatives of MUFAs. The electron multiplier voltage is between 1800 V and 2000 V; the transfer line is kept at 300°C. The GC/MS is periodically autotuned with perfluorotributylamine with an ionization energy of 70 eV.

Calculation

A representative soil extract sample containing all the presented FAMEs is taken as a relative standard. This sample is injected for the SCAN and for the SIM mode. The amounts of each FAME in this soil extract are estimated by SCAN chromatogram using the following relationship:

$$C = \frac{R_f\, k\, m}{R_i\, A\, MW} \times 1000 \qquad (9.4)$$

where C is the content of the compound of interest in soil (nM g^{-1}), R_f and R_i are the mean responses of the compound of interest and internal standard, respectively, k is the factor for the compound of interest against the internal standard in terms of response, m is the amount of internal standard injected (ng), *A* is the amount of soil extract injected (mg) and *MW* is the molecular weight of the compounds of interest.

Discussion

Time can be saved if smaller amounts of soil (1 g) are used by this method and the separation of FAMEs in different functional groups is not carried out (Korner and Laczko 1992).

Mass spectrometer is only necessary for the identification of fatty acids. The use of an FID detector has also been applied (Tunlid and White 1992).

Potential problems of quantifying microbial community structure by analysis of phospholipid fatty acids could result from a shift in fatty acid composition of some monocultures with changes in temperature and media composition (Parker et al 1982). However, there are no studies on phospholipid fatty acids composition of microbial community under natural growth conditions (Tunlid and White 1992). Bacteria growing under natural conditions contain a relatively constant proportion of their biomass as phospholipids (White et al 1979b).

Despite the fact that the analysis of PLFA cannot provide an exact determination of species as such or physiological state of microorganism in a given environment, the analysis provides a quantitative description of the overall microbiota in the particular environment sampled. By use of a statistical analysis, it is possible to estimate the differences among various samples (Tunlid and White 1992).

The minimum detection limit of the presented method is approximately 5 ng g^{-1}.

The separation of saturated from monounsaturated fatty acids by the SPE column allows the determination of cyclopropane fatty acids, despite the similarities of their mass spectra to the MUFAs.

Correlation coefficients larger than 0.97 were obtained in soils between the total amount of PLFA and different classical procedures of biomass determination (Zelles et al 1992).

It has been demonstrated that this method can be used as an indicator for the effect of soil management (Zelles et al 1992).

Diaminopimelic acid

(Grant and West 1986)

m-Diaminopimelic acid (DAP) has been found in the peptidoglycan of most bacterial cell walls (Schleifer and Kandler 1972) and can be a precursor in the biosynthesis of lysine. It plays an important role in the cross-linking of the cell wall; cross-linking takes place between the carboxyl group of the terminal D-alanine of one peptide chain and the free -NH_2 group of the diaminopimelic acid. Instead of M-diaminopimelic acid, LL-diaminopimelic acid has been found in the cell walls of gram-positive bacteria. Little is known about the decomposition of diaminopimelic acid in soils and it has received little attention as an indicator of bacterial biomass (Steubing 1970; Lehninger 1977; Durska and Kaszubiak 1980a, 1980b, 1980c, 1983).

Principle of the method

The method is based on the extraction of DAP from soil, hydrolysis by HCl and quantitative determination after paper chromatography.

Materials and apparatus

Spectrophotometer
Rotary evaporator
Chromatography paper (Whatman No. 3 MM)

Chromatography chamber
Shaker
Water baths adjustable at 25° and 100°C
Filter paper (Whatman GFC)
Centrifuge and centrifuge tubes.

Chemicals and solutions

HCl (12 M)
Trimethylchlorosilan (1% in toluene)
NaOH pellets
Mobile phase I Butan-1-ol/pyridine/acetic acid (concentrated)/water (30:20:6:24 v/v)

Mobile phase II Butan-1-ol/formic acid (concentrated)/water (7:1:v/v)

Ninhydrin reagent consisting of 1 g ninhydrin, 100 mg cadmium acetate, 5 ml acetic acid, 100 ml acetone and 10 ml distilled water
H_2SO_4 (concentrated)
Methanol
DAP standard solution

Procedure

Triplicate soil subsamples (2.5 g fresh weight) are incubated with 20 ml of HCl (12 M) at 25°C for 48 h on a rotary shaker (100 rev min^{-1}), diluted with 20 ml distilled water and hydrolysed at 105°C for 18 h. Hydrolysates are filtered, residues washed with distilled water (2 × 10 ml) and combined filters rotary evaporated (45–50°C) to dryness in flasks which had been silylated to prevent surface adsorption of DAP. Hydrolysates are deacidified by rotary evaporation to dryness (twice) from the aqueous solution, then finally dried *in vacuo* over NaOH. Deacidified hydrolysates are redissolved in distilled water and centrifuged to remove humin, which is then washed and recentrifuged. The combined supernatants are made up to a standard volume. Duplicate aliquots of each hydrolysate and volumes of authentic DAP solution are applied to Whatman paper as 2 cm bands and chromatographed in the mobile phase in descending mode for 72 h. Papers are dried in air then dipped in ninhydrin reagent. Spots, developed within 24 h over H_2SO_4 (concentrated) in darkness, are cut out, eluted with methanol (4 ml) and absorbances measured spectrophotometrically at 500 nm. Blank regions on each pair, parallel and equal in area to the DAP spots, are eluted to correct for background absorbance. To assess homogeneity of the DAP spots from soil hydrolysates, samples are chromatographed in a second dimension using the mobile phase II.

Discussion

The method has a sensitivity of about 10 µg g^{-1}soil.

The losses of DAP during the extraction were up to 38%.

A large proportion of DAP is present in the non-living organic fraction of soil (Grant and West 1986).

Glucosamine

Estimation of glucosamine

(Zelles 1988)

A relatively specific compound found in the prokaryotic wall is peptidoglycan, which consist of a backbone of alternating *N*-acetyl glucosamine (*N*-acetyl-2-amine-2-deoxy-D-glucose) and *N*-acetyl muramic acid. The cell walls of a large number of fungi, including those of the higher orders, contain chitin (Bartnicki-Garcia 1968), which consists predominantly of unbranched chains of *N*-acetyl glucosamine. The exoskeleton of invertebrates also contains chitin. Glucosamine has been recovered from soil samples by strong acid hydrolysis with HCl followed by purification of cation-exchange chromatography (Frankland et al 1978). To omit the cation-exchange step, glucosamine has been determined in soil hydrolysates by paper chromatography (Grant and West 1986). The quantification can be carried out by colorimetry (Zelles et al 1987a, 1987b), but more accurately by gas chromatography (Hicks and Newell 1983) or by HPLC (Zelles 1988).

The glucosamine content of fungi varies over a wide range, depending on species, growth condition and age. Aaronson (1981) reported that the concentration of chitin ranged between 10 and 250 mg g^{-1} dry weight in various fungal species. Hicks and Newell (1984) described that the glucosamine content varied from 8.5 to 92.8 μg mg^{-1} dry weight. Grant and West (1986) measured 505–2109 μg, Zelles et al (1990) *c.* 3500 μg in the organic layer and 137–479 μg in the mineral layers of forest soils. Storage increased the glucosamine content (Zelles et al 1991), or did not cause a clear effect (West et al 1987).

There are two significant problems that occur when using glucosamine analysis to estimate fungal biomass: the contribution of glucosamine from prokaryotes and that from invertebrates. In the first case glucosamine can be accounted for by measuring the amount of muramic acid in a sample and assuming a muramic acid glucosamine molar ratio of 1:1. However, the main problem of the glucosamine method is, as for the muramic acid analysis, the presence of glucosamine in non-living organic material (Tunlid and White 1992). The evaluation of the glucosamine figures on the basis of known soil biomass data indicates that these compounds were largely associated with non-living organic matter (Grant and West 1986).

The method for the determination of muramic acid was also used for determining glucosamine. All the parameters described in the muramic acid section are valid for the glucosamine assay.

Estimation of glucosamine

(Grant and West 1986)

Principle of the method

The method is based on the extraction of glucosamine from soil, hydrolysis by HCl and quantitative determination after paper chromatography.

Materials and apparatus

Spectrophotometer
Rotary evaporator
Chromatography paper
Chromatography chamber
Rotary shaker
Water baths adjustable at 25° and 100°C
Filter paper (Whatman GFC)
Centrifuge and tubes

Chemicals and solutions

HCl (12 M)
Trimethylchlorosilane (1% in toluene)
NaOH pellets
Mobile phase I

Butan-1-ol/pyridine/acetic acid (concentrated)/water (60:40:3:30 v/v)

Mobile phase II

Pyridine/water (4:1 v/v)

Ninhydrin reagent

See page 386

H_2SO_4 (concentrated)
Methanol
Glucosamine standard solution

Procedure

Subsample replication, extraction and assay procedures for soil glucosamine are identical to those for diaminopimelic acid except for the following details. Soil samples are incubated with 10 ml HCl (12 M) at 25°C under N_2 for 48 h, then diluted with distilled water (30 ml) and hydrolysed under N_2 at 100°C for 6 h. Filtrates of hydrolysates are made up to 100 ml with distilled water, 10 ml subsamples deacified and aliquots chromatographed with volumes of authentic glucosamine solution for 48 h on Whatman No. 1 paper with solvent ratios of 60:40:3:30 (v/v). Glucosamine spots developed over 4–5 days. For homogeneity testing, chromatography is carried out in the second dimension with pyridine:water (4:1 v/v).

Discussion

The glucosamine losses during the extraction and analysis accounted for about 8%.

Isolation and identification of DNA from soil

A. Saano
K. Lindström

Most traditional methods developed for the detection of microorganisms in soil suffer from the lack of specificity or sensitivity, or require the isolation of the target organisms. In theory, methods based on isolation and identification of the nucleic acids of target organisms overcome these problems. In practice, however, we still lack the ideal protocol for fully making use of the unique properties of the nucleic acids and of the molecular methods for microbial identification in soil.

Two different techniques for isolation of DNA from soil can be carried out:

1. The cell extraction method.
2. The direct lysis method.

In the first technique, the extraction of microbial cells from soil precedes the DNA extraction, whereas in the second technique DNA is extracted directly from soil. The protocols described in this section are based on direct extraction and we refer to recent reviews (e.g. Saano and Lindström 1990; Trevors 1992) for more information on the different approaches.

All methods aim at getting a high yield of DNA that is pure enough for molecular analysis by the DNA–DNA hybridization, the restriction fragment length polymorphism analysis or the amplification by the polymerase chain reaction. Humic and clay compounds in many soils inhibit these reactions and the presence of colloids in soil render the extraction of pure DNA problematic. For this reason an extensive purification step in the DNA isolation protocols is necessary.

Freshly inoculated as well as indigenous, even unculturable, microorganisms have been detected by means of DNA isolation, amplification and/or hybridization with suitable probes and primers. However, it is important to take into account that freshly inoculated cells, recently cultivated on rich laboratory media, are likely to behave differently from indigenous organisms occupying soil niches.

Isolation of DNA from soil

Three methods based on direct extraction of DNA in soil will be described (Selenska and Klingmüller 1991; Tsai and Olson 1991; Picard et al 1992). These methods represent three different approaches aiming at the same goal: the highest yield of extracted DNA sufficiently pure to allow the identification of its genetic origin.

The isolation of DNA in soil uses the following steps: (1) the lysis of the cells; (2) the separation of DNA from other cell components such as polysaccharides and proteins; (3) the release of DNA from soil particles and purification of the DNA extract from soil constituents; and (4) the precipitation of DNA. Steps 2–4 do not always proceed in succession, and steps 2 and 3 may need to be repeated after step 4.

The method of Selenska and Klingmüller (1991)

Principle of the method

Selenska and Klingmüller (1991) use sodium dodecyl sulphate (SDS) at 70°C with shaking, and repetitive centrifugation for the lysis of the cells, separation of DNA from cell components and for the release of DNA from soil particles. Polyethyleneglycol (PEG) has been used for precipitation of DNA.

Materials and apparatus

Water bath at 70°C, shaking
Cooled centrifuge with the ability to spin volumes of at least 25 ml at 8000*g*
Ultracentrifuge (180,000*g*), at least 5 ml tubes
Dialysis facilities

Chemicals and solutions

Na_2HPO_4 (120 mM), pH 8 with 1% SDS, 5 ml/sample
Na_2HPO_4 (120 mM), pH 8, 12 ml/sample
NaCl 5 M, approximately 3 ml/sample
PEG 6000, approximately 4.5 g/sample
Tris–HCl (10 mM)–EDTA (1 mM) (TE), 3.7 ml/sample; for dialysis a few litres of autoclaved TE
CsCl 4.1 g/sample
Ethidium bromide (5 mg ml^{-1}), 0.5 ml/sample
Phenol, Tris–HCl-saturated, pH 8

Procedure

Suspend 2 g wet weight of soil in 5 ml of 120 mM Na_2HPO_4, pH 8 with 1% SDS.

Incubate in a shaking water bath at 70°C for 1 h.

Centrifuge for 15 min at 2800*g* at 10°C. Store the supernatant at 4°C.

Resuspend the pellet in 6 ml Na_2HPO_4 120 mM, pH 8.

Incubate in a shaking water bath at 70°C for 20 min.

Repeat third to fifth steps. Repeat third step once more.

Collect the three supernatants in one tube at 4°C. Centrifuge 30 min at 8000*g*.

Transfer the supernatant into a fresh tube.

Add 1/10 vol. 5 M NaCl.

Add PEG 6000 to 15%. Mix and leave overnight at 4°C.

Centrifuge for 10 min at 5000*g* at 4°C.

Resuspend the pellet in 3.7 ml TE. Add 4.1 g CsCl, dissolve and add 0.5 ml ethidium bromide 5 mg ml^{-1}.

Centrifuge in 5 ml tubes for 16 h at 180,000*g*.

Collect the DNA band and dialyse it for 2 h against TE at 4°C. Extract ethidium bromide with Tris-saturated phenol and continue the dialysis overnight.

Discussion

Different methods can be used after the penultimate step to purify the DNA pellet, e.g. phenol–chloroform extraction or hydroxyapatite column. Selenska and Klingmüller use the traditional CsCl–ethidium bromide gradient ultracentrifugation (for detailed description of this technique, see Ausubel et al 1989).

If the ultracentrifugation step is omitted, this method is easy to perform, and it does not require many devices and rare chemicals.

Handling of ethidium bromide must be done in a ventilated hood with protective gloves. Retrieval of the DNA band after ultracentrifugation demands special care and some training, and does not always proceed successfully.

The whole procedure takes about 1.5 days, and with ultracentrifugation, 2.5 days.

The procedure gives yields are up to 100 μg DNA per 2 g soil with a molecular weight of about 25 kb, which is susceptible to treatment with at least EcoRI, PstI, BamHI, HindIII and Taq polymerase, as well as being suitable for Southern blotting and hybridization. However, the inhibiting effects of soil constituents have not been sufficiently well studied and may cause unexpected problems. The crude DNA extract should be purified by other means, especially if ultracentrifugation is omitted.

The method of Tsai and Olson (1991)

Principle of the method

Tsai and Olson (1991) use lysozyme at 37°C for the lysis of the cells, SDS and three cycles of freezing and thawing at −70°C/+65°C for the lysis of the cells and for the separation of DNA from cell components, phenol–chloroform treatment for the separation of DNA from the cell components and the release of DNA from soil particles. Elutip is used for the release of DNA from soil particles and isopropanol for the precipitation of DNA.

Materials and apparatus

Shaker at 150 rev min^{-1}
Centrifuge with the ability to spin tubes with volumes of at least 10 ml at 10,000*g*
Water bath adjustable to 37°C and 65°C
Facilities for −70°C (dry ice–ethanol bath recommended)
Elutip-d column (Schleicher & Schüll) attached to a Schleicher & Schuell NA010/27 0.45 μm pore size cellulose prefilter
Vacuum drier

Chemicals and solutions

Na_2HPO_4 (120 mM), pH 8, 4 ml/sample
NaCl (0.15 M), Na_2EDTA, pH 8 with 15 mg lysozyme ml^{-1}, 2 ml/sample
NaCl (0.1 M), Tris–HCl (0.5 M), pH 8, SDS 10%, 2 ml/sample
Phenol, Tris–HCl (pH 8)-saturated, 3.5 ml/sample
Chloroform (chloroform:isoamylalcohol, 24:1), 4 ml/sample
Isopropanol, 2 ml/sample
Tris–HCl (20 mM)–EDTA (1 mM) (TE), pH 8, 100 μl/sample
RNase, A, 20 μg/sample

Procedure

Suspend 1 g wet weight soil in 2 ml of 120 mM Na_2HPO_4, pH 8

Shake for 15 min at 150 rev min^{-1}

Centrifuge for 10 min at 6000*g*

Wash the pellet once more with Na_2HPO_4 and repeat step three

Resuspend the pellet in 2 ml of 0.15 M NaCl, Na_2EDTA, pH 8 with 15 mg lysozyme ml^{-1}, incubate for 2 h in a water bath at 37°C and agitate every 20–30 min

Add 2 ml of 0.1 M NaCl, 0.5 M Tris–HCl, pH 8, SDS 10%, and mix

Run three cycles of freezing and thawing at −70°C in a dry ice–ethanol bath/+65°C water bath.

Add 2 ml of Tris–HCl (pH 8)-saturated phenol and vortex

Centrifuge for 10 min at 6000*g*

Collect 3 ml of the upper aqueous phase, mix with 1.5 ml phenol and 1.5 ml chloroform–isoamylalcohol

Centrifuge for 10 min at 6000*g*

Collect 2.5 ml of the upper aqueous phase and mix with 2.5 ml chloroform–isoamylalcohol

Centrifuge for 10 min at 6000*g*

Collect 2 ml of the upper aqueous phase, mix with 2 ml of cold isopropanol, let DNA precipitate at −20°C for at least 1 h

Centrifuge 10 min at 10,000*g* and dry the pellet in a vacuum at room temperature

Resuspend the pellet in 100 μl of TE, pH 8

Add RNase to a final concentration of 0.2 μg/μl and incubate for 2 h at 37°C

Purify the DNA sample with Elutip-d column according to the instructions of the manufacturer (optional)

Discussion

The method does not require expensive apparatus or rare chemicals and can be run through in 1 day.

Handling of phenol, chloroform and isopropanol must be done in a properly ventilated hood using protective gloves. When the number of samples is high, the successive extraction with phenol and chloroform becomes exhausting.

Yields are easily over 10 μg of DNA g^{-1} of sample and are mostly more than 23 kb in length.

Elutip purification is optional; it may be necessary for enzyme manipulations as well as to improve hybridization efficiency; on the other hand, it may reduce the yield by 40%.

The method of Picard et al (1992)

Principle of the method

Picard et al (1992) use no enzymes or detergents whatsoever, but sonicate, freeze and thaw the samples at −196°C/+100°C for the lysis of the cells, separation of DNA from the cell components and for the release of DNA from soil particles. Elutip is used for the release of DNA from soil particles and ethanol for the precipitation of DNA.

Materials and apparatus

Ultrasonicator with a titanium microtip
Table-top centrifuge (Eppoendorf type) capable of at least 12,000*g*
Microwave oven (optional)
Thermos flask for liquid nitrogen
Stove and a pot for boiling water
Elutip-d columns (Schleicher & Schuell)

Chemicals and solutions

Tris (50 mM), Na_2EDTA (20 mM), pH 8, NaCl (100 mM) and polyvinyl pyrrolidone 1% (w/v, Sigma) (TENP), 1.1 ml/sample
Liquid nitrogen
Ethanol

Procedure

Suspend 100 mg soil in 500 μl TENP.

Sonicate at a power setting of about 15 W with a titanium microtip for 5 min at 50% of active cycles.

Centrifuge for 1 min at 12,000*g*. Remove the supernatant, set aside and store.

Place the tube with the pellet in a microwave oven, heat five times for 1 min at 900 W, then suspend in 100 μl TENP.

Apply three successive thermal shocks with liquid nitrogen/boiling water, for 10 min, to the suspension.

Centrifuge for 1 min at 12,000*g*. Remove the supernatant, set it aside and store.

Resuspend the pellet in 100 μl TENP, then centrifuge for 1 min at 12,000*g*. Remove the supernatant, set aside and store. Repeat this step three more times.

Pool the supernatants and purify with three successive Elutip-d columns as specified by the manufacturer.

Precipitate the DNA with 2 volumes of 94% ethanol.

Discussion

This the easiest of the three methods described here; only the handling of liquid nitrogen and the boiling water requires some care. No harmful detergents are needed.

Using a small sample size (100 mg) allows a greater number of samples to be handled at one time. The availability of more than one ultrasonicator will greatly reduce the time needed for this method.

A yield of about 50 μg DNA/g of soil is high. However, because of the small sample size, DNA from one isolation will not be sufficient to carry out many DNA hybridizations.

Sonication and thermal shock treatments are the efficient parts of this procedure; microwave treatment gives only residual amounts of DNA. Sonication shears DNA down to 500–1000 bp fragments compared with the DNA of over 23–25 kb long resulting from the two other methods. Obviously it is difficult to set up a universal method applicable for every soil and microorganism. Therefore, before making conclusions about the efficiency of a particular method with respect to a certain inoculant, the isolation of DNA should not be started too soon after the inoculation. Selenska and Klingmüller (1991) began the isolation of DNA 70 days after the inoculation of *Enterobacter agglomerans* harbouring Tn5 in its genome. The method, together with subsequent DNA hybridization, managed to reveal the presence of Tn5 in the soil sample when the inoculant itself was no longer detectable with viable count testing. The efficiency of recovery of total DNA of *Pseudomonas luteola* V55 and *Pseudomonas putida* VNM43 from soil samples was originally tested 30 min after inoculation with the method of Tsai and Olson (1991). The applicability of this method for successful isolation of total DNA of *Rhizobium galegae* HAMBI 1174 has been tested for soil samples inoculated 30 min, 6 days and 15 days before the isolation (Saano and Lindström 1992). Picard et al (1992) started the isolation of total DNA from soil samples 1 h after the inoculation with *Agrobacterium tumefaciens*.

The methods presented above are not directly comparable with each other in the sense of quantitative isolation of specific DNA. Before a decision between the three methods is made, it is advised that several factors are considered: the soil type (see the original publications), the size of the soil sample, the desired size of the DNA yielded, the availability of the apparatus and reagents, and the simplicity of the procedure.

In contrast to the direct extraction of DNA presented here, some methods begin with separation of the bacterial fraction from other soil components. Pillai et al (1991) based their technique on sucrose gradient centrifugation, and, Jacobsen and Rasmussen (1992) on using cation-exchange resin.

Methods for the characterization of DNA from soil

Before the characterization of the isolated soil DNA is started, its concentration should be evaluated. Because of the humic compounds present in the DNA solution, the A_{260} spectrophotometric measurement does not give reliable values. Therefore, it is recommended that an agarose gel (0.7–1.2%) electrophoresis with DNA standards should be run. Usually the soil DNA samples can be loaded directly into the wells of the gel; the soil compounds serve both as a load to bring the DNA down to the bottom of the well and as a coloured marker for following the electrophoresis. The humic compounds seem to push ethidium bromide away from DNA in the electric field, so it is better to treat the gel with ethidium bromide after the run rather than to add it into the gel and the running buffer before the electrophoresis.

DNA–DNA hybridization

Soil DNA samples can be tested by dot blotting and Southern blotting, as any other DNA samples [for general outlines, see Ausubel et al (1989) or Sambrook et al (1989)]. The humic compounds present in soil DNA samples seem to block strongly the immobilization on to nylon membranes of DNA obtained from pure cultures. This has been observed for both dot and Southern blotting of both total bacterial and plasmid DNA. This effect, which can decrease the amount of immobilized DNA by more than 100-fold, has been found for both [^{32}P]dCTP-labelled and Digoxigenin-dUTP (Boehringer-Mannheim GmbH, Germany)-labelled DNA. It has also been found that resin purification (Magic DNA Clean-up Columns, Promega, USA) of soil DNA solution improves the immobilization of soil DNA (Saano et al 1993).

Dot blotting

Principle of the method

For dot blots it is recommended that a vacuum device is used to achieve regular rounded dots. Without a device the best results will be produced if a piece of dry Whatman 3MM filter paper is placed beneath the membrane, which should be dampened with 2 × SSC just before pipetting the samples. If the soil DNA sample is heavily contaminated with soil compounds, they may fill the membrane pores and hinder blotting. In this case there is no other way than to purify the sample or to dilute it. Nitrocellulose and nylon membranes of different manufacturers may cause differences in the performance of the blots (Saano and Lindström 1992). No further immobilization step of DNA is needed after the transfer on to positively charged nylon membranes.

Materials and apparatus

Stove and a pot for boiling water
Table-top centrifuge (Eppendorf type)
Mechanical vacuum pump or a water aspirator connected to a sink
Bio-Dot blotting manifold device (Bio-Rad, USA) or an analogue

Chemicals and solutions

2 × SSC (0.3 M NaCl, 30 mM sodium citrate, pH 7)
Nylon membrane 0.45 μm pore size, positively charged
Whatman 3MM filter paper

Procedure

Dampen the membrane in 2 × SSC and place it in the blotting device according to the instructions of the manufacturer.
Denature e.g. 50 ng of the sample DNA in a microtube by boiling for 3 min. Cool the tube for 1 min in an ice-water bath.
Centrifuge for 30 s at high speed to gather the condensate. Apply vacuum to the membrane. Pipette the samples. Wait till all the liquid has been sucked from the wells.
Close the valve between the blotting device and the pump, and shut down the vacuum.
Disconnect the blotting device from the pump and slowly open the valve. Open the blotting device and remove the membrane.
Air-dry the membrane by placing it on a sheet of Whatman 3MM filter paper.
Optional: immobilize the DNA by placing the membrane under 254 nm UV for 1–3 min at a 15 cm distance, or alternatively, wrap the membrane loosely between two sheets of Whatman 3MM filter paper and bake for 2 h at 80°C.

Discussion

Only moderate vacuum should be applied to the membrane so that 20 μl of water pipetted in the bottom of a well of the blotting device disappears within 1–2 s. Stronger vacuum may damage the membrane or lead to uneven dots.

Some of the liquid sucked from the well may run along the membrane surface when the membrane is removed from the device. This can be avoided by lifting the membrane with two forceps from the opposite sides keeping it in a horizontal position. Some DNA may easily disperse with the liquid on the membrane surface and cause hybridization background problems later. Therefore, it is recommended that small sample volumes are used.

Southern blotting

Principle of the method

Southern blotting of soil DNA can be done at least with a vacuum device, but the traditional paper towel capillary blotting as well as electroblotting should also work. At least the electrophoretic mobility of soil DNA does not seem to differ from that of pure culture DNA. Here the vacuum blotting procedure is presented using NaOH-mediated transfer, as it is the easiest and fastest to perform. Prior to blotting, treat the gel with UV light for 2–3 min on a 302 nm UV-transilluminator to improve the transfer of large-sized DNA.

Materials and apparatus

Blotting device (VacuGene, Pharmacia, Sweden, or an analogue)
Mechanical vacuum pump (VacuGene Blotting Pump 2016, Pharmacia, Sweden, or an analogue)

Chemicals and solutions

NaOH, 0.4 M
Nylon membrane 0.45 μm pore size, positively charged
2 × SSC (0.3 M NaCl, 30 mM sodium citrate, pH 7)
Whatman 3MM filter paper

Procedure

Cut a piece of membrane smaller than the gel by at least 0.5 cm from each side. Cut a membrane-sized window in the plastic folio, which will cover the porous support in the blotting device. Check that there are no holes elsewhere in the folio. Prepare the blotting device and adjust it to the pump. Immerse the membrane in 0.4 M NaOH for some seconds, then place it within the window of the plastic folio. Place the gel on to the membrane so that there will be no bubbles under the gel or leakages of air from any of the sides of the gel. The gel wells can be placed within the blotting area if they are not damaged. Apply 60–100 cm H_2O vacuum. If there are leakages, these values cannot be reached and the setup should be checked. After some minutes it should be possible to see (or feel with your fingers), how the gel over the window is sucked downwards forming a flat basin. Pipette 2–10 ml (depending on the size of the window) of 0.4 M NaOH into the basin, filling also the wells of the gel. Blot for 45–60 min adding 0.4 M NaOH every 15 min. Shut down the vacuum. Remove the gel together with the membrane from the blotting support and turn it bottom up. Detach the membrane from the gel, rinse it in 2 × SSC and place on a sheet of Whatman 3 MM filter paper.

Discussion

The bigger the margin between the gel and the membrane size, the easier it will be to place the gel properly on the membrane. Therefore, it is wise to plan the electrophoresis so that the side lanes in the gel are left without important samples.

Traditionally, prior to the transfer of the DNA, the gel has first been treated with 0.25 M HCl for depurination of DNA, then the DNA has been denatured by a 1 h treatment with NaOH, whereafter the gel has been neutralized. The transfer itself has been performed using high salt buffer. However, the NaOH-based transfer presented above is much simpler and less expensive. The only prerequisite is the availability of nylon membrane.

Probe labelling, hybridization (reassociation) and detection

Principle of the method

Non-radioactive hybridization and detection with the Digoxigenin technique can be used for the analysis of soil DNA blots. Soil compounds do not cause non-specific probe-binding problems or non-specific enzyme–substrate binding when the blot is washed properly under high stringency conditions. Only non-radioactive hybridization and detection is described here; for radioactive techniques, the basic manuals should be consulted (Ausubel et al 1989; Sambrook et al 1989).

Materials and apparatus

Oven with a temperature range of 40–80°C
Platform shaker
Thermosealer for plastic folio (optional)

Chemicals and solutions

This list is based on the instructions for a non-radioactive DNA Labelling and Detection Kit (Boehringer-Mannheim GmbH, Germany).

Hexanucleotide mixture, 2 µl/labelling of up to 1 µg of DNA

dNTP labelling mixture, 2 µl/labelling
Klenow enzyme (2 units μl^{-1}), 1 µl/labelling
Anti-digoxigenin-AP Fab fragments (750 units ml^{-1}), 1 µl 10 ml^{-1} of detection buffer 1 (see below)

Nitroblue tetrazolium (NBT) salt 75 mg/ml in 70% (v/v) dimethylformamide, 4.5 µl ml^{-1} of detection buffer 3 (see below)

X-phosphate 50 mg ml^{-1} in 100% dimethylformamide, 3.5 µl ml^{-1} of detection buffer 3 (see below).

Optionally, instead of nitroblue tetrazolium salt and X-phosphate: Lumigen PPD 10 mg ml^{-1} in detection buffer 3 (see below), or Lumi-Phos 530 (ready-to-use mixture).

Blocking reagent 50 g/bottle
20 × SSC: 3 M NaCl, 0.3 M sodium citrate, pH 7 (20°C)

Hybridization solution: 5 × SSC, blocking reagent 1%, *N*-lauroylsarcosine sodium salt 0.1% (w/v), SDS 0.02% (w/v)

Washing solution 1: 2 × SSC, 0.1% SDS (w/v)

Washing solution 2: 0.1 × SSC, 0.1% SDS (w/v)

Detection buffer 1: Tris base 0.1 M, NaCl 0.15 M, pH 7.5 (20°C)

Detection buffer 2: 1% blocking reagent dissolved in detection buffer 1. Adjust pH to 7.5 after adding the blocking reagent

Detection buffer 3: Tris base 0.1 M, NaCl 0.1 M, $MgCl_2$ 0.05 M, pH 9.5 (20°C). Optional for luminescent substrates of alkaline phosphatase only: X-ray film and a cassette, or, for example, Polaroid 55 (low light sensitivity but high resolution, for getting both negatives and positives) with a cassette, or Polaroid 667 (fast as X-ray film, only for positives) with a cassette. For treating X-ray films, developing and fixing solutions are needed; for Polaroid films only Na_2SO_3 18% (w/v) is needed. Cassettes for Polaroid 55 and Polaroid 667 are different from each other.

Procedure

Labelling of the probe DNA

Measure approximately 1 µg of probe DNA in a microtube and dilute with water up to 15 µl. Denature DNA by boiling for 3 min. Chill in an ice-water bath. Collect the condensate by brief centrifugation. Add 2 µl of hexanucleotide mixture and 2 µl of the dNTP labelling mixture. Add 1 µl of Klenow enzyme and mix gently. Incubate for 1–20 h (the longer the better) at 37°C. Pre-hybridize the filter or membrane for 1 h by placing it with the hybridization solution (without a probe) at 68°C (this is high stringency, find the best temperature for the hybridization specificity required).

Hybridization assay

Denature the probe DNA by boiling it for 3 min and chill in an ice-water bath. Collect the condensate by centrifuging briefly. Discard the pre-hybridization solution. Dilute the probe DNA with fresh hybridization solution to about 2.5 ml 100 cm^{-2} filter and place the membrane in the probe DNA solution. Allow to hybridize for 3 h at 68°C. Collect the probe DNA solution and store it at −20°C for possible reuse.

Washing of the blot

Wash twice for 5 min with 2 × SSC, 0.1% SDS at room temperature with shaking and twice for 30 min with 0.1 × SSC, 0.1% SDS at 68°C with shaking.

Detection of the hybrids

Immerse the blot in detection buffer 1 for 1 min and discard the buffer. Incubate the blot in detection buffer 2 for 30 min and discard the buffer. Dilute the anti-digoxigenin-AP in buffer 1 (1:10 000), cover the blot with it (about 2.5 ml 100 cm^{-2}) and incubate for 30 min. Discard the buffer. Wash the blot twice for 15 min in buffer 1 with shaking. Immerse the blot in detection buffer 3 for at least 1 min and discard the buffer. Dilute NBT and X-phosphate in detection buffer 3 (4.5 µl ml^{-1} and 3.5 µl ml^{-1}, respectively), cover the blot with the mixture. Allow to stand in the dark. Check the formation of a coloured precipitate every now and then to stop the reaction at the appropriate moment. Sometimes the results are visible in 20 min, other times after 3 days. Stop the reaction by removing the blot from the colour mixture and by rinsing it with distilled water. If the reaction was complete within a few hours, the colour mixture can be reused within the next 1–2 days.

The membrane is now ready for photography. Alternatively, allow the membrane to dry and store it in a dry (e.g. sealed in a plastic bag) and dark place at room temperature. Optionally, dilute Lumigen PPD in detection buffer 3 (1:100) and incubate the membrane with it for 5 min in the dark. Collect the liquid and store at 4°C in the dark for possible reuse. Place the membrane for a few seconds on a sheet of dry Whatman 3MM filter paper. Seal the membrane in a plastic bag. Pre-incubate for 5–15 min at 37°C. Expose to X-ray or polaroid film.

Discussion

In contrast to the kit instructions, it is not necessary to separate unincorporated nucleotides from the labelled DNA molecules. Practice shows that the binding capacity for nucleotides is weak compared with that for DNA molecules so that Dig-dUTP does not create an unspecific hybridization background, at least not during high stringency hybridization. Omitting separation efforts saves time and gives a higher probe yield.

In contrast to the kit instructions, it is recommended that Tris base is used instead of Tris–HCl for the preparation of the detection buffers. It is cheaper and the pH needed is easily reached with HCl.

The amounts of blocking reagent, 20 × SSC, hybridization solution, washing solutions, or detection buffers depend on the size and amount of blotting membrane treated at the same time. The easiest way of handling the membranes in all the steps is in plastic thermostable boxes with lids. A more economic way is to enclose the membranes in plastic bags corresponding to the size of the largest blot at pre-hybridization and hybridization, as well as for the last step of detection. When a hybridization oven is used, the membranes are enclosed in plastic tubes.

The detection of the hybrids with NBT and X-phosphate, or with the luminescent substrates of alkaline phosphatase Lumigen PPD, or with Lumi-Phos 530 is equally convenient and has approximately the same sensitivity.

NBT and X-phosphate come as ready-to-use vials with the kit, but for frequent use it is cheaper to prepare them individually.

Restriction fragment length polymorphism

Principle of the method

Both Selenska and Klingmüller (1991) and Tsai and Olson (1991) have fragmented their soil DNA with several restriction enzymes. In principle, this might already enable identification of specific soil microorganisms with known restriction patterns in gel electrophoresis restriction fragment length polymorphism (RFLP). However, a much better resolution can be reached if a Southern blot is hybridized with a certain gene probe (Tsai and Olson 1991). This technique produces a pattern of much fewer and clearly distinguishable bands, which can be specific not only for species, but also for strains and even individual genes. There are no specific protocols for RFLP as such, they are just combinations of the usual techniques in molecular biology: restriction enzyme digestion of DNA, agarose gel electrophoresis, visualization of the pattern, Southern blotting of the gel, DNA–DNA hybridization of the blot and detection of the hybrid bands. The following is a description of a simple procedure for RFLP. It is applicable directly to such restriction enzymes as EcoRI and HindIII, which are very broadly used and inexpensive.

Materials and apparatus

Incubator at 37°C
Power supply
Electrophoresis chamber
UV transilluminator (302 nm)
Camera with UV filter (e.g. Kodak Wratten 22)

Chemicals and solutions

Agarose, electrophoresis grade

Electrophoresis buffer, TAE solution

0.04 M Tris acetate, 0.002 M EDTA, pH 8.5;
or TBE solution:
0.089 M Tris base, 0.089 M boric acid, 0.002 M EDTA, pH 8. Prepare 10 × stock solutions for frequent use

Ethidium bromide solution

(0.5 mg ml^{-1}), for gels or stain solutions dilute 1:1000

Lambda DNA digested with BstEII or HindIII (MW marker for electrophoresis runs)
10 × loading buffer

20% Ficoll 400, 0.1 M Na_2EDTA, pH 8, 1% SDS, 0.25% bromophenol blue

Procedure

Measure 1 µg of soil DNA in a tube. Add 2 µl 10 × enzyme incubation buffer. Add about 10 units of the restriction enzyme. Dilute with sterile distilled water up to 20 µl and mix gently. Incubate for at least 1 h at 37°C. Add 3 µl of 10 × loading buffer and 7 µl of water, and mix. Load the lambda DNA marker (100 ng) and the samples into 0.8–1.2% agarose gel. Run the electrophoresis for 1–2 h at 10 V cm^{-1}. Stain the gel for 20 min with ethidium bromide solution. Destain the gel for 15 min with distilled water (ethidium bromide is attached covalently to DNA but not to agarose). Visualize the DNA on a UV transilluminator and document the restriction pattern by taking a photograph. Make a Southern blot, hybridize and detect the hybrid band pattern.

Discussion

Ethidium bromide is a mutagen and potential carcinogen. Ethidium bromide solutions should be handled wearing protective gloves and weighing of the powder should be done in a ventilated hood.

Polymerase chain reaction

Principle of the method

The general principles of the polymerase chain reaction (PCR), developed by Saiki et al (1988), have been described elsewhere (Sambrook et al 1989). PCR can improve the sensitivity of detection by several orders of magnitude. Every particular DNA sequence to be identified with the help of PCR must first be sequenced (at least part from both of its ends), then appropriate oligonucleotide primers should be found (e.g. computer programmes PRIMER of the Whitehead Institute for Biomedical Research, OLIGO 4.1 of Wojciech Rychlik, NBI, or PCGENE of IntelliGenetics Inc.). The optimal annealing temperatures depend on the melting temperature of the primers. DNA denaturation is achieved at about 95°C and strand synthesis (DNA extension) is performed at 72°C according to the temperature optimum of the polymerase.

The other factors in PCR, such as concentrations of the salts and nucleotides as well as the number of cycles, must be taken into account when the PCR protocols are designed. Phenolic and humic compounds present in soil extracts are strong inhibitors of polymerases (Tsai and Olson 1992a), and the presence of large amounts of foreign DNA in the sample may inhibit PCR (Picard et al 1992).

Many different approaches have been invented to avoid the inhibition of polymerase by the soil compounds. Most of these studies have been conducted using known inoculants. Pillai et al (1991, 1992) used a two-step PCR for the identification of a Tn5 mutant of *Rhizobium leguminosarum* bv. phaseoli, where an aliquot from 25 cycles was used as a template for a further 25 cycles of amplification. The limit of the detection sensitivity was below 10^2 colony forming units (CFU). Tsai and Olson (1992b) used Sephadex G-200 spun columns to purify soil DNA extracts and they reached the sensitivity of detection of *Escherichia coli* 35346 of less than 70 CFU g^{-1} soil. Picard et al (1992) found that formamide improves the sensitivity of PCR by improving annealing of the primers. The sensitivity of detection of *Agrobacterium tumefaciens* was 10^4 CFU g^{-1} soil. Smalla et al (1993) used CsCl precipitation and centrifugation, potassium acetate precipitation and centrifugation, isopropanol precipitation, glass milk purification, and spermine, for the purification of the DNA extracts before PCR. They reached a sensitivity of 10^3 inoculant *Pseudomonas fluorescens* R2f (RP4:pat) per gram of soil. PCR can also be used for the identification of naturally occurring microorganisms in soil. Bruce et al (1992) used primers both for "universal" eubacterial 16s rRNA, and for mercury resistance (*mer*) genes. The identity of the amplified fragments was confirmed by Southern blot DNA–DNA hybridization. Reysenbach et al (1992) have found that using primers complementary to universally conserved regions may lead to preferential amplification of some DNA templates. This would lead to a biased view of the community structure. To avoid this, the authors recommend addition of 5% (w/v) acetamide to a PCR mixture. It should be emphasized strongly that the following protocol is just to give the basic idea of a PCR procedure. For each particular template and its primers, the conditions should be optimized.

Materials and apparatus

Thermocycler
Electrophoresis chamber
Power supply
UV transilluminator (302 nm)
Camera with UV filter (e.g. Kodak Wratten 22)

Chemicals and solutions

10 × PCR buffer

100 mM Tris–HCl (pH 8.3), 500 mM KCl, Triton X-100 1%

25 mM $MgCl_2$
10 mM 4dNTP (deoxynucleoside triphosphate mixture)
50 µM primer mixture
Thermostable DNA polymerase
Sterile mineral oil

Procedure

Measure 39.5 µl of sterile, ultrapure water in a PCR tube. Add 5 µl of 10 × PCR buffer, 3 µl $MgCl_2$ (final concentration 1–3 mM), 1 µl 4dNTP (final concentration 200 µM), 0.5 µl primer mixture (final concentration 50 nM), 0.5 µl thermostable DNA polymerase (2.5 units) and 50 µl sterile mineral oil. Heat the mixture up to 85°C ("hot start" to prevent non-specific annealing of the primers). Add 0.5 µl soil DNA into the reaction mixture (see that the tip of the pipette goes through the oil layer) and incubate 3 min at 95°C (initial denaturation), 1 min at the annealing temperature and 1 min at 72°C. Run 25–60 cycles: 1 min at 95°C, 1 min at the annealing temperature and 1 min at 72°C. Cool quickly and store at −20°C. Run an agarose gel electrophoresis of an aliquot. Stain the gel with ethidium bromide, destain with distilled water and visualize the results on an UV transilluminator.

Discussion

It is important to set a negative control sample, for which the whole procedure is run through as described, except that no template DNA is added in the PCR mixture. PCR can be very sensitive to minute amounts of contaminant DNA. A positive control sample, with pure culture DNA as a template, is good for comparison with the bands resulting from the soil samples.

If the sensitivity of ethidium bromide staining is not good enough, a more sensitive alternative is to run a polyacrylamide gel and visualize the bands with silver staining (Bassam et al 1991).

The use of antibody techniques in soil

J.W.L. Van Vuurde
J.M. van der Wolf

Serological assays for the detection of target microorganisms are based on a reaction of one or more antigenic determinants, present on molecules of the target, with antibodies directed against these determinants.

Polyclonal antisera, produced in, for example, rabbits after immunization with the target microorganism, contain a variety of antibodies against typical and less typical determinants of the target. The latter can lead to cross-reactions with non-target organisms. At cell level, the degree of cross-reactivity will depend on the incidence of cross-reacting antibodies in the antiserum and the prevalence of shared determinants on the cross-reacting strains. The degree of cross-reactivity is often below the detection level of the test. Monoclonal antibodies (Goding 1986) are directed against only one antigenic determinant and strongly decrease the risk of cross-reactions with non-target organisms.

Many examples are known for the specific detection of target bacteria below the species level (e.g. Benedict et al 1989). The use of serological techniques in soil studies was most successful for bacteria. Some promising applications for the detection of fungi were reported using monoclonal antibodies (Dewey 1992) but the specificity below the species level was often poor.

For bacteria, the most important serological methods (see Hampton et al 1990; Klement et al 1990) are:

1. Agglutination assays based on an agglutination reaction of the target cells by linkage through antibodies. The detection threshold is a target bacterial density of 10^7 or more cells per millilitre and the specificity is often lower than for other serological assays. The method is suitable for simple and fast screening of colonies, or pure cultures, but positive results should be confirmed with other techniques.
2. Agar diffusion tests are based on the formation of precipitin bands in the agar due to a reaction between agar-diffusible antigens with antibodies against an antigenic determinant of that compound. The detection threshold of Ouchterlony double diffusion (ODD) is *c.* 10^8 cells per millilitre. The banding pattern is often typical for the cross-reacting strains. ODD can be applied for identification purposes and to study serological relationships between immunologically related strains.
3. Immuno(magnetic) trapping can be used for the selective concentration of target bacteria from sample extracts. Immunoisolation is based on selective trapping of target cells onto a solid phase prior to plating. In immunomagnetic separation, paramagnetic beads are coated with antibodies and can be used to concentrate target cells from environmental samples (Morgan et al 1991). The method can also be used as a purification step before PCR, in order to avoid DNA extraction, which is often necessary to remove PCR-inhibiting compounds.
4. Labelled-antibody assays are based on the use of antibodies conjugated with a reporter molecule. Fluorescent and luminescent

dyes, gold beads, biotin and enzymes are mainly used for conjugation. Gold-labelled antibodies are used for light and electron-microscopical detection. Enzyme-linked immunosorbent assay (ELISA) is very suitable for automated routine screening of samples, but has a detection level of *c.* 10^5 cells per millilitre. Coupling peroxidase-labelled specific monoclonal antibodies, as used in the conventional ELISA, with a chemoluminescence reaction, the sensitivity of detection can be increased to 10^3 cells per ml extract (Schloter et al 1992). For detection of target bacteria in soil, assays based on immunofluorescence (IF) microscopy offer the best possibilities for reliable and sensitive detection of target bacteria and will be discussed further.

The sensitivity of IF-cell staining equals that of PCR for the detection of target bacteria in soil extracts at a level of 10^3 to 10^4 cells per ml (Smalla et al 1993). IF can be used for *in situ* studies of target cells, e.g. on roots or substrate particles. For soil extracts, the method is easy and fast; small volumes of samples are dried on a microscope object glass, the bacteria are fixed on the glass and stained with fluorescent antibodies in a direct or an indirect way (see Hampton et al 1990; Klement et al 1990). Under the fluorescence microscope, fluorescent target cells will be visible against a dark background in a microscope field, which may contain thousands of non-target bacteria. The method can be used for quantitative studies, but does not discriminate between culturable and non-culturable cells. A direct confirmation for positive cells is not possible.

Immunofluorescence colony-staining (IFC) has been developed for the detection of culturable bacteria at a detection level between 10^1–10^3 per millilitre of undiluted sample extract (Van Vuurde and Roozen 1990; Leeman et al 1991). IFC allows the direct confirmation of IFC-positive colonies and can be efficiently applied for *in situ* detection of culturable target bacteria, e.g. on roots (Underberg and Van Vuurde 1990). The principle, procedure and applications of IFC are based on Van Vuurde (1990) and Van Vuurde and van der Wolf (1994).

Principle of the method

The advantages of isolation and of serology are combined in IFC. Agar-mixed sample plating, so-called pour plating, allows detection of the target at a 100–1000 times lower detection threshold by reduced interference of background organisms as compared with surface plating. The immunofluorescent staining differentiates the target colonies from the background.

Semi-selective media selected for high recovery rates of the target are recommended for pour plating. Pour plates are incubated for 1–2 days until small colonies are formed in the medium. The detection level is determined by the diameter of the Petri dish and the selectivity of the medium. Wells (16 mm diameter) of 24-well tissue-culture plates are used for routine application of a large series of samples. The detection level of this format is *c.* 10^2 CFU ml^{-1} for bacteria with average colony growth.

Bacterial cells (target or cross-reacting) can be isolated directly from IFC-positive colonies for confirmation by isolation or PCR (see Fig. 9.3). IFC allows the possible elimination of cross-reacting bacteria by making the medium selective against these bacteria.

Materials and apparatus

24-well tissue-culture plates
Ventilated incubator (*c.* 40°C) or warm air drier
Incubator for microbial growth
Microscope with incident blue light [filter block for fluorescein isothiocyanate (FITC)] and low magnification objectives with a high numeric aperture (e.g. 4×/NA 0.12 to 10×/NA 0.22) and a low magnification eyepiece (e.g. 6.3×)

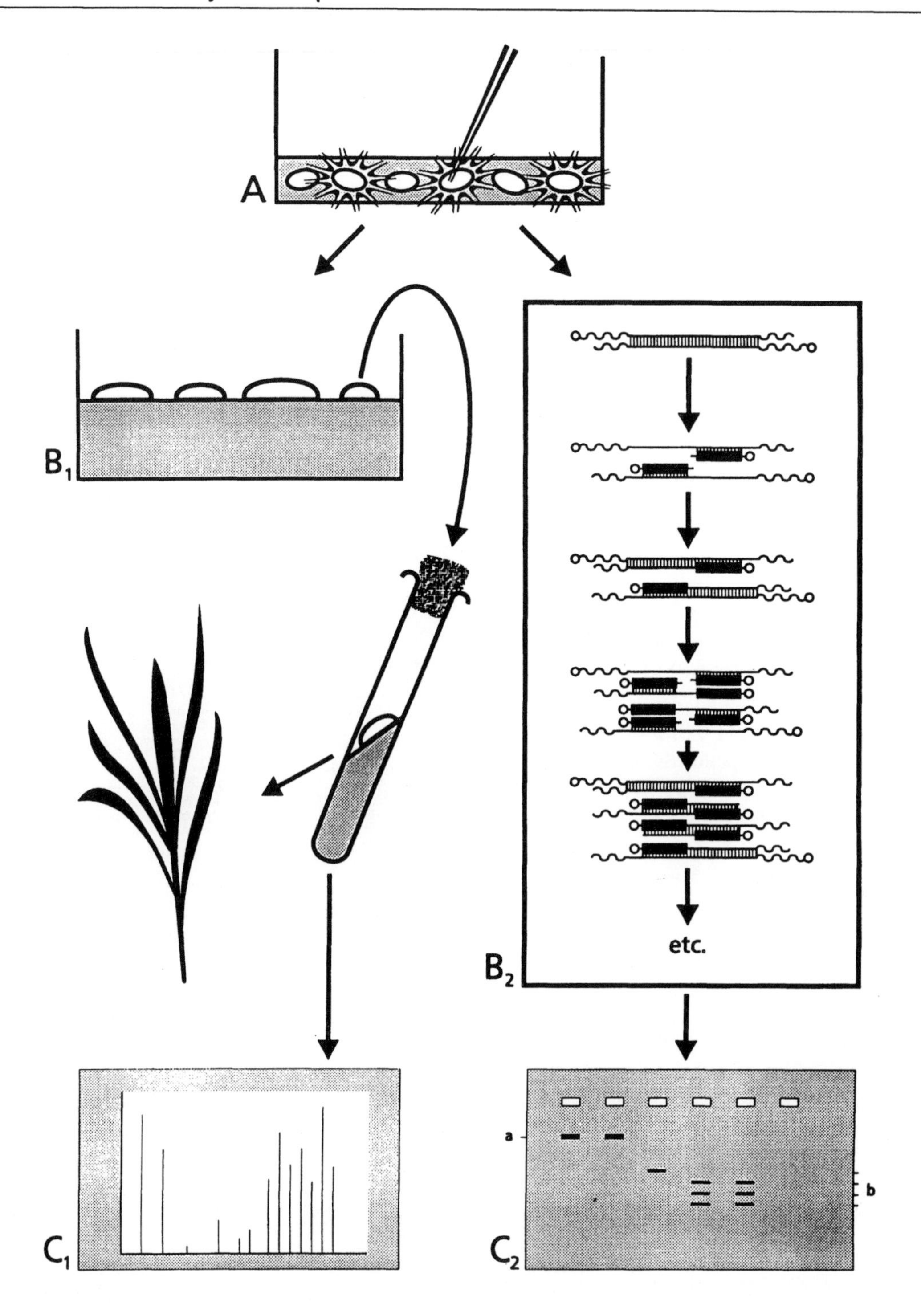

Figure 9.3. Scheme of fluorescent colonies in pour plates (A) and verification by isolation from IFC-positive colonies to obtain a pure culture (B_1), e.g. for testing of ecological behaviour or characterization by fatty-acid profiling (C_1), or to perform PCR (B_2) for characterization by gel electrophoresis (C_2). (Scheme reproduced with permission from IPO-DLO Annual Report, 1992.)

ELISA plate shaker (Titertek, Flow Laboratories)

Chemicals and solutions

Antiserum conjugated with FITC
Agar medium
0.01 M phosphate-buffered saline pH 7.4 with 0.1% Tween 20 (PBST) for conjugate dilution and washing

Procedure for sample extracts

Add each test sample (maximum 100 µl) to a 16 mm diameter well of a 24-well tissue culture plate. Add 300 µl liquified agar medium at 45°C to each well. Swirl the plate gently clockwise and counterclockwise for a homogeneous distribution of the bacteria in the agar, or use an ELISA plate shaker.

Incubate at optimal conditions for the target bacterium. Stop the incubation when pinhead-sized colonies are formed (1 or 2 days for most bacteria).

Open the multiwell plate and dry the agar layer completely by blowing warm (*c.* 40°C) air over the agar surface. A ventilated incubator provides regular drying and so helps to prevent the medium from cracking.

Add 400 µl of diluted FITC-conjugated antiserum to the well and incubate the agar film overnight at room temperature. The conjugate is diluted with PBST.

Remove the non-bound conjugate from the wells. Wash by two 15 min rinses with 1 ml of PBST. Remove excess PBST from the well after the last washing. Removal of liquid from the wells can be done with a pipette or a tube connected to a vacuum aspirator.

Place the well with the stained agar film under the microscope. Inspect the agar film for colonies with green fluorescence.

The conjugate titre will be determined by the experimental conditions such as incubation time and temperature, and by optical conditions such as the light source, the filter system and the numeric aperture (NA) of the objective and eyepiece magnification. The titre and the working dilution of the conjugate for target colonies should be determined using two-fold dilution steps for the microscopical setup to be used. Often a working dilution is found between 50 and 100 times the dilution of the conjugate. As the staining brilliance of colonies of pure cultures of the target is often less intense than that of target colonies against a background of non-target colonies, both preparations should be used for titration.

Incubation with FITC conjugate can often be reduced to 4 h. Diluted conjugate from a stock containing glycerine may result in a weaker staining.

Standard dried reference preparations of target colonies in agar medium at a density of *c.* 10^3 CFU cm^{-2} can be made in 9 cm Petri dishes. Small parts (e.g. 5 × 5 mm) can be cut out and used for antiserum titration, comparative experiments and reference controls. These standard preparations can be stored dry for several months at low temperature in the presence of silica gel.

The following controls are recommended to check the staining: standard dried reference preparations of target or non-target colonies from pure cultures, and from sample extracts spiked with and without target bacteria. Check for autofluorescent colonies in unstained replicate test samples (autofluorescence can in general be eliminated by optimizing the FITC filter system).

Discussion

For confirmation, if needed, isolate from the IF-positive colonies by puncturing the colony under the microscope with a fine needle or glass capillary tube (by hand or with a micromanipulator) and plating on a medium or by PCR (Van Vuurde and Van der Wolf 1995).

The detection level is dependent on the format of the test, and can be varied by using the wells of e.g. a 6, 12, or 48-well tissue-culture plate or Petri dishes with diameters of 60 or 90 mm. A final agar concentration of 0.8–1.0% after mixing with the sample is recommended.

Dried agar films with colonies can be stored for several months in a container with silica gel at 4°C before IFC staining is done.

For IFC with FITC conjugates, objectives of a standard light microscope can be used. The highest colony-fluorescence intensity is obtained for objectives with a high NA and a relatively low eyepiece magnification. Most 4× objectives have a sufficient working distance to look in the well from above. For higher objective magnifications, special long working distance objectives are needed (Nikon 10× and 20× LWD), or the well should be inspected through the bottom of the inverted plate.

Use a 10× objective to check for small fluorescent colonies.

Reading of IFC-stained agar films can best be done directly after staining. However, stained preparations can often be stored dry for several months at 4°C and still show brilliant fluorescence when they are rewetted with PBST before observation.

IFC can be used to discriminate between culturable and non-culturable cells. With IF cell staining, both culturable and non-culturable cells will be stained. When IFC is used for thin agar layers at a magnification level at which single cells can be observed, CFU can be distinguished from non-culturable cells by their ability to form a microcolony. The method can be used for target cells in a sample extract and on thin roots.

Mix the sample and the agar, and prepare a layer *c.* 1 mm thick, e.g. on a microscope slide. Incubate to allow the target bacterium to form microcolonies (5–50 cells per colony). For most plant pathogenic bacteria, a 4–16 h incubation is suitable (check, for example, with phase contrast). Dry the agar layer and perform IFC as described above. Observe the preparation at *c.* 400× magnification. Best results are obtained with a 40× long distance objective with coverglass correction (e.g. Olympus LWDCDPL 40/0.55 with 1.99 mm working distance).

Double staining in IFC. Double staining of two different target bacteria with conjugates with a different fluorescence colour allows their detection in the same preparation. Texas Red is recommended in combination with FITC. A special dual filter block (Omega) is available for simultaneous observation of both targets in the same preparation.

Isolation from IFC-positive colonies for pure culture identification. In general, bacteria survive in the colonies during the preparation and staining. Isolation directly from fluorescent colonies is possible with a needle or fine capillary tube, and pure cultures can be obtained after further purification. This method can be used for reliable confirmation of the target colonies but can also be used for the isolation of cross-reacting bacteria. The contamination with non-target cells often makes a laborious and time-consuming purification of the target bacteria from the IFC-positive colony necessary. Some bacterial species show low recoveries from colonies after IFC (e.g. *Erwinia chrysanthemi*). Recovery can be improved for this bacterium by eliminating the drying step and increasing the washing after staining to two times 2 h.

Confirmation of fluorescent colonies with the PCR. Isolation to obtain pure cultures from IFC-positive colonies is laborious and may fail at high densities of non-target colonies. Therefore, an alternative routine method was developed for rapid characterization of fluorescent colonies based on PCR (Van der Wolf et al 1994).

IFC-positive colonies can be confirmed

with PCR even when the cells sampled from the IFC-positive colony are dead. PCR is a rapid detection method that can be performed in one working day. If specific primers are available, the amplification reaction will not be influenced by contamination of target cells with non-target cells.

The procedure is carried out as follows. Dried agar films should be rewetted in phosphate-buffered saline in advance of the collection of fluorescent colonies from the agar films. Punch the fluorescent colonies from the agar using a Pasteur pipette with the end bent at 90°. Transfer the punches to an Eppendorf reaction tube with demineralized water or a buffer suitable for the PCR. The volume of liquid should be twice the sample volume for the PCR, but at least 15 μl. Boil the punches for 5–15 min. Centrifuge for 3 min in an Eppendorf centrifuge at maximal speed. Use the supernatant for the PCR. The amplification product can be analysed by standard techniques (dot blot, gel electrophoresis, RFLP and Southern blotting).

PCR can be inhibited by non-target colonies in agar punches. This may incidentally cause false-negative reactions. During the incubation with FITC-conjugated antibodies, contamination of the agar surface may occur with target cells or even colonies due to secondary growth. The cells and secondary colonies will only be visible at higher magnification. They can cause positive reactions in PCR when sampling colonies as a negative control in the same preparation. Secondary growth of the target can be reduced by adding a bactericide to the conjugate buffer, e.g. 0.05 mg ml^{-1} NaN_3.

References

Aaronson JM (1981) Cell wall chemistry, ultrastructure, and metabolism. In: Biology of Conidial fungi, vol 2. Cole GT, Kendrick B (eds). Academic Press, New York, pp. 459–507.

Anderson JPE, Domsch KH (1974) Use of selective inhibitors in the study of respiratory activities and shifts in bacterial and fungal populations in soil. Ann Microbiol 24: 189–1947.

Anderson JPE, Domsch KH (1975) Measurement of bacterial and fungal contributions to respiration of selected agricultural and forest soils. Can J Microbiol 21: 314–322.

Anderson JPE, Domsch KH (1982) Decomposition of ^{14}C- and ^{15}N-labelled microbial cells in soil. Soil Biol Biochem 14: 461–467.

Ausubel FM, Brent R, Kingston RE, Moore DD, Seidman JG, Smith JA, Struhl K (1989) Current Protocols in Molecular Biology. Wiley, New York.

Bååth E, Lundgren B, Söderström B (1981) Effects of nitrogen fertilization on the activity and biomass of fungi and bacteria in a podzolic soil. Zbl Bakt Hyg I Abt Orig C2: 90–98.

Balkwill DL, Leach FR, Wilson JT, McNabb JF, White DC (1988) Equivalence of microbial biomass measures based on membrane lipid and cell wall components, adenosine triphosphate, and direct counts in subsurface aquifer sediments. Microb Ecol 16: 73–84.

Bartnicki-Garcia S (1968) Cell wall chemistry, morphogenesis, and taxonomy of fungi. Ann Rev Microbiol 22: 87–108.

Bassam BJ, Caetano-Anolles G, Gresshoff PM (1991) Fast and sensitive silver staining of DNA in polyacrylamide gels. Anal Biochem 196: 80–83.

Benedict AA, Alvarez AM, Berestecky J, Imanaka W, Mizumoto CY, Pollard LW, Mew TW, Gonzalez CF (1989) Pathovar-specific monoclonal antibodies for Xanthomonas campestris pv. oryzae and Xanthomonas campestris pv. oryzicola. Phytopathology 79: 322–328.

Bitton G, Marshall KC (1980) Adsorption of Microorganisms to Surfaces. Wiley, New York.

Bligh EG, Dyer W (1959) Rapid method of rapid lipid extraction and purification. Can J Biochem Physiol 37: 911–917.

Bruce KD, Hiorns, WD, Hobman JL, Osborn AM, Strike P, Ritchie DA (1992) Amplification of DNA from native populations of soil bacteria by using the polymerase chain reaction. Appl Environ Microbiol 58: 3413–3416.

Christie WW (1989) Silver ion chromatography using solid-phase extraction columns packed with a bonded-sulfonic acid phase. J Lipid Res 30: 1471–1473.

Costerton JW, Irvin RT, Cheng KJ (1981) The bacterial glycocalyx in nature and disease. Ann Rev Microbiol 35: 299–324.

Davis MW, Lamar RT (1992) Evaluation of methods to extract ergosterol for quantification of soil fungal biomass. Soil Biol Biochem 24: 189–198.

DeLong EF, Wickham GS, Pace NR (1989) Phylogenetic stains: Ribosomal RNA-based probes for the identification of single cells. Science 241: 1360–1363.

Dewey FM (1992) Detection of plant invading fungi by monoclonal antibodies. In: Techniques for the Rapid Detection of Plant Pathogens. Duncan JM, Torrance L (eds). Blackwell Scientific Publications, Oxford, pp. 47–62.

Durska G, Kaszubiak H (1980a) Occurrence of m-diaminopimelic acid in soil. I. The content of m-diaminopimelic acid in different soils. Pol Ecol Stud 6: 189–193.

Durska G, Kaszubiak H (1980b) Occurrence of m-diaminopimelic acid in soil II. Usefulness of m-diaminopimelic acid determination for calculations of the microbial biomass. Pol Ecol Stud 6: 195–199.

Durska G, Kaszubiak H (1980c) Ocurrence of m-diaminopimelic acid in soil. III. m-Diaminopimelic acid as the nutritional component of the soil microorganisms. Pol Ecol Stud 6: 201–206.

Durska G, Kaszubiak H (1983) Occurrence of bound muramic acid and m-diaminopimelic acid in soil and comparison of their contents with bacterial biomass. Acta Microb Pol 3: 257–263.

Ellwood DC, Melling J, Rutter P (eds) (1979) Adhesion of Microorganisms to Surfaces. New York, Academic Press.

Fazio SD, Mayberry WR, White DC (1979) Muramic acid assay in sediments. Appl Environ Microbiol 38: 349–350.

Fazio SA, Uhlinger DJ, Parker JH, White DC (1982) Estimations of uronic acids as quantitative measure of extracellular polysaccharide and cell wall polymers from environmental samples. Appl Environ Microbiol 43: 1151–1159.

Findlay RH, White DC (1983) Polymeric β-hydroxy alkylonates from environmental samples and *Bacillus megaterium.* Appl Environ Microbiol 45: 71–78.

Findlay RH, Pollard PC, Moriarty DJW, White DC (1985) Quantitative determination of microbial activity and community nutritional status in estuarine sediments: evidence for a disturbance artifact. Can J Microbiol 31: 493–498.

Frankland JC, Lindley DK, Swift MJ (1978) A comparison of two methods for the estimation of mycelial biomass in leaf litter. Soil Biol Biochem 10: 323–333.

Gehorn MJ, Davis JD, Glen AS, White DC (1984) Determination of the gram-positive bacterial content of soils and sediments by analysis of teichoic acid components. J Microbiol Meth 2: 165–176.

Goding JW (1986) Monoclonal Antibodies: Principles and Practice. Academic Press, London.

Goosens H, Rijpstra WIC, Düren RR, De Leeuw JW, Schenck PA (1985) Bacterial contribution to sedimentary organic matter; a comparative study of lipid moieties in bacteria and recent sediments. Adv Org Geochim 10: 683–696.

Glaser L, Burger MM (1964) The synthesis of teichoic acids. III. Glucosylation of polyglycerophosphate. J Biol Chem 239: 3187–3191.

Grant WD, West AW (1986) Measurement of bound muramic acid and m-diaminopimelic acid and glucosamine in soil: evaluation as indicators of microbial biomass. J Microbiol Meth 6: 47–53.

Guckert JB, White DC (1986) Phospholipid, ester-linked fatty acid analysis in microbial ecology. In: Perspectives in Microbial Ecology. Megusar F, Kantar G (eds). Proceedings of the Fourth International Symposium on Microbial Ecology, Ljubljana. American Society for Microbiology, Washington, DC, pp. 455–459.

Hahn D, Dorsch M, Stackebrandt E, Akkermans ADL (1989) Synthetic oligonucleotide probes for identification of *Frankia* strains. Plant Soil 118: 211–219.

Hampton RO, Ball E, De Boer S (eds) (1990) Serological Methods for Detection and Identification of Viral and Bacterial Plant Pathogens. APS Press, Minnesota.

Hicks RE, Newell SY (1983) An improved gas chromatographic method for measuring glucosamine and muramic acid concentrations. Anal Biochem 128: 438–445.

Hicks RE, Newell SY (1984) Comparison of glucosamine and biovolume conversion factors for estimating fungal biomass. Oikos 42: 355–360.

Jacobsen CS, Rasmussen OF (1992) Development and application of a new method to extract bacterial DNA from soil based on separation of bacteria from soil with cation exchange resin. Appl Environ Microbiol 58: 2458–2462.

Jantzen E, Bryn K (1985) Whole-cell and lipopolysaccharide fatty acids and sugars of gram-negative bacteria. In: Chemical Methods in Bacterial Systematics. Goodfellow M, Minnikin DE (eds). Academic Press, London, pp. 145–171.

Jenkinson DS, Ladd NJ (1981) Microbial biomass in soil: measurement and turnover. In: Soil Biochemistry, vol 5. Paul EA, Ladd JN (eds). Marcel Dekker, New York, pp. 415–471.

Johnson BN, McGill WB (1990) Comparison of ergosterol and chitin as quantitative estimates of mycorrhizal infection and *Pinus contora* seedling response to inoculation. Can J Forest Res 20: 1125–1131.

Kates M (1964) Bacterial lipids. Adv Lipid Res 2: 17–90.

Kato T (1986) Sterol-biosynthesis in fungi, a target for broad spectrum fungicides. In: Sterol Biosynthesis Inhibitors and Anti-feeding compounds. Haug G, Hoffman H (eds). Springer Verlag, Berlin, pp. 1–25.

King JD, White DC (1977) Muramic acid as a measure of microbial biomass in estuarine and marine samples. Appl Environ Microbiol 33: 777–783.

Klement Z, Rudolph K, Sands DC (1990) Methods in Phytobacteriology. Akademia Kiado, Budapest.

Korner J, Laczko E (1992) A new method for assessing soil microorganism diversity and evidence of vitamin deficiency in low diversity communities. Biol Fertil Soils 13: 58–60.

Lang WK, Glasey K, Archibald AR (1982) Influence of phosphate supply on teichoic acid and teichuronic acid content of *Bacillus subtilis* cell walls. J Bacteriol 151: 367–375.

Lechevalier MP (1977) Lipids in bacterial taxonomy – a taxonomist view. CRC Crit Rev Microbiol 5: 109–210.

Lee C, Howarth RW, Howes BL (1980) Sterols in decomposing *Spartina alterniflora* and the use of ergosterol in estimating the contribution of fungi to detrital nitrogen. Limonol Ocean 25: 290–303.

Leeman M, Raaijmakers JM, Bakker PAHM, Schippers B (1991) Immunofluorescence colony stainng for monitoring pseudomonads introduced in soil. In: Biotic Interactions and Soil-borne Diseases. Beemster ABR, Bollen BJ, Gerlagh M, Ruissen MA, Schippers B, Tempel A (eds). Elsevier, Amsterdam, pp. 374–380.

Lehninger LL (1977) Biochemie. Verlag Chemie, Weinheim, New York.

Lovley DR, Klug MJ (1982) Intermediary metabolism of organic matter in the sediments of a eutrophic lake. Appl Environ Microbiol 43: 552–560.

Matcham SE, Jordan BR, Wood DA (1985) Estimation of fungal biomass in a solid substrate reactor by three independent methods. Appl Microbiol Biotechnol 21: 108–112.

McKinley VL, Costerton JW, White DC (1988) Microbial biomass, activity, and community structure of water and particulates retrieved by backflow from a waterflood injection well. Appl Environ Microbiol 54: 1383–1393.

Millar WN, Cassida LE (1970) Evidence for muramic acid in the soil. Can J Microbiol 18: 299–304.

Mimura T, Romano J-C (1985) Muramic acid measurements for bacterial investigations in marine environments by high-pressure liquid chromatography. Appl Environm Microbiol 50: 229–237.

Morgan JAW, Winstanley C, Pickup RW, Saunders J (1991) Rapid immunocapture of Pseudomonas putida cells from lake water by using bacterial flagella. Appl Environ Microbiol 57: 503–509.

Moriarty DJW (1975) A method for estimating the biomass of bacteria in aquatic sediments and its application to trophic studies. Oecologia 20: 219–229.

Nichols PD, Guckert JB, White DC (1986) Determination of monounsaturated fatty acid double-bond position and geometry for microbial monocultures and complex consortia by capillary GC-MS of their dimethyl disulphide adducts. J Microb Methods 5: 49–55.

Newell SY, Miller JD, Fallon RD (1987) Ergosterol content of salt-marsh fungi: effect of growth condition and mycelial age. Mycologia 79: 688–695.

Odham G, Tunlid A, Westerdahl G, Marden P (1986) Combined determination of poly-β-hydroxyalkanoic and cellular fatty acids in starved marine bacteria and sewage sludge by gas chromatography with flame ionization or mass spectrometry detection. Appl Environ Microbiol 52: 905–910.

Parker JH, Smith GA, Fredrickson HL, Vestal JR, White DC (1982) Sensitive assay, based on hydroxy fatty acids from lipopolysaccharidelipid A, for gram-negative bacteria in sediments. Appl Environ Microbiol 44: 1170–1177.

Parkinson D (1982) Filamentous Fungi. In: Methods of Soil Analysis Part 2, Chemical and microbiological properties. Page AL, Miller RH, Keeney DR (eds). American Society of Agronomy, Inc., Madison, WI, pp. 949–967.

Picard C, Ponsonnet C, Paget E, Nesme X, Simonet P (1992) Detection and enumeration of bacteria in soil by direct DNA extraction and polymerase chain reaction. Appl Environ Microbiol 58: 2717–2722.

Pillai SD, Josephson KL, Bailey RL, Gerba CP, Pepper IL (1991) Rapid method for processing soil samples for polymerase chain reaction amplification of specific gene sequences. Appl Environ Microbiol 57: 2283–2286.

Pillai SD, Josephson KL, Bailey RL, Pepper IL (1992) Specific detection of rhizobia in root nodules and soil using the polymerase chain reaction. Soil Biol Biochem 24: 885–891.

Ratledge C, Wilkinson SG (1988) Microbial Lipids, vol 1. Academic Press, London.

Reysenbach A-L, Giver LJ, Wickham GS, Pace NR (1992) Differential amplification of rRNA genes by polymerase chain reaction. Appl Environ Microbiol 58: 3417–3418.

Rizza B, Tucker AN, White DC (1970) Lipids of *Bacteroides melaninogenicus*. J Bacteriol 101: 84–91.

Saano A, Kaijalainen S, Lindström K (1993) Inhibition of DNA, immobilization to nylon membrane by soil compounds. Microb Releases 2: 153–160.

Saano A, Lindström K (1990) Detection of rhizobia by DNA–DNA hybridization from soil samples: problems and perspectives. Symbiosis 8: 61–73.

Saano A, Lindström K (1992) Detection of Rhizobium galegae from one-gram soil samples by non-radioactive DNA–DNA hybridization. Soil Biol Biochem 24: 969–977.

Saddler JN, Wardlaw AC (1980) Extraction, distribution and biodegradation of bacterial lipopolysaccharides in estuarine sediments. Antonie van Leeuwenkoek J Microbiol 46: 27–39.

Saiki RK, Gelfand DH, Stoffel S, Scharf SJ, Higuchi R, Horn GT, Mullis KG, Erlich HA (1988) Primer-directed enzymatic amplification of DNA with a thermostable DNA polymerase. Science 239: 487–494.

Sambrook J, Fritsch EF, Maniatis T (1989) Molecular Cloning: A laboratory manual, 2nd edn. Cold Spring Harbor Laboratory, Cold Spring Harbor, NY.

Schleifer KH, Kandler O (1972) Peptidoglycan types of bacterial cell walls and their taxonomic implications. Bacteriol Rev 36: 407–477.

Schloter M, Bode W, Hartmann A, Besse F (1992) Sensitive chemoluminescence-based immunological quantification of bacteria in soil extracts with monoclonal antibodies. Soil Biol Biochem 24: 399–403.

Seitz LM, Mohr HE, Burroughs R, Sauer DB (1977) Ergosterol as an indicator of fungal invasion in grains. Cereal Chem 54: 1207–1217.

Seitz LM, Sauer DB, Burroughs R, Mohr HE, Hubbard ID (1979) Ergosterol as a measure of a fungal growth. Phytopathol 69: 1202–1203.

Selenska S, Klingmüller W (1991) DNA recovery and direct detection of Tn5 sequences from soil. Lett Appl Microbiol 13: 21–24.

Sieburth J McN (1975) Microbial Seascapes. Baltimore, University Park Press.

Singer SJ, Nicolson GL (1972) The fluid mosaic model of the structure of cell membranes. Science 175: 720–725.

Smalla K, Cresswell N, Mendonca-Hagler LC, Wolters A, van Elsas JD (1993) Rapid DNA extraction protocol from soil for polymerase chain reaction-mediated amplification. J Appl Bacteriol 74: 78–85.

Smith GA, Nickels JS, Kerger BD, Davis JD, Collins, SP, Wilson JT, McNabb JF, White DC (1986) Quantitative characterization of microbial biomass and community structure in subsurface material: a prokaryotic consortium responsive to organic contamination. Can J Microbiol 32: 104–111.

Steubing L (1970) Chemische Methoden zur Bewertung des mengenmäßgen Vorkommens von Bakterien und Algen im Boden. Zentralbl. Bakteriol Parasitenk Abt 2 124: 245–249.

Thompson GA (1972) Ether-linked lipids in molluscs.

In: Ether Lipids: Chemistry and Biology. Snyder F (ed.). New York, Academic Press, pp. 321–340.

Tollefson TS, McKercher RB (1983) The degradation of ^{14}C-labelled phosphatidyl choline in soil. Soil Biol Biochem 15: 145–148.

Tonn SJ, Gander JE (1979) Biosynthesis of polysaccharides by prokaryotes. Ann Rev Microbiol 33: 169–199.

Trevors JT (1992) DNA extraction from soil. Microb Releases 1: 3–9.

Tsai Y-L, Olson BH (1991) Rapid method for direct extraction of DNA from soil and sediments. Appl Environ Microbiol 57: 1070–1074.

Tsai Y-L, Olson BH (1992a) Detection of low numbers of bacterial cells in soils and sediments by polymerase chain reaction. Appl Environ Microbiol 58: 754–757.

Tsai Y-L, Olson BH (1992b) Rapid method for separation of bacterial DNA from humic substances in sediments for polymerase chain reaction. Appl Environ Microbiol 58: 2292–2295.

Tunlid A, Godham G (1986) Ultrasensitive analysis of bacterial signatures by gas chromatography/ mass spectrometry. Proceedings of the IVth ISME, pp. 447–454.

Tunlid A, White DC (1992) Biochemical analysis of biomass, community structure, nutritional status, and metabolic activity of microbial communities in soil. In: Soil Biochemistry, vol 7. Stotzky G, Bollag J-M (eds). Marcel Dekker, New York, pp. 229–262.

Tunlid A, Odham G, Findlay RH, White DC (1985) Precision and sensitivity in the measurement of ^{15}N enrichment in D-alanine from bacterial cell walls using positive/negative ion mass spectrometry. J Microbiol Meth 3: 237–245.

Underberg HA, Van Vuurde JWL (1990) In situ detection of Erwinia chrysanthemi on potato roots using immunofluorescence and immunogold staining. In: Klement Z (ed.). Proceedings of the 7th International Conference on Plant Pathogens and Bacteria, Budapest, 1989. Akademiai Kiado, Budapest, pp. 937–942.

Van der Wolf JM, Van Beckhoven JRCM, De Vries PhM, Raaijmakers JM, Bakker PAHM, Bertheau Y, Van Vuurde JWL (1994) Polymerase chain reaction for verification of positive colonies in immunofluorescence colony staining. Appl Environ Microbiol (submitted).

Van Vuurde JWL (1990) Immunofluorescence colony staining. In: Serological Methods for Detection and Identification of Viral and Bacterial Plant Pathogens. Hampton R, Ball E, De Boer S (eds). APS Press, St Paul, pp. 299–305.

Van Vuurde JWL, Roozen NJM (1990) Comparison of immunofluorescence colony-staining in media, selective isolation on pectate medium, ELISA and immunofluorescence cell staining for detection of Erwinia carotovora subsp. atroseptica and E. chrysanthemi in cattle manure slurry. Neth J Plant Path 96: 75–89.

Van Vuurde JWL, Van der Wolf JM (1995) Immunofluorescence colony-staining (IFC). In: Molecular Microbial Ecology Manual. Akkermans ADL, Van Elsas JD, De Bruin FJ (eds). Kluwer Academic Publishers, Dordrecht (in preparation).

Vestal JB, White DC (1989) Lipid analysis in microbial ecology. Bioscience 39: 535–541.

Ward TE, Frea JI (1979) Determining the sediment distribution of methanogenic bacteria by direct fluorescence antibody methodology. In: Methodology for Biomass Determinations and Microbial Activities in Sediments. Lichtfield CD, Seyfried PL (eds). ASTM STP 673. American Society for Testing and Materials, Philadelphia, pp. 75–86.

Ward JB (1981) Teichonioc acid and teichuronic acids: biosynthesis, assembly and location. Microbiol Rev 45: 211–243.

Wardle, Parkinson (1990) Soil Biol Biochem

Weete JD (1974) Fungal Lipid Biochemistry: Distribution and Metabolism. Plenum Press, New York.

Weete JD (1989) Structure and function of sterols in fungi. Adv Lipid Res 23: 115–167.

West (1986) Improvement of the selective respiratory inhibition technique to measure eukaryote: prokaryote ratios in soils. J Microbiol Methods 5: 125–138.

West AW, Grant WD, Sparling GP (1987) Use of ergosterol, diaminopimelic acid and glucosamine content of soils to monitor changes in microbial populations. Soil Biol Biochem 19: 607–612.

Wilkinson SG (1988) Gram-negative bacteria. In: Microbial Lipids, vol 1. Ratledge C, Wilkinson SG (eds). Academic Press, London, pp. 299–457.

White DC (1983) Analysis of microorganisms in terms of quantity and activity in natural environments. In: Microbes in Their Natural Environments. Slater JH, Whittenbury R, Wimpenny JWT (eds). Society of General Microbiology Symposium 34, pp. 37–66.

White DC, Tucker AN (1969) Phospholipid metabolism during bacterial growth. J Lipid Res 10: 220–233.

White DC, Davis WM, Nickels JS, King JD, Bobbie RJ (1979a) Determination of the sedimentary microbial biomass by extractable lipid phosphate. Oecologia 40: 51–62.

White DC, Bobbie RJ, King JD, Nickels JS, Amoe P (1979b) Lipid analysis of sediments for microbial biomass and community structure. In: Methodology for Biomass Determinations and Microbial Activities in Sediments. Lichtfield CD, Seyfrieds PL (eds). ASTM STP 673. American Society for Testing and Materials, Philadelphia, pp. 87–103.

Wollenweber H-W, Rietschel ET (1990) Analysis of lipopolysaccharide (lipid A) fatty acids. J Microbiol Meth 11: 195–211.

Zelles L (1988) The simultaneous determination of

muramic acid and glucosamine in soil by high-performance liquid chromatography with per-column fluorescence derivatization. Biol Fertil Soils 6: 125–130.

Zelles L, Bai QY (1993) Fractionation of fatty acids derived from soil lipids by solid phase extraction and their quantitative analysis by GC-MS. Soil Biol Biochem (in press).

Zelles L, Hund K, Stepper K (1987a) Methoden zur relativen Quantifizierung der pilzlichen Biomasse im Boden Z Pflanzenernäh Bodenkd 150: 249–252.

Zelles L, Hund K, Stepper K (1987b) Differenzierte Erfassung der Bioaktivität von Pilzen und Bakterien zweier Böden unter Fichte. Z Pflanzenernähr Bodenkd 150: 253–257.

Zelles L, Stepper K, Zsolnay A (1990) The effect of lime on microbial activity in spruce (*Picea abies L.*) forest. Biol Fertil Soils 9: 78–82.

Zelles L, Adrian P, Bai QY, Stepper K, Adrian MV, Fischer K, Maier A, Ziegler A (1991) Microbial activity measured in soils stored under different temperature and humidity conditions. Soil Biol Biochem 23: 955–962.

Zelles L, Bai QY, Beck T, Beese F (1992) Signature fatty acids in the phospholipids and lipopolysaccharides as indicators of microbial biomass and community structure in agricultural soils. Soil Biol Biochem 24: 317–323.

10 Field methods

The measurements of microbial biomass and activity under laboratory conditions supply useful information on the physiological state of microbial populations in soil. Such measurements, however, do not reflect the real situation under natural conditions. Generally, the objectives of field measurements are to quantify mineralization processes, and thereby to gain insight into how the nutrient minerals and organic matter of the soil can be more efficiently utilized and conserved. Furthermore, field methods provide information on the effect of environmental factors on soil microflora. In this chapter, up-to-date field methods are presented and discussed.

Methods in Applied Soil Microbiology and Biochemistry
ISBN 0–12–513840–7

Estimation of soil respiration

K. Alef

Measurement of CO_2 evolution rates (long-term assay)

(Anderson 1982)

Principle of the method

The method is based on the determination of CO_2 evolved from undisturbed soils. The NaOH solution is placed in an open glass jar above the soil surface and the area to be measured is covered with a metal cylinder closed at the upper end. After the incubation, the NaOH solution is removed and the CO_2 concentration is measured by titration.

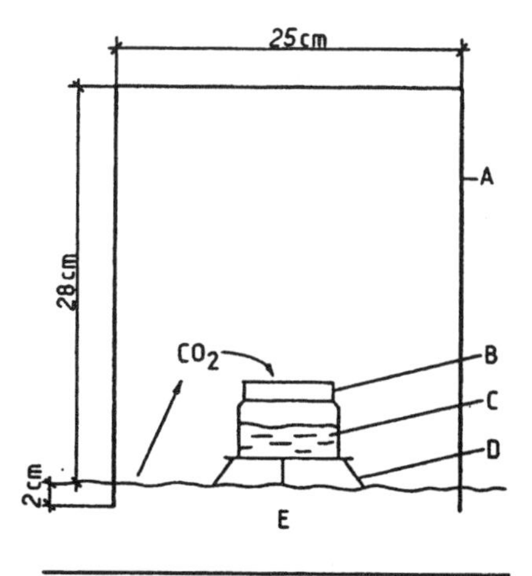

Figure 10.1. Long-term estimation of CO_2 production from the soil surface (Alef 1991; Anderson 1982): (A) Metal cylinder; (B) glass jar with screw cap; (C) NaOH solution; (D) tripod and (E) soil.

Materials and apparatus

Metal cylinders with one sealed end (Fig. 10.1)
Screw-capped glass jars (7 cm high, 6.5 cm diameter)
Tripods made of heavy metal or plastic

Chemicals and reagents

NaOH solution (1.0 M)
Barium chloride ($BaCl_2$, 3 M)
HCl (1.0 M)
Phenolphthalein indicator

1 g of phenolphthalein is dissolved in about 80 ml ethanol (95%) and brought up with ethanol (95%) to 100 ml.

Procedure

Pipette 20 ml of NaOH solution into a glass jar and place on a tripod located on the soil surface of the selected site. Immediately place the metal cylinder over the NaOH solution and press the edges about 2 cm into the surface of the soil. The system should be shielded from direct sunlight (e.g. using aluminium foil). After an incubation time of 24 h or more, the glass jar is removed (cap the jar for transport to the laboratory) and the NaOH solution is titrated as described earlier (see Chapter 5). Controls are performed by incubating the jars (containing NaOH solution) in the field in completely sealed metal cylinders.

Calculation

The CO_2 evaluation rates can be calculated as follows:

$$\text{C or } CO_2 \text{ (mg)} = (B - V)\, NE \quad (10.1)$$

where B is the HCl (ml) needed to titrate the NaOH solution from the control, V is the

HCl (ml) needed to titrate the NaOH solution in the jars exposed to the soil atmosphere, N = 1.0 (HCl normality) and E is the equivalent weight (22 for CO_2 and 6 for C). The data are expressed as milligrams of CO_2 per square metre per hour.

Discussion

A full standardization of this method has not yet been achieved. Results obtained by this method are controversial due to many modifications used in the assay. Underestimations and overestimations of the amount of CO_2 evolved have been reported (Wanner 1970; Kucera and Kirkhham 1971; Edwards and Sollins 1973; Edwards 1974).

The CO_2 evolution rates depend very strongly on the physical and chemical properties of the studied soils. Furthermore, the temperature as well as the water content affected these rates.

CO_2 evolution in the field assay is due to the respiration of microorganisms, animals, plant roots and abiotic CO_2 production.

Destruction of soil structure should be avoided. This causes temporary increases in CO_2 evolution.

Measurement of CO_2 evolution rates (short-term assay)

(Anderson 1982)

Principle of the method

The method is based on the collection of CO_2 evolved from known areas of the soil surface by the means of an apparatus (Fig. 10.2; Richter 1972), which consists of four small gas collectors, a gas analysis tube and a small bellows pump. In the gas

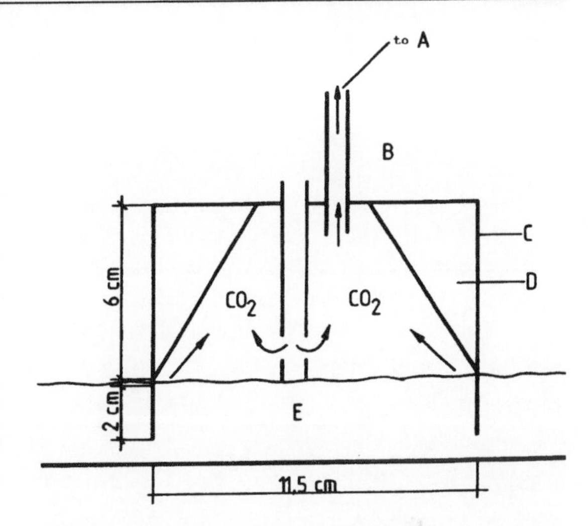

Figure 10.2. Short-term estimation of CO_2 production from the soil surface (Alef 1991; Anderson 1982): (A) pump; (B) copper tubes; (C) metal can; (D) plastic funnel; (E) soil.

analysis tube, CO_2 reacts with a hydrazine compound ($CO_2 + N_2H_2 \rightarrow NH_2COOH$), whose consumption is shown by a change in the redox indicator.

Materials and apparatus

Gas collectors

The four gas collectors cover a surface area of 400 cm^2 and have a total volume of about 1.1 l. The cans of each collector are shown in Fig. 10.2

Hand-operated bellows pump (100 ml volume, Drägerwerk AG Lübeck, Germany)
Stopwatch
Thermometer

Procedure

After selecting an appropriate site, press the edges of the four gas collectors 1–2 cm into the soil, then draw 1 l of the air (10 strokes of the pump) through the system. Break off the fused ends of a gas analysis tube, and insert it between the collectors and the pump. An arrow on each tube shows the proper direction of the gas flow. Draw 2 l of air through the system (20 strokes, about 400 s), read the volume of CO_2 present in the 2 l of gas directly from the tube by comparing the leading edge of coloration with the markings on the tube's surface. Then repeat the process to determine the volume of the ambient air. To do this disconnect the gas collectors and, using a fresh gas analysis tube, draw an air sample (2 l) from a height of 1 m above the soil surface. During the measurement, record the air temperature.

Calculation

Example

A 2 l sample of soil air and ambient air drawn within 400 s at 20°C contained 0.15% of CO_2 by volume. Under the same conditions, 2 l of ambient air contains 0.07% of CO_2 by volume. The total surface under the four collectors is 400 cm^2 (100 cm^2 under each collector).

Then 0.08% of CO_2 by volume or a total 0.0008×2000 ml = 1.6 ml CO_2 400 s^{-1} was evolved from the soil. This means 14.4 ml CO_2 h^{-1} 400 cm^{-2} or 360 ml CO_2 m^{-2} h^{-1}. With the use of simple gas laws, this value can be corrected to standard temperature (0°C) and pressure (101.3 kPa = 760 Torr) as follows:

$$\frac{(360 \text{ ml } CO_2 \text{ m}^{-2} \text{ h}^{-1}) \times (101.3 \text{ kPa}) \times (273°\text{C})}{(101.3 \text{ kPa}) \times (273°\text{C} + 20°\text{C})} = 335.4 \text{ ml } CO_2 \text{ m}^{-2} \text{ h}^{-1} \quad (10.2)$$

Since at standard conditions, 1 ml of CO_2 is equal to 1.96 mg of CO_2

$$(335.4 \text{ ml } CO_2 \text{ m}^{-2} \text{ h}^{-1}) \times \frac{1.196 \text{ mg } CO_2}{1 \text{ ml } CO_2} = 657 \text{ mg } CO_2 \text{ m}^{-2} \text{ h}^{-1} \quad (10.3)$$

Discussion

The measurements can be carried out with minimal disturbance of soil and within a short time.

Automatic bellows pumps are commercially available (Dräger Werk AG, Lübeck, Germany).

Measurement of CO_2 and O_2 concentrations at various soil depths

(Richter 1972; Anderson 1982)

Principle of the method

The method is based on the gas chromatographic analysis of small samples of the soil atmosphere drawn from the desired depths by means of gas sampling probes and gas-tight syringes.

Materials and apparatus

Sampling probe
Gas-tight syringes (5 ml); needles and rubber stoppers into which the needles can be inserted
Gas chromatograph equipped with TCD; molecular sieve of 5 Å column (2 m) for analysis of O_2; Porapak R column (2 m) for analysis of CO_2.

Procedure

The probe is pushed into the soil to the desired depth. Two millilitres of air are withdrawn and discarded by using a syringe (Richter 1972; Anderson 1982) without a needle inserted into the opening of the cannula within the probe. An additional 2 ml of air are withdrawn and the needle is fitted to the syringe. Immediately after that a 1 ml sample of air is discharged (to flush out the ambient air from the needle) and the needle is sealed by inserting it deep into a rubber stopper. The analysis of soil atmospheric samples is carried out by a gas chromatograph according to standard analytical

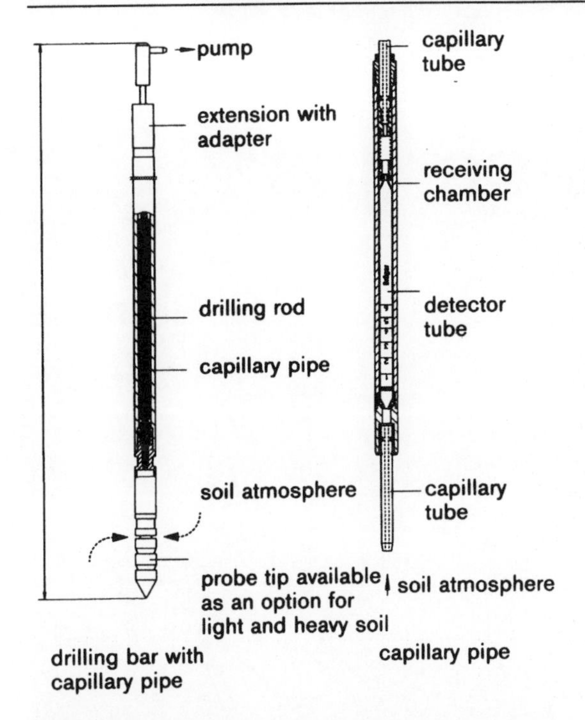

Figure 10.3. Dräger probe for collecting gas samples from soil.

procedures (He as carrier gas at flow rate of 60 ml min^{-1}, column temperature 25°C, detector temperature 50°C). A standard curve can be prepared from standard gases containing either 0.1 or 1% of CO_2 by volume or 10% of O_2 by volume.

Similarly, a probe (Fig. 10.3, Dräger, Lübeck, Germany) can also be used. Instead of the syringe, air samples are withdrawn using a small manual pump. The Dräger probe is usually applied for collecting air samples in contaminated sites. Chemical compounds present in the air sample are adsorbed on an active coal column. These compounds can then be analysed in the laboratory.

Discussion

The most difficult part of this technique is to obtain samples from the desired depths without contaminating them with the atmosphere from other depths or the surface.

In compacted soils a pilot hole should be made (use the slide hammer) and the probe inserted into the hole as quickly as possible to prevent excess atmospheric air from entering the hole (Roulier et al 1974).

The gas chromatographic determinations of CO_2 and O_2 are very sensitive.

Automated monitoring of biological trace gas production and consumption

R. Brumme
F. Beese

Soil acts as source or sink of many trace gases and plays an important role in global atmosphere/biosphere interactions (Andrae and Schimel 1989; Bouwman 1990). The quantification of the contribution of soil to the biogeochemical cycling of different trace gases requires automatic monitoring of the fluxes (Loftfield et al 1992), because temporal (diurnal as well as seasonal) and spatial variations prevent reasonable estimates of overall annual fluxes with simple approaches (Brumme and Beese 1992). Another purpose of automated field measurements is to improve or verify our knowledge about factors controlling trace gas emission.

Principle of the method

The method is based on the estimation of the gas concentration under field conditions in a covered soil (boxes, Fig. 10.4). The increase or decrease in the gas concentrations during the closure of the boxes can be used to calculate gas fluxes. An automated system, in which a personal computer controls the closing and opening of the chamber, the gas sampling in the chambers, the analysis of CO_2, N_2O and CH_4, and the calculation of the gas fluxes, enables the monitoring of diurnal, day-to-day and seasonal fluctuations.

Materials and apparatus

The chamber

The double-walled chamber is shown in

Figure 10.4. Double-walled chamber with motor driven lid (Loftfield et al 1992).

Fig. 10.4. It is made from Plexiglass (10 and 6 mm thick), and consists of an inner and outer part, and a lid. The area of each part is 0.25 m^2 corresponding to a gas volume of 50 l each. The inner part is the zone where the measurements are made. The outer part acts as a buffer zone to reduce disturbances of the gas

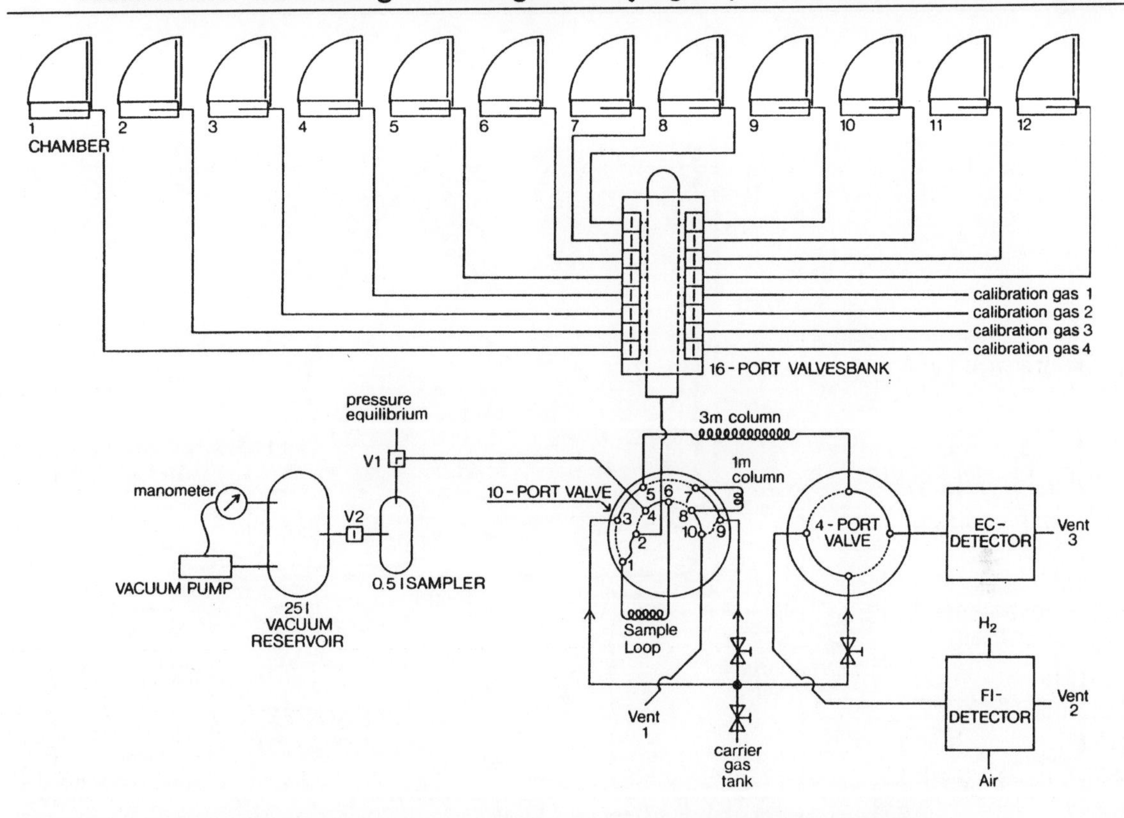

Figure 10.5. Gas plumbing for chamber sampling and gas chromatography analysis.

fluxes caused by wind action (Matthias et al 1980). To avoid damage to the root system, the chambers are placed on top of the forest floor or only inserted deep enough to achieve the horizontal placement required for the 2 cm water trap to seal the lid. Driven by a 12 V motor, the lid can be closed and opened at given time intervals. The lid has an orifice to ensure that no pressure or vacuum can be built up during the process of closing and opening. A magnetic contact controls the movement of the lid (see Fig. 10.6). Twelve chambers are connected to a gas chromatograph by heatable sampling tubes over a 16-port valve bank (Fig. 10.5). They consist of an inner Teflon tube (3 mm i.d.) and an outer polyethylene hose (9 mm i.d.). A heating wire is wound around the Teflon tube to prevent condensation of water.

Sixteen-port valve bank

Besides the ports of the 12 chambers, four ports for calibration gases are linked to the 16-port valve bank (Fig. 10.5).

Gas sampling

The gas in the chambers is sampled in a 0.5 l flask by opening a valve of the 16-port valve bank. The flask has been evacuated before by the opening of V2. In this way, 0.5 l of gas are drawn into the flask over a 5 ml sample loop within the gas chromatograph. This amount is enough to exchange the gas in the tube system five times. After the sample loop has been filled, the tension is equalized to the atmosphere by turning V1. The gas sample now is introduced into the carrier gas stream of the gas chromatograph by shifting the 10-port valve.

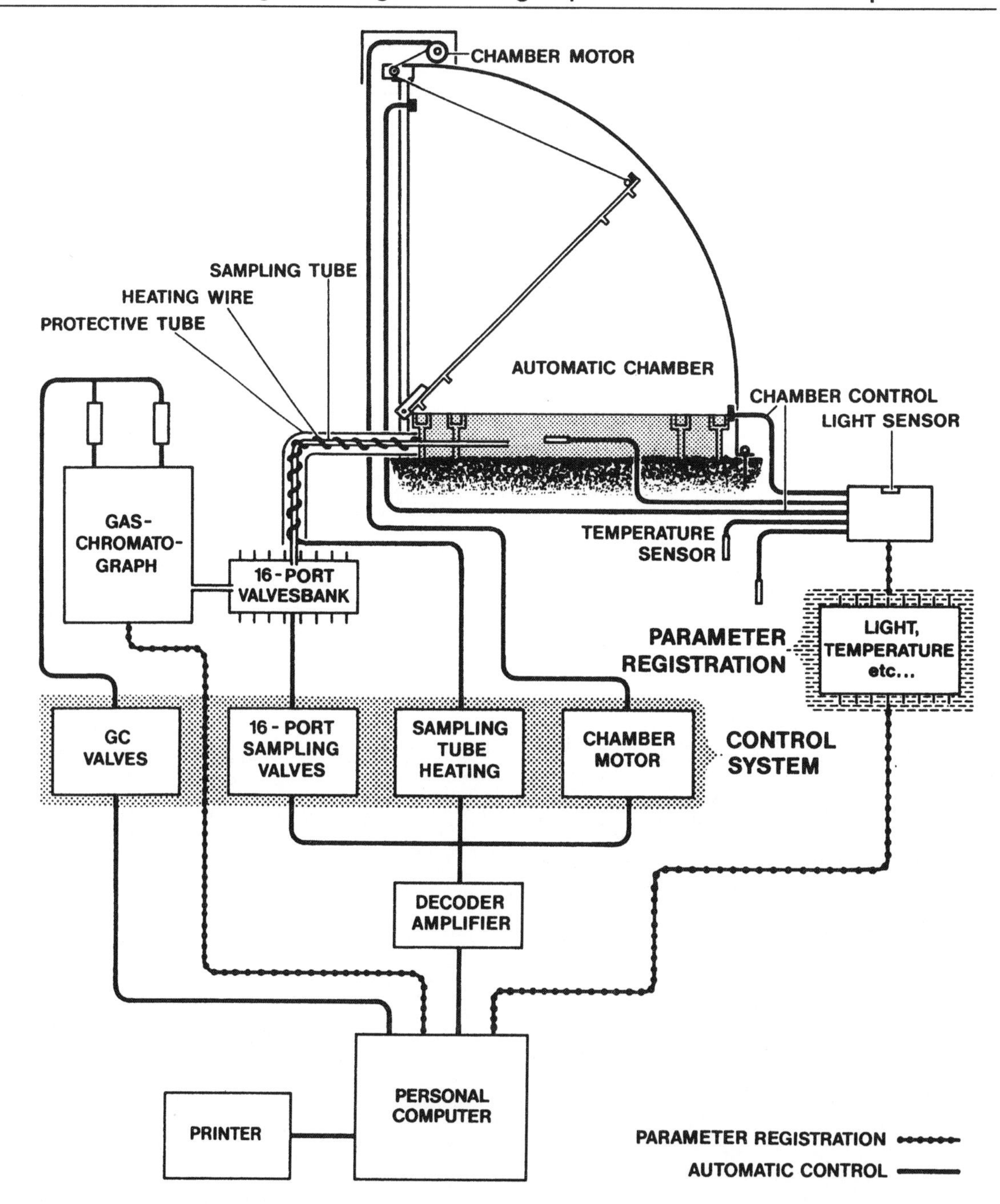

Figure 10.6. Schematic diagram of the personal computer supported control and data acquisition. ●-●-●, parameter registration; ———, automatic control (Loftfield et al 1992).

Gas chromatography

The air samples are analysed by a gas chromatograph, equipped with a pneumatic 10- and 4-port valve, a ^{63}N electron capture detector (ECD) and a flame ionization detector (FID) (Fig. 10.5). A plumbing system (1/16 inch stainless steel) described by Mosier and Mack (1980) was chosen because it has the advantage of removing dirt and vapour by a precolumn back-flush-system. Both columns, the precolumn (1 m) and the analytical column (3 m), are filled with 150–200 µm Porapak Q (Millipore, Milford, MA). The electron capture detector, the valves and the column oven temperatures are 280, 65 and 65°C, respectively. The flow rate of the carrier gas was 18 ml min^{-1}. In contrast to Mosier and Mack (1980), N_2O, CO_2 and CH_4 are measured in the same gas sample.

AT-compatible personal computer and printer

The interface was achieved by using a 16-channel A/D card (no. 2814) and a 32-port I/O card (no. 2817), obtained from Data Translations (Marlborough, MA).

Hardware

All chambers are equipped with one light and three temperature sensors (Fig. 10.6). One temperature sensor is located in the chamber and the other two are in the soil. Hardware was developed for connecting the PC with the GC valves, the 16-port valve bank, the 12 sampling tube heatings and the 12 chamber motors.

Software

The system is controlled by a personal computer (Fig. 10.6). Special software, written in Turbo Pascal (Umweltfreundliche Energieanlagen, Göttingen, Germany), operates the 16-port valve bank, the V1 and V2 valves for collecting the samples, the 12 chamber motors to close and open the chambers, the 12 sampling tube heatings and the 10- and 4-port valves of the gas chromatograph. The gas analysis sequence for the calibration gases and the air from the chambers is variable and can be adapted to the problems under investigation. All important operations, errors in operation and results are displayed both on the screen and on the printer. In addition to the control of the sampling operations, the PC monitored the GC detector signals, the lid position of each chamber, the light intensity and the temperatures.

Gases

Nitrogen 5.0, ECD quality
Hydrogen 5.0, ECD quality
Hydrocarbon-free air
Different concentrations of calibration gases

Procedure

The operation of the system is described by a typical analysis cycle (Table 10.1). First the heating for the sampling tubes is turned on. Next, the 0.5 l flask, the 16-port valve bank, the sample lines and the GC sample loop are evacuated by opening valve V2. Following this step, valve V2 is closed and a single port of the 16-port valve bank is opened to clean the system. A gas sample (495 ml) from one of the open chambers is drawn into the evacuated system and the 5 ml GC sample loop of the 10-port valve (solid lines in Fig. 10.5) is filled. Thereafter, the sample valve is closed and the whole procedure is repeated for gas analysis with two extensions. Before closing the valve V2 and the sample valve, the pressure is measured and a malfunction is noted if the given values were not obtained (Table 10.1). Afterwards the lid is closed and the tube heater is turned off before the sample valve is closed. Next, the system is vented to the atmosphere (valve V1) and the pressure is measured for calculation of the sample volume.

The 10-port valve then acts (dotted lines in Fig. 10.5) for the injection of a 5 ml gas sample into the carrier gas stream. After N_2, O_2, CH_4, CO_2 and N_2O have left the precolumn, the 10-port valve turns back to the initial position (solid lines in Fig. 10.5). Water vapour remains on the precolumn

Table 10.1. Sequence of timed events for a typical operation cycle of the automated soil gas chamber system.

Time (s)	*Event*
0	Turn on tube heater
30	Open valve V2
90	Close valve V2
91	Open sample valve (V1–V16) (for cleaning)
151	Close sample valve (V1–V16)
152	Open valve V2
210	Measure pressure If > 1 kPa, note *error*, continue If < 1 kPa, continue
212	Close valve V2
213	Open sample valve (V1–V16) (for 1st gas analysis)
273	Close lid Is lid closure verified? No: Note *error*, turn motor off, turn heater off, continue Yes: turn off lid motor, continue
274	Measure pressure If < 80 kPa, note *error*, continue If > 80 kPa, continue
275	Turn tube heater off
276	Close sample valve (V1–V16)
277	Open valve V1 to equalize pressure in the system
287	Measure pressure for sample volume calculation
289	Close valve V1
293	Actuate 10-port valve to "inject"
368	Reset 10-port valve to "load" (solid lines in Fig. 10.5)
370	Open sample valve (V1–V16) (for 2nd gas analysis)
387	Open integration window for FID signal
421	Close window, turn 4-port valve to ECD (solid lines in Fig. 10.5), integrate CH_4 signal
440	Open integration window for ECD signal
540	Close window, return 4-port valve to FID (dotted lines in Fig. 10.5), integrate CO_2 and N_2O signal

and is displaced by the back flush. At the same time the sample loop is refilled with the next sample while the gases are separated in the analytic column. CH_4 is led to the FID (4-port valve, dotted lines in Fig. 10.5) and, after closing the 4-port valve, CO_2 and N_2O are led to the ECD. Then the 4-port valve is reset. The automatic registration and analysis of the gas measurements are performed by an integration subroutine. The chromatograms are saved for a later control as well as the calculated peak areas. Errors during the measurements are indicated and noted. One cycle lasted 6 min and was a part of an overall cycle in which all 12 chambers were closed for 1 h (variable) and samples were taken and analysed at 0, 30 and 60 min (variable) after closure of the chambers. Calibration gases and samples from the chambers are analysed alternatively. The order of gas analysis has to be chosen at the beginning. The temperature and light intensities were collected at 5-min intervals and recorded as hourly averages. During the opening and closing of the chambers a control subroutine is activated. If the operation is not finished within a fixed time of 50 s (variable), the system indicates a malfunction and the chamber is taken out of operation until the next check up.

Discussion

The automatic chamber system has been developed by two manufacturers, Umweltfreundliche Energieanlagen (UfE) in Göttingen and Loftfields Analytische Lösungen (LAL) in Neu Eichenberg, Germany.

An adjustable chamber system has been developed to measure gas fluxes from crops and soil. With increasing height of the crop, additional extensions can be installed between the basal chamber and an upper chamber with the lid.

The GC system has to be installed in an isothermal room because of the extreme sensitivity of the detector signal to changes in temperature.

By using N_2 as carrier gas, the calibration curve for CH_4 and CO_2 is not always linear. Therefore it is recommended that four different calibration gases are used.

Quantification of total denitrification losses from undisturbed field soils by the acetylene inhibition technique

G. Benckeiser
H.J. Lorch
J.C.G. Ottow

Denitrification (nitrate respiration) is probably the major source of nitrogen loss from terrestrial and aquatic ecosystems (Benckiser and Syring 1992; Körner et al 1993). Because of the lack of reliable direct measurements under natural conditions, the estimates of total global denitrification losses range from 83 to 390 g N a^{-1}. Such estimates have little significance so far, because basic data on denitrification measurements from soils and aquatic ecosystems are essentially missing and insufficient to allow reliable calculations.

Denitrification:

$$2NO_3^- \xrightarrow[\text{[H]}]{\text{ATP}} 2NO_2^- \xrightarrow[\text{[H]}]{\text{ATP}} [2NO]\uparrow \xrightarrow[\text{[H]}]{\text{ATP}} N_2O\uparrow\downarrow \overset{C_2H_2}{\dashv\!\!\rightarrow} N_2 \quad (10.4)$$

Denitrification is an aerobic energy-conserving process (ATP synthesis by means of cytochromes). It may occur alternatively to or simultaneously with respiration and nitrification in a great number and variety of taxonomically aerobic microorganisms (1–10% of the culturable bacteria) (Burth et al 1982; Ottow and Fabig 1985; Abou Seada and Ottow 1985, 1988; Schmider and Ottow 1986; Hooper et al 1990). The need to use nitrate (or nitrite and N_2O) alternatively or even simultaneously with O_2 occurs when the demand for electron acceptors cannot be met during intensive mineralization processes in the water films of so-called hot spots (Ottow 1992; Simarmata et al 1993). Consequently, denitrification losses can be recorded even at high oxygen partial pressures in the soil air (Prade and Trolldenier 1990; Benckiser 1994; Schwarz et al 1994). All soil treatments (such as using organic manure, soil tillage, fertilization, grassland conversion, etc.), which enhance microbial activity will stimulate nitrate respiration, especially at high soil moisture, temperature (5–65°C) and nitrate availability or supply (von Bischopinck and Ottow 1985; Benckiser and Warneke-Busch 1990). At the various microsites in the field the specific favourable conditions for denitrification may change rapidly, diurnally, and locally as well as temporarily (Arah 1990; Aulakh et al 1991; Benckiser and Syring 1992; Ottow 1992). For every field method developed to quantify total denitrification losses during a longer (e.g. vegetation) period, consideration should be taken of the large temporal and spatial variability of dentitrification caused by the inhomogeneous distribution of the carbon sources, diffusional constraints of nitrate, O_2 and/or N_2O imposed by the ever-changing soil water content as well as by the delayed N_2O fluxes as a result of its entrapping in pores and water films, and adsorption to soil colloids (Becker et al 1990; Rolston 1990; Benckiser 1994). Thus, the spatial and temporal variability of denitrification in the field can vary by up to 300% and more (Benckiser et al 1986, 1987; Rolston 1990; Smith 1990, Aulakh et al 1991). This high variability, however, provides the statistical framework explaining both the interactions of essential ecological factors controlling denitrification as well as the wide range of gaseous nitrogen losses recorded in the field. The reliability of field measurements is consequently limited by the previously mentioned variability in denitrification. The results obtained by the acetylene inhibition techniques can only be considered as estimates of the magnitude of gaseous nitrogen losses.

Principle of the method

The acetylene inhibition technique (AIT), used to quantify total denitrification losses from soils, is based on a complete blockage of the N_2O reductase activity by HC≡CH, which has a structure similar to that of N=N−O (Federova et al 1973; Balderston et al 1976; Yoshinari and Knowles 1976; Yoshinari et al 1977; Ryden et al 1979; Ryden and Dawson, 1982; Benckiser et al 1986; Kapp et al 1990). Concentrations in the range of 0.2–1.0% v/v exhibited complete inhibition of N_2O reduction in denitrifying organisms in soils (Balderston et al 1976; Yoshinari and Knowles 1976; Yoshinari et al 1977; Kapp et al 1990).

Consequently, the N_2O fluxes collected from C_2H_2-treated soils can provide a measure of the overall denitrification rate, which is equal to the N_2O-N and N_2 release (Benckiser et al 1986; Klemedtsson and Hansson 1990; Kapp et al 1990, 1993). At present, there are two AIT-based approaches to quantify total denitrification losses from undisturbed soils:

1. Soil core incubation method (Parkin et al 1985; Robertson et al 1987; Ryden et al 1987; Aulakh et al 1991).

2. Flow-through soil cover method (Ryden et al 1979; Ryden and Dawson 1982; Benckiser et al 1986; Kapp et al 1990, 1991; Lehn-Reiser et al 1990; Schwarz et al 1994).

Materials and apparatus

Soil core incubation method

Airtight anaerobic jars (Fig. 10.7)
Glass bridges with tubes, airtight double-way glass cocks, rubber ball and one-way valves
Rubber septum seals fitting to the tubes of the glass bridges
Core samplers (varying volumes)
Airtight syringes fitted with a lock.

Flow-through soil cover method

Self-constructed PVC soil covers (of variable dimensions) equipped with a sharpened steel frame at the bottom, an air inlet and outlet (PVC tubes, 10 mm i.d.), and fixed or removable Plexiglass lids
Glass or polyethylene columns for CO_2 and H_2O trapping
Glass tubes (straight, 20 cm length, 2 cm i.d., or U-shaped, 50 cm length, 2 mm i.d., for trapping N_2O) + fitting rubber stoppers
Glass tubes (0.5 mm i.d.)
Rubber tube (0.5 mm i.d.) flow meter (0–70 l min^{-1})
Membrane pump
PVC tubes (1 m length, 6 mm i.d.) for C_2H_2 contribution
Polyethylene tube (6 and 10 mm i.d.)
Electric borer
Borer (10 mm × 60 cm)
Erlenmeyer flasks with tubes and fitting rubber stoppers
Glass containers with glass tubes as outlets and fitting rubber stoppers
Tube-clamps and double-way cocks

Rubber septum seals fitting to the necks of the Erlenmeyer flasks
Airtight syringes fitted with a lock
PVC or steel tubes (1 m length, 6–10 mm i.d.) with soil air inlets and outlets for soil air sampling
Micro-PVC tube (0.5 mm i.d.)
Septum seals (8–12 mm i.d.)
Double-sided needles
Evacuated vacucontainer (5 ml)

Gas chromatography

A gas chromatograph equipped with an ECD and/or a thermal conductivity detector (TCD) is required. N_2O can be separated from CO_2 and C_2H_2 using a Porapak Q column (120 mesh, 2 m long), and N_2 from O_2 using a 0.5 mm molecular sieve column (60/80 mesh, 2 m long). For the detection of low N_2O-N concentrations an ECD detector is essential (carrier gas N_2, flow rate 30 ml min^{-1}, make up gas at 40 ml min^{-1}; instead of N_2, a mixture of Ar and CH_4 at 95:5% can be used). Proper conditions for separating and quantifying N_2O are an oven, injector and ECD temperature of 40°C, 60°C and 300°C, respectively. The range of detection is 0.1–400 ng N_2O-N. Higher N_2O-N concentrations as well as CO_2, C_2H_2, N_2 and O_2 can be quantified by a TCD detector (carrier gas He, flow rate 30 ml min^{-1}). Proper conditions for separating and quantifying N_2, O_2, CO_2 and C_2H_2 are an oven, injector and TCD temperature of 40°C, 60°C and 150°C, respectively. The range of N_2O-N detection is 0.1–50 µg N_2O-N. The high sensitivity for N_2O detection by GC makes the AIT a favourable tool for an accurate determination of denitrification losses. The heterogeneity of soils and soil samples, as well as the high spatial and temporal variability in denitrifying activities are the major restrictions.

Gases and chemicals

Acetylene (technical grade)

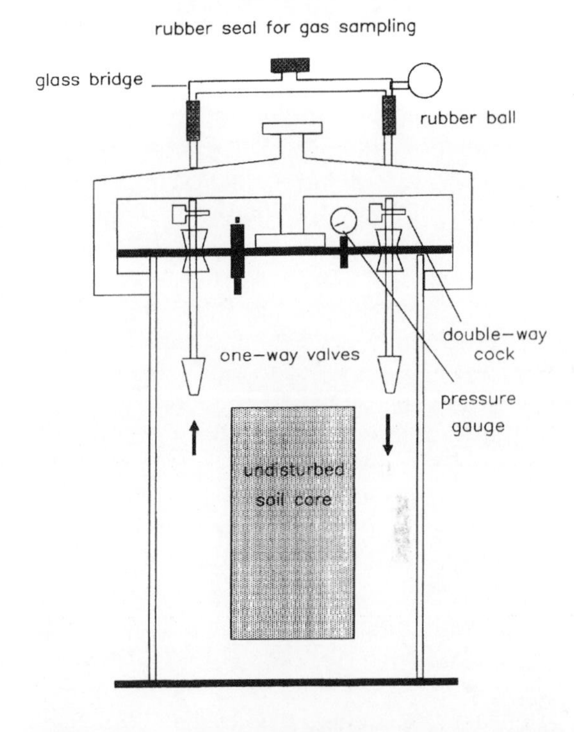

Figure 10.7. A modified anaerobic jar (4.5 l) used to incubate intact soil cores in the presence of acetylene for total denitrification measurements. Commercially available anaerobic jars (Gössner Company, Hamburg, Germany) are modified by introducing a glass bridge, gas taps, suba seal and rubber ball (Ottow et al 1985; Abou Seada and Ottow 1985, 1988).

Nitrogen or Ar and CH_4 (ECD grade)
Helium (purity 99.996% v/v)
N_2O diluted in N_2 (100 parts per million by volume; purity 99.995 v/v, calibration gas), CO_2, O_2 and N_2O (purity 99.995 v/v, calibration gases)
Granulated $CaCl_2$
NaOH + $Ca(OH)_2$ (tablets with indicator)
Molecular sieve, 0.5 nm (2 mm pellets)
Concentrated H_2SO_4
Distilled water

Procedure

The soil core incubation method

Using a rubber hammer, steel corers (metal rings generally used by soil physicists, 6–11 cm diameter × 7.5–20 cm) are randomly and carefully driven in the soil in order to obtain almost undisturbed soil cores. Cores damaged by the presence of stones or organic debris are discarded. Some compacting of about 10% of the volume may occur during sampling (Ryden et al 1987). The collected cores (at least five replicates) are placed immediately into modified, *ca.* 4.5 l anaerobic jars (Fig. 10.7). The jars and the cores are flushed with air or He (Ar, N_2) gas (for about 15 min), depending on whether aerobic or anaerobic incubation is desired. After syringe injection of C_2H_2, using airtight PVC syringes (Becton and Dickinson, Germany), through the seal, the total air phase is carefully mixed using the rubber ball (Fig. 10.7) to give a final acetylene concentration of 5–10% v/v. The jars are incubated in darkness (20–25°C) or at the actual soil temperature directly in the field in prepared holes (Ryden et al 1987). Daily (up to 14 days) 1 ml gas samples are taken after gas mixing from the jar with a gas-tight syringe, and the N_2O, CO_2 and O_2 concentrations are determined by GC. The respired O_2 from the atmosphere of the jars is adjusted to 20% v/v by adding O_2 with a syringe through the rubber seal and mixed with the gas phase using the rubber ball. Samples may be run without acetylene, if the N_2O part of total denitrification is of interest. This soil core method is suitable for evaluating soil denitrification rates of almost undisturbed soil under nearly natural conditions. Denitrification losses are generally expressed in kg N ha^{-1} day^{-1}.

In situ determination of total denitrification losses

Self-constructed (Fig. 10.8) open cover boxes (50 cm × 10 cm × 15 cm), equipped with a sharpened steel base and a removable Plexiglass lid (to enable photosynthesis) are inserted carefully about 5 cm deep into the tilled soil between the plant rows or into the sward (four replicates). Six holes (*ca.* 60 cm deep and 10 mm diameter) are established by an electric borer around the cover boxes. One day before measuring the N_2O fluxes from the soil, each box is flushed (without pressure) for 4 h with 40 l of acetylene (Messer-Griesheim, Germany) through six perforated probes (PVC tubes, 8 mm in diameter; Benckiser et al 1986; Lehn-Reiser et al 1990; Schwarz et al 1994). Three hours prior to the N_2O collection, an additional 10 l of acetylene is introduced into the soil for 2 h. The acetylene distribution and concentration in the soil air are checked by special gas sampling tubes inserted carefully in prebored holes (Fig. 10.8). Appropriate acetylene concentrations of 0.2–1.0% (v/v) should be present in the soil air and are easily ascertained even at relatively high water tensions up to 6 kPa (*ca.* field capacity) (Benckiser et al 1986; Kapp et al 1990). In order to eliminate acetone contamination of the soil, acetylene is passed through flasks containing concentrated sulphuric acid before introducing it into the soil (Benckiser et al 1986). During the equilibration period (2 h), the gas lids of the cover boxes are removed to avoid the accumulation of acetylene gas and an increase in temperature of the chamber. During the actual period of N_2O sampling, the open chambers are flushed continuously with an air stream of 20 l h^{-1} using a vacuum pump (Vacubrand M2), needle valves and flow meters (Platon, Germany; Fig. 10.8). The N_2O released in the air stream is led through a CO_2 and H_2O filter (a flask containing granules of $CaCl_2$ and NaOH) and collected in three traps (19 cm long, 2.5 cm in diameter) filled with 0.5 nm molecular sieve pellets (2 mm, Merck, Germany). N_2O absorbed over 4–8 h periods in the traps is transported to the laboratory. Concomitantly, with the same procedure, the N_2O in the surrounding air used for flushing the cover boxes is collected. In order to

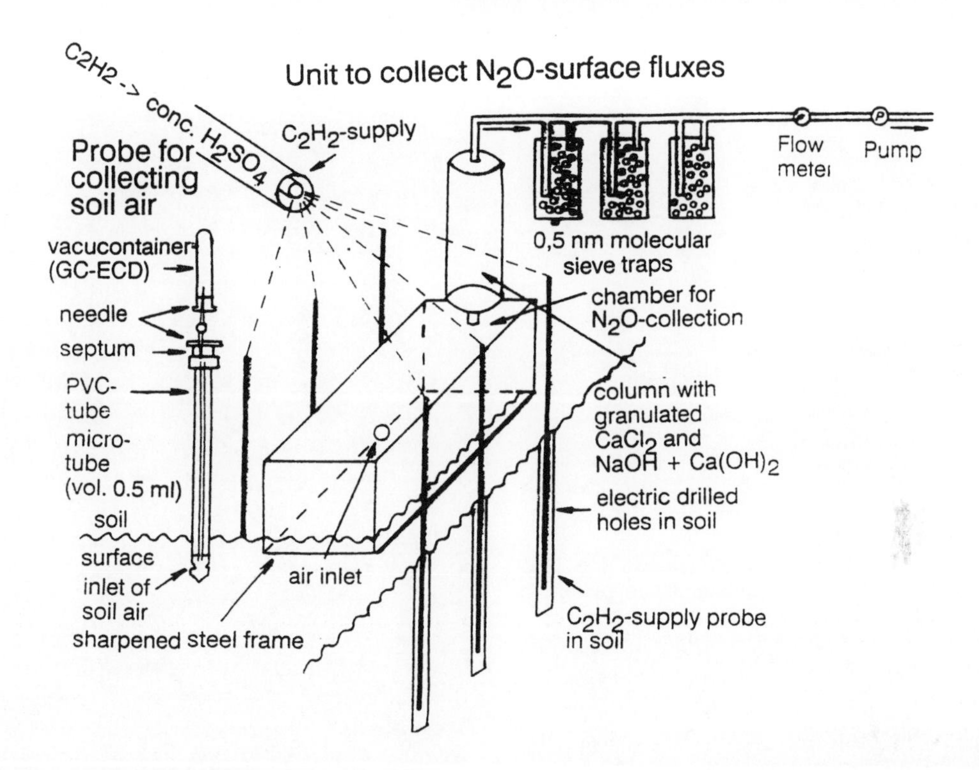

Figure 10.8. Schematic diagram of the open soil cover method to quantify total denitrification losses in undisturbed fields with the acetylene inhibition technique (Benckiser et al 1986; Kapp et al 1990; Lehn-Reiser et al 1990; Schwarz et al 1994).

avoid microorganisms adapting to acetylene as a substrate or bypassing the inhibition effect, the sampling sites are changed every 4 days. The N_2O absorbed on the 105 g of the 0.5 nm molecular sieve is liberated from the pellets in the evacuated Erlenmeyer flasks (*ca.* 1 l) containing 150 ml water (Fig. 10.9); Ryden and Dawson 1982; Benckiser et al 1986; Kapp et al 1990). After 2 h the vacuum is replaced by air, and the N_2O concentration in each flask is determined by GC. The amount of N_2O dissolved in water is calculated as proposed by Moraghan and Buresh (1977). After subtraction of the N_2O in the air used for flushing the cover boxes, the N_2O surface fluxes are calculated in g N_2O-N ha^{-1} day^{-1}. Before starting AIT measurements, the actual nitrate concentrations in the soil should be determined because soils are sinks rather than N_2O sources, if the nitrate concentration is low with respect to the amount of easily decomposable carbon (Knowles 1990; Simarmata et al 1993).

Discussion

Estimates of total denitrification losses suffer from high variability even within short distances of a field. A temporarily high denitrification activity at the microsite level is essentially determined by the amount and availability of the easily decomposable organic matter in the soils (Ottow 1992; Benckiser 1994; Schwarz et al 1994). The higher the inhomogeneity of the soil with respect to the distribution of the organic debris, the greater is the variability in denitrification. Homogeneously tilled (arable) soils with a low tortuosity will give a

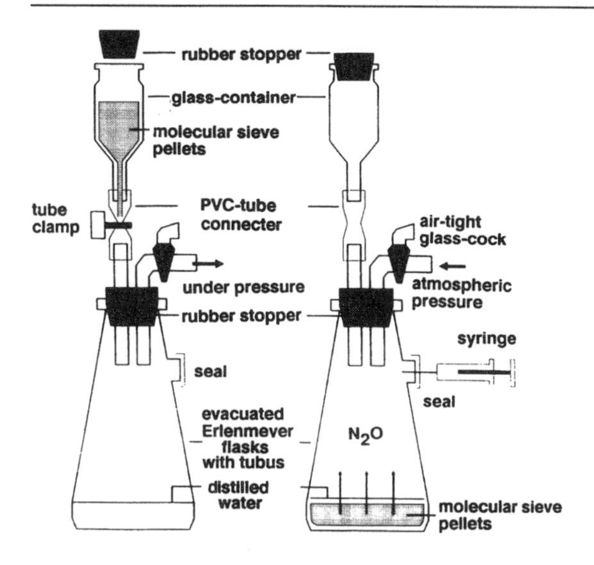

Figure 10.9. Units required to release the N_2O absorbed to the 0.5 nm molecular sieve pellets in the field (Ryden and Dawson 1982; Benckiser et al 1986; Kapp et al 1990; Lehn-Reiser et al 1990; Schwarz et al 1994).

uniform diffusion and the lowest variability, and thus the most reliable results with the AIT (Benckiser et al 1986, 1987). At present, the AIT is the most suitable and probably the most reliable method for routine determinations of total denitrification losses in undisturbed fields (Knowles 1990).

The main advantages of the AIT are:

1. The application in undisturbed as well as in disturbed, fertilized ecosystems.
2. The relatively simple and accessible equipment required.
3. The high sensitivity of the N_2O measurements by gas chromatography.

Another approach, the ^{15}N method, is mainly restricted to the evaluation of fertilizer nitrogen losses and is less reliable for measurements of total denitrification losses from undisturbed soils (grassland, forests and zero-tilled arable land). The use of ^{15}N needs (a) a homogeneous incorporation of ^{15}N in the (top) soil, and (b) expensive materials and equipment (mass spectrometer). If denitrification is calculated indirectly by difference, all ^{15}N pools have to be considered (^{15}N mass balances; Aulakh et al 1991). Denitrification losses deriving from the nitrogen out of the native organic nitrogen pool are not considered sufficiently. Despite this, there is good agreement between the denitrification losses estimated by the AIT and the ^{15}N method at comparable field conditions (Parkin et al 1985; Aulakh et al 1991). The major restriction of the AIT is caused by the inhibition of nitrification even in the presence of small C_2H_2 concentrations (Mosier 1980; Knowles 1990; Mosier and Schimel 1993). Consequently, during the denitrification measurements, the nitrate supply by nitrification is interrupted and total nitrogen losses by nitrate respiration refer only to the amount of nitrate present in the soil at the moment of measurement. Frequent changes of the measuring site is one way to try to minimize this limitation. Further, at low nitrate concentrations and a relatively high availability of easily decomposable organic matter, N_2O will be respired in soils to N_2 even in the presence of C_2H_2 (Simarmata et al 1993). Therefore, frequent nitrate monitoring during denitrification measurements by the AIT is essential for reliable interpretations of field data. Additional effects on denitrification by the metabolism of C_2H_2 under aerobic and particularly under anaerobic conditions may occur, but should be neglected in all soils containing 1–2% of total carbon. However, C_2H_2 may increase the rate of soil carbon mineralization or added glucose, if nitrate is not limiting (Haider et al 1983). This mechanism remains to be clarified. Acetylene concentrations of 1–2% (v/v) should be present in the soil air to establish C_2H_2 gradients that allow the N_2O reductase activity in all aggregates to be blocked sufficiently (Becker et al 1990). Even at such high concentrations acetylene does not affect the general soil metabolism (Kapp et al 1990; Simarmata et al 1993).

P. Burauel
W. Steffen
F. Führ

Lysimeter

Principle of the lysimeter

The lysimeter type that is described here enables experiments to be carried out under natural environmental conditions. Questions concerning nutrient turnover as well as the fate of xenobiotics can be studied in the soil plant system. This also includes research activities in soil microbiology such as questions about microbial ecology and soil biochemistry. The relationship between laboratory and field studies can be drawn using data from lysimeter experiments. The lysimeter units in use are composed of a square or round cylindrical casing (0.5 or 1 m^2 surface area) filled with an undisturbed soil monolith of 1.10 m of depth. The casing, which is placed in a rack with a perforated bottom, is inserted in a completely watertight container embedded in the ground. All equipment is made from stainless steel (Führ et al 1976, 1991; Führ 1985; Steffens et al 1990, 1992; Brumhard 1991; Pütz 1992).

The lysimeter station

The lysimeter station at the Institute of Radioagronomy was designed and constructed to perform experiments closely matching natural environmental conditions to evaluate, for example, the fate of ^{14}C-labelled pesticides in the soil/plant system in accordance with good agricultural practice. Using the lysimeter as a test system, mass balances can be drawn

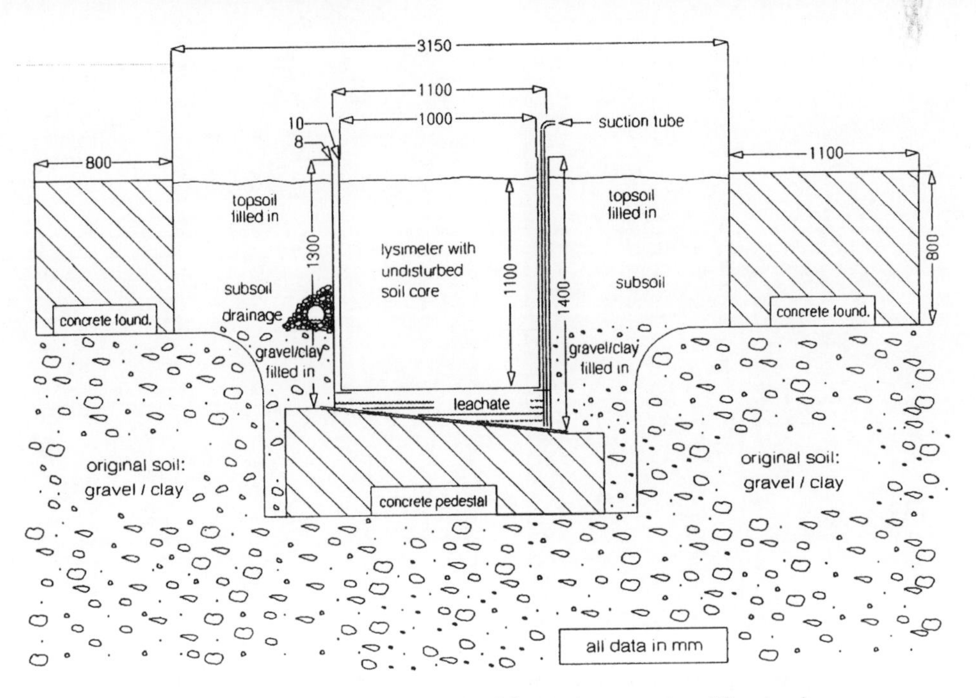

Figure 10.10. Cross-section of the lysimeter system (dimensions are in millimetres).

up considering almost all processes of dissipation. A cross-section of the lysimeter installation is shown in Fig. 10.10. A concrete bed serves as a basement to place the lysimeter unit in the ground. The lysimeter casing itself stands in a second watertight container on four arms fastened to the walls at a depth of 120 cm. This container is attached to the concrete basement. In total it is 140 cm deep so that there is enough storage space to collect water percolating through the soil monolith. The percolating drainage water is sampled by a suction tube inserted into a pipe fixed to the higher wall of the container. Figure 10.11 shows a single lysimeter embedded in a control plot to minimize possible side effects. Fifty lysimeters (20 of 0.5 m^2 and 30 of 1.0 m^2) are distributed in 10 strips of five lysimeters, each covering in total approximately 1000 m^2.

Filling of the lysimeters

The filling of a lysimeter is an important procedure in order to obtain undisturbed soil monoliths. To collect an undisturbed soil monolith, the lysimeter casing is covered at the top with a steel plate (30 mm thick). At the bottom, the walls of the casing (8–10 mm thick) have sharpened edges for cutting the soil. The casing is pressed into the soil by the shovel of an excavator to a depth of 110 cm (Fig. 10.12).

This technique guarantees that the soil monolith is pressed close to the walls of the casing and that there are no gaps between the walls and the soil. This is very important when studying the movement of xenobiotics in the soil. After the casing is pressed into the soil, the surrounding soil is removed and the bottom of the rack is pushed under the casing, cutting the soil with the sharpened edge at the front of the

Figure 10.11. The lysimeter station at the Institute of Radioagronomy.

Figure 10.12. Filling of the lysimeter with an undisturbed soil monolith.

Figure 10.13. Placement of the lysimeter casing in the lysimeter station.

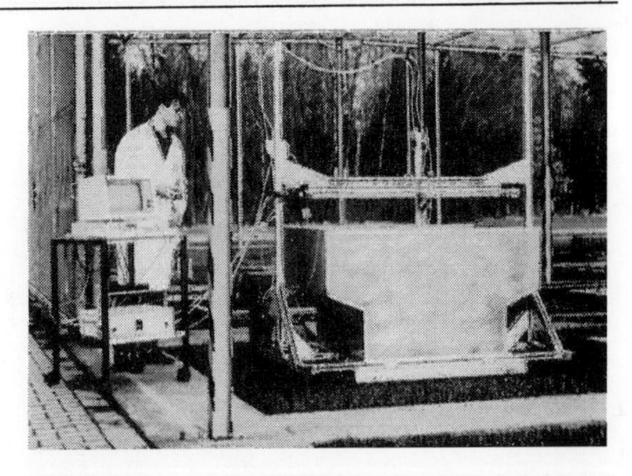

Figure 10.14. Automatic spraying apparatus for applying ^{14}C-labelled agrochemicals in lysimeter experiments.

base (Fig. 10.12). Then the lysimeter casing with the undisturbed soil monolith standing in the rack is lifted from the hole and transported to the lysimeter station. There the lysimeter casings are inserted into the containers, which are permanently installed in the ground (Fig. 10.13). With this technique, up to 15 lysimeters can be filled per day (9 h). In principle, any soil with a deep profile can be used. However, it will be very difficult, if not impossible, to collect undisturbed monoliths from heavy clays and stony soils using the technique described above.

Application of the lysimeter

Lysimeter experiments offer the opportunity to run experiments under farming conditions with the ability to change the type of crops, soil type and tillage. In the case of rotations, the crop treated with a ^{14}C-labelled pesticide is always tested first. If appropriate, intermediate crops such as mustard plants, phacelia or clover are also grown, Fertilization matches agricultural practice and questions of nutrient supply strategies can also be included in the design of the study, as long as sampling dates and procedures do not interfere with each other. In order to control weeds, fungal pests and insects within the lysimeter, and the surrounding plots, agrochemicals are applied in accordance with integrated pest control management. A meteorological station registers the air temperature and humidity, precipitation and wind velocity – all factors governing the fate of a pesticide in the system. Temperature and water content in the soil are also recorded due to their effects on the microbial activity of soil. The homogeneous application of the labelled pesticide or fertilizer is an important step in lysimeter experiments, especially for sampling reasons. Realistic rates of the tested pesticide and spray solution volumes (20–40 ml m^{-2}) as in agricultural practice should be followed to provide results that are transferrable to the field. Two different spraying techniques can be used: a hand-operated garden sprayer and an especially designed automatic spraying apparatus with nozzles, used in agricultural practice (Fig. 10.14). Before spraying, the lysimeter is surrounded by thin aluminium plates covered with foil in order to avoid contamination of the surrounding area and to control for balance purposes – the amount of the ^{14}C-labelled pesticide not reaching the lysimeter area. Several treatments have shown that up to 20% of the pesticide is lost during application and it mainly remains attached to the tin foil (Mittelstaedt et al 1992). This has to be taken into account when planning the application of definite rates of a particular agrochemical.

Extraction of soil solution with porous suction cups

H. Deschauer

In environmental studies the importance of the soil solution has continued to increase over the last decade. By reflecting the dynamic processes in the soil system, the soil solution is one of the most important components of the ecosystem used for studying the temporal and spatial distribution of nutrients and pollutants, the biological availability of all soil constituents, and the biological and chemical turnover of mineral and organic soil material. Several techniques are used to extract the soil solution. Extracts from bulk soil material are made by centrifugation, percolation, with immiscible liquids or by pressure filtration. All these methods produce a severe disturbance of the physicochemical properties of the soil as well as of the soil sampling site, and are not suitable for long-term and continuous investigations. *In situ* methods used are extraction with porous suction cups or sampling by resin bags. Because of their easy handling and the suitability for continuous sampling, suction cups are the most widely used method for ecological studies (Liator 1988).

Principle of the method

For sampling of the soil solution, suction is applied to a porous suction cup, which is in contact with the soil material. In this way capillary-bound soil water is extracted by the induced pressure gradient when the suction generated is lower than the soil water potential. The solution is stored in a sample chamber or in a collection vessel.

Three collection strategies are used (Grossmann and Udluft 1991)

The soil solution can be extracted continuously over a long period (1 week) with an electrical pump; or discontinuous sampling (regular or episodic) can be done either using a hand pump or an electrical pump. While the continuous sampling integrates over a long period, both discontinuous methods sample a solution from a single water-transport event.

Materials and apparatus

Many materials have been tested for their suitability for collecting soil solution. The hydrological and physical properties, the chemical resistance, and the exchange capacity as well as the hydrophilic properties of a material are relevant to its use in long-term field experiments. Table 10.2 summarizes the properties of the materials most frequently used for suction cups.

Field installation and maintenance

Installation procedure is similar to soil tensiometers. Installation is possible horizontally as well as vertically (Fig. 10.15). All compartments of the suction system containing soil solution should be placed below ground to prevent freezing. For most suction cup materials and with simple electrical pumps a maximum suction of

Table 10.2. Comparison of different materials tested as suction cups.

	Pore size (μm)	*Air entry (kPa)*	*Physical resistance*	*Chemical resistance*	*Exchange capacity*	*Hydrophilia*
P80 ceramic[1]	1	400	++	+	+	++
Frited glass[2, 4]	10–16	200–300	–	++	++	++
PTFE[3, 4]	2–10		+	++	++	–
Stainless steel[4]	0.5*	300	++	+	+	++
Sintered nickel[1]	n.a.	100	++	–	–	++
PVDF[5]	0.22	350	+	+	n.a.	+
Nylon[5]	0.45	210	++	++	n.a.	–

References: [1]Hetsch et al (1978); [2]Long (1978); [3]Beier and Hansen (1992); [4]Köhler (1993); [5]Grossmann et al (1985).
n.a., no data available; –, poor; + good; ++, very good; * irregular porosity.

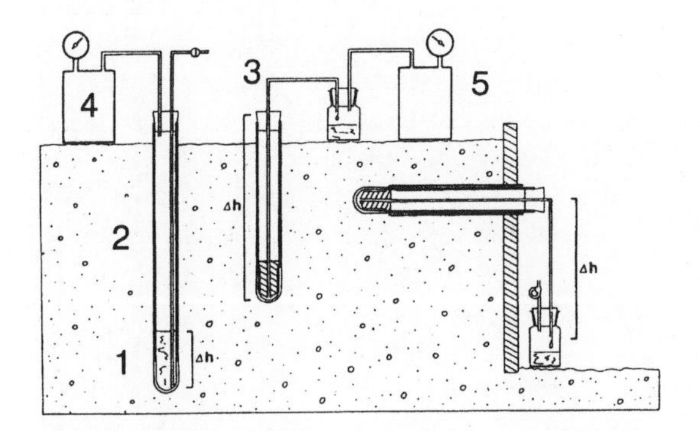

Figure 10.15. Different arrangements of soil solution samplers (Grossmann 1988): (1) suction cup; (2) hard or flexible cover; (3) capillary; (4) sampling vessel; (5) pump.

about 80 kPa may be reached, resulting in the extraction of water preferentially bound in mesopores. The equipment used is summarized in Grossmann and Udluft (1991) and shown in Fig. 10.12. The suction cup itself consists of the porous cup (1), a hard or flexible cover (2) and a capillary (3) for sampling soil solution. The system is completed by a sampling vessel (4) and a pump (5). Suction cup covers and capillaries are mostly made of PVC and polyethylene, but stainless steel, Teflon and glass are also used. The energy supply is provided either by batteries or power cables. Small membrane pumps are often used because of the low energy requirement.

For some of the materials mentioned (P-80, sintered Ni), a prewash with dilute acid and demineralized water is recommended to prevent bleeding of metals. Teflon cups are produced using fatty acids and they should therefore be cleaned with an organic solvent before use. Furthermore, all materials should be preconditioned with a solution comparable to the soil solution. In laboratory experiments, sorption of trace metals has been shown for many materials (P-80, PVDF, sintered nickel), which can be reduced by preconditioning with soil solution. Guggenberger and Zech (1992) showed a marked effect from preconditioning on the sorption of dissolved organic carbon (DOC) with new P-80 suction cups. The sorption of P and NO_3^- has also been reported for ceramic suction cups.

Besides the chemical sorption of trace metals, P and nitrate, a filter effect has also been found. Colloidal or particular organic constituents of the soil solution may be filtered or precipitated on the surface of the suction cup. For sintered nickel cups, the precipitation of nickel phosphates is discussed as a reason for the reduced permeability after percolation of the soil solution. On the other hand, P-80 cups have been used for several years in field experiments with no effects on permeability.

Sorption of mineral or organic substances on the suction cup surface or filter effects may change the surface properties of the cup material. Köhler and Deschauer (unpublished results) found an increased sorption of hydrophobic organic compounds on several suction cup materials after preconditioning with dissolved organic material. Hence, the results of laboratory investigations on sorption or filtering on cup surfaces should be evaluated critically and, after installation, the suction cups should be conditioned in the field for several weeks.

In fine and medium-textured soils under unudic moisture regime, the suction cups are not influenced by the climatic conditions over a period of several years. In winter, suction cups may be destroyed by mechanical forces due to soil freezing and thawing. Therefore, all constituents of the suction cup system containing soil solution should be placed below ground to prevent freezing. In coarse-textured soils or under dry climatic conditions, permanent contact of the suction cup with the soil may be disturbed by drying and shrinking of the soil (clay) or the silty material used for installation. Both problems require re-installation of the suction cups. The system should be tested for leaks at intervals of several months. For suction systems with a battery supply, fast depletion of the batteries is an indicator of small leaks.

Discussion

The use of suction cups for the extraction of soil solution is a well-established method in long-term field investigations. Despite their suitability for many experiments and questions, some problems with the method should be taken into consideration. Due to the suction generated, cups sample the soil solution from mesopores and fail to include the water transported in the macropores and fine pores. Hence, in structured soils with large and deep soil cores a combination of suction cups and suction-free lysimeters is necessary to collect a "real" soil solution. Due to the size of the cups, only a small part of the soil is in contact with the suction cups, resulting in a large number of replications being required to cover the soil heterogeneity. Depending on the number of replicates and the soil under investigation, a large error must be accepted. Differences in the inflow rate of suction cups, and the problems of filtering and adsorption increase this error.

Water volumes sampled by suction cups are not applicable for calculating the mass of water transported in the soil system. Hence a separate determination of soil water transport (soil tensiometers) is necessary to calculate water and substance transport. Besides the problems of adsorption and filtering, degassing of the soil solution is further influenced by the suction cups. The result of degassing by suction generation is changes in the buffer properties and the pH of the soil solution as well as losses of volatile organic compounds.

Considering the problems of suction soil solution samplers, the user has to solve this problem by choosing the appropriate suction cup with respect to the following questions

1. Which constituents of the soil solution are of particular interest for the investigation?
2. What level of precision is needed?
3. Which physical properties of the soil are to be expected?
4. How much will it cost?

As stated by Liator (1988), "no single and simple solution to soil solution collection at most soil conditions" is available.

Litterbag method

H.A. Verhoef

The litterbag method is commonly used to study the decomposition of organic matter in terrestrial ecosystems. In this method a large number of bags is placed in the field and a randomly chosen set of replicate bags is retrieved at predetermined time intervals, and analysed for loss of mass and/or changes in the chemical composition of the organic matter. The method was firstly used by Falconer et al (1933) and is often attributed to Bocock and Gilbert (1957). The litterbag is an open structure that allows a free exchange of air, water and solutes. The mesh size of (parts of) the bags is essential. By using bags with different mesh sizes, particular groups of organisms (soil animals) can be excluded, while those entering the bags can be extracted, and an idea of the relative contribution of various animal groups can be obtained.

An interesting modification of the traditional litterbag methodology of putting all the bags in the soil simultaneously, followed by subsequent sampling, is the method of Herlitzius (1983). The aim of this method is to elucidate whether, and to what extent, the decomposition of leaf litter is controlled by the date or duration of the exposure of the material in the soil.

Recently, the information available from using litterbags has been increased by the use of stratified litterbag sets (Faber and Verhoef 1991). These have been applied in stratified organic soil layers typical of coniferous forest soils.

Principle of the method

This is the determination of the temporal changes in a specific amount of soil organic matter, concerning mass and chemical composition, often combined with determination of temporal changes in root biomass, and soil biota densities.

Materials and apparatus

The material for the construction of litterbags can be nylon, fibre glass, polyester, polyvinyl and Terylene. Some bag types have closed plastic walls. These have the advantage of preventing compression of the enclosed litter and keeping the enclosed litter in a more or less fixed position. It has the disadvantage that vertical abiotic and biotic influences have greater impact than horizontal ones.

Procedure

For pine forests the procedure presented here discriminates for L, F and H layers, including fresh litter. Fresh litter is sampled at abscission time from the tree. Differentiation between the L, F and H layers is made in the field, and all organic material is taken to the laboratory and air-dried at room temperature for 1 week. To reduce heterogeneity, the L layer is sorted by hand for intact needles, material from the F layer is sieved until 2–5 mm in length, whereas the H layer is sieved with a 0.35 mm mesh sieve.

All material is stored at room temperature until required for replacement. One week before introduction the water content of the different organic layers is adjusted to the water content of the specific layer at the period of introduction. The weight and thickness of the L, F and H litterbags is in agreement with the thickness of the organic layer in the field, in such a way that the litterbags can be fitted into the organic layer. The "fresh" layer is put on top of the organic layers, and the amount is comparable to the needle fall at that time of the year. The size of the litterbags is 10 × 10 cm but may vary depending on the size of the experimental plot. The mesh sizes are different for the different layers. To achieve unrestricted movement of biota through the bags for a mass loss as natural as possible, a mesh size of 350 μm is used for H for the bottom side and 1.0 mm for the top, 2 mm for F and 4 mm for L and fresh litter. These mesh sizes also minimize losses of material through handling. If it is the intention to exclude specific groups of organisms, the mesh size should be adapted to the body width of the animals to be excluded.

The litterbags are then placed as stratified sets in the field, preferentially in a random fashion. The number of replicates depends on the heterogeneity of the study site, with a minimal number of 5. The total number of bags depends on the frequency of recollection which depends on the rate of decomposition. The higher the rate of decomposition, the shorter the interval.

Anderson and Ingram (1989) suggest that the sampling interval should be dictated by the requirement for at least four sets of samples before 50% of the original mass is lost.

For ecosystems without stratified organic layers, such as deciduous forests, grasslands and agricultural systems, the bags include the whole organic layer and the mesh sizes should be adapted to the body width of the animals present. As mentioned earlier, it is possible by using a specific experimental layout, to increase the information from the litterbag experiments, giving information on the effect of date and duration of exposure of the material on the breakdown of organic matter in soil (Herlitzius 1983). Therefore, we can distinguish three exposure treatments:

1. In the first treatment the litterbags are exposed for a fixed period (e.g. 1 month) and replaced month by month at the same spot for a certain period (e.g. 1 year).

2. In the second treatment, the so-called "from sequence", all bags are placed in soil and then sequentially removed after 1, 2, 3, . . . 11 months; each bag removed is immediately replaced at the same spot by a bag of the so-called "to sequence". All "to sequence" bags are removed after 1 year, and form the chronological mirror image of the "from sequence" treatment, providing a series that not only experiences different periods of exposure but also different dates of exposure.

3. In the third treatment, samples are exposed for a full year. The fixed sampling interval (1 month) and the total experimental period (1 year) can be adapted to the material used and the environmental conditions.

After exposure, the bags should be carefully lifted in the field to reduce losses of particulate materials and be placed in closed containers for transport to the laboratory. After extraction of fauna (e.g. using a Tullgren apparatus), roots and soil are removed. The material is then air-dried and subsamples are taken for oven dry weight corrections.

Calculation

The difference between the original weight and the weight after exposure gives information about the decomposition rate. The decomposition can be expressed as percentage mass loss or percentage remaining weight. Further analysis of the data is quite variable among researchers. There are two general analytical approaches to the examination of decomposition data. If a comparison of the treatments is needed, ANOVA is useful. If the intent is to determine decomposition rate constants, then fitting mathematical models to the data is the more appropriate analysis. Single and double exponential models best describe the loss of mass over time (see Wieder and Lang 1982). The experimental layout described above, based on Herlitzius (1983), can be analysed in accordance with Gunadi (1993).

The combination of the data from the "from" and the "to sequence" treatment in one figure gives two possible mirror symmetries: (1) the axis of symmetry can be parallel to the *y*-axis, which means that the duration or the length of the exposure is important in relation to litter quality; and (2) the axis of symmetry can be parallel to the *x*-axis, which means that the date or seasonality effects are more important in the decomposition process. To test whether the axis of symmetry can be found parallel to the *x*- or *y*-axis, one should first change the direction of the "to sequence" treatment to become one direction with the "from sequence" treatment. If the two-way ANOVA result is not significant in the first-order interaction and in the main effect, then the duration or length of the exposure is important. Then one should change the value of the "to sequence" treatment to become negative. If the ANOVA result is not significant in the first-order interaction, the data of exposure or seasonality is important.

Discussion

The litterbag method is commonly used to study the decomposition of organic matter at different sites or under different treatments.

Apart from changes in weight and chemical composition of organic matter, one can use litterbags to follow root ingrowth and population dynamics of the soil inhabitants.

Using the "to" and "from sequence" method one can get information about the importance of the date or duration of exposure of the material for its breakdown.

References

Abou Seada MNI, Ottow JCG (1985) Effect of increasing oxygen concentration on total denitrification and nitrous oxide release from soil by different bacteria. Biol Fertil Soils 1: 31–38.

Abou Seada MNI, Ottow JCG (1988) Einfluβ chemischer Bodeneigenschaften auf Ausmaβ und Zusammensetzung der Denitrifikationsverluste drei verschiedener Bakterien. Z Pflanzenernähr Bodenkd 151: 109–115.

Alef K (1991) Methodenhandbuch Bodenmikrobiologie. Ecomed, Landberg/Lech, Deutschland.

Anderson JM, Ingram JSI (eds) (1989) Tropical Soil Biology and Fertility: A handbook of methods. CAB International.

Anderson JPE (1982) Soil respiration. In: Methods of Soil analysis, part 2. Chemical and microbiological properties, 2nd edn. Page AL, Miller RH, Keeney DR (eds). Agronomy 9/2. American Society of Agonomy, Madison, WI, pp. 831–871.

Andrae MO, Schimel DS (eds) (1989) Exchange of Trace Gases between Terrestrial Ecosystems and the Atmosphere. Wiley, New York.

Arah JRM (1990) Steady-state denitrification: An adequate approximation. In: International Workshop on Denitrification in Soil Rhizosphere and Aquifer. Mitteilgn Dtsch Bodenkdl Gesellsch 60: 413–418.

Aulakh MS, Doran JW, Mosier AR (1991) Field evaluation of four methods for measuring denitrification. Soil Sci Soc Am J 55: 1332–1338.

Balderston WL, Sherr B, Payne WJ (1976) Blockage by acetylene of nitrous oxide reduction in Pseudomonas perfectomarinus. Appl Environ Microbiol 31: 504–508.

Becker KW, Höpper H, Meyer B (1990) Rates of denitrification under field conditions as indicated by the acetylene inhibition technique – a critical review. In: International Workshop on Denitrification in Soil, Rhizosphere and Aquifer. Mitteilgn Dtsch Bodenkdl Gesellsch 60: 25–30.

Beier C, Hansen K (1992) Evaluation of porous cup soil water samplers under controlled field conditions: comparison of ceramic and PTFE cups. J Soil Sci 43: 261–271.

Benckiser G (1994) Relationships between field-measured denitrification losses, CO_2-formation and diffusional constraints. Soil Biol Biochem 26: 891–899.

Benckiser G, Syring K-M (1992) Denitrifikation in Agrarstandorten – Bedeutung, Quantifizierung und Modellierung. BioEngin 8: 46–52.

Benckiser G, Warneke-Busch G (1990) Denitrifikation mineralisch und/oder klärschlammgedüngter Agrarökosysteme in Abhängigkeit von Kohlenstoff, Wasser- und Nitratversorgung sowie der Bodentemperatur. Forum Städte-Hyg 41: 157–160.

Benckiser G, Haider K, Sauerbeck D (1986) Field measurements of gaseous nitrogen losses from an Alfisol planted with sugar beets. Z Pflanzenernähr Bodenkd 149: 249–261.

Benckiser G, Gaus G, Syring K-M, Haider K, Sauerbeck D (1987) Denitrification losses from an Inceptisol field treated with mineral fertilizer or sewage sludge. Z Pflanzenernähr Bodenkd 150: 241–248.

Bischopinck von KU, Ottow JCG (1985) Einfluβ der Temperatur auf Kinetik und Gaszusammensetzung der Denitrifikation in einem sandigen Lehm. Mitteilgn Dtsch Bodenkdl Gesellsch 43: 537–542.

Bocock KL, Gilbert OJ (1957) The disappearance of leaf litter under different woodland conditions. Plant and Soil 9: 179–185.

Bouwman, AF (ed.) (1990) Soils and the Greenhouse Effect. Wiley, New York.

Brumhard B (1991) Lysimeterversuche zum Langzeitverhalten der Herbizide Metamitron (GoltixR) und Methabenzthiazuron (TribunilR) in einer Parabraunerde mit besonderer Berncksichtigung der Transportund Verlagerungsprozesse unter Einbeziehung von Detailuntersuchungen. Ph.D. thesis, University of Bonn, Juel-Report 2465.

Brumme R, Beese F (1992) Effects of liming and nitrogen fertilization on emission of CO_2 and N_2O from a temporate forest. J Geophys Res 97: 851–858.

Burth I, Benckiser G, Ottow JCG (1982) N_2O-Freisetzung aus Nitrit (Denitrifikation) durch ubiquitäre Pilze unter aeroben Bedingungen. Die Naturwiss 69: 598–599.

Edwards NT (1974) A moving chamber desing for measuring soil respiration rates. Oikos 25: 97–101.

Edwards NT, Sollins P (1973) Continuous measurement of carbon dioxide evolution from partitioned forest floor. Soil Sci Soc Am Proc 39: 361–365.

Faber JH, Verhoef HA (1991) Functional differences between closely related soil arthropods with respect to decomposition and nitrogen mobilization in a pine forest. Soil Biol Biochem 23: 15–23.

Falconer GJ, Wright JW, Beall HW (1933) The decomposition of certain types of fresh litter under field conditions. Am J Bot 20: 196–203.

Federova R, Milekhina EI, Il'Yukhina NI (1973) Evaluation of the method of "gas metabolism" for detecting extraterrestrial life. Identification of nitrogen-fixing microorganisms. Izv Akad Nauk SSSR Ser Biol 6: 797–806.

Führ F (1985) Application of ^{14}C-labelled herbicides

in lysimeter studies. Weed Sci 33 (Suppl. 2): 11–17.

Führ F, Cheng HH, Mittelstaedt W (1976) Pesticide balance and metabolism studies with standardized lysimeters. Landwirtschaftliche Forsch SH 32: 272–278.

Führ F, Steffens W, Mittelstaedt W, Brumhard B (1991) Lysimeter experiments with ^{14}C-labelled pesticides – an agroecosystem approach. Pesticide Chem 14: 37–48.

Grossmann J (1988) Physikalische und chemische Prozesse bei der Probenahme von Sickerwasser mittels Saugsonde. Dissertation, TU, München.

Grossmann J, Udluft P (1991) The extraction of soil water by the suction cup method: a review. J Soil Sci 42: 83–93.

Grossmann J, Freitag G, Merkel B (1985) Eignung von Nylon-und Polyvinylidenfluoridmembranfiltern als Materialien zum Bau von Saugkerzen. Z Wasser-Abwasser-Forsch 18: 187–190.

Guggenberger G, Zech W (1992) Sorption of dissolved organic carbon by ceramic P80 suction cups. Z Pflanzenernähr Bodenkd 155: 1–5.

Gunadi B (1993) Decomposition and nutrient flow in a pine forest plantation in Central Java. Ph.D. thesis. Vrije Universiteit, Amsterdam.

Haider K, Mosier AR, Heinemeyer O (1983) Side effects of acetylene on the conversion of nitrate in soil. Z Pflanzenernähr Bodenkd 146: 623–633.

Herlitzius H (1983) Biological decomposition efficiency in different woodland soils. Oecologia 57: 78–97.

Hetsch W, Beese F, Ulrich B (1979) Die Beeinflussung der Bodenlösung durch Saugkerzen aus Ni-Sintermetall und Keramik. Z Pflanzenernähr Bodenk 142: 29–38.

Hooper AB, Arciero DM, DiSpirito AA, Fuchs J, Johnson M, LaQuier F, Mundfrom G, McTavish H (1990) Production of nitrite and N_2O by the ammonia-oxidizing nitrifiers. In: Nitrogen Achievements and Objectives. Gresshoff PM, Roth LE, Stacey G, Newton WE (eds). Chapman and Hall, New York, pp. 387–392.

Kapp M, Schwarz J, Benckiser G, Ottow JCG, Daniel P, Opitz von Boberfeld W (1990) Der Einsatz der Acetylen-Inhibierungstechnik zur Quantifizierung von Denitrifikationsverlusten in unterschiedlich gedüngten Weidelgrasbeständen. Forum Städte-Hyg 41: 168–172.

Kapp M, Schwarz J, Benckiser G, Ottow JCG, Daniel P, Opitz von Boberfeld W (1991) Einfluβ von Mineral- und Gülledüngung auf Mikroflora und Denitrifikationsverluste einer Weidelgrasmonokultur (Lolium perenne). Verh Ges Ökol 19: 375–384.

Kapp M, Schwarz J, Benckiser G, Ottow JCG (1993) Quantifizierung und Modellierung von Denitrifikationsverlusten eines Grünlandstandortes nach mehrjähriger Gülledüngung. Mitteilgn Dtsch Bodenkdl Gesellsch 72: 735–738.

Klemedtsson L, Hansson GI (1990) Methods to separate N_2O produced from denitrification and nitrification. In: International Workshop on Denitrification in Soil, Rhizosphere and Aquifer. Mitteilgn Dtsch Bodenkdl Gesellsch 60: 19–24.

Knowles R (1990) Acetylene inhibition technique: Development, advantage and potential problems. In: Denitrification in Soil and Sediment. Revsbech NP, Sorensen J (eds). Plenum Press, New York, pp. 151–165.

Körner R, Benckiser G, Ottow JCG (1993) Quantifizierung der Lachgas (N_2O)-Freisetzung aus Kläranlagen unterschiedlicher Verfahrensführung. Korresp Abwasser 40: 514–525.

Kucera CL, Kirkham DR (1971) Soil respiration studies in tallgrass prairie in Missouri. Ecology 52: 912–915.

Lehn-Reiser M, Munch JC, Chapot JY, Ottow JCG (1990) Field measured denitrification losses from a calcareous Inceptisol after green manure. Forum Städte-Hyg 41: 164–167.

Liator MI (1988) Review of soil solution samplers. Water Res 24: 727–733.

Loftfield NS, Brumme R, Beese F (1992) Automated monitoring of nitrous oxide and carbon dioxide flux from forest soils. Soil Sci Soc Am J 56: 1147–1150.

Long LFA (1978) A glass filter soil solution sampler. Soil Sci Soc Am J 42: 834–835.

Matthias AD, Blackmer AM, Bremner JM (1980). A simple chamber technique for field measurement of emission of nitrous oxide from soil. J Environ Qual 9: 251–256.

Mittelstaedt W, Führ F, Zohner A (1992) Abbau und Verlagerung von [pyridazin-4,5-^{14}C]Pyridat in drei Ackerboeden – Ergebnisse von Lysimeterstudien 1987–1991. Tagungsband der 48. Deutschen Pflanzenschutztagung, Göttingen, 5–8 October 1992 (in preparation).

Moraghan JL, Buresh RJ (1977) Chemical reduction of nitrite and nitrous oxide by ferrous iron. Soil Sci Soc Am J 41: 47–50.

Mosier AR (1980) Acetylene inhibition of ammonium oxidation in soil. Soil Biol Biochem 12: 443–444.

Mosier AR, Mack L (1980) Gas chromatography system for precise, rapid analysis of nitrous oxide. Soil Sci Soc Am J 44: 1121–1123.

Mosier AR, Schimel DS (1993) Nitrification and denitrification. In: Nitrogen Isotope Techniques. Knowles R, Blackburn TH (eds). Academic Press, New York, pp. 181–208.

Ottow JCG (1992) Denitrifikation, eine kalkulierbare Grösse in der Stoffbilanz von Böden? Wasser Boden 9: 578–581.

Ottow JCG, Fabig W (1985) Influence of oxygen aeration on denitrification and redox level in different batch cultures. In: Planetary Ecology.

Caldwell DE, Brierly JA, Brierly CL (eds). Van Nostrand Reinhold Co, New York, pp, 101–120.

Ottow JCG, Burth-Gebauer I, El Demerdash ME (1985) Influence of pH and partial oxygen pressure on the N_2O-N to N_2 ratio of denitrification. In: Denitrification in the Nitrogen Cycle. Goltermann HL (ed.). Nato Conference Series I Ecology. Plenum Press, New York, pp. 101–120.

Parkin TB, Sexstone AL, Tiedje JM (1985) Comparison of field denitrification rates by acetylene-based soil core and nitrogen-15 methods. Soil Sci Soc Am J 49: 94–99.

Prade K, Trolldenier G (1990) Denitrification in the rhizosphere of rice and wheat seedlings as influenced by the K status of the plants. In: International Workshop on Denitrification in Soil, Rhizosphere and Aquifer. Mitteilgn Dtsch Bodenkdl Gesellsch 60: 121–125.

Pütz T (1992) Gegenberstellende Lysimeteruntersuchungen zum Abbau- und Verlagerungsverhalten von [^{14}C]Metabenzthiazuron und ^{14}C-markiertem Haferstroh in einer landwirtschaftlich genutzten Parabraunerde unter Berncksichtigung vergleichender Messungen des Bodenwasserhaushaltes, der Bodentemperatur und weiterer klimatischer Parameter in Lysimetern sowie im Feld. Ph.D. thesis, University of Bonn, Juel-Report (in preparation).

Richter J (1972) Zur Methodik des Bodengashaushaltes. II. Ergebnisse und Diskussion. Z Pflanzenernähr Bodenkd 132: 220–239.

Robertson GP, Vitousek PM, Matson PA, Tiedje JM (1987) Denitrification in a clearcut loblolly pine (Pinus taeda L.) plantation in the southeastern US. Plant and Soil 97: 119–129.

Rolston DE (1990) Modeling of denitrification: Approaches, successes and problems. In: International Workshop on Denitrification in Soil, Rhizosphere and Aquifer. Mitteilgn Dtsch Bodenkdl Gesellsch 60: 397–402.

Roulier MH, Stolzy LH, Szuskiewcz TE (1974) An improved procedure for sampling the atmosphere of field soils. Soil Sci Am Proc 38: 687–689.

Ryden JC, Dawson KP (1982) Evaluation of the acetylene inhibition technique for the measurement of denitrification in grassland soils. J Sci Food Agric 33: 1197–1207.

Ryden JC, Lund LJ, Letey J, Focht DD (1979) Direct measurements of denitrification loss from soils: II. Development and application of field methods. Soil Sci Soc Am J 43: 110–118.

Ryden JC, Skiner JH, Nixon DJ (1987) Soil core incubation system for the field measurement of denitrification using acetylene inhibition. Soil Biol Biochem 19: 753–757.

Schmider F, Ottow JCG (1986) Charakterisierung der denitrifizierenden Mikroflora in den verschiedenen Reinigungsstufen einer biologischen Kläranlage. Arch Hydrobiol 106: 497–512.

Schwarz J, Kapp M, Benckiser G, Ottow JCG (1994) Evaluation of denitrification losses by the acetylene inhibition technique in a ryegrass field (Lolium perenne L.) fertilized with animal slurry or ammonium nitrate. Biol Fertil Soils 18: 327–333.

Simarmata T, Benckiser G, Ottow JCG (1993) Effect of an increasing carbon:nitrate ratio on the reliability of acetylene in blocking the N_2O-reductase activity of denitrifying bacteria in soil. Biol Fertil Soils 15: 107–112.

Smith KA (1990) Denitrification losses from manured and fertilized soils, and the problems of measurement. In: International Workshop on Denitrification in Soil, Rhizosphere and Aquifer. Mitteilgn Dtsch Bodenkdl Gesellsch 60: 153–158.

Steffens, W, Führ F, Mittelsteadt W (1990) Lysimeter studies on long-term fate of pesticides: the experimental design. Seventh International Congress of Pesticide Chemistry, Hamburg, Abstracts Book 3, p. 78.

Steffens, W, Mittelstaedt W, Stork A, Führ F (1992) The lysimeter station at the Institute of Radioagronomy of the Research Centre Jülich GmbH. Lysimeter Studies of Pesticides in Soil, BCPC MONO 53: 21–34.

Wanner (1970) Soil respiration, litter fall and productivity of tropical rain forest. J Ecol 58: 543–547.

Wieder RK, Lang GE (1982) A critique of the analytical methods used in examining decomposition data obtained from litterbags. Ecology 63: 1636–1642.

Yoshinari T, Knowles R (1976) Acetylene inhibition of nitrous oxide reduction by denitrifying bacteria. Biochem Biophys Res Commun 69: 705–710.

Yoshinari, T, Hynes R, Knowles R (1977) Acetylene inhibition of nitrous oxide reduction and measurement of denitrification and nitrogen fixation in soil. Soil Biol Biochem 9: 177–183.

Bioremediation of soil

11

Bioremediation is a technology using microorganisms to clean up chemically contaminated soil. This technology is based on the activation of microbial degradation of pollutants in contaminated sites by optimizing environmental factors, such as nutrient concentrations, water content, pH, oxygen supply and availability of contaminants to microorganisms and temperature (Wolf et al 1988; Arndt et al 1990; DECHEMA 1992a; Alef et al 1993; MacDonald and Rittmann 1993; Müller et al 1993; Skladany and Metting 1993). In comparison with chemical, physical and thermal remediation techniques, the microbiological soil remediation has mostly proved to be ecological, economical and the most favourable technique for cleaning up soils contaminated with oil. The most important advantages of bioremediation are: (1) it is a natural process; (2) the remediation products are mostly harmless; and (3) the decontaminated soil can be recultivated.

Bioremediation of soil can, under special circumstances, be conducted without the excavation of soil, and implemented below and around existing buildings and paved surfaces. Furthermore, bioremediation is one of the few treatment processes currently accepted by public opinion. The major disadvantage of this technology is that not all chemical compounds can be degraded by microorganisms and the accumulation of toxic metabolites during the remediation of soil can not be excluded. Treatment problems may also arise at sites dealing with mixed wastes, such as soils contaminated with organic compounds and heavy metals.

Despite the expanding market for this technology and its many legitimate applications, however, bioremediation still requires standardization.

In this chapter, up-to-date methods in the microbial decontamination of soils are presented and discussed. The chapter deals with many aspects of bioremediation including security, microbiological and chemical analysis, *in situ* and *ex situ* techniques, as well as the reuse of soils after decontamination.

Methods in Applied Soil Microbiology and Biochemistry
ISBN 0-12-513840-7

Technical safety and guidelines

H. Burmeier

The problem

Together with the level of danger present in any construction venture, working in contaminated areas involves an additional risk potential due to contaminants. In addition, the use of microbiological remediation results, at least for the period of regeneration itself, in possible additional risks due to the use of microorganisms themselves. To control these health risks, safety measures must be used to protect the workers and the neighbourhood (work safety and neighbourhood safety measures). Of course, the protective measures depend on the nature of the contaminants and the biological remediation process to be used.

The *in situ* remediation results in a lower number of safety measures than on-site and off-site remediation because there is less chance of coming into contact with the contaminated materials. On/off-site remediations are equal with regard to the amount of safety technology required.

There are no general technical and safety guidelines used worldwide. Guidelines differ widely among different countries. Here German technical safety and guidelines will be presented as an example to show the different aspects of this topic. The reader is referred to Burmeier et al (1990), Rumler (1989) and Stegmann et al (1992).

Statutory basis

Because of the complexity of the problems related to soil contamination and land reclamation, the regulations for work safety and protection that are applied against emissions are drawn from various disciplines. When the Dangerous Material Regulations (German Chemicals Act) were in the process of being drawn up, the remediation of contaminated land was not yet of any importance, so this problem was not taken into consideration. In spite of this, there are a variety of regulations and annexes, as well as technical rules for handling hazardous material, which can be extended to cover measures for soil remediation.

The guidelines for protection against harmful effects on the environment provided by the German Federal Law, which is applied to neighbourhood protection, are to be considered for the microbiological treatment of contaminated media.

In order to bring together all the provisions relating to soil remediation, the sub-committee on the reclamation of derelict land of the specialist committee for civil engineering in the main association of industrial liability insurance associations drew up "Guidelines for Work in Contaminated Areas" (ZH 1/183), which were adopted and came into force on 1 April 1992. These guidelines contain the relevant safety standards for microbiological remediation work as well, and are to be used alongside the accident prevention regulations "Biotechnology" (VBG 102).

Accidents

There has not yet been a statistically significant number of accidents concerning work in contaminated areas. On the other hand, there have been incidents in nearly every remediation site in which the health of workers has been impaired.

The following causes, resulting from examinations of the accidents and observations of the incidents, have been identified:

- Lack of awareness concerning the existence of the problem;
- Ignorance of the potential danger;
- Insufficient/unplanned safety measures;
- Untrained personnel;
- Inadequate test methods;
- Non-existent or unused safety equipment for workers;
- Poor standards of hygiene;
- Unexpected events;
- Incomplete invitations to bid;
- Inadequate safety during the work for reasons of cost.

This list of reasons has a qualitative and indicative value only, but is nevertheless still suitable for drawing conclusions with regard to safety measures.

Determination of danger

The fundamental precondition for the determination of safety is the careful examination of the site and the determination of the danger level to the customer before any work is started. Of particular importance for the requirements of work safety and protection from emissions is information on the reactivity and mobility of the contaminants on contact with the atmosphere. For this aspect, a quantitative assessment of the release of the contaminants into the atmosphere will become important.

Along with volatilization processes, mechanical movements of the contaminated material, and the formation of dusts and aerosols are also responsible for charging the atmosphere with dangerous material. For this reason, there is need to carry out a gas chromatograph–mass spectrometry (GC-MS) screening of the existing dangerous material in order to assess the risk potential. The determined risk potential of the existing or expected dangerous material is then assessed in a further step with regard to the danger facing persons working in the microbiological remediation process, if necessary by including chemists, toxicologists, industrial doctors and other specialists.

The assessment of the risk potential that might emanate from the applied microorganisms must be carried out by the commissioner for biological safety (paragraph 16 UVV "Biotechnology"). Special attention should be paid here to pathogenic microorganisms and whether these can be enriched during the remediation. These examinations should be carried out in particular when enriched special cultures are to be used for the remediation. Commercially obtainable preparations contain an extremely high concentration of bacteria, which can easily be mobilized. Safety measures are then based on the results of these examinations. In the case of the accumulation of toxic metabolites during the remediation, special safety engineering measures are also necessary. In this context, the availability of transformation products is of great importance. This is based among others on the physical and chemical properties of the product. Decisive importance is attached to its interaction with the soil matrix. This interaction should be quantified in the framework of the preliminary examinations. With this data, assessments of the mass flow density from the soil can be carried out. Only then can a practical evaluation of transformation products with regard to the safety of engineering measures be carried out. Corresponding to the risk potential, which has been determined, and the knowledge of the dispersion and absorption behaviour of the dangerous material, the personal safety equipment, the necessary technical safety measures, the accompanying measuring standards programme to be carried out during the work and the range of necessary industrial medical examinations of the employees can all now be determined. It has proved to be of great value to pay strict attention to the planned work steps, and to describe in detail a safety plan for the site and the safety measures to be taken for each of these steps.

Planning and preparing for work

Before commencing the work, the customer should draw up work plans (safety plans), which clearly regulate the behaviour of those employed on the site. These plans should be drawn up with the help of industrial doctors, toxicologists, specialists in biological safety, safety engineers and other safety engineering experts. The plans should contain the following points:

- The working area, activity, and designation/ assessment of danger of the contaminants;
- Dangers for people and the environment;
- Technical, organizational and personal protective measures, and action in case of danger;
- Decontamination measures, first aid, and an industrial medical examination programme;
- Type and extent of safety coordination measures;
- Instructions.

In addition, the following points should be taken into consideration during preparation for work:

- Careful selection of suitable staff;
- Drawing up emergency plans;
- Training measures for employees (first aid, measuring technology, respiratory protection, and where necessary, a course on "Safety at work in contaminated sites");
- Procurement in good time of suitable measuring appliances, construction machines, technical ventilation equipment and personal safety equipment.

Site equipment

The site should be fenced in such a way that it cannot be entered by trespassers, and "locks" should be installed for employees and for

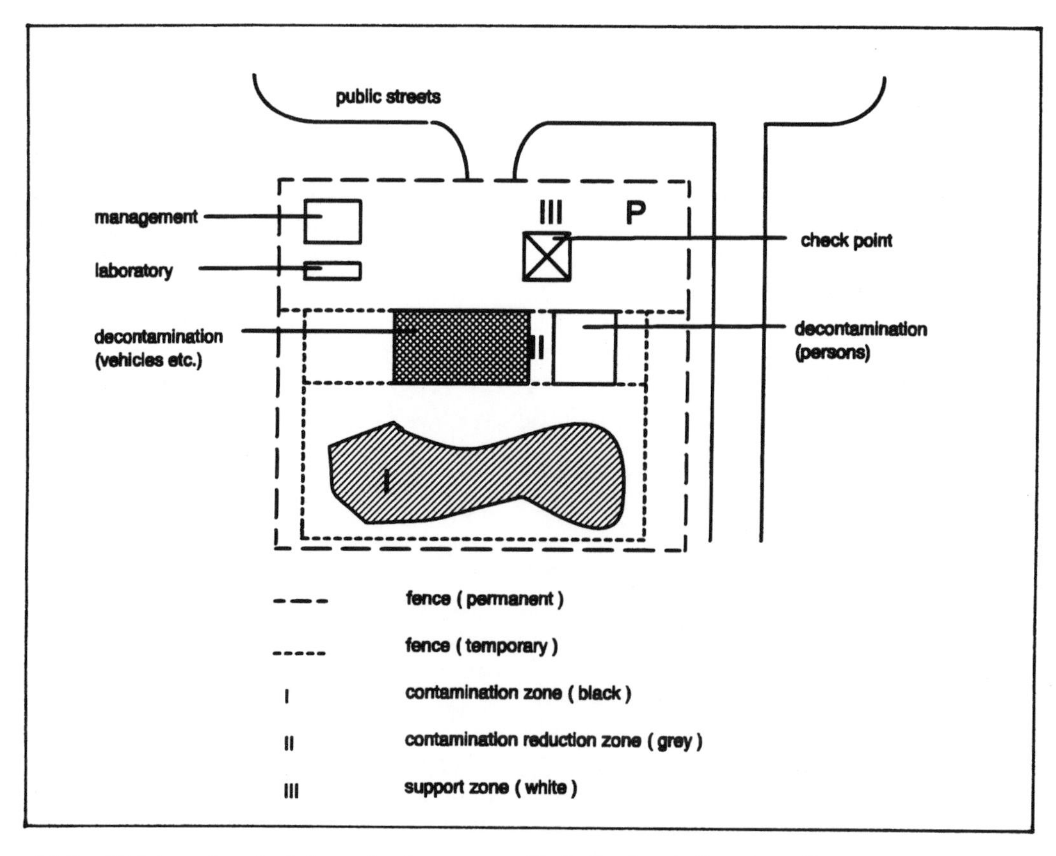

Figure 11.1. Division of a site into protective areas with locks for staff and appliances. — —, permanent fence; – – –, temporary fence; I, contamination zone (black); II, contamination reduction zone (grey); III, support zone (white).

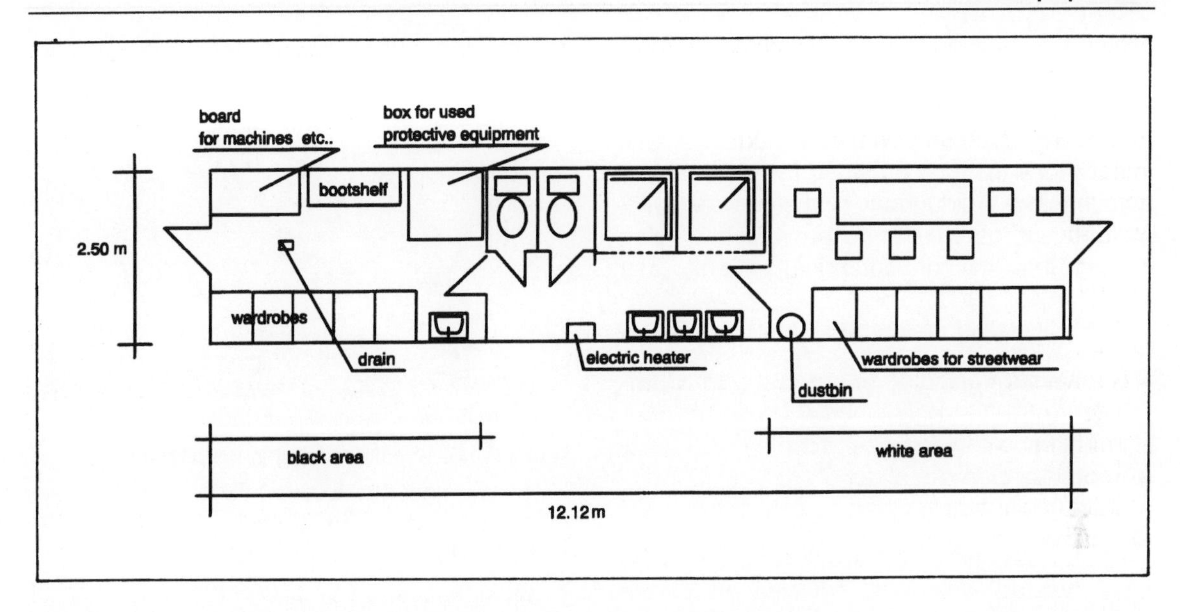

Figure 11.2. Basic sketch of a black–white installation as a 40 ft container.

appliances in such a way that workers on the site are only able to enter and leave through these "locks" (Fig. 11.1).

Along with securing the site by means of a fence, suitable rooms must be set up for the hygienic needs of the employees. A decontamination unit is to be equipped and maintained as changing rooms, toilets and wash rooms for the employees. A decontamination unit usually consists of three connected cells, whereby the cell facing the entrance is used for changing out of and into working clothes, and for breaks from work, the middle cell is provided with showers, wash rooms and toilets, and the third cell, which faces the site, is used for changing into and out of work clothes, and for storing them. Figure 11.2 shows this type of decontamination unit for six people on a site. The decontamination unit must correspond with the factory regulations and must be thoroughly cleaned every working day. The atmosphere in the rooms, the water supply and the furniture should be as pleasant as possible, so that the installation is acceptable to the employees.

In special cases, it may be necessary to have a first-aid post, and this should comply with the provisions of the "Code of practice for first-aid posts and first-aid containers in factories" (ZH 1/507). The permanent presence of a company paramedic will also be required in special cases. The first-aid material, which has to be kept on site, under paragraph 6 of the VBG 109 (Accident Prevention Regulations First Aid) must be supplemented in accordance with the dangers on the site in question. Eyewashes and protective skin creams should be available at all times (Fig. 11.3).

A special room must be planned in the decontamination area for cleaning contaminated work clothes and other safety equipment; this room may only be entered wearing

Figure 11.3. Mobile decontamination unit, integrated into the fence, with site manager's office.

protective clothing and it contains cleaning and disinfecting agents.

Immediately in front of the entrance to the decontamination unit on the site side, suitable installations must be provided for the preliminary cleaning of dirty work clothes, in particular of footwear, as well as to prevent dirt being carried into the decontamination unit (Fig. 11.4).

These installations consist of, for example:

- Showers for cleaning returnable work clothing;
- Footwear cleaning apparatus;
- Footwear cleaning area;
- Store-room for footwear

Appliances and vehicles that are used in the contaminated area must be cleaned before leaving the site, before being repaired or maintained, and at regular intervals. Suitable decontamination installations must be provided for this cleaning work and these are listed below:

Figure 11.4. Footwear cleaning installation.

Figure 11.5. Vehicle cleaning installation.

- Vehicle and tyre wash;
- Reinforced, where possible enclosed, washing area with separator for cleaning vehicles and appliances;
- Special lock for the decontamination of workers' protective clothing and of tools;
- Containers for collecting and transporting dangerous material, liquids or objects.

As water is contaminated by work with water jets or steam, it is necessary for the cleaning staff to be equipped with waterproof, chemical-resistant protective clothing and face masks. Wearing respiratory protective devices or protective suits with a line oxygen supply will depend on the circumstances of individual cases (Fig. 11.5).

If required decontamination installations are not yet available at the start of a construction measure, care must be taken that the appliances and parts of the equipment used are decontaminated at a suitable place.

Before any work starts, a plan must be drawn up for emergencies together with the local fire brigade; this plan must contain details of the steps to be taken in the case of fire. Fire extinguishers must be provided that are capable of extinguishing fires caused by the existing dangerous materials, or by those whose presence is suspected.

It is recommended that a weather station be set up to document the weather; the station should be able to record the atmospheric pressure, humidity, temperatures, rainfall, wind direction and wind velocity.

Protective measures

The protective measures that have to be taken in individual cases can be determined on the basis of the expected danger potential and the expected path taken by the dangerous material.

In all measures for work safety and protection from emissions, it is necessary that:

1. Remediation and reclamation procedures are used that cause the least possible amount of hazardous emissions, or that avoid these entirely;
2. Technical steps should be taken to ensure that dangerous matter is not released;
3. Organizational steps should ensure that workers are not exposed to any danger by emissions of dangerous material;
4. Personal safety equipment should be used to ensure that workers taking part in handling dangerous materials are reliably protected against any danger.

After the remediation process to be implemented has been selected, technical protective measures always take precedence over organizational and personal protective measures.

Reclamation and regeneration measures

Gaseous, liquid and even dusty emissions can be released by air and water having access to the contaminated ground. For this reason, *in situ* remediation processes are most suitable for the workers on the site as well as for the immediate neighbourhood. On the other hand, contaminated water and earth can often only be treated by on-site or off-site processes, whereby the safety requirements in these cases must be placed behind the remediation/reclamation requirements. If possible, it is recommended to expose and work on small areas.

When choosing the procedures for extracting, loading, transporting, unloading and depositing the material, methods giving lower emissions of dangerous material are preferred (Fig. 11.6). These include:

- Suction systems with integrated filling stations for dusty media;
- Suction systems for liquid and pulp materials;
- Transport systems consisting of liquid-proof, gas-proof or dust-tight containers;
- Vehicle superstructures with sealed loading areas.

Technical protective measures

Technical protective measures are needed in the setting up of appliances, tools and machines to be used, for example, in areas in which an explosive atmosphere cannot be excluded. In addition, these measures cover technical ventilation equipment used to dilute concentrations of gas, and casings and coverings used to prevent the release of dangerous material from the site to the atmosphere.

Technical ventilation is extremely important in biological on-/off-site remediation, where treatment is carried out in temporary buildings or enclosed halls. Along with conventional influencing factors (the process, the number and the output of the diesel engines used for

Figure 11.6. Ro-ro container with cover for transporting solid and paste-like waste matter.

Figure 11.7. Driver's cab with filter unit.

vehicles and appliances, the number of staff involved), it is necessary to suit the dimensions of the ventilator to the dangerous material. The efficiency of the technical ventilation system must be monitored constantly by measuring the respirable atmosphere.

Protective equipment for construction machines, appliances and tools

Earth-moving machines and other special machines for use in civil engineering, as well as other vehicles (e.g. dumpers) should be equipped with driver's cabs with filter units, according to the provisions of the "Code of practice for driver's cabs with filter units in earth-moving machines and other special civil engineering vehicles" (ZH 1/184). It is not always necessary to use the active carbon filter unit contained in the filter cartridge (Fig. 11.7).

Technical ventilation

If the measurements carried out to monitor the site show that dangerous material is present in harmful concentrations, suitable technical ventilation measures must be implemented. In the case of gaseous matter, it is recommended that ventilation is used that feeds fresh air to the area where work is being carried out (blowing ventilation). The location for the suction device of the air inlet should be placed at a sufficient distance from the source of the emissions, taking the prevailing wind into consideration, so that gases are not sucked into the system from the area near to the surface. Where a suction ventilation system is used, the rapid mixing, thinning and transportation of harmful gases, which a blowing ventilation system produces, is not achieved. In addition, there is a danger that harmful gases and vapours will be expelled in increased volumes, and in the worst possible case the ventilator can become a possible source of ignition.

A suction ventilation system should be used for dust-borne dangerous material to prevent clouds of dust being blown about. When a suction ventilation system is used, the behaviour of the dusts in fire and explosions should be examined.

In order to determine whether the ventilation measures are sufficient, repeated single measurements of the concentrations of dangerous material must be carried out. In addition, continuous measurements must be carried out for monitoring the oxygen content as well as the explosive atmosphere.

Enclosing

Measures have to be carried out together with the customer and the planning authorities that reliably prevent the transportation of dangerous material from contaminated areas into the neighbourhood of the construction sites, or which are necessary for the implementation of an optimum microbiological decomposition of the dangerous material. These measures include:

- Enclosing whole sites;
- Enclosing highly contaminated site areas, such as centres of contamination, contaminated buildings etc.;
- Enclosing individual areas such as, for example, in the disposal of dusts containing dioxin from derelict land;
- Enclosing biobeds;

Figure 11.8. Microbiological treatment of soil under the protection of an enclosure.

- Construction of temporary buildings in which the microbiological process is carried out.

Enclosures offer effective protection against dangerous material being carried into the ambient atmosphere of a site but they have the disadvantage that, because of the lack of natural ventilation, higher concentrations of the dangerous material develop inside the area in which work is being carried out. With the large volumes of air involved, technical ventilation systems can only be realized with a limited output. Because of the downstream filter units, this means that increased requirements must be met for protecting workplaces in the interior of the enclosure (Fig. 11.8).

The exhausted gases from all the construction machines used within an enclosure must be purified by suitable technical measures, such as catalytic converters, soot particle filters or an exhaust gas scrubber. Petrol engines should not be used in enclosed areas because of their high level of pollutant emissions. Even with the use of appliances driven by electricity, a technical ventilation system must be installed. The dimensions of this ventilation system should depend on each individual case.

Securing areas near the surface of contaminated sites

If unforeseen events bring work on the site to a standstill, or if contaminated excavated soil has to be stored temporarily, uncovered sections or intermediate storage areas for contaminated soil, as well as exposed drilled material, must be covered temporarily with suitable foil. It is recommended that gaseous emissions from the covered sections be extracted by a suction system and led off over an active carbon filter unit (Fig. 11.9).

Success in securing exposed sections can also be achieved by spreading foam, by using cohesive material and by covering the surface with concrete. Dust generation in contaminated areas should be prevented by wetting roads and paths, or by other means.

Organizational protective measures

Along with the technical protective measures, organizational protective measures are also important (Fig. 11.10). The arguments cited above are also essential for organizational protective measures. In addition, the following should be noted:

Figure 11.9. Temporary covering for contaminated material with integrated gas extraction.

Figure 11.10. Workplace monitoring with measuring equipment.

- Instructions should be given to employees regarding the danger on the site (dangerous material, biological agents);
- Before work begins and then at regular intervals (e.g. monthly), instructions should be given to employees regarding the use of protective clothing and equipment, and possibly first aid;
- Employees should have an emergency identity card, which shows the doctor or hospital to be informed;
- Listing of the names of the employees involved in the remediation work, and recording of this information together with the data on the hazards involved, as well as data on the biological agents used;
- Implementation of controlling investigations to accompany the decontamination work to assess whether the microbiological regeneration process has caused decomposition products to be generated in concentrations that could endanger health;
- Working out a measurement standards monitoring programme for workplaces in compliance with the technical regulations for dangerous material TRGS 402. The materials matrix is decisive for the use of the measuring appliances. Appliances that indicate directly and can be used in the field are to be preferred. In the case of work in sections in which, e.g. methane occurs in relevant volumes, it is essential that measuring appliances are used that work continuously, are equipped with visual and acoustic alarm functions, and that are able to monitor the lower explosion limits of combustible gases and vapours, as well as carbon monoxide content. These measures are supplemented by individual measuring but these should be coordinated by experienced technicians.

Personal safety equipment

The choice of personal safety equipment that has to be supplied in individual cases is dependent on:

- The nature and amount of dangerous material;
- Its concentration and mobility;
- The planned duties of the employee.

It is imperative that the relevant technical regulations for the quality of the equipment are observed, as well as those for the professional and health requirements of the users. To avoid buying unsuitable equipment, equipment should be purchased that carries the GS sign to show that it has been tested under the Appliance Safety Act. Where there is any doubt, the trade supervisory bodies or professional associations should be consulted.

Personal protection

As most contaminated areas contain matter that is absorbed via the skin, and hygiene

Figure 11.11. Basic personal safety equipment.

measures in the case of microbiological remediation and reclamation are essential, the following should be regarded as basic equipment for the prevention of skin contact (Fig. 11.11):

- Rubber safety footwear (anti-static/chemical resistant);
- Disposable protective clothing (with active breathing membrane);
- Chemical-resistant, tear-resistant gloves (gauntlet type), if possible with vinyl or cotton gloves underneath;
- Protective helmets;
- Face masks to protect against splashing.

Disposable protective clothing can be obtained in various qualities. The decision to buy material that is permeable to air, breathes or is coated with polyethylene depends on the expected dangers. The manufacturers or distributors of the protective clothing should be informed of the danger against which the material has to protect the wearer. In addition, care should be taken that, where possible, the material should be tear-resistant because of the mechanical stresses the suits will face. Where levels of pollution are low, the clothing may be worn more than once, if it shows no signs of damage.

If intensive skin contact with dangerous material is to be expected, or if highly toxic matter can be found on the site, a protective suit against chemicals, grade 1, DIN 32 762 (complete protection suit) with an integrated respiratory protective system should be worn.

Figure 11.12. Protective suit against chemicals. DIN 36762, grade 1, with a respiratory protective device independent of the ambient atmosphere.

The limits on the length of wearing the suit, TRgA 415, must be observed (Fig. 11.12).

Respiratory protection

If the development of gaseous and dusty emissions, and the formation of aerosols and vapours cannot be prevented by safety measures, suitable respiratory protective devices should be provided. If the workplace measurements, which must be carried out in all cases, show an excess of 10% of the relevant threshold limit value (lower toxic limit) or the TRK value, workers must use suitable respiratory protective devices. The provisions of the code of practice on respiratory protection must be observed. Along with the physical fitness of the wearers (determined by means of an occupational medical examination), they must be professionally qualified as well. These professional qualifications must be proven by a period of practical and theoretical training.

The decision to use filter devices (independent of the ambient atmosphere) or insulating devices (dependent on the ambient atmosphere), depends on both the degree of concentration of the expected dangerous materials, and the oxygen content of the ambient atmosphere. With an oxygen content of lower than 19% by volume the use of filter appliances is forbidden.

Respiratory protective devices should be

used for short periods only, where possible. The provisions of the TRgA 415, which lays down the restrictions on the periods in which respiratory protective devices may be worn, must be observed and complied with. Where temperatures are high and heavy physical work must be carried out, these devices should, where possible, not be used.

Fan-supported respiratory protective systems, which reduce breathing resistance as far as possible, and ventilated suits offering complete protection should be used to reduce the physical stress on the wearers of respiratory protective devices.

However, experience gained with a great number of applications of this kind of respiratory protective system has shown that, where the ambient temperature is less than 10°C, workers wearing respiratory protective devices catch colds. For this reason, fan-supported respiratory protective systems should only be used at these temperatures when the respiratory air supply and the air for the body has been prewarmed.

Discussion

Biological remediation is in itself not usually characterized by a particular danger potential. Appreciable accidental danger is mainly caused by the accompanying work, such as handling the contaminated material that is to be remediated or reclaimed. This potential danger can be reduced if, for example, volatile substances can be removed before the actual microbiological treatment takes place. However, a residual risk still remains, which can only be countered by the use of preventive measures. Careful monitoring of the microbiological parameters is, therefore, of central importance because this is the only way of determining the status of the biological remediation process. Only with knowledge of these data is it possible to judge whether the reduction of dangerous material is taking place in the desired manner. Only then can the formation of undesirable transformation products be excluded. Careful monitoring of the decomposition process of the dangerous material is therefore of fundamental importance, in particular as the metabolite question still has not been satisfactorily answered.

There is at present little research being carried out into the effects of mixtures of material on humans. So it is all the more astonishing in many cases just how carelessly customers and contractors act on these sites. Minimizing the problems involved helps as little as exaggerating them.

Measures in the field of work safety and protection from emissions, which should always be taken together, are preventive measures that contribute to the trouble-free running of the construction measures where they are applied with reference to the individual case and with corresponding knowledge. For the determination of the protective measures, the principle applies that, when assessing the expected risks for each case, the least favourable case should be used as a standard for the protective measures.

Extensive knowledge of safety regulations and of how to put these regulations into practice is necessary for the effective protection of workers on the site and the inhabitants of the neighbourhood. Often, those involved in this type of building measure lack this type of knowledge.

This deplorable state of affairs can only be remedied if sufficient attention is paid to the requirements of "safety", not only by the customer but also by the contractor, by training their employees. It is to be hoped that the standards with regard to invitations to tender for such work and the award of contracts, which at present are non-existent, will be drawn up as soon as possible.

With the compilation of the "Guidelines for work in contaminated areas", the reclamation of derelict land sub-committee in the civil engineering committee has drawn up a set of rules that will contribute to the trouble-free running of a building site, where they are applied with reference to the individual case.

Microbiological characterization of contaminated soils

K. Alef

Besides chemical, geological and hydrological studies of the risk potential of contaminated sites, the estimation of microbial activity and the quantification of microbial populations supports information on the physiological state of soil microflora or whether microorganisms do exist in the contaminated soil. In this chapter soil microbiological methods used in the remediation practice are presented and discussed. Methods for identification and counting of soil bacteria are also presented.

Sampling, transport and storage of soil

Planned soil sampling is very important for chemical and biological characterization of the contaminated site, particularly as achieving a representative soil sampling is nearly impossible in remediation practice. Soil sampling is strongly dependent on the distribution of the pollutants in the site. However, the statistical design of soil sampling strategy is of great importance. For experimental details the reader is referred to Chapter 2 (see also Liphard 1992). Soil samples should be transported in sterilized plastic bags (cooled, if possible) to the laboratory (DECHEMA 1992b). There are no general rules for the storage of soils; measurements should be carried out on the same day, otherwise samples can be stored at −20°C (Alef 1991; DECHEMA 1992b). For the determination of water-holding capacity (WHC), pH and water content of soils, the reader is referred to Chapter 3.

Estimation of microbial activity

Methods presented in Chapter 5 can be used for estimating microbial activity in contaminated soil; the estimation of soil respiration (O_2 consumption or CO_2 production) is the most frequently used method in remediation practice (DECHEMA 1992b).

Estimation of soil respiration in the presence of substrate

Principle of the method

The method is based on the determination of O_2 consumption or CO_2 production in contaminated soil mixed with glucose and incubated at 22°C for 24 h.

Materials and apparatus

Respirometer or other apparatus described in Chapter 5.

Chemicals and solutions

Glucose
NH_4Cl
K_2HPO_4

Procedure

One hundred grams of moist sieved (2 mm, 40–60% WHC) soil is mixed with 1 g glucose, 20 mg K_2HPO_4, 150 mg NH_4Cl and placed in the respirometer vessel. All samples are incubated at 22°C for 24 h and the O_2 consumption is recorded. The samples can be incubated for up to 7 days if the metabolism is low or the growth rate has a long lag phase (see also Chapter 5).

Calculation

The results are expressed in mg O_2 100 g^{-1} dry wt day^{-1}.

Estimation of soil respiration in the absence of substrate

Principle of the method

The method is based on the determination of O_2 consumption or CO_2 production in contaminated soil incubated at 22°C for up to 7 days.

Materials and apparatus

See Chapter 5.

Procedure

Up to 200 g of sieved moist soil (2 mm, 40–60% WHC) is placed in the respirometer vessel and incubated at 22°C. The incubation time should not exceed 7 days (see also Chapter 5).

Calculation of results

See Chapter 5

Discussion

The following points are valid for the measurement of soil respiration in the presence and absence of substrate.

1. Abiotic CO_2 production can be quantitatively important, especially in soils containing large amounts of carbonate. In this case a definitive determination of the activity can only be accomplished by measuring O_2 consumption.

2. The water content of the soil has a great effect on the measured activity, therefore the WHC should be adjusted to 50% to obtain comparable results.

Evaluation of results

The results of the tests describe the physiological state of the microorganisms in contaminated soil. In the case of the contaminants producing an inhibitory effect on the microorganisms, microbiological remediation may be limited, if no definitive respiratory activity can be found. The soil respiration in the absence of substrate should exceed 0.4 mg O_2 or 0.5 mg CO_2 (100 g^{-1} day^{-1}) and that in the presence of substrate 4 mg O_2 or 5 mg CO_2 (100 g^{-1} day^{-1} (Dott 1992; DECHEMA 1992b).

Quantification of microbial populations in contaminated soil

A lack of respiratory activity in the soil would provide evidence that there are no microorganisms or that there is a lack of nutrients (carbon source) in the sample. In order to remove this doubt, it is necessary to quantify microbial populations in the contaminated soils. For the enrichment and counting of viable microorganisms, media described in Chapter 4 can be used. By using media with different types and levels of nutrients, the range of species identified may be completely different. Also the use of different extractants or extraction methods can give different results (Alef 1991; DECHEMA 1992b). It is important that the medium used should facilitate the growth of the largest possible range of microorganisms.

Principle of the method

The method is based on the enrichment and counting of microorganisms from contaminated soil.

Materials and apparatus

See Chapter 4

Solutions and media

Tetrasodium pyrophosphate (0.2%)
NaCl (0.9%, sterile)
KCl (0.9%, sterile)
NaH_2PO_4 (1 M, pH 7.0, sterile)
Complex medium (g l^{-1} distilled H_2O)

- 0.50 g Glucose
- 0.50 g Yeast extract
- 0.50 g Proteose pepton
- 0.50 g Casamino acids
- 0.50 g Soluble starch
- 0.30 g Sodium pyruvate
- 0.30 g K_2HPO_4
- 0.05 g $MgSO_4.7H_2O$
- 15.0 g Agar

Adjust the pH to 7.2 by adding K_2HPO_4 or KH_2PO_4 before adding agar. The medium is autoclaved at 121°C for 20 min. After cooling to 45°C, the agar plates are prepared as described in Chapter 4.

Procedure

Contaminated soil (10 g) is suspended under aseptic conditions in 100 ml tetrasodium pyrophosphate solution. Sterile demineralized water can also be used. After shaking for 30 min (150 rev min^{-1}) at room temperature, the supernatant is transferred to a sterile measuring flask and allowed to stand for 2–5 min. An aliquot of the clear supernatant is then removed with a sterile pipette and a dilution series is prepared in decimal steps in a NaCl, KCl or NaH_2PO_4 solution. Aliquots of 0.1 ml of each dilution step (up to 10^{-7}) are spread on the agar plates and incubated at 20°C for 10 days. At time intervals (4, 7 and 10 days) the bacterial growth is evaluated by counting the bacterial colonies.

Evaluation of the results

An inhibitory effect of the contaminants in soil may occur if the detected vial bacterial count is less than 10^3 g^{-1} dry wt (DECHEMA 1992; Dott 1992).

Basic chemical analysis in contaminated soils

The determination of hydrocarbon content by infrared spectroscopy (according to DIN 38409, part 18)

H. Platen

Principle of the method

The method is based on the treatment of naturally moist soil samples with anhydrous sodium sulphate, followed by extraction of hydrocarbons and other lipophilic compounds with the solvent 1, 1, 2-trichlorotrifluoroethane.

Amphiphilic compounds are subsequently removed from the extract by passing it through an aluminium oxide-filled column. Hydrocarbons are detected with an infrared spectrophotometer at wave numbers ranging from 3200 to 2800 cm^{-1}. Carbon–hydrogen valance bonds of methyl, methylene and aromatic groups show maximum absorption at 2959, 2924 and 3030 cm^{-1}, respectively.

The hydrocarbon content is calculated by using standard calculation equations or by an external calibration. The detection limit for hydrocarbons is about 20 mg kg^{-1}. Fundamentals of IR spectroscopy are given by Coates (1990) and Peters et al (1976).

Materials and apparatus

Laboratory balance
Spoon
Erlenmeyer flasks with ground joints (NS29/32) (300 ml) with conical joint stoppers (NS29/32) and clips
Glass rod (20 cm × 5 mm)
Measuring cylinder (50 ml)
Rotary shaker
Glass columns (Fig. 11.13)
Funnel (glass or polypropylene)
Cellulose filters
Volumetric flasks (50 ml) with conical joint socket (NS14/23) and corresponding conical stoppers
Pasteur pipettes and pipette teats
2 quartz cuvettes (1 cm light path) with conical Teflon stoppers, transparent for infrared light
IR spectrophotometer
Microlitre syringes (10–25 µl)

Chemicals and solutions

Sodium sulphate, anhydrous
1,1,2-trichlorotrifluoroethane ($C_2Cl_3F_3$)
Aluminium oxide (neutral, 100–125 mesh, activity I)
Reference substances: petrol (gasoline), Diesel fuel, motor oil

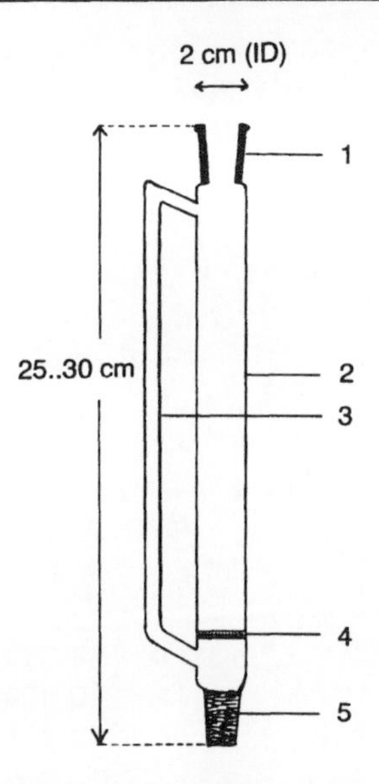

Figure 11.13. Glass column for the aluminium oxide treatment of extracts: (1) conical joint socket NS 14/23; (2) glass tubing 25 cm × 2 cm (i.d.); (3) glass tubing 7 mm (5 mm i.d.); (4) glass filter disc, pore size P1 (100–160 μm); (5) conical joint cone NS 14/23.

Procedure

Soil samples are directly analysed or stored at 4°C. Approximately 20 g of naturally moist soil are placed in an Erlenmeyer flask and the exact weight of soil is noted. After adding about 10 g of anhydrous sodium sulphate, the flask is closed with a glass stopper fastened with clip and shaken vigorously. If necessary, lumps should be removed with a glass rod; 50 ml of $C_2Cl_3F_3$ are then added, the flask is immediately closed and incubated for 30–60 min on a rotary shaker at 100 rev min^{-1}. The total content of the flask is filtered through a cellulose filter, which is situated in a funnel on a column containing 8 g of aluminum oxide (approximately 3.5 cm high). Once the solvent has passed through the filter, the funnel is removed and the column closed with a conical stopper. The extract is collected in a volumetric flask until all of it has passed through the column. Infrared transmission spectra of the extract are recorded against $C_2Cl_3F_3$ at wave numbers ranging from 3200 to 2800 cm^{-1} (scan speed of 500 cm^{-1} min^{-1}).

If transmission at the wave numbers 3030, 2959 or 2924 cm^{-1} is lower than 20%, the extract should be diluted as follows. The dry weight of the empty cuvette including the stopper is noted. The cuvette is then filled with $C_2Cl_3F_3$ to about 4/5 of its volume, closed with stopper and the weight determined. After the addition of a small amount of the extract, the cuvette is immediately closed and weighed.

The dilution factor (mfd) is calculated according to the following equation:

$$\text{Dilution factor} = \frac{(C) - (A)}{(C) - (B)} \tag{11.1}$$

where A is the dry weight of the cuvette, including stopper in grams, B is the weight of the cuvette, including stopper and $C_2Cl_3F_3$ in grams and C is the weight of the cuvette, including stopper, $C_2Cl_3F_3$ and extract in grams.

Calculation

Calculation by the standard equation

The following equations contain the assumption that various mineral oils consist of approximately constant ratios of methyl, methylene and aromatic groups. Extinction values at 2924, 2959 and 3030 cm^{-1} are calculated from corresponding transmission values by the following equation:

$$E = -\log(0.01\ T) \tag{11.2}$$

where E is the extinction and T is the transmission (%)

(A) Calculation of mineral oil content in soil in the absence of aromatic compounds (no peak at 3030 cm^{-1}). The equation is only valid for 1 cm cuvettes.

Mineral oil content (mg kg^{-1} dry soil) =

$$\frac{1.4 \times 10^5 \times V_{sol}}{m \times dsc}\left(\frac{E_{2959}}{8.3} + \frac{E_{2924}}{5.4}\right) mfd \tag{11.3}$$

where V_{sol} is the volume of the solvent used for extraction in millilitres, m is the weight of natural wet soil used for the extraction in grams, dsc is the dry substance of soil (%), E_{2959} is the extinction at 2959 cm^{-1} and E_{2924} is the extinction at 2924 cm^{-1}. The values 1.4×10^5, 8.3 and 5.4 are factors calculated from the specific absorption of methyl and methylene groups.

(B) Calculation of mineral oil content includes aromatic compounds (peak at 3030 cm^{-1}). The equation is only valid for 1 cm cuvettes.

Mineral oil content (mg kg^{-1} dry soil) =

$$\frac{1.3 \times 10^5 \times V_{sol}}{m \times dsc} \times \left(\frac{E_{2959}}{8.3} + \frac{E_{2924}}{5.4} + \frac{E_{3030}}{0.9} \right) mfd \qquad (11.4)$$

The values 1.3×10^5, 8.3, 5.4 and 0.9 are factors calculated from specific absorption of methyl and methylene groups.
Calculation by external calibration

A calibration curve is prepared by the dilution of 2, 4, 6, . . . 20 µl of mineral oil with 25 ml $C_2Cl_3F_3$. Mass concentration is calculated by multiplying the volume with its specific weight.

The extinction values obtained are plotted against the concentrations of the mineral oil in the solvent and the response factor is then determined from the slope of the regression line. Mineral oil content is calculated according to the following equation:

Mineral oil content (mg kg^{-1} dry soil) =

$$\frac{E_{2959} \times V_{sol} \times 100 \times mfd}{\text{Response factor} \times m \times dsc} \qquad (11.5)$$

where 100 is the correction factor of the dry soil substance and the response factor = E_{2959} / (mg l^{-1}).

This method provides good results if the particular type of soil contamination is known. In this case, calibration can be accomplished by utilizing the same type of mineral oil. The concentration of reference solutions given in microlitres per millilitre should be calculated in milligrams per millilitre using the specific weight of mineral oil.

Discussion

Soil samples (undried soil) are treated with anhydrous sodium sulphate to bind soil water. This enhances the quality of hydrocarbon extraction and prevents the disruption of IR measurement by traces of water. A weight of 10 g of anhydrous sodium sulphate is sufficient to treat 20 g of soil with a water content up to 25%.

Representative results can be obtained if the soil is homogenized prior to the analysis. This may lead to the loss of a great amount of volatile hydrocarbons. On the other hand large variability in results are to be expected, when heterogeneous soil is analysed. The analysis of samples depends, therefore, on the experiences of the laboratory personnel. Nevertheless, the reasons for the renunciation of homogenization should be noted in the analysis reports.

Dry weight of the soil is estimated by the incubation of the moist soil at 105°C for 24 h.

To increase the extraction yield of hydrocarbons, some analysts prefer to sonify soil suspensions for 15–30 min in an ultrasonic water bath.

A high loss of solvent and volatile hydrocarbons may occur during the slow filtration of the soil suspension. To avoid this, soil suspension is allowed to stand for several minutes and the supernatant is then applied directly to the aluminium oxide column. Experience, however, shows that the amount of extract obtained is not sufficient to perform the IR measurements.

A clear-cut transmission reduction at 3030 cm^{-1}, as described, was never observed. A

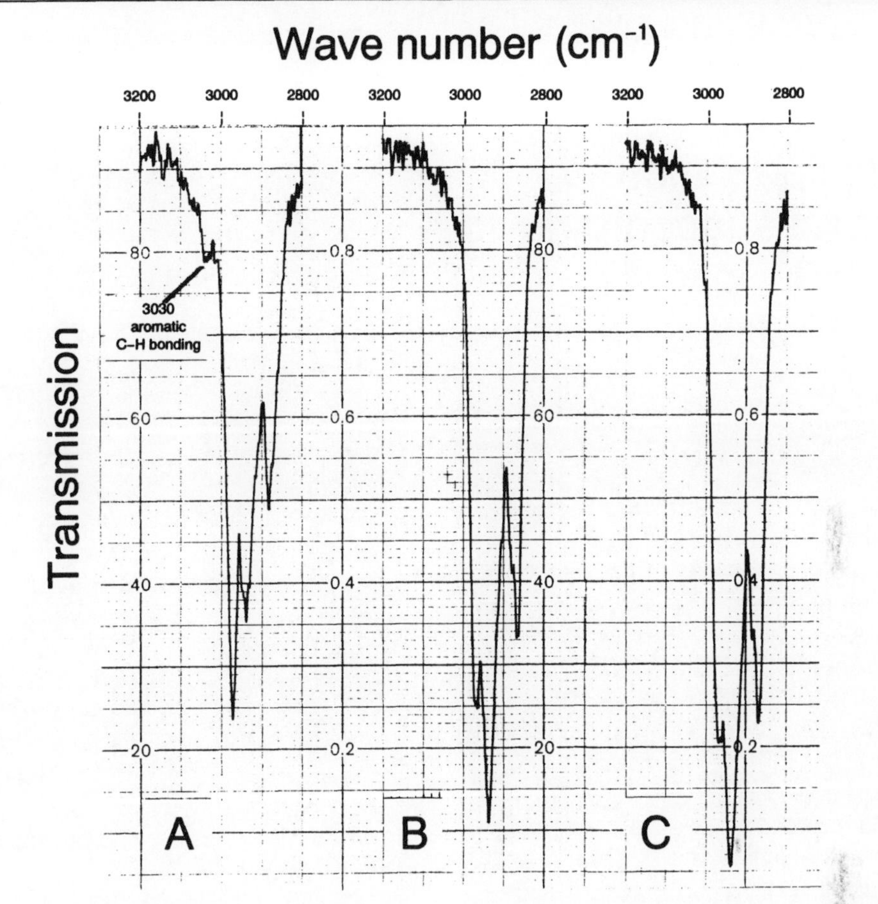

Figure 11.14. Reference absorption spectra of petrol (A), Diesel fuel (B), and motor oil (C), each 10 µl/25 ml dissolved in $C_2F_3Cl_3$, recorded with a Perkin Elmer type 283 IR spectrophotometer at a scan speed of 500 cm^{-1} min^{-1} using a 1 cm cuvette. The spectra of Diesel fuel and motor oil are very similar (peak at 2924 cm^{-1}), whereas petrol shows a significant shoulder at 3030 cm^{-1} and the main peak at 2959 cm^{-1}, due to its higher content of methyl groups.

typical spectrum for petrol is shown in Fig. 11.14; only a shoulder or a small peak could be detected in the region of 3030 cm^{-1}, when soil extracts are analysed. This could be due to the adsorption of hydrocarbons by soil particles and to the hydrocarbon mobility through soil (Bundesminister der Innern 1979).

The dilution procedure described can be used for small amounts of extract. Conventional procedures using dilutions in volumetric flasks (25 or 50 ml) require considerably larger amounts of both solvent and extract.

Some chemical compounds other than hydrocarbons (e.g. long-chain aliphatic alcohols, ethers, siloxanes, etc.) are retained in the solvent phase after the aluminium oxide treatment. Such compounds interfere with the estimation of hydrocarbons in the soil.

Soils contaminated with oil have a typical smell due to the presence of compounds such as aldehydes, ketones and ketone ethers.

The gravimetric determination of non-volatile lipophilic substances (according to DIN 38409, part 17)

H. Platen

Principle of the method

The gravimetric method allows the determination of the total content of non-volatile lipophilic compounds with a boiling point greater than 250°C (e.g., motor oil, fats, lipophilic polymerization products, etc.). The method is based on the treatment of soil samples with anhydrous sodium sulphate, followed by an extraction with the solvent $C_2Cl_3F_3$. After evaporating the solvent, the lipophilic compound content in the residue is determined gravimetrically. The detection limit of this method is about 60 mg kg^{-} in dry soil.

Materials and apparatus

Laboratory balance (accuracy to 0.1 g)
Laboratory analytical balance (accuracy to 0.1 mg)
Spoon
Erlenmeyer flasks with ground joints (NS29/32) (300 ml), conical joint stoppers (NS29/32) and clips
Glass rod (20 cm × 5 mm)
Measuring cylinder (50 ml)
Rotary shaker
Funnel (glass or polypropylene)
Aluminium foil
Cellulose filters
Pasteur pipettes and pipette teats
Drying oven
Heating plate

Chemicals and solutions

Sodium sulphate, anhydrous
Solvent: 1,1,2-trichlorotrifluoroethane ($C_2Cl_3F_3$)

Procedure

Soil samples are directly analysed or stored at 4°C. Approximately 50 g of naturally moist soil are placed in an Erlenmeyer flask and the exact weight of the soil is noted. After adding about 25 g of anhydrous sodium sulphate, the flask is immediately closed and shaken vigorously. If necessary, lumps should be removed using a glass rod. Solvent (50 ml) is added, the flask is immediately closed with a glass stopper and fastened with a clip. The flask is then incubated for 30–60 min on a rotary shaker at 100 rev min^{-}. After shaking, the flask is allowed to stand for several minutes and the supernatant is applied to a cellulose filter which is situated in a funnel on a measuring cylinder. The recovered volume of the extract is noted.

The total volume of the extract is transferred into a boat made of aluminium foil (10 cm × 10 cm × 1 cm) with a known weight. For greater accuracy, the use of boats with a weight of about 800 mg is recommended. After evaporating most of the solvent on a heating plate at 70°C (for about 20 min), the extract is then dried to a constant weight at 105°C (for about 10 min). The weight of aluminium foil with residual oil is noted (Fig. 11.15).

Calculation

Non-volatile lipophilic substances (mg kg^{-1} dry soil) =

$$\frac{nvl \times 5 \times 10^6}{V_{rec} \times m \times dsc} \qquad (11.6)$$

where *nvl* is the recovery of non-volatile lipophilic substances on aluminium foil in

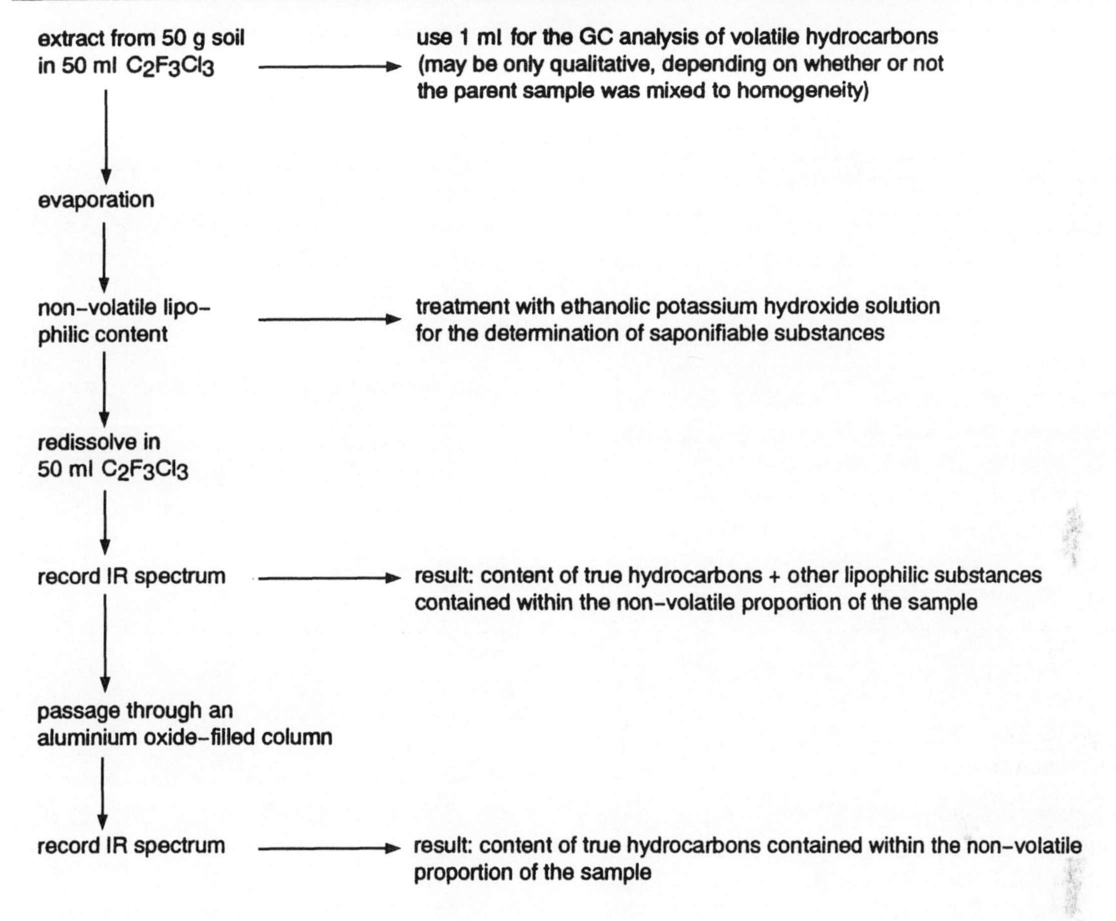

Figure 11.15. Analytical sequence scheme for the further analysis of non-volatile lipophilic substances, obtained by the method described.

milligrams, V_{rec} is the volume of solvent recovered after extraction in millilitres, m is the dry weight of moist soil used for extraction in grams, dsc is the dry substance of soil (%) and 5×10^6 is the factor to convert the result in milligrams per kilogram dry soil.

Discussion

If only non-volatile substances are to be determined, soil samples should be mixed to homogeneity prior to the analysis.

The dry weight of a soil sample should be determined by incubating the sample at 105°C for 24 h.

Twenty-five grams of sodium sulphate are sufficient to absorb water from a soil sample (50 g) with water content up to 25%.

To increase the extraction yield, some analysts prefer sonifying the soil suspension for 15–30 min in an ultrasonic water bath. In specific cases, where a higher temperature is necessary for the complete extraction of hydrocarbons (i.e. lipophilic substances such as tar, which are tightly bound to soil particles, and therefore not dissolved by the described extraction procedure), the Soxleth extraction method can be used, probably in combination with another solvent like toluene (Liphard 1987).

High losses of solvent may occur during filtration, therefore evaporation of the supernatant is recommended. During the sample preparation, sunlight should be avoided to minimize the evaporation.

Because of the sensitivity of gravimetric estimations, the use of round-bottomed flasks for the rotary evaporator is not recommended.

This method has not until now been used very commonly in environmental analyses. However, its determination is required by German law. The acceptable lipophilic content of soil and that considered to be in excess, are 600 mg kg^{-1} and 8000 mg kg^{-1}, respectively (Hessisches Ministerium für Umwelt-, Energie- und Bundesangelegenheiten 1992).

Pentane, hexane or toluene can also be used as solvents. However, they should not be evaporated as described in the procedure.

Analysis of the non-volatile substances can be performed by IR spectroscopy, after saponifiation with ethanolic potassium hydroxide solution (Hansen 1973) or by gas chromatography.

The gas chromatographic analysis of BTX (benzene, toluene and xylenes) and other aromatic and aliphatic volatile hydrocarbons (According to DIN 38407, part 9)

H. Platen

Principle of the method

The method is based on the treatment of soil sample with anhydrous sodium sulphate, followed by extraction with the solvent pentane. The extract is analysed directly by a gas chromatograph fitted with flame ionization detector (FID). If the elimination of oxidation products of hydrocarbons is desired, the extract can be purified prior to the analysis by passing it through an aluminium oxide- or florisil-filled column. The detection limit ranges from 0.1 to 1.0 mg kg^{-1}, for a pure single hydrocarbon and from 5 to 50 mg kg^{-1} for hydrocarbon mixtures containing about 50 or more compounds.

Materials and apparatus

Laboratory balance
Spoon
300 ml Erlenmeyer flasks, stoppers and clips
Glass rod (20 cm × 5 mm)
Measuring cylinder (50 ml)
Rotary shaker
Pasteur pipettes and pipette teats
Gas chromatograph fitted with an FID
Two or more different fused silica capillary columns. The results presented in this chapter are obtained with a fused silica capillary column DB624 (30 m, i.d. 320 µm; film thickness 1.8 µm; J&W Scientific, USA).
Various microlitre syringes (5–1000 µl volume)
Sample vials with Teflon septa

Chemicals and solutions

Sodium sulphate, anhydrous
Pentane
Reference substances

Petrol (gasoline), Diesel fuel, or pure substances such as benzene, toluene, xylenes or decane

Reference stock solutions (volume concentration: 1 ml l^{-1})

Pipette 50 µl of a reference substance into a 50 ml volumetric flask containing 40 ml of pentane and bring up with pentane to 50 ml. The concentration is calculated by considering the density value of the reference material.

Reference standard solutions

Dilute the stock solutions with pentane to obtain the following solutions: 100 μl l^{-1}, 10 μl l^{-1} and 1 μl l^{-1}.

Procedure

Soil samples are analysed directly or stored at 4°C. Hydrocarbons are extracted from soil samples with the solvent pentane (20 ml per 20 g moist soil). The soil suspension is allowed to stand for several minutes and approximately 1 ml of the clear supernatant is then added to the vial. One microlitre of the supernatant is used for gas chromatography analysis. The following GC operation conditions are recommended:

- Injection volume 1 μl (split ratio of SSL injector 1:50)
- Injector temperature: 325°C
- Detector temperature: 325°C
- Carrier gas: nitrogen at a flow rate of 1 ml min^{-1}
- Oven temperature programme: constant temperature of 50°C for 5 min; heating to 120°C at a rate increase of 4°C min^{-1}; heating to 160°C at a rate increase of 10°C min^{-1}; heating at a rate increase of 50°C min^{-1} (ballistically) to 260°C; maintain at 260°C for 21 min.

Chromatograms can be quantified if the peaks do not exceed the linear range of the

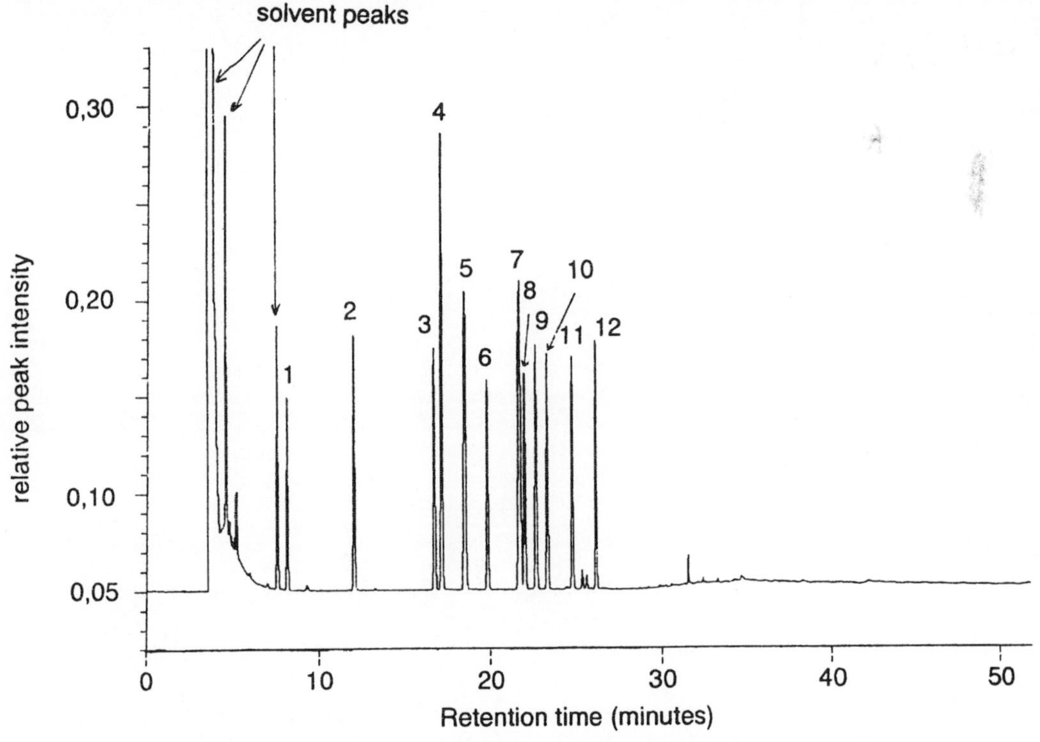

Figure 11.16. Chromatogram of aromatic hydrocarbon standard mixture (BTX and other aromatic derivatives, 100 μl l^{-1} each in pentane): (1) benzene; (2) toluene; (3) ethylbenzene; (4) *m*- and *p*-xylene; (5) *o*-xylene and styrene; (6) isopropylbenzene; (7) 3-ethyltoluene and 4-ethyltoluene; (8) 1,3,5-trimethylbenzene; (9) 2-ethyltoluene; (10) 1,2,4-trimethylbenzene; (11) 1,2,3-trimethylbenzene; (12) 1,2-diethylbenzene. The gas chromatographic conditions are described in the text. A relative peak sensitivity value of 0.1 corresponds to 100 μV.

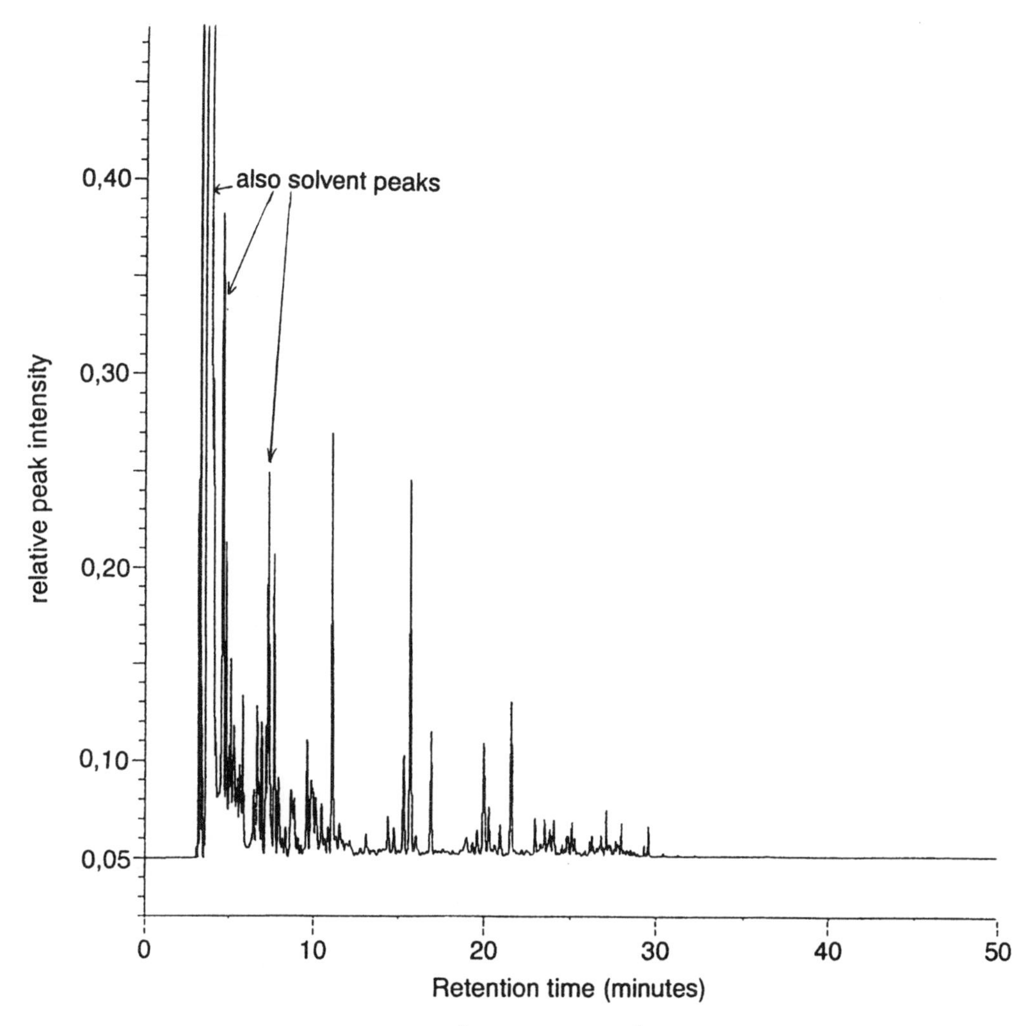

Figure 11.17. Chromatogram of petrol [5 ml l^{-1} (= 3590 mg l^{-1}) in pentane]. The arrows indicate peaks derived from the solvent. Pentane and the two other identified impurities are also compounds that naturally occur in petrol. It has to be decided on an individual basis, if this negatively influences the sample examination. The gas chromatographic conditions are described in the text. A relative peak sensitivity value of 0.1 corresponds to 100 µV.

detector. If necessary the supernatant should be diluted with pentane.

The dilution factor (*mfd*) can be calculated as follows:

$$mfd = V_d/V_{Ex} \qquad (11.7)$$

where V_d is the total volume after dilution and V_{Ex} is the volume of the original extract to be diluted.

Evaluation and calculation

Qualitative evaluation of a chromatogram

Sample identification is accomplished by comparing the peak pattern with those of the single substances, like BTX (Fig. 11.16) and/or mixtures like petrol and Diesel fuel (Figs 11.17–11.19).

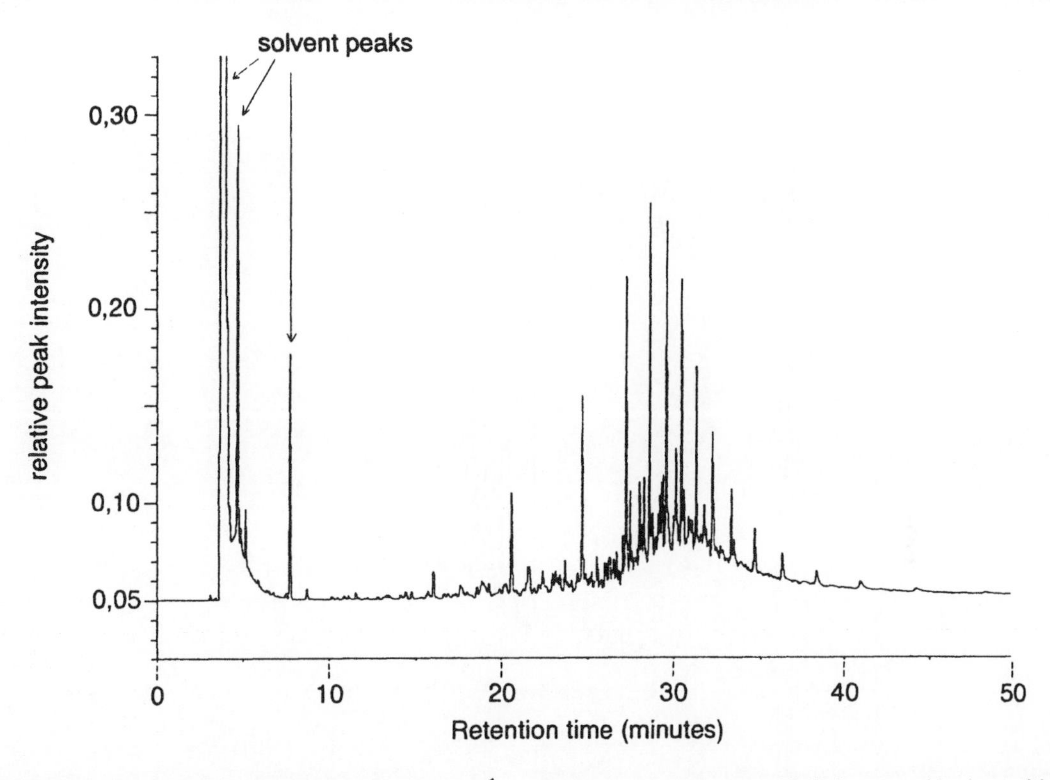

Figure 11.18. Chromatogram of Diesel fuel [5 ml l^{-1} (= 4130 mg l^{-}) in pentane L]. Solvent and its impurities (marked by arrows) do not negatively influence the evaluation of the chromatogram. The gas chromatographic conditions are described in the text. A relative peak sensitivity value of 0.1 corresponds to 100 μV.

Calculation

Solutions of single substances containing 1–100 mg l^{-1} of benzene, toluene, *o*-xylene, decane, etc., or solutions of complex hydrocarbons mixtures, such as petrol or Diesel fuel, containing 50–50,000 mg l^{-1}, are used for calibration. Peak areas (μV s^{-1}) are plotted against concentrations (mg l^{-1}) and the response factor is read from the slope of the plot (μV s^{-1}/mg l^{-1}). The mineral oil content in soil is calculated according to the following equation:

Mineral oil content (mg kg^{-1} dry soil) =

$$\frac{PA \times V_{sol} \times 100}{RF \times m \times dsc} \times mfd \qquad (11.8)$$

where *PA* is the peak area (μV s^{-1}), *RF* is the response factor (μV s^{-1}/mg l^{-1}), V_{sol} is the volume of solvent used for extraction (ml), *m* is the weight of naturally moist soil used for extraction (g), *dsc* is the dry substance of soil (%), *mfd* is the dilution factor and 100 is the correction factor (to calculate as soil dry weight). Response factors obtained for various hydrocarbons are very similar (Schomburg 1990) and, therefore, results can be calculated using the response factor of a single substance.

The use of this calculation method should, however, be noted in the analysis report as presented in the following example:

"Total hydrocarbon content (calculated for xylene as a reference) is 588 mg kg^{-1} dry soil."

Qualitative and quantitative results within the analytical report, including those for single substances, if they are identified in

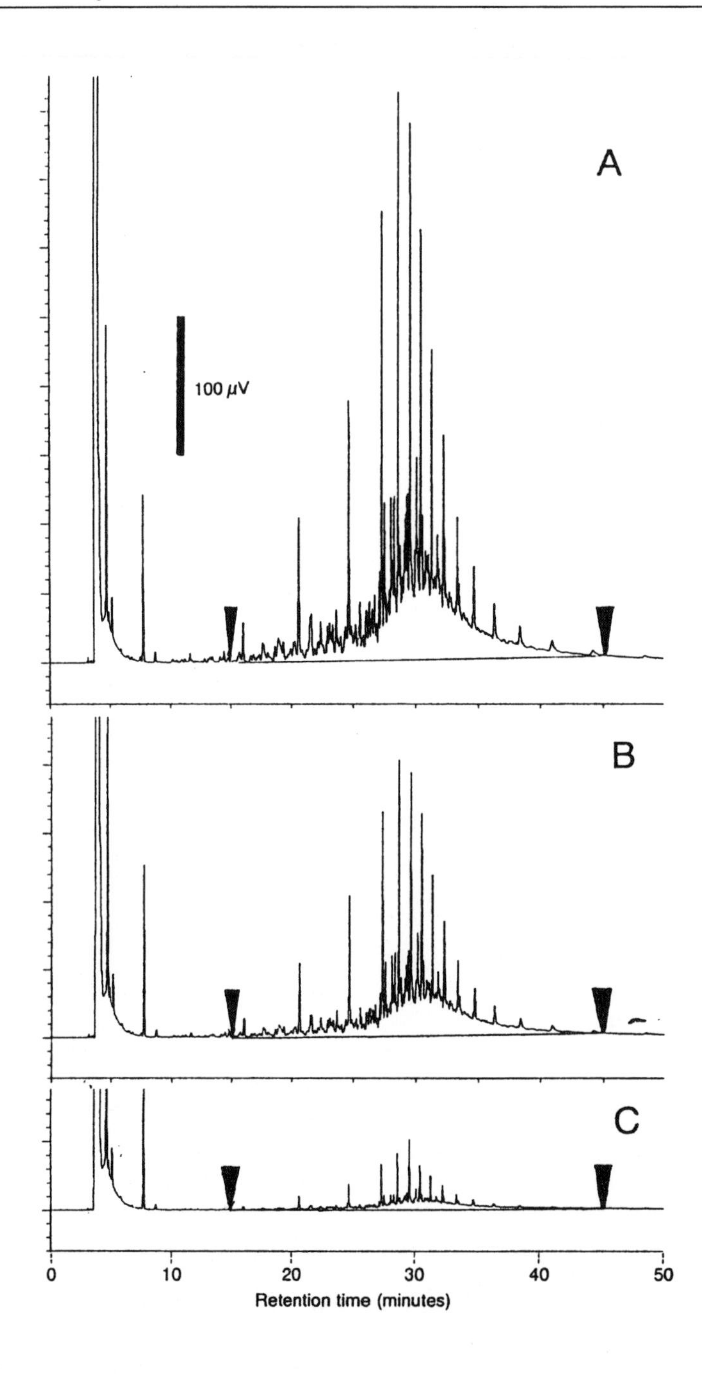

Figure 11.19. Calibration by peak grouping integration, determined using three different concentrations of Diesel fuel in pentane. The sum of all peaks area is calculated in the range between 15 and 45 min. (A) 1 ml l^{-1} (= 826 mg l^{-1}); (B) 5 ml l^{-1} (= 4130 mg l^{-1}); (C) 10 ml l^{-} (8260 mg l^{-1}). Integration of grouped peaks gave the following results (μV s): 4,497,000, 22,620,000 and 42,730,000, respectively. Response factors (μV s/mg l^{-1}) calculated for these results were 5444, 5477 and 5173, respectively (mean value: 5365; standard deviation: 3.1%). The black triangle marks the integration time frame, the horizontal line marks the baseline of the integration. The gas chromatographic conditions are described in the text.

complex mixtures, should be noted as follows:

"The peak pattern of the gas chromatogram indicates petrol contamination. The total hydrocarbon content (calculated for xylene as reference) is 588 mg kg^{-1} dry soil. This includes the following single substances (mg kg^{-1} dry weight): benzene 44; toluene 53; *o*-xylene 34".

or

"The peak pattern of the gas chromatogram is not similar to those of petrol or Diesel fuel. The total hydrocarbon content (calculated with the use of xylene) is 799 mg kg^{-1} dry soil. The single substances benzene, toluene, *o*-, *m*- and *p*-xylene, and ethylbenzene were not identified. The source of soil contamination is, therefore, unknown; the solvent mixture requires further examination."

Discussion

The dry weight of the soil should be estimated by incubating the sample at 105°C for 24 h.

Ten grams of sodium sulphate are sufficient to absorb water from 20 g soil with a water content of up to 25%.

Trichlorotrifluoroethane or tetrachloromethane can be used as extractants. A suitable solvent should be selected.

To increase the extraction yield, some analysts prefer to sonify the soil suspension for 15–30 min.

Substances that interfere with the hydrocarbons estimation can be removed from the extract by filtering it through an aluminium oxide column (e.g. a Pasteur pipette containing approximately 2 g of aluminium oxide or florisil). However, in practice, this purification step can be omitted because the use of overlays of sample chromatograms in conjunction with reference chromatograms provides an indication of possible changes that have occurred to the original pollutant.

The use of internal calibration is not recommended because internal reference substances are often coeluted with soil contaminants. This makes an accurate calculation of the hydrocarbon recovery rates very difficult.

A great variety of columns and temperature programmes are available as alternatives for performing such analysis (Farwell 1990; Schomburg 1990). The sensitivity of the analysis depends on the split ratio, as well as on other typical GC characteristics.

The identification of a specific substance should be confirmed by performing a second analysis with a different column.

Motor oil and hydrocarbons with a boiling point greater than 450°C can not be detected by this method.

Results should be stated in analytical reports as follows: "The sample chromatogram is absolutely identical to the reference chromatogram for Diesel fuel" or "The sample chromatogram is similar to the reference chromatogram for petrol". The composition of mineral oils in soils can be altered over time. This would lead to an altered peak pattern in the original contaminant (Steiof et al 1993). This fact should be considered in the interpretation of the results.

Analysis of volatile halogenated hydrocarbons in soil

H.M. Berstermann

Volatile halogenated hydrocarbons are often found as contaminants in industrial sites (cleaning, degreasing). They include fluorinated, chlorinated, brominated or iodated compounds with 1–6 carbon atoms and a boiling point range from 20 to 180°C.

These substances can be detected in soil by either:

1. Extraction followed by capillary gas chromatography (DIN 38407 F4, 1993), or
2. Headspace-coupled capillary gas chromatography (DIN 38407–F5, 1993).

A comparison between the methods shows that headspace chromatography yields higher recovery rates than Soxhlet extraction or extraction by shaking with pentane (Hagendorf et al 1987). Soil air can also be analysed by collecting the air on charcoal, followed by extraction and capillary gas chromatography. This presents a fast and cheap method for the analysis of volatile halogenated hydrocarbons in soil by headspace-coupled capillary gas chromatography.

Principle of the method

A soil sample is transferred into a headspace vial. Heating the vial causes the volatile substances to be distributed between the solid and gas phases. An aliquot of the gas phase is injected into a gas chromatograph via a heated transfer line, separated on a suitable capillary column and detected by a mass spectrometric or electron capture detector. According to DIN 38407–F5, it is necessary to analyse two columns of different polarity, when working with a GC-ECD. When using the two column-technique, splitting of the sample after the injection port is recommended. In our experience it is possible to use only one column in the routine analysis of samples of lower concentration.

Range of application

All types of soils. If hydrocarbon content is higher than 10%, see "Calibration" (p. 514).

Materials and apparatus

Headspace vials
Automatic headspace sample changer (e.g. DANI-Headspace 3950)
Capillary gas chromatograph with mass spectrometric detection (GC-MS) or election capture detector (GC-ECD)
Capillary column (30–50 m), film thickness 0.5–1.0 cm, i.d. 0.32 cm, e.g. Durabond 1 (DB1), Durabond 5 (DB5), SE 54 CB, Durabond 1701 (DB 1701)
Computer or integrator for data collection and processing
Balance (10 mg precision)
Calibration vessels, volumetric flasks etc.

Chemicals and reagents

Double-distilled water

Water for calibration mixtures and blanks has to be analysed before use and must be free of halogenated hydrocarbons. Otherwise it should be cleaned by flushing with nitrogen (180 ml min^{-1}) for 1 h at 60°C in a clean (free of halogenated hydrocarbons) 2 l glass bottle. Flushing with nitrogen should also be performed while cooling down to room temperature.

Reference substances of high purity grade

Internal standards: 1-bromo-3-chloropropane, 1-bromo-2-chloroethane, bromotrichloromethane. The calibration mixture is prepared by weighing 100 mg each of the following chlorinated hydrocarbons (analysis according to the German drinking water rules, TVO) into 50 ml of methanol:

Dichloromethane
Trichloromethane
Tetrachloromethane
1,1,1-Trichloroethane
Trichloroethene
Tetrachloroethene

The spectrum of analysed substances can be extended by adding additional halogenated hydrocarbons to the calibration mixture. This stock solution can be stored in a dark cool place for several weeks. For GC-MS analysis the stock solution is diluted 1:10, for GC-ECD analysis 1:200 (2 dilution steps) with methanol.

The standards are prepared by pipetting 10, 50, 100 and 250 μl of the calibration mixture into 1 l of unloaded water. The water is stirred well with a Teflon stirrer whilst injecting the calibration mixture under the water surface.

The calibration samples should have the following concentrations

GC-MS analysis: 2, 10, 20 and 100 $\mu g\ l^{-1}$

GC-ECD analysis: 0.1, 0.5, 1 and 5 $\mu g\ l^{-1}$

The analysis of blanks (unloaded water used for the preparation of calibration samples) as well as calibration mixtures are carried out in triplicate.

Calibration using unloaded soil samples is also possible. In this case 10 g of unloaded moist soil are weighed into headspace vessels and then spiked with the calibration mixtures. The calibration mixtures have to be diluted 1:100.

Procedure

Sampling and preparation of the samples

The sample bottles, headspace vials and all other vessels should be heated for 1 h at 150°C immediately before use. The soil samples have to be transferred into the sample bottles directly after sampling and must be stored at 4°C. The time between soil sampling and filling of the headspace vial as well as that between filling of the headspace vial and analysis should not exceed 48 h. The samples should not be homogenized, dried or sieved. Ten grams of sample are weighed into the headspace vial, which is immediately sealed (before placing the vials inside the thermostat, try to rotate the aluminium cap seal by hand; if loose, put a thick septum on the cap and press again with the hand crimper).

Analysis

The headspace vials are heated to 80°C for at least 1 h. An aliquot of the gas phase is then automatically injected into the gas chromatograph by the automated sampling system (Figs 11.20 and 11.21). The parameters of the headspace sampler (DANI HSS 3950) are: single sampling, pressurization for 10 s, filling the loop for 5 s and injection for 10 s.

Time (s)	Function
0	The probe needle penetrates the vial
1	Pressurization starts
11	Pressurization stops
12	Filling of the loop starts
17	Filling of the loop stops
18	GC injection starts, GC programme starts
28	GC injection stops
29	The probe moves up

Bath temperature:	80°C
Manifold temperature:	100°C
Transfer tube temperature:	150°C
Sampling interval:	20 min
Sample loop:	1 ml

Parameters of the gas chromatograph (VARIAN 3400):

Carrier gas: hydrogen
Carrier gas pressure: 14 psi
Carrier gas flow: approximately 3 ml min^{-1}
Split flow: 15–20 ml min^{-1}
Injector temperature: 200°C
Column: Macherey + Nagel, Permabond SE-54-54-DF-0.50; 50 m × 32 mm i.d.

Oven programme

Start temperature (°C)	End temperature (°C)	Rate (°C min^{-1})	Time (min)
32	32	0.0	3.3
32	60	8.0	3.5
60	60	0.0	3.0

Parameters of the Finnigan ITS40 mass spectrometer:

Low mass: 30 amu
High mass: 250 amu
Background mass: 29 amu
Scan rate: 1000 ms
Scan mode: Electron ionization with automatic gain control

Calibration

For calibration an unloaded water sample is spiked with standards. If the soil is highly loaded with hydrocarbons or with oil, the calibration must be carried out using an internal standard (e.g. bromodichloromethane).

In our experience, it is possible to work with external standards, especially when large series are analysed. To control the precision of the external calibration, it is necessary to analyse a calibration sample at least for every 20 samples in a series.

Evaluation

Halogenated hydrocarbons are identified by their retention times and mass spectra in

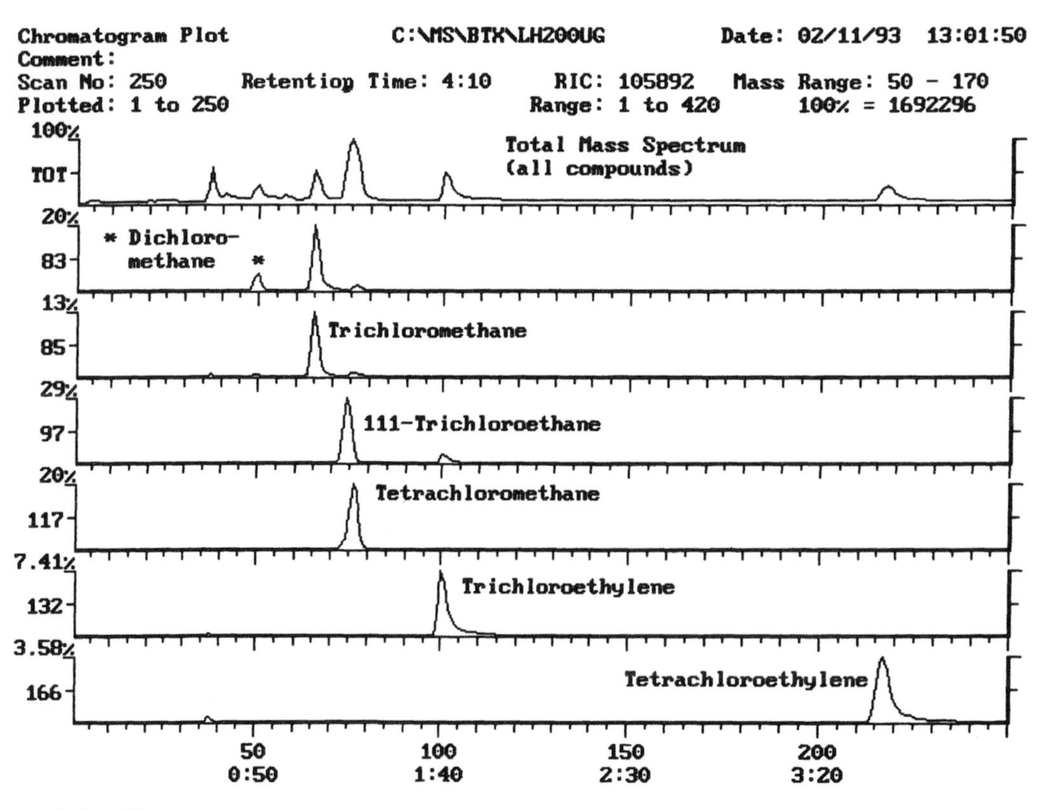

Figure 11.20. The total mass spectrum (top) and the spectra of the characteristic masses of the different volatile halogenated hydrocarbons.

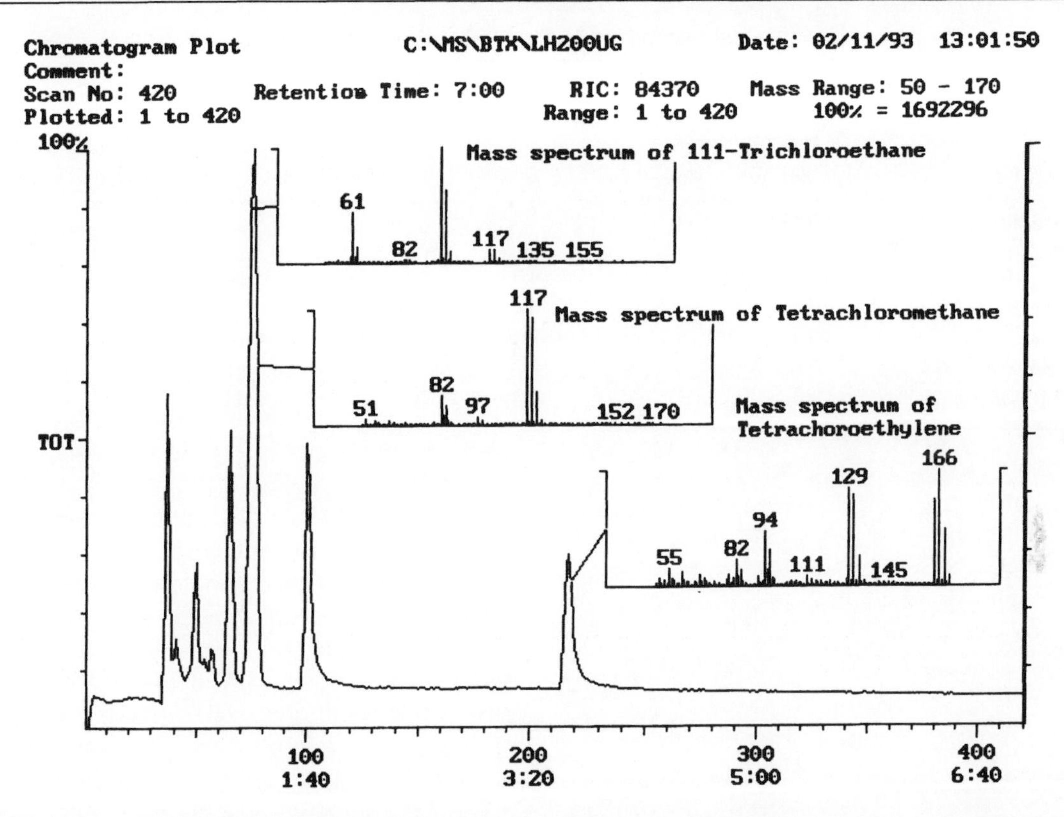

Figure 11.21. Mass spectra of some compounds. The large peak with retention time of 1:18 min is due to 1,1,1,-trichloroethane and tetrachloromethane. It is possible to separate the substances by their mass spectra (see also Fig. 11.20) and in this way to quantify each substance.

GC-MS analysis, which are characteristic for halogenated substances (see Fig. 11.20). Quantification is carried out by integration of the characteristic mass.

In GC-ECD analysis the halogenated hydrocarbons are only identified by retention times on one or two columns. Quantitative evaluation is done by a regression curve evaluated by the programme for data collection and data processing of the gas chromatograph, or the mass spectrometer (Finnigan ITS 40).

Discussion

The equilibration time in the bath of the headspace changer could last several hours. To determine the optimum equilibration time, an aliquot of the sample has to be analysed at different times during the incubation. A plot of peak area versus equilibration time shows the optimum time when the curve reaches a plateau (Technical bulletin Fa. DANI, "Headspace sample unit 3950").

Using water as a calibrant matrix is faster and provides better correlation coefficients than using unloaded soil. In our experience there is no significant difference between using water or soil. The differences between two soils (e.g. sand and clay) are often greater than those between water and soil. It is possible to test the suitability of the calibration curve by spiking a contaminated sample.

Manual injection of a gas-phase aliquot using gas-tight syringes is not suitable for quantitative analysis. It is possible to use this technique for semi-quantitative analysis.

Table 11.1. Soil contamination values according to the "Netherlands list".

Substance	*Reference value (mg kg^{-1} TS)*	*Levels suggesting further investigation necessary (mg kg^{-1} dry weight)*	*Sanitation limit value (mg kg^{-1} TS)*
Naphthalene	0.1	5	50
Anthracene	0.1	10	100
Phenanthrene	0.1	10	100
Fluoranthene	0.1	10	100
Pyrene	0.1	10	100
3,4-Benzpyrene	0.05	1	10
PAH (total)	0.1	20	20

Contamination of calibration solution or samples by halogenated solvents normally used in laboratory is a serious problem when using a GC-ECD. Dichloromethane, especially, is a frequent contaminant in the laboratory.

Some testing protocols recommend addition of water to the sample in the headspace vial, to increase the precision and the sensitivity of the analysis. In our laboratory, the effect of water addition to naturally moist soil has always been negligible. It may be a possible source of sample contamination and, for this reason, it should be avoided. Very dry samples (< 5% water) have to be moistened, however, to obtain a water-saturated gas-phase in the heated headspace vessel.

The standard deviation has been found to range from 2% to 5%, and the extraction recovery from 81% to 91%; measurement sensitivity on GC-MS is usually equal to 2 µg kg^{-1}.

Analysis of polycyclic aromatic hydrocarbons in soil

B. Schieffer

Because of the carcinogenic potential of some species, the analysis of polycyclic aromatic hydrocarbons (PAH) in contaminated soils has become increasingly important during the last few years. PAHs originate from the incomplete combustion of organic matter, e.g. in heating installations, in power plants or in engines. At present, PAHs are found in air, water, sediment and soil. Because of their relatively slow degradation, they accumulate in the food chain.

The Environmental Protection Agency (EPA 1976) put 16 PAHs on to the priority pollutant list. These PAHs are either highly distributed in the environment or represent a high carcinogenic potential.

For drinking water, the limits for six PAHs are defined, while for soil contamination (Table 11.1), there are only the limits provided by the "Netherlands list".

Testing methods for drinking water in Germany are:

GC-FID (VDI 3872)
HPTLC (DIN 38407-F7)
HPLC (DIN 38407-F8)
TLC (DIN 38409-F13)

An analysing method using capillary gas chromatography and mass spectrometric detection (GC-MS) is recommended by the EPA. There are no regulations for the analysis of solid matter or soil samples.

There are several methods available for preparing contaminated soil samples. Extraction can be performed in an ultrasonic bath or in a

Soxhlet extractor (Brilis and Marsen 1990). There is a wide variety of extractants with satisfactory recovery rates (Novotny et al 1974; Björseth 1989).

After sample drying, homopolar solvents like cyclohexane, hexane, dichloromethane or toluene can be used for extraction. Moist samples can be extracted without drying using methanol or methanol/benzene (1:1). Cleaning and enrichment of PAHs is normally done with solid-phase columns (Sephadex, Silica or C18).

A fast and precise method suitable for a routine laboratory with high throughput of samples will be presented and discussed.

Principle of the method

The moist soil is extracted several times with methanol, and the dissolved PAHs are enriched on C18 solid-phase extraction columns and eluted with dichloromethane. Analysis is carried out by capillary gas chromatography and mass spectrometric detection.
Determination limits are

5 µg kg^{-1} (dry matter) for phenanthrene, fluoranthene and pyrene and 10 µg kg^{-1} (dry matter) for the other EPA-PAHs.

Range of application

Soils with water concentration under 40%. At higher water concentrations, the sample must be dried by mixing it with sodium sulphate.

Materials and apparatus

Screw cap vessels (100 ml)
Dispenser 5–50 ml
Ultrasonic bath
Centrifuge
Pipettes (up to 1000 µl)
Capillary gas chromatograph with mass spectrometric detection (GC-MS) (e.g., Finnigan ITS 40)

Chemicals and solutions

Methanol, nanograde
Dichloromethane, nanograde
Internal standard solution I (stock solution)

Deuterated PAHs (Naphthaline d8, Acenaphthene d10, Phenanthrene d10, Crysene d12, Perylene d12) – 4 mg of each substance is dissolved in 1 ml of a mixture of Dichloromethane d2 (80%) and Benzene d6 (20%) (Cambridge Isotope Laboratories)

Internal standard solution II

Deuterated PAHs, each 100 mg l^{-1} in methanol

Internal standard solution III

Deuterated PAHs, each 10 mg l^{-1} in methanol

Calibration solution I (stock solution)

PAH-Mix (EPA) each 100 mg l^{-1} in toluene (Fa. Promochem)

Calibration solutions II, III, IV, V

10, 100, 500 and 1000 µl of calibration solution I in 10 ml

Methanol produces

Calibration solution II: 0.1 mg l^{-1} of each PAH
Calibration solution III: 1.0 mg l^{-1} of each PAH
Calibration solution IV: 5.0 mg l^{-1} of each PAH
Calibration solution V: 10.0 mg l^{-1} of each PAH

Octadecyl (C18) solid phase extraction column 6 ml, Fa. Baker.
Bidistilled water

Procedure

Preparation of the samples:

Extraction

Depending on the expected

concentrations, up to 10 g of a homogenized moist sample are weighed (10 mg precision) into a screw cap vessel. The internal standard (0.5 ml of internal standard solution III) is added to the sample. The sample is extracted three times with 10 ml methanol in an ultrasonic bath (20 min). The temperature should not exceed 25°C. Following each extraction, the sample is centrifuged and the extracts are combined in another screw cap vessel.

Enrichment

Conditioning of the solid-phase extraction (columns) is performed by washing them twice with water and once with bidistilled water (5 ml each). The combined extracts are spiked with 5 ml internal standard solution III (5 µg of each substance) and diluted with 30 ml bidistilled water. This water–methanol solution is sucked over a solid-phase extraction column. The loaded column is dried by a nitrogen stream for 30 min, followed by centrifugation at 4000 rev min^{-1} and 30 min drying in an N_2 stream. The organic substances, collected from the column are then eluted with 1.5 ml dichloromethane into the autosampler vial. The vial is sealed and the sample analysed.

Analysis

Parameters of the gas chromatograph (VARIAN 3400):
Carrier gas: hydrogen
Carrier gas pressure: 14 psi
Carrier gas flow: 3 ml min^{-1}
Split flow: 15–20 ml min^{-1}
Injector (PTV)
Start: 60°C
End: 300°C
Rate: 300°C min^{-1}
Transfer line: 240°C

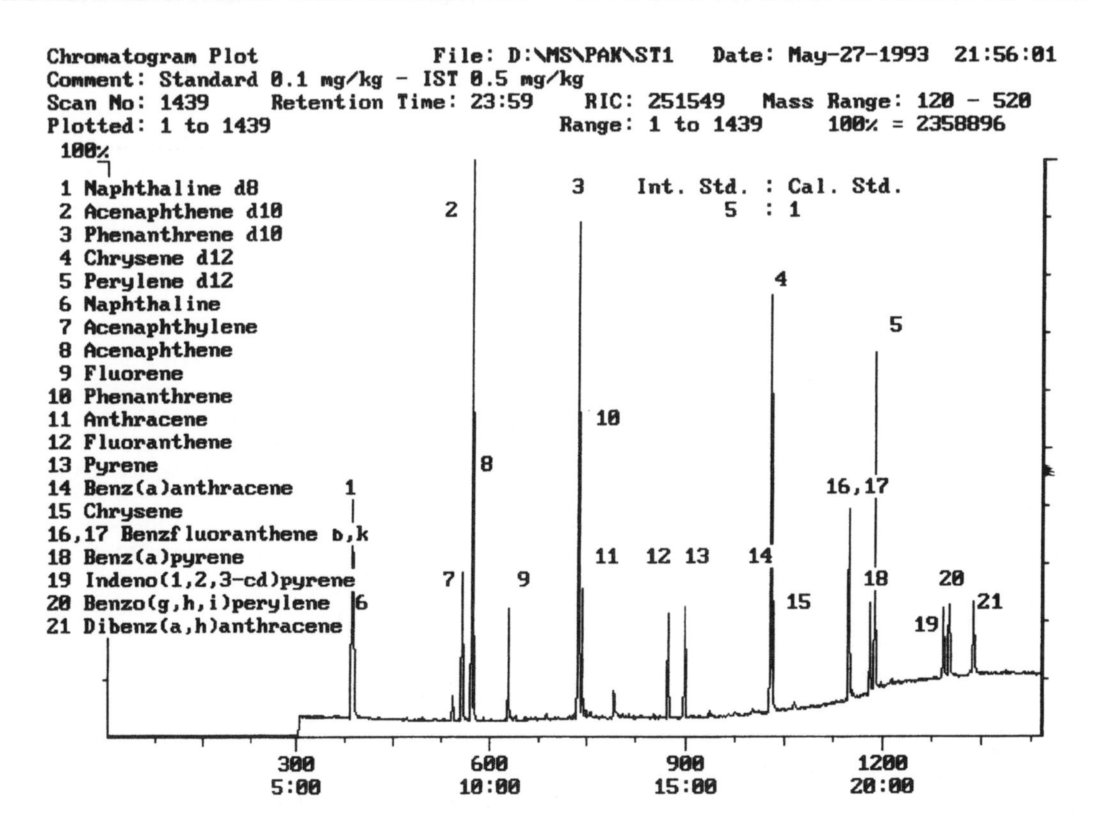

Figure 11.22. Retention of PAHs according to EPA method together with those of the internal standard.

Column: HT 8 (SGE GmbH), i.d. 0.32 mm, film thickness 0.25 μm
Carrier gas: hydrogen
Column pressure: 1.2 bar

Column table:

Start temperature (°C)	End temperature (°C)	Rate (°C min^{-1})	Time (min)	Total time (min)
60	60	0.0	2.0	2.0
60	330	15.0	18.0	20.0
330	330	0.0	6.0	26.0

Parameters of the Finnigan ITS40 mass spectrometer:

Low mass: 120 amu
High mass: 520 amu
Background mass: 119 amu
Scan rate: 1000 ms
Scan mode: electron ionization with automatic gain control
Acquire time: 24 min

Retention of the PAHs according to the EPA method together with those of the internal standards is shown in Fig. 11.22. The characteristic mass spectrum of a PAH is nearly that of its molecular mass.

Calibration

Calibration is performed using internal standards with the following deuterated substances

Naphthaline d8; acenaphthene d10; phenanthrene d10; chrysene d12; and perylene d12. To obtain data for the calibration curve, 10 g samples of unloaded soil are spiked with 1 ml of each of the calibration mixtures II, III, IV and V, respectively. The concentrations of the calibrants are: 10, 100, 500 and 1000 μg kg^{-1}. Calibrants are treated in the same way as analytical samples.

Quantification is done by integration of the characteristic mass, as shown in Table 11.2.

Table 11.2. Characteristic mass for various polycyclic aromatic hydrocarbons (OAHs).

PAH	*Characteristic mass*	*Deuterated PAH used for calibration*
(1) Naphthaline d8	136	
(2) Acenaphthene d10	162	
(3) Phenanthrene d10	182	
(4) Chrysene d12	240	
(5) Perylene d12	264	
(6) Naphthalene	128	Naphthaline d8
(7) Acenaphthylene	152	Acenaphthene d10
(8) Acenaphthene	152,153	Acenaphthene d10
(9) Fluorene	165	Acenaphthene d10
(10) Phenanthrene	178	Phenanthrene d10
(11) Anthracene	178	Phenanthrene d10
(12) Fluoranthene	202	Phenanthrene d10
(13) Pyrene	202	Phenanthrene d10
(14) Benz(a)anthracene	228	Chrysene d12
(15) Chrysene	228	Chrysene d12
(16,17) Benzofluoranthene (b,k)	252	Chrysene d12
(18) Benz(a)pyrene	252	Chrysene d12
(19) Indeno (1,2,3-cd) pyrene	276	Perylene d12
(20) Benzo (g,h,i) perylene	276	Perylene d12
(21) Dibenz (a,h) anthracene	278	Perylene d12

Evaluation

Quantitative evaluation is done by using a regression curve evaluated by a programme for data collection and data processing of the mass spectrometer (Finnigan ITS 40). Alternatively the evaluation of the regression curve can be done by a separate programme or according to the German standard methods DIN 38402–A51 and DIN 51405.

Discussion

It is useful to test loads of unknown samples by a semiquantitative test to estimate possible dilution steps. Ten microlitres of the combined methanolic extract are tipped on a thin layer chromatography plate (0.1 mm/acetylated cellulose). The intensity of fluorescence is estimated using UV light. For comparison, two calibration solutions (10 μl of 1 mg l^{-1} and 5 mg l^{-1}) are treated in the same way. The standard deviation ranges from 3% to 8% and the measurement sensitivity from 2 to 10 μg kg^{-1}; the extraction recovery ranges from 67% to 89%.

Because of their similar retention times on the chosen column, the benzofluoroanthenes (b,j,k), which have identical molecular weights, cannot be separated by the mass spectrometer. In this case the sum of their retention time is considered.

If the soil sample is highly contaminated, cleaning of the dichloromethane extract can be performed by washing it over benzenesulphonic acid with hexane as an eluent.

This method is not suitable for soils with hydrocarbon concentrations higher than 100 mg kg^{-1} (the Netherlands list), because of decreasing the recovery rate.

Determination of the extractable halogenated hydrocarbons in soil samples

I. Lorenz

Detailed analytical investigation of organic compounds in soil samples requires equipment and is time consuming. Because of the great variety of organic pollutants, the determination of single components is often inefficient. Therefore, the first step should be a screening method.

For the detection of halogenated organic compounds in many contaminated sites, the EOX method is considered to be useful. The following method is described according to German standard method DIN 38414 part 17 (1989) for the examination of sludge and sediments (group S).

Principle of the method

The soil sample is extracted in a Soxhlet apparatus with a suitable solvent. A certain amount of the extract is combusted in a H_2/O_2 flame (Wickbold apparatus). The elements (halogens) are directly determined as ions in a hydrous solution. The combustion product (condensate) contains halides, which are determined as chloride with suitable analytical methods, e.g. potentiometry or ion chromatography.

Range of application

The method described below is suitable for the determination of the total content of non-volatile halogenated organic compounds in contaminated soil at the level of about 1 mg kg^{-1}.

Materials and apparatus

Soxhlet extractor
Water bath
Rotary evaporator

Hydrogen–oxygen combustion apparatus (e.g. Wickbold apparatus) with combustion burner for liquids
Halogen detection apparatus (e.g. Potentiometer, ion chromatograph)
Calibration vessels
Pipettes

Chemicals and solutions

All substances and chemicals must be of high grade purity (pro analysi, p.a.). The purity of the chemicals and gases should be tested by blank values
Sodium sulphate, Na_2SO_4, 1 h at 600°C
Hexane, C_6H_{14}
Oxygen
Hydrogen
NaOH (100 mM)

Dissolve 4 g NaOH in about 800 distilled water and bring up with distilled water to 1000 ml

p-Chlorophenol, C_6H_5ClO
Chlorophenol stock solution, c(Cl) = 100 mg l^{-1}: 36.3 mg chlorophenol in 100 ml hexane (this can be stored for 1 month)
Chlorophenol standard solution, c(Cl) = 10 mg l^{-1}

Dilute chlorophenol stock solution 1:10 with hexane (this can be stored for 1 week)

Procedure

Fresh soil sample (equivalent to 20 g dry weight) is mixed with approximately 10 g Na_2SO_4. The mixture is extracted in a Soxhlet apparatus with 150 ml hexane for 20 h. The solution is then concentrated in a rotary evaporator to less than 50 ml, put in a 50 ml calibration vessel and filled up with hexane to 50 ml.

The extract determined (25 ml) is combusted in a H_2/O_2 flame (>2000°C). The mineralization products are washed into a calibrated 100 ml receptacle with sodium hydroxide solution (0.1 M). The chloride is determined by ion chromatography (detection limit 0.2 mg l^{-1}). To obtain blank values and to detect possible contaminations, pure hexane is combusted after every sample.

The combustion of one sample lasts about 15 min. The flame must not be smoking during combustion.

Evaluation

The chloride concentration in the combustion condensate is converted to mg Cl in 100 ml, which is the amount of Cl in the combusted part of the extract. With reference to the whole extract, the mass percentage of extractable organically bound halogens in 20 g of dry soil is found.

Chloride from combusted extract in 100 ml condensation product (mg)

$$\frac{(N - N_o) \times 100}{1000} \times \frac{V_{ge}}{V_e} \qquad (11.9)$$

$$\text{EOX (mg kg}^{-1}) = \frac{\dfrac{(N - N_o) \times 100}{1000} \times \dfrac{V_{ge}}{V_e}}{m_{soil}} \qquad (11.10)$$

where N is the chloride concentration in condensation product in milligrams per litre, N_0 is the blank value of the condensation product in milligrams per litre, V_{ge} is the volume of the whole extract, V_e is the volume of combusted extract and m_{soil} is the dry matter of the weighed soil sample in kilograms.

Discussion

It is possible to use other suitable solvents instead of hexane for the Wickbold combustion apparatus. Recovery rates depend on the solvent and on the extraction method. Friege et al (1987) reported a good yield when toluene was used as solvent. They also found out that the Soxhlet extraction was the best extraction method for non-volatile halocarbons (Friege et al 1987).

A *p*-chlorophenol standard can be used to determine recovery rates for the combustion. Adding a standard to a soil is a good method to control the accuracy of analysis and extraction yields.

The chloride blank value obtained by the combustion of pure hexane can be subtracted. To do this, the combusted hexane volume and combustion time must be the same as in contaminated samples.

Attention must be paid to the air in the laboratory as a possible source of contamination.

Besides the Wickbold apparatus there are other combustion apparatus for the analysis of halogenated organic compounds, which allow combustion in an oxygen stream at 1000°C. Both methods show comparable results. The Wickbold method has the advantage of combusting greater sample volumes and, therefore, of minimizing the effect of possible errors (Stachel et al 1984).

Degradation parameters

W. Müller-Markgraf

The objective of optimization studies

It is necessary to carry out detailed studies on the influence of different parameters on the degradation rate of xenobiotics present in a contaminated soil matrix, prior to any large-scale bioremediation procedure. Since this is apparent from recent publications and conference contributions regarding various scientific aspects of microbial soil remediation techniques, it seems even more surprising to observe, in conjunction with the practical approach, the widespread reluctance to perform optimization investigations. Although part of this may be due to the additional costs of a comprehensive pre-investigation programme, there might be yet another threshold, due to the profound uncertainty about the rational of a parameter optimization programme that goes well beyond a simple initial assessment of the biodegradability of the contaminants in a specific soil sample. In many cases, the decision concerning how far a given polluted soil can be treated using biotechnological methods is merely based on the knowledge of the nature and concentration of the contaminant by chemical analysis. Slightly more thorough approaches rely on viable cell counts and measurements of the respiratory activity. Although, the latter type of investigation provides answers to the key questions, of whether an indigenous microorganism population exists and how far the microorganisms have an active metabolism (or whether such a metabolism can be induced), this "minimum feasability study" does not guarantee that the contaminant is actually being degraded in the soil matrix.

Experience from practical approaches shows that, despite the fact that the pollutant is readily degradable, the degradation rate levels off after a relatively short initial period, causing the system to remain at high concentration values for longer times than are commensurate with efficient soil remediation processes. In other cases a significant reduction of the xenobiotics can not be observed at all, even though the simple feasibility study, as mentioned above, suggests that a microbial remediation of the soil is promising.

Consequently, a more refined pre-investigation programme, finely tuned to the specific circumstances of a particular contaminated site, can actually help to reduce the costs for remediation in all but the very simplest cases. Furthermore, as will be shown later, a detailed knowledge of the influence of the degradation parameters (to be defined) on the decontamination process is a necessary prerequisite for the development of process control strategies required for specific remediation technology.

The following discussion of the optimization studies does not include the "initial investigation" (see earlier section on microbiological characterization of contaminated soils). Rather we assume that simple tests suggesting that an indigenous population of microorganisms with the desired metabolic capabilities exists in soil have already been performed. Starting from this point, we are able to develop a systematic programme for the investigation of the degradation parameters with the goal of answering the following questions:

1. What are the limiting factors responsible for the diminishing degradation rate observed *in situ* (i.e. at the undisturbed contaminated site), although an active microbial population is present?
2. What factors must be optimized to remove the inhibition of the natural degradation process, thus achieving substantial biodegradation rates of the contaminants?

3. How can the kinetics of degradation be optimized, in order to reach a throughput per unit time and volume that allows for an economic application of a biotechnological approach to soil remediation?
4. How can the extent of pollutant degradation be maximized in order to comply with the relevant environmental regulations?

It is important to distinguish between kinetic considerations and mass balance (biodegradation yield). Whereas the latter plays an important role with respect to the assessment of the success of the technical bioremediation procedure, the former are vital to the question of how to reach the required residual concentration levels as fast as possible and how to control the process whenever an unwanted deviation from the regular pathway of pollutant depletion occurs. It must be noted, however, that most experimental parameter optimization programmes tend to deal only with the mass-balance aspect of the problem. This means, typically, that a soil sample is being incubated for a specified period of time; for instance, in a fixed-bed soil column where nutrients, oxygen (air) and water are provided. The final contaminant concentration is then compared with the initial concentration and the instantaneous levels of biological activity of soil before and after the incubation are measured. Clearly, an unambiguous conclusion on the fate of contaminants, nutrients and biomass between initial and final measurement is not possible. It is even more misleading if an extrapolation concerning the time-scale of the system is inferred from this sparse data. Nevertheless, it can be observed in many instances that predictions about the time required to reach a given residual concentration are made solely on the basis of the "before and after" experiment. One goal of the following sections is to illustrate some of the possible pitfalls that might be encountered by following this approach. As an alternative, a set of optimizing experiments is suggested that are capable of yielding mass-balance and kinetic information on the following degradation parameters:

- Inhibition of the microflora by unknown toxic substances present in soil.
- Nutrient salt demand.
- Aeration, oxygen demand and alternative electron acceptors.
- Adsorption, desorption and bioavailability.
- Temperature and pH.
- Co-substrates.
- Special microorganisms and bioaugmentation.

The biochemical and microbiological aspects of each parameter will not be discussed in great detail. Instead attention is focused primarily on very practical aspects of the investigation techniques, in conjunction with the degradation parameters, using examples whenever possible. Thus, the importance of each parameter and its influence on the overall depletion rate of the pollutants will be demonstrated in terms of its effect in a specific degradation experiment. The examples will deal with aliphatic hydrocarbons such as Diesel oil, heating oil and gasoline (petroleum hydrocarbons, PH), as well as with polyaromatic hydrocarbons (PAH). However, the scope of the methodology can easily be extended to other organic contaminants, such as monoaromatic compounds, partially halogenized constituents, nitroaromatics, etc.

Respirometry

Although it appears highly unlikely that, say, a PH-contaminated soil matrix exhibits a deficiency in mineral-oil degrading bacteria (since these species are ubiquitous in the environment), a fast, simple and cost-effective respirometric test is still indispensable. By measuring the oxygen consumption and/or the CO_2 formation of an incubated soil sample containing the pollutant, many decisions can be made at a very early stage of the pre-investigation programme, especially the decision as to whether it is reasonable to continue with the more costly and time-consuming optimization studies. A variety of commercially available respirometers, originally intended for waste-water treatment experiments, can be used. In the present example, an electrolytic respirometer is employed (Voith Sapromt D12 with control unit and microcomputer for data acquisition, Voith GmbH Heidenheim, Germany, see Chapter 5). Basically, this instrument comprises a closed reaction vessel of different possible sizes, which is connected to a simple pressure detector (contact manometer) that controls an electrolytic oxygen generator. Furthermore, a small vial with soda lime is incorporated in the sealed reaction vessel. As the biochemical oxidation of the hydrocarbons proceeds in the reactor flask, the liberated CO_2 will be adsorbed by the soda lime, thereby leading to a slight pressure drop in the system. The latter will be detected by the pressure detector used to trigger the oxygen generator for a certain period of time. This process will go on until the pressure drop, caused by the respiratory activity of the microflora present in the reactor, is replenished by the oxygen produced by the electrolytic cell.

Counting the oxygen production cycles provides a simple means to register the integral, cumulative biological oxygen demand (BOD) as a function of time.

Since most of the bioremediation applications in the field tend to supplement nitrogen and phosphorus as critical limiting factors to biodegradation under *in situ* conditions, these elements are also provided in a typical respirometer experiment. For a very simple assay, a standardized amount of contaminated soil (10 g) with a maximum grain size of 2 mm is stirred into 250 ml of a buffered minimal medium to form a thin slurry. The medium is prepared by dissolving the following in 800 ml of distilled water:

0.78 g KH_2PO_4
1.83 g K_2HPO_4
11.00 g $(NH_4)_2SO_4$
0.85 g $MgSO_4.7H_2O$
0.17 g $FeSO_4.7H_2O$

The pH is then adjusted to 7.0 (NaOH or H_2SO_4) and the volume is brought up with distilled water to 1000 ml. The reactor vessel

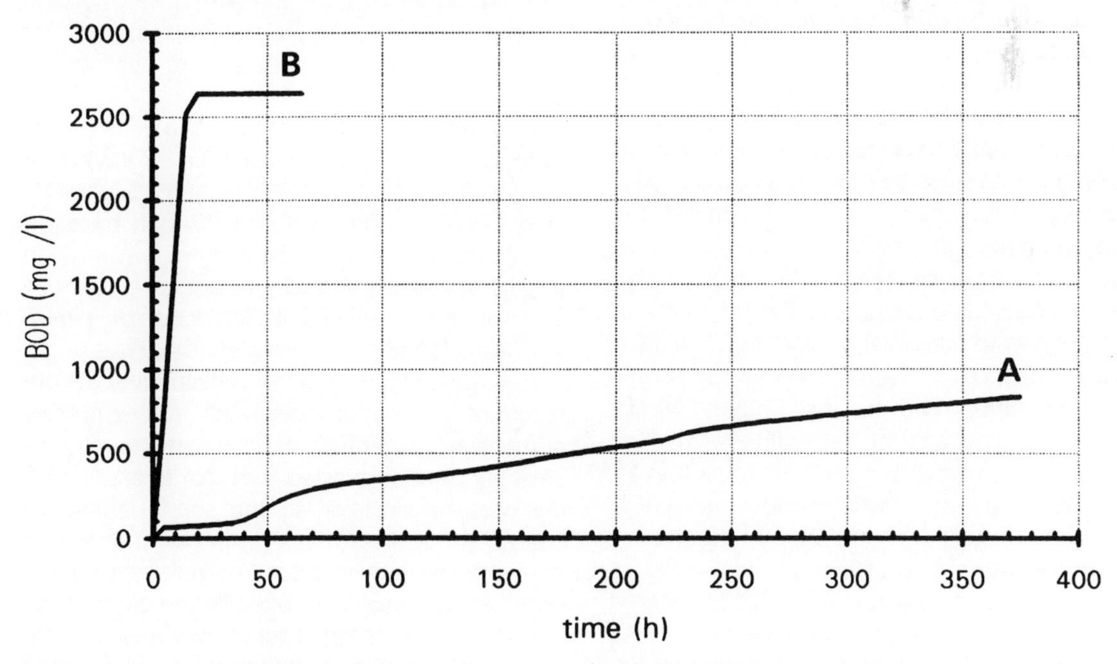

Figure 11.23. Time dependency of the cumulative biological oxygen demand (BOD) for two mineral oil contaminated soil samples. Ten grams of soil (< 2 mm) with a concentration of total petroleum hydrocarbons of 60 g kg^{-1} dry weight) was used in both cases. B was inoculated with 1 ml activated sludge from a municiple wastewater treatment plant.

is then incubated at 25°C for at least 200 h while the cumulative BOD (mg O_2 l^{-1} suspension) is recorded.

Figure 11.23 shows two examples of BOD functions obtained by this simple and straightforward experiment. These examples are chosen to demonstrate that the observation of certain features in the course of the BOD functions can suggest preliminary conclusions about the actual behaviour of the entire soil contaminant/microbial population system. "Preliminary conclusions" is used here because we are not observing the decay of the different constituents of the contamination but an indirect parameter (BOD) possibly linked to the desired contaminant mineralization only to a certain degree. To illustrate this, a short discussion of the curves measured in Fig. 11.23 may be useful.

Curve A shows the respirometric signal of a PH-contaminated soil with a known total aliphatic hydrocarbon concentration, which accounts for almost the entire total organic carbon (TOC) of the sample. This BOD versus time function is characterized by: (1) immediate oxygen consumption after the beginning of the incubation; (2) several periods of BOD increase (often, as in this example, in steps) that subsequently level off to gradually decreasing respiration rates (as indicated by decreasing slopes as the incubation time increases). The lack of a pronounced lag time, the period of time needed by the microbial population to reach the exponential growth phase, is indicative of the presence of a well-adapted microbial population. The subsequent periods of steady increase can tentatively be interpreted as the phases where different constituents of the pollutant mixture are utilized, whereas the smooth transition to the final phase of smaller slopes marks the onset of maintenance respiration of the microorganisms, after most of the substances have been depleted. A qualitative test of this preliminary hypothesis is to correlate the "final value" after the relaxation of the BOD function into the maintenance phase, with the approximate overall conversion of substrate, inorganic nutrients and biomass as given by the stoichiometry of the overall biochemical reaction equation (Einsele 1985):

$$2nCH_2 + 2nO_2 + 0.19nNH_4^+ \rightarrow n(CH_{1.7}\,O_{0.5}\,N_{0.19}) + nCO_2 + 1.5nH_2O + n \times 48{,}000\ \text{kJ} \qquad (11.11)$$

This equation provides a first approximation of the oxygen demand per "aliphatic hydrocarbon" to be oxidized, assuming uninhibited biomass growth. On the basis of this assumption, one would expect about 2.3 g O_2 to be utilized to oxidize 1.0 g of a common Diesel oil. Conversely, if the biomass does not grow, due to unknown growth-limiting factors prevailing in the soil, one would observe a consumption of approximately 3.4 g O_2 l g^{-1} petroleum hydrocarbons. The latter value corresponds to the mere "chemical oxidation" of the hydrocarbon, without any sizeable amount of carbon ending up in the biomass during the course of the reaction. The final O_2/hydrocarbon ratio can range from 2.3 to 3.4, depending on the conditions given by the soil/contaminant system and its interaction with the microbial population. This again holds true only under the assumption that the total PH content accounts for the observed BOD signal and no other utilizable substrates are available in significant amounts.

How the different features of the BOD versus time function, such as lag time (if any), rate constant describing the exponential increase (if observable), final value, etc., can be used to set up an entire classification system with regard to biodegradability of a vast amount of xenobiotics, is discussed in detail by Urano and Kato (1986).

However, since drawing conclusions on the basis of a respirometer study alone carries numerous potential pitfalls, it must be strongly recommended not to overinterpret the data. In addition to the semiquantitative character of the mass balance mentioned above, more subtle misleading results are possible, as can easily be exemplified by investigating curve B of Fig. 11.23. After a very short initial adaptation period (not shown in the figure due to the resolution of the plot), an instantaneous and rather steep jump to a constant final level can be observed. Qualitatively, this behaviour appears quite satisfactory at a first glance; however, a closer data evaluation reveals that the respiratory rate is exactly zero after the initial rapid increase. Clearly, this means that the microorganisms are not active and this result may have practical consequences. A respiration rate (i.e. the slope of the cumulative BOD function) of zero means that there is no maintenance respiration. The very fast respiratory rate for a short period of time at the beginning may be due to the presence of chemical compounds in the sample (for instance, 2, 4-dinitrophenol) that have the potential to uncouple the phosphorylation step in the respiratory chain. It is important to mention that the sample in the BOD measurement, represented by curve B in Fig. 11.23, was initially inoculated with a small amount of activated sludge; otherwise, no response whatsoever would be seen.

Inhibition test

Given the simple respirometer test as described in the former section, it is rather straightforward to design an experiment that allows for the determination of the relative inhibition of the biochemical oxidation reaction by the contaminant itself, or by metabolites formed during the biodegradation. An inhibition test based on respirometric data must not be confused with the measurement of the BOD time evolution of a single soil sample. Xenobiotics such as PAH can exhibit a marked toxicity on the soil microflora and at the same time they can act as a substrate to be metabolized. For this reason, a single BOD time function is not sufficient to assess possible inhibition effects.

A BOD function, like the one from the example in curve A of Fig. 11.23, could have shown a faster rise and an even higher final level if there were no inhibiting effects slowing down the overall process. On the other hand, a detailed knowledge of the inhibiting or even toxic effects is very important when it comes to the assessment of the prospects of a specific remediation technology for a given contaminated site. To overcome this difficulty the relative "depression" of the BOD function of a readily degradable substrate, caused by the presence of the polluted soil sample, versus a theoretical "best case" curve, can be employed. The theoretical curve is simply constructed by addition of the BOD-functions of a separate aliquot of the soil and of the easily degradable substrate ("reference BOD"), the latter with a small inoculum of soil bacteria to start the reaction. Separate controls consisting of soil sample alone and reference BOD alone are necessary because the pollutants in the soil sample will contribute to the measured BOD of the mixture of soil sample and reference BOD, although they might act as an inhibiting factor at the same time. The standard laboratory procedure can be summarized as follows.

A total of three respirometer reaction flasks are prepared. The first contains 10 g of the contaminated soil (< 2 mm) in 250 ml buffered minimal salt medium ("sample"). In the second, the same amount of soil is treated with 0.2 g of peptone (peptone from casein, trypsin-digested, Merck no. 7213) ("sample/peptone"). The third reactor flask contains just 0.2 g of peptone in minimal salt medium and an inoculum of a few microlitres of the supernatant of a soil suspension from the contaminated soil under investigation ("peptone"). These three flasks are then incubated at 20°C for at least 200 h and the individual BOD time traces are recorded. The procedure is similar to that used by Offhaus (1973) to assess wastewater toxicity. However, in order to obtain the kinetic information, the data evaluation has been modified slightly. For practical reasons it is better to calculate a time-dependent dimensionless quantity, "relative inhibition" (RI), following the simple relationship:

$$RI = f(t) = (BOD_{th} - BOD_{sample/peptone}) / BOD_{th} \quad (11.12)$$

where RI is the relative inhibition, BOD_{th} is the theoretical biological oxygen demand, $BOD_{sample/peptone}$ is the rue biological oxygen demand in the sample containing peptone and $f(t)$ is the function (time). The theoretical BOD is equal to the BOD of peptone and that of the sample:

$$BOD_{th} = BOD_{peptone} + BOD_{sample} \quad (11.13)$$

Clearly, the value for RI at a given time does not necessarily provide any information about the absolute toxicity of the contaminated soil, which can be relevant to the technical remediation procedure. To overcome this problem, a relatively large amount of contaminated soil (10 g) is chosen. Presuming that we are dealing with pollutants with a low solubility in water, one can expect that we are likely to obtain the thermodynamic equilibrium concentration while still being far from depleting the adsorbed phase of PH or PAH on the surface of the soil particles. Following the hypothesis that the bacteria are able to convert only the dissolved hydrocarbons (Breure et al 1992), it is possible to conclude that bacteria encounter the same effective pollutant concentration in laboratory experiments as in any technical remediation process that involves the participation of an aqueous phase. At least during the initial stages of the degradation process this concentration always corresponds to the maximum solubility. Therefore, the effective toxicity with respect to the microbial population should roughly be the same in the respirometer experiment and in soil reclamation in the field.

Even in cases where the latter assumption is not valid, one can at least compare the RI functions of different polluted soils in a semi-quantitative way because the assay is always done in the same standardized manner. Figure 11.24 compares the time evolution of the relative inhibition (which can assume values between 0.0 and 1.0) for a PH-contaminated and a PAH-contaminated soil. The sum of the concentrations of the 16 PAH species, considered by the US Environmental Protection Agency (EPA) to be harmful to the environment, was 730 mg kg^{-1} dry weight in the PAH sample (determination by HPLC, equipped with a fluorescence detector). The total PH concentration in the oil contaminated sample was 5.5 g kg^{-1} dry weight.

The two curves representative for the PAH-

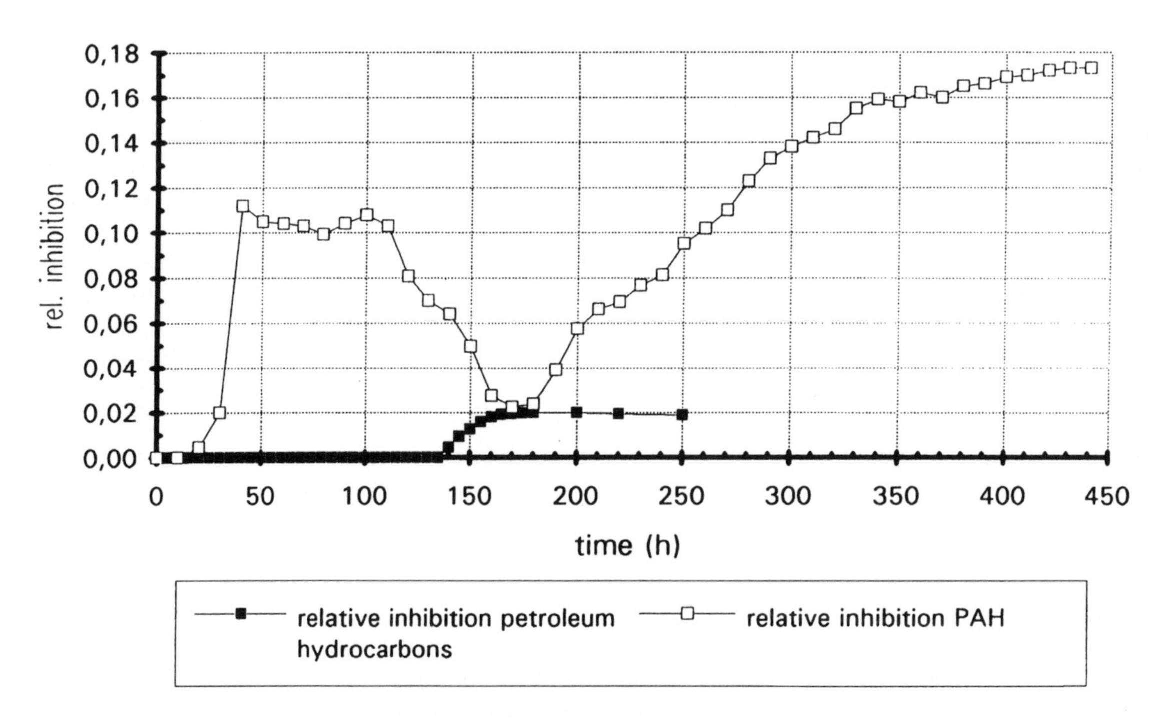

Figure 11.24. Comparison of the time evolution of the relative inhibition of soil contaminated with petroleum hydrocarbon (■) and soil contaminated with a polyaromatic hydrocarbon (□).

and PH-polluted soils, respectively, show distinct differences. The relative inhibition (after minor corrections for an artifact that is introduced by the comparatively high inoculation density in the "sample/peptone", run close to $t = 0$) in the case of the PH contamination is essentially zero until 135 h when a slight increase in the toxicity is shown. Conversely, the inhibition caused by the presence of PAH shows a steep rise at the beginning and reaches a plateau at approximately 0.11. After dropping to a minimum at 170 h, a steady rise to even higher RI values can be observed on a longer time-scale (up to $t = 450$ h). It should be emphasized at this point, that the latter time evolution "pattern" could roughly be reproduced for any PAH that has been studied so far.

Although a scientifically sound interpretation using the respirometry behaviour alone is not possible, one can take advantage of this information, together with some insight derived from the studies described below, to arrive at a reasonable hypothesis. In the case of PH-polluted soil it is possible to hypothesize that substrate toxicity is low and that the bioconversion may give rise to a minor inhibition, potentially caused by metabolites, during the later stages of the process. On the contrary, the presence of PAH causes a distinct substrate inhibition that is being diminished gradually as the bioconversion of the PAH proceeds. During the further course of the reaction, the accumulation of toxic metabolites probably occurs due to an increasing inhibitory effect that reaches even higher values as compared to the initial level. At this point, it is important to issue another warning about the possibly misleading interpretations of respirometric data. If a given sample, regardless of the kind of contamination being analysed, shows very little toxicity, either one of the following conclusions can be true:

1. The constituents of the total contamination are not toxic with respect to the soil microflora.
2. The microbes do not experience any pollutant, i.e. strong binding forces between the organic pollutant molecules and the soil matrix are responsible for a diminishing bioavailability, which is one of the most important degradation parameters.

The second conclusion can easily be verified by following up investigations as described in the next sections. In conclusion, caution is required in the decision as to whether a contaminated site can be reclaimed using biotechnology, when pre-investigations using respirometry alone are considered. Respirometry is a fast, cost-effective method to address (but not to optimize) the degradation parameters affecting inhibition of the microflora and bioavailability (the latter in conjunction with other studies). In principle, it can also be used to study the effect of special organisms/bioaugmentation and co-substrates just by supplementing these factors and observing the respiratory response of the system as compared to the control experiment without any amendment.

Closed-loop soil-slurry reactor

Technical aspects

The use of respirometry with soil slurries can give results that are only very loosely related to the actual decay of the contaminants. To overcome this problem, many bench-scale degradation experiments have been developed (Irvine et al 1992). They are based on the use of a fermenter, set up or modified to handle suspensions of fine soil particles. Both the solid soil matrix and the aqueous supernatant can be sampled from these reactors, and then analysed off-line, using different laboratory techniques. However, most soil slurry reactors are designed for degradation parameter optimization studies rather than for the development of technical soil-biorector concepts. Therefore, they suffer from the drawbacks of limited versatility and the inability to measure the time evolution of several important parameters simultaneously. Since degradation studies involving soil matrices that contain old pollutants tend to take a long time for each individual experimental run (sometimes weeks), it is especially important to obtain as much information from a single experiment as possible.

The following modular soil-slurry reactor was especially designed to facilitate studies of the response of the relevant parameters (such as BOD, depletion of total and individual contaminants, nutrient consumption, bioactivity parameters, etc.) to process control measures, such as the addition of co-substrates, supplementation of specialized biomass with specific metabolic capabilities (bioaugmentation), and change in the amount and composition of the gas phase, etc. Furthermore, this experimental approach allows the degradation experiment to be run in a totally closed system, if necessary, thus no corrections, e.g. for stripping effects caused by the aeration, are necessary.

A scheme of the apparatus is given in Fig. 11.25. The central part comprises a 4 l fermenter, agitated by a magnetically driven stirrer and equipped with an automatic pH and temperature controller. The aeration system is carried out as a closed loop: the air or oxygen-enriched air enters the reactor driven by a membrane pump. Several optional spargers are available to distribute the air in the slurry. As the air leaves the reactor, it first passes through a large (5 l) trap to allow any foam, which has been driven out of the headspace, to collapse. Then the air is allowed to pass through a second trap, filled with 5% NaOH, to absorb the carbon dioxide formed during the aerobic bioreaction. By connecting the exit of the carbon dioxide absorber flask back to the inlet of the membrane pump, a closed system is established.

As an option, a paramagnetic oxygen sensor can be installed in a bypass, close to the reactor's gas outlet. This has proved to be advantageous in a series of experiments where the air inside the closed apparatus was enriched with oxygen, to maintain a level of 60–90% O_2, simply by purging with pure O_2 prior to the experiment. These experiments were conducted with the intention of studying the influence of higher O_2 concentrations on the degradation rate of PH in a soil slurry. As soon as the oxygen concentration inside the apparatus is significantly greater than 21%, it has to be recorded and, if necessary, replenished by additional purging during the experiment, since diffusion across the tubing cannot be neglected. Furthermore, one of the inlet ports is equipped with a pH electrode, which is connected to an automatic pH regulator, comprising controller and peristaltic pumps for acid (0.5M H_2SO_4) or base titration (1M NaOH).This is essential, because some of the slurries under investigation showed a tendency to drift towards pH values as high as 8.3, due to high carbonate content of the soil. A substantial decrease in both the rates of respiration and hydrocarbon depletion was observed as the pH slowly drifted towards higher values, for this reason pH values must be kept constant at approximately 7.0. On the other hand, the microbial growth and substrate utilization is frequently accompanied by a pH drift towards more acidic values (depending, of course, on the pH and buffer capacity of the soil), which has to be compensated as well, in order to maintain optimum conditions for the bacterial activity.

The rest of the apparatus strongly resembles an electrolytic respirometer, in that the reactor headspace is connected through silicon tubing to a contact manometer that registers the pressure drop, caused by the absorbed CO_2 produced by the oxidation of the contaminants. However, in contrast to most of the commercial respirometers, where an external clock cycle is defining periodic time slices, and where the biological system may or may not demand oxygen production, in this bioreactor the microbial population inside the apparatus is totally free to trigger O_2 production at any time. In addition, the larger O_2 production capacity of the electrolytic oxygen generator allows oxygen to be supplied to a much larger biomass/soil slurry volume compared to the analytical respirometer.

The respirometer comprises an input signal detector/amplifier to evaluate the signals of the contact manometers, an adjustable timer (monoflop) to switch on the current through the oxygen generator for a fixed time (30–

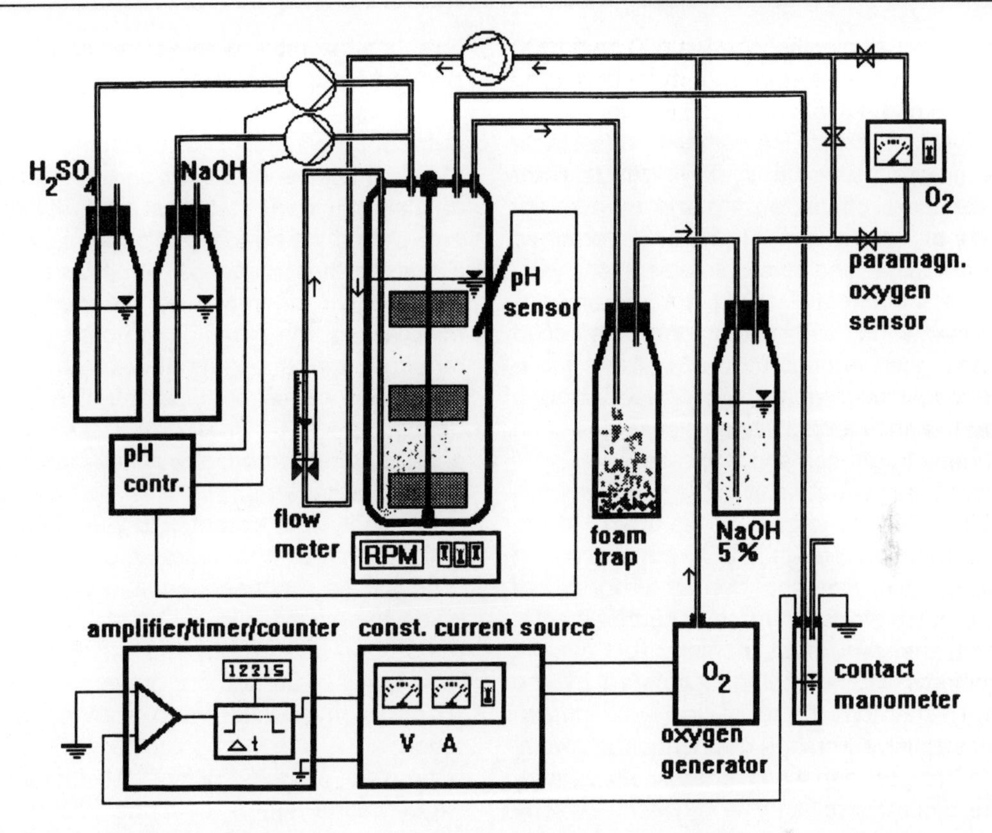

Figure 11.25. Schematic diagram of the closed-loop soil-slurry bioreactor for degradation parameter studies.

120 s) and a constant source of current for the O_2 generator.

As the contact manometer triggers the monoflop via the input amplifier, a constant current between (typically) 1.0 and 2.5 A is allowed to flow through the electrolytic cell containing a platinum anode and a $CuSO_4$/Cu cathode, thereby producing O_2 at the anode for a period of time that can be defined using the monoflop. According to Farady's law, the mass of oxygen produced per "on" cycle can be adjusted according to the size of the fermenter system and the differential oxygen demand simply by fine-tuning *I* and *t*.

$$mO_2 = 8 \times I \times t / F = 8.29 \times 10^{-5} \times I \times t \qquad (11.14)$$

where mO_2 is the mass of oxygen produced in grams, *I* is the current in amperes, *t* is the duration of the current flow as defined by the monoflop setting in seconds and *F* is Faraday's constant (9.6480×10^4 As mol^{-1}).

Since the "on"-time is fixed to a specific value prior to an experimental run, a simple counter module can be used to count the oxygen production cycles. This translates directly into the amount of oxygen consumed, thus leading to a cumulative BOD versus time function similar to the one measured in the analytical respirometer. Oxygen production rates of the order of 1 g h^{-1} or more can easily be attained, if necessary. The use of a microcomputer (while not essential) for data acquisition and processing will greatly enhance the system's capabilities, especially when several parallel units are operated simultaneously.

This bench-scale experimental setup is superior to other, similarly designed degradation experiments for the following reasons:

1. The bioreactor contains a sufficient amount

of material (typically between 500 and 1000 g of moist soil, giving a slurry with a total volume of 3.5 l) to allow sampling for off-line analysis, without imposing a detectable perturbation on the system. All relevant parameters characterizing the momentary state of the experiment can be determined by chemical and microbiological analysis. These include the concentration of contaminants in the liquid and the solid phase, the concentration of nutrient salts, the viable cell counts, the protein content and the enzyme activities of enzymes such as dehydrogenase and esterase.

2. Processes taking place in the soil slurry will immediately be shown by the shape of the BOD function. From the BOD curve one can quickly ascertain the correct progress of aerobic degradation of the contaminants. Thus, one can react instantly to possible problems without being hampered by the time delay that is inevitable if one has to rely exclusively on off-line chemical analysis.
3. The oxygen supply is entirely automatic. The biological system is receiving exactly as much oxygen as needed.
4. The effect of the suplementation of a critical factor (for instance, co-substrates, surfactants for the increase of bioavailability, etc.) can be observed directly using the BOD signal. Furthermore, the relaxation of the BOD signal after the addition of such factors can be followed. This allows for a repeated study of the system's response to the supplementation within one single degradation experiment, since the beginning and the end of each "sub-experiment" can clearly be resolved, thereby avoiding unintended overlapping that would obscure the results.

Nutrient salt demand

The nutrient salt demand for the mineralization of hydrocarbons is often defined using the C:N:P ratio, which should be approximately 100:5:1 (in weight units). Although similar nutrient ratios are frequently reported in the literature, its exact meaning is frequently not precisely understood.

Firstly, it is not sufficient to provide the right nutrient ratio initially. The possible implications of the C:N:P ratio should be understood as a critical constraint that must be satisfied at any time during the course of biochemical reaction. Since in a batch process the carbon content of the soil will decrease as the pollutants are mineralized, the minimum amount of N and P required must be monitored and supplemented if necessary (i.e. the C:N:P ratio is defining a kinetic criterion for the absolute N and P concentration, and is not merely a mass balance argument).

Secondly, it is essential to keep in mind that higher nitrogen concentrations may inhibit the process. As a consequence, it can be appropriate to stay initially well below the N concentration "prescribed" by the C:N:P ratio, when high levels of carbon are present.

Thirdly, the C:N:P ratio should be understood as a rule that may be subject to the optimized process using, for instance, the experimental setup described above. Also, the desired amount of "major bioelements" can be influenced by totally different considerations, e.g. the appropriate buffer capacity of the soil slurry system. For instance, it may be necessary in some cases to add more HPO_4^{2-} than given by the C:N:P ratio for the phosphorus concentration, because the phosphate buffer is stabilizing the otherwise critical pH value.

Furthermore, the C:N:P ratio gives the preferred concentration of available N and P. Adding NH_4^+ to the soil, however, frequently leads to a substantial initial loss of ammonia due to its fixation on the soil. Therefore the free ammonia concentration after saturation of the surface sites with ion-exchanging capabilities should be monitored and adjusted if necessary. Alternatively, the ion exchanging surface sites can be deactivated in some cases by adding Ca or Mg ions prior to the amendment of nutrient salts.

Last, but not least, one should consider "minor bioelements" and trace elements like Fe, K, Ca, Na, Zn, Mn, Mo, Se, Co, Cu, Ni, W, etc. If a deficiency in some of the necessary trace elements must be suspected from the chemical analysis of the soil, it might be extremely useful to measure the response of the pollutant elimination by increasing the critical factor in a degradation experiment using the soil slurry bioreactor, while keeping all other important factors beyond their limiting threshold.

Aeration, oxygen demand and alternative electron acceptors

The closed loop soil slurry biorector is well suited to conduct investigations of the influence of the oxygen content of the gas phase on the degradation kinetics of the contaminants.

Several studies of the aliphatic hydrocarbon mineralization by the indigenous microflora in PH-contaminated soils were carried out in the past using this technique. As a general result, it could be observed that the respiration rate (the on-line BOD signal) shows a marked and persistent increase immediately after the increase of the O_2 concentration in the gas phase of the reaction system from 21 to 85%. This jump in the respiration rate was accompanied by a sharp decline in the total PH-concentration level (measured off-line by spectroscopy using samples drawn from the reactor). So far, no inhibitory impact of high oxygen concentrations in the gas phase on the mineralization process has been observed. Probably the limited solubility of O_2 in water mitigates its toxic effect or the actual oxygen concentration in the aqueous phase is very low because the specific rate of the consumption by microbial oxidation of the hydrocarbons is much higher than the rate coefficient for the gas–liquid transfer. Unfortunately, the direct measurement of the dissolved oxygen concentration using, e.g. a Clark electrode is not feasible due to the properties of the soil slurry.

Several pre-investigations of both PAH- and mineral oil-contaminated soils have shown that nitrification occurred in the presence of NH_4^+ as a nitrogen source and in soils with a high carbonate content. The formation of nitrite from ammonia and the subsequent oxidation to nitrate using CO_2 as a carbon source is accompanied by a stagnation of the hydrocarbon degradation, due to a shift in the population distribution towards bacteria with chemolithotrophic metabolism. If this shift in population distribution occurs, it has, of course, enormous implications for the design of a technical soil remediation process. The failure to detect this unwanted effect in appropriate pre-investigations can result in a disastrous effect, that is, the contaminants are not being depleted at all while at the same time huge amounts of ammonia are converted into nitrite and eventually accumulated as nitrate.

Fortunately, once the situation is known, one can correct this effect by stopping the ammonia input and the aeration so as to force the system into denitrification, where the nitrate is used as an alternative electron acceptor to oxidize the hydrocarbons. Nitrate can even be deliberately added to the system, especially in cases where anaerobic (micro-) zones in the soil matrix occur, due to the specific properties of the soil (e.g. large inner surfaces holding a substantial amount of the contaminants).

In any case, however, it is essential to be aware of these complications before a technical process is designed and built, because the cost of a later change of the technology tends to be high, if not prohibitive.

Adsorption, desorption and bioavailability

So far the discussion of the different degradation parameters has been conducted under the implicit assumption that factors like nutrient salt amendment and aeration, etc. are the rate-limiting steps in the oxidative degradation of hydrocarbons. While this usually holds in the liquid–gas system (e.g. fermentation), the additional solid phase (soil particles) changes the situation profoundly (Autrey and Ellis 1992). Whenever the observed degradation rate of the contaminants is too slow, while at the same time all other known factors (such as nutrient salts, aeration, pH, etc.) are in the non-limiting regime, one can already suspect that this depends on the bioavailability. So far it can be shown in almost every case that the desorption process of the organic pollutants from an adsorbed to a solubilized state (which is considered bioavailable) is the actual limiting rate, at least in certain phases of the overall reaction. Likewise, different contaminated soils only tend to vary in the percentage of non-bioavailable pollutants and in the "apparent binding force" between contaminant molecules and the soil matrix. Depending on the properties of soil (adsorption on to inner surfaces, dissolution in natural organic soil constituents such as humic acids, interaction of organic substances with mineral surfaces, etc.) and the tolerable residual concentration of PAH or mineral oil, it may be necessary in many cases to increase the mass transfer rate coefficient k_{des} for the "initial reaction":

$$\text{Contaminant (adsorbed)} \xrightarrow{k_{des}} \text{Contaminant (solubilized)} \qquad (11.15)$$

How this can be accomplished is the subject of numerous endeavours undertaken at present in many research groups and cannot be discussed in great detail at this point. However, using the soil slurry bioreactor, a variety of optimization studies can be carried out, with the objective of monitoring the response of the depletion kinetics to the addition of phase-transfer mediating agents. The latter can be representative of totally different chemical groups. A few possibilities are for instance:

- Surfactants, preferentially anionic and/or non-ionic.
- Biosurfactants, optionally produced *in situ* by the indigenous microflora.
- Long-chain alcohols (especially for PAH desorption).
- Selected aliphatic hydrocarbons (for PAH desorption).
- Substances capable of forming hydrophilic complexes with the otherwise non-polar contaminant (e.g. cyclodextrines/PAH).
- Inorganic compounds as, for instance, pyrophosphate.

Each of these options comes with its own problems that need to be addressed during the pre-investigations. For example, most of these agents can act as an additional carbon source, thereby giving rise to an unwanted diauxie, that is, the contaminant utilization rate is decreasing while the added compound is being oxidized. Conversely, the new compound could also act as a co-substrate in cases where the microflora cannot yield enough energy from the utilization of the contaminant (which is true for most PAH). Another problem is that the desorption-mediating substance must be biodegradable itself but not too fast compared to the "host-molecule" (e.g. PAH), since readsorption of the contaminants could otherwise occur.

Biodegradable surfactants, while well suited for hydrophilization of non-polar substances such as the hydrocarbons, exhibit yet another constraint: for a stable colloidal solution of the contaminants in water, a minimum surfactant concentration given by the critical micellar concentration (CMC) is necessary. On the other hand, higher concentrations of surfactants are believed to cause damage to some of the microbial membranes. The bench-scale soil slurry bioreactor is especially suited for systematic investigations of the different additives for the increase of bioavailability, since the fate of the supplemented compounds can be followed by means of the on-line BOD signal. The field of microbial soil remediation is unique in biotechnology in that mass transfer processes tend to play the major role as the kinetic "bottleneck" of the overall reaction, rather than the biochemical reaction itself. At present, the merely chemicophysical effects of adsorption/desorption and mass transfer should be given great consideration because they bear the highest potential for improvements to microbial soil remediation and detoxification. It should be kept in mind, however, that many of the bioavailability studies are still related to basic research and to the development of future technologies like technical-scale soil slurry bioreactors. In order to optimize pre-investigations, it is only important to see whether the lack of bioavailability causes a contamination to be persistent or what other factor must be supplemented.

Temperature and pH

The role of temperature can be studied as well, using the bench-scale experimental setup. However, from practical experience it is usually given less attention compared to the other degradation parameters. The reason for this is that the temperature optimum is well known from both a variety of our own degradation experiments and the literature (Autry and Ellis 1992). Although the temperature optimum for the growth of most PH degraders seems to be around 27°C, one should consider how far this temperature can be maintained in a technical soil detoxification process. Lower temperatures will result in lower depletion rates, but the soil slurry bioreactor experiment is not recommended any way for drawing conclusions on the absolute degradation rate in a field remediation using land farming or a related technology. The results of a soil slurry reactor experiment can only be used in a meaningful way, when an assessment of the expected degradation kinetics in a technical scale bioreactor is required.

The importance of the pH has already been pointed out. The pH in a soil slurry system can be adjusted freely, so an optimization can easily be performed. It was found, that a pH fluctuation between approximately 6.5 and 7.3 did not result in any significant slow-down or acceleration of the depletion rate during PAH and/or PH degradation, although very rapid changes should be avoided. Also of considerable importance is the observation that the natural pH evolution of a soil with approximately 65% carbonate in the grain size fraction < 2 mm during a degradation study in the agitated bioreactor was totally different with respect to the pH behaviour of the process water of a lysimeter experiment using the same soil. This will be discussed comprehensively in the following section.

Co-substrates

Co-substrates play an important role in the oxidation of substrates that do not deliver sizeable amounts of energy for the microbial cell, although the enzymatic prerequisites are present or can be induced. While metabolizing the co-substrate that ideally should bear a (sub-) structure analogue to the substrate (the contaminant), the organism yields energy and the persistent contaminant is being degraded along with the co-substrate. It is known, for instance, that some organisms are not capable of growing on PAH as a sole carbon source but can at least hydroxylate polycyclic aromatics in the presence of a co-substrate (e.g. naphthalene or cyclohexane, although the structural analogy in this latter example is not entirely self-explanatory). Again, a thorough treatise of the whole concept of co-metabolism is a subject of basic research, rather than practical optimization studies.

Systematic studies of different co-substrates in soil slurries are in a very early stage. This is primarily due to the complexity of the system. In "clean", sterile model experiments, as conducted typically in basic research, it is fairly easy to attribute a certain observed effect in the degradation yield or the growth rates of the organisms to the critical factor "co-substrate". Unfortunately, this is rarely possible in real contaminated soils because changing one parameter inevitably changes other parameters at the same time. For instance, due to their chemical structure, many co-substrates have a potential to increase the bioavailability as well, so it becomes harder to distinguish between the different effects. In these soil-slurry systems one has frequently to rely merely on indirect conclusions that might not be unique. Nevertheless, in some cases the influence of acetate on the biodegradation of old mineral oil-contaminated soil has been studied in the soil-slurry reactor. So far, no acceleration greater than approximately 5–10% was observed. The relatively small effect does not come as a surprise, however: the well-adapted microbial population is able to draw energy from the aliphatic hydrocarbon oxidation anyway, so there is no great potential for improvements left. Analogous experiments with PAH, where the contaminants are more persistent, were obscured by the abovementioned superposition of the different contributions, thus making unique conclusions impossible.

Special organisms (bioaugmentation)

The addition of acclimated microorganisms to contaminated soil, generally referred to as bioaugmentation, is currently subject to a major controversy (Atlas 1991). Bioaugmentation experiments in the closed-loop soil-slurry bioreactor showed very little effect in accelerating the degradation rate of PH pollutants. A maximum of about a 10% increase in degradation yield could be achieved using an inoculum of acclimated activated sludge. In a PAH-degradation experiment, a significant degradation could be observed after the addition of a special mixed culture of bacteria capable of mineralizing monoaromatics and alkyl-substituted aromatic hydrocarbons (Dr Bronnenmier, Linde AG, Höllriegelskreuth, Germany, personal communication). In this case, the control experiment, which exclusively took advantage of the indigenous population but was otherwise conducted under identical conditions, showed no contaminant elimination at all. However, from a variety of observations it was inferred that most of the induction of the PAH depletion in this example must be attributed to the production of phase-transfer mediating biosurfactants by the special biomass. In addition to these findings, basic studies of the microbial population naturally occurring in that particular PAH-contaminated soil revealed that these bacteria already had a higher metabolic activity with respect to PAH elimination than the supplemented specialists. This tends to imply again that only an insufficient bioavailability was responsible for the observed deficiency in the degradation potential of the autochthonous microflora. Despite these limitations, the importance of bioaugmentation as a degradation parameter can be summarized as follows:

1. Bioaugmentation studies during parameter optimization can be useful when specific requirements regarding single constituents out of the total spectrum of pollutants must be met. If, for instance, the indigenous population of a PAH-contaminated site is capable of degrading most of the PAH species with the exception of, e.g. benzo(a)pyrene, and no other limiting factors are responsible, it can be highly desirable to add some specialists after the elimination of the "easily" degradable substances by the naturally occurring bacteria. If the specialists were added from the beginning, the effect would be negligible, since the readily utilizable carbon sources would most likely prevent the oxidative depletion of the "problem compound". The development of such bioaugmentation strategies can be the task of bench-scale experimental optimization studies.
2. If, for whatever reason, the contaminated matrix is treated in a bioreactor system rather than with more conservative techniques such as land-farming (and related methods), it is important to develop technologies that allow a sufficient throughput. One possible approach to the desired increase of the time/volume-specific degradation performance of a soil-slurry bioreactor design is the continuous or periodic inoculation with a special biomass.

Soil columns and lysimeters

The closed-loop soil-slurry bioreactor shown in Fig. 11.25 and its variations is an invaluable instrument for the assessment of biodegradability, the identification of the degradation parameters and the optimization of the contaminant depletion rates. However, the majority of all large-scale soil remediation processes work with a variety of fixed-bed bioreactors such as lysimeters, "bio-heaps" and land farming. Although the well-stirred soil-slurry reactor offers a relatively well-defined system that provides the best access to the relevant parameters, it does not necessarily reflect the behaviour and interaction of contaminant, soil matrix and microbial ecology in a fixed bed of soil particles. This is especially true for the prediction of achievable depletion rates. In order to simulate the soil remediation in fixed-bed techniques, the soil column is frequently used. This comprises a vertical tube of variable size that is filled with the soil sample. This soil column is either submersed in water that must be pumped continuously through the column and saturated with oxygen externally after passing the soil, or the water containing nutrients and oxygen is percolated through the sample. Thus the soil column can be viewed as a vertical section out of a bio-heap, i.e. a large body of contaminated soil that is typically aerated and supplemented with nutrients and water. However, due to its limited size the soil column suffers from several drawbacks:

1. It does not allow samples to be drawn during the course of a degradation experiment without disturbing the structure of the matrix. As opposed to the well-stirred homogeneous bioreactor, the "soil percolator" or the submersed soil column is subject to transport processes that eventually result in a spatial variation of the variables that are characterizing the system (for example, the concentration and distribution of contaminants and nutrients, microorganisms, dissolved oxygen content, etc.).
2. Due to the large surface to volume ratio of a column compared to a bio-heap, the results can not be easily transferred to the large-scale remediation process.

Although these criteria are clearly not prohibitive, it has proved to be advantageous to extend the underlying idea of soil columns to larger units that would avoid these disadvantages. Degradation parameter optimization studies in the context of fixed-bed treatment of contaminated soils are conducted preferentially in lysimeters containing approximately 1.0 m^3 of the polluted matrix. These lysimeters are simply rectangular containers equipped with watering and aeration systems, a gravel drainage layer and a process-water recirculation system. The latter comprises the drainage layer underneath the soil body, a water collection container and a time-cycle controlled pump for the redistribution of the nutrient-containing process water over the surface of the soil.

A schematic diagram of one variety of these lysimeters is given in Fig. 11.26. In this particular example the aeration of the soil bed is accomplished by perforated tubing placed close to the bottom and directly on top of the drainage layer. This is connected to a membrane pump that causes an air flow from the surface of the soil bed through the pores of the matrix into the tubing. Thus, volatile compounds potentially present in the contaminated soil matrix will not be lost to the environment but will be adsorbed on to an activated carbon filter attached to the pump's outlet. Nutrient salt amendment, pH control, etc. can be done in the process water collector. Possible variations of the concept include heating the soil bed and injecting pure oxygen by special tubing inside the lysimeter, replacing the aeration manifold. The injection of pure oxygen is done by time control at a rate and duty cycle commensurate with the momentary BOD which, in turn, can be derived from measurements of the O_2 and CO_2 content in the pore volume of the soil matrix. The latter concentrations are measured by probes driven into the soil bed to specific depths. The gases are analysed using a paramagnetic oxygen sensor or an automatic process gas chromatograph for CO_2, respectively.

Samples for off-line analysis of pollutant content and biological parameters can be drawn with spatial resolution by augers. Since the sample volume is very small compared with the total volume of the soil bed, the disturbance of the system by sampling can be neglected.

In particular, the following paramaters are monitored periodically:

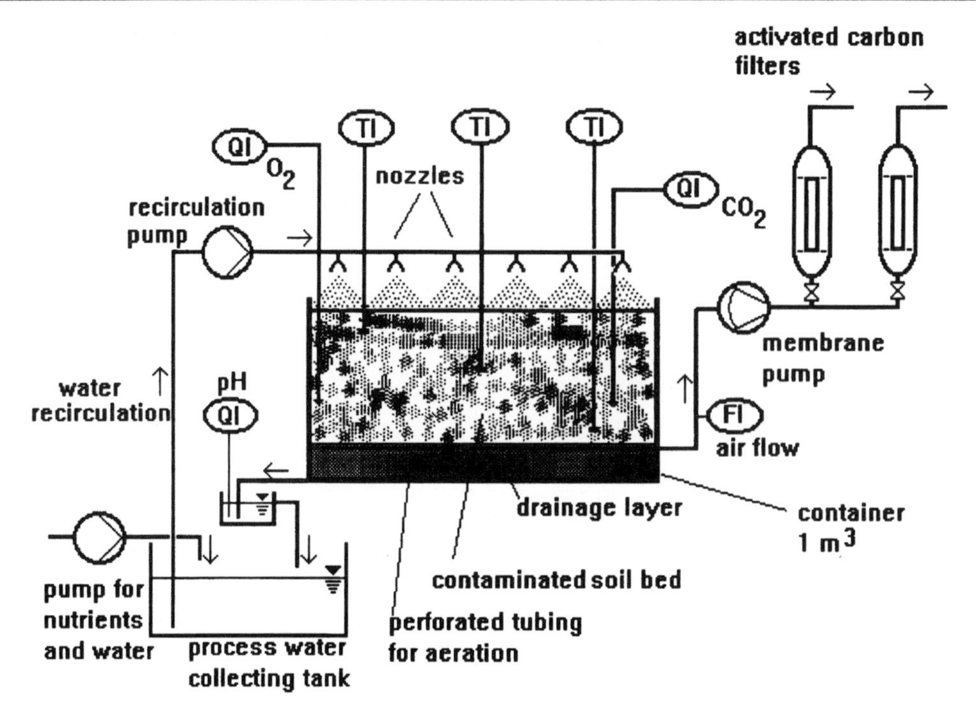

Figure 11.26. Schematic diagram of the 1 m^3 semi-technical-scale lysimeter setup. Only the aerated version is shown here as an example.

- The contaminant concentration and composition at different locations in the soil bed.
- The biological parameters in the solid matrix:
  - Esterase activity.
  - Dehydrogenase activity.
  - Protein content.
  - Viable cell counts.
  - Characterization of the microbial population in the circulating process water.
- The composition of the gas in the soil's pore volume.
- The temperature at different sites in the soil bed.
- The pH in the process water.
- The concentration and composition of the nutrient salts in the process water.

The relevant process parameters can be summarized as follows:

- Mass of contaminated soil matrix: approximately 1.8 metric tons.
- Aeration rate: 42–100 l h^{-1} continuously, depending on the pore gas composition (version "aerated").
- Oxygen injection: approximately 12 l h^{-1} for 30 s, every 2 h. The duty cycle can be adjusted according to the pore gas composition ("oxygen injection" version).
- Water recirculation: 15–30 l day^{-1}.
- Temperature: unheated (18°C) and heated (27–30°C).

The containers for the soil bed and the periphery such as pumps, manifolds, nozzels, etc. are reusable. Because the cost of the entire optimization programme is largely determined by the cost of labour (analysis, maintenance, data acquisition and evaluation, reports, etc.), the cost-effectiveness of the whole system, despite its size, is comparable to that of bench-scale laboratory soil column experiments. However, contrary to most similar laboratory investigations, the system yields data that can be used directly for scaling up to the full technical scale.

To illustrate the potential of this semi-technical experimental setup for pre-investigations, a short summary of some results is given. The soil used for the investigations in this particular example was highly contaminated with old Diesel oil. The oilspill was 11–27 years old, meaning that the oil has been spilled more or less continuously for 16 years, with the last fresh Diesel oil contamination occurring 11 years ago. As inferred from this history, GC-FID analysis revealed that a major part of the *n*-alkanes had already evaporated or degraded, thereby leading to a relative accumulation of more persistent constituents of the oil, as cyclic and branched alkanes (pristane and phytane) and substituted aromatic hydrocarbons. The total PH concentration was determined to be about 10 g kg^{-1} soil (dry matter) in the soil fraction lower than 2 mm (IR spectroscopy). The soil showed a very high hydraulic conductivity. The specific water conductivity coefficient of a saturated soil column was determined to be higher than 10^{-3} m s^{-1}. The pH of the soil was 8.3, due to a very high carbonate content (about 60–65%) in the soil fraction lower than 2 mm.

A total of four parallel lysimeter experiments were carried out, comparing the following process parameter options:

1. Aeration at 18°C.
2. Injection of pure oxygen at 18°C.
3. Injection of pure oxygen at 28°C.
4. Control experiment: no aeration, no oxygen, water circulation without nutrient salts, ambient temperature (average 18°C).

Pre-investigations were conducted in the closed-loop soil-slurry reactor. On the basis of the results, it was concluded that there was no need to use special microorganisms. The most important aspects of the observations and results can be summarized as follows.

A PH concentration level lower than 500 mg kg^{-1} in the soil fraction lower than 2 mm was achieved within 90 days from the start of the degradation process in all experiments except in the control run (analysis: total PH by extraction and IR spectrometry). The relative depletion of all GC-detectable oil constituents after 125 days exceeded 98% (capillary gas chromatography, flame ionization detector). Traces of pristane (2, 6, 10, 14-tetramethylpentadecane) could be detected in the residual contamination.

As opposed to the corresponding bench-scale slurry bioreactor experiments, no pH drift towards higher pH values was observed in the process water. On the contrary, the start of the microbial degradation was marked by a slight tendency to drift towards acidic pH, which is commonly observed in "bio-heaps" during depletion of mineral oil.

In this particular case, the two lysimeters featuring pure oxygen injection were not superior to the "simple" aeration version. This may be due to the fact that oxygen consumption was not the rate-limiting factor. Another hint is the relatively large amount of CO_2 (up to 8%) that was measured in the pore-volume gas in the O_2-injected lysimeters as compared to the "aerated" version (always smaller than 1%). The latter variety was obviously stripping out the CO_2 from the mineralization of the aliphatic hydrocarbons by means of the comparatively large air flow through the soil bed, thus eliminating the carbon dioxide as a potentially inhibiting factor. It must be emphasized, however, that oxygen injection may prove advantageous in other cases, e.g. when the oxygen concentration was clearly identified as a rate-limiting factor.

The lysimeters with oxygen injection, especially, showed a pronounced inhibition of the PH depletion shortly after the start. The degradation stopped at a very high concentration level (> 70% of the initial pollutant concentration). This effect was clearly caused by nitrification. If no steps are taken to control the relative development of the microbial population, the contaminant mineralization will eventually come to a total halt. However, the nitrate formed by nitrification can advantageously be used, as reported before, to degrade the hydrocarbons during a denitrification phase. It should be noted, however, that the sole use of nitrate as an alternative electron acceptor is not feasible, since elementary oxygen is mandatory for the first oxidation step of an aliphatic hydrocarbon chain. Stopping the ammonia input suffices under most circumstances to force the system into denitrification. However, stopping the aeration or even purging with nitrogen can accelerate the process. The nitrate respiration then continues, even at oxygen concentrations in the pore-volume gas as high as 65% (in the lysimeters with oxygen injection).

Since the bacteria are only capable of using either the nitrate or molecular oxygen as an electron acceptor for hydrocarbon utilization, it must be concluded that anaerobic centres are present in the soil matrix, even under well-aerated conditions. These (micro-) centres are only accessible for the nitrate dissolved in the process water. This leads to the conclusion that a fixed-bed remediation technique, as opposed to a bioreactor, should take advantage of the denitrification for the oxidation of mineral oil hydrocarbons in almost any case, at least when strict requirements concerning the tolerable residual concentrations must be met. This implies, of course, that the momentary state of the microbial consortium during a remediation process must be monitored in order to make reasonable control strategies possible.

The recycled process water shows an increase in the PH concentration immediately after the start of the reaction. The concentration can grow even beyond the limits of physical solubility of aliphatic hydrocarbons, indicating emulsification of the hydrophobic hydrocarbons due to the bacterial activity. As the oil concentration in the solid soil matrix approaches its final low value, the concentration in the process water also decreases. Other important parameters, like relative toxicity, concentration of known metabolites, total organic carbon (TOC), etc. are shown qualitatively in the same time evolution. As a consequence, a soil remediation procedure should not be considered as completed successfully unless the mentioned parameters measured in the process water have reached their natural background values. The residual contaminant content in the solid phase is not a sufficient criterion.

Conclusions

The last example illustrated very clearly how important a parameter optimization programme can be. Although the necessary prerequisites for a successful bioremediation were given, the degradation would have stopped completely when the critical degradation parameters (in this case, nitrification) had gone undetected. Many steps of the comprehensive programme outlined here can be replaced by mere experience. If, for instance, there is a high chance that bioavailability is not the rate-limiting criterion, the pre-investigation programme should primarily focus on the optimization of nutrient demand and aeration. In this case, extensive soil-slurry experiments might not be necessary and it could be more appropriate to switch directly to process design studies using lysimeters. The objective of the description of methods and examples was to point out that a parameter optimization investigation does not simply reduce to the adjustment of nutrient salt concentrations and oxygen consumption. Also, the pre-investigation of a contaminated soil has to be done with respect to the planned technology for bioremediation. As can be seen from the examples given, a particular contaminated soil may behave totally differently, depending on whether it is treated in a slurry bioreactor or in a large fixed-bed bioreactor, i.e. a "bioheap". Therefore, the investigation and optimization of the degradation parameters includes not only the detection of the "kinetic bottleneck" and the development of problem-orientated control strategies, but also the correct choice of the appropriate technology for the large-scale bioremediation process.

Microbiological decontamination of soils

"On-site" and "off-site" techniques

R. Eisermann

The treatment of contaminated soil can be performed routinely by applying several different technologies that are currently available (Lee et al 1987; Hülscher 1993; McDonald and Rittmann 1993; Skladany and Metting 1993). The German market offers services for the decontamination of soils with over a hundred companies. However, it should be noted that their experience can vary enormously, especially if such projects need to be performed for larger amounts of contaminated soil (> 2500 tons). The established companies are able to remediate 20,000–250,000 tons per year of contaminated ground.

There are two possible methods for decontamination: the "on-site" procedure and the "off-site" procedure. The "on-site" procedure is used for the treatment of contaminated and excavated ground on the property. In contrast, the "off-site" procedure is the process of decontamination executed in specific plants away from the property. In both cases the treatment for bioremediation will be substantially the same.

In Germany, the official requirements for licensing are dependent upon the procedure to be used in decontamination. The process of getting approval for all "off-site" and "on-site" projects exceeding 6 months is to be done according to a procedure called "Planfeststellung", which is based on the German rules ("Bundes-Immissionsgesetz, BImSchG"). If the decontamination is "on-site" and can be finished in less than 6 months, the application for approval will be handled according to a procedure called "Plangenehmigungsverfahren". In special cases, the authorization for an "on-site" project can be evaluated based on the water budget law ("Wasserhaushaltsgesetz, WHG") and the waste law ("Abfallgesetz, AbfG") or will be given by the official instruction of the appropriate authorities.

Principles and general remarks

The remediation of soil is based on either a static or a dynamic process. For the static method, at first, the contaminated soil will be mixed with nutrients and supplements. It will then be piled up into a stack or a heap that can be equipped with a system for drainage and a system for ventilation. The whole process of bioremediation can be performed through these systems. Oxygen, nutrients and supplements can be added without further mixing of the soil.

In contrast, the dynamic method requires extensive turning of the soil throughout the bioremediation process. In some cases, the procedure will be initiated by the homogenization of the soil. During this step

Table 11.3. Features relevant to on- and off-site remediation processes.

	Static procedure	*Dynamic procedure*
Processing	No processing during bioremediation	Periodic processing with special equipment
Process controlling and regulation; application of nutrients	Inexact	Exact
Application of supplements	Necessary; up to 10 vol%	Not necessary
Microbial degradation process	Catalysed by bacteria and fungi	Mainly catalysed by bacteria
Average duration of the bioremediation process	> 12 months	3–12 months

supplements, nutrients and biomass may be added to the soil matrix. The cultivation of the ground is carried out by periodic, intensive mechanical treatment. The decontamination process will be supervised by the analysis of samples taken after each step of cultivation.

The features relevant to both methods are summarized in Table 11.3. However, it should be noted that there is already a tendency for the dynamic procedure for soil decontamination to be used in preference.

Practical approaches

In this section the principles of the dynamic procedure will be described. The process consists of three distinct steps, the excavation, the classification and homogenization, and the remediation of soil. The excavation of soil and the problems relevant to this step will not be considered further here. For a more detailed discussion of this topic, the reader should refer to the manual of regulations for working with contaminated soil and to the methods of civil engineering.

Preparation of contaminated soils

The first important step for bioremediation is the classification and the pretreatment of the soil. The contaminated soil will be transferred continuously to a screening installation where it will be separated according to particle size.

Screening installations with a relatively low capacity are highly mobile and are ideal for "on-site" projects of up to 5000 m^3. The most common models are drum screen installations, which are set up on a chassis. The soil can be moved directly by an excavator into a feed hopper, which can accommodate 3 m^3. An infinitely variable scraper conveyer transports the earth to a screening drum, which is 3 m long and has a diameter of 1.3 m. The electricity is provided by a Diesel-driven power supply. In this step the soil is separated into two fractions: particles smaller than 35 mm and those larger than 35 mm.

The larger fraction will be applied to a sizing machine to achieve the required particle size (i.e. less than 35 mm). Small twin-shaft sizing machines are often used for this purpose. In contrast to other sizing machines, they are relatively easy to operate. Their high efficiency, based on a combination of shearing and tensile forces results in less accumulation of the fine fraction. An external generator supplies the necessary power at about 130 kW.

Slight modifications to the sizing

Figure 11.27. Basic scheme for an on-site bioremediation plant. Stacks with contaminated soil are treated by a stack processor. Elution of contaminants is prevented by an industrial shed and sealing the ground.

machine allow fertilizer, supplements (compost, straw, etc.) and biomass to be added. The dosage of fertilizer containing nitrogen and phosphate should be chosen to achieve concentrations in the soil of 30–60 mg l^{-1}, 10–20 mg l^{-1} and 1–2 mg l^{-1} for nitrate, ammonium and orthophosphate, respectively. Potassium may also need to be added. In practice, the use of commercial liquid fertilizers has proved to be worthwhile. Elementary compounds such as potassium nitrate (KNO_3), ammonium nitrate (NH_4NO_3), ammonium phosphate ($(NH_4)_3PO_4$), orthophosphate (P_2O_5), etc., can be included to fulfil the specific requirements. The addition of supplements and biomass can be carried out through an injection system connected to the screening and sizing unit. During this process, the inclusion of semi-ripe compost made from foliage, straw and bark mould ($p = 0.3$ g cm^{-3}) to a final concentration of 10% of the soil (by volume) can improve the bioremediation.

Performing and controlling the bioremediation process

An important aspect of the decontamination is to avoid the release of harmful substances into the environment. In "off-site" plants, this is ensured by sealing the ground, where drains for routine examination are installed. Pollution of the air is prevented by an appropriate filter system (a biofilter and/or an active charcoal filter).

Similar demands should also be addressed to "on-site" projects. However, it should be noted that the necessary safety equipment can only be used once. In the specific case of residual pollution of soil contaminated with oil, volatile compounds are often evaporated before the decontamination begins. Therefore, an installation for air cleaning may be unnecessary. The basic scheme for an "on-site" project is shown in Fig. 11.27.

The contaminated soil will be arranged in stacks 3.5 m wide and approximately 1.8 m high. If necessary, layers of compost can be included (10% by volume; p = 0.3 g cm^{-3}). The complete mixing of contaminated soil and compost will then be performed throughout the first treatment of bioremediation. The periodic cultivation will be carried out with a modified compost stack processor. These machines are altered especially for handling soil and strengthened in the essential construction elements. Some suppliers offer compost stack processors with an additional tank for the injection of fertilizer and biomass suspensions. The basic models are equipped with a 160 kW Diesel engine and are able to process about 100–150 m of soil stacks per hour. The driver's cabin of these machines should contain an appropriate air-conditioning unit with an active charcoal filter to guarantee safety during the operation.

The process of remediation will be performed by a weekly or monthly mixing of the soil, where all parameters will be checked and controlled concomitantly. It is imperative that, at the beginning of the treatment, the cultivation is performed in short cycles to supply the soil with oxygen and to adjust nutrient levels. After this initial phase the amount of contamination usually decreases rapidly if oxygen has been applied to an optimal concentration. Furthermore, it is essential to provide a sufficient and homogeneous supply of nutrients and biomass during the initial phase. Turning over the stacks results in the soil matrix being broken up. The overall pore volume in the soil increases and facilitates the diffusion of gases, so that the aerosol biomass reaches all the contaminated particles.

After every treatment step in

Table 11.4. Process parameters and their variability during the control of the process.

Parameter	*Optimal range*	*Regulation by*	*Remarks*
Nutrients			
Nitrate	10–50 mg l^{-1}	Application of fertilizer	Adaptation of applied biomass
Ammonium	10–20 mg l^{-1}	Application of fertilizer	Adaptation to applied biomass
Phosphate	1–2 mg l^{-1}	Application of fertilizer	Adaptation to applied biomass
Other parameters			
Biomass	10^7–10^8 CFU g^{-1} soil	Inoculation with starter cultures isolated from the habitat	Optional
pH value	6.5–8.5	Fertilizers or lime	Facultative
Temperature	18–25°C	Only possible in off-site remediation plants	Minimum soil temperature 10°C
Humidity	30–50% of maximum water retention capacity	Application of water	

bioremediation, samples will be collected systematically and analysed. The data of each parameter will be compared with the data from the preliminary decontamination experiment, which will have been performed earlier to classify the optimal level of nutrients, seeding of biomass, humidity, etc. The different parameters and their variability during the controlling of the process are shown in Table 11.4.

The nutrient concentrations listed in Table 11.4 are only valid at the beginning of a decontamination process and should be less than 10 mg l^{-1} for all nutrients in the advanced state of remediation (organic hydrocarbons concentrations lower than 1.300 mg kg^{-1}). It should be mentioned that there are new synthetic compounds available that can replace the application of inorganic nitrogen and phosphate. In contrast to the inorganic compounds listed above, they have no effect on water quality. Their efficiency, however, remains to be established, since, at present, the relevant field experiments have not yet been completed.

The significance of the application of specifically cultivated biomass is still disputed. Numerous remediation projects have been successful without the use of biomass. Also, the inclusion of different laboratory bacterial strains has been unsuccessful because these bacteria were unable to supersede the microbial flora of the autochthonous biocoenosis. The reinoculation of soil with selected autochthonous microorganisms appears to be the successful alternative to colonizing the soil particles homogeneously. Comparative studies performed by different companies have shown the advantage of this strategy. Freshly prepared cultures of autochthonous bacteria were applied of a concentration of 10^{11} colony forming units (CFU) per millilitre of stock solution; 50 l of this stock solution are sufficient for inoculation of about 1000 m^3 of contaminated soil. The soil moisture also influences the progress of decontamination. The optimal bioremediation of harmful compounds is significantly reduced if the soil moisture drops below a threshold of 30% of the maximum water holding capacity (WHC). If the value exceeds 60% of the maximum WHC, the exchange rate of gases in the soil matrix becomes critical and the availability of oxygen for microorganisms becomes insufficient. Some soils will also become "paste-like" and, at this consistency, are unsuitable for bioremediation.

Conclusions

The procedure for the dynamic treatment is applicable to the bioremediation of soils contaminated with oil or other fuels, as well as for the degradation of non-halogenic organic acids, alcohols and amines. It is not suitable for the elimination of halogenated organic compounds if polycyclic hydrocarbons are the sole carbon source for microorganisms in the polluted soil.

The periodic treatment of soil during the process of decontamination is more personnel- and cost-intensive than the static procedure. Nevertheless, the advantage of shorter times for dynamic redevelopment procedures makes this method profitable. In addition, the static procedure does not always ensure the homogeneous decontamination of the soil. The problems are linked to an increased water flow around the drainage and sprinkler system, and to soil compression in other areas. The effect is an insufficient supply of nutrients and oxygen to some parts of the soil stack.

The unrestricted application of supplements can also exhibit disadvantages. The addition of compost increases the gas exchange in the soil matrix. However, after decontamination, the soil often displays reduced ability for compression. Therefore, it becomes unsuitable for use in civil engineering. The only possible use for this recycled soil is as top soil. The market, however, for this type of soil is very limited.

The procedures described above, "on-site" and "off-site", for large-scale remediation of oil-contaminated soils are important for bioremediations, particularly when these techniques are optimized fo finely pored soils containing more than 65% of silt. They offer an inexpensive alternative compared with procedures performed in bioreactors, to procedures based on a thermal treatment and to washing procedures. Finally, these simple techniques are quite ecologically safe and harmless to the environment.

Decontamination in bioreactors

U. Gauglitz

In classical methods of biological soil decontamination, the problem of homogenizing the soil adequately during treatment is often encountered. Further difficulties include the introduction of additives and the breaking up of local concentrations of pollutants. These problems can be significantly reduced by the use of bioreactors, where the material is mixed more thoroughly. This allows more reliable sampling and can measure the success of a decontamination operation.

Bioreactors are usually completely enclosed systems. This makes it easier to control emissions such as process water and fumes. The control of emissions, in particular, should have a positive effect on the duration of approval procedures. Compared with regeneration heaps and land farming, decontamination in bioreactors offers a range of advantages such as improvements in:

- Equalization of the pollutants.
- Control of the degradation of volatile compounds.
- Bioavailability.
- Process control (pH value, temperature, moisture, moisture content, etc.).
- Control of pollutant degradation.
- Incorporation of additives.
- Treatment of soils with a large fraction of fine particles.

Furthermore, the use of bioreactors permits a free choice between aerobic and anaerobic conditions. Because of the advantages described above, over the past few years, a number of types of bioreactors have been tested. These are described below.

General principle

Most types of reactor will take excavated soil with particle sizes up to about 100 mm without prior classification. Only if larger particles are present is it necessary to go to the trouble of classifying and breaking up the material before treatment. Solid and liquid additives (water, air, bacterial cultures, surfactants, acids, alkalis, co-substrates, etc.) are best put directly into the reactor. Depending on the type of bioreactor, the contaminated soil can be mixed continuously or discontinuously. If the bioreactor is equipped for forced ventilation, the contaminants can be removed by controlled stripping as well as by biological degradation.

The pollutants are eliminated exclusively in the bioreactor. Once the reactor has been filled, there is no direct contact between the contents and the environment, which is the major advantage in terms of environmental and work safety.

Types of bioreactor

Various types of bioreactor are currently being tested for different applications. For treating contaminated soil without prior removal of the fine fraction, it is best to use a horizontal bioreactor. Washing the soil produces a highly contaminated suspension of the fine fraction. Vertical bioreactors are most suitable for decontaminating these suspensions.

All types of reactor incorporate a mixing device, which ensures that the material to be decontaminated is thoroughly homogenized. It is also possible to mix in additives to adjust such parameters as the water content, oxygen and nutrient supply, pH value, foaming, etc. Some reactors can be heated to accelerate the degradation process. The heating occurs either directly by an immersed heat exchanger or indirectly from outside by a heated enclosure.

Horizontal bioreactors

Horizontal bioreactors are particularly suitable for treating material with a high solid content (dry reactor). This provides a higher capacity for a given reactor volume, which can have a capacity of several hundred cubic metres.

Vertical bioreactors

Vertical bioreactors are preferred for treating suspended soil (wet reactor). As thorough homogenization is desirable, they can only operate with a solid content of 20–40%. Further, the particle size of the soil must not exceed about 100 µm. Because of these requirements, vertical bioreactors are mainly used in combination with soil washing systems.

The efficiency of the process can be increased by connecting several reactors in series. Such cascade reactors can have a height of up to 20 m and each reactor can exceed 400 m^3 in capacity (Fig. 11.28; Luyben and Kleijntjens 1992).

Mixing of the soil

All the reactor types have means of ensuring that the contents are thoroughly mixed. With horizontal bioreactors, there are two possible designs.

Rotary-drum reactors

The entire reactor rotates around its axis and the mixing blades fixed to the inside mix the contents. This type of reactor is usually pear-shaped, like a concrete mixer. However, our investigations have shown that a series of modifications are necessary before they can be used as bioreactors. Improved mixing and a system for introducing oxygen are required. The largest rotary reactors that have been built up to now have a capacity of only 12 m^3. The effective capacity of this type of reactor is about 60% of its total volume.

Fixed-drum reactors

Here the reactor is fixed and contains a central rotary shaft fitted with mixing and cutting blades. This design has the disadvantage that repairs are very difficult when the reactor is full and the shaft jams; the reactor can not readily be emptied without turning over the mixing shaft and the air inside is usually heavily contaminated, sometimes with flammable vapours.

The advantage of the fixed- over the rotary-drum reactor lies in its size; pilot plants with a total volume of 420 m^3 have already been constructed. The effective capacity of a fixed-drum reactor is about 40% of its total volume.

With vertical reactors, the contents are mixed with a helical mixer and by the introduction of large volumes of air through the bottom of the reactor. These two methods of mixing can be used alone or in combination. If compressed air is used for mixing, the bioreactor can operate simultaneously as a stripping reactor.

Mode of operation

Batch operation is the simplest mode of operation for bioreactors and is used in most of the processes that have been developed so far. The bioreactor is filled with contaminated soil and completely

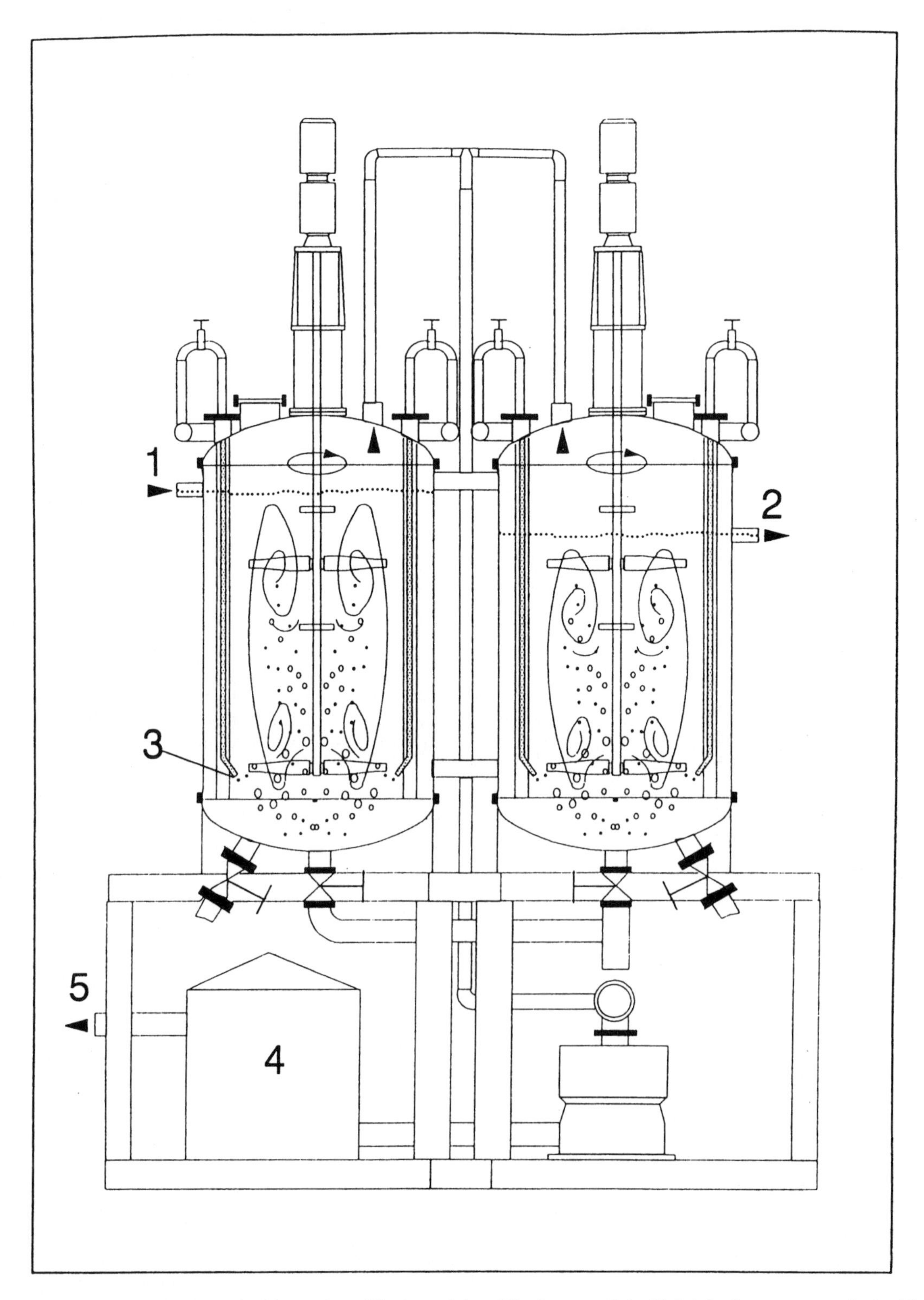

Figure 11.28. Vertical cascade bioreactor: (1) slurry inlet; (2) slurry outlet; (3) inlet of compressed air; (4) activated carbon filter; (5) outlet of filtered air.

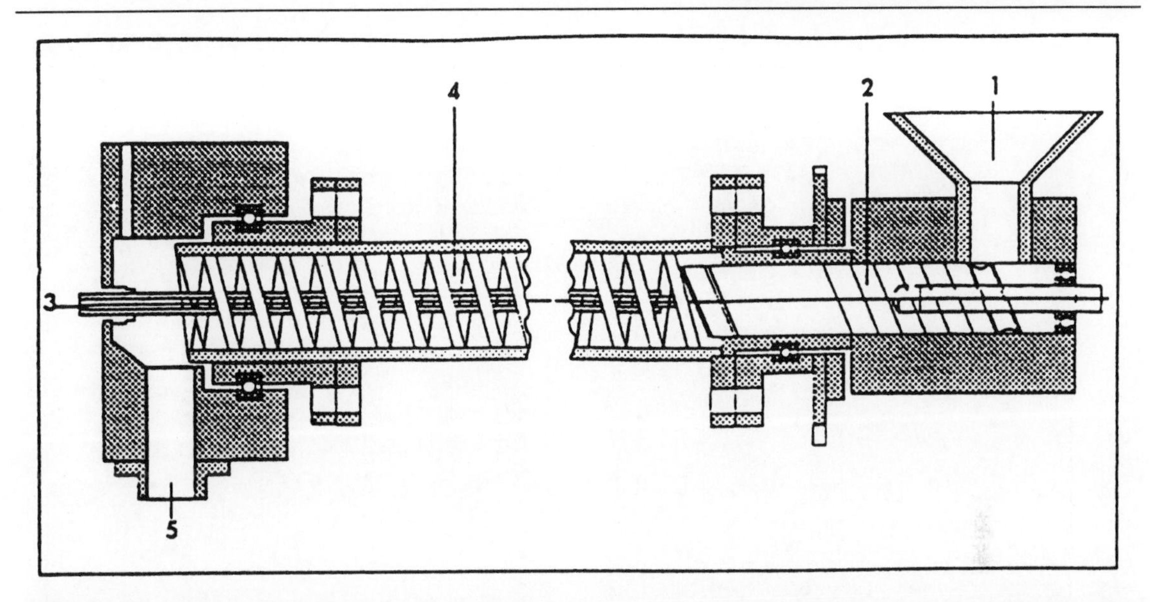

Figure 11.29. Mode of operation of rotary cylinder reactor: (1) feeder for contaminated soil into the charging hopper; (2) transport of contaminated soil into the reactor by a conveyor screw; (3) charging of microorganisms (mixed populations), nutrient salts and water through a central feed point; (4) decomposition of the contaminants during passage through the rotating cylinder; (5) discharge of the decontaminated soil.

emptied only after the contaminants have been removed.

Some vertical bioreactors are designed to operate as a soil-slurry sequencing bath reactor (SS-SBR; Irvine et al 1992; Kleijntjens et al 1992; Nitschke et al 1992). This means, that when the reactor is emptied, a quantity of the treated slurry is retained as biologically adapted material that serves to inoculate the next cycle.

A working group in Berlin is currently developing a rotary cylinder reactor for continuous operation, which is shown in Fig. 11.29 (Hülscher 1993; Kiehne et al 1993). It consists of a rotating cylinder with a helical screw on the inside. Its rotation conveys the soil through the reactor, constantly turning it over. The mixing action takes place only within one segment, so that the time spent by the soil in the reactor is defined exactly. Initial trials have shown that this type of reactor can be operated only using soil with a low moisture content; if the soil contains too much water, it forms lumps, which are much less readily penetrated by oxygen, thus reducing the rate of degradation of the contaminants. However, under continuous operation, it must be expected that the soil matrix will change constantly so as to make control of the process difficult.

Moisture content

Horizontal bioreactors can be operated with soil with a very low moisture content. This has the advantage that the available capacity can be used more effectively. However, as water is needed to allow the contact between the pollutant and microorganisms, a minimum moisture content is nevertheless required.

A certain level of soil moisture is also required in order to mix the soil to be treated. Generally, the lowest soil moisture content in practice is 15%. Because of their design, vertical bioreactors usually require a water content in excess of 60% to give a soil suspension that is easily controlled.

If water is added, this can exceed the water absorption capacity of soil, particularly in vertical bioreactors. It is, therefore, necessary to dewater the soil after it has been treated. This final dewatering must be carried out either by mechanical means, e.g. filter presses, vacuum drum filters, centrifugation, or by natural evaporation and percolation. The cost of dewatering is very high and can account for 40% of the total decontamination costs. The water that is obtained can either be reused again in the same process, or it can be discarded. As a rule, it is necessary to purify the water.

Nutrient supply

As with regeneration heaps, it is also necessary to add nutrients, mainly nitrogen and phosphorus, to bioreactors. Co-substrates may also be required to bring about the degradation of certain contaminants. The types and quantities of nutrients to be added are determined in trials for each case of pollution. The nutrients can be added either in dissolved form or as solids, with controlled-release characteristics if required. The use of bioreactors makes it possible to distribute the nutrients in the soil more quickly and more evenly, which accelerates the degradation process.

Microorganisms

Prior to the bioreactor treatment, the microbial activity of autochthonous microorganisms in the soil is determined in the laboratory. As a rule the microorganisms are sufficiently active in older contaminated soil and their activity can be enhanced in the bioreactor by optimizing the medium. In exceptional cases it may be advisable to add starter cultures (allochtonic microorganisms). In our experience, it is better to enrich or isolate the microorganisms already present in the contaminated soil than to use starter cultures from elsewhere. However, the usefulness of starter cultures in decontaminating soil is controversial. Some commercially available starter cultures have even been shown to contain large numbers of pathogenic and facultative pathogenic bacteria (Dott et al 1989). As described under "Mode of operation", it is also possible to use a portion of the freshly decontaminated soil or the extracted waste water to inoculate the next batch.

Additives

Various additives can be used, particularly for horizontal bioreactors, and mainly serve as carrier materials for the microorganisms, although they also assist in homogenizing the soil. Additives can include the bark of trees, compost, mineral granules and similar materials. The proportion of additives should not exceed 20% as larger quantities make it difficult to reprocess the purified material. It must, therefore, always be considered, whether it is really necessary to use additives. It can be useful to add surfactants to accelerate the treatment of soil containing pollutants not readily soluble in water. These emulsify pollutants such as hydrocarbons, thereby increasing their surface area. Since pollutants can only be decomposed at the surface, this method effectively increases their bioavailability. However, it is important to use the correct surfactant. Readily biodegradable surfactants can be metabolized before they are able to emulsify pollutants completely. The use of non-biodegradable surfactant is not possible because it would tend to replace one type of organic pollution with another.

Oxygen supply

As already described, molecular oxygen is required for the mineralization of many contaminants. In dry reactors, the oxygen is introduced into the soil by the mixing action. In wet reactors, it is better to introduce oxygen by blowing compressed air into the soil slurry (see Fig. 11.28). It can be worth introducing pure oxygen into a bioreactor to minimize the volume of waste gas. However, it must be considered that an explosive gas mixture can be formed in the presence of volatile contaminants.

Treatment of the waste air

The use of bioreactors provides full control of the emissions produced by the treatment. This applies particularly when forced ventilation is used. The waste air can be treated in activated carbon filters, biofilters, by incineration, etc.

Treatment time

For the reasons given above, it requires much less time to decontaminate soil in bioreactors than in regeneration heaps. The time required depends on factors such as the type of contaminants, their concentration, the type of soil matrix, etc. A sandy soil with only a few hundred parts per million of easily degradable aromatic compounds can be decontaminated in a bioreactor in a few days. The decontamination of clay or silt soil containing polycyclic aromatic hydrocarbons can be expected to require several weeks.

Treatment costs

The cost of treating contaminated soil in bioreactors depends on a number of factors. Some of these factors are mentioned above under "Treatment time". The type of bioreactor used also affects the overall cost, which includes investment costs, operating costs, etc. Currently, the average rate lies in the range of 200–300 US dollars per ton. In particular cases the price can be considerably higher, so that, from an economic point of view, other methods of decontaminating the soil can be more favourable.

Assessment

Compared with regeneration heaps, the advantages of bioreactors for treating polluted soil are obvious. Nevertheless, it must be checked from case to case which is the most suitable process, as there is no universal process for the bioremediation of the polluted soil.

In situ bioremediation of the saturated soil

T.H. Held, G. Rippen

In contrast to the well-established biological on-site techniques, most of the *in situ* remediation techniques are still in the phase of research and development. The few advantages of this technique (e.g. no necessity to excavate the soil or reduced occupational safety measures) are still outweighed by difficulties in process engineering (Lee et al 1987). The main field of application is the microbiological removal of mineral and waste oil contaminations.

Under optimized conditions, pollutants can be degraded to CO_2 and H_2O by microorganisms. However, metabolites (toxic or harmless) can be accumulated during the remediation process. Since the degradation of hydrocarbons, the main fraction of oil and waste oil, is a process with high oxygen consumption (e.g. 3.5 kg O_2 kg^{-1} hydrocarbons), mineralization requires, besides nutrient salts (PO_4^{3-}, NH_4^+), large amounts of oxygen donors (electron acceptors) such as O_2, H_2O_2 or KNO_3 (Morgan and Watkinson 1991). Trace elements are provided by the fine fraction of soil. In addition, environmental factors such as soil pH, content of substrates, etc. should be checked and optimized. Enhancing temperature (heating to 20–30°C) to increase biological activity and solubility of pollutants is often not economical since the degradation process also takes place at 10 °C, which is near to the constant temperature of groundwater in Europe.

Microbial cultures have also been applied to increase the degradation rates. However, the results are still controversial.

In this section the *in situ* remediation process will be presented and discussed. The reference for this section is a study of a complex waste oil pollution site at Pintsch, near

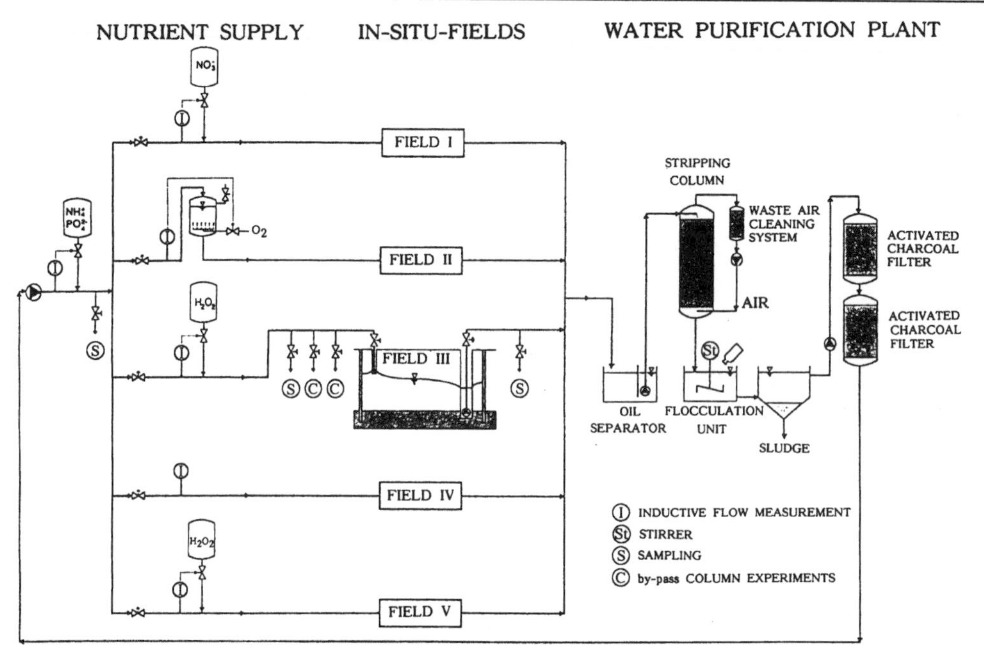

Figure 11.30. Process scheme of the *in situ* fields.

Hanau, Central Germany, which is accompanied by a research and development project supported by the German Federal Ministry of Research and Technology (BMFT).

General principles

The technique is based on the slow circulation of groundwater using infiltration and exfiltration tubes, which are installed to a specific depth in the contaminated site. Environmental conditions are also optimized to obtain the highest microbial activity. A common *in situ* remediation process is comparable to that occurring in a fixed-bed reactor (here the soil) with the groundwater being the transport medium. This process is mediated by the following steps (Fig. 11.30):

1. Microbial metabolism in groundwater and soil.
2. Chemical dissolution of pollutants and transport via the percolation of groundwater into the water purification plant.
3. Hydromechanical output of pollutants and metabolites in the form of emulsions or particles.
4. Chemical/physical clean-up in the water purification plant.

Location, and geological and hydrological aspects

Based upon the fact that up to now promising process engineering has been lacking, the application of *in situ* techniques is restricted to those sites where the following boundary conditions are given:

- A favourable geological structure: only sand or gravel exhibit permeability factors sufficient for hydraulic measures.
- The biodegradability of all pollutants.
- There is no biocidal activity in the environment.
- There is sufficient bioavailability of the pollutant.

Prior to the realization of microbial *in situ* remediation the following steps are necessary:

PINTSCH SITE, HANAU

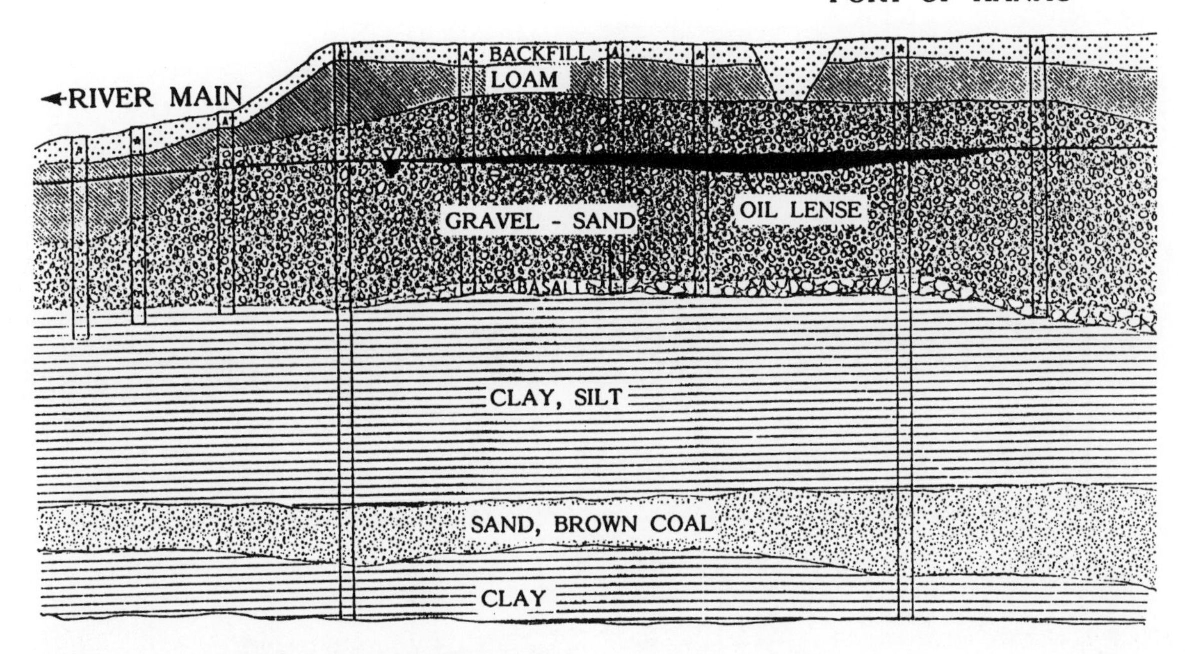

Figure 11.31. Geological survey of the Pintsch site.

1. A historical exploration in order to supply information about the nature, amount and distribution of the expected pollutants.
2. A geological and hydrological investigation, which is necessary to choose a feasible technique of remediation.
3. Drillings for soil and groundwater sampling, which should be planned and carried out in order to obtain undisturbed samples (e.g. drillings with in-liners).

The former waste oil refinery, Pintsch Oil, is situated on a peninsula between the river Main and the port of Hanau. The groundwater is overlaid by an oil lens of more than 70 cm thickness in the centre of a field covering almost the whole site of 25,000 m^2. The water level is about 4 m below surface, corresponding to the water level of the river Main (Fig. 11.31). The soil consists of Main terrace gravel with an overlay of 1 m of loamy material and 1 m of backfill material. At a depth of about 10 m dense clay can be found. In 1989, the Pintsch site was surrounded by a slurry wall and can, therefore, be regarded as a bioreactor isolated from the surrounding groundwater. Since the oil-saturated zone of former groundwater fluctuations is not bioremediable and since the oil phase provides a constant supply of soluble pollutants into the water phase by means of diffusion, not only the unsaturated layer but also 0.5 m of the saturated soil are planned to be treated using soil-washing processes.

Loamy material and backfill material is to be remediated microbiologically in on-site piles, and finally the remaining saturated layer is assigned to be cleaned *in situ*. To elaborate the process engineering, a research and development project was started in 1989 that included five *in situ* test fields 5–10 m in size. Each field was surrounded by a slurry wall to separate the sites (Figs 11.30 and 11.32).

FORMER:

INFILTRATION ABOVE THE AQUIFER

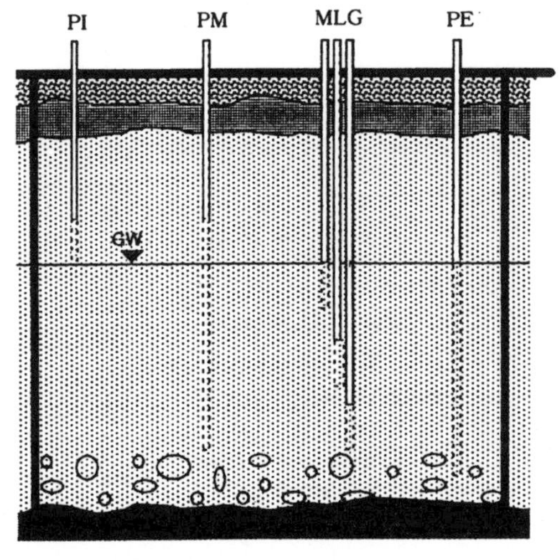

TODAY:

INFILTRATION WITHIN THE AQUIFER WITH ADDITIONAL SPRINKLING

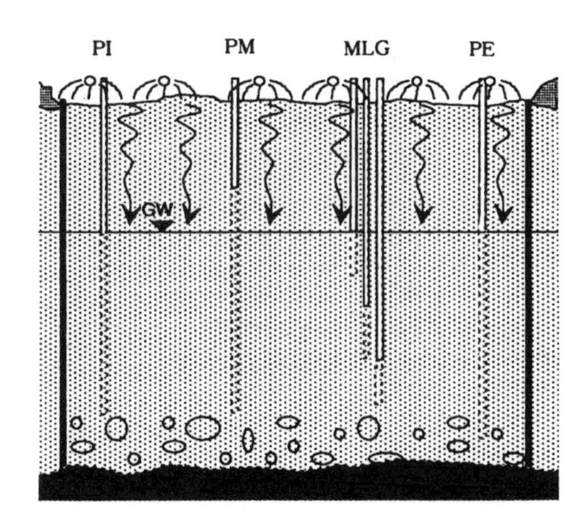

PIEZOMETER TUBE OF INFILTRATION PI
MULTILEVEL GAUGES MLG
PM PIEZOMETER TUBE FOR MEASUREMENT
PE PIEZOMETER TUBE OF EXFILTRATION

LOAM
BACKFILL
MAIN TERRACE GRAVEL
TERTIAR

Figure 11.32. Cross-section of the *in situ* fields. (a) The former system of infiltration above the aquifer; (b) the present system of infiltration within the aquifer with additional sprinkling. GW, ground water; PI, piezometer tube for infiltration; PM, piezometer tube for measurement; MLG, multi-level gauges; PE, piezometer tube for exfiltration.

Field work

The field experiments at Pintsch, Hanau, started in 1989 (Riss et al 1991). Besides piezometer tubes for infiltration and exfiltration, multi-level gauges were installed in the centre of the field to receive water samples from specific depths (Fig. 11.32). The exfiltration tubes exhibited filter areas over the entire depth of the unsaturated layer. Since the infiltration tubes were destroyed by corrosion and the oil layer was spread into the aquifer by hydraulic measures, in the spring of 1990 the infiltration technique was changed. Instead of infiltrating into the unsaturated soil, the piezometer tubes were lengthened to the field bottom thus providing a filter area over the whole saturated layer. For additional treatment of the unsaturated soil, the loamy material and backfill material were removed in the autumn of 1990 to treat the soil by sprinkling with nutrient-supplied water. Contrary to expectation, the preferred flow direction of the process water was along the bottom of the field (8–10 m depth) in both cases of infiltration, as shown by tracer experiments. Samples from the groundwater and soil were removed, and used for chemical and microbiological analyses.

Chemical analysis

On the basis of the existing information (e.g. nature of pollutants, pollution due to an accident or due to the operation of an industrial plant, geological and hydrogeological profiles, etc.) the concentrations of the different pollutants must be determined in the whole area for balancing the total amount of pollutants. In addition, the concentrations of the most important nutrients (ammonium, phosphate, nitrate) and eventually biological inhibitors have to be determined so as to permit the maximal microbial degradation of pollutant.

The site investigated at Pintsch, Hanau, is different from other oil-contaminated locations because, in addition to a massive pollution of about 900 t hydrocarbons (HC) in the saturated soil layer, it is characterized by the presence of a number of highly toxic chemicals such as volatile chlorinated hydrocarbons (VCH), benzene/toluene/ xylenes (BTX), polychlorinated biphenyls (PCB) and polychlorinated dibenzodioxins and dibenzofurans (PCDD/F).

Microbiological assays

Samples of groundwater and soil are removed at the start, at given time intervals and at the end of the remediation; the samples are then microbiologically characterized. The consumption of O_2, H_2O_2 and nitrate is monitored in the process water. CO_2 and and O_2 are determined in ground air by using a special instrument (Binos, Rosemount). Pre-investigations showed that the Pintsch site is characterized by a high level of colonization with indigenous microorganisms adapted to pollution and environmental factors. The distribution of species is site specific (Dott et al 1989; Kämpfer et al 1991).

Starting and controlling of the *in situ* remediation

The remediation process is started by infiltrating and pumping out the process water (Fig. 11.30); the given water column provided sufficient pressure to penetrate the soil after infiltration. The exfiltrated water is cleaned in the water purification plant including the following steps:

1. The separation of oil.

2. The stripping of volatile compounds, which are then immobilized by an activated charcoal filter (waste air cleaning system).

3. A flocculation unit to remove iron.

4. The removal of sludge;

5. The use of activated charcoal filters for final cleaning of the process water. The resulting water purity allowed excess water to be led off into the river Main.

Subsequently the infiltration water is supplied with nutrient salts (ammonium, phosphate) via five single plastic conductors. These are controlled by an inductive flow measurement, which gives a signal to the respective pumps supplying the electron acceptors. The piezometer tubes are made of plastic to avoid their destruction by corrosion and ochre formation. At Pintsch, Hanau, fields I–IV were run with 0.5 $m^3 h^{-1}$ process water (0.4 $m^3 h^{-1}$ by infiltration; 0.1 $m^3 h^{-1}$ by sprinkling) containing 4 mg l^{-1} NH_4^+ and 4 mg l^{-1} PO_4^{3-}. Field I was additionally supplied with NO_3^- (200 mg l^{-1}), field II with oxygen (40 mg l^{-1}), field III with H_2O_2 (500 mg l^{-1}). Field IV was the reference field receiving water with 6–8 mg l^{-1} O_2. Field V was run with 1.75 $m^3 h^{-1}$ of infiltration water containing 100 mg l^{-1} H_2O_2. The final concentrations of nutrients and electron acceptors in the process water were calculated depending on the initial concentrations in the site (groundwater and soil), laboratory optimization experiments and previous remediation cases. Many problems of process engineering could be solved. The main problem influencing the assays was the oil phase on top of the groundwater. This phase at any time supplies the aquifer with pollutants by diffusion and, therefore, serves as a constant source of recontamination. Because of this problem, microbiological activities can only be monitored by indirect parameters, such as CO_2 formation in ground air, nitrate or O_2 consumption, and determination of bacterial numbers and specific degradation activities of soil. The concentrations of the main pollutants are significantly reduced by hydraulic and microbial processes. The volatile chlorinated hydrocarbon concentrations decreased from about 70 to 0.2 mg mi^{-1} (sum of VCH), and that of BTX from 6 to 0.2 mg ml^{-1}. The hydrocarbons in solution were reduced from about 10 to 1 mg l^{-1} and soil HC from about 4300 to 3200 mg kg^{-1} dry weight. Furthermore, high bacterial counts and a significant potential of bacterial degradation was observed.

The removal of hydrocarbon content from the soil using a water purification plant was less than 1% within 3 years. The degradation of hydrocarbon content is mainly performed by microorganisms, when a sufficient amount of electron acceptors were used. After 1300 days, most of the pollution parameters are nearly reduced to the target values defined for groundwater (Held and Rippen 1994). At this time, a breakthrough of all electron acceptors occurred in the respective fields. This indicates that the soil located near the infiltration area in the main water flow direction might have been cleaned to a certain extent and the diffusion of hydrocarbons might have become the limiting factor (Grathwohl 1992).

The concentration of the pollutants in groundwater is rather low at the end of the remediation while the concentration in soil remains high. This may depend on the reduced bioavailability of the remaining pollutants due to:

1. The integration of the pollutants into or strong adsorption on the soil matrix.
2. The presence of non-degradable fractions.
3. The unfavourable chemical/physical conditions (very low solubility).
4. The formation of "dead-end" metabolites (the degradation of some pesticides, e.g. results in the formation of toxic polymers) (Stegmann and Franzius 1991).

Evaluation of the bioremediation success

Target values for *in situ* bioremediation

Usually, a bioremediation is considered successful if predefined target values are achieved. Since pollutants are heterogeneously distributed even over a small area, testing the soil will not provide a satisfactory measure of success. Therefore, target values were fixed for groundwater only. The Pintsch specific target values are 500 µg l^{-1} of hydrocarbons, 30 µg l^{-1} of VCH (3 µg l^{-1} of vinyl chloride) and 50 µg l^{-1} of BTX (10 µg l^{-1} of benzene) (Held and Rippen 1994). However, it is important to state as reported before that, in addition to pollutant concentrations, the formation of metabolites and the "tailing" of pollutant concentrations in groundwater have to be considered.

Metabolites of degradation

The detection of toxic metabolites during the remediation process using the DIN 38409–H18 method is not possible (Stegmann and Franzius 1991). Therefore, it is recommended that (eco)toxicological tests are performed with the decontaminated groundwater and soil.

As shown by the investigation of the bioremediated Pintsch soil in the on-site process, the remaining fraction comprises 30% of the starting hydrocarbon concentration (volatile compounds have been stripped out) and consists of branched C–C chain frameworks with more than 20 carbon atoms (Wanior and Ripper 1993), which are insoluble and are not (eco)toxicologically relevant (Rippen et al 1992). This fraction is supposed to be non-bioavailable.

Besides the VCH and some hydrocarbons, most metabolites are more polar and water soluble and, therefore, more mobile. They can be efficiently washed out and removed by the water purification plant. In addition, the formation of dead-end metabolites was not expected considering the specific distribution of pollutants for Pintsch.

Whether or not metabolites are chemically immobilized during the humus formation and the nature of their fate after eventual humus degradation is the subject of current discussion.

Tailing of pollutants

Applying the remediation processes discussed above, a large tailing of pollutants in groundwater has been found (about 1.0 mg HC l^{-1}; about 0.2 mg VCH l^{-}; about 0.2 mg BTX l^{-1}). This tailing depends on several factors. Firstly, the oil phase serves as a constant supply of pollutants via diffusion. Secondly, heterogeneously distributed layers of silt consist of larger amounts of pollutants with a lower availability for remediation processes and, therefore, the silt layers release pollutants over an extremely long period of time by diffusion. The same effect is achieved if the hydraulic measures and the support with electron acceptors are too heterogeneous due to the geological structure of the ground.

In the earlier stages of the bioremediation the degradation rate is limited by the amount of available electron acceptors but the degradation is later reduced due to lower concentrations of the pollutants. These substrates have to diffuse to the bacterial cell where they are metabolized according to the Michaelis–Menten kinetics resulting in lower cell growth rates (i.e. biocatalysts) and an additional reduction in degradation rate (kinetic limitation).

Discussion

The data presented are not pointing directly to a particular decontamination method; they are sufficient to design a suitable and practicable large-scale process of *in situ* bioreclamation.

The removal of oil phases is absolutely necessary to exclude recontamination and to lower the tailing of pollutants.

The concentrations of the nutrients (NH_4^+, PO_4^{3-}) and the oxygen donors (O_2, H_2O_2) should be chosen according to the field experiments. Usually large amounts of oxygen donors are required; therefore, the degradation rate is mainly limited by the transport rate of the electron acceptors to the place of pollution, and the transport is influenced by the maximum water flow within the soil matrix given by the permeation factor between narrow limits. The oxygen donors used besides molecular oxygen are hydrogen peroxide and nitrate. Oxygen can be supplied in a solvated form (Biox® process; Müller 1992). The oxygen supply by gas lances has only been reported for the unsaturated soil layer (Lund 1991). A large amount from 30% (author's results) to 60% (Barenschee and Bochem 1990)] of the supplied oxygen remains in the system and is used by the biomass, which is partially leached out with the exfiltrated groundwater.

Independently of the used oxygen donor, the most evident problem of the *in situ* remediation is provided by the process engineering, i.e. the transport of nutrients, the characteristics of groundwater, the flow in oxygen bubbles produced by O_2 and H_2O_2, or formation of ochre under oxidative conditions. In this context it should be noted that oxygen input via gas lances within the saturated soil might lead to stripping effects of the volatile compounds and to contamination of the unsaturated soil.

Nitrate can also be used as an alternative electron acceptor. The denitrification of nitrate during the remediation process may be considered as a technical advantage.

To our knowledge, the application of microbial cultures by the *in situ* remediation processes has not yet been successful, probably because of the adsorption of the cells at the place of inoculation and due to their inability to succeed over the indigenous flora. Therefore, inoculation of soil with special pregrown bacterial cultures is not recommended.

(Eco)toxicological tests are necessary. They are recommended at the beginning and at the end of the remediation.

Reuse of decontaminated soils

Whenever possible, the decontaminated soil is returned to its original location. However, the material has often been excavated to make way for building, so that it cannot be returned. Such decontaminated soil is then used, for instance, in road construction, in noise-deflecting embankments and dikes, landscaping and other construction work. The reuse of soil depends on its physical and mechanical properties. Before the soil is decontaminated, it should, therefore, be considered to what extent any additives might affect its reuse.

References

Alef K (1991) Methodenhandbuch Bodenmikrobiologie, Aktivität, Biomass, Differenzierung. Ecomed-Verlagsgesellschat, Landsberg.

Alef K, Fiedler H, Hutzinger O (1993) Bodensanierung, Bodenkontamination, Verhalten und Ökologische Wirkung von Umweltchemikalien im Boden. Ecoinforma "1992", vol 2, Ecoinforma Press, Bayreuth.

Arndt F, Hinsenveld M, Van den Brink WJ (1990) Contaminated Soil 90. Third International KfK/TNO Conference on Contaminated Soil. Kluwer Academic Publishers, Dordrecht.

Atlas RM (1991) In: In-Situ Bioreclamation Applications and Investigations for Hydrocarbon and Contaminated Site Remediation, 1st edn. Butterworth-Heinemann, Stoneham, MA, pp. 14–33.

Autry AR, Ellis GM (1992) Bioremediation: an Effective remedial alternative for Petroleum Hydrocarbon-Contaminated Soil. Envir Progr 11: 318.

Barenschee E-R, Bochem P (1990) Laboruntersuchungen zur in situ-Bodensanierung mit Wasserstoffperoxid. WLB Wasser Luft Boden 11/12: 86–89.

Björseth A (1989) Handbook of Polycyclic Aromatic Hydrocarbons, Marcel Decker, New York.

Breure AM, Sterkenburg A, Volkering F, van Andel JG (1992) Contribution to the International Symposium "Soil Decontamination Using Biological Processes", Karlsruhe, Germany, Book of Preprints, pp. 147.

Brilis GM, Marsen PJ (1990) Comparative evaluation of Soxhlet and sonication extraction in the determination of polynuclear aromatic hydrocarbons (PAHs) in soil. Chemosphere 21: 91–98.

Bundesminister der Innern (1979) Beurteilung und Behandlung von Mineralölschadensfällen im Hinblick auf den Grundwasserschutz. Teil 3: Analytik. Umweltbundesamt (ed.), Berlin.

Burmeier H, Dreschmann P, Egermann R, Ganse J, Rumler R (1990) "Sicheres Arbeiten auf Altlasten" "focon", Aachen.

Coates J (1990) Instrumentation for infrared spectroscopy. In: Analytical Instrumentation Handbook. Ewing GW (ed.). Marcel Dekker Inc., New York, pp. 233–279.

DECHEMA (ed.) (1992a) Soil Decontamination using Biological Processes. Frankfurt, Germany.

DECHEMA (1992b) Laboratory Methods for the Evaluation of Biological Soil Cleanup Processes. Deutsche Gesellschaft für Chemisches Apparatewesen, Chemisches Technik und Biotechnologie, Frankfurt.

DIN 38407 part 9 Bestimmung von Benzol und einigen Derivaten mittel Gaschromatographie. In: Deutsche Einheitsverfahren zur Wasser-, Abwasser- und Schlammuntersuchung, 25. Lieferung, VCH Verlagsgesellschaft, Weinheim, 1991.

DIN 38409 part 18: Bestimmung von Kohlenwasserstoffen. In: Deutsche Einheitsverfahren zur Wasser-, Abwasser- und Schlammuntersuchung, 9. Lieferung, VCH Verlagsgesellschaft, Weinheim, 1981.

DIN 38409 part 17: Bestimmung von schwerflüchtigen lipophilen Stoffen. In: Deutsche Einheitsverfahren zur Wasser-, Abwasser- und Schlammuntersuchung, 10. Lieferung, VCH Verlagsgesellschaft, Weinheim, 1981.

DIN 38407 Teil 4–F4, Grouppe F: Bestimmung von leichflüchtligen halogenierten Substanzen. In: Deutschen Einheitsverfahren zur Wasser-, Abwasser- und Schlammuntersuchung, VCH Verlagsgesellschaft, Weinheim, 1988.

DIN 38407 Teil 5–F5, Gruppe F: Bestimmung von leichtflüchtligen Substanzen mit Headspace-Chromatographie. In: Deutschen Einheitsverfahren zur Wasser-, Abwasser- und Schlammuntersuchung, VCH Verlagsgesellschaft Weinheim, 1991.

DIN 38407 Teil 7–F7: Bestimmung von polycyclischen aromatischen Kohlenwasserstoffen. In: Deutschen Einheitsverfahren zur Wasser-, Abwasser- und Schlammuntersuchung, VCH Verlagsgesellschaft, Weinheim, 1991.

DIN 38407 Teil 8–F8: Bestimmung von polycyclischen aromatischen Kohlenwasserstoffen. In: Deutschen Einheitsverfahren zur Wasser-, Abwasser- und Schlammuntersuchung, VCH Verlagsgesellschaft, Weinheim, 1981.

DIN 38409 Teil 13–H13: Bestimmung von polycyclischen aromatischen Kohlenwasserstoffen. In: Deutschen Einheitsverfahren zur Wasser-, Abwasser- und Schlammuntersuchung, VCH Verlagsgesellschaft, Weinheim, 1981.

DIN 51405: Prüfung von Mineralöl-Kohlenwasserstoffen, verwandten Flüssigkeiten und Lösemitteln für Lacke und Anstrichstoffe. Gaschromatgraphische Analyse-Allgemeine Arbeitsbedingungen. In: Deutsche Einheitsverfahren zur Wasser-, Abwasser- und Schlammuntersuchung, VCH Verlagsgesellschaft, Weinheim.

Dott W, Feidicker B, Kämpfer P, Schleibinger H, Strechel S (1989) Comparison of autochthones bacteria and commercially available cultures with respect to their effectiveness in fuel oil degradation. J Bact Microbiol 4: 365–374.

Dott W (1992) Pre-investigation of biological decontamination potential assessment. In: Soil Decontamination Using Biological Processes.

DECHEMA (ed.), Frankfurt, Germany, pp. 133–135.

Farwell SO (1990) Modern Gas Chromatographic Instrumentation. In: Analytical Instrumentation Handbook. Ewing GW (ed.). Marcel Dekker Inc., New York, pp. 673–744.

Friege H, Jörissen U, Darskus R, Schlesing H (1987) An EOX method for the detection of halogenated hydrocarbons in soil. Fresenius Z Anal Chem 326: 154–155.

Grathwohl P (1992) Die molekulare Diffusion als limitierender Faktor bei der Sanierung von Boden- und Grundwasserkontaminationen. UWSF Z Umweltchem Ökotox 4: 231–236.

Hagendorf U, Leschber R, Neger M, Rotard W (1987) Bestimmung leichtflüchtiger chloroeter kohlenwasserstoffe in Bodenproben. Fresenius Z Anal Chem 326: 33–39.

Hansen NW (1973) The determination of unsaponifiable matter in oils and fats. In: Official, Standardized and Recommended Methods of Analysis. The Society for Analytical Chemistry, London, pp. 169–172.

Held T, Rippen G (1993) Sanierungszielwerte für die Sanierung einer Altöl-Kontaminiesten Industrie fläche. Altlaster-Spektrum (4), 209–216.

Hessisches Ministerium für Umwelt-, Energie- und Bundesangelegenheiten (1992) Entsorgung von belastetem Boden. In: Staatsanzeiger für das Land Hessen vom 1.2.1993, pp. 331–341.

Hülscher M (1993) Similarity study on scaling of rotary reactors for biotechnical soil clean-up. Chem -Ing -Tech 4: 436–440.

Irvine RL, Earley JP, Yocum PS (1992) Slurry reactors for assessing the treatability of contaminated soil. In: Soil Decontamination Using Biological Processes. DECHEMA (ed.), Frankfurt, Germany, pp. 187–194.

Kämpfer P, Steiof M, Dott W (1991) Microbial characterization of a fuel-oil contaminated site including numerical identification of heterotrophic water and soil bacteria. Microbial Ecol 21: 227–251.

Kiehne M, Müller-Kuhrt L, Berghof C, Buchholz R (1993) Kontiuierlicher Drehrohrreaktor zur Bodensanierung. Chem -Ing -Tech 11: 1343–1345.

Kleijntjens RH, Van der Lans RG, Luyben KC (1992) Design of a three phase slurry reactor for soil processing. Process Safety Environ Prot 70: 84–92.

Lee MD, Wilson JT, Ward CH (1987): *In-situ* restoration techniques for aquifers contaminated with hazardous wastes. J Haz Mat 14: 71–82.

Liphard KG (1987) Bestimmung von polycyclischen aromatischen Kohlenwasserstoffen im Boden. Gewässerschutz Wasser Abwasser 99: 205–226.

Liphard KG (1992) Analysis of organic contaminants in soil. How reliable are data? In: Soil Decontamination Using Biological Processes. DECHEMA (ed.). Frankfurt, Germany, pp. 213–218.

Lund NC (1991) Beitrag zur biologischen in situ-Reinigung kohlenwasserstoffbelasteter Böden. In: Gudehus G, Natau O (eds). Veröffentl. d. Institutes für Bodenmechanik und Felsmechanik der Universität Fridericiana in Karlsruhe, Heft 119.

Luyben KC, Kleijntjens RH (1992) Bioreactor design for soil decontamination. In: Soil Decontamination Using Biological Processes DECHEMA (ed.). Frankfurt, Germany, pp. 195–204.

McDonald JA, Rittmann B (1993) Performance standards for *in-situ* bioremediation. Environ Sci Technol 27: 1974–1979.

Morgan P, Watkinson RJ (1991) Factors limiting the supply and efficiency of nutrient and oxygen supplements for the in situ biotreatment of contaminated soil and groundwater. Water Res 26, 73–78.

Müller D (1992) Sanierungsprojekt ehemalige Erölraffinerie Speyer. TerraTech 1: 44–46.

Mueller JG, Lantz SE, Ross D, Colvin RJ, Middaugh DP, Pritchard PH (1993) Strategy using bioreactors and specially selected microorganisms for bioremediation of groundwater contaminated with creosote and pentachlorophenol. Environ Sci Technol 27: 691–698.

Nitschke V, Beyer M, Klein J (1992) Biologische Behandlung von Böden in einem suspensionsreaktor (DMT-Biodyn-Verfahren). In: Soil Decontamination Using Biological Processes. DECHEMA (ed.). Frankfurt, Germany, pp. 201.

Novotny MV, Lee ML, Bartle KD (1974) Methods for fractionation analytical separation and identification of polyaromatic hydrocarbons in complex mixture. J Chromatogr Sci 12: 606–612.

Offhaus K (1973) Münchner Beiträge zur Abwasser-, Fischerei- und Flußbiologie, Bd. 24, p. 169.

Peters DG, Hayes JM, Hieftje GM (1976) A Brief Introduction to Modern Chemical Analysis. WB Saunders Co., Philadelphia.

Rippen G, Ripper J, Ripper P (1992) Derivation of guideline values for the microbial reclamation of a complex waste oil pollution site. In: International Symposium "Soil Decontamination Using Biological Processes". Preprints. Karlsruhe, 6–9 December 1992, pp. 251–253.

Riss A, Bareusche ER, Helmsling O, Ripper P (1991) Einsatz von wassostoffpuoxid zum mikrobiologischen kohlenwassostofftabban. Lubur und Feldunsache in-situ. gwf Wasser Abwasser 132: 115–126.

Rumler R (1989) "Arbeitsmedizinische Aspekte bei der Sanierung von Altlasten" "Die Tiefbau-Berufsgenossenschaft" -Amtliches Mitteilungsblatt, Heft 1/1989.

Schomburg G (1990) Gas Chromatography–A practical course. VCH Verlagsgesellschaft, Weinheim.

Skladany GJ, Metting FB Jr (1993) Bioremediation of contaminated soil. In: Soil Microbial Ecology.

Metting FB Jr (ed) Marcel Dekker, New York, pp. 483–513.

Stachel B, Lahl, U, Schröer W, Zeschmar B (1984) Determination of the summation parameter organically bound halogen in water samples. Studies on mineralization. Chemosphere 13: 703–714.

Stegmann R, Franzius V (1991) Biologische Verfahren der Altlastensanierung. Ökologische Briefe no. 43: 6–8.

Stegmann R, Franzius V, Wolf K (1992) Handbuch der Altlastensanierung. R. v. Decker's Verlag, G. Schenk, Heidelberg.

Steiof M, Heinz H, Tautorat G, Dott W (1993) Investigation on the biodegradability of mineral-oil-hydrocarbons with different boiling ranges. In: Soil Decontamination using Biological Processes. DECHEMA (ed.), Frankfurt, Germany. pp. 400–407.

Urano K, Kato Z (1986) A method to classify biodegradability of organic compounds. J Haz Mat 13: 135–145, 147–159; Evaluation of biodegradation ranks of priority organic compounds J Haz Mat 13: 147–159.

Wanior J, Ripper J (1993) GC, 13C–NMR and IR study of a mixture of waste oil from the old refinery site Pintsch-Oil GmbH i.L. in Hanau, Germany. Fresenius Z Anal Chem 347: 423–429.

Wolf K, Van den Brink WJ, Colon FJ (1988) Contaminated Soil 88. Second International TNO-BMFT-Conference on Contaminated Soil, Kluwer Academic Publishers, Dordrecht.

Index

Printed and bound by CPI Group (UK) Ltd, Croydon, CR0 4YY
11/06/2025
01899189-0020